MECHANICAL AND STRUCTURAL VIBRATIONS

MECHANICAL AND STRUCTURAL VIBRATIONS

Demeter G. Fertis
Department of Civil Engineering
University of Akron

A WILEY-INTERSCIENCE PUBLICATION

JOHN WILEY & SONS, INC.

New York   •   Chichester   •   Brisbane   •   Toronto   •   Singapore

Copyright © 1995 by John Wiley & Sons, Inc.

All rights reserved. Published simultaneously in Canada.

This publication is designed to provide accurate and
authoritative information in regard to the subject
matter covered. It is sold with the understanding that
the publisher is not engaged in rendering legal, accounting,
or other professional services. If legal advice or other
expert assistance is required, the services of a competent
professional person should be sought.

Library of Congress Cataloging in Publication Data:

Fertis, Demeter G.
  Mechanical and structural vibrations / Demeter G. Fertis.
    p.   cm.
  Includes bibliographical references and index.
  ISBN 0-471-10600-3 (cloth: acid-free paper)
  1. Vibration. I. Title.
TA355.F36 1995
620.3—dc20   94-33121

Printed in the United States of America

10 9 8 7 6 5 4 3 2 1

I dedicate this book to my wife, Anna,
for her valuable help and participation
in the creation of this work

PREFACE

The vibration and dynamics analysis of structural and mechanical systems is becoming increasingly complex today, and it will continue to do so in order to meet the challenges and demands of 21st century technology. To meet such demands, students, as well as practicing engineers, should be trained in a way that will provide them with a better knowledge and physical understanding of the vibration and dynamic response of both present and future structural and mechanical systems. In both academia and industry, emphasis should be placed on understanding the physical behavior of such systems, and then mathematical models can be provided that incorporate all the essential elements needed for system design and analysis. It should be pointed out, however, that the solution of such mathematical models should be kept as simple as possible in order to cope with the pressing demands of cost reduction. Utilization of equivalent and idealized systems often provides a simpler and very effective solution of very complex problems. Such equivalences and idealizations should continue to be employed today and should become more sophisticated while retaining the original philosophy of simplicity and accuracy. This procedure enables the engineer to effectively design structural and mechanical systems with responses that are becoming increasingly nonlinear.

By following this line of reasoning, this book can serve both academia and industry. In an engineering college, this book can be used as a text for two semesters of undergraduate courses on vibrations, which are regularly offered as part of the civil, mechanical, mechanics, and aeronautical and aerospace engineering departments. The material is class tested and is organized in a way that permits the instructor to select the topics that are most appropriate to the students' level of instruction. The variety of topics is wide, and the selection of such topics may include subjects that are not included in other texts, which prepare the student for more advanced work in the areas of vibration and dynamics. This book can also be used as a first-year graduate text, for students who take a course in vibrations for the first time, or as a graduate text by selecting the more advanced topics or chapters.

This book can also be used by professional civil, mechanical, aeronautical, and polymer engineers in industry and government. It emphasizes both theory and practical applications, and it contains valuable new methods and concepts that are not available in other texts dealing with this subject, in order to help both the student and the practicing engineer to cope with the demands of advanced technology. Nonlinear analysis and the inelastic response of structural and mechanical components of uniform and variable stiffness are examples of such topics.

Chapter 1 is devoted to the fundamentals of vibratory motions of both linear and nonlinear systems, as well as an introduction to chaos and chaotic systems. Differential equations of motion of various continuous and spring–mass systems are derived. Chapter 2 deals with the vibration analysis of simple spring–mass systems and systems with continuous mass and elasticity. Various aspects of damping are also examined. This chapter provides a fair exposure to the general vibration problem and prepares the reader for the more advanced work that follows. Chapter 3 deals with the forced vibration of simple damped and undamped systems by using both rigorous and numerical methods. The dynamic response of idealized beams and frames, as well as the dynamic response of elastoplastic systems, is included. Such topics are particularly useful when we are dealing with the dynamic response of systems that are subjected to earthquake excitations and blast overpressures.

Chapter 4 is devoted primarily to the vibration and dynamic response of systems with continuous mass and elasticity, such as beams and plates, beam columns with axial compressive or tensile loads, and elastically supported beams with axial restraints. Static and flutter instabilities are also examined. This chapter incorporates the unique concept of the dynamic hinge and dynamically equivalent systems, as developed by the author and his collaborators. This concept may be used as a diagnostic tool to investigate and improve the accuracy of mode shapes that are obtained by using the various computer codes or available methodologies. It also permits one to determine directly a frequency of vibration by using a small portion of a structure and constructing a dynamically equivalent system. This procedure does not require knowledge of the preceding frequencies of vibration.

Commonly used methods of vibration analysis are discussed in Chapter 5. They include the methods of Rayleigh, Stodola, and Myklestad, as well as an extensive treatment of transfer matrices. A discussion of the application of the dynamic hinge concept for mode shape diagnosis is found in this chapter. The very extensively used finite element method is covered in Chapter 6. Various aspects regarding the use of this method are discussed, and element stiffness and mass matrices for various beam and plate elements are derived. Vibration and dynamic response is obtained by using direct integration, modal analysis, exact solution, and the Newmark–Beta method. Element completeness and compatibility, as well as higher order effects, are also examined.

The utilization and development of Lagrange's equation, as well as the utilization of modal analysis for the dynamic response of various systems subjected to external dynamic excitations, are discussed in Chapter 7. Modal equations for spring–mass systems and systems with continuous mass and elasticity, such as beams and plates, are derived. These equations are then used to determine the dynamic response of such systems, when they are subjected to dynamic external excitations. The analysis also includes the development of modal equations for earthquake response and their application to various practical problems.

Chapter 8 provides an extensive treatment regarding the linear and nonlinear responses of members with continuously varying stiffness along their lengths. Static and vibration response of such members is determined by including both small and large deformations by utilizing the author's method of equivalent systems. This method, as developed by the author and his collaborators, provides unique, rigorous solutions of very complicated problems by using exact, or very accurate, simplified equivalent systems. The approach provides very accurate and reliable solutions to very complicated linear and nonlinear engineering problems, by using relatively simple equivalent mathematical models. This methodology also includes the vibration response of inelastic members.

Chapter 9 is devoted to the dynamic response of systems with continuous mass and elasticity, which can be idealized as spring–mass systems, by including the effects of damping. Fourier and Laplace transforms are used in the analysis of such systems. The analysis also includes machine foundations that are subjected to external excitations and viscous damping. This work, together with the convolution integral, prepares the groundwork for the utilization of stochastic approaches.

The development and application of variational and stochastic methods of analysis are discussed in Chapter 10. Both methodologies are used extensively in the solution of numerous complicated engineering problems and deserve appropriate attention. Necessary conditions for functionals with fixed and movable boundaries are examined, with application to practical engineering problems. Basic aspects of random vibrations, expectations, and correlation functions, as well as power spectra analysis, are included in this chapter. It also covers the response of various systems to random excitations and the utilization of probability in design processes. The last part of the chapter includes the utilization of power spectra analysis for concrete material response by using the recorded acoustic emissions of concrete material specimens as they are gradually loaded to failure. Such methodologies may be applied to many types of materials and can be used to evaluate existing materials, as well as materials that need to be developed for advanced future applications.

The last chapter is devoted to dimensional and model analyses. It includes principles of dimensional analysis, the Pi theorem, and the prediction equation, as well as the application of the prediction equation and the preparation of appropriate scale models, which can accurately predict the response of the prototype in the required aspect. Such models, when possible, are often used to determine the dynamic and vibration response of complicated systems, and the student, as well as the practicing engineer, should be made familiar with the advantages, or sometimes the disadvantages, resulting from the utilization of scale models. Suitable practical applications are included, in order to pinpoint the merits and importance of such models and their analysis.

Several appendices are also included at the end of the text, which complete or facilitate the comprehension and application of the various theories included in the main text. They include graphs, tables, and mathematical treatments, as well as basic principles of mechanics and computer programs. Answers to selected problems for the ones listed at the end of each chapter are also included at the end of the text.

The material presented in this text is fully class-tested through the many years of the author's teaching experience at Wayne State University, the University of Iowa, and the University of Akron. It also represents the author's long and extensive experience with both academia and industry and incorporates valuable suggestions from both graduate and undergraduate students, in addition to suggestions from practicing engineers in industry and government who collaborated with the author on various projects. Since industry and government employ a good portion of graduating engineers, the universities should train the engineering students in a way that meets the demands of these sectors.

I wish to thank my graduate and undergraduate students for their help in locating printing errors and for their valuable suggestions regarding the material included in the text. I am indebted to Dr. Paul A. Bosela, of Cleveland State University, for his major contributions to the preparation and writing of Chapter 6. My special thanks and gratitude go to my wife, Anna, for her constant encouragement during the writing of this text and for her assistance in the typing of the manuscript. Thanks are also due to the Dean of the College of Engineering, Dr. Nicholas Sylvester, and the University of Akron, for approving a semester faculty improvement leave that allowed me to devote all my time to the writing of the text. Special thanks are also due to Drs. M. Adams, Professor of Mechanical and Aerospace Engineering at Case Western Reserve University; J. Padovan, Distinguished Professor of Mechanical Engineering at the University of Akron; and F. Shaker, Deputy Chief of Engineering Directorate of NASA Lewis Research Center, for their valuable suggestions and discussions during the writing of the manuscript. Last but not least, I wish to thank John Wiley & Sons for making my work available to the academic and professional audience. In particular, I wish to thank Bob Argentieri, Engineering Editor; Donna Conte, Associate Managing Editor; Minna Panfili, Editorial Assistant; and their staff, for their fine work in the copyediting and general handling of the manuscript.

DEMETER G. FERTIS

Akron, Ohio
March 1994

CONTENTS

MECHANICAL AND STRUCTURAL VIBRATIONS

1 Fundamentals of Vibratory Motions

1.1 INTRODUCTION

The analysis and design of structural and mechanical systems to resist the effects of vibratory motions and time-dependent forces dated back to the 17th century [1], when great inquirers like Galileo, Huygens, and Newton proposed relationships and laws that ultimately formed the basis of the subject. The laws of motion, as stated by Newton in 1687 in his *Philosophiae Naturalis Principia Mathematica*, formalized the thoughts of the times on the subject and constituted its basic concepts.

The dynamics of periodic motion and mechanical vibrations was treated in its most rudimentary form when Galileo studied the motion of the pendulum and observed certain relationships between the period and amplitude of vibration. In the 19th century, problems of vibrating shafts in ships indicated the need for more extensive knowledge in the theory of periodic motion, in order to provide adequate design information for the shipbuilders.

The advent of the 20th century with its great technological advances, and the ever evolving challenges for the development of new technologies that meet the needs of the 21st century, provided the necessary impetus to study in greater detail the phenomena of both linear and nonlinear vibratory motions and their effect on both earth and space structures and machines. The solution of the vibratory problems associated with the interaction of the solar arrays of the space station Freedom and its main truss and the design of the overall space station system control are examples of such challenges.

In this chapter, fundamental aspects regarding the vibratory motions of both linear and nonlinear systems are examined, in order to prepare the reader for the more advanced topics that are treated in following chapters of the text.

1.2 FUNDAMENTAL ASPECTS OF PERIODIC MOTIONS OF LINEAR SYSTEMS

The field of engineering vibrations can be thought of as a special application of the principles of dynamics to bodies performing some kind of repetitive

1

motion. Therefore, Newton's laws of motion can be used to study such vibratory motions. Vibration, in a general sense, can be classified as either free or forced. A *free vibration* could be defined as a repetitive motion of a body that takes place about its static equilibrium position. The motion is performed under the action of forces inherent in the system itself and without the benefit of fluctuating external forces of any kind. The force that tends to move the mass of the body toward the neutral position is called the *restoring force*. Therefore, it becomes clear that any freely vibrating system must possess both mass and some kind of a restoring force. *Forced vibration*, on the other hand, is a vibratory motion imposed on a body or system by the action of an external excitation. In this section, fundamental aspects of free vibrations are discussed.

Any irregular motion of a particle about some fixed position of equilibrium may be thought of as a vibration. The displacements associated with such types of vibratory motions are time dependent, and they are measured from the static equilibrium position of the particle. If these displacements are repetitive and their repetition is executed at equal intervals of time with respect to an equilibrium position, the resulting motion is said to be *periodic*. The time τ required for the motion to be repeated is called the *period*. The *frequency of vibration*, f, is defined as the number of cycles per unit of time. In equation form, we write

$$f = \frac{1}{\tau} \tag{1.1}$$

A *cycle* is defined as a complete motion during the period τ of vibration. Also, the frequency of vibration is often designated as the number of radians per unit of time, and it is usually represented by the Greek letter ω. On this basis, we can relate f and ω by the equation

$$f = \frac{\omega}{2\pi} \tag{1.2}$$

and τ and ω by the expression

$$\tau = \frac{1}{f} = \frac{2\pi}{\omega} \tag{1.3}$$

Fortunately, the free vibration of many engineering structural and mechanical systems is usually harmonic. The assumption of harmonic vibration permits a problem to be expressed mathematically in terms of sine and

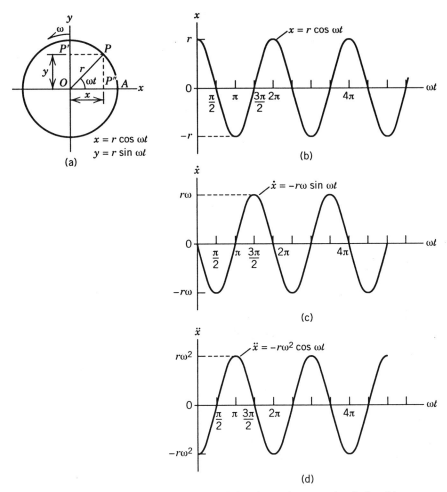

FIGURE 1.1. (a) Particle P moving around the circumference of a circle with constant angular velocity ω. (b) Displacement curve. (c) Velocity curve. (d) Acceleration curve.

cosine functions, which greatly simplifies its solution. A harmonic motion is characteristically periodic. However, the opposite is not true, because periodic motions do not have to be harmonic.

Harmonic motion can be described by the motion of the projection P' of the particle P, on the vertical diameter of the circle in Fig. 1.1a, moving at constant angular velocity ω around the circumference of the circle of radius r. Similar observations can be made by considering the projection P'' of P on the horizontal diameter of the circle. If at any time $t = 0$ the particle P

is at A, the angular velocity θ at any time t is $\theta = \omega t$. Hence, the displacement y of the projection P', measured from the origin O, is

$$y = r \sin \omega t \tag{1.4}$$

The linear velocity $\dot{y}$ of P' is

$$\dot{y} = \frac{dy}{dt} = r\omega \cos \omega t \tag{1.5}$$

and the linear acceleration $\ddot{y}$ is

$$\ddot{y} = \frac{d^2 y}{dt^2} = -r\omega^2 \sin \omega t = -\omega^2 y \tag{1.6}$$

or

$$\ddot{y} + \omega^2 y = 0 \tag{1.7}$$

Equation (1.7) is the differential equation of motion of P' along the vertical diameter of the circle with the specified initial conditions. The general solution y of this equation is

$$y = A \sin \omega t + B \cos \omega t \tag{1.8}$$

The constants of integration A and B may be determined by using the initial conditions of the motion. Such considerations are examined in detail later in this text.

The displacement, velocity, and acceleration of P', as given by Eqs. (1.4), (1.5), and (1.6), are plotted, respectively, in Figs. 1.1b, 1.1c, and 1.1d. The shapes of the three curves in Fig. 1.1 are exactly the same, except that they are displaced relative to each other along the ωt axis. This relative difference is called the *phase angle* between these three curves, and it is designated by the Greek letter ϕ. The curves show that the velocity $\dot{y}$ is 90° ($\phi = 90°$) ahead of the displacement vector, and the acceleration $\ddot{y}$ is 180° ($\phi = 180°$) ahead of the displacement vector. For example, at $\omega t = 90°$, Eqs. (1.4), (1.5), and (1.6) yield $y = r$, $\dot{y} = 0$, and $\ddot{y} = -\omega^2 r$, which verifies the preceding statement. Similar conclusions can be made by considering the projection $x = r \cos \omega t$ of P on the horizontal diameter of the circle in Fig. 1.1a.

Equation (1.8) may also be written in terms of a constant C and the phase angle ϕ by substituting for A and B the expressions

$$A = C \cos \phi \quad \text{and} \quad B = C \sin \phi \tag{1.9}$$

On this basis, Eq. (1.8) yields

$$y = C(\cos \phi \sin \omega t + \sin \phi \cos \omega t)$$

or

$$y = C \sin(\omega t + \phi) \tag{1.10}$$

We also have

$$\tan \phi = \frac{B}{A} \tag{1.11}$$

and

$$A^2 + B^2 = C^2(\cos^2 \phi + \sin^2 \phi)$$

or

$$C = \sqrt{A^2 + B^2} \tag{1.12}$$

Differentiation of Eq. (1.10) with respect to time t yields the following expressions for the velocity $\dot{y}$ and acceleration $\ddot{y}$:

$$\dot{y} = C\omega \cos(\omega t + \phi) \tag{1.13}$$

$$\ddot{y} = -C\omega^2 \sin(\omega t + \phi) \tag{1.14}$$

1.3   DEGREES OF FREEDOM AND MODES OF VIBRATION

The *degrees of freedom* of a freely vibrating body may be defined by the number of independent coordinates that are required to identify its displacement configuration during vibration. For example, a rigid block supported by a linear spring of stiffness k can vibrate in six different ways; consequently, it can have six degrees of freedom. It can translate in the direction of its three x, y, and z principal axes, and it can also rotate about the same three axes. Therefore, we need three displacement coordinates associated with the x, y, and z directions and three angular displacements with respect to the same axes in order to be able to define the position of the particle at any time t during vibration.

Each of these possibilities of free vibration is defined as a *mode of vibration*, and the natural frequency associated with each mode is independent of the frequency of the other modes. On this basis, it can be stated that the *natural frequencies* of a freely vibrating body are equal in number to its degrees of freedom and that there is a mode shape associated with each frequency.

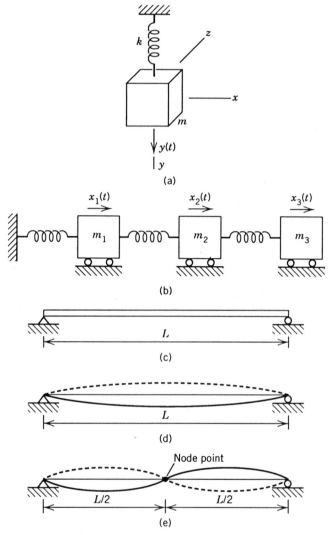

FIGURE 1.2. (a) Spring–mass system. (b) Spring–mass system restricted to move in the horizontal direction only. (c) Simply supported beam. (d) Fundamental mode of vibration of the simply supported beam. (e) Second mode of vibration of the simply supported beam.

If the mass of the spring–mass system in Fig. (1.2a) is restricted to move in the vertical direction only, then this system has one degree of freedom, and the single coordinate $y(t)$ shown in the figure is sufficient to define completely the position of the mass m during vertical vibration. The spring–mass system in Fig. 1.2b, which is restricted to move in the horizontal direction only, has three degrees of freedom, because the displacements $x_1(t)$, $x_2(t)$, and $x_3(t)$ must be known in order to determine its displacement configuration during motion.

An elastic body, such as the simply supported beam in Fig. 1.2c, can have infinite modes of vibration and consequently an infinite number of degrees of freedom. In each mode, all particles making up the elastic body vibrate at the same free frequency of vibration. Since the particles are infinite in number, it requires an infinite number of coordinates to determine the position of each particle during vibration. The modes of vibration are also infinite in number, because there is an infinite number of ways in which these particles can line themselves up during free vibration and form a mode shape configuration. Figures 1.2d and 1.2e show the shapes of the first two modes of vibration in the transverse direction for the simply supported beam in Fig. 1.2c. The lowest frequency of vibration, represented by the mode shape in Fig. 1.2d, is called the *fundamental frequency*, and the higher frequencies, such as the second one that is represented by the mode shape in Fig. 1.2e, are termed *harmonics*. In many problems, especially those dealing with structures, the fundamental mode of vibration is of particular importance because the amplitudes of vibration would be the largest, and it is often used to define the flexibility of the structure. The rigidity of the structure is a function of its free frequency of vibration. The larger the frequency the stiffer the structure would be.

A *node* is defined as any point having zero amplitude on the unsupported length of a vibrating elastic body. An elastic body vibrating at its fundamental frequency, such as the one in Fig. 1.2d, has no nodes. The second mode of vibration, like the one in Fig. 1.2e, has one node; the third mode of vibration has two nodes, and so on. As the number of nodes increases, the amplitude of vibration decreases. As the number of nodes becomes infinitely large, the amplitude approaches zero in the limit.

1.4   DAMPED FREE VIBRATIONS

Damping may be thought of as a force that resists the motion at all times, and such forces are always present in practical systems. The amplitude of vibration of a damped structure, vibrating freely, may be attenuated by such resisting forces developing during motion, since it is not subject to external exciting forces to sustain the vibration. The resisting forces dissipate energy, and in time the vibrations dies out. This phenomenon is known as damping. The causes for such actions are many, but the associated energy losses are often small and they could be neglected in the analysis of many engineering structures.

The three most common types of damping are *viscous*, *Coulomb*, and *hysteresis*. Viscous damping is fairly common and occurs, for example, when bodies are vibrating in a fluid medium at rather low velocities. The resisting

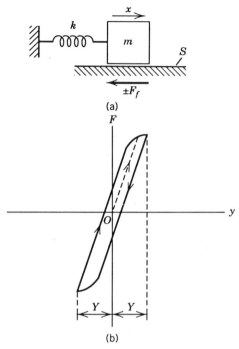

FIGURE 1.3. (a) Spring–mass system vibrating on a dry surface S. (b) Force–displacement diagram (hysteresis loop) for a spring–mass system.

force $F_d(t)$ that is developed from such damping is considered to be proportional to the velocity of the oscillatory motion. In mathematical form,

$$F_d(t) = -c\frac{dx}{dt} = -c\dot{x} \tag{1.15}$$

In the above equation, $\dot{x}$ is the velocity of the body, and c is the damping constant, which is defined as the resisting force that is developed per unit of velocity.

Coulomb damping, which is also known as dry friction damping, is the result of rubbing and sliding of vibrating bodies on dry surfaces. For example, the vibratory motion in the horizontal direction of the spring–mass system in Fig. 1.3a is under the influence of Coulomb damping, because during vibration the mass m is rubbing against the dry surface S. The resisting force F_f that is developed from such damping is given by the expression

$$F_f = \pm\mu N \tag{1.16}$$

where N is the force that is normal to the surface S, and μ is the coefficient of kinetic friction of the material of the surface S.

Hysteresis damping, which is also known as solid damping or structural damping, is the result of internal friction in the material of the body during its oscillatory motion, that is, when the elastic body deforms during its oscillatory motion, frictional forces are developing inside the body from the friction between internal planes that slip and slide during deformation. This type of resistance is independent of the frequency of vibration, but it is considered to be approximately proportional to the amplitude of the deformed elastic body.

By considering a cycle of motion of a vibrating single-degree-of-freedom spring–mass system and plotting the force–displacement diagram shown in Fig. 1.3b, the area within the loop of this diagram represents the amount of energy ΔU transformed into heat, per cycle of motion, due to the internal friction in the material. In this figure, F is the spring force, y is the displacement of the mass, and Y is the amplitude of its vibration. Experiments have shown that ΔU could be obtained from the approximate expression

$$\Delta U = k \pi c_0 Y^2 \tag{1.17}$$

In this expression, c_0 is the dimensionless constant of the material for solid damping, and k is the spring constant, which is defined as the required force to deflect the spring by an amount equal to unity.

Each of the above types of damping occurs in nature in an approximate way, but, by a judicious combination of these three types, it is often possible to get a better representation of actual phenomena than by using only one. The amplitudes of free vibration under the influence of viscous damping die out exponentially. With constant friction, the reduction in amplitude is linear and the system will come to rest in finite time. When the free vibration is under the influence of hysteresis damping, the rate of the reduction of its amplitude depends on the size of the area within the hysteresis loop. Rubber-type materials, for example, exhibit a much wider loop, and consequently a much larger area, when they are compared to metallic materials such as steels. This is one good reason why rubber materials are often used in practice as dampers.

1.5   FORCED VIBRATION AND FORCE FUNCTIONS

It was stated in Section 1.2 that vibrations are usually classified as free or forced. In the case of forced vibration, the body is subjected to external force functions that make it vibrate with the frequency of the exciting force. An alternating external force system may arise as a consequence of many natural phenomena such as waves, sound, blast, earthquake, and heavy vehicular traffic on highway pavements and bridges, as well as from any

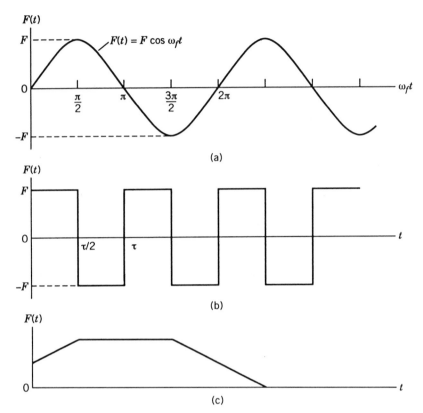

FIGURE 1.4. (a) Periodic and harmonic forcing function. (b) Periodic forcing function. (c) Nonperiodic forcing function.

mechanically produced causes. In each case the wave motion of the disturbance will vibrate a structure at the frequency of the oscillating force. A condition of *resonance* will occur if the frequency of the applied force system coincides with one of the natural free frequencies of the body. At the resonant condition, the amplitude of vibration will approach infinity with time. In practical situations, however, the amplitude of vibration may exceed allowable values in a short period of time, with the subsequent loss of structural integrity.

In a general sense, forcing functions may be classified as *periodic, nonperiodic, random,* or *impulsive.* A periodic force is one that repeats itself at equal intervals of time. A special type of periodic force is a harmonic one, where its time variation may be represented by a sine or a cosine function. An example of such a force is plotted in Fig. 1.4a, where the forcing function

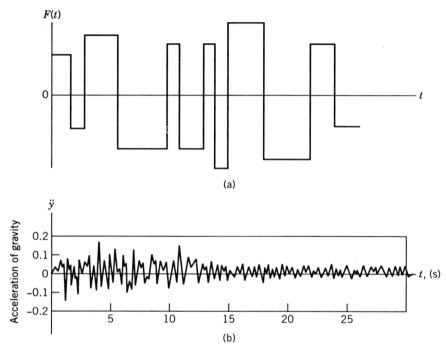

FIGURE 1.5. (a) Random forcing function. (b) Random ground earthquake acceleration.

is $F(t) = F \sin \omega_f t$ and ω_f is the frequency of the force in radians per second (rps). A periodic force that is not harmonic is plotted in Fig. 1.4b, where τ is the time required for the force to complete one cycle of motion.

A nonperiodic force is a force of some general type that does not repeat itself with time. The force plotted in Fig. 1.4c is a nonperiodic forcing function.

Since "random" implies unpredictability, a random forcing function would be one of a very irregular form in terms of both time and amplitude. These types of excitation involve the element of prediction, and they are usually called *nondeterministic*. Random excitations produced by earthquake motions, blast, or wind gust are nondeterministic. A simple type of random force is shown in Fig. 1.5a. A plot of the random ground accelerations produced by a major earthquake is shown in Fig. 1.5b.

An impulsive force $F(t)$ can be thought of as an instantaneously applied force of large magnitude and having only an instantaneous time duration. A unit impulse is obtained if the area within the force–time curve is unity. Mathematically, a unit impulse is represented by the delta Dirac function

$\delta(t)$, defined as

$$\delta(t) = 0 \quad \text{for } t \neq 0 \tag{1.18}$$

$$\int_{-\infty}^{\infty} \delta(t)\, dt = 1 \tag{1.19}$$

The Dirac function may be visualized geometrically as a spike of unlimited height and vanishing width, occurring at the origin of its argument t. The width vanishes and the height grows in a relative way so that the bounded area is constant and equal to unity. For example, a rectangle of width a and height $1/a$ becomes a Dirac function in the limit $a \to 0$.

1.6   VIBRATIONAL ASPECTS OF NONLINEAR SYSTEMS

The discussion in the preceding five sections was limited to linear systems, where vibration analysis of such systems may be carried out by linear analysis. For linear systems, such as the ones shown in Fig. 1.2, it is always assumed that the restoring force is proportional to the resulting deformation. For example, if the mass of the spring–mass system in Fig. 1.2a vibrates in its vertical y direction, the force in the spring is proportional to the deformation of the spring. In Eq. (1.15), the resisting force $F_d(t)$ due to viscous damping was assumed to be proportional to the velocity $\dot{x}$ of the oscillatory motion. For such systems, the differential equations of motion are linear, and linear analysis may be used to determine their linear response. However, there are physical situations in which linear differential equations are not sufficient to describe the vibrational behavior of such systems, and nonlinear differential equations of motion must be developed.

For example, the free vibration of a flexible cantilever beam, such as the one shown in Fig. 1.6a, may be characterized as a nonlinear vibration problem that requires nonlinear methods of analysis to determine its vibration response (see also ref. 2). In this figure, y_s is the static equilibrium position of the flexible member, and y_d represents the amplitudes of vibration. Since the member vibrates in its transverse direction with respect to its static equilibrium position y_s, nonlinear analysis is required to determine its vibratory motion. Since the static deformation y_s is large, its magnitude is a function of the large deformation of the member, and, consequently, nonlinear analysis is required to determine y_s. If, on the other hand, the vibration amplitudes y_d are also large, then the frequencies of vibration are functions of y_d and they are not harmonic—a situation that further complicates the solution of such problems. For purposes of illustration, the computed first three mode shapes of this beam are shown in Fig. 1.6b. In the solution of this problem, the static amplitudes y_s are large, but the vibration amplitudes

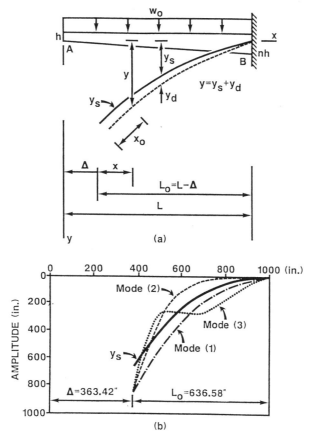

FIGURE 1.6. (a) Deformation configuration of a flexible cantilever beam undergoing free vibration. (b) First three modes of vibration of the cantilever beam.

y_d are assumed to be small. It should be pointed out here that the principle of superposition, which is commonly used for linear systems, does not apply for nonlinear systems. For example, the static displacement y_s in Fig. 1.6a is not proportional to the attached weight w_0.

1.7 DIFFERENTIAL EQUATIONS OF MOTION OF LINEAR SPRING–MASS SYSTEMS

Consider the linear spring-mass system shown in Fig. 1.7a, whose mass m is restricted to move in the vertical direction only and under the influence of viscous damping. The time-varying force acting on m is $F(t)$, k is the spring constant defined as the force required to stretch the spring by an amount equal to unity, y is the vertical displacement of m measured from the equilib-

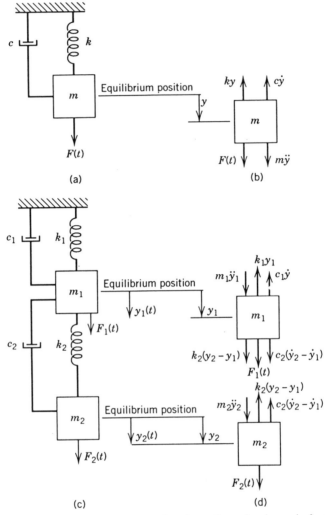

FIGURE 1.7. (a) Spring–mass system under the action of a dynamic force $F(t)$ and viscous damping. (b) Free-body diagram of mass m. (c) Two-degree spring–mass system under the action of forces $F_1(t)$ and $F_2(t)$ and viscous damping. (d) Free-body diagrams of masses m_1 and m_2.

rium position, and c is the damping constant. The position of m at any time t is completely determined if $y(t)$ is known, and, consequently, this system has only one degree of freedom.

The differential equation of motion for this system can be derived by using the free-body diagram in Fig. 1.7b and applying Newton's second law of motion. In this free-body diagram, ky is the force in the spring, $c\dot{y}$ is the resisting force due to viscous damping, $m\ddot{y}$ is the inertia force, and $F(t)$ is the applied force. By applying Newton's second law of motion, we have

$$m\ddot{y} = F(t) - ky - c\dot{y} \tag{1.20}$$

or

$$m\ddot{y} + c\dot{y} + ky = F(t) \tag{1.21}$$

where $\ddot{y}$ denotes the second derivative with respect to time. Equation (1.21) is a linear, nonhomogeneous, second-order differential equation with constant coefficients. Its solution will yield the displacement function $y(t)$.

If the time-varying force $F(t)$ is zero, Eq. (1.21) yields

$$m\ddot{y} + c\dot{y} + ky = 0 \tag{1.22}$$

which is a second-order, linear, homogeneous differential equation with constant coefficients, and represents the free vibration of the one-degree spring–mass system under the influence of viscous damping.

In the case where the viscous damping force $c\dot{y}$ is also zero, we have

$$m\ddot{y} + ky = 0 \tag{1.23}$$

This homogeneous differential equation of motion represents the undamped free vibration of the spring–mass system.

The spring–mass system in Fig. 1.7c, which is restricted to move in the vertical direction only, has two degrees of freedom, because it requires the two independent coordinates y_1 and y_2 to locate the positions of masses m_1 and m_2, respectively, during motion. The system is again assumed to be under the influence of viscous damping, where c_1 and c_2 are the damping constants. The free-body diagrams in Fig. 1.7d depict the forces acting on masses m_1 and m_2. The coordinates y_1 and y_2 are again measured from the equilibrium positions of m_1 and m_2, respectively. By using the indicated free-body diagrams of m_1 and m_2 and applying Newton's second law of motion, we have

$$m_1\ddot{y}_1 = -k_1 y_1 - c_1\dot{y}_1 + k_2(y_2 - y_1) + c_2(\dot{y}_2 - \dot{y}_1) + F_1(t) \tag{1.24}$$
$$m_2\ddot{y}_2 = -k_2(y_2 - y_1) - c_2(\dot{y}_2 - \dot{y}_1) + F_2(t) \tag{1.25}$$

or

$$m_1\ddot{y}_1 + k_1 y_1 + c_1\dot{y}_1 - k_2(y_2 - y_1) - c_2(\dot{y}_2 - \dot{y}_1) = F_1(t) \tag{1.26}$$
$$m_2\ddot{y}_2 + k_2(y_2 - y_1) + c_2(\dot{y}_2 - \dot{y}_1) = F_2(t) \tag{1.27}$$

Equations (1.26) and (1.27) are the differential equations of motion of the two-degree spring–mass system. Simultaneous solution of these two equations will yield the displacement functions $y_1(t)$ and $y_2(t)$ for masses m_1 and m_2, respectively.

The free vibration under the influence of viscous damping is obtained when the time-varying forces $F_1(t)$ and $F_2(t)$ are zero. In this case, the two homogeneous differential equations representing this vibratory motion are as follows:

$$m_1 \ddot{y}_1 + k_1 y_1 + c_1 \dot{y}_1 - k_2(y_2 - y_1) - c_2(\dot{y}_2 - \dot{y}_1) = 0 \qquad (1.28)$$

$$m_2 \ddot{y}_2 + k_2(y_2 - y_1) + c_2(\dot{y}_2 - \dot{y}_1) = 0 \qquad (1.29)$$

If, on the other hand, the damping forces are zero, Eqs. (1.28) and (1.29) yield

$$m_1 \ddot{y}_1 + k_1 y_1 - k_2(y_2 - y_1) = 0 \qquad (1.30)$$

$$m_2 \ddot{y}_2 + k_2(y_2 - y_1) = 0 \qquad (1.31)$$

Equations (1.30) and (1.31) represent the free undamped vibration of the two-degree spring–mass system.

In the above derivations, it was assumed that the elastic springs connecting the masses are linear and of negligible mass compared to the heavy rigid masses m, m_1, and m_2. In addition, it was also assumed that the time-varying forcing functions acting on the masses are low-frequency forcing functions. If, for example, the frequency of $F(t)$ is of high magnitude that can excite the natural free modes of the spring itself, the system in Fig. 1.7a can no longer be analyzed as a one-degree spring–mass system. The distributed mass and elasticity of the spring should be taken into consideration in the analysis.

The differential equations of motion derived above could also be obtained by using *D'Alembert's principle*. This principle was introduced by D'Alembert in 1743, and it is used extensively in the fields of vibration and dynamics. This principle will be applied to the spring–mass system in Fig. 1.8a. The mass m of the spring is subjected to the harmonic force $F(t) = F \cos \omega_f t$, where ω_f is the frequency of the force, and it moves under the influence of viscous damping of damping constant c as shown. For comparison purposes, both Newton's second law of motion and D'Alembert's principle will be applied for the derivation of the differential equation of motion.

At an amplitude y from the equilibrium position, the free-body diagram of m is shown in Fig. 1.8b. Since the mass m is assumed to be moving downward with a downward acceleration $\ddot{y}$, the inertia force $m\ddot{y}$ would also be acting downward. The differential equation of motion may be derived by equating the inertia force $m\ddot{y}$ to the sum of the forces acting in the direction of $m\ddot{y}$, as before, which is Newton's second law of motion. This yields

$$m\ddot{y} + c\dot{y} + ky = F \cos \omega_f t \qquad (1.32)$$

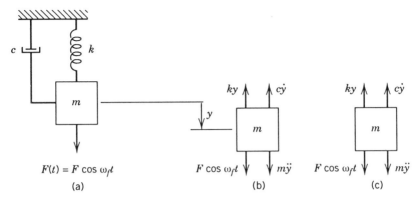

FIGURE 1.8. (a) Spring–mass system under the action of a harmonic force and viscous damping. (b) Free-body diagram of mass m in accordance with Newton's second law of motion. (c) Free-body diagram of mass m in accordance with D'Alembert's principle.

If the sense of the inertia force $m\ddot{y}$ is reversed, as shown in Fig. 1.8c, then D'Alembert's principle of dynamic equilibrium may be applied to determine the differential equation of motion of the spring–mass system. This is done in the same way as statics, by summing up all the forces acting in the vertical direction on the mass m in Fig. 1.8c and setting them equal to zero. If upward forces are assumed positive, the following result is obtained:

$$m\ddot{y} + c\dot{y} + ky - F\cos\omega_f t = 0 \qquad (1.33)$$

or

$$m\ddot{y} + c\dot{y} + ky = F\cos\omega_f t \qquad (1.34)$$

which is identical to Eq. (1.32).

By using Newton's second law of motion or D'Alembert's principle, the differential equations of motion of other types of linear spring–mass systems may be obtained. The above examples serve only as guides for the application of the general methodology.

1.8 DIFFERENTIAL EQUATIONS OF MOTION FOR TORSIONAL VIBRATIONS OF LINEAR DISK–SHAFT SYSTEMS

In practical situations, there is a large number of structural elements and machine parts that are subjected to torsional oscillatory motions. For

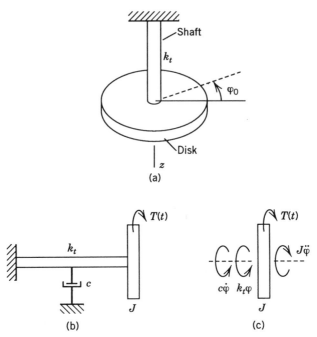

FIGURE 1.9. (a) Disk–shaft system. (b) Disk–shaft system under the influence of viscous damping. (c) Free-body diagram of the disk.

example, the disk–shaft system in Fig. 1.9a will vibrate freely about its centroidal z axis if the disk is rotated about this axis by an amount φ_0 and then released suddenly. In the figure, k_t denotes the *torsional spring constant* of the shaft, defined as the twisting moment required to twist the shaft an angular twist φ equal to unity, say, 1 radian. Hence, the units of k_t would be in·lb/rad for the English system and N·m/rad for the metric system of units. In the following discussion, differential equations of motion for disk–shaft systems are derived.

Consider the disk–shaft system in Fig. 1.9b and assume that it is subjected to an oscillatory torque $T(t)$ as shown. The motion of the shaft is under the influence of viscous damping of torsional viscous damping constant c, where c is defined as the resisting torque per unit velocity. In the figure, the symbol J is used to denote the mass moment of inertia of the disk about the axis of rotation, which in this case is the centroidal axis of the shaft.

The free-body diagram of the disk is shown in Fig. 1.9c, where φ is the angular twist of the disk and shaft, $\dot{\varphi}$ and $\ddot{\varphi}$ are the first and second derivatives of φ with respect to time t, $J\ddot{\varphi}$ and $c\dot{\varphi}$ are the inertia and damping torques, respectively, and $k_t\varphi$ is the shaft torque. By applying Newton's second law of motion, we find

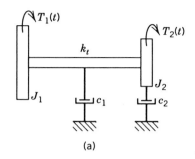

(a)

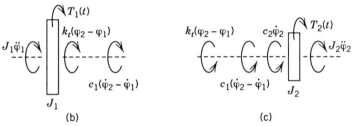

(b)   (c)

FIGURE 1.10. (a) Disk–shaft system under the influence of viscous damping and torques $T_1(t)$ and $T_2(t)$. (b) Free-body diagram of disk of mass moment of inertia J_1. (c) Free-body diagram of disk of mass moment of inertia J_2.

$$J\ddot{\varphi} = -k_t\varphi - c\dot{\varphi} + T(t) \tag{1.35}$$

or

$$J\ddot{\varphi} + c\dot{\varphi} + k_t\varphi = T(t) \tag{1.36}$$

Equation (1.36) is the linear differential equation of motion of the disk–shaft system, and its solution will yield the angular time displacement $\varphi(t)$. If the applied torque $T(t)$ is zero, Eq. (1.36) yields

$$J\ddot{\varphi} + c\dot{\varphi} + k_t\varphi = 0 \tag{1.37}$$

which is the differential equation of motion for the free vibration of the system under the influence of viscous damping.

If there is no damping, $c\dot{\varphi}$ will be equal to zero, and Eq. (1.37) yields

$$J\ddot{\varphi} + k_t\varphi = 0 \tag{1.38}$$

This equation represents the free, undamped, torsional vibration of the disk–shaft system.

Consider now the system in Fig. 1.10a, which consists of two disks of mass moments of inertia J_1 and J_2, joined together by a shaft of torsional stiffness k_t. The disks are subjected to oscillatory torques $T_1(t)$ and $T_2(t)$ as shown. The motion of the shaft is under the influence of viscous damping of damping constant c_1, and the viscous damping constant for the disk with inertia J_2 is c_2. Since the system is composed of two disks, where each disk can assume an angular position that is independent of the other, we can say that the disk–shaft system in Fig. 1.10a has two degrees of freedom.

The angular displacements of the disks of mass moments of inertia J_1 and J_2 are denoted as φ_1 and φ_2, respectively. If φ_2 is assumed to be larger than φ_1, the angular twist of the shaft is $\varphi_2 - \varphi_1$. On this basis, the free-body diagrams of the disks would be as shown in Figs. 1.10b and 1.10c. By using the disk in Fig. 1.10b and applying Newton's second law of motion, we have

$$J_1\ddot{\varphi}_1 = T_1(t) + c_1(\dot{\varphi}_2 - \dot{\varphi}_1) + k_t(\varphi_2 - \varphi_1) \tag{1.39}$$

or

$$J_1\ddot{\varphi}_1 - c_1(\dot{\varphi}_2 - \dot{\varphi}_1) - k_t(\varphi_2 - \varphi_1) = T_1(t) \tag{1.40}$$

where $J_1\ddot{\varphi}_1$ is the inertia torque of the J_1 disk.

In a similar manner, by using the free-body diagram in Fig. 1.10c, we obtain

$$J_2\ddot{\varphi}_2 = T_2(t) - c_1(\dot{\varphi}_2 - \dot{\varphi}_1) - c_2\dot{\varphi}_2 - k_t(\varphi_2 - \varphi_1) \tag{1.41}$$

or

$$J_2\ddot{\varphi}_2 + c_1(\dot{\varphi}_2 - \dot{\varphi}_1) + c_2\dot{\varphi}_2 + k_t(\varphi_2 - \varphi_1) = T_2(t) \tag{1.42}$$

where $J_2\ddot{\varphi}_2$ is the inertia torque of the J_2 disk.

Equations (1.40) and (1.42) are the differential equations of motion of the two-degree disk–shaft system in Fig. 1.10a, and their simultaneous solution will yield the angular time displacements $\varphi_1(t)$ and $\varphi_2(t)$. If $T_1(t)$ and $T_2(t)$ are both zero, then Eqs. (1.40) and (1.42) yield the following pair of equations:

$$J_1\ddot{\varphi}_1 - c_1(\dot{\varphi}_2 - \dot{\varphi}_1) - k_t(\varphi_2 - \varphi_1) = 0 \tag{1.43}$$
$$J_2\ddot{\varphi}_2 + c_1(\dot{\varphi}_2 - \dot{\varphi}_1) + c_2\dot{\varphi}_2 + k_t(\varphi_2 - \varphi_1) = 0 \tag{1.44}$$

These two homogeneous differential equations represent the free torsional vibration of the disk–shaft system that oscillates under the influence of viscous damping.

The free, undamped, torsional vibration of the disk–shaft system may be represented by Eqs. (1.43) and (1.44) by making the damping constants c_1 and c_2 equal to zero. This yields the following pair of homogeneous differential equations:

$$J_1\ddot{\varphi}_1 - k_t(\varphi_2 - \varphi_1) = 0 \tag{1.45}$$

$$J_2\ddot{\varphi}_2 + k_t(\varphi_2 - \varphi_1) = 0 \tag{1.46}$$

The differential equations of motion of other similar types of disk–shaft problems may be obtained by following similar reasoning. However, it should be pointed out here that the above theory is reasonably accurate if the mass of each disk is large compared to the mass of the shaft — a condition often encountered in practical situations. If this condition cannot be satisfied, then the shaft should be treated as an elastic body with infinite degrees of freedom in order to obtain a satisfactory solution.

1.9 DIFFERENTIAL EQUATIONS OF MOTION FOR CONTINUOUS LINEARLY ELASTIC SYSTEMS

Systems with continuous mass and elasticity, such as beams, columns, frames, and shafts, have an infinite number of degrees of freedom because, during motion, an infinite number of coordinates are required to locate the position of the particles making up the deformed elastic body. In this section, differential equations of motion for continuous linearly elastic systems will be derived.

Beams and shafts, for example, can vibrate longitudinally as well as transversely, and they can also be subjected to torsional vibration. Longitudinal vibrations occur along the horizontal axis of the member and they can be either free or forced. For example, free longitudinal vibrations can be produced by the sudden release of an axially applied force, and a forced vibration is produced by the application of an axial harmonic force.

Transverse vibrations are the ones occurring in the transverse direction of a beam or shaft and about an equilibrium position. If an initial vertical displacement is applied to a member and then it is suddenly released, the member will vibrate freely in the transverse direction about its static equilibrium position. A harmonic forced vibration in the transverse direction is obtained by the application of a vertical harmonic force.

By similar reasoning, it may be stated that beams and shafts can also be subjected to torsional free vibration about their centroidal axis by the sudden release of an initial angular rotation about the centroidal axis, and they can be subjected to a forced torsional vibration by the application of a harmonic torque.

1.9.1 Free Longitudinal Vibration of Beams and Shafts

Consider the free-free prismatic member in Fig. 1.11a of length L, and let us determine the differential equation of motion of its free, undamped, longitudinal vibration. The cross-sectional area of the member is A, its modulus of elasticity is E, and ρ is the mass per unit volume. It is also assumed

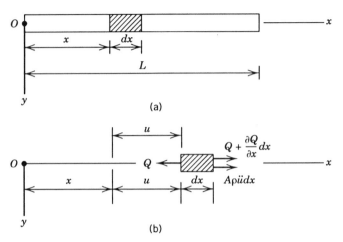

FIGURE 1.11. (a) Uniform elastic beam subjected to longitudinal vibration. (b) Free-body diagram of an element of the beam.

that the cross-sectional dimensions of the member are small compared to the wavelength of the longitudinal vibration and that there is no lateral deformation.

At a distance x from the origin O, an element of length dx of the member is shown in Fig. 1.11a. The free-body diagram of this element at a time t is shown in Fig. 1.11b. In this figure, u is the longitudinal displacement of the cross section measured from the static equilibrium position, Q and $Q + (\partial Q/\partial x)dx$ are the axial forces acting on the element, and $A\rho\ddot{u}\,dx$ is the inertial force. By applying Newton's second law of motion, we find

$$A\rho\ddot{u}\,dx = Q + \frac{\partial Q}{\partial x}dx - Q \qquad (1.47)$$

or

$$A\rho\ddot{u} = \frac{\partial Q}{\partial x}$$

Referring to the mechanics of solids, we note that the strain ε_x and stress σ_x of the member in the x direction are given as follows:

$$\varepsilon_x = \frac{\partial u}{\partial x} \quad \text{and} \quad \sigma_x = E\frac{\partial u}{\partial x} \qquad (1.48)$$

Therefore, the axial force Q can be written

$$Q + A\sigma_x = AE\frac{\partial u}{\partial x} \tag{1.49}$$

By substituting Eq. (1.49) into Eq. (1.47), we find

$$A\rho\ddot{u} = AE\frac{\partial^2 u}{\partial x^2}$$

or

$$E\frac{\partial^2 u}{\partial x^2} - \rho\frac{\partial^2 u}{\partial t^2} = 0 \tag{1.50}$$

Equation (1.50) is the differential equation of motion for the free, undamped, longitudinal vibration of the member. Its solution will yield the required information regarding its vibratory motion.

1.9.2 Transverse Vibration of Beams and Shafts

It was stated earlier that transverse vibrations of beams and shafts can be either free or forced. Their differential equations of motion are derived here by considering the member in Fig. 1.12a of length L, and an element of length dx located at a distance x from the origin O.

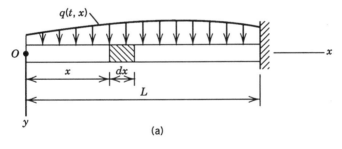

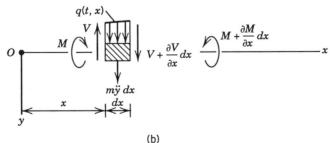

FIGURE 1.12. (a) Uniform elastic beam subjected to transverse motions. (b) Free-body diagram of an element of the beam.

To be somewhat more general, the exciting force on the beam is designated as $q(t, x)$, and it can vary with time as well as with x.

The free-body diagram of the element is shown in Fig. 1.12b. In this figure, M and $M + (\partial M/\partial x)\, dx$ are the dynamic moments acting on the sides of the element, V and $V + (\partial V/\partial x)\, dx$ are the dynamic shear forces, y is the vertical dynamic displacement, $\ddot{y}$ is the acceleration in the y direction, m is the mass per unit length of the member, and $m\ddot{y}\, dx$ is the inertia force.

By applying Newton's second law of motion, we find

$$m\ddot{y}\, dx = q(t, x)dx - V + V + \frac{\partial V}{\partial x}\, dx$$

or

$$m\ddot{y} - \frac{\partial V}{\partial x} = q(t, x) \tag{1.51}$$

By using the expression

$$V = \frac{\partial M}{\partial x} = -\, EI\frac{\partial^3 y}{\partial x^3} \tag{1.52}$$

and substituting into Eq. (1.51), we find

$$m\ddot{y} - \frac{\partial}{\partial x}\left(-\, EI\frac{\partial^3 y}{\partial x^3}\right) = q(t, x)$$

or

$$EI\frac{\partial^4 y}{\partial x^4} + m\frac{\partial^2 y}{\partial t^2} = q(t, x) \tag{1.53}$$

Equation (1.53) is the differential equation of motion of the member. For a given force function $q(t, x)$, the solution of this equation yields the vertical time displacements $y(t, x)$ of the member. If $q(t, x)$ is zero, Eq. (1.53) yields the expression

$$EI\frac{\partial^4 y}{\partial x^4} + m\frac{\partial^2 y}{\partial t^2} = 0 \tag{1.54}$$

which is the differential equation of the free undamped vibration of the member.

1.9.3 Torsional Vibration of Circular Beams or Shafts

Consider the circular shaft in Fig. 1.13a and an element of length dx located at a distance x from the origin O. The length of the shaft is L, φ is the angle of twist, and G is the shear modulus of its material.

The free-body diagram of the element is shown in Fig. 1.13b. In this figure, T and $T + (\partial T/\partial x)dx$ are the dynamic torques, J is the mass moment of inertia of the element about the axis of rotation, and $J\ddot{\varphi}$ is the inertia torque. By applying Newton's second law of motion, we find

$$J\ddot{\varphi} = T + \frac{\partial T}{\partial x} dx - T$$

or

$$J\ddot{\varphi} - \frac{\partial T}{\partial x} dx = 0 \qquad (1.55)$$

From the mechanics of materials we know that

$$\frac{\partial \varphi}{\partial x} = \frac{T}{GI_p} \qquad (1.56)$$

where I_p is the polar moment of inertia of the circular cross section. We also

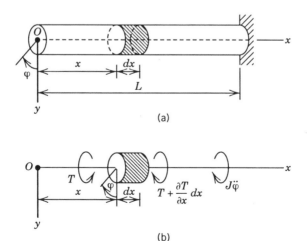

FIGURE 1.13. (a) Uniform elastic circular shaft subjected to torsional motion. (b) Free-body diagram of an element of the shaft.

know that

$$J = \rho \, dx \, I_p \qquad (1.57)$$

where ρ is the mass per unit volume.

By using Eqs. (1.56) and (1.57) and substituting into Eq. (1.55), we obtain

$$\rho \, dx \, I_p \ddot{\varphi} - \frac{\partial^2 \varphi}{\partial x^2} GI_p \, dx = 0$$

or

$$\frac{\partial^2 \varphi}{\partial x^2} - \frac{\rho}{G} \frac{\partial^2 \varphi}{\partial t^2} = 0 \qquad (1.58)$$

Equation (1.58) is the differential equation of motion for the free, un-damped, torsional vibration of the shaft.

The differential equations of motion for other types of problems having continuous mass and elasticity may be derived by using similar reasoning.

1.10 DIFFERENTIAL EQUATIONS OF MOTION FOR NONLINEAR SYSTEMS

In Section 1.6, the basic aspects of nonlinear systems and their nonlinear characteristics were discussed. In order to get a better understanding of the difficulties involved in the analysis of nonlinear systems, the differential equations of motion of some of those systems will be introduced in this section, and the difficulties associated with their solution will briefly be explained.

A classic example of a nonlinear system would be a spring–mass system with a nonlinear spring. In this case, the restoring force in the spring is not linearly proportional to its displacement. The load–displacement curve of such a restoring force may be symmetric with respect to the origin, or it may be unsymmetric. Therefore, we can have either a symmetric restoring force or an unsymmetric one. For the case of a symmetric nonlinearity, and conse-quently a symmetric restoring force, the differential equation of motion is as follows:

$$\ddot{x} + c\dot{x} + \omega_0^2 x \pm \alpha x^3 = F \cos \omega_f t \qquad (1.59)$$

which is known as *Duffing's equation*, after the mathematician who studied the equation. It represents the differential equation of motion of a damped,

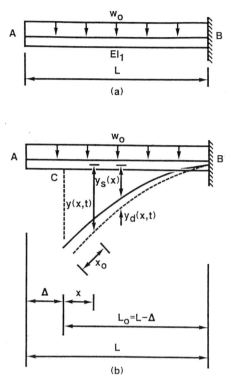

FIGURE 1.14. (a) Flexible cantilever beam of constant stiffness EI. (b) Large deformation configuration of the member.

harmonically excited, single-degree-of-freedom system with a nonlinear spring. The $\pm$ sign signifies a hardening or softening spring.

In Eq. (1.59) we may assume that $\omega_0^2 = k/m$ and that α is a function of the amplitude. If the frequency of the force is the same as the output frequency response, then resonance will occur. For problems of this type, the amplitude experiences a sudden discontinuous jump near resonance, known as the *jump phenomenon*, and a region of instability in the amplitude–frequency plot is observed. The extent of this instability region would depend on the amount of damping present in the nonlinear system and on the rate of change of exciting frequency, as well as on other reasons. More detailed discussion of this subject is provided later in the text.

Another very interesting nonlinear problem is the free vibration of a flexible member, such as the one shown in Fig. 1.14. The large deflection configuration of the member is shown in Fig. 1.14b, where $y_s(x)$ denotes the static equilibrium position of the member, $y_d(x, t)$ is the amplitude of vibration, $y(x, t)$ is the amplitude of the member from its initial straight position, and w_0 is the weight of the beam per unit length.

For flexible members, the static equilibrium position is associated with large static amplitudes, and the differential equation of motion that expresses the free vibration of the member becomes nonlinear. As stated earlier, free vibrations, in general, are taking place from the static equilibrium position of the member. If the vibrational amplitudes $y_d(x, t)$ are small, the free frequencies of vibration are independent of the amplitude of vibration, but they are dependent on the static amplitude $y_s(x)$ that defines the static equilibrium position. If, on the other hand, the frequency amplitudes are also large, the free frequencies of vibration would be dependent on both static and vibrational amplitudes.

The nonlinear differential equation of motion representing the free vibration of flexible members, where both static and vibrational amplitudes are large (see also ref. 2 for details), is given as follows:

$$\frac{d^2}{dx^2}\left(E_x I_x \frac{y''}{[1 + (y')^2]^{3/2}}\right) + [1 + (y')^2]^{1/2} m(x_0) \frac{d^2 y}{dt^2} = 0 \qquad (1.60)$$

Equation (1.60) takes into consideration that the modulus of elasticity E_x of the member, as well as its moment of inertia I_x, may vary along the length of the member. In this equation, we also note that the mass $m(x_0)$ per unit length of the member is a function of the large deformation of the member, where x_0 defines points along the arch length of the member in Fig. 1.14b.

Equation (1.60) cannot be solved by separation of variables, because the free frequencies of vibration are amplitude dependent. The difficulties associated with the solution of this equation may be reduced by assuming that the amplitudes $y_d(x, t)$ of the free vibration are small. This is a situation that exists in many practical flexible beam problems. The problem, however, is still nonlinear because the static displacement $y_s(x)$ is large. Extensive work on the subject has been done by the author [2] and his collaborators. In this case, the differential equation of motion representing the free undamped vibration of a member with small vibration amplitudes $y_d(x, t)$ is as follows:

$$\frac{d^2}{dx^2}\left(E_x I_x \frac{y_d''(x)}{[1 + (y_s')^2]^{3/2}}\right) - \{[1 + (y_s')^2]^{1/2} m(x_0)\}\omega^2 y_d(x, t) = 0 \quad (1.61)$$

The solution of Eq. (1.61) may be simplified by using equivalent systems as discussed in Chapters 4 and 5 of ref. 2.

1.11 CHAOS AND CHAOTIC SYSTEMS

An interesting development in the areas of nonlinear dynamics and nonlinear vibration is a rather newly formulated idea of chaos and chaotic dynamics [3–6]. Although *chaos* is a Greek word that is used to describe an event that has no physical or scientific explanation, the term *chaos* here is reserved for the class of deterministic problems that are free of random or unpredictable input parameters. Poincaré has stated that prediction becomes impossible when small differences in the initial conditions produce great differences in the final output, which means that a small error in the former will produce an extremely large error in the latter. In the current literature, chaos is associated with the class of motions in deterministic physical and mathematical systems whose time history has a sensitive dependence on initial conditions.

Researchers in the field of chaos want us to believe that engineers always knew about chaos when they used the terms noise or turbulence and inserted safety factors to design around unknowns that appeared in the design of structural systems and devices. It is believed, however, that a chaotic system must have nonlinear elements or properties. A linear system cannot exhibit chaotic vibration. What is the basic difference, if any, between chaotic systems as newly defined and the classical definition of nonlinear systems? It is believed that chaotic vibrations occur when some strong nonlinearity exists in the system. What are the criteria, if any, that help us to define when the response is nonlinear, in the classical sense, and not chaotic?

Much of the hope in establishing the idea of chaos is concentrated on turbulence, because it is one of the few remaining unsolved problems of classical physics, and on the discovery of deterministic systems that exhibit chaotic oscillations.

It is believed by researchers [3–6] that there is an underlying structure to chaotic dynamics, which can be determined by searching the order in phase space (position versus velocity). On this basis, one will find that chaotic motion exhibits a new geometric property known as *fractal structure*. Fractals are geometric structures that appear at many scales, and one goal [3] is to discover the fractal structure in chaotic vibration, as well as to measure the loss of information in these random-like structures.

A partial list of physical systems that are believed to exhibit chaotic vibrations includes buckled elastic structures, flow-induced or aeroelastic problems, large, three-dimensional vibrations of structures such as beams and shells, systems with sliding friction, and feedback control systems.

The search for chaotic responses continues, because researchers believe that chaotic dynamics are inherent in all, nonlinear physical phenomena, and this belief has created a sense of revolution in physics today.

1.12   EQUIVALENT SPRINGS

In many practical situations, the vibration and general dynamic analysis of elastic systems becomes convenient if the actual system is replaced by an equivalent one (or an idealized one) that meets design requirements in a

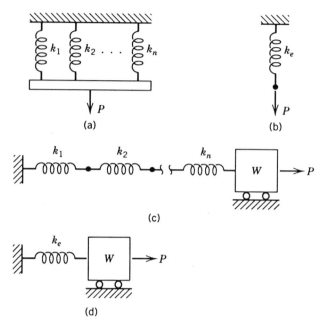

FIGURE 1.15. (a) Springs in parallel arrangement and (b) equivalent spring of constant k_e. (c) Springs in series arrangement and (d) equivalent spring of constant k_e.

given situation. For example, the springs in Fig. 1.15a are in parallel, and they can be replaced by an equivalent system, such as the one shown in Fig. 1.15b. The equivalent spring constant k_e may be determined by using basic principles of mechanics, and it can be obtained from the expression

$$k_e = k_1 + k_2 + \cdots + k_n \tag{1.62}$$

or

$$k_e = \sum_{i=1}^{n} k_i \qquad i = 1, 2, \ldots, n \tag{1.63}$$

The derivation of the above equation is based on the assumption that the bar that connects the springs $k_1, k_2, \ldots, k_n$ remains horizontal during vertical motion.

The springs in Fig. 1.15c are in series, and they can be replaced by the equivalent one in Fig. 1.15d, which has an equivalent constant k_e. This constant may be determined from the equation

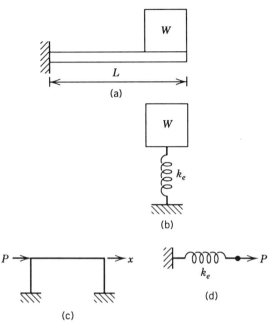

FIGURE 1.16. (a) Uniform cantilever beam supporting a heavy weight W at the free end. (b) Idealized spring–weight system of constant k_e. (c) Steel frame subjected to a horizontal load at the position shown. (d) Equivalent spring of constant k_e.

$$\frac{1}{k_e} = \frac{1}{k_1} + \frac{1}{k_2} + \cdots + \frac{1}{k_n} \tag{1.64}$$

In the above two cases of equivalency, the deflection of the attached bar, or weight W, of the equivalent spring of stiffness k_e, would be identical to the corresponding deflection of the original system when a force P is applied as shown.

The cantilever beam in Fig. 1.16a that supports a concentrated weight W, which is considered to be large compared to the weight of the beam, can be replaced by the idealized one-degree spring–mass system shown in Fig. 1.16b. The spring constant k_e represents the transverse stiffness of the cantilever beam at its free end. If the weight of the beam is neglected as being small compared to the weight W, the deflection of the weight W in Fig. 1.16b, which is caused by the action of the weight W, would be identical to the deflection at the free end of the cantilever beam in Fig. 1.16a under the action of the same load.

The spring constant k_e represents the value of a vertical concentrated load P applied at the free end of the cantilever beam and causing a vertical deflection δ at this end that is equal to unity. From the mechanics of solids, we know that δ is given by the expression

$$\delta = \frac{PL^3}{3EI} \tag{1.65}$$

where L is the length of the member, E is the modulus of elasticity of its material, and I is the cross-sectional moment of inertia. When $\delta = 1$, we have

$$P = k_e = \frac{3EI}{L^3} \tag{1.66}$$

In Fig. 1.16c the horizontal deflection x of the top of the frame, due to a concentrated load P acting as shown, may be represented by the deflection of the spring of constant k_e (Fig. 1.16d), when the load P is acting as shown. The spring constant k_e represents the value of the force P that deflects the top of the frame in the horizontal direction by an amount equal to unity. The value of k_e can be determined by applying methods of analysis that can be found in books on the theory of structures and mechanics of solids [27, 71].

Other practical situations may be handled in a similar manner.

Example 1.1. The spring assembly in Fig. 1.17a is loaded by a concentrated load $P = 120$ kips as shown. By assuming that the bar connecting the springs k_1, k_2, and k_3 remains horizontal during vetical movement, determine its vertical deflection δ by using an equivalent spring. The values of k_1, k_2, and k_3 are 10, 30, and 20 kips/in., respectively.

Solution. The springs in Fig. 1.17a are in parallel. Thus, by using Eq. (1.63), we find

$$k_e = \sum_{i=1}^{3} k_i = k_1 + k_2 + k_3$$

$$= 10 + 30 + 20$$

$$= 60 \text{ kips/in.}$$

The vertical deflection δ of the bar is

$$\delta = \frac{P}{k_e} = \frac{120}{60} = 2.0 \text{ in.}$$

Example 1.2. The uniform steel beam in Fig. 1.17c supports the heavy weight W located at the overhang point C of the member, as shown in the figure. Determine an idealized spring–weight system whose deflection due to W is

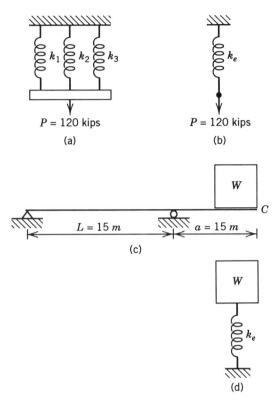

FIGURE 1.17. (a) Horizontal bar supported by three springs and acted on by a vertical load P. (b) Equivalent spring of constant k_e. (c) Uniform steel beam supporting a heavy weight W as shown. (d) Idealized spring–weight system of spring constant k_e.

identical to the vertical deflection of the beam at point C. The cross-sectional moment of inertia I of the beam is 40×10^3 cm^4, and the modulus of elasticity $E = 206.84 \times 10^9$ Pa. Neglect the weight of the member.

Solution. The idealized spring–weight system is shown in Fig. 1.17d. The spring constant k_e represents the stiffness of the member at point C, which is defined as the value of the vertical load P applied at point C and producing a vertical deflection δ at the point equal to unity. From the mechanics of solids, the deflection δ is given by the equation

$$\delta = \frac{Pa^2}{3EI}(L + a) \tag{1.67}$$

When $\delta = 1$, we have

$$P = k_e = \frac{3EI}{a^2(L + a)}$$

This yields

$$k_e = \frac{(3)(206.83 \times 10^9)(40 \times 10^3)(1/100)^4}{(5)^2(15 + 5)}$$

$$= 496.416 \times 10^3 \text{ N/m}$$

The spring–weight system in Fig. 1.17d would have a spring deflection that is identical to the vertical deflection at point C of the beam in Fig. 1.17c, provided that $k_e = 496.416 \times 10^3$ N/m.

PROBLEMS

1.1 In Fig. 1.1a, the particle P moves with a constant angular velocity $\theta = \omega t$ around the circumference of a circle of radius r. If P starts to move from point A, determine the differential equation of motion of its projection on the horizontal diameter of the circle.

1.2 The particle P in Fig. 1.1a starts from point A and moves with constant angular velocity $\theta = \omega t$ around the circumference of the circle as shown. Write the expressions for the displacement, velocity, and acceleration of its projection on the horizontal diameter of the circle. Plot the variation of these quantities with respect to ωt and discuss important characteristics of the motion.

1.3 With C and ϕ as constants, plot Eq. (1.10) for various values of ωt. Discuss important aspects of the graph.

1.4 Draw a spring–mass system consisting of two springs and one mass and having one degree of freedom. Explain the conditions involved.

1.5 Draw a spring–mass system consisting of three springs and two masses and having two degrees of freedom. Explain the conditions.

1.6 Define viscous, Coulomb, and hysteresis damping. Give some practical examples of each kind of damping.

1.7 Make a plot of a harmonic excitation and explain the fundamental aspects of the graph.

1.8 Draw on graph paper two examples of periodic nonharmonic excitations and two examples of nonperiodic excitations.

1.9   Draw on graph paper two examples of random forcing functions, and explain the random nature of the excitation.

1.10   Sketch the geometric representation of the delta Dirac function.

1.11   Explain the difference between D'Alembert's principle and Newton's second law of motion. Use appropriate sketches.

1.12   For each of the spring–mass systems shown in Fig. P1.12, draw appropriate free-body diagrams for the masses involved and derive the differential equations of motion by applying Newton's second law of motion. Also, write equations for the free damped and free undamped vibrations.

1.13   Repeat Problem 1.12 by applying D'Alembert's principle of dynamic equilibrium.

1.14   Repeat Problem 1.12 by using the values $c = 0.10 \, \text{lb·s}^2/\text{in.}$ and $m_3 = 0.15 \, \text{kip·s}^2/\text{in.}$

1.15   For each of the spring-mass systems in Fig. P1.15, determine the differential equation of motion by using appropriate free-body diagrams and applying Newton's second law of motion.

1.16   Repeat Problem 1.15 by using D'Alembert's principle of dynamic equilibrium.

1.17   Determine the differential equations of motion for the spring–mass systems in Fig. P1.17 by using appropriate free-body diagrams and applying Newton's second law of motion.

1.18   Repeat Problem 1.17 by using D'Alembert's principle of dynamic equilibrium.

1.19   For each of the disk–shaft systems in Fig. P1.19, draw the appropriate free-body diagrams for the disks, and derive the differential equations of motion by applying Newton's second law of motion. Also, write the equations for the free damped and free undamped vibrations of the disk–shaft systems.

1.20   Repeat Problem 1.19 by applying D'Alembert's principle of dynamic equilibrium.

1.21   Repeat Problem 1.19 by using the values $c_1 = 0.10 \, \text{in.·lb·s/rad}$, $c_2 = 0.50 \, \text{in.·lb·s/rad}$, $k_{t_1} = 2 \times 10^6 \, \text{in.·lb/rad}$, $k_{t_2} = 3 \times 10^6 \, \text{in.·lb/rad}$, $k_{t_3} = 6 \times 10^6 \, \text{in.·lb/rad}$, $J_1 = 80 \, \text{in.·lb·s}^2$, $J_2 = 120 \, \text{in.·lb·s}^2$, and $J_3 = 200 \, \text{in.·lb·s}^2$.

1.22   The beams in Fig. P1.22 are of uniform cross section and of moment of inertia $I = 1500 \, \text{in.}^4$ and modulus of elasticity $E = 30 \times 10^6 \, \text{psi}$. For

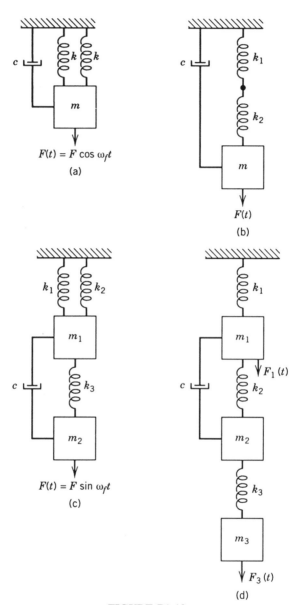

$$F(t) = F \cos \omega_f t$$
(a)

$$F(t)$$
(b)

$$F(t) = F \sin \omega_f t$$
(c)

(d)

FIGURE P1.12

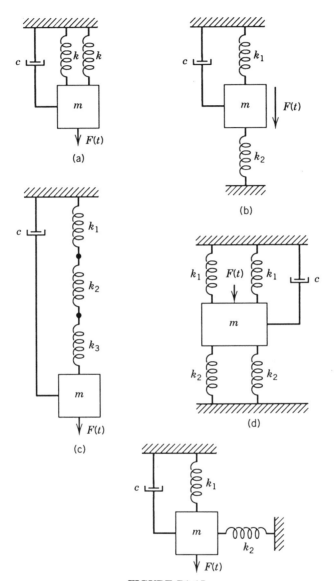

FIGURE P1.15

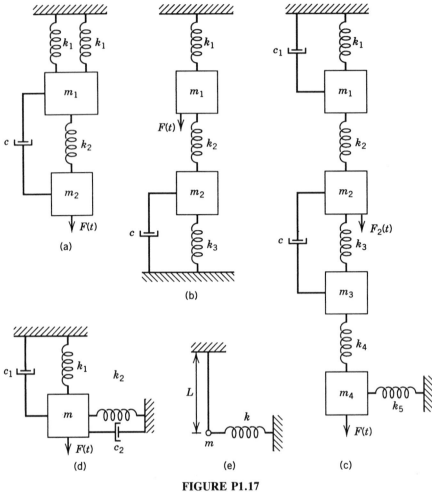

FIGURE P1.17

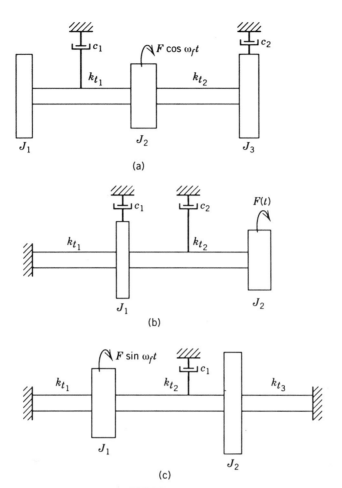

(a)

(b)

(c)

FIGURE P1.19

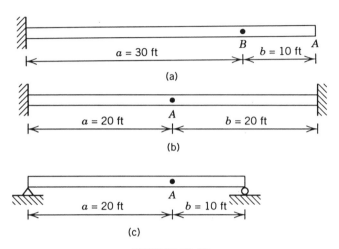

FIGURE P1.22

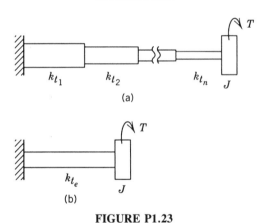

FIGURE P1.23

each beam, determine the spring stiffness in the transverse direction for points A and B. Neglect the weight of the beams.

1.23 In Fig. 1.15a, the linear springs $k_1, k_2, \ldots, k_n$ are in parallel, and the equivalent spring of constant k_e is shown in Fig. 1.15b. The corresponding torslonal systems are shown in Figs. P1.23a and P1.23b, respectively. Prove that k_{t_e} is given by the expression

$$\frac{1}{K_{t_e}} = \frac{1}{K_{t_1}} + \frac{1}{k_{t_2}} + \cdots + \frac{1}{k_{t_n}} \tag{1.68}$$

1.24 Assume that the torsional system in Fig. P1.23a consists of the torsional springs k_{t_1} and k_{t_2}, of values $k_{t_1} = 6 \times 10^6$ in.·lb/rad and $k_{t_2} = = 2 \times 10^6$ in.·lb/rad. Determine the equivalent torsional spring constant k_{t_e} by using the expression given in Problem 1.23.

2 Vibration Analysis of Simple Systems

2.1 INTRODUCTION

The differential equations of motion that have been derived in the preceding chapter will be solved in this chapter in order to determine the vibration responses of simple structural and mechanical systems. Systems with continuous mass and elasticity, as well as systems with a limited number of degrees of freedom, are investigated. The vibration response is determined here by assuming that the elastic system is vibrating freely, that is, without any oscillatory external forces acting on it, but the effects of various types of damping are considered.

The types of problems discussed in this chapter include spring–mass systems; disk–shaft systems; torsional, flexural, and longitudinal vibrations; and the wave equation.

2.2 FREE UNDAMPED VIBRATION OF ONE-DEGREE SPRING–MASS SYSTEMS

In Section 1.7, the differential equations of motion describing the free vibration of spring–mass systems have been derived. In this section, we consider the spring–mass system in Fig. 1.7a; we assume that the dynamic force $F(t)$ and damping coefficient c are zero. It is also assumed that the mass m is large compared to the mass of the spring and that it moves only in the vertical direction y. With these restrictions in mind, if the mass m is displaced from its static equilibrium position by an amount y_0 and then released, the mass m will vibrate about its static equilibrium position with an amplitude y_0, for an indefinite amount of time. As stated in Section 1.7, Eq. (1.23), this vibratory motion is represented by the following differential equation of motion:

$$m\ddot{y} + ky = 0 \qquad (2.1)$$

The vibratory motion represented by Eq. (2.1) is harmonic, and the dynamic amplitude y at any time t may be expressed by the equation

$$y = y_0 \cos \omega t \tag{2.2}$$

where y_0 is the maximum amplitude of vibration, and ω is the free, un-damped, natural frequency of the spring–mass system.

By substituting Eq. (2.2) into Eq. (2.1), we find

$$-m\omega^2 y_0 \cos \omega t + k y_0 \cos \omega t = 0$$

or

$$\omega = \sqrt{\frac{k}{m}} \tag{2.3}$$

By using Eq. (2.3), we can write Eq. (2.1) as

$$\ddot{y} + \omega^2 y = 0 \tag{2.4}$$

Note that $k = \omega^2 m$.

Equation (2.4) is a second-order, homogeneous, differential equation; its solution $y(t)$ is well known and is given by the equation

$$y(t) = A \sin \omega t + B \cos \omega t \tag{2.5}$$

The constants A and B may be determined from the initial conditions of the motion. If, for example, the motion starts at time $t = 0$ with an initial displacement y_0 and an initial velocity $\dot{y}_0$, Eq. (2.5) yields the expressions

$$y_0 = A \sin(0) + B \cos(0) \tag{2.6}$$
$$\dot{y}_0 = A\omega \cos(0) - B\omega \sin(0) \tag{2.7}$$

The simultaneous solution of Eqs. (2.6) and (2.7) yields

$$A = \frac{\dot{y}_0}{\omega} \tag{2.8}$$

$$B = y_0 \tag{2.9}$$

With A and B known, Eq. (2.5) yields the following general solution:

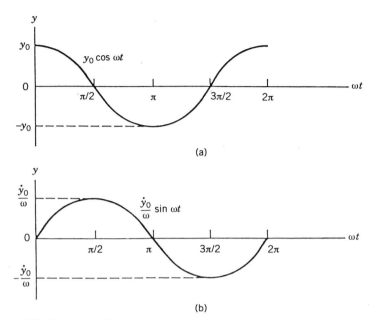

FIGURE 2.1. (a) Plot of term $y_0 \cos \omega t$. (b) Plot of term $(\dot{y}_0/\omega) \sin \omega t$.

$$y(t) = \frac{\dot{y}_0}{\omega} \sin \omega t + y_0 \cos \omega t \qquad (2.10)$$

If the initial velocity $\dot{y}_0$ is zero, then Eq. (2.10) yields

$$y(t) = y_0 \cos \omega t \qquad (2.11)$$

This equation is identical to the expression given by Eq. (2.2).

In Eq. (2.10) we note that the vibration consists of two parts; one part is proportional to $\cos \omega t$ and depends on the initial condition y_0, while the other part is proportional to $\sin \omega t$ and depends on the initial velocity $\dot{y}_0$. The curves in Figs. 2.1a and 2.1b provide a graphical representation of the two parts of Eq. (2.10).

The solution given by Eq. (2.5) may become somewhat more general if the initial conditions of the motion are defined with respect to some initial time $t = t_0$; that is, at time $t = t_0$, the displacement and velocity are y_0 and $\dot{y}_0$, respectively. By using Eq. (2.5) and applying these two initial conditions at an initial time $t = t_0$ and solving for A and B, we find

$$A = y_0 \sin \omega t_0 + \frac{\dot{y}_0}{\omega} \cos \omega t_0 \qquad (2.12)$$

$$B = y_0 \cos \omega t_0 - \frac{\dot{y}_0}{\omega} \sin \omega t_0 \tag{2.13}$$

On this basis, by using Eqs. (2.12) and (2.13), we find that Eq. (2.5) yields the following expression:

$$y(t) = y_0 \cos \omega(t - t_0) + \frac{\dot{y}_0}{\omega} \sin \omega(t - t_0) \tag{2.14}$$

Differentiation with respect to time yields the velocity $\dot{y}(t)$ at any time t, as follows:

$$\dot{y}(t) = -y_0 \omega \sin \omega(t - t_0) + \dot{y}_0 \cos \omega(t - t_0) \tag{2.15}$$

The format of Eqs. (2.5) and (2.14) can be rearranged differently by using the expressions

$$A = y_0 = Y \sin \phi \quad \text{and} \quad B = \frac{\dot{y}_0}{\omega} = Y \cos \phi \tag{2.16}$$

where ϕ is the phase angle and Y is a constant. On this basis, Eq. (2.14) yields

$$y(t) = Y \sin[\omega(t - t_0) + \phi] \tag{2.17}$$

and

$$\dot{y}(t) = Y\omega \cos[\omega(t - t_0) + \phi] \tag{2.18}$$

where

$$Y = \sqrt{A^2 + B^2} = \sqrt{y_0^2 + \left(\frac{\dot{y}_0}{\omega}\right)^2} \tag{2.19}$$

and

$$\tan \phi = \frac{A}{B} = \frac{Y \sin \phi}{Y \cos \phi} = \frac{\omega y_0}{\dot{y}_0} \tag{2.20}$$

Equation (2.17) is plotted in Fig. 2.2.

It should be noted here that the free undamped frequency of vibration of the spring–mass system in radians per second (rps) is given by Eq. (2.3). In cycles per second (cps) or hertz (Hz), after the German physicist Hertz, the free vibration f is given by the expression

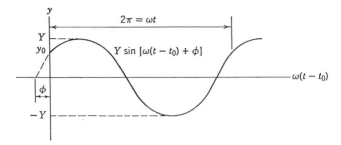

FIGURE 2.2. Graphic representation of Eq. (2.17).

$$f = \frac{\omega}{2\pi} = \frac{1}{2\pi}\sqrt{\frac{k}{m}} \tag{2.21}$$

The period τ of vibration, defined as the time the spring–mass system requires to complete one cycle of motion, is given by the equation

$$\tau = \frac{1}{f} = 2\pi\sqrt{\frac{m}{k}} \tag{2.22}$$

Example 2.1. A weight $W = 80\,$lb is suspended by a spring of spring constant $k = 100\,$lb/in. The vibratory motion of the weight has a maximum velocity $\dot{y}_{max} = \dot{Y} = 30\,$in./s. Determine the natural frequency and period of vibration of the system, the maximum amplitude of the vibratory motion, and the maximum acceleration.

Solution. The mass m of the spring–mass system is

$$m = \frac{W}{g} = \frac{80}{386} = 0.208\,\frac{\text{lb}\cdot\text{s}^2}{\text{in.}}$$

where $g = 386\,$in./s^2 is the acceleration of gravity. From Eq. (2.3), we find

$$\omega = \sqrt{\frac{k}{m}} = \sqrt{\frac{100}{0.208}} = 21.90\,\text{rps}$$

and from Eq. (2.21), we obtain

$$f = \frac{\omega}{2\pi} = \frac{21.90}{2\pi} = 3.49 \text{ Hz}$$

From Eq. (2.22), the period of vibration τ is

$$\tau = \frac{1}{f} = \frac{1}{3.49} = 0.287 \text{ s}$$

By differentiating Eq. (2.2) with respect to time t, we find

$$\dot{y} = -y_0\omega \sin \omega t$$
$$\ddot{y} = -y_0\omega^2 \cos \omega t$$

Therefore, $\dot{y}_{max} = \dot{Y} = -y_0\omega$ and $\ddot{y}_{max} = \ddot{Y} = -y_0\omega^2$. On this basis, the maximum amplitude of vibration y_{max} is

$$y_{max} = y_0 = -\frac{\dot{Y}}{\omega} = -\frac{30}{21.90} = -1.37 \text{ in.}$$

The maximum acceleration $\ddot{y}_{max}$ is

$$\ddot{y}_{max} = \ddot{Y} = -y_0\omega^2 = -(-1.37)(21.90)^2 = 657.07 \text{ in./s}^2$$

Example 2.2. A uniform simply supported steel beam supports a weight W of 27.0 kN located at the center of the member. The length L of the member is 8.0 m, the width $b = 10.0$ cm, the depth $h = 20.0$ cm, and the modulus of elasticity $E = 206.843$ MPa. By neglecting the mass of the beam and using an idealized one-degree spring–mass system, determine the free undamped frequency of the beam and its period of vibration.

Solution. The cross-sectional moment of inertia I of the member is

$$I = \frac{bh^3}{12} = \frac{(0.10)(0.20)^3}{12} = 66.67 \times 10^{-6} \text{ m}^4$$

The stiffness k of the member, at its center, is the value of a vertical load P applied at the center and producing a vertical deflection y equal to unity. From the mechanics of materials, the deflection y at the center of the member due to P is given by the expression

$$y = \frac{PL^3}{48EI} \qquad (2.23)$$

For $y = 1$, we have

$$P = k = \frac{48EI}{L^3}$$

$$= \frac{(48)(206.843 \times 10^6)(66.67 \times 10^{-6})}{(8.0)^3}$$

$$= 1292.83 \text{ kN/m}$$

The mass m at the center of the member is

$$m = \frac{27,000 \text{ N}}{9.81 \text{ m/s}^2} = 2752.29 \text{ kg}$$

The symbol kg is used to denote 1 kilogram of mass in SI units. The free undamped frequency of vibration ω of the idealized one-degree spring–mass system is

$$\omega = \sqrt{\frac{k}{m}}$$

$$= \sqrt{\frac{1292.83 \times 10^3}{2752.29}}$$

$$= 21.67 \text{ rps}$$

or, in hertz,

$$f = \frac{\omega}{2\pi} = \frac{21.67}{2\pi} = 3.45 \text{ Hz}$$

The period of vibration τ is

$$\tau = \frac{1}{f} = \frac{1}{3.45} = 0.29 \text{ s}$$

Example 2.3. A steel cantilever beam of length $L = 18$ ft supports a weight

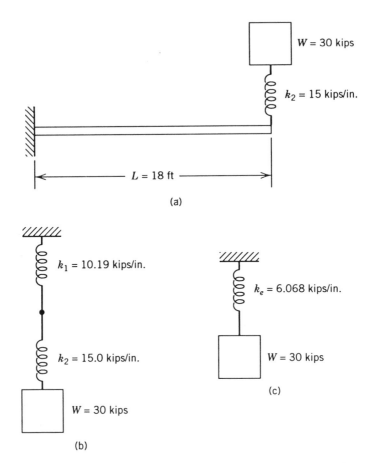

FIGURE 2.3. (a) Steel cantilever beam supporting a weight W at its free end. (b) Idealized one-degree system consisting of weight W and spring constants k_1 and k_2 in parallel. (c) Idealized one-degree system with weight W and equivalent spring of constant k_e.

W of 30 kips at its free end as shown in Fig. 2.3a. The beam has a W21 × 55 cross section with moment of inertia $I = 1140.7\,\text{in}^4$. In addition, a linear spring of stiffness $k = 15.0\,\text{kips/in.}$ is attached at the free end of the member as shown in the figure. By neglecting the weight of the steel cantilever beam, determine the idealized one-degree spring–mass system consisting of one mass and an equivalent spring constant k_e. Also, determine its free frequency of vibration and its period of vibration. The modulus of elasticity $E = 30 \times 10^3$ ksi, where ksi denotes kips per square inch.

Solution. The idealized spring–mass system consisting of two springs of constants k_1 and k_2 is shown in Fig. 2.3b. The spring constant k_1 represents the stiffness of the cantilever beam at its free end, which can be determined from the equation

$$k_1 = \frac{3EI}{L^3} = \frac{(3)(30 \times 10^3)(1140.7)}{(18 \times 12)^3}$$

$$= 10.19 \text{ kips/in.}$$

Since the two springs in Fig. 2.3b are in parallel, the equivalent constant k_e may be determined from Eq. (1.64). That is,

$$\frac{1}{k_e} = \frac{1}{k_1} + \frac{1}{k_2} = \frac{1}{10.19} + \frac{1}{15}$$

or, by solving for k_e, we find

$$k_e = 6.068 \text{ kips/in.}$$

The idealized one-degree spring–mass system consisting of the weight $W = 30$ kips and the equivalent spring of constant $k_e = 6.068$ kips/in. is shown in Fig. 2.3c.

The free frequency of vibration ω may be determined from the following equation:

$$\omega = \sqrt{\frac{k_e}{m}} = \sqrt{\frac{6.068}{(30/386)}}$$

$$= 8.836 \text{ rps}$$

The period of its free vibration τ is

$$\tau = \frac{1}{f} = \frac{2\pi}{\omega} = \frac{2\pi}{8.836} = 0.711 \text{ s}$$

2.3   FREE VIBRATION WITH VISCOUS DAMPING

We examine again the one-degree spring–mass system in Fig. 1.7a, by assuming that the dynamic force $F(t)$ is zero. The mass m is restricted to move only in the vertical direction y under the influence of viscous damping, where c is the viscous damping coefficient. With these conditions in mind, if the mass m is displaced vertically by an initial displacement y_0, it will vibrate in the vertical direction with respect to the equilibrium position and it will come to rest with time, because the spring–mass system moves under the influence of viscous damping. The time it takes for the mass m to come to rest depends on the amount of damping present in the system.

The differential equation of motion for this system is derived in Section 1.7, and it is as follows:

$$m\ddot{y} + c\dot{y} + ky = 0 \tag{2.24}$$

The solution $y(t)$ of Eq. (2.24) may be written

$$y(t) = Ae^{pt} \tag{2.25}$$

where A and p are constants.

By substituting Eq. (2.25) into Eq. (2.24) and carrying out the required differentiations, we find

$$mp^2 + cp + k = 0 \tag{2.26}$$

The solution of this quadratic equation yields the following two roots:

$$p_{1,2} = -\frac{c}{2m} \pm \sqrt{\left(\frac{c}{2m}\right)^2 - \frac{k}{m}} \tag{2.27}$$

Therefore, the solution given by Eq. (2.25) may be written

$$y(t) = A_1 e^{p_1 t} + A_2 e^{p_2 t} \tag{2.28}$$

where A_1 and A_2 are constants. The roots p_1 and p_2 are given by Eq. (2.27).

The constants A_1 and A_2 may be determined from the initial conditions of the vibratory motion. For example, if at $t = 0$ the initial displacement is y_0 and the initial velocity is $\dot{y}_0$, then Eq. (2.28) yields

$$A_1 + A_2 = y_0 \tag{2.29}$$
$$A_1 p_1 + A_2 p_2 = \dot{y}_0 \tag{2.30}$$

By solving these two equations simultaneously for A_1 and A_2, we find

$$A_1 = y_0 - \frac{\dot{y}_0 - y_0 p_1}{p_2 - p_1} \tag{2.31}$$

$$A_2 = \frac{\dot{y}_0 - y_0 p_1}{p_2 - p_1} \tag{2.32}$$

By substituting Eq. (2.27) into Eq. (2.28), we find

$$y(t) = e^{-(c/2m)t}(A_1 e^{[\sqrt{(c/2m)^2 - k/m}]t} + A_2 e^{-[\sqrt{(c/2m)^2 - k/m}]t}) \qquad (2.33)$$

The factor $e^{-(c/2m)t}$ in Eq. (2.33) is an exponentially decaying function of time, which indicates that the damped vibratory motion has an exponentially decaying amplitude with time. The constants A_1 and A_2 are given by Eqs. (2.31) and (2.32), respectively.

By examining the terms under the radical in Eq. (2.27), three important conclusions can be drawn:

1. If the sum of the terms under the radical is equal to zero, then

$$\left(\frac{c}{2m}\right)^2 = \frac{k}{m} = \omega^2 \qquad (2.34)$$

where ω is the undamped natural frequency of vibration of the spring–mass system. Under this condition, Eq. (2.27) yields $p_{1,2} = -c/2m$. This type of damping represents a transition from oscillatory to nonoscillatory configuration and is known as the *condition of critical damping*; that is, if the motion starts from the position y_0 with velocity $\dot{y}_0$, the mass will return to rest without oscillation. This situation usually does not occur in practice.

2. If the sum of the terms under the radical is positive, we have the inequality

$$\left(\frac{c}{2m}\right)^2 > \omega^2 \qquad (2.35)$$

which is known as the *condition of overdamping*. In this case, the motion is aperiodic, and p_1 and p_2 are always real values and negative.

3. The *condition of underdamping* occurs when the sum of the terms under the radical is negative. In this case, we have

$$\left(\frac{c}{2m}\right)^2 < \omega^2 \qquad (2.36)$$

Under this condition, the spring–mass system will vibrate with decreasing amplitude. Most structural and mechanical systems are under the influence of light damping.

For the underdamped condition, with $i = \sqrt{-1}$, the application of Euler's expression

$$e^{\pm i\theta} = \cos \theta \pm i \sin \theta \tag{2.37}$$

leads to the equation

$$e^{\pm [i\sqrt{k/m - (c/2m)^2}]t} = \cos\left[\sqrt{\frac{k}{m} - \left(\frac{c}{2m}\right)^2}\,\right]t \pm i \sin\left[\sqrt{\frac{k}{m} - \left(\frac{c}{2m}\right)^2}\,\right]t \tag{2.38}$$

By using Eq. (2.38), we may write the trigonometric form of Eq. (2.33) as follows:

$$y(t) = e^{-(c/2m)t}\left\{ c_1 \cos\left[\sqrt{\frac{k}{m} - \left(\frac{c}{2m}\right)^2}\,\right]t + C_2 \sin\left[\sqrt{\frac{k}{m} - \left(\frac{c}{2m}\right)^2}\,\right]t\right\} \tag{2.39}$$

The constants C_1 and C_2 would have to be determined from the initial conditions of the motion.

The three conditions outlined above may be expressed in a somewhat more convenient way by introducing the *critical damping factor* c_c. The critical damping factor is defined as the value of c, in Eq. (2.27), which makes the algebraic sum of the terms under the radical equal to zero. Thus, from Eq. (2.34), we find

$$c_c = 2\sqrt{km} = 2m\omega \tag{2.40}$$

From this point of view, the damping of the spring–mass system may be specified in terms of c_c and the damping ratio ζ that is given by the expression

$$\zeta = \frac{c}{c_c} \tag{2.41}$$

Therefore, we can write

$$\frac{c}{2m} = \zeta\omega \tag{2.42}$$

$$\frac{k}{m} - \left(\frac{c}{2m}\right)^2 = \omega^2(1 - \zeta^2) \tag{2.43}$$

On this basis, Eq. (2.27) may be written

$$p_{1,2} = (-\zeta \pm \sqrt{\zeta^2 - 1})\omega \tag{2.44}$$

This equation shows that the condition of critical damping occurs when $\zeta = 1$, overdamping will occur when $\zeta > 1$, and the spring–mass system will be under the influence of light damping when $\zeta < 1$.

By using Eq. (2.44), we can write Eq. (2.39) in terms of the damping ratio ζ as follows:

$$y(t) = e^{-\zeta \omega t}[C_1 \cos(\omega \sqrt{1 - \zeta^2})t + C_2 \sin(\omega \sqrt{1 - \zeta^2})t] \tag{2.45}$$

From Eq. (2.45), we can conclude that the damped frequency ω_d of the spring–mass system may be obtained from the equation

$$\omega_d = \omega \sqrt{1 - \zeta^2} \tag{2.46}$$

The damped period of vibration τ_d is given by the expression

$$\tau_d = \frac{2\pi}{\omega_d} = \frac{2\pi}{\omega \sqrt{1 - \zeta^2}} \tag{2.47}$$

The constants C_1 and C_2 in Eq. (2.45) may be determined by applying the initial conditions of the motion. For example, if at $t = 0$ the initial displacement is y_0 and the initial velocity is $\dot{y}_0$, application of Eq. (2.45) yields

$$C_1 = y_0 \tag{2.48}$$

$$C_2 = \frac{\dot{y}_0 + \zeta \omega y_0}{\omega_d} \tag{2.49}$$

By using Eqs. (2.48) and (2.49), we can write Eq. (2.45) as follows:

$$y(t) = e^{-\zeta \omega t}\left(y_0 \cos \omega_d t + \frac{\dot{y}_0 + \zeta \omega y_0}{\omega_d} \sin \omega_d t\right) \tag{2.50}$$

This shows that the first term in Eq. (2.50) is proportional to $\cos \omega_d t$ and depends only on the initial displacement y_0. The second term of the same equation is proportional to $\sin \omega_d t$ and depends on both the initial displacement y_0 and the initial velocity $\dot{y}_0$.

Equation (2.50) can also be written in terms of a constant Y and the

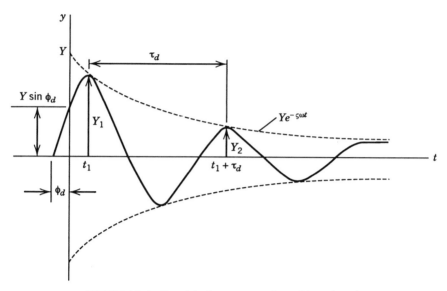

FIGURE 2.4. Graphical representation of Eq. (2.51).

phase angle ϕ_d in the following manner:

$$y(t) = Ye^{-\zeta\omega t}\sin(\omega_d t + \phi_d) \tag{2.51}$$

where

$$Y = \sqrt{C_1^2 + C_2^2} = \sqrt{y_0^2 + \frac{(\dot{y}_0 + \zeta\omega y_0)^2}{\omega_d^2}} \tag{2.52}$$

$$\phi_d = \tan^{-1}\left(\frac{C_2}{C_1}\right) = \tan^{-1}\left(\frac{\dot{y}_0 + \zeta\omega y_0}{\omega_d y_0}\right) \tag{2.53}$$

Equation (2.51) may be thought of as representing a pseudoharmonic motion with an exponentially decaying amplitude $Ye^{-\zeta\omega t}$ and a phase angle ϕ_d. Its damped period of vibration τ_d may be obtained from Eq. (2.47). A graph of this motion is shown in Fig. 2.4. Since $Ye^{-\zeta\omega t}$ depends on the damping ratio ζ, the rate of decay of the amplitude would depend on the amount of damping present in the spring–mass system.

Example 2.4. A weight $W = 40$ lb is suspended by a linear spring of constant $k = 60$ lb/in. The vibratory motion is under the influence of viscous damping with damping constant $c = 1.60$ lb·s/in. Determine the damped and un-damped free frequencies of the spring–mass system, the damped and un-damped periods of vibration, and the critical damping factor.

Solution. The undamped frequency ω is

$$\omega = \sqrt{\frac{k}{m}} = \sqrt{\frac{(60)(386)}{40}}$$

$$= 24.0 \text{ rps}$$

From Eq. (2.40), the critical damping factor c_c is as follows:

$$c_c = 2m\omega = \frac{(2)(40)(24)}{386}$$

$$= 4.97 \text{ lb·s/in.}$$

Therefore, from Eq. (2.41), the damping factor ζ is

$$\zeta = \frac{c}{c_c} = \frac{1.60}{4.97} = 0.323$$

The damped free frequency of vibration ω_d may be obtained from Eq. (2.46):

$$\omega_d = \omega\sqrt{1 - \zeta^2} = 24\sqrt{1 - (0.323)^2}$$

$$= 22.75 \text{ rps}$$

The undamped period of vibration τ is

$$\tau = \frac{2\pi}{\omega} = \frac{2\pi}{24} = 0.262 \text{ s}$$

and the damped period τ_d is

$$\tau_d = \frac{2\pi}{\omega_d} = \frac{2\pi}{22.75} = 0.276 \text{ s}$$

The above results show that ω_d is about 5.21% lower than ω. This result indicates that it takes a comparatively large amount of damping ($\zeta = 0.323$) to reduce ω_d by 5.21% as compared to ω. In practice, since damping is rather light, it is often assumed in the analysis that the damped and undamped frequencies maintain the same value. The amplitude of vibration, however, is much more sensitive to damping, as can be observed by examining the term $e^{-\zeta\omega t}$ in Eq. (2.50). For example, if $\zeta = 0$ and $t = 0$, we have $e^{-\zeta\omega t} = 1.0$; and $e^{-\zeta\omega t} = 1/e^{(0.1)(24)(0.262)} = 0.533$ when $\zeta = 0.10$ and $t = \tau =$

0.262 s. This indicates that 10% damping produces a 46.7% reduction in amplitude from pick-to-pick during one cycle of motion. For $\zeta = 0.323$ and $t = \tau = 0.262$ s, we have $e^{-\zeta\omega t} = 0.131$, indicating that the pick-to-pick reduction in amplitude is 86.9%. The system comes to rest much quicker when damping is higher.

2.4 LOGARITHMIC DECREMENT

Consider the solution given by Eq. (2.51), which represents the free vibration of a one-degree spring–mass system under the influence of viscous damping. This equation is written again for convenience:

$$y(t) = Ye^{-\zeta\omega t}\sin(\omega_d t + \phi_d) \tag{2.54}$$

The decay of this oscillatory motion is shown schematically in Fig. 2.4.

If we consider any two successive amplitudes, say, the amplitudes y_1 and y_2 at times t_1 and $t_1 + \tau_d$, respectively, as shown in Fig. 2.4, we may write the following expressions for y_1 and y_2:

$$y_1 = Ye^{-\zeta\omega t_1}\sin(\omega_d t_1 + \phi_d) \tag{2.55}$$

$$y_2 = Ye^{-\zeta\omega(t_1 + \tau_d)}\sin[\omega_d(t_1 + \tau_d) + \phi_d] \tag{2.56}$$

The logarithmic decrement δ is defined as the natural logarithm of the ratio of any two successive amplitudes. Thus, for the successive amplitudes given by Eqs. (2.55) and (2.56), we have

$$
\begin{aligned}
\delta &= \ln\frac{y_1}{y_2} \\
&= \ln\frac{e^{-\zeta\omega t_1}\sin(\omega_d t_1 + \phi_d)}{e^{-\zeta\omega(t_1 + \tau_d)}\sin[\omega_d(t_1 + \tau_d) + \phi_d]} \\
&= \ln\frac{e^{-\zeta\omega t_1}}{e^{-\zeta\omega(t_1 + \tau_d)}} \\
&= \ln e^{\zeta\omega\tau_d}
\end{aligned}
\tag{2.57}
$$

or

$$\delta = \zeta\omega\tau_d \tag{2.58}$$

In Eq. (2.57), the sines have been canceled out because their values are equal at times t_1 and $t_1 + \tau_d$.

From Eq. (2.47), we have

$$\tau_d = \frac{2\pi}{\omega_d} = \frac{2\pi}{\omega\sqrt{1 - \zeta^2}} \tag{2.59}$$

By substituting Eq. (2.59) into Eq. (2.58), we find

$$\delta = \frac{2\pi\zeta}{\sqrt{1 - \zeta^2}} \tag{2.60}$$

The logarithmic decrement δ given by Eq. (2.60) defines the rate of decay of the oscillation. This is easily observed by comparing Eq. (2.58) with the term $Ye^{-\zeta\omega t}$ in Fig. 2.4, which represents the amplitude decay of the motion. Since $\zeta\omega = \delta/\tau_d$, we note that $Ye^{-\zeta\omega t} = Y/e^{\delta t/\tau_d}$, which prescribes the influence of δ on the oscillatory motion of consecutive amplitudes Y_1, Y_2, Y_3, and so on.

For very light damping, the value of the damping factor ζ is small, and the denominator in Eq. (2.60) is approximately equal to unity. Therefore, in this case, we have

$$\delta = 2\pi\zeta \tag{2.61}$$

For values of $\zeta < 0.30$, Eq. (2.61) yields reasonable results for practical purposes.

It can easily be deduced that the ratio of any two successive amplitudes is constant. See, for excample, the ratio of the amplitudes y_1 and y_2 given by Eqs. (2.55) and (2.56), respectively. On this basis, the logarithmic decrement δ can also be calculated from the ratio of amplitudes several cycles apart. If y_n is the amplitude n cycles after the amplitude y_0, we can prove that the logarithmic decrement δ may be determined from the equation

$$\delta = \frac{1}{n}\ln\left(\frac{y_0}{y_n}\right) \tag{2.62}$$

Equation (2.62) becomes very useful and convenient when determining the required number of cycles for a system to reach a specified reduction in amplitude.

By using Eq. (2.60) and solving for ζ, we may also write the equation

$$\zeta = \frac{\delta}{\sqrt{(2\pi)^2 + \delta^2}} \tag{2.63}$$

which expresses ζ as a function of δ.

Example 2.5. A weight $W = 150 \text{ lb}$ is suspended by a spring of constant $k =$

300 lb/in. The spring–mass system is under the influence of viscous damping with damping constant $c = 4.65$ lb·s/in. Determine the logarithmic decrement δ by using Eqs. (2.60) and (2.61) and compare the results.

Solution. The free undamped frequency of vibration ω is

$$\omega = \sqrt{\frac{k}{m}} = \sqrt{\frac{(300)(386)}{150}} = 27.8 \text{ rps}$$

The critical damping factor c_c is

$$c_c = 2m\omega = \frac{(2)(150)(27.8)}{386} = 21.6 \text{ lb·s/in.}$$

Thus, the damping ratio ζ is

$$\zeta = \frac{c}{c_c} = \frac{4.65}{21.60} = 0.215$$

From Eq. (2.60), we find

$$\delta = \frac{2\pi\zeta}{\sqrt{1 - \zeta^2}} = \frac{2\pi(0.215)}{\sqrt{1 - (0.215)^2}} = 1.385$$

From Eq. (2.61), we find

$$\delta = 2\pi\zeta = 2\pi(0.215) = 1.350$$

The above results show that the approximate expression given by Eq. (2.61) yields a value of δ that is 2.53% lower than the value obtained by using the exact expression represented by Eq. (2.60). This is an interesting observation, because in many practical situations such errors are easily tolerated. On the other hand, in many practical problems, damping is very light, usually much less than the value of 21.5% found above, and the error would be much smaller, as can be observed by examining Eq. (2.60).

2.5   FREE VIBRATION WITH COULOMB DAMPING

A short discussion regarding Coulomb damping was given in Section 1.4, and the associated damping force F_f is given by Eq. (1.16). An example of a free vibratory motion that is under the influence of Coulomb damping is

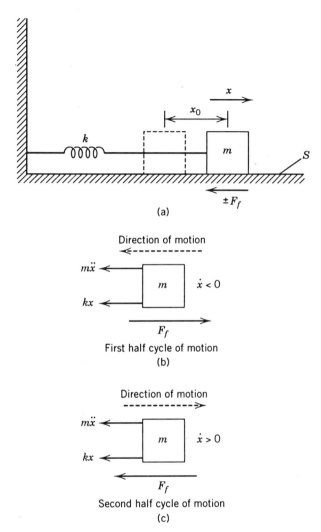

FIGURE 2.5. (a) Spring–mass system under the influence of Coulomb damping. (b) Free-body diagram of m during the first half-cycle of motion. (c) Free-body diagram of m during the second half-cycle of motion.

shown in Fig. 2.5. From the undeformed position of the spring of stiffness k, an initial displacement x_0 is given to the mass m as shown and then the mass is released. The resisting force F_f, which is produced from the friction between m and the surface S, is assumed to be constant. In other words, the difference in value between starting and moving conditions is neglected. In practical situations, however, the roughness of the surface S may not be uniform, and, consequently, the frictional coefficient may not be constant.

During the first half-cycle of motion after release, the free-body diagram of the mass m is shown in Fig. 2.5b. During this time, the velocity $\dot{x}$ is negative and the mass m is moving to the left. On this basis, the differential equation of motion representing the first half-cycle of motion may be written

$$m\ddot{x} + kx = F_f \qquad \dot{x} < 0 \tag{2.64}$$

When the mass m reaches the extreme position to the left, the velocity $\dot{x}$ becomes zero and the mass will be moving to the right during the second half-cycle of motion. During the second half-cycle of motion, the velocity $\dot{x}$ is positive and the free-body diagram of the mass m would be as shown in Fig. 2.5c. The differential equation representing this motion is

$$m\ddot{x} + kx = -F_f \qquad \dot{x} > 0 \tag{2.65}$$

Equations (2.64) and (2.65) are nonhomogeneous differential equations of motion, and their solution can be obtained by superimposing the homogeneous (complementary) solution $x_c(t)$ and the particular solution $x_p(t)$.

The homogeneous solution $x_c(t)$ for Eq. (2.64), which is the solution of $m\ddot{x} + kx = 0$, is similar to the one given by Eq. (2.5); that is,

$$x_c(t) = A \sin \omega t + B \cos \omega t \tag{2.66}$$

The particular solution $x_p(t)$ of Eq. (2.64), since the damping force F_f is constant, would be of the form

$$x_p = C \tag{2.67}$$

where C is a constant that needs to be determined. By substituting Eq. (2.67) into Eq. (2.64), we find $kC = F_f$, or

$$C = \frac{F_f}{k} \tag{2.68}$$

Therefore,

$$x_p = \frac{F_f}{k} \tag{2.69}$$

On this basis, the complete solution $x(t)$ of Eq. (2.64) is

$$x(t) = x_c(t) + x_p(t)$$

or

$$x(t) = A \sin \omega t + B \cos \omega t + \frac{F_f}{k} \qquad (2.70)$$

The constants A and B can be determined by applying the initial conditions of the motion. By using Eq. (2.70) and applying the initial conditions $x(0) = x_0$ and $\dot{x}(0) = 0$ at $t = 0$, we find

$$A = 0 \qquad (2.71)$$

$$B = x_0 - \frac{F_f}{k} \qquad (2.72)$$

Therefore, the complete solution $x(t)$ given by Eq. (2.70) yields

$$x(t) = \left(x_0 - \frac{F_f}{k} \right) \cos \omega t + \frac{F_f}{k} \qquad (2.73)$$

$$\dot{x}(t) = \left(x_0 - \frac{F_f}{k} \right) \sin \omega t \qquad (2.74)$$

Equation (2.74) shows that, at the end of the first half-cycle—that is, when $\omega t = \pi$—the velocity of the mass m is zero, and the mass will start moving to the right. From Eq. (2.73), we find that the displacement at $\omega t = \pi$, or $t = \pi/\omega$, is

$$x_{t = \pi/\omega} = -\left(x_0 - \frac{2F_f}{k} \right) \qquad (2.75)$$

By following a similar procedure, we find that the complete solution of Eq. (2.65), which represents the second half-cycle of motion, is

$$x(t) = A_1 \sin \omega t + B_1 \cos \omega t - \frac{F_f}{k} \qquad (2.76)$$

The constants A_1 and B_1 may be determined by using Eq. (2.76) and applying the following initial conditions:

$$x_{t = \pi/\omega} = -\left(x_0 - \frac{2F_f}{k} \right) \qquad (2.77)$$

$$\dot{x}_{t = \pi/\omega} = 0 \qquad (2.78)$$

This yields

$$A_1 = 0 \qquad (2.79)$$

$$B_1 = \left(x_0 - \frac{3F_f}{k}\right) \qquad (2.80)$$

On this basis, Eq. (2.76) is written as follows:

$$x(t) = \left(x_0 - \frac{3F_f}{k}\right)\cos \omega t - \frac{F_f}{k} \qquad (2.81)$$

and

$$\dot{x}(t) = -\omega\left(x_0 - \frac{3F_f}{k}\right)\sin \omega t \qquad (2.82)$$

From Eq. (2.82), we observe again that the velocity at the end of the second half-cycle, that is, at $\omega t = 2\pi$, is zero; and from Eq. (2.81), we find that the displacement at $t = 2\pi/\omega$ is

$$x_{t = 2\pi/\omega} = x_0 - \frac{4F_f}{k} \qquad (2.83)$$

This shows that, in each cycle of motion, the amplitude is reduced by the amount $4F_f/k$. This quantity represents the rate of amplitude decay for Coulomb damping. This decay is linear, and it is shown schematically in Fig. 2.6. When the amplitude becomes less than F_f/k, the spring force is no longer able to overcome the static friction force and the motion will cease,

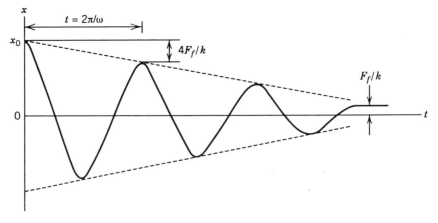

FIGURE 2.6. Rate of decay of amplitude for a free vibration with Coulomb damping.

as shown in the figure. The oscillatory motion is harmonic, and the frequency of oscillation ω is the same as the free undamped frequency of the spring–mass system; that is, $\omega = \sqrt{k/m}$.

Example 2.6. A weight $W = 800$ N is attached to a spring of constant $k = 60.0$ kN/m and slides back and forth on a dry surface. The coefficient of friction μ between the weight and the surface is 0.30, and the motion is initiated after the weight W was displaced by an initial displacement $x_0 = 20.0$ cm. Determine the displacement of the weight W at the end of the fourth cycle and the tenth cycle of motion.

Solution. The frictional force F_f is

$$F_f = \mu W = (0.30)(800) = 240 \text{ N}$$

Per each cycle of motion, the amplitude is reduced by the amount

$$\frac{4F_f}{k} = \frac{(4)(240)}{(60 \times 10^3)} = 0.016 \text{ m}$$
$$= 1.60 \text{ cm}$$

At the end of the fourth cycle of motion, the displacement x of the weight may be determined from the equation

$$x = x_0 - (4)\frac{4F_f}{k}$$
$$= 20.0 - (4)(1.60) = 13.60 \text{ cm}$$

At the end of the tenth cycle of motion, we have

$$x = 20.0 - (10)(1.60) = 4.0 \text{ cm}$$

The number of cycles n required for the amplitude to be reduced to zero may be obtained from the equation

$$x_0 - n\frac{4F_f}{k} = 0 \tag{2.84}$$

or

$$n = \frac{x_0 k}{4F_f} \tag{2.85}$$

This yields

$$n = \frac{(0.20 \text{ m})(60 \times 10^3) \text{ N/m}}{(4)(240 \text{ N})} = 12.50 \text{ cycles}$$

However, since the spring force has to overcome the static friction force $F_f = 240.0$ N, the weight W would retain the displacement $F_f/k = 0.40$ cm when it comes to rest.

2.6 FREE VIBRATION WITH HYSTERESIS DAMPING

Basic aspects of hysteresis damping were discussed in Section 1.4, where Eq. (1.17) provides the energy loss per cycle of motion. The magnitude of the energy loss is dependent on the size of the area within the hysteresis loop shown plotted in Fig. 1.3b. For many rubber-type materials, the area within the loop is large when it is compared to metallic materials such as steel. For many engineering structures, however, the amount of hysteresis damping is rather small and is often neglected in the analysis.

In the case of one-degree-of-freedom spring–mass systems, the effects of hysteresis damping could be taken into consideration by determining an equivalent viscous damping ratio ζ_e and an equivalent viscous damping factor c_e. On this basis, the vibration response of the spring–mass system that incorporates hysteresis damping may be determined from the following expression:

$$y(t) = e^{-\zeta_e \omega t}[A_1 \cos(\omega\sqrt{1 - \zeta_e^2})t + A_2 \sin(\omega\sqrt{1 - \zeta_e^2})t] \qquad (2.86)$$

where A_1 and A_2 are constants that can be determined from the initial conditions of the motion. Equation (2.86) is similar to Eq. (2.45), which was derived in Section 2.3. In this case, the damped frequency of vibration ω_d for hysteresis damping may be obtained from the equation

$$\omega_d = \omega\sqrt{1 - \zeta_e^2} \qquad (2.87)$$

where ω in this equation is the free undamped frequency.

If at time $t = 0$ the initial displacement and initial velocity are y_0 and $\dot{y}_0$, respectively, we find

$$A_1 = y_0 \qquad (2.88)$$

$$A_2 = \frac{\dot{y}_0 + \zeta \omega y_0}{\omega_d} \tag{2.89}$$

and Eq. (2.86) may be written

$$y(t) = e^{-\zeta_e \omega t} \left[y_0 \cos \omega_d t + \frac{\dot{y}_0 + \zeta \omega y_0}{\omega_d} \sin \omega_d t \right] \tag{2.90}$$

The damped frequency ω_d in Eq. (2.90) may be obtained from Eq. (2.87). Therefore, when an equivalent damping ratio ζ_e and an equivalent viscous damping factor c_e are used, the single-degree spring–mass system can be analyzed as in Section 2.3, where viscous damping is discussed.

The equivalent damping ratio ζ_e and equivalent viscous damping factor c_e may appropriately be determined by considering the consecutive picks A, B, and C in Fig. 2.7 and applying energy principles. For example, if the energy loss per quarter cycle is assumed to be $k\pi c_0 y^2/4$, the energy equation for the half-cycle between A and B in Fig. 2.7 may be written

$$\frac{kY_1^2}{2} - \frac{k\pi c_0 Y_1^2}{4} - \frac{k\pi c_0 Y_2^2}{4} = \frac{kY_2^2}{2} \tag{2.91}$$

With some rearrangements and mathematical manipulations, Eq. (2.91) may

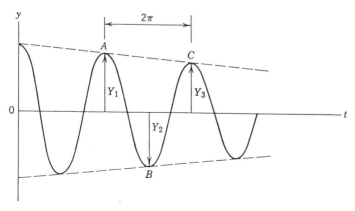

FIGURE 2.7

be written

$$\left(\frac{Y_1}{Y_2}\right)^2 = \frac{2 + \pi c_0}{2 - \pi c_0} \tag{2.92}$$

Similarly, for the next half-cycle between B and C in Fig. 2.7, we have

$$\left(\frac{Y_2}{Y_3}\right)^2 = \frac{2 + \pi c_0}{2 - \pi c_0} \tag{2.93}$$

If the hysteresis damping constant c_0 is considered to be small, then the product of Eqs. (2.92) and (2.93) yields the approximate expression

$$\frac{Y_1}{Y_3} \approx 1 + \pi c_0 \tag{2.94}$$

Since the logarithmic decrement is defined as the natural logarithm of any two successive amplitudes Y_1 and Y_3 shown in Fig. 2.7, we find

$$\delta = \ln\left(\frac{Y_1}{Y_3}\right) \approx \pi c_0 \tag{2.95}$$

By using Eqs. (2.61) and (2.95), we find

$$2\pi\zeta_e \approx \pi c_0 \tag{2.96}$$

and

$$\zeta_e = \frac{c_0}{2} \tag{2.97}$$

By using Eqs. (2.40), (2.41), and (2.97), we may also write

$$c_e = c_c\zeta_e = c_0\sqrt{km} = \frac{c_0 k}{\omega} \tag{2.98}$$

Equations (2.97) and (2.98) provide the means to determine ζ_e and c_e, which may be used to transform hysteresis damping into an equivalent viscous damping. In this manner, the vibration analysis may be carried out by using the procedures established in Section 2.3.

In terms of the phase angle ϕ_d and a constant Y, the response $y(t)$ of the spring–mass system under the influence of hysteresis damping may be written:

$$y(t) = Ye^{-\zeta_e\omega t}\sin[(\omega\sqrt{1 - \zeta_e^2})t] \tag{2.99}$$

The damping ratio ζ_e for many practical problems is often small, and it may be possible to assume that the damped and undamped frequencies have the same magnitude $\omega_d = \omega = (k/m)^{1/2}$. In some cases, this assumption would facilitate the vibration analysis of the system without appreciable loss of accuracy.

Example 2.7. A weight $W = 120$ lb is suspended from a spring of constant $k = 80$ lb/in. Experimental results show that, during vibration, the ratio of two successive amplitudes, such as Y_1 and Y_3, in Fig. 2.7, is equal to 1.106. Determine the equivalent damping ratio ζ_e, the equivalent damping factor c_e, and the energy loss ΔU per cycle of motion when the amplitude $Y = 1.2$ in.

Solution. From Eq. (2.94), we find

$$1.106 = 1 + \pi c_0$$

or

$$c_0 = \frac{1.106 - 1}{\pi} = 0.03374$$

Thus, from Eq. (1.17), we find that the loss of energy ΔU per cycle of motion when $Y = 1.2$ in. is

$$\Delta U = (80)(0.0338)(1.2)^2 \pi$$

$$= 12.2112 \text{ in.·lb}$$

and from Eq. (2.97), we find

$$\zeta_e = \frac{c_0}{2} = 0.01687$$

The undamped frequency ω of the spring–mass system is

$$\omega = (k/m)^{1/2} = [(80)(386)/120]^{1/2}$$

$$= 16.04 \text{ rps}$$

Therefore, from Eq. (2.98), we obtain

$$c_e = \frac{c_0 k}{\omega} = \frac{(0.03374)(80)}{16.04}$$

$$= 0.1683 \text{ lb·s/in.}$$

With known ζ_e, the general vibration response of the spring–mass system is represented by Eq. (2.90), or Eq. (2.99), which represents an equivalent viscously damped motion.

2.7 TORSIONAL VIBRATION OF DISK–SHAFT SYSTEMS

Consider the one-degree disk–shaft system in Fig. 1.9a. The differential equation of its free undamped vibration is derived in Section 1.8 and is given by Eq. (1.38). We write this equation again:

$$J\ddot{\varphi} + k_t\varphi = 0 \tag{2.100}$$

where φ is the angular twist of the disk and shaft, k_t is the torsional spring constant of the shaft, and J is the mass moment of inertia of the disk about the axis of rotation.

The format of Eq. (2.100) is similar to the one given by Eq. (2.1). Since the vibration is harmonic, we may express φ as

$$\varphi = A \sin \omega t \tag{2.101}$$

where A is a constant. By substituting Eq. (2.101) into Eq. (2.100) and solving for the torsional frequency ω, we find

$$\omega = \sqrt{\frac{k_t}{J}} \tag{2.102}$$

In hertz, the torsional frequency f is

$$f = \frac{\omega}{2\pi} = \frac{1}{2\pi}\sqrt{\frac{k_t}{J}} \tag{2.103}$$

and the period of vibration τ is

$$\tau = \frac{1}{f} = 2\pi\sqrt{\frac{J}{k_t}} \tag{2.104}$$

The general solution of Eq. (2.100) is of the form

$$\varphi(t) = C_1 \sin \omega t + C_2 \cos \omega t \tag{2.105}$$

where C_1 and C_2 are constants that can be determined from the initial conditions of the motion. For example, if the initial angular displacement at $t = 0$ is φ_0, and the initial velocity is $\dot{\varphi}_0$, we find

$$\varphi(0) = \varphi_0 = C_1 \sin(0) + C_2 \cos(0)$$

or

$$C_2 = \varphi_0 \tag{2.106}$$

By using the velocity initial condition at $t = 0$ and the first derivative of Eq. (2.105) with respect to time, we find

$$\dot{\varphi}(0) = \dot{\varphi}_0 = C_1 \omega \cos(0) - C_2 \omega \sin(0)$$

or

$$C_1 = \frac{\dot{\varphi}_0}{\omega} \tag{2.107}$$

On this basis, Eq. (2.105) yields

$$\varphi(t) = \frac{\dot{\varphi}_0}{\omega} \sin \omega t + \varphi_0 \cos \omega t \tag{2.108}$$

The solution of Eq. (2.100) may also be given in terms of a constant Y and a phase angle ϕ_t, as was done in Section 2.2 for Eq. (2.1). This solution may be written

$$\varphi(t) = Y \sin(\omega t + \phi_t) \tag{2.109}$$

which is similar in format to Eq. (2.17) when the initial time t_0 is zero. The phase angle ϕ_t may be determined from the equation

$$\tan \phi_t = \frac{\omega \varphi_0}{\dot{\varphi}_0} \tag{2.110}$$

which is similar to Eq. (2.20).

If the disk–shaft system in Fig. 2.8 is under the influence of viscous

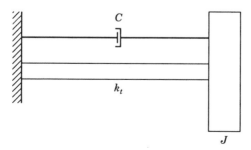

FIGURE 2.8. Disk–shaft system under the influence of viscous damping.

damping with damping constant c, its differential equation of motion is given by the following expression:

$$J\ddot{\varphi} + c\dot{\varphi} + k_t\varphi = 0 \qquad (2.111)$$

Equation (2.111) is identical to Eq. (1.37) in Section 1.8.

The form of Eq. (2.111) is similar to the one given by Eq. (2.24) in Section 2.3, and the solution procedure used in that section may also be applied here. By using this procedure, the solution of Eq. (2.111) in terms of the damping ratio ζ is

$$\varphi(t) = e^{-\zeta\omega t}[C_1 \cos(\omega\sqrt{1 - \zeta^2})t + C_2 \sin(\omega\sqrt{1 - \zeta^2})t] \qquad (2.112)$$

where ω in this equation is the free, undamped, torsional frequency of the disk–shaft system, and it can be determined from Eq. (2.102). The viscously damped torsional frequency ω_d of the disk–shaft system may be obtained from the equation

$$\omega_d = \omega\sqrt{1 - \zeta^2} \qquad (2.113)$$

The constants C_1 and C_2 in Eq. (2.112) may be determined from the initial conditions of the vibratory motion. For example, if at $t = 0$ the initial torsional displacement is φ_0 and the initial torsional velocity is $\dot{\varphi}_0$, application of Eq. (2.112) yields

$$C_1 = \varphi_0 \qquad (2.114)$$

$$C_2 = \frac{\dot{\varphi}_0 + \zeta\omega\varphi_0}{\omega_d} \qquad (2.115)$$

and, consequently,

$$\varphi(t) = e^{-\zeta\omega t}\left(\varphi_0 \cos \omega_d t + \frac{\dot{\varphi}_0 + \zeta\omega\varphi_0}{\omega_d} \sin \omega_d t\right) \qquad (2.116)$$

Again, we observe here that the first term of Eq. (2.116) is proportional to $\cos \omega_d t$ and depends only on the initial torsional displacement φ_0. The second term of the same equation is proportional to $\sin \omega_d t$ and depends on both the initial torsional displacement φ_0 and the initial torsional velocity $\dot{\varphi}_0$.

Equation (2.116) may also be written in terms of a constant Φ and the phase angle ϕ_d as follows:

$$\varphi(t) = \Phi e^{-\zeta\omega t} \sin(\omega_d t + \phi_d) \qquad (2.117)$$

where

$$\Phi = \sqrt{C_1^2 + C_2^2} = \sqrt{\varphi_0^2 + \frac{(\dot{\varphi}_0 + \zeta\omega\varphi_0)^2}{\omega_d^2}} \qquad (2.118)$$

and

$$\phi_d = \tan^{-1}\left(\frac{C_2}{C_1}\right) = \tan^{-1}\left(\frac{\dot{\varphi}_0 + \zeta\omega\varphi_0}{\omega_d\varphi_0}\right) \qquad (2.119)$$

The damped torsional period of vibration τ_d may be determined from the equation

$$\tau_d = \frac{1}{f_d} = \frac{2\pi}{\omega_d} = \frac{2\pi}{\omega\sqrt{1 - \zeta^2}} \qquad (2.120)$$

The following examples illustrate the application of the above methodologies.

Example 2.8. An aluminum disk is shrunk onto a steel shaft to form the arrangement shown in Fig. 1.9a. The length L of the shaft is 80 in., the diameter d of the shaft is 2 in., the steel shear modulus G is 12×10^6 psi, and the shaft is assumed to be weightless. If the mass moment of inertia J of the disk is 2.8 in.·lb·s^2, determine the torsional frequency of the one-degree disk–shaft system and its period of vibration.

Solution. From the mechanics of solids, the angle of twist φ of the shaft is given by the expression

$$\varphi = \frac{TL}{GI_p} \qquad (2.121)$$

where T is the applied torque at the free end of the shaft, and I_p is the polar moment of inertia of the shaft's cross section. The torsional stiffness k_t of the shaft is the value of the torque T that makes φ equal to unity: that is, 1 radian. Therefore, for $\varphi = 1$ radian, we have

$$T = k_t = \frac{GI_p}{L} \qquad (2.122)$$

The polar moment of inertia I_p of the shaft is

$$I_p = \frac{\pi d^4}{32} = \frac{\pi(2)^4}{32} = 1.57 \text{ in}^4.$$

Thus, from Eq. (2.122), we find

$$k_t = \frac{(12 \times 10^6)(1.57)}{80} = 235,500 \text{ lb·in./rad}$$

From Eq. (2.102), we find

$$\omega = \sqrt{\frac{235,500}{2.8}} = 290.0 \text{ rps}$$

In hertz, we have

$$f = \frac{\omega}{2\pi} = 46.16 \text{ Hz}$$

The period of vibration τ is

$$\tau = \frac{1}{f} = \frac{1}{46.16} = 0.0217 \text{ s}$$

Example 2.9. The disk–shaft system in Fig. 2.8 is subjected to a torsional oscillatory motion that is under the influence of viscous damping. If $k_t = 20,000 \text{ N·m/rad}$ and $J = 12.6 \text{ N·m·s}^2$, determine the value of the viscous damping constant c so that the damped torsional frequency ω_d of the disk–shaft system is 80% of its free, undamped, torsional frequency ω. Also, calculate the values of ω, ω_d, τ, and τ_d.

Solution. By using Eq. (2.102), we find that the free, undamped, torsional frequency ω of the disk–shaft system is

$$\omega = \sqrt{\frac{k_t}{J}} = \sqrt{\frac{20,000}{12.6}}$$

$$= 39.84 \text{ rps}$$

Therefore, the damped torsional frequency ω_d is

$$\omega_d = 0.80\omega = (0.80)(39.84)$$
$$= 31.87 \text{ rps}$$

From Eq. (2.113), when $\omega_d = 0.80\omega$, we find

$$0.80\omega = \omega\sqrt{1 - \zeta^2}$$

or

$$\zeta = 0.6$$

The damping ratio is $\zeta = c/c_c$, where c_c is the critical viscous damping factor for free torsional vibration. The critical damping factor c_c is given by the equation

$$c_c = 2J\omega \qquad (2.123)$$

On this basis, we have

$$c = \zeta c_c = 2J\omega\zeta$$
$$= (2)(12.6)(39.84)(0.6)$$
$$= 602.38 \text{ N} \cdot \text{m} \cdot \text{s/rad}$$

The values of τ and τ_d are

$$\tau = \frac{2\pi}{\omega} = \frac{2\pi}{39.84} = 0.1577 \text{ s}$$

$$\tau_d = \frac{2\pi}{\omega_d} = \frac{2\pi}{31.87} = 0.1972 \text{ s}$$

The above results indicate that ζ, which is the ratio of c/c_c, must be equal to 0.6 in order to be able to reduce the free, undamped, torsional frequency ω by 20%. This is an enormous amount of damping required to accomplish the task, because in many practical problems the amount of damping is small.

It should also be noted that Eq. (2.123) is identical to Eq. (2.40) in Section 2.3 if m is replaced by J.

2.8   FREE UNDAMPED VIBRATION OF SPRING–MASS SYSTEMS WITH TWO OR MORE DEGREES OF FREEDOM

Consider the spring–mass system in Fig. 2.9a, which consists of two masses, m_1 and m_2, and three springs with constants k_1, k_2, and k_3. If masses m_1 and m_2 are restricted to move in the vertical direction only, the system will have two degrees of freedom. The free-body diagrams of m_1 and m_2 are shown in Fig. 2.9b. By using the free-body diagrams of masses m_1 and m_2 and applying Newton's second law of motion, we find the following two differential equations of motion:

$$m_1\ddot{y}_1 - k_2(y_2 - y_1) + k_1 y_1 = 0 \tag{2.124}$$

$$m_2\ddot{y}_2 + k_2(y_2 - y_1) + k_3 y_2 = 0 \tag{2.125}$$

Equations (2.124) and (2.125) represent the free, undamped, vibratory motion of the spring–mass system. Since the vibration is harmonic, the displacements y_1 and y_2 of masses m_1 and m_2, respectively, may be expressed as follows:

$$y_1 = Y_1 \cos \omega t \quad \text{and} \quad y_2 = Y_2 \cos \omega t \tag{2.126}$$

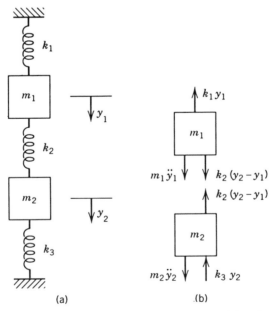

(a)   (b)

FIGURE 2.9. (a) Spring–mass system with two degrees of freedom. (b) Free-body diagrams of masses m_1 and m_2.

where Y_1 and Y_2 are the amplitudes of vibration. By substituting Eq. (2.126) into Eqs. (2.124) and (2.125) and carrying out the required mathematics and rearrangements, we find

$$(-m_1\omega^2 + k_1 + k_2)Y_1 - k_2Y_2 = 0 \tag{2.127}$$

$$-k_2Y_1 + (-m_2\omega^2 + k_2 + k_3) = 0 \tag{2.128}$$

Equations (2.127) and (2.128) are the two homogeneous auxiliary equations of the spring–mass system. They can be satisfied with any values of the amplitudes Y_1 and Y_2 other than $Y_1 = Y_2 = 0$ only if the determinant of the coefficients of Y_1 and Y_2 is zero; that is, we must have

$$\begin{vmatrix} -m_1\omega^2 + k_1 + k_2 & -k_2 \\ k_2 & -m_2\omega^2 + k_2 + k_3 \end{vmatrix} = 0 \tag{2.129}$$

Equation (2.129) is known as the *frequency determinant*. The *frequency equation* may be obtained by expanding the determinant in Eq. (2.129) in accordance with the rules of mathematics. This yields the following frequency equation:

$$m_1m_2\omega^4 - (m_1k_2 + m_1k_3 + m_2k_1 + m_2k_2)\omega^2 + k_1k_2 + k_1k_3 + k_2k_3 = 0 \tag{2.130}$$

Equation (2.130) is a biquadratic equation, and its solution yields four values of the frequency ω of the spring–mass system. Two of these roots are positive, while the other two are negative. The two negative values of the frequency ω have no physical significance and can be ignored.

In terms of ω^2, the solution of Eq. (2.130) yields the following expression for the roots ω_1^2 and ω_2^2:

$$\omega_{1,2}^2 = \frac{(m_1k_2 + m_1k_3 + m_2k_1 + m_2k_2)}{2m_1m_2}$$
$$\pm \frac{[(m_1k_2 + m_1k_3 + m_2k_1 + m_2k_2)^2 - 4m_1m_2(k_1k_2 + k_1k_3 + k_2k_3)]^{1/2}}{2m_1m_2} \tag{2.131}$$

When numerical values are assigned for the m's and k's in the above equation, the values for ω_1^2 and ω_2^2 are obtained. The four roots of ω are $\pm\omega_1$ and $\pm\omega_2$, but the two negative values, $-\omega_1$ and $-\omega_2$, are usually of no physical significance.

When the spring–mass system vibrates with frequencies ω_1 and ω_2, there is a definite relationship between the amplitudes Y_1 and Y_2 in each case. The amplitude relationship between Y_1 and Y_2 for each positive value of ω

is known as the mode shape, and it can be determined by using either Eq. (2.127) or Eq. (2.128). For example, from Eq. (2.127), we find

$$Y_1 = \frac{k_2}{(-m_1\omega^2 + k_1 + k_2)} Y_2 \qquad (2.132)$$

For $\omega = \omega_1$, Eq. (2.132) yields

$$Y_1^{(1)} = \frac{k_2}{(-m_1\omega_1^2 + k_1 + k_2)} Y_2^{(1)} \qquad (2.133)$$

For $\omega = \omega_2$, we find

$$Y_1^{(2)} = \frac{k_2}{(-m_1\omega_2^2 + k_1 + k_2)} Y_2^{(2)} \qquad (2.134)$$

The superscripts (1) in Eq. (2.133) and (2) in Eq. (2.134) are used to denote first and second mode, respectively. The values of ω_1 and ω_2 are determined from Eq. (2.131). Since only the ratio of the amplitudes Y_1 and Y_2 is used to describe the modes of vibration, it is customary to assign a value of unity to one of these amplitudes. On this basis, the modes are said to be normalized to unity. This procedure is permissible because the frequencies ω_1 and ω_2 are dependent only on the shape of the mode and not on the actual amplitudes of the mode shape.

Spring–mass systems with more than two degrees of freedom can be treated in a similar manner. The computations, however, become increasingly tedious, and the use of computers or other methods of analysis would be more appropriate for such cases. These situations are discussed in detail later in this text.

The same expressions and conclusions regarding the frequencies and mode shapes of the two-degree spring–mass system in Fig. 2.9a can be drawn, by assuming that the solutions of Eqs. (2.124) and (2.125) are given by the expressions

$$y_1 = Y_1 e^{\psi t} \quad \text{and} \quad y_2 = Y_2 e^{\psi t} \qquad (2.135)$$

and proceeding as above. Thus, by substituting the expressions in Eq. (2.135) into Eqs. (2.124) and (2.125) and carrying out the required mathematical manipulations, we obtain

$$(m_1\psi^2 + k_1 + k_2)Y_1 - k_2 Y_2 = 0 \qquad (2.136)$$

$$-k_2 Y_1 + (m_2\psi^2 + k_2 + k_3)Y_2 = 0 \qquad (2.137)$$

By comparing Eqs. (2.136) and (2.137) with the auxiliary equations given by Eqs. (2.127) and (2.128), we can conclude that $\psi^2 = -\omega^2$, and, consequently, the four values of ψ would be $\psi_1 = i\omega_1$, $\psi_2 = -i\omega_1$, $\psi_3 = i\omega_2$, and $\psi_4 = -i\omega_2$, where $i = \sqrt{-1}$. Therefore, by using these values of ψ and the amplitude relationships given by Eqs. (2.133) and (2.134), the displacement functions y_1 and y_2 in Eq. (2.135) yield the following expressions:

$$y_1 = (Y_1)_1 e^{i\omega_1 t} + (Y_1)_2 e^{-i\omega_1 t} + (Y_1)_3 e^{i\omega_2 t} + (Y_1)_4 e^{-i\omega_2 t} \tag{2.138}$$

$$y_2 = C_1[(Y_1)_1 e^{i\omega_1 t} + (Y_1)_2 e^{-i\omega_1 t}] + C_2[(Y_1)_3 e^{i\omega_2 t} + (Y_1)_4 e^{-i\omega_2 t}] \tag{2.139}$$

where

$$C_1 = \frac{k_2}{(-m_1\omega_1^2 + k_1 + k_2)} \tag{2.140}$$

$$C_2 = \frac{k_2}{(-m_1\omega_2^2 + k_1 + k_2)} \tag{2.141}$$

By applying Euler's relation, the trigonometric expressions of Eqs. (2.138) and (2.139) are written as follows:

$$y_1 = A_1 \cos \omega_1 t + A_2 \sin \omega_1 t + A_2 \cos \omega_2 t + A_4 \sin \omega_2 t \tag{2.142}$$

$$y_2 = C_1(A_1 \cos \omega_1 t + A_2 \sin \omega_1 t) + C_2(A_3 \cos \omega_2 t + A_4 \sin \omega_2 t) \tag{2.143}$$

Equations (2.142) and (2.143) show that the motion is composed of two harmonic motions corresponding to the free frequencies ω_1 and ω_2 of the spring–mass system. When the spring–mass system vibrates with frequency ω_1 only, Eq. (2.143) yields the amplitude relationship

$$y_2 = C_1 y_1 \tag{2.144}$$

From the same equation, we find that

$$y_2 = C_2 y_1 \tag{2.145}$$

when the spring–mass system vibrates with frequency ω_2 only.

Similar conclusions regarding mode shapes may also be drawn here. The constants A_1, A_2, A_3, and A_4 in Eqs. (2.142) and (2.143) may be determined by applying the initial conditions of the motion. For example, we may have

$$\text{at } t = t_0 \quad \begin{aligned} y_1 &= y_1^0 & \dot{y}_1 &= \dot{y}_1^0 \\ y_2 &= y_2^0 & \dot{y}_2 &= \dot{y}_2^0 \end{aligned} \tag{2.146}$$

The above initial conditions can be modified and written in terms of initial conditions that take into account the normal modes of the spring–mass system. From Eqs. (2.142) and (2.143), we note that

$$y_1 = y_1^{(1)} + y_1^{(2)} \tag{2.147}$$

$$y_2 = y_2^{(1)} + y_2^{(2)} = C_1 y_1^{(1)} + C_2 y_1^{(2)} \tag{2.148}$$

where the superscripts (1) and (2) denote the first and second mode, respectively. Therefore, the initial conditions in Eq. (2.146) may be written as follows:

$$
\begin{aligned}
\text{at } t = t_0 \qquad y_1 &= y_1^0 = y_1^{0(1)} + y_1^{0(2)} \\
y_2 &= y_2^0 = y_2^{0(1)} + y_2^{0(2)} = C_1 y_1^{0((1)} + C_2 y_1^{0(2)} \\
\dot{y}_1 &= \dot{y}_1^0 = \dot{y}_1^{0(1)} + \dot{y}_1^{0(2)} \\
\dot{y}_2 &= \dot{y}_2^0 = \dot{y}_2^{0(1)} + \dot{y}_2^{0(2)} = C_1 \dot{y}_1^{0(1)} + C_2 \dot{y}_1^{0(2)}
\end{aligned}
\tag{2.149}
$$

Simultaneous solution of the above four expressions in Eq. (2.149) will yield the four initial conditions $y_1^{0(1)}$, $y_1^{0(2)}$, $\dot{y}_1^{0(1)}$, and $\dot{y}_1^{0(2)}$ for the two modes of vibration of the spring–mass system.

Example 2.10. For the two-degree spring–mass system in Fig. 2.10a, determine the free undamped frequencies of vibration and the corresponding mode shapes. The spring–mass system is restricted to move in the vertical direction only.

Solution. The values of the free undamped frequencies of vibration ω_1^2 and ω_2^2 may be determined from Eq. (2.131). This equation yields

$$\omega_{1,2}^2 = \frac{17}{6}\left(\frac{k}{m}\right) \pm \frac{\sqrt{157}}{6}\left(\frac{k}{m}\right)$$

Therefore

$$\omega_1^2 = \frac{17}{6}\left(\frac{k}{m}\right) - \frac{\sqrt{157}}{6}\left(\frac{k}{m}\right) = 0.75\,\frac{k}{m}$$

$$\omega_2^2 = \frac{17}{6}\left(\frac{k}{m}\right) + \frac{\sqrt{157}}{6}\left(\frac{k}{m}\right) = 4.93\,\frac{k}{m}$$

and

$$\omega_1 = 0.866\,\sqrt{\frac{k}{m}}$$

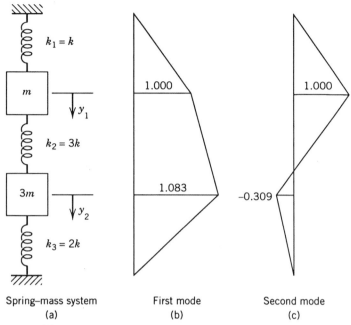

Spring–mass system
(a)

First mode
(b)

Second mode
(c)

FIGURE 2.10. (a) Spring–mass system consisting of two masses and three springs. (b) Shape of the first mode. (c) Shape of the second mode.

$$\omega_2 = 2.220 \sqrt{\frac{k}{m}}$$

The smallest value of ω, which in this case is $\omega_1 = 0.866 \sqrt{k/m}$, is called the *fundamental frequency* of the free undamped vibration, and the corresponding mode of vibration is called the fundamental mode.

The two modes of vibration may be determined from Eqs. (2.133) and (2.134). By using Eq. (2.133), the fundamental mode of vibration is represented by the relation

$$Y_1^{(1)} = \frac{3k}{[-m(0.75)(k/m) + k + 3k]} Y_2^{(1)}$$

or

$$Y_1^{(1)} = 0.923 Y_2^{(1)}$$

The second mode shape may be obtained by using Eq. (2.134). This equation yields

$$Y_1^{(2)} = \frac{3k}{[-m(4.93)(k/m) + k + 3k]} Y_2^{(2)}$$

or

$$Y_1^{(2)} = -3.226 Y_2^{(2)}$$

If Y_1 is assumed equal to unity, the normalized shapes of the two modes are as follows:

First Mode	Second Mode
$Y_1 = 1.000$	$Y_1 = 1.000$
$Y_2 = 1.083$	$Y_2 = -0.310$

The shapes of the normalized first and second modes of the free undamped vibration of the two-degree spring–mass system are shown in Figs. 2.10b and 2.10c, respectively. In the first mode, both masses are vibrating up and down in the same direction, while, in the second mode of vibration, the two masses are moving up and down but out of phase with each other.

Example 2.11. The disk in Fig. 2.11a is suspended from a vertical spring of spring constant k, and it can roll in the vertical direction about its static equilibrium position without slipping, as shown. The mass of the disk is m, its weight is W, the radius is r, and its mass moment of inertia about the centroid G is J_G. Determine the expressions for the vertical displacement y and the angular rotation θ.

Solution. If an initial displacement y_0 in the vertical direction is given to the disk and the disk is then released, the disk will vibrate in the vertical direction about its static equilibrium position, and it will also rotate about an axis vertical to the plane of the page and passing through the centroid G of the disk. If at any time t the vertical displacement y from the static equilibrium position is assumed to be small, then $y = r\theta$, and the inertia torque $J_G\ddot{\theta} = -J_G\omega^2\theta$ becomes equal to $-J_G\omega^2 y/r$.

The free-body diagram of the disk at any vertical displacement y from the static equilibrium position is shown in Fig. 2.11b. In this diagram, the inertia force $m\ddot{y}$ and the inertia torque $J_G\ddot{\theta}$ are directed in accordance with D'Alembert's principle for dynamic equilibrium. By taking moments about the instantaneous center of rotation B in Fig. 2.11b, the following equation is obtained:

$$-kyr + m\omega^2 yr + J_G\omega^2\theta = 0 \tag{2.150}$$

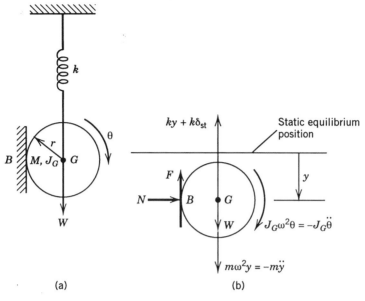

FIGURE 2.11. (a) Disk–spring system. (b) Free-body diagram of the disk.

In this equation, ω is the free undamped frequency of vibration in radians per second.

By substituting y/r for θ, Eq. (2.150) yields

$$(mr^2 + J_G)\omega^2 y - kr^2 y = 0 \tag{2.151}$$

By solving for ω, we obtain

$$\omega = \sqrt{\frac{kr^2}{mr^2 + J_G}} \tag{2.152}$$

The expression for the vertical displacement $y(t)$ may be obtained from Eq. (2.10) by substituting for ω the value obtained from Eq. (2.152). This yields

$$y(t) = y_0 \cos\left(\sqrt{\frac{kr^2}{mr^2 + J_G}}\right) t + \frac{\dot{y}_0}{\sqrt{kr^2/(mr^2 + J_G)}} \sin\left(\sqrt{\frac{kr^2}{mr^2 + J_G}}\right) t \tag{2.153}$$

In this equation, the initial displacement and initial velocity at $t = 0$ are taken as y_0 and $\dot{y}_0$, respectively.

Since for a disk its mass moment of inertia J_G about G is $mr^2/2$, Eq.

(2.153) yields

$$y(t) = y_0 \cos\left(\sqrt{\frac{2k}{3m}}\right)t + \frac{\dot{y}_0}{\sqrt{2k/3m}} \sin\left(\sqrt{\frac{2k}{3m}}\right)t \qquad (2.154)$$

The equation for θ is given by the following expression:

$$\theta(t) = \frac{y(t)}{r} = \frac{y_0}{r} \cos\left(\sqrt{\frac{2k}{3m}}\right)t + \frac{\dot{y}_0}{(\sqrt{2k/3m})r} \sin\left(\sqrt{\frac{2k}{3m}}\right)t \qquad (2.155)$$

The peculiarity of this problem is that the vertical and rotational displacements are coupled, and that the time required to complete a rotational cycle of motion is equal to the time required to complete a vertical translational cycle of motion. Therefore, the rotational and translational periods of the free undamped vibration are equal.

2.9   TWO-DEGREE SPRING–MASS SYSTEMS WITH VISCOUS DAMPING

In the preceding section, the spring–mass system was assumed to vibrate freely and there was no damping in the system to eventually eliminate the vibratory motion. Under such conditions, the vibratory motion will continue for an indefinite period of time. In real situations, however, some kind of damping is always present and the free vibration will die out with time. The presence of damping, as it will be shown below, usually complicates the analysis for dynamic and vibration response because of the many unknown factors and conditions involved in the analysis. It is fortunate, however, that, for many structural and mechanical problems, experience has shown that damping is usually low, at least during the time when the response of the system is maximum, and it can be neglected.

Let it be assumed in this section that the two-degree spring–mass system in Fig. 2.12a is under the influence of viscous damping as shown. The free-body diagrams in Fig. 2.12b depict the forces acting on masses m_1 and m_2. By applying Newton's second law of motion, the two differential equations of motion are as follows:

$$m_1 \ddot{y}_1 + c_1 \dot{y}_1 - c_2(\dot{y}_2 - \dot{y}_1) + k_1 y_1 - k_2(y_2 - y_1) = 0 \qquad (2.156)$$

$$m_2 \ddot{y}_2 + c_2(\dot{y}_2 - \dot{y}_1) + k_2(y_2 - y_1) = 0 \qquad (2.157)$$

The motion to be examined here is the one where y_1 and $\dot{y}_1$ are assumed to be smaller than y_2 and $\dot{y}_2$, respectively. This is only an assumption, because

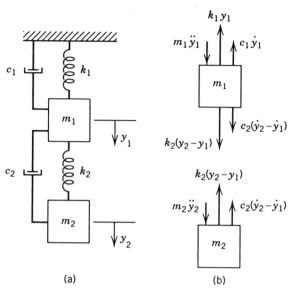

FIGURE 2.12. (a) Two degree spring–mass system under the influence of viscous damping. (b) Free-body diagrams of masses m_1 and m_2.

the equations would apply even if the opposite occured. With this in mind, the terms in Eqs. (2.156) and (2.157) are rearranged as follows:

$$m_1\ddot{y}_1 + (c_1 + c_2)\dot{y}_1 - c_2\dot{y}_2 + (k_1 + k_2)y_1 - k_2 y_2 = 0 \qquad (2.158)$$

$$m_2\ddot{y}_2 - c_2\dot{y}_1 + c_2\dot{y}_2 - k_2 y_1 + k_2 y_2 = 0 \qquad (2.159)$$

The solutions for the displacements y_1 and y_2 of masses m_1 and m_2, respectively, may be assumed to be of the form

$$y_1 = A_1 e^{\psi t} \qquad (2.160)$$

$$y_2 = A_2 e^{\psi t} \qquad (2.161)$$

where ψ, A_1, and A_2 are constants that need to be determined.

By substituting the solutions given in Eqs. (2.160) and (2.161) into Eqs. (2.158) and (2.159) and by rearranging the results in terms of A_1 and A_2, we find

$$[m_1\psi^2 + (c_1 + c_2)\psi + k_1 + k_2]A_1 - (c_2\psi + k_2)A_2 = 0 \qquad (2.162)$$

$$-(c_2\psi + k_2)A_1 + (m_2\psi^2 + c_2\psi + k_2)A_2 = 0 \qquad (2.163)$$

For a solution other than the trivial one, the determinant of the coefficients of A_1 and A_2 must be zero. This yields the following frequency determinant:

$$\begin{vmatrix} m_1 \psi^2 + (c_1 + c_2)\psi + k_1 + k_2 & -(c_2 \psi + k_2) \\ -(c_2 \psi + k_2) & m_2 \psi^2 + c_2 \psi + k_2 \end{vmatrix} = 0 \quad (2.164)$$

The characteristic equation may be obtained by expanding the determinant. This yields

$$m_1 m_2 \psi^4 + [m_1 c_2 + (c_1 + c_2)m_2]\psi^3 + [m_1 k_2 + (k_1 + k_2)m_2 + c_1 c_2]\psi^2$$
$$+ (c_1 k_2 + c_2 k_1)\psi + k_1 k_2 = 0 \quad (2.165)$$

This is an equation of the fourth degree, and, consequently, there are four values of ψ that satisfy this equation. When these values of ψ are appropriately substituted into Eqs. (2.160) and (2.161), the general solutions of Eqs. (2.158) and (2.159) may be obtained.

An examination of Eq. (2.165) reveals that the coefficients of ψ^4, ψ^3, ψ^2, and ψ, as well as the product $k_1 k_2$, are all positive. Therefore, a solution that would satisfy Eq. (2.165) is the case in which all four roots are complex. Since the two-degree spring–mass system is under the influence of viscous damping, the real part of each root is negative, and they can be written as follows:

$$\psi_1 = -a_1 + ib_1 \quad (2.166)$$

$$\psi_2 = -a_1 - ib_1 \quad (2.167)$$

$$\psi_3 = -a_2 + ib_2 \quad (2.168)$$

$$\psi_4 = -a_2 - ib_2 \quad (2.169)$$

In the above four equations, the quantities a_1, a_2, b_1, and b_2 are positive, and b_1 and b_2 provide the values of the two frequencies of vibration of the two-degree spring–mass system.

By substituting the above four values of ψ into the solutions given by Eqs. (2.160) and (2.161), we find

$$y_1 = (A_1)_1 e^{\psi_1 t} + (A_1)_2 e^{\psi_2 t} + (A_1)_3 e^{\psi_3 t} + (A_1)_4 e^{\psi_4 t} \quad (2.170)$$

$$y_2 = (A_2)_1 e^{\psi_1 t} + (A_2)_2 e^{\psi_2 t} + (A_2)_3 e^{\psi_3 t} + (A_2)_4 e^{\psi_4 t} \quad (2.171)$$

where ψ_1, ψ_2, ψ_3, and ψ_4 are given by Eqs. (2.166), (2.167), (2.168), and (2.169), respectively. By applying Euler's relation and simplifying, Eqs. (2.170) and (2.171) may be written in trigonometric form as follows:

$$y_1 = e^{-a_1 t}(C_1 \cos b_1 t + C_2 \sin b_1 t) + e^{-a_2 t}(C_3 \cos b_2 t + C_4 \sin b_2 t) \quad (2.172)$$

$$y_2 = e^{-a_1 t}(C_5 \cos b_1 t + C_6 \sin b_1 t) + e^{-a_2 t}(C_7 \cos b_2 t + C_8 \sin b_2 t) \quad (2.173)$$

Equations (2.172) and (2.173) show that the decay of the motion is exponential and is defined by the real parts a_1 and a_2 of the roots of ψ in Eqs. (2.166) through (2.169). The vibration of the spring–mass system, however, is characterized by the imaginary parts $\pm ib_1$ and $\pm ib_2$ in Eqs. (2.166) through (2.169), where b_1 and b_2 are the free damped frequencies of vibration of the spring–mass system. If damping is very light, a_1 and a_2 will be very small, and b_1 and b_2 will approach in the limit the free undamped frequencies ω_1 and ω_2 of the system.

There are eight constants in Eqs. (2.170) and (2.171), or in Eqs. (2.172) and (2.173), that need to be determined. The first step for the calculation of these eight constants includes the use of Eqs. (2.166) through (2.169) and the auxiliary equation (2.162) or (2.163). For example, by substituting the value of ψ_1 from Eq. (2.166) into Eq. (2.163), the ratio $(A_2)_1/(A_1)_1$ is determined. In a similar manner, the ratios $(A_2)_2/(A_1)_2$, $(A_2)_3/(A_1)_3$, and $(A_2)_4/((A_1)_4$ are obtained by using the values of ψ_2, ψ_3, and ψ_4, respectively, and Eq. (2.163). From these ratios, the ratios C_5/C_1, C_6/C_2, C_7/C_3, and C_8/C_4 may also be obtained, if desired, because Eqs. (2.172) and (2.173) were derived from Eqs. (2.170) and (2.171), respectively, by applying Euler's relation. The remaining four conditions that are needed to complete the computation of all eight constants are the initial conditions of the motion; that is,

$$\text{at } t = t_0 \qquad \begin{array}{cc} y_1 = y_1^0 & \dot{y}_1 = \dot{y}_1^0 \\ y_2 = y_2^0 & \dot{y}_2 = \dot{y}_2^0 \end{array} \qquad (2.174)$$

From Eq. (2.165), an additional set of four roots of ψ may be determined by substituting $-\psi$ for ψ in Eq. (2.165). Under this condition, the solution will yield four negative real roots of ψ. By analyzing this motion, as was done previously, we conclude that the motion is aperiodic, and it represents the case of a large amount of damping in the spring–mass system; that is, if the spring–mass system is subjected to a disturbance, it will return to its equilibrium position without oscillation.

The final possible set of four roots of ψ that could be obtained from Eq. (2.165) would consist of two real negative roots and a complex conjugate pair. In this case, for each mass of the spring–mass system, an aperiodic part of the motion is superimposed on a damped vibration. In practice, however, the amount of damping on structures and machines is usually light, and only the first case, if any, is usually of concern.

Example 2.12. The two-degree spring–mass system in Fig. 2.12a is vibrating freely under the influence of viscous damping, with damping coefficients $c_1 = 2.0\,\text{lb}\cdot\text{s/in.}$ and $c_2 = 1.0\,\text{lb}\cdot\text{s/in.}$ The spring constants are $k_1 = 500\,\text{lb/in.}$ and $k_2 = 300\,\text{lb/in.}$; the masses are $m_1 = 0.90\,\text{lb}\cdot\text{s}^2/\text{in.}$ and $m_2 = 0.30\,\text{lb}\cdot\text{s}^2/\text{in.}$ Determine the free damped frequencies of vibration of the spring–mass system.

Solution. By using the given values for m_1, m_2, k_1, k_2, c_1, and c_2 and the characteristic equation given by Eq. (2.165), we find

$$0.27\psi^4 + 1.80\psi^3 + 512.00\psi^2 + 900.00\psi + 150{,}000 = 0$$

or

$$\psi^4 + 6.6667\psi^3 + 1896.2963\psi^2 + 3333.3333\psi + 555{,}555.55 = 0 \quad (2.175)$$

This is a quartic equation of the form

$$\psi^4 + \alpha\psi^3 + \beta\psi^2 + \delta\psi + \gamma = 0 \tag{2.176}$$

The quartic equation (2.176) has the resolvent cubic equation

$$y^3 - \beta y^2 + (\alpha\delta - 4\gamma)y - \alpha^2\gamma + 4\beta\gamma - \delta^2 = 0 \tag{2.177}$$

which can be reduced to the form

$$w^3 + aw + b = 0 \tag{2.178}$$

if we substitute for y the value $w - (-\beta/3)$. The values of the constants a and b in Eq. (2.178) are

$$a = \tfrac{1}{3}[3\alpha\delta - 12\gamma - \beta^2]$$
$$= \tfrac{1}{3}[3(6.6667)(3333.3333) - 12(555{,}555.55) - (1896.2963)^2]$$
$$= -3{,}398{,}646.42$$
$$b = \tfrac{1}{27}[-2\beta^3 - 9(4\beta\gamma - \beta\alpha\delta) + 27(4\beta\gamma - \alpha^2\gamma - \delta^2)]$$
$$= \tfrac{1}{27}\{-2(1896.2963)^3 - 9[4(1896.2963)(555{,}555.55)$$
$$\quad -(1896.2963)(6.6667)(3333.3333)] + 27[4(1896.2963)(555{,}555.55)$$
$$\quad -(6.6667)^2(555{,}555.55) - (3333.3333)^2]\}$$
$$= 2{,}282{,}463{,}145.8$$

To determine the roots of w, we let

$$A = \left[-\frac{b}{2} + \sqrt{\frac{b^2}{4} + \frac{a^3}{27}} \right]^{1/3}$$

$$= \sqrt[3]{-\frac{2{,}282{,}463{,}145.8}{2} + \sqrt{\frac{(2{,}282{,}463{,}145.8)^2}{4} + \frac{(-3{,}398{,}646.42)^3}{27}}} \quad \text{(a)}$$

$$= \sqrt[3]{-1{,}141{,}231{,}572.9 + i389{,}302{,}320.0}$$

where $i = \sqrt{-1}$. We also have

$$B = \left[-\frac{b}{2} - \sqrt{\frac{b}{4} + \frac{a^3}{27}} \right]^{1/3}$$

$$= \sqrt[3]{-1{,}141{,}231{,}572.9 - i389{,}302{,}320.0} \quad \text{(b)}$$

Since A and B are complex numbers, the cube roots of A and B are determined by using known rules of complex algebra. They are:

$$A_1 = 629.8003 + i858.0407$$
$$A_2 = -1057.9851 + i116.4027$$
$$A_3 = 428.1849 - i947.4433$$
$$B_1 = 428.1849 + i974.4433$$
$$B_2 = -1057.9851 - i116.4027$$
$$B_3 = 629.8003 - i858.0407$$

With A and B known, the values of w may be determined from the equations

$$w_1 = A_3 + B_1$$
$$= 856.3698$$
$$w_2 = -\frac{A_3 + B_1}{2} + \frac{A_3 - B_1}{2}\sqrt{-3}$$
$$= 1259.6004$$
$$w_3 = -\frac{A_3 + B_1}{2} - \frac{A_3 - B_1}{2}\sqrt{-3}$$
$$= -2115.9702$$

The roots y_1, y_2, and y_3 of Eq. (2.177) may now be determined from the equation

$$y = w + \frac{\beta}{3} \tag{2.179}$$

That is,

$$y_1 = w_1 + \frac{\beta}{3}$$

$$= 856.3698 + \frac{1896.2963}{3} = 1488.4686$$

$$y_2 = w_2 + \frac{\beta}{3}$$

$$= 1259.6004 + \frac{1896.2963}{3} = 1891.6992$$

$$y_3 = w_3 + \frac{\beta}{3}$$

$$= -2115.9702 + \frac{1896.2963}{3} = -1483.8714$$

By following the rules regarding the solution of the quartic equation, which has the form given by Eq. (2.176), any one of the three roots of y may be used to determine the constant R, which is given by the following equation:

$$R = \sqrt{\frac{\alpha^2}{4} - \beta + y_{1,\,2,\,\text{or}\,3}} \tag{2.179a}$$

It would be advisable, when possible, to select the root of y that makes R a real number in order to reduce the complexity of the solution. In our case here, we could use $y_2 = 1891.6992$, which yields

$$R = \sqrt{\frac{(6.6667)^2}{4} - 1896.2963 + 1891.6992}$$

$$= 2.5523$$

Since R is different from zero, the required values of the parameters D and E may be determined from the following equations:

$$D = \sqrt{\frac{3\alpha^2}{4} - R^2 - 2\beta + \frac{4\alpha\beta - 8\delta - \alpha^3}{4R}} \qquad (2.179b)$$

$$E = \sqrt{\frac{3\alpha^2}{4} - R^2 - 2\beta - \frac{4\alpha\beta - 8\delta - \alpha^3}{4R}} \qquad (2.179c)$$

On this basis, Eqs. (2.179b) and (2.179c) yield the following values for D and E:

$$D = \left[\frac{(3)(6.6667)^2}{4} - (2.5523)^2 - (2)(1896.2963)\right.$$

$$\left. + \frac{(4)(6.6667)(1896.2963) - (8)(3333.3333) - (6.6667)^3}{(4)(2.5523)}\right]^{1/2}$$

$$= (-3765.7731 + 2312.1486)^{1/2}$$

$$= i38.1264$$

$$E = (-3765.7731 - 2312.1486)^{1/2}$$

$$= i77.9610$$

The four roots ψ_1, ψ_2, ψ_3, and ψ_4 of ψ are as follows:

$$\psi_1 = -\frac{\alpha}{4} + \frac{R}{2} + \frac{D}{2}$$

$$= -\frac{6.6667}{4} + \frac{2.5523}{2} + \frac{i38.1264}{2}$$

$$= -0.3905 + i19.0632$$

$$\psi_2 = -\frac{\alpha}{4} + \frac{R}{2} - \frac{D}{2}$$

$$= -0.3905 - i19.0632$$

$$\psi_3 = -\frac{\alpha}{4} - \frac{R}{2} + \frac{E}{2}$$

$$= -2.9429 + i38.9805$$

$$\psi_4 = -\frac{\alpha}{4} - \frac{R}{2} - \frac{E}{2}$$

$$= -2.9429 - i38.9805$$

The two damped frequencies of vibration ω_1 and ω_2 are

$$\omega_1 = 19.0632 \text{ rps}$$

$$\omega_2 = 38.9805 \text{ rps}$$

If damping were neglected, we would have $\omega_1 = 19.09$ rps and $\omega_2 = 39.0444$ rps. The difference is very small, which means that the amount of damping present in the two-degree spring–mass system does not alter appreciably the values of the free undamped frequencies. On the other hand, the vibration will die out after a definite period of time because the spring–mass system is vibrating under the influence of viscous damping. This is an important practical benefit that we get when damping is present.

2.10   TORSIONAL VIBRATION OF DISK–SHAFT SYSTEMS WITH TWO OR MORE DEGREES OF FREEDOM

Consider the two-degree disk–shaft system in Fig. 1.10a of Section 1.8. If the damping constants c_1 and c_2 and the exciting torques $T_1(t)$ and $T_2(t)$ are all zero, the case of free, undamped, torsional vibration is obtained. The differential equations of motion for the disk–shaft system are derived in Section 1.8: Eqs. (1.45) and (1.46). These two homogeneous differential equations are written again:

$$J_1\ddot{\varphi}_1 - k_t(\varphi_2 - \varphi_1) = 0 \tag{2.180}$$

$$J_2\ddot{\varphi}_2 + k_t(\varphi_2 - \varphi_1) = 0 \tag{2.181}$$

By assuming harmonic torsional vibration, the angular twists φ_1 and φ_2 of the two disks may be written

$$\varphi_1 = \Phi_1 \sin \omega t \tag{2.182}$$

$$\varphi_2 = \Phi_2 \sin \omega t \tag{2.183}$$

By substituting Eqs. (2.182) and (2.183) into Eqs. (2.180) and (2.181) and performing the required mathematical manipulations, the following two equations are obtained:

$$(k_t - J_1\omega^2)\Phi_1 - k_t\Phi_2 = 0 \tag{2.184}$$

$$-k_t\Phi_1 + (k_t - J_2\omega^2)\Phi_2 = 0 \tag{2.185}$$

For a solution other than the trivial one, the determinant of the coefficients of Φ_1 and Φ_2 must be zero; that is,

$$\begin{vmatrix} k_t - J_1\omega^2 & -k_t \\ -k_t & k_t - J_2\omega^2 \end{vmatrix} = 0 \tag{2.186}$$

Equation (2.186) is the frequency determinant. By expanding this determinant in accordance with the mathematical rules for determinants, we obtain the following frequency equation:

$$(k_t - J_1\omega^2)(k_t - J_2\omega^2) - k_t^2 = 0$$

or

$$J_1 J_2 \omega^4 - k_t(J_1 + J_2)\omega^2 = 0 \tag{2.187}$$

The solution of Eq. (2.187) yields the following two free frequencies of torsional vibration:

$$\omega_1^2 = 0$$

$$\omega_2^2 = k_t \frac{(J_1 + J_2)}{J_1 J_2}$$

This shows that the only nonzero positive torsional frequency is

$$\omega_2 = \left(k_t \frac{J_1 + J_2}{J_1 J_2}\right)^{1/2} \tag{2.188}$$

Consider now the disk–shaft arrangement shown in Fig. 2.13. By proceeding as in Section 1.8, the differential equations of motion are derived as follows:

$$J_1\ddot{\varphi}_1 + k_{t_1}\varphi_1 + k_{t_2}(\varphi_1 - \varphi_2) = 0 \tag{2.189}$$

$$J_2\ddot{\varphi}_2 + k_{t_2}(\varphi_2 - \varphi_1) = 0 \tag{2.190}$$

where φ_1 and φ_2 are the angular twists of the disks of mass moments of inertia J_1 and J_2, respectively, and k_{t_1} and k_{t_2} are the torsional spring stiffnesses of the two shafts as shown in Fig. 2.13.

The torsional frequencies of free torsional vibration and the corresponding mode shapes may be obtained from the solution of Eqs. (2.189) and (2.190),

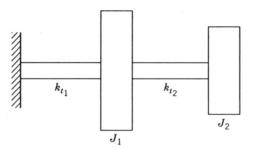

FIGURE 2.13. Disk–shaft arrangement.

as was done earlier. That is, by assuming harmonic torsional vibration, we express φ_1 and φ_2 as

$$\varphi_1 = \Phi_1 \sin \omega t \tag{2.191}$$

$$\varphi_2 = \Phi_2 \sin \omega t \tag{2.192}$$

By substituting Eqs. (2.191) and (2.192) into Eqs. (2.189) and (2.190) and performing the required mathematical manipulations, we obtain

$$(-J_1\omega^2 + k_{t_1} + k_{t_2})\Phi_1 - k_{t_2}\Phi_2 = 0 \tag{2.193}$$

$$-k_{t_2}\Phi_1 + (-J_2\omega^2 + k_{t_2})\Phi_2 = 0 \tag{2.194}$$

The frequency determinant may be obtained by using the coefficients of Φ_1 and Φ_2, as was done earlier. For a solution other than the trivial one, we set this determinant equal to zero; that is,

$$\begin{vmatrix} -J_1\omega^2 + k_{t_1} + k_{t_2} & -k_{t_2} \\ -k_{t_2} & -J_2\omega^2 + k_{t_2} \end{vmatrix} = 0 \tag{2.195}$$

Expansion of the above determinant yields the following frequency equation of torsional vibration:

$$J_1J_2\omega^4 - (k_{t_2}J_1 + k_{t_1}J_2 + k_{t_2}J_2)\omega^2 + k_{t_1}k_{t_2} = 0 \tag{2.196}$$

The roots ω_1^2 and ω_2^2 of Eq. (2.196) are

$$\omega_{1,2}^2 = \frac{k_{t_2}J_1 + k_{t_1}J_2 + k_{t_2}J_2}{2J_1J_2}$$

$$\pm \frac{[(k_{t_2}J_1 + k_{t_1}J_2 + k_{t_2}J_2)^2 - 4J_1J_2k_{t_1}k_{t_2}]^{1/2}}{2J_1J_2} \qquad (2.197)$$

The mode shapes corresponding to the two positive values ω_1 and ω_2 of the torsional frequencies ω may be obtained by using either Eq. (2.193) or Eq. (2.194). By using Eq. (2.194), we find

$$\Phi_2 = \frac{k_{t_2}}{k_{t_2} - J_2\omega^2}\Phi_1 \qquad (2.198)$$

For $\omega^2 = \omega_1^2$, the relationship of the angular amplitudes Φ_1 and Φ_2 of the fundamental mode of vibration may be obtained from Eq. (2.198) as follows:

$$\Phi_2^{(1)} = \frac{k_{t_2}}{k_{t_2} - J_2\omega_1^2}\Phi_1^{(1)} \qquad (2.199)$$

The second mode relationship for the amplitudes Φ_1 and Φ_2 may be obtained again from Eq. (2.198) by substituting ω_2^2 for ω^2. This yields

$$\Phi_2^{(2)} = \frac{k_{t_2}}{k_{t_2} - J_2\omega_2^2}\Phi_1^{(2)} \qquad (2.200)$$

The superscripts (1) in Eq. (2.199) and (2) in Eq. (2.200) denote the first and second mode, respectively, and the values of ω_1^2 and ω_2^2 in these equations may be determined from Eq. (2.197).

We note here that the general format of the above equations for torsional vibration of two-degree disk–shaft systems is similar to the one obtained for two-degree spring–mass systems discussed in Section 2.8. For example, Eq. (2.197) can be obtained directly from Eq. (2.131) by making $k_3 = 0$ and replacing the k's with k_t's and the m's with J's.

Example 2.13. Determine the free, undamped, torsional frequencies of vibration and the corresponding mode shapes of the disk–shaft system in Fig. 2.13. The shaft spring constants are $k_{t_1} = 30{,}000\ \text{N·m/rad}$ and $k_{t_2} = 20{,}000\ \text{N·m/rad}$; the mass moments of inertia of the disks are $J_1 = 16\ \text{N·m·s}^2$ and $J_2 = 10.5\ \text{N·m·s}^2$.

Solution. The free, undamped, torsional frequencies of vibration ω_1 and ω_2

may be determined from Eq. (2.197); that is,

$$\omega_{1,2}^2 = [(20{,}000)(16) + (30{,}000)(10.5) + (20{,}000)(10.5)]/(2)(16)(10.5)$$
$$\pm \{[(20{,}000)(16) + (30{,}000)(10.5) + (20{,}000)(10.5)]^2$$
$$- (4)(16)(10.5)(30{,}000)(20{,}000)\}^{1/2}/(2)(16)(10.5)$$
$$= 2514.88 \pm 1659.28$$

The fundamental torsional frequency of free vibration ω_1 is given by

$$\omega_1 = (2514.88 - 1659.28)^{1/2}$$
$$= 29.25 \text{ rps}$$

and the higher frequency of vibration ω_2 is given by

$$\omega_2 = (2514.88 + 1659.28)^{1/2}$$
$$= 64.61 \text{ rps}$$

The mode shapes corresponding to ω_1 and ω_2 may be determined from Eqs. (2.199) and (2.200). These two equations yield

$$\Phi_2^{(1)} = \frac{20{,}000}{20{,}000 - (10.5)(29.25)^2} \Phi_1^{(1)}$$
$$= 1.815\Phi_1$$

and

$$\Phi_2^{(2)} = \frac{20{,}000}{20{,}000 - (10.5)(64.61)^2} \Phi_1^{(2)}$$
$$= -0.839\Phi_1$$

If we make the assumption that $\Phi_1 = 1.000$, then the normalized torsional mode shapes corresponding to ω_1 and ω_2 are as follows:

Fundamental Mode Shape	Second Mode Shape
$\Phi_1 = 1.000$	$\Phi_1 = 1.000$
$\Phi_2 = 1.815$	$\Phi_2 = -0.839$

We note here that, in the fundamental mode of vibration, both disks are vibrating harmonically clockwise and counterclockwise, while, in the second mode of vibration, the two disks vibrate in opposite rotational directions.

For example, if the disk of mass moment of inertia J_1 is moving clockwise, the disk of mass moment of inertia J_2 is moving counterclockwise.

2.11   FREE TRANSVERSE VIBRATION OF UNIFORM BEAMS

The differential equation of motion for the free undamped vibration of elastic beams is derived in Section 1.9 and is given by Eq. (1.54). We rewrite this equation here as follows:

$$EI\frac{\partial^4 y}{\partial x^4} + m\frac{\partial^2 y}{\partial t^2} = 0 \qquad (2.201)$$

Equation (2.201), if modified appropriately, may be also used for elastic beams where the moment of inertia I and modulus of elasticity E can vary along the length of the member [1, 2, 7]. In this section, the case of uniform stiffness EI is considered. Beams with variable stiffness EI are considered in Chapter 8. It should be noted here that shear and rotational inertia effects are also neglected, but they are taken into consideration in Chapter 4.

The solution $y(t, x)$ of Eq. (2.201) may be assumed to be composed of a function $Y(x)$ that varies only with x, and a function $f(t)$ that varies only with time t; that is,

$$y(x, t) = Y(x)f(t) \qquad (2.202)$$

By substituting Eq. (2.202) into Eq. (2.201) and performing the required mathematical operations, we find

$$\frac{EI}{m}\frac{\partial^4 Y(x)/\partial x^4}{Y(x)} = -\frac{\partial^2 f(t)/\partial t^2}{f(t)} \qquad (2.202a)$$

In Eq. (2.202a), we note that the left-hand side is arranged so that it contains only functions of x, and the right-hand side contains only functions of time t.

Equation (2.202a) can be satisfied for all values if, and only if, each side of the equation is equal to the same constant. If this constant is taken to be equal to ω^2, we find

$$\frac{EI}{m}\frac{\partial^4 Y(x)/\partial x^4}{Y(x)} = \omega^2$$

$$-\frac{\partial^2 f(t)/\partial t^2}{f(t)} = \omega^2$$

or

$$\frac{\partial^4 Y(x)}{\partial x^4} - \lambda^4 Y(x) = 0 \qquad (2.203)$$

$$\frac{\partial^2 f(t)}{\partial t^2} + \omega^2 f(t) = 0 \qquad (2.204)$$

where

$$\lambda^4 = \frac{m\omega^2}{EI} \qquad (2.205)$$

Equation (2.203) is a fourth-order partial differential equation, and its solution yields the expression for the function $Y(x)$, which defines the shapes of the modes corresponding to the free undamped frequencies ω of an elastic beam. Since a member with continuous mass and elasticity has infinite degrees of freedom, the frequencies of vibration ω will be infinite in number; consequently, we are going to have an infinite number of functions $Y(x)$ that define the corresponding mode shapes. Equation (2.204) has the same form as Eq. (2.4) in Section 2.2, which deals with free, undamped, harmonic vibration of one-degree spring–mass systems. The solution is similar to the one given by Eq. (2.5) or (2.17). These solutions are written below as follows:

$$f(t) = A \sin \omega t + B \cos \omega t \qquad (2.206)$$

$$f(t) = C \sin(\omega t + \phi) \qquad (2.207)$$

In the above two equations, the initial time t_0 is taken as equal to zero; ϕ is the phase angle; and A, B, and C are constants. The pair of ϕ and C, or the pair of A and B, may be determined by applying the initial conditions of the vibratory motion.

The solution $Y(x)$ of Eq. (2.203) may be assumed as

$$Y(x) = Ce^{\psi x} \qquad (2.208)$$

where C and ψ are constants.

By substituting Eq. (2.208) into Eq. (2.203) and performing the required mathematics, we find

$$\psi^4 = \lambda^4 \qquad (2.209)$$

which yields the following roots of ψ:

$$
\begin{array}{ll}
\psi_1 = \lambda & \psi_3 = i\lambda \\
\psi_2 = -\lambda & \psi_4 = -i\lambda
\end{array}
\qquad (2.210)
$$

In Eq. (2.210), $i = \sqrt{-1}$. By using the roots of ψ given by Eq. (2.210), we find that Eq. (2.208) may be written

$$Y(x) = C_1 e^{\lambda x} + C_2 e^{-\lambda x} + C_3 e^{i\lambda x} + C_4 e^{-i\lambda x} \qquad (2.211)$$

By using the Eulerian relations

$$e^{\pm \lambda x} = \cosh \lambda x \pm \sinh \lambda x \qquad (2.212)$$

$$e^{\pm i\lambda x} = \cos \lambda x \pm i \sin \lambda x \qquad (2.213)$$

the trigonometric form of Eq. (2.211) is obtained:

$$Y(x) = A_1 \cos h \lambda x + A_2 \sinh \lambda x + A_3 \cos \lambda x + A_4 \sin \lambda x \qquad (2.214)$$

Equation (2.214) is the general solution of Eq. (2.203), and it may be used for beams of uniform mass m and stiffness EI along the length of the member. The constants A_1, A_2, A_3, and A_4 and the values of λ may be determined by using Eq. (2.214) and applying the beam's boundary conditions. The function $Y(x)$, which depends only on x, represents the mode shape of the beam, and it can be determined within one constant, as will be shown later in this section. For example, three boundary conditions may be used to determine three constants, say, A_1, A_2, and A_3, and the fourth boundary condition can be used to determine λ. Since λ in Eq. (2.205) is a function of the free frequency ω, for every value of λ we can determine a value of ω from Eq. (2.205).

By substituting now the solutions given by Eqs. (2.206) and (2.214) into Eq. (2.202), the general solution of Eq. (2.201) may be written as follows:

$$y(x, t) = (A_1 \cosh \lambda x + A_2 \sinh \lambda x + A_3 \cos \lambda x$$
$$+ A_4 \sin \lambda x)(A \sin \omega t + B \cos \omega t) \qquad (2.215)$$

or, if Eqs. (2.207) and (2.214) are used, we may write

$$y(x, t) = (A_1 \cosh \lambda x + A_2 \sinh \lambda x + A_3 \cos \lambda x$$
$$+ A_4 \sin \lambda x) \sin(\omega t + \phi) \qquad (2.216)$$

The constant C of Eq. (2.207) is absorbed by the constants A_1, A_2, A_3, and A_4.

Equation (2.215), or Eq. (2.216), represents the general solution of the free, undamped, transverse vibration of beam spans. The constants A_1, A_2, A_3, A_4, λ, A, and B in Eq. (2.115), or the constants A_1, A_2, A_3, A_4, ω, and ϕ in Eq. (2.216), may be determined by using the four boundary conditions of the beam span and the two initial time conditions of the motion, such as initial displacement and velocity at an initial time $t = 0$, or $t = t_0$. Note that, from Eq. (2.205), we find that the frequencies ω, in radians per second, may be determined from the following expression:

$$\omega = \lambda^2 \sqrt{\frac{EI}{m}} \qquad (2.217)$$

The corresponding period of vibration τ is

$$\tau = \frac{2\pi}{\omega} = \frac{2\pi}{\lambda^2} \sqrt{\frac{m}{EI}} \qquad (2.218)$$

The following example illustrates the application of the above theory.

Example 2.14. Determine the free undamped frequencies of vibration and the corresponding mode shapes of a simply supported beam of length L, uniform stiffness EI along its length, and uniform mass m per unit length. Also, write the general expression for the free undamped vibration of the beam.

Solution. The boundary conditions of the simply supported beam are as follows:

$$\text{at } x = 0 \qquad Y = 0 \qquad (2.219)$$

$$\frac{\partial^2 Y}{\partial x^2} = 0 \qquad (2.220)$$

$$\text{at } x = L \qquad Y = 0 \qquad (2.221)$$

$$\frac{\partial^2 Y}{\partial x^2} = 0 \qquad (2.222)$$

By using Eq. (2.214) and the two boundary conditions given by Eqs. (2.219) and (2.220), we find

$$A_1 + A_3 = 0$$

$$A_1 - A_3 = 0$$

Since A_1 cannot be equal to A_3 and $-A_3$, we must have

$$A_1 = A_3 = 0 \qquad (2.223)$$

Again, by using Eq. (2.214) and the two boundary conditions given by Eqs. (2.221) and (2.222), we find

$$A_2 \sinh \lambda L + A_4 \sin \lambda L = 0 \qquad (2.224)$$

$$A_2 \sinh \lambda L - A_4 \sin \lambda L = 0 \qquad (2.225)$$

where A_1 and A_3 are considered to be equal to zero. For a solution other than the trivial one—that is, $A_1 = A_2 = A_3 = A_4 = 0$—the determinant of the coefficients of A_2 and A_4 in Eqs. (2.224) and (2.225) must be zero; that is,

$$\begin{vmatrix} \sinh \lambda L & \sin \lambda L \\ \sinh \lambda L & -\sin \lambda L \end{vmatrix} = 0 \qquad (2.226)$$

Expansion of the above determinant yields the following frequency equation:

$$\sinh \lambda L \sin \lambda L = 0 \qquad (2.227)$$

In Eq. (2.227), $\sinh \lambda L$ will go to zero only when $\lambda L = 0$, which is the case of no vibration. Therefore, we must have

$$\sin \lambda L = 0 \qquad (2.228)$$

The roots of λL that satisfy Eq. (2.228) are

$$\lambda L = n\pi \qquad n = 1, 2, 3, \ldots \qquad (2.229)$$

or

$$\lambda = \frac{n\pi}{L} \qquad n = 1, 2, 3, \ldots \qquad (2.230)$$

The frequencies of vibration ω, in radians per second, can be determined from Eq. (2.217) by substituting for λ the expression given by Eq. (2.230). This yields

$$\omega_n = \frac{n^2 \pi^2}{L^2} \sqrt{\frac{EI}{m}} \qquad n = 1, 2, 3, \ldots \qquad (2.231)$$

In hertz, the frequencies of vibration f_n are given by the expression

$$f_n = \frac{\omega}{2\pi} = \frac{n^2\pi}{2L^2}\sqrt{\frac{EI}{m}} \qquad n = 1, 2, 3, \ldots \qquad (2.232)$$

The fundamental frequency f_1 has the smallest value, and it corresponds to $n = 1$ in Eq. (2.232).

Addition of Eqs. (2.224) and (2.225) yields the expression

$$A_2 \sinh \lambda L = 0 \qquad (2.233)$$

Again, in this case, we must have $A_2 = 0$, because $\sinh \lambda L$ cannot be equal to zero for the reasons stated earlier. Thus, with $A_1 = A_2 = A_3 = 0$, Eq. (2.214) yields

$$Y(x) = A_4 \sin \lambda x \qquad (2.234)$$

or, by using Eq. (2.230), we write

$$Y_n(x) = A_{4n} \sin \frac{n\pi x}{L} \qquad n = 1, 2, 3, \ldots \qquad (2.235)$$

Equation (2.235) provides the mode shapes of the simply supported beam with A_{4n} as constant. For $n = 1$ we obtain the fundamental mode of vibration $Y_1(x)$ corresponding to the fundamental frequency ω_1. Since the frequencies of vibration are mode shape dependent only and do not depend on the actual amplitudes of the mode, the constant A_{4n} may be taken as equal to unity. On this basis, the expression $\beta_n(x)$ for the mode shapes may be written

$$\beta_n = \sin \frac{n\pi x}{L} \qquad n = 1, 2, 3, \ldots \qquad (2.236)$$

By using Eq. (2.216), the general expression $y(x, t)$ for the free vibration of the member may be written

$$y(x, t) = \sum_{n=1}^{\infty} A_{4n} \sin \frac{n\pi x}{L} \sin(\omega_n t + \phi_n) \qquad (2.237)$$

which represents the superposition of the amplitudes of all modes of vibration.

The first three mode shapes of the simply supported beam are shown in

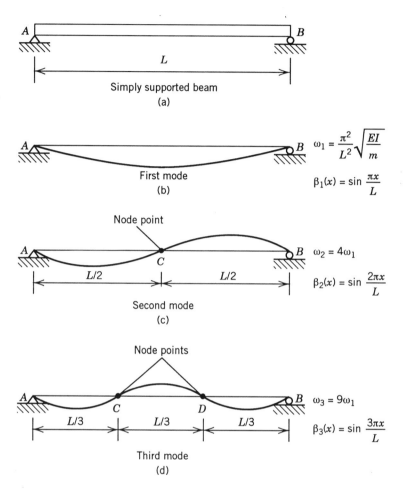

FIGURE 2.14. (a) Uniform simply supported beam. (b) First mode of vibration. (c) Second mode of vibration. (d) Third mode of vibration.

Figs. 2.14b, 2.14c, and 2.14d. The fundamental mode of vibration does not have any node points, the second node has a node point at $x = L/2$, the third mode has two node points at $x = L/3$ and $x = 2L/3$, and so on.

2.12 LONGITUDINAL VIBRATION OF UNIFORM BEAMS

The differential equation of motion for the free longitudinal undamped vibration of beams has been derived in Section 1.9, and it is given by Eq.

(1.50). We write this equation again here as follows:

$$E\frac{\partial^2 u}{\partial x^2} - \rho\frac{\partial^2 u}{\partial t^2} = 0 \tag{2.238}$$

By following a procedure similar to the one used in the preceding section, we may assume that the solution for the longitudinal displacement $u(x, t)$ is given by the equation

$$u(x, t) = U(x)f(t) \tag{2.239}$$

where $U(x)$ is only a function of x, and $f(t)$ depends only on the time t. By substituting Eq. (2.239) into Eq. (2.238) and carrying out the required mathematical manipulations, we find

$$\frac{E}{\rho}\frac{\partial^2 U(x)/\partial x^2}{U(x)} = \frac{\partial^2 f(t)/\partial t^2}{f(t)} \tag{2.240}$$

Equation (2. 240) will be satisfied for all values of x and t if each side of the equation is equal to the same constant. Note that the right-hand side of the equation is only a function of time t and the left-hand side of the equation depends only on x. If the constant is assumed to be equal to $-\omega^2$, the following differential equations may be obtained:

$$\frac{\partial^2 U(x)}{\partial x^2} + \left(\frac{\omega}{c}\right)^2 U(x) = 0 \tag{2.241}$$

$$\frac{\partial^2 f(t)}{\partial t^2} + \omega^2 f(t) = 0 \tag{2.242}$$

where

$$c = \sqrt{\frac{E}{\rho}} \tag{2.243}$$

The solution of Eq. (2.241) is given by the expression

$$U(x) = A \sin\frac{\omega x}{c} + B \cos\frac{\omega x}{c} \tag{2.244}$$

and the solution of Eq. (2.242) is

$$f(t) = C \sin(\omega t + \phi) \tag{2.245}$$

The constants A and B and the frequencies ω in Eq. (2.244) may be determined by using the boundary conditions of the member. The constants C and ϕ in Eq. (2.245) may be determined from the initial time conditions of the motion. Note that Eq. (2.245) is similar to Eq. (2.207) in Section 2.11. If preferred, the form of the solution given by Eq. (2.206) could be used.

By using Eqs. (2.244) and (2.245), the general solution given by Eq. (2.239) may now be written

$$u(x, t) = \left(A \sin \frac{\omega x}{c} + B \cos \frac{\omega x}{c} \right) \sin(\omega t + \phi) \tag{2.246}$$

where the constant C in Eq. (2.245) is absorbed by the constants A and B of Eq. (2.244).

The unknowns in Eq. (2.246) are the two constants A and B, the phase angle ϕ, and the free, undamped, longitudinal frequencies ω. They can be determined by using the two boundary conditions of the member and the two initial time conditions of the motion, such as displacement and velocity at an initial time $t = t_0$. The following example illustrates the application of the general methodology.

Example 2.15. For a uniform member of length L and fixed at both ends, determine the free, undamped, longitudinal frequencies of vibration and the corresponding mode shape functions. Also, write the general solution $u(x, t)$ of its longitudinal vibration.

Solution. The boundary conditions of the member are as follows:

$$\text{at } x = 0 \qquad u(0, t) = U(0) = 0 \tag{2.247}$$

$$x = L \qquad u(L, t) = U(L) = 0 \tag{2.248}$$

By using Eq. (2.244) and applying the boundary condition given by Eq. (2.247), we find that the constant B is equal to zero. By using the same equation and applying the boundary condition given by Eq. (2.248), we find

$$A \sin \frac{\omega L}{c} = 0 \tag{2.249}$$

If the constant A in Eq. (2.249) is zero, the member will not vibrate longitudinally, because both A and B are zero in Eq. (2.244). Therefore, in order to have vibration, we must have

$$\sin \frac{\omega L}{c} = 0 \tag{2.250}$$

Equation (2.250) suggests that the values of $\omega L/c$ that make $\sin(\omega L/c) = 0$ should be as follows:

$$\frac{\omega L}{c} = n\pi \qquad n = 1, 2, 3, \ldots \tag{2.251}$$

By using Eqs. (2.243) and (2.251), we find

$$\omega_n = \frac{n\pi}{L}\sqrt{\frac{E}{\rho}} \qquad n = 1, 2, 3, \ldots \tag{2.252}$$

In hertz, the longitudinal frequencies f_n of the member are

$$f_n = \frac{\omega_n}{2\pi} = \frac{n}{2L}\sqrt{\frac{E}{\rho}} \qquad n = 1, 2, 3, \ldots \tag{2.253}$$

By using Eqs. (2.252) and (2.244) and remembering that the constant B is zero, we may write the mode shapes of the free, undamped, longitudinal vibration as follows:

$$U_n(x) = A_n \sin\frac{n\pi x}{L} \qquad n = 1, 2, 3, \ldots \tag{2.254}$$

For $n = 1$, Eq. (2.252) will yield the fundamental frequency of the longitudinal vibration and Eq. (2.254) will yield the corresponding mode shape. Note that the mode shape in Eq. (2.254) is determined within a constant A_n. Since the longitudinal vibrational frequency ω_n depends only on the shape of the corresponding mode $U_n(x)$, the constant A_n in Eq. (2.254) may be assumed to be equal to unity, as was done in the preceding section.

By using Eqs. (2.246) and (2.254), the general solution of the longitudinal vibration of the member given by Eq. (2.239) may be written

$$u(x, t) = \sum_{n=1}^{\infty} A_n \sin\frac{n\pi x}{L} \sin(\omega_n t + \phi_n) \tag{2.255}$$

where ω_n in Eq. (2.255) is given by Eq. (2.252). The constant A_n and phase angle ϕ_n in Eq. (2.255) may be determined by using the initial conditions of the longitudinal motion.

2.13   ORTHOGONALITY PROPERTIES OF NORMAL MODES

Consider a system that consists of n masses and has n degrees of freedom. If the system vibrates at a mode i, the inertia force acting on a mass m_r is $-m_r\omega_i^2 y_r^{(i)}$. The superscript (i) is used here to denote the displacement $y_r^{(i)}$ of the mass m_r in the ith mode. When the system vibrates in the jth mode, the inertia force acting on the mass m_r is $-m_r\omega_j^2 y_r^{(j)}$. The superscript (j) is used here to denote the amplitude $y_r^{(j)}$ of the mass m_r in the jth mode.

By using D'Alembert's principle of dynamic equilibrium, it can easily be shown that, if all the inertia forces of a mode i are applied statically to the system, the displacement configuration of the ith mode will be obtained.

Let it now be assumed that the inertia forces of the ith mode are permitted to move through a virtual displacement corresponding to the shape of the jth mode. The work done by the inertia forces $m_r\omega_i^2 y_r^{(i)}$ of the ith mode moving through the displacements $y_r^{(j)}$ of the jth mode is

$$\sum_{r=1}^{n} (m_r\omega_i^2 y_r^{(i)}) y_r^{(j)} \tag{2.256}$$

If the reverse is now permitted to occur, the work done by the inertia forces $m_r\omega_j^2 y_r^{(j)}$ of the jth mode moving through the displacements $y_r^{(i)}$ of the ith mode is

$$\sum_{r=1}^{n} (m_r\omega_j^2 y_r^{(j)}) y_r^{(i)} \tag{2.257}$$

By virtue of the well-known Betti's law [8], the work given by Eq. (2.256) must be equal to the work represented in Eq. (2.257). Thus, we have

$$\sum_{r=1}^{n} (m_r\omega_i^2 y_r^{(i)}) y_r^{(j)} = \sum_{r=1}^{n} (m_r\omega_j^2 y_r^{(j)}) y_r^{(i)} \tag{2.258}$$

or

$$(\omega_i^2 - \omega_j^2) \sum_{r=1}^{n} m_r y_r^{(i)} y_r^{(j)} = 0 \tag{2.259}$$

Since $\omega_i^2 - \omega_j^2$ in Eq. (2.259) cannot be equal to zero because ω_i^2 and ω_j^2 are supposed to have different distinct and positive values, we must have

$$\sum_{r=1}^{n} m_r y_r^{(i)} y_r^{(j)} = 0 \tag{2.260}$$

Equation (2.260) is known as the *orthogonality property* or *orthogonality condition* of normal modes. It is an extremely important property, and it has found many applications in the vibration analysis of structural and mechanical systems, as will be shown in later parts of the text. The orthogonality property is valid for any pair combination of modes i and j of a system that has two or more degrees of freedom.

From a mathematical point of view, another possibility in Eq. (2.259) is to have $i = j$. In this case, $\omega_i = \omega_j$, and, consequently,

$$\sum_{r=1}^{n} m_r (y_r^{(i)})^2 = \text{any arbitrary constant} \qquad (2.261)$$

If the absolute values of $y_r^{(i)}$ in Eq. (2.261) are selected so that the sum of this equation is equal to unity, then the shape of the ith mode is said to be *normalized*, and Eq. (2.261) is the *normalizing condition*.

2.14 TORSIONAL VIBRATION OF UNIFORM BEAMS

The differential equation of motion for the free, undamped, torsional vibration of circular beams or shafts is derived in Section 1.9 and is given by Eq. (1.58). This equation is written again below:

$$\frac{\partial^2 \varphi}{\partial x^2} - \frac{\rho}{G} \frac{\partial^2 \varphi}{\partial t^2} = 0 \qquad (2.262)$$

In this equation, φ is the angle of twist, G is the shear modulus of the material, and ρ is the mass per unit volume.

The form of Eq. (2.262) is identical to the one in Eq. (2.238), which represents the free, undamped, longitudinal vibration of beams. Therefore, its solution $\varphi(x, t)$ may be expressed as

$$\varphi(x, t) = \Phi(x) f(t) \qquad (2.263)$$

By proceeding as in Section 2.13, we find the following expressions for $\Phi(x)$ and $f(t)$:

$$\Phi(x) = A \sin \frac{\omega x}{a} + B \cos \frac{\omega x}{a} \qquad (2.264)$$

$$f(t) = C \sin(\omega t + \phi) \qquad (2.265)$$

where

$$a = \sqrt{\frac{G}{\rho}} \qquad (2.266)$$

The constants A and B and the frequency ω (in radians per second) in Eq. (2.264) may be determined by using the boundary conditions of the member. The constant C and phase angle ϕ in Eq. (2.265) may be determined by using the initial conditions of the motion.

By using Eqs. (2.264) and (2.265), the general solution given by Eq. (2.263) may be written

$$\varphi(x, t) = \left(A \sin \frac{\omega x}{a} + B \cos \frac{\omega x}{a} \right) \sin(\omega t + \phi) \qquad (2.267)$$

where the expression for a is given by Eq. (2.266). The following example illustrates the application of the methodology.

Example 2.16. The ends of a uniform rod of length L are free. If the rod is subjected to free, undamped, torsional oscillation, determine its torsional frequencies of vibration and the corresponding mode shapes. Also, write the general expression $\varphi(x, t)$ for the torsional vibratory motion of the rod.

Solution. The partial derivative of Eq. (2.267) with respect to x is

$$\frac{\partial \varphi}{\partial x} = \frac{\omega}{a} \left(A \cos \frac{\omega x}{a} - B \sin \frac{\omega x}{a} \right) \sin(\omega t + \phi) \qquad (2.268)$$

At the free ends of the rod, the torque is zero, and, consequently, $\partial \varphi / \partial x = 0$. By applying the boundary condition that at $x = 0$ we have $\partial \varphi / \partial x = 0$, we find that $A = 0$. The boundary condition of $\partial \varphi / \partial x = 0$ at $x = L$ yields

$$B \frac{\omega}{a} \sin \frac{\omega L}{a} = 0 \qquad (2.269)$$

In order to have vibration, the constant B in Eq. (2.269) must be different from zero. Therefore, in order to satisfy Eq. (2.269), we must have

$$\sin \frac{\omega L}{a} = 0 \qquad (2.270)$$

The values of $\omega L / a$ that make $\sin(\omega L / a) = 0$ are

$$\frac{\omega L}{a} = n\pi \qquad n = 1, 2, 3, \ldots \qquad (2.271)$$

Solving for ω, we find

$$\omega_n = \frac{n\pi a}{L} \qquad n = 1, 2, 3, \ldots \qquad (2.272)$$

or, by using Eq. (2.266),

$$\omega_n = \frac{n\pi}{L}\sqrt{\frac{G}{\rho}} \qquad n = 1, 2, 3, \ldots \qquad (2.273)$$

In hertz, the torsional frequencies f_n are given by

$$f_n = \frac{\omega_n}{2\pi} = \frac{n}{2L}\sqrt{\frac{G}{\rho}} \qquad (2.274)$$

By using Eq. (2.264), the corresponding mode shapes are given by the expression

$$\Phi_n(x) = B_n \cos\frac{n\pi x}{L} \qquad n = 1, 2, 3, \ldots \qquad (2.275)$$

For $n = 1$, Eq. (2.273) yields the fundamental frequency of vibration, and Eq. (2.275) yields the corresponding mode shape of the fundamental frequency.

By using Eqs. (2.263), (2.267), and (2.275), the general expression $\varphi(x, t)$ for the free, undamped, torsional vibration of the rod may be written

$$\varphi(x, t) = \sum_{n=1}^{\infty} B_n \cos\frac{n\pi x}{L}\sin(\omega_n t + \phi_n) \qquad (2.276)$$

In Eq. (2.276), ω is given by Eq. (2.273), and ϕ_n are the phase angles. The constant B_n and phase angles φ_n may be determined by applying the initial time conditions of the torsional oscillatory motion.

2.15   THE WAVE EQUATION

Mathematically, the differential equations of motion for longitudinal and torsional vibrations of beams or shafts as given by Eqs. (2.238) and (2.262), respectively, are both of the same form. They involve some general function $\psi(x, t)$ that can be determined from a differential equation of motion, which can be written in the following form:

$$c^2 \frac{\partial^2 \psi}{\partial x^2} = \frac{\partial^2 \psi}{\partial t^2} \tag{2.277}$$

Equation (2.277) is known as the *wave equation*. A way to solve the differential equation given by Eq. (2.277), other than using the method of separation of variables, is to introduce the new variables

$$u = x - ct \tag{2.278}$$

$$v = x + ct \tag{2.279}$$

By substituting Eqs. (2.278) and (2.279) into Eq. (2.277) and carrying out the required mathematical manipulations, we find

$$\frac{\partial^2 \psi}{\partial u \, \partial v} = 0 \tag{2.280}$$

The solution of Eq. (2.280), in its most general form, is

$$\psi(x, t) = g_1(u) + g_2(v) \tag{2.281}$$

or

$$\psi(x, t) = g_1(x - ct) + g_2(x + ct) \tag{2.282}$$

Equation (2.282) represents a wave motion in both negative and positive x directions, where a wave can move in these directions with velocity c, and a shape that remains unchanged during travel.

The constant c in the two problems discussed in Sections 2.12 and 2.14 is known as the *velocity of propagation*. Therefore, for longitudinal stress waves, the velocity of propagation c is equal to $(E/\rho)^{1/2}$, as can be observed from Eq. (2.238). For shear stress waves, Eq. (2.262) in Section 2.14 indicates that $c = (G/\rho)^{1/2}$.

PROBLEMS

2.1 The differential equation of motion for the free undamped vibration of a single-degree spring–mass system is given by Eq. (2.1). By assuming that the solution $y(t) = Ce^{\psi t}$, where C and ψ are constants, and applying the Eulerian relation $e^{\pm i \psi t} = \cos \psi t \pm i \sin \psi t$, where $i = \sqrt{-1}$, derive the solution given by Eq. (2.5).

2.2 A weight $W = 550 \, \text{N}$ is suspended by a spring of spring constant $k = 20,000 \, \text{N/m}$. If the maximum velocity of the free vibratory motion of

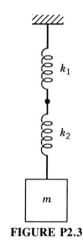

FIGURE P2.3

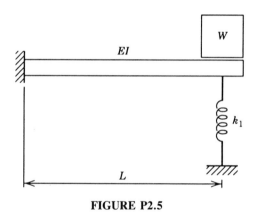

FIGURE P2.5

W is 1.20 m/s^2, determine the free frequency of vibration of the spring–mass system, the maximum amplitude of the vibratory motion, and the maximum acceleration.

2.3 The spring–mass system in Fig. P2.3 has a free frequency of vibration $f = 15 \text{ Hz}$. If $k_1 = 60{,}000 \text{ N/m}$ and $k_2 = 40{,}000 \text{ N/m}$, determine the value of the mass m.

2.4 Repeat Problem 2.3 by assuming that $k_1 = 2.20 \text{ kips/in.}$ and $k_2 = 1.65 \text{ kips/in.}$

2.5 The steel cantilever beam in Fig. P2.5 is restrained at the free end by a linear spring of constant $k_1 = 120 \text{ kips/in.}$ If the beam supports a weight $W = 60 \text{ kips}$ at the position shown in the figure, determine the free frequency of vibration by using an idealized one-degree spring–

mass system of spring constant k_e. The stiffness EI of the member is 30×10^6 kips·in.^{2} and the length $L = 20$ ft.

2.6 Determine the constant Y and the phase angle ϕ_d in Eq. (2.51) by assuming that the initial displacement and the initial velocity at $t = 0$ are y_0 and $\dot{y}_0$, respectively.

2.7 If the damping ratio ζ for a spring–mass system is 0.10 and the initial displacement and initial velocity at $t = 0$, are y_0 and $\dot{y}_0$, respectively, write the differential equation of its free damped vibration.

2.8 A spring–mass system consists of a weight $W = 5.0$ lb that hangs by a linear spring of constant $k = 8.0$ lb/in. and a dashpot with a damping constant $c = 0.06$ lb·s/in. Determine the critical damping factor c_c, the damping ratio ζ, and the damped and undamped free frequencies of vibration of the spring–mass system.

2.9 A spring–mass system consists of a weight $W = 3500.0$ N, a linear spring of spring constant $k = 210,000$ N/m, and a dashpot of viscous damping constant c. Determine the value of c such that the damped free frequency ω_d is 90% of the undamped frequency ω of the spring–mass system.

2.10 A weight $W = 180.0$ N is suspended from a spring of constant $k = 28,000.0$ N/m. A dashpot is attached between the weight and the ground and has a resistance of 0.9 N at a velocity of 0.08 m/s. Determine the free damped frequency of vibration of the system and the critical damping factor of the dashpot.

2.11 Determine the constants C_1 and C_2 in Eq. (2.45) by using the initial conditions $y = y_0$ and $\dot{y} = 0$ at time $t = 0$. Show all required mathematical manipulations.

2.12 A one-degree spring–mass system is vibrating under the influence of viscous damping with damping constant $c = 0.20$ lb·s/in. The mass m that is suspended by a linear spring of constant $k = 50$ lb/in. is equal to 0.055 lb·s^2/in. Determine the logarithmic decrement by using Eqs. (2.60) and (2.61) and compare results. Also, determine the ratio of any two successive amplitudes.

2.13 A vibrating system consists of a mass of weight $W = 14.0$ lb, a spring of spring constant $k = 30.0$ lb/in., and a dashpot that produces a velocity of 1.057 in./s for a force of 1.0 lb. Determine the damping ratio ζ and the logarithmic decrement δ. Also, determine the ratio between two consecutive amplitudes.

2.14 Prove that the logarithmic decrement δ can also be determined from the expression

$$\delta = \frac{1}{n} \ln\left(\frac{y_0}{y_n}\right)$$

where y_n is the amplitude after the completion of n cycles of motion.

2.15 A weight $W = 35$ lb is suspended by a spring of constant $k = 80$ lb/in. The motion is under the influence of viscous damping with damping constant $c = 0.16$ lb·s/in. Determine the damping ratio, the logarithmic decrement, and the ratio of any two successive amplitudes.

2.16 A weight $W = 1100$ N is attached to a spring of constant $k = 59{,}500$ N/m and slides back and forth on a dry surface. The motion is initiated after the weight W is displaced 30.0 cm. If the initial velocity is zero and the amplitude at the end of the sixth cycle of motion is 12.70 cm, determine the coefficient of friction μ.

2.17 A weight $W = 40$ lb is attached to a spring of constant $k = 120$ lb/in. and slides back and forth on a dry surface with coefficient of friction $\mu = 0.4$. The initial velocity is zero and the weight is released when the spring is stretched 10 in. What is the displacement of the weight at the end of the second and fifth cycles of motion measured from the undeformed position of the spring?

2.18 The sliding weight in Fig. (2.5a) has a weight of 450 N and the coefficient of friction μ between the weight and the surface is 0.22. The spring constant $k = 7000$ N/m, and the weight is initially released with zero velocity when the spring is stretched 14.0 cm. Determine the displacement of the weight at the end of the first and third cycles and the position at which the weight will stop. Measure these displacements from the undeformed position of the spring.

2.19 If the ratio Y_1/Y_3 in Fig. 2.7 is measured experimentally and is found to be equal to 2.8, determine the equivalent viscous damping factor ζ_e and the equivalent viscous damping constant c_e.

2.20 If the ratio Y_1/Y_3 in Fig. 2.7 is 1.078, determine the equivalent viscous damping ratio ζ_e and the equivalent viscous damping constant c_e, and write the expression for the solution $y(t)$ of the one-degree spring–mass system.

2.21 An aluminum disk is shrunk onto a steel shaft to form the arrangement shown in Fig. 1.9a. The length L of the shaft is 3 m and its diameter is 0.70 cm. Determine the polar moment of inertia of the disk if the torsional period of vibration of the disk is 1.25 s. The shear modulus G of the steel shaft is 80×10^9 N/m^2.

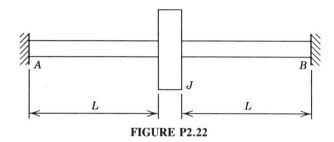

FIGURE P2.22

2.22 An aluminum disk is shrunk onto a steel shaft to form the arrangement shown in Fig. P2.22. The two portions of the shaft have the same length $L = 60$ in. and the same diameter $d = 2$ in. The steel shear modulus $G = 12 \times 10^6$ psi, and the shaft is assumed to be weightless. If the mass moment of inertia J of the disk is 2.8 in.·lb·s^2, determine the torsional frequency of vibration of the one-degree disk–shaft system and its period of vibration.

2.23 Solve Problem 2.21 when the disk–shaft system is under the influence of 20% viscous damping and the damped torsional period of vibration of the disk is 0.95 s.

2.24 Solve Problem 2.22 by assuming that the disk–shaft arrangement is under the influence of 30% viscous damping.

2.25 The disk–shaft arrangement in Fig. 2.8 is subjected to a torsional oscillatory motion that is under the influence of viscous damping with damping constant $c = 450.0$ N·m·s/rad. If the length of the shaft is 4.0 m and $J = 14.50$ N·m·s^2, determine the diameter d of the shaft when the damped torsional frequency of vibration is 32.0 rps. The shear modulus G of the steel shaft is 80×10^9 N/m^2.

2.26 For the two-degree spring–mass system in Fig. 2.9a, determine the free undamped frequencies of vibration and the corresponding mode shapes when $k_1 = k$, $k_2 = 3k$, $k_3 = 0$, $m_1 = m$, and $m_2 = 3m$. The spring–mass system is restricted to move in the vertical direction only. Compare the results with the ones obtained in Example 2.10.

2.27 Derive the differential equations of motion for the free undamped vibration of the spring–mass systems shown in Fig. P2.27. The spring–mass systems are restricted to move in the vertical direction only.

2.28 Determine the free undamped frequencies of vibration and the corresponding mode shapes of the spring–mass systems in Fig. P2.27 if $m_1 = m_2 = m_3 = m$ and $k_1 = k_2 = k_3 = k_4 = k$.

2.29 Determine the frequencies of vibration and the corresponding mode

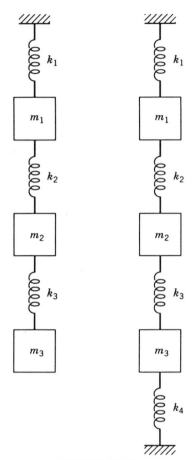

FIGURE P2.27

shapes for the spring–mass system in Fig. P1.17a when $k_1 = k_2 = k$, $m_1 = 3m$, $m_2 = m$, and $c = F(t) = 0$.

2.30 Rework Example 2.12 with $c_1 = 2.0 \, \text{lb·s/in.}$ and $c_2 = 3.0 \, \text{lb·s/in.}$ All other parameters remain the same. Write the expressions for the displacements y_1 and y_2 of masses m_1 and m_2, respectively.

2.31 Determine the damped frequencies of vibration for the spring–mass system in Fig. P1.17b if $k_1 = 87{,}500 \, \text{N/m}$, $k_2 = 45{,}000 \, \text{N/m}$, $k_3 = 30{,}000 \, \text{N/m}$, $m_1 = 150.0 \, \text{N·s}^2/\text{m}$, $m_2 = 50.0 \, \text{N.s}^2/\text{m}$, $c = 400 \, \text{N·s/in.}$, and $F(t) = 0$. Write the expressions for the displacements y_1 and y_2 of masses m_1 and m_2, respectively.

2.32 By using the results obtained in Problem 2.30 and with the initial conditions of the motion at time $t = 0$ being $y_1 = 6.0 \, \text{in.}$, $y_2 = 8.0 \, \text{in.}$,

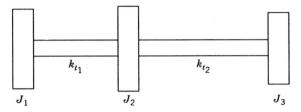

FIGURE P2.33

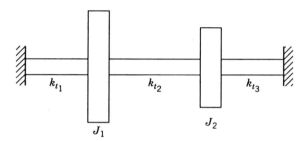

FIGURE P2.34

and $\dot{y}_1 = \dot{y}_2 = 0$, determine the constants $C_1, C_2, C_3, \ldots, C_8$ in the expressions for the displacements y_1 and y_2 of masses m_1 and m_2, respectively, of the spring–mass system.

2.33 For the disk–shaft system in Fig. P2.33, determine (a) the differential equations of motion for its free, undamped, torsional vibration and (b) the natural frequencies of its torsional vibration and the corresponding mode shapes. The pertinent quantities are $k_{t_1} = 600$ kips·ft/rad, $k_{t_2} = 200$ kips·ft/rad, $J_1 = 300$ in.·lb·s², $J_2 = 600$ in.·lb·s², and $J_3 = 200$ in.·lb·s².

2.34 For the disk–shaft system in Fig. P2.34, determine (a) the differential equations of motion for its free, undamped, torsional vibration and (b) the free torsional frequencies of vibration and the corresponding mode shapes. The pertinent quantities are $k_{t_1} = 30,000$ N·m/rad, $k_{t_2} = 16,000$ N·m/rad, $k_{t_3} = 22,000$ N·m/rad, $J_1 = 5.0$ N·m·s², and $J_2 = 2.0$ N·m·s².

2.35 Rework Example 2.13 when $k_{t_1} = 80,000$ N·m/rad, $k_{t_2} = 50,000$ N·m/rad, $J_1 = 16$ N·m·s², and $J_2 = 10.5$ N·m·s² and compare the results.

2.36 Prove that the mode shapes $\beta_n(x)$ of a uniform cantilever beam of

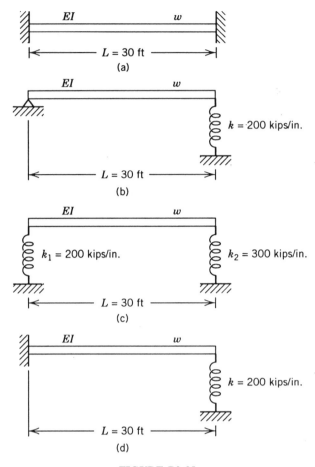

FIGURE P2.38

length L may be obtained from the following expression:

$$\beta_n(x) = C_n\bigg((\cos \lambda_n x - \cosh \lambda_n x)$$
$$- \frac{\cosh \lambda_n L + \cos \lambda_n L}{\sinh \lambda_n L + \sin \lambda_n L}(\sin \lambda_n x - \sinh \lambda_n x)\bigg)$$

2.37 For the uniform cantilever beam in Problem 2.36, prove that the values of λ can be determined from the following frequency equation:

$$\cosh \lambda L \cos \lambda L + 1 = 0$$

2.38 For the uniform single-span beams supported as shown in Fig. P2.38, determine their free, undamped, transverse frequencies of vibration and their corresponding mode shapes by following the procedure discussed in Section 2.11. The stiffness EI is constant and equal to 30×10^6 kip·in.^{2}, and the weight w per unit length of the member is 0.6 kips/ft.

2.39 By using the frequency equation given in Problem 2.37, determine the first three frequencies of vibration of the beam. If preferred, these results may be obtained graphically by using values of λ.

2.40 The ends of a uniform bar of length L are both free. Determine its free longitudinal frequencies of vibration and its corresponding mode shapes. Note that the stress $E \, \partial u/\partial x$ and strain $\partial u/\partial x$ at the free ends of the bar are both zero.

2.41 Determine the longitudinal frequencies of vibration and the corresponding mode shapes for a cantilever bar of length L. Note that the strain $\partial u/\partial x$ at the free end is zero.

2.42 Repeat Problem 2.40 when both ends of the bar are fixed.

2.43 A uniform rod of length L is fixed at the one end and it is free at the other end. By following the procedure discussed in Section 2.14, determine its free torsional frequencies of vibration and the corresponding mode shapes.

2.44 Repeat Problem 2.43 when both ends of the uniform bar are fixed.

2.45 The ends of a uniform aluminum bar in free torsional oscillation are both free. The solid circular cross section of the bar has a radius $r = 1.5$ in., its length $L = 30$ ft, the shear modulus $G = 4 \times 10^6$ psi, and the unit weight of the material is 0.10 lb/in.^{3}. Determine its first three free frequencies of vibration and the corresponding mode shapes.

2.46 Determine the velocity of the longitudinal waves of a uniform aluminum cantilever bar. The cross section of the bar is a square of side $a = 0.7$ in., its length $L = 10.0$ ft, the modulus of elasticity $E = 10^7$ psi, and the unit weight of the material is 0.10 lb/in.^{3}

2.47 Determine the velocity of the shear waves of the aluminum bar in Problem 2.45.

3 Forced Vibration of Simple Systems

3.1 INTRODUCTION

In the preceding chapter, the vibration responses of spring–mass systems, disk–shaft systems, and systems with continuous mass and elasticity, such as beams, were examined in detail. In many cases, the influence of various kinds of damping was also investigated in detail. There are, however, situations in which these systems could be subjected to external excitations that arise from many practical situations and unforeseen events. Operating machinery, gusty winds, vehicular traffic on highway bridges, blast effects from various types of conventional and nuclear explosions, and earthquake ground motions are examples of such sources of excitation.

In Chapter 1, it was shown how external excitations can be incorporated into the differential equations of motion of various types of linear and nonlinear systems. Spring–mass systems and disk–shaft systems, as well as systems with continuous mass and elasticity, were included. In the present chapter, the solution of these differential equations of motion will be carried out in detail by using both rigorous and numerical methods of analysis. It will also be shown how systems with infinite degrees of freedom can be idealized, or reduced, to systems with one or more degrees of freedom, which provide an accurate solution for practical applications. Both elastic and elastoplastic responses will be considered, and the effect of various types of dynamic excitations will be investigated.

3.2 UNDAMPED HARMONIC FORCES

Harmonic forces, as stated in Section 1.5, are represented by force functions where the time variation is a sine or cosine function. For example, $F(t) = F \cos \omega_f t$ and $F(t) = F \sin \omega_f t$ are harmonic forces. In these expressions, F is the maximum magnitude of the force and ω_f is its forced frequency.

Consider, for example, the spring–mass system in Fig. 3.1a, which is subjected to the harmonic force

$$F(t) = F \cos \omega_f t \tag{3.1}$$

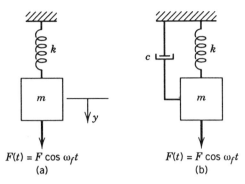

$F(t) = F \cos \omega_f t$
(a)

$F(t) = F \cos \omega_f t$
(b)

FIGURE 3.1. (a) One-degree spring–mass system subjected to a harmonic force. (b) One-degree spring–mass system subjected to a harmonic force and under the influence of visous damping.

The differential equation of motion of the spring–mass system is derived as discussed in Section 1.7, and it is as follows:

$$m\ddot{y} + ky = F \cos \omega_f t \tag{3.2}$$

The solution $y(t)$ of Eq. (3.2) consists of the complementary or homogeneous solution $y_c(t)$ and the particular solution $y_p(t)$; that is,

$$y(t) = y_c(t) + y_p(t) \tag{3.3}$$

The solution $y_c(t)$ represents the solution of the homogeneous equation $m\ddot{y} + ky = 0$, representing the free undamped vibration of the spring–mass system, and it is given by Eq. (2.5) in Section 2.2; that is,

$$y_c(t) = A \sin \omega t + B \cos \omega t \tag{3.4}$$

where A and B are constants and ω is the free undamped frequency of the spring–mass system.

The particular solution $y_p(t)$ is the one resulting from the application of the harmonic force $F \cos \omega_f t$. Since the force is harmonic, it would be reasonable to assume that $y_p(t)$ is also harmonic, and it can be represented by the expression

$$y_p(t) = Y \cos \omega_f t \tag{3.5}$$

where Y is the maximum amplitude.

By substituting Eq. (3.5) into Eq. (3.2) and simplifying, we find

$$-m\omega_f^2 Y + kY = F$$

which yields

$$Y = \frac{F}{k - m\omega_f^2} \tag{3.6}$$

By performing a few simple manipulations, we may write Eq. (3.6) as follows:

$$Y = \frac{F}{k} \frac{1}{[1 - (\omega_f/\omega)^2]} \tag{3.7}$$

By using Eqs. (3.3), (3.4), (3.5), and (3.7), the complete solution $y(t)$ of the spring–mass system is given as

$$y(t) = y_c(t) + y_p(t)$$

$$= A \sin \omega t + B \cos \omega t + \frac{F}{k} \frac{1}{[1 - (\omega_f/\omega)^2]} \cos \omega_f t \tag{3.8}$$

The constants A and B can be determined from the initial conditions of the motion. If the initial displacement and initial velocity at time $t = 0$ are y_0 and $\dot{y}_0$, respectively, then, by Eq. (3.8), we find;

$$A = \frac{\dot{y}_0}{\omega} \tag{3.9}$$

$$B = y_0 - \frac{F}{k} \frac{1}{[1 - (\omega_f/\omega)^2]} \tag{3.10}$$

Thus by substituting Eqs. (3.9) and (3.10) into Eq. (3.8), we find

$$y(t) = \frac{\dot{y}_0}{\omega} \sin \omega t + \left\{ y_0 - \frac{F}{k} \frac{1}{[1 - (\omega_f/\omega)^2]} \right\} \cos \omega t$$

$$+ \frac{F}{k} \frac{1}{[1 - (\omega_f/\omega)^2]} \cos \omega_f t \tag{3.11}$$

The first two terms of Eq. (3.11) prescribe the free vibration of the one-degree spring–mass system, and the third term gives the effect of the harmonic force $F \cos \omega_f t$. It should be noted, however, that, when the force frequency ω_f becomes equal to the free undamped frequency ω of the system, the second and third terms of Eq. (3.11) become infinite. This is known as the phenomenon of *resonance*. In such a case, the amplitude $y(t)$ increases gradually with time and becomes infinite as the time t approaches infinity ($t \to \infty$). In practical situations, however, the material of the engineering element will fail long before the amplitude becomes infinite.

Example 3.1. The one-degree spring–mass system in Fig. 3.1a has a spring constant $k = 500.0$ kN/m. The mass $m = 500$ kg, $F = 8000$ N, and the forced frequency $\omega_f = 18$ rps. Determine the maximum displacement of the mass m due to the harmonic force. Also, determine the displacement of the mass m when $\omega_f = 20, 22, 25, 28, 30,$ and 31.62 rps. The spring–mass system was at rest before the application of the harmonic force.

Solution. The particular solution $y_p(t)$ of Eq. (3.2) is often termed the *steady-state solution*, and the complementary one $y_c(t)$ is usually called the *transient solution*. The transient solution is neglected in many cases because it usually dies out with time when the system is under the influence of some kind of damping. The steady-state response represents the direct effect of the harmonic force, and this is the one to consider here.

The free undamped frequency ω of the spring–mass system is

$$\omega = \sqrt{\frac{k}{m}} = \sqrt{\frac{500,000}{500}}$$

$$= 31.62 \text{ rps}$$

The maximum displacement y_{max} of the mass m can be determined from Eq. (3.7); that is,

$$y_{max} = Y = \frac{8000}{500,000} \cdot \frac{1}{[1 - (18.00/31.62)^2]}$$

$$= 0.02367 \text{ m}$$

$$= 2.367 \text{ cm}$$

Equation (3.7) may also be used to determine y_{max} for $\omega_f = 20, 22, 25,$ 28, 30, and 31.62 rps. For example, when ω_f 28 rps, we have

$$y_{max} = Y = \frac{8000}{500,000} \cdot \frac{1}{[1 - (28.00/31.62)^2]}$$

$$= 0.07407 \text{ m}$$

$$= 7.407 \text{ cm}$$

TABLE 3.1. Variation of the Maximum Displacement of the Mass m with Increasing Ratio ω_f/ω

ω (rps)	ω_f (rps)	ω_f/ω	$y_{max} = Y$ (cm)
31.62	18.00	0.569	2.367
31.62	20.00	0.633	2.667
31.62	22.00	0.696	3.101
31.62	25.00	0.791	4.278
31.62	28.00	0.886	7.407
31.62	30.00	0.949	16.162
31.62	31.62	1.000	∞

The results are shown in Table 3.1. Note how the maximum displacement Y is increasing as the ratio ω_f/ω is approaching unity. The material, however, will fail long before infinite amplitude is attained.

3.3 HARMONIC FORCE WITH VISCOUS DAMPING

Consider now the one-degree spring–mass system in Fig. 3.1b, which is subjected to the harmonic force $F \cos \omega_f t$ and is oscillating under the influence of viscous damping with damping constant c. The differential equation of motion is derived as discussed in Section 1.7, and it is as follows:

$$m\ddot{y} + c\dot{y} + ky = F \cos \omega_f t \qquad (3.12)$$

The solution of this equation consists of the complementary solution $y_c(t)$ and the particular solution $y_p(t)$; that is,

$$y(t) = y_c(t) + y_p(t) \qquad (3.13)$$

By using Eq. (2.51) and replacing the constant Y by the constant C, the complementary solution $y_c(t)$ may be written

$$y_c(t) = Ce^{-\zeta\omega t} \sin(\omega_d t + \phi_d) \qquad (3.14)$$

In Eq. (3.14),

$$\omega_d = \omega\sqrt{1 - \zeta^2} \qquad (3.15)$$

is the damped frequency of vibration of the system, ω is its undamped frequency, C is a constant, and ϕ_d is the phase angle.

The particular solution $y_p(t)$ is based on the forcing function, and it can have the form

$$y_p(t) = A \cos \omega_f t + B \sin \omega_f t \qquad (3.16)$$

or the form

$$y_p(t) = Y \cos(\omega_f t - \phi) \qquad (3.17)$$

In Eq. (3.17), the constant Y and the phase angle ϕ are related to the constants A and B of Eq. (3.16) as follows:

$$Y = \sqrt{A^2 + B^2} \qquad (3.18)$$

$$\tan \phi = \frac{B}{A} \qquad (3.19)$$

By using the solution given by Eq. (3.16) and substituting it into Eq. (3.12), the following expression may be obtained after performing the required mathematical operations and rearranging terms:

$$(-mA\omega_f^2 + c\omega_f B + kA) \cos \omega_f t + (-mB\omega_f^2 - c\omega_f A + kB) \sin \omega_f t$$
$$= F \cos \omega_f t \qquad (3.20)$$

This equation will be satisfied for all values of time t if we match the expressions of the coefficients of both sine and cosine terms on the right-hand and left-hand sides of the equation. This mathematical operation yields

$$-mA\omega_f^2 + c\omega_f B + kA = F \qquad (3.21)$$
$$-mB\omega_f^2 - c\omega_f A + kB = 0 \qquad (3.22)$$

Equation (3.21) matches the coefficients of $\cos \omega_f t$ on each side of Eq. (3.20), and Eq. (3.22) matches the coefficients of $\sin \omega_f t$.

The simultaneous solution of Eqs. (3.21) and (3.22) yields the following expressions for the constants A and B:

$$A = \frac{F(k - m\omega_f^2)}{(k - m\omega_f^2)^2 + (c\omega_f)^2} \qquad (3.23)$$

$$B = \frac{Fc\omega_f}{(k - m\omega_f^2)^2 + (c\omega_f)^2} \qquad (3.24)$$

Thus, by substituting Eqs. (3.23) and (3.24) into Eqs. (3.18) and (3.19), we find

$$Y = \frac{F}{\sqrt{(k - m\omega_f^2)^2 + (c\omega_f)^2}} \tag{3.25}$$

$$\tan \phi = \frac{c\omega_f}{k - m\omega_f^2} \tag{3.26}$$

By using known relationships and performing the required manipulations, Eqs. (3.25) and (3.26) may also be written as follows:

$$Y = \frac{F/k}{\sqrt{[1 - (\omega_f/\omega)^2]^2 + (2\zeta\omega_f/\omega)^2}} \tag{3.27}$$

$$\tan \phi = \frac{2\zeta\omega_f/\omega}{1 - (\omega_f/\omega)^2} \tag{3.28}$$

By using Eqs. (3.17) and (3.27), the particular solution $y_p(t)$ may be written

$$y_p(t) = \frac{F/k}{\sqrt{[1 - (\omega_f/\omega)^2]^2 + (2\zeta\omega_f/\omega)^2}} \cos(\omega_f t - \phi) \tag{3.29}$$

Therefore, the complete solution of Eq. (3.12) is

$$y(t) = y_c(t) + y_p(t)$$
$$= Ce^{-\zeta\omega t} \sin(\omega_d t + \phi_d) + Y \cos(\omega_f t - \phi) \tag{3.30}$$

The constant Y and phase angle ϕ in this expression are given by Eqs. (3.27) and (3.28), respectively. The constant C and the phase angle ϕ_d can be determined from the initial conditions of the motion.

The term F/k in Eq. (3.27) is usually called the *static deflection*, Y_{st}, due to the force F, because it is assumed to represent the deflection of the mass m due to F, when F is applied gradually as a static load at zero frequency. The ratio Y/Y_{st} is termed the *magnification factor*, and it is denoted by the Greek letter Γ. Thus,

$$\Gamma = \frac{Y}{Y_{st}} = \frac{1}{\sqrt{[1 - (\omega_f/\omega)^2]^2 + (2\zeta\omega_f/\omega)^2}} \tag{3.31}$$

The magnification factor Γ yields the value by which Y_{st} should be multiplied in order to obtain Y; Γ depends only on the damping ratio ζ and the ratio ω_f/ω. For a given value of ζ, the magnification factor Γ can be plotted against various values of the ratio ω_f/ω. The curve so obtained illustrates

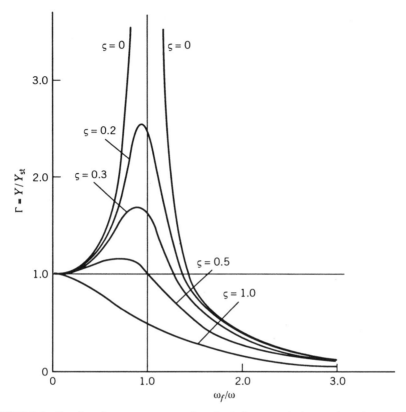

FIGURE 3.2. Family of curves representing the influence of viscous damping on the magnification factor Γ for $\zeta = 0, 0.2, 0.3, 0.5$, and 1.00.

the influence of ζ at resonance, where $\omega_f/\omega = 1$. By using various values of ζ e.g., ($\zeta = 0, 0.05, 0.10, \ldots, 1$), a family of curves may be obtained, which illustrates the degree of influence for the various amounts of damping. Such a family of curves is shown in Fig. 3.2, where ζ is assumed to take values of $0, 0.2, 0.3, 0.5$, and 1.00.

Example 3.2. For the one-degree spring–mass system in Fig. 3.1b, we assume that the spring constant $k = 400$ lb/in., the mass $m = 0.40$ lb $\cdot$ s²/in., the maximum value F of the harmonic force is 200 lb, the forced frequency $\omega_f = 20$ rps, and the damping ratio $\zeta = 0.30$. Determine (a) the maximum amplitude of the steady-state motion, (b) the amplitude Y when ω_f is equal to the free undamped frequency ω of the spring–mass system, (c) the damping factor c, and (d) the phase angle ϕ.

Solution. The free undamped frequency ω of the one-degree spring–mass system is

$$\omega = \sqrt{\frac{k}{m}} = \sqrt{\frac{400}{0.40}} = 31.60 \text{ rps}$$

From Eq. (3.27), the maximum amplitude Y of the steady-state motion is

$$Y = \frac{200/400}{\sqrt{[1 - (20/31.60)^2]^2 + [(2)(0.30)(20)/31.60]^2}}$$

$$= 0.706 \text{ in.}$$

When ω_f is equal to the free undamped frequency ω of the spring–mass system, the phenomenon of resonance occurs. Thus, from Eq. (3. 27), the resonant amplitude is

$$Y = \frac{200/400}{\sqrt{(1 - 1)^2 + [(2)(0.30)(1.0)]^2}}$$

$$= 0.833 \text{ in.}$$

The critical damping factor c_c is

$$c_c = 2m\omega = (2)(0.40)(31.60)$$

$$= 25.20 \text{ lb} \cdot \text{s/in.}$$

Thus

$$c = \zeta c_c = (0.30)(25.20)$$

$$= 7.55 \text{ lb} \cdot \text{s/in.}$$

From Eq. (3.28), the phase angle ϕ is

$$\tan \phi = \frac{(2)(0.30)(20)/31.60}{1 - (20/31.60)^2}$$

$$= 0.635$$

and

$$\phi = 32°25'$$

3.4 HARMONIC FORCE WITH COULOMB DAMPING

An approximate solution to this problem can be obtained by introducing an equivalent viscous damping constant c_e. This constant can be determined by computing the energy absorption ΔU_v per cycle of viscous damping and equating it to the corresponding ΔU_f for Coulomb damping; that is,

$$\Delta U_v = \Delta U_f \tag{3.32}$$

Consider the single-degree spring–mass system in Fig. 3.3, which is acted on by the harmonic force $F \sin \omega_f t$ and moves under the influence of Coulomb damping with friction force F_f as shown. If the friction is not large, it would be reasonable to assume that the steady-state motion is approximately harmonic, and, consequently, it could adequately be defined by the expression

$$x = X \sin(\omega_f t - \phi) \tag{3.33}$$

This equation is similar to the one used in the preceding section for the case of harmonic force with viscous damping. If the friction is large, an exact solution to the problem should be obtained. For most practical situations, however, the friction is usually small. The energy absorption ΔU_v per cycle of equivalent viscous damping can be obtained from the expression

$$\Delta U_v = \int c_e \dot{x} \, dx \tag{3.34}$$

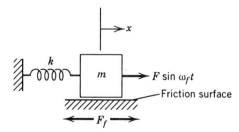

FIGURE 3.3.   One-degree spring–mass system subjected to a harmonic force and moving under the influence of Coulomb damping.

where x is given by Eq. (3.33). Thus, by substitution, Eq. (3.34) yields

$$\Delta U_v = c_e \int_0^{2\pi/\omega_f} [\omega_f X \cos(\omega_f t - \phi)][\omega_f X \cos(\omega_f t - \phi)] \, dt$$

$$= c_e(\omega_f X)^2 \int_0^{2\pi/\omega_f} \cos^2(\omega_f t - \phi) \, dt \tag{3.35}$$

$$= \pi c_e \omega_f X^2$$

The energy ΔU_f per cycle of Coulomb damping is

$$\Delta U_f = 4F_f X \tag{3.36}$$

where F_f is the constant friction force. By substituting Eqs. (3.35) and (3.36) into Eq. (3.32), we find

$$c_e = \frac{4F_f}{\pi \omega_f X} \tag{3.37}$$

From Eq. (2.42), we find that the damping ratio ζ is

$$\zeta = \frac{c}{2m\omega} \tag{3.38a}$$

or

$$\frac{2\zeta}{\omega} = \frac{c}{m\omega^2} = \frac{c}{k} \tag{3.38b}$$

Thus, by Eq. (3.38b), the steady-state solution for viscous damping given by Eq. (3.31) can be written

$$X = \frac{X_{st}}{\sqrt{[1 - (\omega_f/\omega)^2]^2 + (c\omega_f/k)^2}} \tag{3.39}$$

Equation (3.39) can be modified by replacing c with the equivalent viscous damping constant c_e given by Eq. (3.37). On this basis, the following expression for the amplitude X is obtained:

$$X = \frac{X_{st}}{\sqrt{[1 - (\omega_f/\omega)^2]^2 + [4F_f/\pi k X]^2}} \tag{3.40}$$

Since on both sides of Eq. (3.40) we have X, the solution of Eq. (3.40) for X yields the following expression for the steady-state amplitude X of the spring–mass system in Fig. 3.3:

$$X = X_{st} \frac{\sqrt{1 - (4F_f/\pi F)^2}}{1 - (\omega_f/\omega)^2} \tag{3.41}$$

where F in this expression is the maximum amplitude of the harmonic force in Fig. 3.3.

A real value for the amplitude X can be obtained from Eq. (3.41), provided that $4F_f < \pi F$, or $F_f < \pi F/4$. If $F_f > \pi F/4$, an exact solution should be used to determine X.

Example 3.3. The weight W of the mass m in Fig. 3.3 is 600 lb, and it rubs against the dry surface during the application of the harmonic force. The coefficient of kinetic friction $\mu = 0.25$, the spring constant $k = 1200$ lb/in., and $\omega_f = 20$ rps. Determine the steady-state amplitude of the motion when $F = 1400$ lb.

Solution. The values of ω, X_{st}, and F_f are as follows:

$$\omega = \sqrt{\frac{k}{m}} = \sqrt{\frac{(1200)(386)}{600}} = 27.8 \text{ rps}$$

$$X_{st} = \frac{F}{k} = \frac{1400}{1200} = 1.165 \text{ in.}$$

$$F_f = \mu W = (0.25)(600) = 150 \text{ lb}$$

Thus, from Eq. (3.41), we have

$$X = (1.165) \frac{\sqrt{1 - [(4)(150)/\pi(1400)]^2}}{1 - (20/27.8)^2}$$

$$= (1.165) \frac{\sqrt{1 - 0.0187}}{1 - 0.52}$$

$$= 2.41 \text{ in.}$$

3.5 DYNAMIC RESPONSE DUE TO A FORCE OF GENERAL TYPE

In this section, the response of a single-degree spring–mass system subjected to a force $F(t)$ of arbitrary time variation will be examined. The undamped as well as the damped cases will be investigated. The solution, however, is based on the idea of dividing the force–time function into an infinite number of impulses and adding up the responses of the system due to these impulses. Therefore, before we proceed with this solution, the response of the system due to an impulse and a unit impulse will first be determined. Some discussion regarding impulses and unit impulses was already given in Section 1.5, and it would be helpful if reference is made to this discussion.

3.5.1 Impulse Response

Assume that the impulse is designated by the symbol F_{imp}, and that it is applied suddenly to the undamped, one-degree spring–mass system at time $t = t_0$, when the system is in its equilibrium position. We are concerned with the response of the system at times $t \geq t_0$, because the impulse acts for an instantaneous time only.

By equating the linear momentum change $m\dot{y}_0$ to the linear impulse F_{imp} that is applied at $t = t_0$—that is, $m\dot{y}_0 = F_{imp}$—we find

$$\dot{y}_0 = \frac{F_{imp}}{m} \tag{3.42}$$

The response $y(t)$ due to the impulse can be determined from Eq. (2.14) by substituting $y_0 = 0$ and $\dot{y}_0 = F_{imp}/m$. This yields

$$y(t) = \frac{F_{imp}}{m\omega} \sin \omega(t - t_0) \qquad t \geq t_0 \tag{3.43}$$

where t_0 is the time of application of the impulse as stated earlier.

If the impulse is a unit impulse, then F_{imp} is unity, and Eq. (3.43) yields

$$y(t) = \frac{1}{m\omega} \sin \omega(t - t_0) \qquad t \geq t_0 \tag{3.44}$$

Equation (3.44) gives the response of an undamped, single-degree spring–mass system due to a unit impulse.

If the one-degree spring–mass system is also subjected to viscous damping, then, by Eq. (2.50), we find

$$y(t) = \frac{F_{inp}}{m\omega_d} e^{-\zeta\omega(t-t_0)} \sin \omega_d(t - t_0) \qquad t \geq t_0 \tag{3.45}$$

where $\omega_d = \omega\sqrt{1 - \zeta^2}$ is the damped frequency of the system, and t_0 is the time of application of the impulse F_{imp}. For a unit impulse,

$$y(t) = \frac{1}{m\omega_d} e^{-\zeta\omega(t - t_0)} \sin \omega_d(t - t_0) \qquad t \geq t_0 \qquad (3.46)$$

Note that Eq. (3.46) incorporates the time t_0, which is the time of application of F_{imp}.

3.5.2 Dynamic Force of Arbitrary Time Variation

A dynamic force of arbitrary time variation is shown in Fig. 3.4. The force is assumed to be divided into an infinite number of impulses. One such impulse is represented by the shaded area in Fig. 3.4, and it is assumed to be applied at time $t_0 = T$. The value of this impulse is $F(T)\,dT/m$, and, by using Eq. (3.43), its dynamic response for the undamped case is

$$\frac{F(T)\,dT}{m\omega} \sin \omega(t - T) \qquad t \geq T \qquad (3.47)$$

The total response $y(t)$ due to the general dynamic force $F(t)$ is obtained by summing up the responses of all impulses. Thus, for an interval of time between zero and t, the total response $y(t)$ may be written

$$y(t) = \frac{1}{m\omega} \int_0^t F(T) \sin \omega(t - T)\,dT \qquad (3.48)$$

Equation (3.48) is also known as *Duhamel's integral* or the *convolution integral*. The complete solution, however, should include the transient response, which in this case is given by Eq. (2.10). Hence, the complete

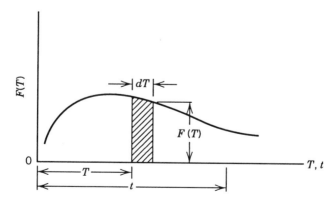

FIGURE 3.4. Dynamic force $F(t)$ of arbitrary time variation.

solution $y(t)$ is

$$y(t) = \frac{\dot{y}_0}{\omega} \sin \omega t + y_0 \cos \omega t + \frac{1}{m\omega} \int_0^t F(T) \sin \omega(t - T) \, dT \qquad (3.49)$$

For the case where the one-degree spring–mass system is also subjected to viscous damping, the response $y(t)$ due to the dynamic force $F(t)$ is

$$y(t) = \frac{1}{m\omega_d} \int_0^t F(T) e^{-\zeta\omega(t - T)} \sin \omega_d(t - T) \, dT \qquad (3.50)$$

If the transient response, which is given by Eq. (2.50), is included, the complete solution $y(t)$ would be as follows:

$$y(t) = e^{-\zeta\omega t} \left(y_0 \cos \omega_d t + \frac{\dot{y}_0 + \zeta\omega y_0}{\omega_d} \sin \omega_d t \right)$$

$$+ \frac{1}{m\omega_d} \int_0^t F(T) e^{-\zeta\omega(t - T)} \sin \omega_d(t - T) \, dT \qquad (3.51)$$

In practice, however, we are concerned mostly with the steady-state response that is caused by the application of the dynamic force. The transient response usually dies out because of damping, and it is often neglected. The following example illustrates the application of the theory.

Example 3.4. The single-degree spring–mass system in Fig. 3.5a is subjected to a dynamic force that varies as shown in Fig. 3.5b. This is a force of magnitude F that is applied suddenly to the system and lasts for an indefinite period of time. Determine the response of the spring–mass system due to this force (a) by considering damping and (b) by neglecting the effects of damping. The initial velocity $\dot{y}_0$ and the initial displacement y_0 are both zero.

Solution. The force $F(T)$ is constant and equal to F at all times. The response $y(t)$ due to this force, for the viscously damped condition, can be determined from Eq. (3.50). This equation yields

$$y(t) = \frac{1}{m\omega_d} \int_0^t Fe^{-\zeta\omega(t - T)} \sin \omega_d(t - T) \, dT$$

$$= \frac{Fe^{-\zeta\omega t}}{m\omega_d} \int_0^t e^{\zeta\omega T}(\sin \omega_d t \cos \omega_d T - \cos \omega_d t \sin \omega_d T) \, dT$$

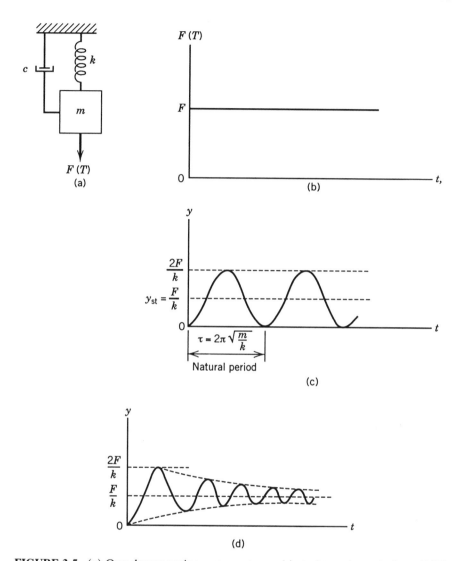

FIGURE 3.5. (a) One-degree spring–mass system subjected to a dynamic force $F(T)$ and moving under the influence of viscous damping. (b) Time variation of $F(T)$. (c) Plot of the response $y(t)$ for the undamped case. (d) Plot of the response $y(t)$ for the viscously damped case.

$$= \frac{Fe^{-\zeta\omega t}}{m\omega_d}\left(\sin\omega_d t \int_0^t e^{\zeta\omega T}\cos\omega_d T\,dT - \cos\omega_d t \int_0^t e^{\zeta\omega T}\sin\omega_d T\,dT\right)$$

$$= \frac{Fe^{-\zeta\omega t}}{m\omega_d}\left\{\sin\omega_d t\left[\frac{e^{\zeta\omega T}}{(\zeta\omega)^2 + \omega_d^2}(\zeta\omega\cos\omega_d T + \omega_d\sin\omega_d T)\right]_0^t\right.$$

$$-\cos \omega_d t \left[\frac{e^{\zeta \omega T}}{(\zeta \omega)^2 + \omega_d^2} (\zeta \omega \sin \omega_d T - \omega_d \cos \omega_d T) \right]_0^t \Bigg\}$$

Or, by carrying out the required mathematical manipulations, we find

$$y(t) = \frac{F}{m[(\zeta \omega)^2 - \omega_d^2]} - \frac{Fe^{-\zeta \omega t}}{m \omega_d [(\zeta \omega)^2 + \omega_d^2]} (\zeta \omega \sin \omega_d t + \omega_d \cos \omega_d t)$$

$$(3.52)$$

The solution for the undamped case can be obtained directly from Eq. (3.52) by making $\zeta = 0$ and $\omega_d = \omega$. This yields

$$y(t) = \frac{F}{m \omega^2} - \frac{F}{m \omega^2} \cos \omega t \qquad (3.53)$$

$$= \frac{F}{k} (1 - \cos \omega t)$$

The same solution may be obtained by using Eq. (3.48). Note that F/k in Eq. (3.53) is what was termed in Section 3.3 the static displacement y_{st}. Therefore the magnification factor Γ for the undamped case is given by the following equation

$$\Gamma = \frac{y}{y_{st}} = (1 - \cos \omega t) \qquad (3.54)$$

When $\omega t = 0$, we have $\Gamma = 0$; and when $\omega t = \pi$, we find that $\Gamma = 2$. Since the maximum value of Γ is 2, we conclude here that the maximum dynamic displacement y is equal to $2y_{st}$.

The graph of Eq. (3.53) is shown in Fig. 3.5c, which illustrates that the force F will make the spring–mass system vibrate about the equilibrium position $y(t) = y_{st}$. If the spring–mass system is under the influence of viscous damping, as shown in Fig. 3.5a, the vibration will eventually die out, and the spring–mass system will come to rest at $y(t) = y_{st}$, as shown in Fig. 3.5d.

The magnification factors for various types of time-dependent dynamic forces are tabulated in Appendix A for easy reference.

3.6 ONE-DEGREE SPRING–MASS SYSTEMS SUBJECTED TO SUPPORT MOTION

Consider now the situation where the support point of a one-degree spring–mass system is excited harmonically by the support displacement y_s in Fig. 3.6a, which has the following form:

$$y_s = y_{s0} \sin \omega_f t \tag{3.55}$$

In this equation, y_{s0} is the maximum displacement and ω_f is the frequency of the support motion. In addition, the spring–mass system is under the influence of viscous damping, as shown in the figure.

By applying Newton's second law of motion to the free-body diagram of the mass m in Fig. 3.6b, we find

$$m\ddot{y} = -k(y - y_s) - c(\dot{y} - \dot{y}_s)$$

or

$$m\ddot{y} + c(\dot{y} - \dot{y}_s) + k(y - y_s) = 0 \tag{3.56}$$

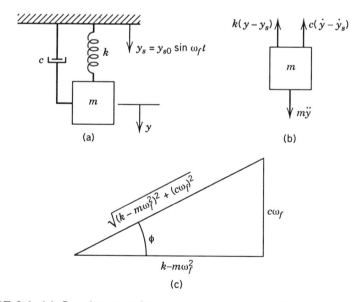

FIGURE 3.6. (a) One-degree spring–mass system subjected to a harmonic support motion. (b) Free-body diagram of mass m. (c) Triangle used to eliminate $\sin \phi$ and $\cos \phi$ from Eq. (3.66).

In Eq. (3.56), the symbol y is used to denote the absolute displacement of the mass m, and $y - y_s$ is the spring deflection, which in reality is the displacement of the mass relative to the support.

By adding to each side of Eq. (3.56) the term $-m\ddot{y}_s$ and rearranging, we have

$$m(\ddot{y} - \ddot{y}_s) + c(\dot{y} - \dot{y}_s) + k(y - y_s) = -m\ddot{y}_s \tag{3.57}$$

From Eq. (3.55), by differentiation with respect to time, we obtain

$$\ddot{y}_s = -\omega_f^2 y_{s0} \sin \omega_f t \tag{3.58}$$

Thus, by using Eqs. (3.57) and (3.58), we obtain

$$m(\ddot{y} - \ddot{y}_s) + c(\dot{y} - \dot{y}_s) + k(y - y_s) = m\omega_f^2 y_{s0} \sin \omega_f t \tag{3.59}$$

or

$$m\ddot{y}_r + c\dot{y}_r + ky_r = Q \sin \omega_f t \tag{3.60}$$

where

$$y_r = (y - y_s) \tag{3.61}$$
$$Q = m\omega_f^2 y_{s0} \tag{3.62}$$

Equation (3.60) is a nonhomogeneous differential equation of motion, and its solution is composed of the homogeneous solution and a particular solution that satisfies the excitation $Q \sin \omega_f t$. In this section, we are interested about the particular solution, which represents the effect of the support harmonic excitation. The particular solution is assumed to be of the form

$$y_r = Y_r \sin(\omega_f t - \phi) \tag{3.63}$$

This is a harmonic motion of maximum amplitude Y_r that has the same frequency as the excitation and lags by a phase angle ϕ.

By substituting Eq. (3.63) into Eq. (3.60) and carrying out the required mathematical operations, we find

$$-m\omega_f^2 Y_r \sin(\omega_f t - \phi) + c\omega_f Y_r \cos(\omega_f t - \phi)$$
$$+ kY_r \sin(\omega_f t - \phi) = Q \sin \omega_f t \tag{3.64}$$

By expanding the sine and cosine terms and rearranging, we find

$$\{Y_r[(k - m\omega_f^2)\cos\phi + c\omega_f\sin\phi] - Q\}\sin\omega_f t$$
$$+ \{Y_r[(k - m\omega_f^2)\sin\phi - c\omega_f\cos\phi]\}\cos\omega_f t = 0 \qquad (3.65)$$

Equation (3.65) will be satisfied if the coefficients of $\sin\omega_f t$ and $\cos\omega_f t$ vanish independently. This procedure yields

$$Y_r = \frac{Q}{(k - m\omega_f^2)\cos\phi + c\omega_f\sin\phi} \qquad (3.66)$$

$$\tan\phi = \frac{c\omega_f}{k - m\omega_f^2} \qquad (3.67)$$

The quantities $\sin\phi$ and $\cos\phi$ in Eq. (3.66) can be eliminated by using Fig. 3.6c. This approach yields

$$Y_r = \frac{Q}{\sqrt{(k - m\omega_f^2)^2 + (c\omega_f)^2}} \qquad (3.68)$$

Further modifications of Eqs. (3.67) and (3.68) can be obtained by using known expressions from preceding sections. Thus, by using Eq. (3.62) and known expressions from preceding sections, Eqs. (3.67) and (3.68) may be written in terms of the damping ratio ζ and the ratio ω_f/ω as follows:

$$\tan\phi = \frac{2\zeta(\omega_f/\omega)}{1 - (\omega_f/\omega)^2} \qquad (3.69)$$

$$Y_r = \frac{y_{s0}(\omega_f/\omega)^2}{\sqrt{[1 - (\omega_f/\omega)^2]^2 + (2\zeta\omega_f/\omega)^2}} \qquad (3.70)$$

where ω is the free undamped frequency of the spring–mass system.

With Y_r and ϕ determined, the solution given by Eq. (3.63) is completely defined. By using Eq. (3.61), the absolute displacement y of the mass m is determined:

$$y = y_r + y_s \qquad (3.71)$$
$$= Y_r\sin(\omega_f t - \phi) + y_{s0}\sin\omega_f t$$

where Y_r and ϕ are given by Eqs. (3.70) and (3.69), respectively.

Example 3.5. For the spring–mass system in Fig. 3.6a, the mass $m = 0.4\ \text{lb·s}^2/\text{in.}$, $k = 500\ \text{lb/in.}$, and the damping ratio $\zeta = 0.10$. The maximum amplitude y_{so} of the support motion is 1.2 in., and $\omega_f = 20$ rps. Determine the displacement y of the mass m at time $t = 0.3$ s.

Solution. The free undamped frequency ω of the system is

$$\omega = \sqrt{\frac{k}{m}} = \sqrt{\frac{500}{0.4}}$$

$$= 35.4 \text{ rps}$$

Thus, from Eq. (3.69),

$$\tan \phi = \frac{(2)(0.10)(20/35.4)}{1 - (20/35.4)^2}$$

$$= 0.1665$$

$$\phi = 9°27' = 0.165 \text{ rad}$$

From Eq. (3.70), we find

$$Y_r = \frac{(1.2)(20/35.4)^2}{\sqrt{[1 - (20/35.4)^2]^2 + [(20)(0.10)(20/35.4)]^2}}$$

$$= 0.538 \text{ in.}$$

From Eq. (3.71), the displacement y of the mass m is as follows:

$$y = 0.538 \sin[(20)(0.3) - 0.165] + 1.2 \sin[(20)(0.3)]$$

$$= 0.538 \sin(334°) + 1.2 \sin(344°)$$

$$= -0.567 \text{ in.}$$

3.7   FORCED VIBRATION OF TWO-DEGREE SPRING–MASS SYSTEMS

In this section, the forced vibration response of the two-degree spring–mass system in Fig. 1.7c will be considered. It is assumed here that the spring–mass system is acted on by the harmonic forces $F_1(t) = F_1 \cos \omega_f t$ and $F_2(t) = F_2 \cos \omega_f t$ and that the motion is under the influence of viscous damping with damping coefficients c_1 and c_2, as shown in the figure.

The differential equations of motion are identical to Eqs. (1.26) and (1.27) in Section 1.7 if $F_1(t)$ and $F_2(t)$ are replaced by $F_1 \cos \omega_f t$ and $F_2 \cos \omega_f t$, respectively, where ω_f is the frequency of the harmonic forces. We write these equations again as follows:

$$m_1\ddot{y}_1 + k_1 y_1 + c_1\dot{y}_1 - k_2(y_2 - y_1) - c_2(\dot{y}_2 - \dot{y}_1) = F_1 \cos \omega_f t \quad (3.72)$$

$$m_2\ddot{y}_2 + k_2(y_2 - y_1) + c_2(\dot{y}_2 - \dot{y}_1) = F_2 \cos \omega_f t \quad (3.73)$$

The steady-state response of the two-degree spring–mass system is obtained here. This response is assumed to be harmonic and has the same frequency ω_f as the harmonic forces. It is assumed also that the harmonic forces $F_1 \cos \omega_f t$ and $F_2 \cos \omega_f t$ are represented by the following expressions:

$$F_1(t) = F_1 e^{i\omega_f t} \quad (3.74)$$

$$F_2(t) = F_2 e^{i\omega_f t} \quad (3.75)$$

If the response lags the harmonic force by a phase angle ϕ, the solutions y_1 and y_2 of Eqs. (3.72) and (3.73) can be written

$$y_1 = Y_1 e^{i(\omega_f t - \phi_1)} \quad (3.76)$$

$$y_2 = Y_2 e^{i(\omega_f t - \phi_2)} \quad (3.77)$$

In Eqs. (3.74) through (3.77), the harmonic forces and the harmonic motion are defined by the real parts of these equations.

Equations (3.76) and (3.77) may be rewritten:

$$y_1 = \bar{Y}_1 e^{i\omega_f t} \quad (3.78)$$

$$y_2 = \bar{Y}_2 e^{i\omega_f t} \quad (3.79)$$

where

$$\bar{Y}_1 = Y_1 e^{-i\phi_1} \quad (3.80)$$

$$\bar{Y}_2 = Y_2 e^{-i\phi_2} \quad (3.81)$$

are the complex amplitudes of y_1 and y_2.

By using Eqs. (3.72) and (3.73) and substituting the expressions given by Eqs. (3.74), (3.75), (3.78), and (3.79), the following system of two equations is obtained:

$$(-m_1\omega_f^2 + ic_1\omega_f + ic_2\omega_f + k_1 + k_2)\bar{Y}_1 - (i\omega_f c_2 + k_2)\bar{Y}_2 = F_1 \quad (3.82)$$

$$-(i\omega_f c_2 + k_2)\bar{Y}_1 + (-m_2\omega_f^2 + i\omega_f c_2 + k_2)\bar{Y}_2 = F_2 \quad (3.83)$$

By applying Cramer's rule, the solution of Eqs. (3.82) and (3.83) yields the following expressions for $\bar{Y}_1$ and $\bar{Y}_2$:

$$\bar{Y}_1 = Y_1 e^{-i\phi_1} = \frac{\begin{vmatrix} F_1 & -(i\omega_f c_2 + k_2) \\ F_2 & -m_2\omega_f^2 + i\omega_f c_2 + k_2 \end{vmatrix}}{\begin{vmatrix} -m_1\omega_f^2 + ic_1\omega_f + ic_2\omega_f + k_1 + k_2 & -(i\omega_f c_2 + k_2) \\ -(i\omega_f c_2 + k_2) & -m_2\omega_f^2 + i\omega_f c_2 + k_2 \end{vmatrix}} \tag{3.84}$$

$$\bar{Y}_2 = Y_2 e^{-i\phi_2} = \frac{\begin{vmatrix} -m_1\omega_f^2 + ic_1\omega_f + ic_2\omega_f + k_1 + k_2 & F_1 \\ -(i\omega_f c_2 + k_2) & F_2 \end{vmatrix}}{\begin{vmatrix} -m_1\omega_f^2 + ic_1\omega_f + ic_2\omega_f + k_1 + k_2 & -(i\omega_f c_2 + k_2) \\ -(i\omega_f c_2 + k_2) & -m_2\omega_f^2 + i\omega_f c_2 + k_2 \end{vmatrix}} \tag{3.85}$$

In Eqs. (3.84) and (3.85), the quantities ϕ_1 and ϕ_2 are the phase angles of the complex amplitudes $\bar{Y}_1$ and $\bar{Y}_2$, respectively.

If we expand the determinants in Eqs. (3.84) and (3.85), we will obtain a format of the form $(a + ib)/(c + id)$. In accordance with the rules of complex numbers, we have the following identities:

$$\frac{A + iB}{C + iD} = \sqrt{\frac{A^2 + B^2}{C^2 + D^2}} \cdot e^{i(-\phi)} \tag{3.86}$$

$$\tan(-\phi) = \frac{BC - AC}{AC + BD} \tag{3.87}$$

If we follow these rules, we will obtain the following expressions for the amplitudes Y_1 and Y_2.

$$Y_1 = \sqrt{\frac{A_1^2 + B_1^2}{C_1^2 + D_1^2}} \tag{3.88}$$

$$Y_2 = \sqrt{\frac{A_2^2 + B_2^2}{C_2^2 + D_2^2}} \tag{3.89}$$

where

$$A_1 = -m_2\omega_f^2 F_1 + F_1 k_2 + F_2 k_2 \tag{3.90}$$

$$B_1 = c_2\omega_f F_1 + c_2\omega_f F_2 \tag{3.91}$$

$$C_1 = m_1 m_2 \omega_f^4 - k_1 m_2 \omega_f^2 - k_2 m_2 \omega_f^2 - c_1 c_2 \omega_f^2 - k_2 m_1 \omega_f^2 + k_1 k_2 \tag{3.92}$$

$$D_1 = k_1 c_2 \omega_f + k_2 c_1 \omega_f \tag{3.93}$$

$$A_2 = -F_2 m_1 \omega_f^2 + F_2 k_1 + F_2 k_2 + F_1 k_2 \tag{3.94}$$

$$B_2 = F_2 c_1 \omega_f + F_2 c_2 \omega_f + F_1 c_2 \omega_f \tag{3.95}$$

$$C_2 = C_1 \tag{3.96}$$

$$D_2 = D_1 \tag{3.97}$$

By considering the real parts of Eqs. (3.76) and (3.77), the steady-state solutions y_1 and y_2 are given by the following expressions:

$$y_1 = Y_1 \cos(\omega_f t - \phi_1) \tag{3.98}$$

$$y_2 = Y_2 \cos(\omega_f t - \phi_2) \tag{3.99}$$

where Y_1 and Y_2 may be obtained from Eqs. (3.88) and (3.89), respectively. The forcing functions that produce these types of motion are $F_1 \cos \omega_f t$ and $F_2 \cos \omega_f t$. The amplitudes y_1 and y_2 are not in phase during motion.

The phase angles ϕ_1 and ϕ_2 may be obtained by using the general expression given by Eq. (3.87). For the problem at hand, ϕ_1 and ϕ_2 may be determined from the following equations:

$$\tan(-\phi_1) = \frac{B_1 C_1 - A_1 D_1}{A_1 C_1 + B_1 D_1} \tag{3.100}$$

$$\tan(-\phi_2) = \frac{B_2 C_2 - A_2 D_2}{A_2 C_2 + B_2 D_2} \tag{3.101}$$

The quantities A_1, B_1, C_1, D_1, A_2, B_2, C_2, and D_2 may be determined from Eqs. (3.90), (3.91), (3.92), (3.93), (3.94), (3.95), (3.96), and (3.97), respectively.

If damping is neglected, $c_1 = c_2 = 0$; and Eqs. (3.82) and (3.83) yield the following two equations:

$$(k_1 + k_2 - m_1 \omega_f^2) Y_1 - k_2 Y_2 = F_1 \tag{3.102}$$

$$-k_2 Y_1 + (k_2 - m_2 \omega_f^2) Y_2 = F_2 \tag{3.103}$$

The simultaneous solution of Eqs. (3.102) and (3.103) yields the following expressions for the maximum amplitudes Y_1 and Y_2 of masses m_1 and m_2, respectively:

$$Y_1 = \frac{F_2 k_2 + F_1(k_2 - m_2 \omega_f^2)}{(k_1 + k_2 - m_1 \omega_f^2)(k_2 - m_2 \omega_f^2) - k_2^2} \tag{3.104}$$

$$Y_2 = \frac{F_1 k_2 + (k_1 + k_2 - m_1 \omega_f^2)}{(k_1 + k_2 - m_1 \omega_f^2)(k_2 - m_2 \omega_f^2) - k_2^2} \tag{3.105}$$

In Eqs. (3.104) and (3.105), we note that their denominators are identical. If these denominators are individually equal to zero, the amplitudes Y_1 and Y_2 become infinite in value. This condition occurs when the frequency ω_f of the harmonic forces coincides with any one of the two free frequencies of vibration of the two-degree spring–mass system. This is again what we have termed earlier as the phenomenon of resonance. The values of ω_f that produce resonance may be found from the following equation:

$$(k_1 + k_2 - m_1 \omega_f^2)(k_2 - m_2 \omega_f^2) - k_2^2 = 0 \tag{3.106}$$

By carrying out the required mathematical operations, Eq. (3.106) may be written in the following form:

$$\omega_f^4 - \left(\frac{k_1 + k_2}{m_1} + \frac{k_2}{m_2}\right) \omega_f^2 + \frac{k_1 k_2}{m_1 m_2} = 0 \tag{3.107}$$

Equation (3.107) is identical to Eq. (2.130) in Section 2.8 if the spring constant k_3 is zero. Its solution will provide the two positive values of ω_f, which will be identical in value to the two free undamped frequencies of vibration of the two-degree spring–mass system.

3.8 ACCELERATION IMPULSE EXTRAPOLATION NUMERICAL METHOD

An interesting numerical method was developed at the Massachusetts Institute of Technology (MIT) during the late 1950s, in order to facilitate the dynamic analysis required for a successful moon landing [1, 7, 9, 10]. The method is based on simple principles of dynamic analysis, and it can be applied with great ease and accuracy to simple as well as very complicated dynamics problems.

In the preceding sections, closed-form solutions were used for the solution of the differential equations of motion for the problem at hand. Such solutions, however, can become extremely complicated when the dynamic loading conditions and the geometry of the system that needs to be analyzed dynamically are represented by complex mathematical functions. This could create a serious limitation for practicing engineers, because they are usually interested in an approximate solution to the problem. A convenient numerical method, such as the acceleration impulse extrapolation method, can

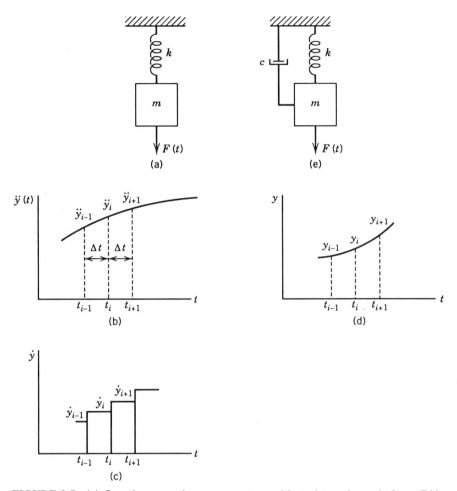

FIGURE 3.7. (a) One-degree spring–mass system subjected to a dynamic force $F(t)$. (b) Acceleration–time curve. (c) Velocity–time curve. (d) Displacement–time curve. (e) One-degree spring–mass system subjected to a dynamic force $F(t)$ and moving under the influence of viscous damping.

provide a reasonably accurate solution for simpler, as well as for very complicated, dynamics problems [1, 7, 10].

In order to illustrate the fundamental aspects of the acceleration impulse extrapolation method, we consider the differential equation of motion for a one-degree spring–mass system, such as the one in Fig. 3.7a:

$$m\ddot{y} + ky = F(t) \tag{3.108}$$

By solving Eq. (3.108) for the acceleration $\ddot{y}$, we obtain

$$\ddot{y} = \frac{F(t) - ky}{m} \tag{3.109}$$

If we seek a closed-form solution of Eq. (3.108), the displacement y of the mass m in Fig. 3.7a will be obtained as a continuous function of time. In a numerical method, such as the acceleration impulse extrapolation method, the value of the displacement y is obtained at discrete values of time.

In the acceleration impulse extrapolation method, the acceleration curve in Fig. 3.7b is replaced by a series of equally spaced impulses occurring at times $t_0, t_1, \ldots, t_{i-1}, t_i, t_{i+1}, \ldots, t_n$. At a time station t_i, the magnitude of the acceleration impulse is

$$\ddot{y}_i(\Delta t) \tag{3.110}$$

where Δt is the time interval between time stations, as shown in Fig. 3.7b. The velocity $\dot{y}$ is constant between time stations and the displacement y is linear, as shown in Figs. 3.7c and 3.7d, respectively.

The solution of Eq. (3.108) by the acceleration impulse extrapolation method involves a step-by-step procedure that starts at $t = 0$, where the displacement and velocity are known, and then the displacement is extrapolated from one time station to the next. For example, if the displacements $y^{(i-1)}$ and $y^{(i)}$ at time stations $i - 1$ and i, respectively, are known from previous computations, then we can determine the acceleration $\ddot{y}^{(i)}$ at station i from Eq. (3.109). Between stations i and $i + 1$, the average velocity $\dot{y}_{av}$ may be obtained from the following approximate equation:

$$\dot{y}_{av} = \frac{y^{(i)} - y^{(i-1)}}{\Delta t} + \ddot{y}^{(i)}(\Delta t) \tag{3.111}$$

At time station $i + 1$, the displacement $y^{(i+1)}$ can be obtained from the expression

$$\begin{aligned} y^{(i+1)} &= y^{(i)} + \dot{y}_{av}(\Delta t) \\ &= 2y^{(i)} - y^{(i-1)} + \ddot{y}^{(i)}(\Delta t)^2 \end{aligned} \tag{3.112}$$

Equation (3.112) may be used to determine the displacement at time station $i + 1$, provided that the displacements $y^{(i)}$ and $y^{(i-1)}$ at time stations i and $i - 1$, respectively, and the acceleration $\ddot{y}^{(i)}$ at time station i are known. Therefore, by using Eq. (3.112), the displacement y from one time station to the other may be obtained rather conveniently. This equation, however, cannot be applied for the first time station, because at $t = 0$ the value of

$y^{(i-1)}$ does not exist. This difficulty can be eliminated if, for the first time station, the displacement $y^{(1)}$ is calculated from one of the following two equations:

$$y^{(1)} = \tfrac{1}{6}(2\ddot{y}^{(0)} + \ddot{y}^{(1)})(\Delta t)^2 \tag{3.113}$$

$$y^{(1)} = \frac{\ddot{y}^{(0)}}{2}(\Delta t)^2 \tag{3.114}$$

In Eq. (3.113), the acceleration during the first time interval is assumed to vary linearly, while in Eq. (3.114), for the same time interval, the acceleration is assumed to be constant and equal to the initial value. Therefore, for the first time interval, either one of Eqs. (3.113) and (3.114) may be used. Equation (3.113), however, must be used for the first time interval if the dynamic force $F(t)$ is zero at $t = 0$, because in this case we have $\ddot{y}^{(0)} = 0$, and Eq. (3.114) does not apply.

Consider now the case where the one-degree spring–mass system is under the influence of viscous damping, as shown in Fig. 3.7e. In this case, the differential equation of motion representing the spring–mass system is

$$m\ddot{y} + c\dot{y} + ky = F(t) \tag{3.115}$$

and

$$\ddot{y} = \frac{F(t) - c\dot{y} - ky}{m} \tag{3.116}$$

In this case, since $\ddot{y}$ in Eq. (3.116) depends on the velocity $\dot{y}$, the above numerical procedure can be applied only if $\dot{y}$ is known. An approximate expression has been developed to determine $\dot{y}^{(i)}$ at a time station i, which is written as follows:

$$\dot{y}^{(i)} = \frac{y^{(i)} - y^{(i-1)}}{\Delta t} + \ddot{y}^{(i)}\left(\frac{\Delta t}{2}\right) \tag{3.117}$$

In Eq. (3.117), the second term on the right-hand side of the equation provides an estimate of the amount by which the average velocity of the preceding time interval should be increased, in order to obtain the velocity at the next time station.

If Eq. (3.117) is substituted into Eq. (3.116), the following expression is obtained:

$$\ddot{y}^{(i)} = \frac{2F(t)}{2m + c(\Delta t)} - \frac{2c(y^{(i)} - y^{(i-1)})}{(\Delta t)[2m + c(\Delta t)]} - \frac{2ky^{(i)}}{2m + c(\Delta t)} \tag{3.118}$$

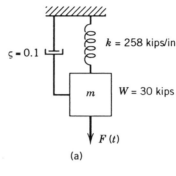

(a)

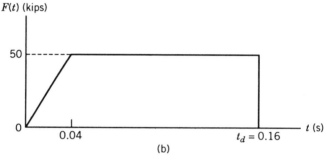

(b)

FIGURE 3.8. (a) One-degree spring–mass system subjected to a dynamic force $F(t)$ and moving under the influence of viscous damping. (b) Time variation of $F(t)$.

In this case, Eq. (3.118) can be used to determine the acceleration $\ddot{y}^{(i)}$ at a station i; Eq. (3.112) may be used to determine the displacement $y^{(i+1)}$ at station $i + 1$; and Eq. (3.113) or (3.114) may be applied to calculate the displacement $y^{(1)}$ at the first time station.

The following example illustrates the application of the numerical method to a one-degree spring–mass system. In the following section, the acceleration impulse extrapolation method will be applied to spring–mass systems with two or more degrees of freedom.

Example 3.6. The one-degree spring–mass system in Fig. 3.8a is subjected to a dynamic force $F(t)$ of general type, which has the time variation shown in Fig. 3.8b. Determine the maximum displacement of the weight W by using the acceleration impulse extrapolation method. The spring–mass system is under the influence of viscous damping, with damping ratio $\zeta = 0.10$.

Solution. The critical damping constant c_c is

$$c_c = 2m\omega = 2m\sqrt{\frac{k}{m}}$$

$$= \frac{(2)(30)}{386} \sqrt{\frac{(258)(386)}{30}}$$

$$= 8.951 \text{ kips·s/in.}$$

The damping constant c is

$$c = \zeta c_c = (0.10)(8.951) = 0.8951 \text{ kips·s/in.}$$

We also have

$$F(t) = \frac{50}{0.04} t = 1250t \qquad 0 \le t \le 0.04$$

$$F(t) = 50 \qquad 0.04 \le t \le 0.16$$

$$F(t) = 0 \qquad t > 0.16$$

$$m = \frac{W}{g} = 0.07764 \text{ kips·s}^2/\text{in.}$$

By using Eq. (3.118) and a time interval $\Delta t = 0.01$ s, the acceleration $\ddot{y}^{(i)}$ at any time station i is as follows:

$$\ddot{y}^{(i)} = \frac{2}{2(0.07764) + (0.8951)(0.01)} F(t)$$

$$- \frac{2(0.8951)[y^{(i)} - y^{(i-1)}]}{(0.10)[2(0.07764) + (0.8951)(0.01)]}$$

$$- \frac{2(258)y^{(i)}}{2(0.07764) + (0.8951)(0.01)}$$

$$= 12.18F(t) + 1090.05y^{(i-1)} - 4231.97y^{(i)}$$

Since the force $F(t)$ is zero at $t = 0$, we must use Eq. (3.113) for the computation of the displacement $y^{(1)}$ at time station 1. This equation yields the following results:

$$y^{(1)} = \frac{1}{6}(2\ddot{y}^{(0)} + \ddot{y}^{(1)})(\Delta t)^2$$

$$= \frac{1}{6}\{2(0) + [12.18F(t) + 1090.05y^{(0)} - 4231.97y^{(1)}]\}(0.01)^2$$

At the first time station, where $t = 0.01$ s, we find from Fig. 3.8b that $F(t) = 12.50$ kips. We also know that the displacement $y^{(0)}$ at $t = 0$ is zero.

Therefore,

$$y^{(1)} = \tfrac{1}{6}[(12.18)(12.50) - 4231.97)y^{(1)}](0.01)^2$$

This equation may be solved for $y^{(1)}$. The result is

$$y^{(1)} = 0.002370 \text{ in.}$$

For the remaining intervals, the displacement $y^{(i+1)}$ at any time station $i + 1$ may be obtained from Eq. (3.112).

The results are tabulated in Table 3.2. The first column in this table indicates the time station i; the second column gives the time t at each time station; the third column gives the values of $F(t)$; columns (4), (5), and (6) are the three terms of the equation for $\ddot{y}^{(i)}$; the seventh column yields $\ddot{y}^{(i)}$; column (8) yields the values of the third term in Eq. (3.112) and the ninth column provides the displacements $y^{(i)}$ at each time station i. This table shows that the maximum displacement $y_{\max}$ of the weight W occurs at $t = 0.07$ s, and it is equal to 0.305246 in. Note in this table how y varies with time. Also, note that at $t > 0.16$ s the force $F(t)$ is zero. Therefore, for $t > 0.16$ s, the force $F(t)$ in column (3) should be marked zero, if it is desired to continue the table beyond 0.16 s and to determine subsequent peaks of the displacement y. Table 3.2 contains only one peak at $t = 0.07$ s.

In general, the maximum displacement of W could be reached at any time $t < t_d$, or at any time $t > t_d$, where t_d is the time duration of the dynamic force. Therefore, some care must be taken in selecting the time interval Δt. The time interval Δt selected for this example problem should provide very accurate results.

3.9 ACCELERATION IMPULSE EXTRAPOLATION METHOD FOR SPRING–MASS SYSTEMS WITH TWO OR MORE DEGREES OF FREEDOM

The acceleration impulse extrapolation method will be applied here for the dynamic response of spring–mass systems with two or more degrees of freedom. The dynamic force acting on the spring–mass system can have any arbitrary time variation, and the spring–mass system is moving under the influence of viscous damping. A spring–mass system that is subjected to these types of force and damping conditions is shown in Fig. 1.7c in Section 1.7. The differential equations of motion for masses m_1 and m_2 are given by Eqs. (1.26) and (1.27), respectively. We write these two equations again

TABLE 3.2. Variation of the Displacement y with Time t

(1) i	(2) t	(3) $F(t)$	(4) $12.18F(t)$	(5) $1090.05y^{(i-1)}$	(6) $4231.97y^{(i)}$	(7) $\ddot{y}^{(i)}$	(8) $\ddot{y}^{(i)}(\Delta t)^2$	(9) $y^{(i)}$ in.
0	0	0	0	0	0	0	0	0
1	0.01	12.5	152.25	0	10.029769	142.220231	0.014222	0.002370[a]
2	0.02	25.0	304.50	0	80.246615	226.836804	0.022684	0.018962
3	0.03	37.5	456.75	2.583419	246.457975	230.961553	0.023096	0.058238
4	0.04	50.0	609.00	20.669528	510.417902	162.064430	0.016206	0.120610
5	0.05	50.0	609.00	63.482332	842.957640	−102.486709	−0.010249	0.199188
6	0.06	50.0	609.00	131.470931	1132.123918	−305.999039	−0.030600	0.267517
7	**0.07**	50.0	609.00	217.124879	1291.791915	−391.185009	−0.039119	**0.305246**
8	0.08	50.0	609.00	291.606906	1285.909476	−344.176074	−0.034418	0.303856
9	0.09	50.0	609.00	332.733402	1134.371095	−194.152862	−0.019415	0.268048
10	0.10	50.0	609.00	331.218233	900.669015	0.516707	0.000051	0.212825
11	0.11	50.0	609.00	292.185722	667.182766	173.807125	0.017381	0.157653
12	0.12	50.0	609.00	231.989891	507.252388	273.597265	0.027360	0.119862
13	0.13	50.0	609.00	171.849653	463.108709	276.546864	0.027655	0.109431
14	0.14	50.0	609.00	130.655573	536.000160	192.285102	0.019229	0.126655
15	0.15	50.0	609.00	119.285262	690.268163	56.792120	0.005679	0.163108
				138.060283				

[a]Equation (3.113) is used here to obtain $y^{(1)}$ at the first time station $i = 1$.

149

here as follows:

$$m_1\ddot{y}_1 + k_1 y_1 + c_1 \dot{y}_1 - k_2(y_2 - y_1) - c_2(\dot{y}_2 - \dot{y}_1) = F_1(t) \qquad (3.119)$$

$$m_2\ddot{y}_2 + k_2(y_2 - y_1) + c_2(\dot{y}_2 - \dot{y}_1) = F_2(t) \qquad (3.120)$$

By solving Eqs. (3.119) and (3.120) for $\ddot{y}_1$ and $\ddot{y}_2$, respectively, we find

$$\ddot{y}_1 = \frac{F_1(t) - k_1 y_1 - c_1 \dot{y}_1 + k_2(y_2 - y_1) + c_2(\dot{y}_2 - \dot{y}_1)}{m_1} \qquad (3.121)$$

$$\ddot{y}_2 = \frac{F_2(t) - k_2(y_2 - y_1) - c_2(\dot{y}_2 - \dot{y}_1)}{m_2} \qquad (3.122)$$

By following the procedure discussed in Section 3.8, the velocities $\dot{y}_1^{(i)}$ and $\dot{y}_2^{(i)}$ at time station i may be expressed as

$$\dot{y}_1^{(i)} = \frac{y_1^{(i)} - y_1^{(i-1)}}{\Delta t} + \ddot{y}_1^{(i)}\left(\frac{\Delta t}{2}\right) \qquad (3.123)$$

$$\dot{y}_2^{(i)} = \frac{y_2^{(i)} - y_2^{(i-1)}}{\Delta t} + \ddot{y}_2^{(i)}\left(\frac{\Delta t}{2}\right) \qquad (3.124)$$

If the expressions for $\dot{y}_1^{(i)}$ and $\dot{y}_2^{(i)}$ in Eqs. (3.123) and (3.124), respectively, are substituted into Eqs. (3.121) and (3.122), the final expressions for $\ddot{y}_1$ and $\ddot{y}_2$ at any time station i may be obtained.

The displacements $y_1^{(1)}$ and $y_2^{(1)}$ for the first time station may be obtained from the following equations:

$$y_1^{(1)} = \tfrac{1}{6}(2\ddot{y}_1^{(0)} + \ddot{y}_1^{(1)})(\Delta t)^2 \qquad (3.125)$$

$$y_1^{(1)} = \frac{\ddot{y}_1^{(0)}}{2}(\Delta t)^2 \qquad (3.126)$$

$$y_2^{(1)} = \tfrac{1}{6}(2\ddot{y}_2^{(0)} + \ddot{y}_2^{(1)})(\Delta t)^2 \qquad (3.127)$$

$$y_2^{(1)} = \frac{\ddot{y}_2^{(0)}}{2}(\Delta t)^2 \qquad (3.128)$$

In Eqs. (3.125) and (3.127), the acceleration is assumed to be linear during the first time interval; in Eqs. (3.126) and (3.128), the acceleration is assumed to be constant during the first time interval and equal to the acceleration at time $t = 0$.

The displacements $y_1^{(i+1)}$ and $y_2^{(i+1)}$ at station $i + 1$ may be determined from the equations

$$y_1^{(i+1)} = 2y_1^{(i)} - y_1^{(i-1)} + \ddot{y}_1^{(i)}(\Delta t)^2 \tag{3.129}$$

$$y_2^{(i+1)} = 2y_2^{(i)} - y_2^{(i-1)} + \ddot{y}_2^{(i)}(\Delta t)^2 \tag{3.130}$$

If damping is zero, Eqs. (3.121) and (3.122) may be written

$$\ddot{y}_1 = \frac{F_1(t) - k_1 y_1 + k_2(y_2 - y_1)}{m_1} \tag{3.131}$$

$$\ddot{y}_2 = \frac{F_2(t) - k_2(y_2 - y_1)}{m_2} \tag{3.132}$$

These two equations provide the acceleration $\ddot{y}_1$ and $\ddot{y}_2$ of the two-degree undamped system, and Eqs. (3.125) through (3.128) may be used to determine the displacement at the first time station. At any other station $i + 1$, Eqs. (3.129) and (3.130) must be used to determine the displacements.

The following example illustrates the application of the methodology.

Example 3.7. Consider the two-degree spring–mass system in Fig. 3.9a, which is acted on by the dynamic forces $F_1(t) = F(t)$ and $F_2(t) = 0.7F_1(t)$ as shown; it moves under the influence of viscous damping with damping constants $c_1 = 0.288$ kips·s/in. and $c_2 = 0.419$ kips·s/in. Determine the maximum amplitudes of masses m_1 and m_2 by applying the acceleration impulse extrapolation method. The spring constants are $k_1 = 9.52$ kips/in. and $k_2 = 32.10$ kips/in.

Solution. From Eqs. (3.121) and (3.122), we obtain

$$\ddot{y}_1 = \frac{F(t) - k_1 y_1 - c_1 \dot{y}_1 + k_2(y_2 - y_1) + c_2(\dot{y}_2 - \dot{y}_1)}{m_1} \tag{3.133}$$

$$\ddot{y}_2 = \frac{0.7F(t) - k_2(y_2 - y_1) - c_2(\dot{y}_2 - \dot{y}_1)}{m_2} \tag{3.134}$$

By substituting Eqs. (3.123) and (3.124) into Eq. (3.133) and performing the required mathematical manipulations, we obtain

$$\left(\frac{2m_1 + (c_1 + c_2)\,\Delta t}{2m_1}\right)\ddot{y}_1^{(i)} = \frac{1}{m_1}\left[F(t) + (y_2 - y_1)\right.$$

$$+ c_2\left(\frac{y_2^{(i)} - y_2^{(i-1)}}{\Delta t} + \frac{\Delta t}{2}\ddot{y}_2^{(i)} - \frac{y_1^{(i)} - y_1^{(i-1)}}{\Delta t}\right)$$

$$\left. - k_1 y_1 - c_1 \frac{y_1^{(i)} - y_1^{(i-1)}}{\Delta t}\right] \tag{3.135}$$

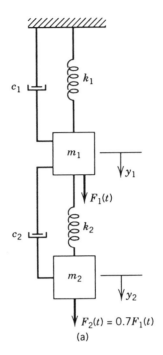

(a)

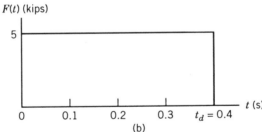

(b)

FIGURE 3.9. (a) Two-degree spring–mass system acted on by the dynamic forces $F_1(t)$ and $F_2(t)$ and moving under the influence of viscous damping. (b) Time variation of the dynamic forces.

In a similar manner, by substituting Eqs. (3.123) and (3.124) into Eq. (3.134) and simplifying, we obtain

$$\ddot{y}_2^{(i)} = \frac{2}{2m_2 + c_2\,\Delta t}\left[0.7F(t) - k_2(y_2 - y_1)\right.$$

$$\left. - c_2\left(\frac{y_2^{(i)} - y_2^{(i-1)}}{\Delta t} - \frac{y_1^{(i)} - y_1^{(i-1)}}{\Delta t} - \frac{\Delta t}{2}\ddot{y}_1^{(i)}\right)\right] \qquad (3.136)$$

By substituting Eq. (3.136) into (3.135) and performing the required mathematical operations, we find

$$
\ddot{y}_1^{(i)} = \left(\frac{2m_1 + (c_1 + c_2)\,\Delta t}{2} - \frac{c_2^2(\Delta t)^2}{2(2m_2 + c_2\,\Delta t)}\right)^{-1}\left(\frac{2m_2 + 1.7c_2\,\Delta t}{2m_2 + c_2\,\Delta t}F(t)\right.
$$

$$
+ \frac{2m_2}{2m_2 + c_2\,\Delta t}k_2(y_2 - y_1) - k_1 y_1 - c_1\frac{y_1^{(i)} - y_1^{(i-1)}}{\Delta t} \tag{3.137}
$$

$$
\left. + \frac{2m_2}{2m_2 + c_2\,\Delta t}\cdot\frac{y_2^{(i)} - y_2^{(i-1)}}{\Delta t}\cdot c_2 - \frac{2m_2}{2m_2 + c_2\,\Delta t}\cdot\frac{y_1^{(i)} - y_1^{(i-1)}}{\Delta t}\cdot c_2\right)
$$

By using the given values for m_1, m_2, c_1, and c_2 and a time interval $\Delta t = 0.03$ s, Eqs. (3.137) and (3.136) yield

$$
\ddot{y}_1^{(i)} = 4.526F(t) + 193.390y_2^{(i)} - 277.349y_1^{(i)} - 58.633y_2^{(i-1)}
$$

$$
- 100.788y_1^{(i-1)} \tag{3.138}
$$

$$
\ddot{y}_2^{(i)} = 5.098F(t) + 310.201y_1^{(i)} - 313.895y_2^{(i)} - 93.313y_1^{(i-1)}
$$

$$
+ 95.168y_2^{(i-1)} \tag{3.139}
$$

Equations (3.126) and (3.128) may be used here to obtain the displacements $y_1^{(1)}$ and $y_2^{(1)}$ at the first time station 1. These equations yield the following results:

$$
y_1^{(1)} = \frac{(4.526)(5)}{2}(0.03)^2
$$

$$
= 0.010184\text{ in.} \tag{3.140}
$$

$$
y_2^{(1)} = \frac{(5.098)(5)}{2}(0.03)^2
$$

$$
= 0.011471\text{ in.} \tag{3.141}
$$

From Fig. 3.9b, we note that for the interval $0 \leq t \leq 0.4$ s the force $F(t) = 5$ kips, and it is equal to zero for $t > 0.4$ s.

The results are tabulated in Tables 3.3a and 3.3b. Since the displacements $y_1^{(1)}$ and $y_2^{(1)}$ are determined and are given by Eqs. (3.140) and (3.141), respectively, the values of $y_1^{(i+1)}$ and $y_2^{(i+1)}$ at any time station $i + 1$ may be determined from Eqs. (3.129) and (3.130), respectively. Note that the format used to set up Table 3.2 is also used here to prepare Tables 3.3a and 3.3b. These two tables must be prepared simultaneously in order to determine $y_1^{(i)}$ and $y_2^{(i)}$ in the tenth column of these tables.

From Table 3.3a, we note that the maximum displacement $y_{1\text{max}}$ of mass

TABLE 3.3a. Variation of the Displacement y_1 of Mass m_1 with Time t

(1)	(2)	(3)	(4)	(5)	(6)	(7)	(8)	(9)	(10)
i	t (s)	$4.526F(t)$	$193.390y_2^{(i)}$	$277.349y_1^{(i)}$	$58.633y_2^{(i-1)}$	$100.788y_1^{(i-1)}$	$\ddot{y}_1^{(i)}$	$\ddot{y}_1^{(i)}(\Delta t)^2$	$y_1^{(i)}$ (in.)
0	0	22.630	0	0	0	0	22.630000	0.020367	0.010184[a]
1	0.03	22.630	2.218377	2.824522	0	0	22.023855	0.019821	0.040189
2	0.06	22.630	8.796344	11.146379	0.672579	1.026425	20.633811	0.018570	0.088764
3	0.09	22.630	19.519626	24.618607	2.666922	4.050569	18.914666	0.017023	0.154362
4	0.12	22.630	34.056559	42.812146	5.918063	8.946346	16.902696	0.015212	0.235172
5	0.15	22.630	51.970468	65.224719	10.325447	15.557837	14.608139	0.013147	0.329129
6	0.18	22.630	72.741908	91.283599	15.756681	23.702516	12.034144	0.010831	0.433917
7	0.21	22.630	95.795737	120.346446	22.054275	33.172254	9.197270	0.008278	0.546983
8	0.24	22.630	120.527997	151.705188	29.043857	43.733627	6.142579	0.005528	0.665577
9	0.27	22.630	146.326996	184.597115	36.542314	55.129323	2.946890	0.002652	0.786823
10	0.30	22.630	172.587231	218.224572	44.364190	67.082175	−0.289356	−0.000260	0.907809
11	0.33	22.630	198.717701	251.779918	52.325907	79.302317	−3.455807	−0.003110	1.025685
12	0.36	22.630	224.146552	284.472709	60.248281	91.496253	−6.448185	−0.005803	1.137758
13	0.39	22.630	248.325137	315.556044	67.957934	103.376740	−9.182101	−0.008264	1.241567
14	0.42	0	270.735364	344.347366	75.288525	114.672353	−34.228174	−0.030805	1.314571
15	0.45	0	286.463386	364.594952	82.082975	125.135055	−35.079486	−0.031572	1.356003
16	0.48	0	295.239617	376.086076	86.851480	132.492982	−35.204957	−0.031684	1.365751
17	**0.51**	0	296.982448	378.789674	89.512304	136.668830	−34.650700	−0.031186	**1.365751**
18	0.54	0	291.804431	372.843866	90.040705	137.651312	−33.428828	−0.030086	1.344313

[a]This is the value of $y_1^{(1)}$ given by Eq. (3.140).

154

TABLE 3.3b. Variation of the Displacement y_2 of Mass m_2 with Time t

(1) i	(2) t (s)	(3) $5.098F(t)$	(4) $310.201y_1^{(i)}$	(5) $313.895y_2^{(i)}$	(6) $93.313y_1^{(i-1)}$	(7) $95.158y_2^{(i-1)}$	(8) $\ddot{y}_2^{(i)}$	(9) $\ddot{y}_2^{(i)}(\Delta t)^2$	(10) $y_2^{(i)}$ (in.)
0	0	25.490	0	0	0	0	25.490000	0.022941	0
1	0.03	25.490	3.159087	3.601722	0	0	25.047365	0.022543	0.011471[a]
2	0.06	25.490	12.466668	14.281608	0.950300	1.091672	23.816432	0.021435	0.045485
3	0.09	25.490	27.534682	31.691762	3.750156	4.328716	21.911480	0.019720	0.100934
4	0.12	25.490	47.883247	55.293700	8.282835	9.605687	19.402399	0.017462	0.176103
5	0.15	25.490	72.950590	84.378445	14.403981	16.759370	16.417534	0.014776	0.268734
6	0.18	25.490	102.096145	108.102632	21.944605	25.374877	13.113785	0.011802	0.376141
7	0.21	25.490	134.601487	155.532470	30.712014	35.796587	9.643590	0.008679	0.495350
8	0.24	25.490	169.674674	195.687383	40.490097	47.141469	6.128663	0.005516	0.623238
9	0.27	25.490	206.462651	237.574238	51.040625	59.312314	2.650102	0.002385	0.756642
10	0.30	25.490	244.073281	280.209948	62.106987	72.008106	−0.745548	−0.000671	0.892431
11	0.33	25.490	281.603260	322.634973	73.420815	84.930873	−4.031655	−0.003628	1.027549
12	0.36	25.490	318.168513	363.920860	84.710381	97.789783	−7.182945	−0.006465	1.159039
13	0.39	25.490	352.933669	403.176835	95.709744	110.303423	−10.159487	−0.009144	1.284064
14	0.42	0	385.135325	439.561731	106.167612	122.201803	−38.392215	−0.034553	1.399945
15	0.45	0	407.781239	465.097503	115.854342	133.229966	−39.940640	−0.035947	1.481273
16	0.48	0	420.633487	479.346456	122.666564	140.969789	−40.409744	−0.036369	1.526654
17	**0.51**	0	423.657326	482.176089	126.532708	145.288608	−39.762863	−0.035787	**1.535666**
18	0.54	0	417.007237	493.769141	127.442323	146.146262	−38.057965	−0.034252	1.508891

[a]This is the value of $y_2^{(1)}$ given by Eq. (3.141).

155

m_1 occurs at $t = 0.51$ s, and it is equal to 1.365751 in. The maximum displacement $y_{2\text{max}}$ of mass m_2 (from Table 3.3b) occurs at $t = 0.51$ s, and it is equal to 1.535666 in.

3.10   FOURIER SERIES ANALYSIS

In many cases, the dynamic forces acting on an engineering system may be periodic but not harmonic. Some examples of periodic force functions are shown in Fig. 3.10. The French mathematician Jean Baptiste Joseph Fourier (1768–1830) proved [11] that any periodic force function can be expressed in terms of combinations of sine and cosine functions. For example, a periodic force function $F(t)$ that repeats itself every T seconds can be written as an infinite sum of sines and cosines. The component harmonic terms will have the frequencies $\omega_f = 2\pi/T$, $2\omega_f, \ldots, n\omega_f$. In mathematical form, we have the expression

$$F(t) = \frac{a_0}{2} + a_1 \cos \omega_f t + a_2 \cos 2\omega_f t + \cdots + a_n \cos n\omega_f t$$

$$+ \cdots + b_1 \sin \omega_f t + b_2 \sin 2\omega_f t + \cdots + b_n \sin n\omega_f t \quad (3.142)$$

or

$$F(t) = \frac{a_0}{2} + \sum_{n=1}^{\infty} (a_n \cos n\omega_f t + b_n \sin n\omega_f t) \quad (3.143)$$

The evaluation of the constants a_0, a_n, and b_n can be carried out by using the following equations:

$$a_0 = \frac{2}{T} \int_d^{d+T} F(t)\, dt \quad (3.144)$$

$$a_n = \frac{2}{T} \int_d^{d+T} F(t) \cos(n\omega_f t)\, dt \quad (3.145)$$

$$b_n = \frac{2}{T} \int_d^{d+T} F(t) \sin(n\omega_f t)\, dt \quad (3.146)$$

The quantity d in the intervals of the above three equations will be equal to either zero or $-T/2$. The constant a_0 provides an estimate of the offsetting of the remaining terms of the Fourier series.

The dynamic response of a structural or a mechanical system that is subjected to a periodic force function $F(t)$ may be carried out by using the Fourier series expansion of $F(t)$, which is given by Eq. (3.143). When the

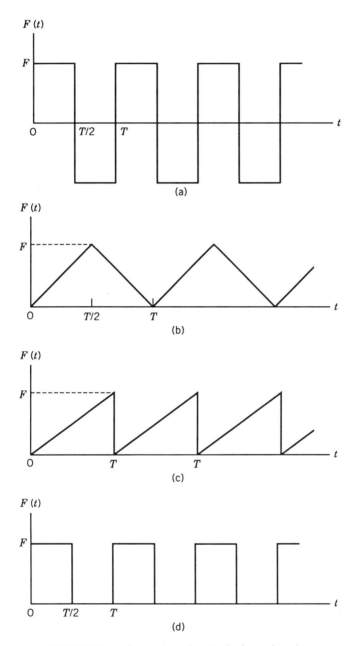

FIGURE 3.10. Examples of periodic force functions.

constants a_0, a_n, and b_n of the series are determined by using Eqs. (3.144), (3.145), and (3.146), the dynamic response may be obtained by applying individually each term of the Fourier series expansion and superimposing the obtained results. An accurate solution may be obtained by using the first three terms of the Fourier series.

A function $F(t)$ may be odd, even, or neither odd nor even. If the function defined in the interval $(-T, T)$ is odd, its Fourier coefficients are as follows:

$$a_0 = 0 \tag{3.147}$$

$$a_n = 0 \tag{3.148}$$

$$b_n = \frac{2}{T} \int_d^{d+T} F(t) \sin(n\omega_f t)\, dt \qquad n = 1, 2, 3, \ldots \tag{3.149}$$

In this case, its Fourier series is reduced to the following expression:

$$F(t) = \sum_{n=1}^{\infty} b_n \sin(n\omega_f t) \tag{3.150}$$

This means that the graph of the function is symmetrical with respect to the origin, and the integral from $-T$ to T is zero. For example, the function in Fig. 3.10a is odd.

If the function is even, the Fourier coefficients and the series are as follows:

$$a_0 = \frac{2}{T} \int_d^{d+T} F(t)\, dt \tag{3.151}$$

$$b_n = 0 \tag{3.152}$$

$$a_n = \frac{2}{T} \int_d^{d+T} F(t) \cos(n\omega_f t)\, dt \qquad n = 1, 2, 3, \ldots \tag{3.153}$$

$$F(t) = \frac{a_0}{2} + \sum_{n=1}^{\infty} a_n \cos(n\omega_f t)\, dt \tag{3.154}$$

In this case, the plot of the function is symmetrical with respect to the ordinate. The function shown in Fig. 3.10b is even.

The Fourier series expansion of $F(t)$ may also be written in exponential form as follows:

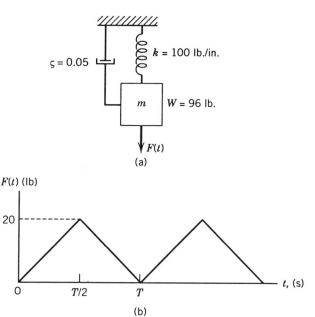

FIGURE 3.11. (a) One-degree spring–mass system acted on by a dynamic force $F(t)$ and moving under the influence of viscous damping. (b) Time variation of $F(t)$.

$$F(t) = \sum_{n=-\infty}^{\infty} c_n e^{in\omega_f t} \tag{3.155}$$

The constant c_n is given by the expression

$$c_n = \frac{1}{T} \int_{-T/2}^{T/2} F(t) e^{-in\omega_f t} \, dt \tag{3.156}$$

where i in this equation denotes the imaginary part of a complex number. Equation (3.155) shows that a periodic function contains all frequencies, both negative and positive, that are harmonically related to the fundamental frequency ω_f.

The following example illustrates the application of the theory.

Example 3.8. The spring–mass system in Fig. 3.11a is acted on by the periodic force $F(t)$ shown in Fig. 3.11b. The maximum magnitude of the force is 20 lb, $T/2 = 0.10$ s, and the motion of the spring–mass system is under the influence of viscous damping with damping ratio $\zeta = 0.05$. Determine the steady-state response of the spring–mass system by applying Fourier series analysis.

Solution. The first step would be the derivation of the Fourier series expansion of $F(t)$. The force function, as stated earlier, is an even function of t and we only have cosine terms in the series. The fundamental frequency $\omega_f = 2\pi/0.2 = 10\pi$. The time variation of $F(t)$ is as follows:

$$F(t) = \frac{40t}{T} \qquad 0 \le t \le T/2 \tag{3.157}$$

$$F(t) = \frac{40(T-t)}{T} \qquad T/2 \le t \le T \tag{3.158}$$

By using Eqs. (3.151) and (3.153), the constants a_0 and a_n are determined as follows:

$$a_0 = \frac{2}{T} \int_d^{d+T} F(t)\, dt$$

$$= \frac{2}{T} \int_0^{T/2} \frac{40t}{T}\, dt + \frac{2}{T} \int_{T/2}^{T} \left[40 - \frac{40t}{T} \right] dt$$

$$= \frac{2}{T} \left(\frac{40t^2}{2T} \right)_0^{T/2} + \frac{2}{T} [40t]_{T/2}^{T} - \frac{2}{T} \left[\frac{40t^2}{2T} \right]_{T/2}^{T}$$

$$= 20$$

$$a_n = \frac{2}{T} \int_d^{d+T} F(t) \cos(n\omega_f t)\, dt$$

$$= \frac{2}{T} \int_0^{T/2} \frac{40t}{T} \cos(n\omega_f t)\, dt + \frac{2}{T} \int_{T/2}^{T} 40 \cos(n\omega_f t)\, dt$$

$$- \frac{2}{T} \int_{T/2}^{T} \frac{40t}{T} \cos(n\omega_f t)\, dt$$

$$= \frac{80}{T^2} \left[\frac{1}{n^2 \omega_f^2} \cos(n\omega_f t) + \frac{t}{n\omega_f} \sin(n\omega_f t) \right]_0^{T/2} + \frac{80}{T} \left[\frac{1}{n\omega_f} \sin(n\omega_f t) \right]_{T/2}^{T}$$

$$- \frac{80}{T^2} \left[\frac{1}{n^2 \omega_f^2} \cos(n\omega_f t) + \frac{t}{n\omega_f} \sin(n\omega_f t) \right]_{T/2}^{T}$$

$$= \frac{20}{n^2 \pi^2} [2\cos(n\pi) - \cos(2n\pi) - 1]$$

or

$$a_n = -\frac{80}{n^2 \pi^2} \qquad n = 1, 3, 5, \ldots \tag{3.159}$$

Since the force function is an even one, we have $F(-t) = F(t)$ and, conse-quently, $b_n = 0$. Thus, the Fourier series expansion of $F(t)$ is

$$F(t) = \frac{a_0}{2} + \sum_{n=1}^{\infty} a_n \cos(n\omega_f t)$$

$$= 10 - \frac{80}{\pi^2} \sum_{n=1,3,5,\ldots} \frac{1}{n^2} \cos(n\omega_f t) \tag{3.160}$$

$$= 10 - \frac{80}{\pi^2} [\cos(10\pi t) + \tfrac{1}{9}\cos(30\pi t) + \tfrac{1}{25}\cos(50\pi t) + \cdots]$$

Since now the Fourier series expansion of $F(t)$ is completed, the dynamic analysis may begin by considering individually each term of the series in Eq. (3.160) and superimposing the results. We consider the first term of the series, which is the constant $a_0/2 = 10$ lb. This term produces a displacement that is equal to $a_0/2k = 20/200 = 0.10$ in.

The fundamental frequency $\omega_f = 2\pi/T = 2\pi/0.20 = 10\pi$. The free un-damped frequency of vibration ω is

$$\omega = \sqrt{\frac{k}{m}} = \sqrt{\frac{(100)(386)}{96}}$$

$$= 20.05 \text{ rps}$$

From Eqs. (3.29) and (3.31) in Section 3.3, the steady-state response $y(t)$ due to the harmonic force $a_n \cos(n\omega_f t)$ $(n = 1, 3, 5, \ldots)$ may be written as follows:

$$y(t) = \frac{a_n}{k} \cdot \frac{1}{\sqrt{[1 - (\omega_f/\omega)^2]^2 + (2\zeta\omega_f/\omega)^2}} \cos(\omega_f t - \phi_n) \tag{3.161}$$

For each term of the Fourier cosine series, the constant a_n is given by Eq. (3.159) for $n = 1, 3, 5, \ldots$. The phase angle ϕ may be determined from the equation

$$\tan \phi_n = \frac{2\zeta\omega_f/\omega}{1 - (\omega_f/\omega)^2} \tag{3.162}$$

which is identical to Eq. (3.28) in Section 3.3. By using the values of $n = 1$, 3, and 5, the following solutions for these three terms are obtained:

n = 1

$$y_{st} = -\frac{80}{100\pi^2} = -0.0811 \text{ in.}$$

$$\frac{\omega_f}{\omega} = \frac{10\pi}{20.05} = \frac{\pi}{2}$$

$$\Gamma = \frac{1}{\sqrt{[1-(\omega_f/\omega)^2]^2 + (2\zeta\,\omega_f/\omega)^2}}$$

$$= \frac{1}{\sqrt{[1-(10\pi/20.05)^2]^2 + [(2)(0.05)\,(10\pi/20.05)]^2}} = 0.6776$$

$$\tan \phi_1 = \frac{2\zeta\,\omega_f/\omega}{1-(\omega_f/\omega)^2} = \frac{(2)(0.05)\,(10\pi/20.05)}{1-(10\pi/20.05)^2}$$

$$= -0.1071$$

$$\phi_1 = -6.11° \quad \text{or} \quad \phi_1 = 173.89°$$

n = 3

$$y_{st} = -\frac{80}{(9)(100)\,\pi^2} = -0.00901 \text{ in.}$$

$$\frac{\omega_f}{\omega} = \frac{30\pi}{20.05} = 4.7124$$

$$\Gamma = \frac{1}{\sqrt{[1-(4.7124)^2]^2 + [(2)(0.05)(4.7124)]^2}}$$

$$= 0.0471$$

$$\tan \phi_3 = \frac{(2)(0.05)(4.7124)}{1-(4.7124)^2} = -0.02222$$

$$\phi_3 = -1.27° \quad \text{or} \quad \phi_3 = 173.73°$$

n = 5

$$y_{st} = -\frac{80}{(25)(100)\,\pi^2} = -0.00324 \text{ in.}$$

$$\frac{\omega_f}{\omega} = \frac{50\pi}{20.05} = 7.854$$

$$\Gamma = \frac{1}{\sqrt{[1-(7.854)^2]^2 + [(2)(0.05)(7.854)]^2}}$$

$$= 0.0165$$

TABLE 3.4. Dynamic Response Due to $a_0/2$ and $n = 1, 3,$ and 5

Term	y_{st} (in.)	ω_f/ω	Γ	ϕ_n (degrees)	$y(t) = y_{st}\Gamma\cos(n\omega_f t - \phi_n)$
$a_0/2$	0.10000				
$n = 1$	-0.08110	$\pi/2$	0.6776	173.89	$-0.055\cos(10\pi t - 173.89°)$
$n = 3$	-0.00901	4.7124	0.0471	178.73	$-0.00042\cos(30\pi t - 178.73°)$
$n = 5$	-0.00324	7.8540	0.0165	179.26	$-0.0000535\cos(50\pi t - 179.26°)$

$$\tan\phi_5 = \frac{(2)(0.05)(7.854)}{1 - (7.854)^2} = -0.01294$$

$$\phi_5 = -0.74° \quad \text{or} \quad \phi_5 = 179.26°$$

The results of the responses for $a_0/2$ and $n = 1, 3,$ and 5 are tabulated in Table 3.4. The results show that the contributions from $a_0/2$ and $n = 1$ are most important for practical purposes. The contributions from $n = 3$ and $n = 5$ are rather small, and they could be neglected.

3.11   DYNAMIC RESPONSE OF IDEALIZED BEAMS AND FRAMES

The dynamic analysis of many structural and mechanical elements with continuous mass and elasticity can be analyzed with sufficient accuracy for practical purposes by replacing the initial infinite-degree-of-freedom system by a simpler mathematical model, commonly known as an *idealized system*, which has fewer degrees of freedom. The purpose of an idealized mathematical model is to simplify the initial complex problem in a way that will provide the practicing engineer with the information required for the design of an engineering system.

Consider, for example, the cantilever beam in Fig. 3.12a, which is assumed to support the heavy weight W at its free end. In addition, the dynamic force $F(t)$ is also acting at the free end of the beam. If it is required to design this beam so that displacement and stress criteria are satisfied, the idealized mathematical model in Fig. 3.12b may be used. In this manner, the initial infinite-degree-of-freedom system in Fig. 3.12a is replaced by the idealized one-degree spring–mass system in Fig. 3.12b. The spring constant k is the stiffness of the cantilever beam at its free end, which is defined as the vertical load P at the free end of the beam, which produces a vertical displacement

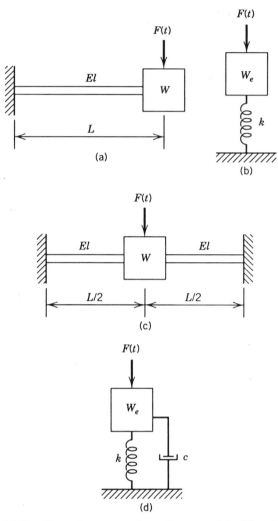

FIGURE 3.12. (a) Cantilever beam carrying a heavy weight W at its free end and subjected to a dynamic force $F(t)$. (b) Idealized one-degree spring–mass system for the cantilever beam. (c) Fixed–fixed beam carrying a heavy weight W and subjected to a dynamic force $F(t)$. (d) Idealized one-degree spring–mass system for the fixed–fixed beam.

y equal to unity; that is,

$$\frac{PL^3}{3EI} = y = 1$$

or

$$P = k = \frac{3EI}{L^3} \qquad (3.163)$$

The weight W_e of the idealized mathematical model would include the weight W acting at the free end of the cantilever beam and some portion of the weight of the cantilever beam. The dynamic force $F(t)$ acting on the idealized model is the force $F(t)$ acting at the free end of the cantilever beam. The idealized system in Fig. 3.12b is a reasonable mathematical model for the design of the cantilever beam, because the maximum displacement occurs at the free end and the maximum stress will occur at the fixed end of the cantilever beam. The idealized spring–mass system in Fig. 3.12b will provide this maximum displacement quite accurately, and it will also provide the dynamic force that produces the maximum stress at the fixed end of the beam. This force is equal to ky_{max}, where y_{max} is the maximum displacement of the spring (or W_e) in Fig. 3.12b, which is produced by the dynamic force $F(t)$. If we apply statically the force ky_{max} at the free end of the beam in Fig. 3.12a, the maximum bending moment at the fixed end may be obtained statically and, consequently, the maximum bending stress may be obtained.

The dynamic analysis of the idealized model in Fig. 3.12b may be carried out by following the methodologies discussed in preceding sections for one-degree spring–mass systems. This procedure would be very accurate if the weight W at the free end of the beam is large compared to the total weight of the beam itself. In many cases, the weight of the beam itself is neglected in the analysis and, consequently, $W_e = W$. This would not create prohibitive loss of accuracy for practical applications.

The fixed–fixed beam in Fig. 3.12c may be idealized by the one-degree spring–mass system in Fig. 3.12d. The weight W_e contains the weight W and about half the total weight of the beam. The spring constant k represents the stiffness of the fixed–fixed beam at its center. By applying the equation

$$\frac{PL^3}{192EI} = y = 1$$

we find

$$P = k = \frac{192EI}{L^3} \qquad (3.164)$$

If we wish to design this beam for maximum displacement and maximum stress, the idealized one-degree spring–mass system in Fig. 3.12d would be sufficient for practical purposes. The maximum displacement of the beam occurs at its center, and the maximum bending stresses occur at the fixed ends of the member and also at its center. The maximum displacement may be determined accurately by using the idealized system in Fig. 3.12d, and the same system may be used to determine the maximum force ky_{max}. When

ky_{max} is applied statically at the center of the beam in Fig. 3.12c, the maximum bending stresses may be determined quite conveniently. If the beam moves under the influence of viscous damping, the idealized one-degree spring–mass system in Fig. 3.12d will also be under the influence of viscous damping with damping coefficient c as shown in the same figure.

Consider now the two-story frame in Fig. 3.13a, which is acted on by the dynamic forces $F_1(t)$ and $F_2(t)$ as shown. The frame may be idealized by the two-degree spring–mass system shown in Fig. 3.13b or in Fig. 3.13c. Both idealized systems are exactly the same, except that engineers may prefer one way over the other. The horizontal members of the frame are often called *girders*, while the vertical members are the *columns* of the frame. In many practical situations, the girders are much stiffer than the columns, and it would be reasonable to assume that they are infinitely stiff in comparison to the stiffness of the columns. The assumption of infinite rigidity for the girders simplifies considerably the dynamic analysis without prohibitive loss of accuracy for practical applications. In other situations, however, the stiffness of the girders may not be sufficiently larger than the stiffness of the columns, and some appreciable error may be introduced in the analysis. In this section, we will assume that the girders are infinitely stiff as compared to the columns.

On this basis, the mass m_1 of the idealized two-degree spring–mass system in Fig. 3.13b or Fig. 3.13c would include the mass of all the girders of the frame at level 1 in Fig. 3.13 plus half the mass of all first-story and second-story columns. Mass m_2 would include the mass of all girders at level 2 of the frame in Fig. 3.13a plus half the mass of all second-story columns. The spring constant k_1 is the sum of the column stiffnesses of the first-story columns, and the spring constant k_2 is the sum of all column stiffnesses of the second story. Since the girders at levels 1 and 2 of the frame are assumed to have infinite rigidity, the rotations of the columns at the girder levels would be zero. Therefore, columns MN, HG, BC, ND, LE, and GF would behave like fixed–fixed beams, as shown in Fig. 3.13e. The stiffness k_i of such a column represents the amount of shear that is developing at the fixed end N, when this end is displaced by a horizontal displacement $x = 1$ as shown. On this basis, we find

$$k_i = \frac{12(EI)_i}{L_i^3} \tag{3.165}$$

For columns AB and IL, the stiffness k_i represents the amount of shear developing at the fixed end B in Fig. 3.13f, when this end is displaced horizontally by an amount $x = 1$. This yields

$$k_i = \frac{3(EI)_i}{L_i^3} \tag{3.166}$$

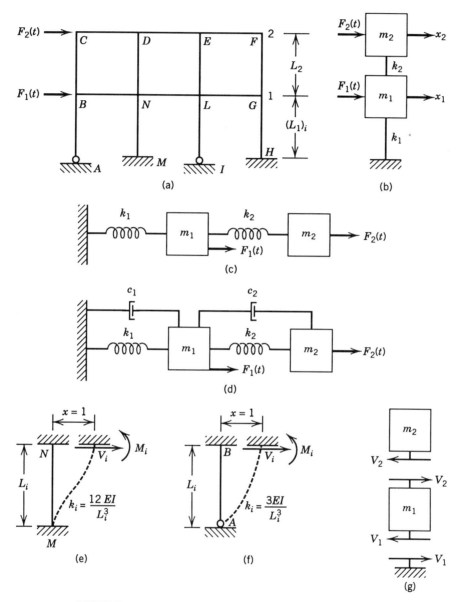

FIGURE 3.13. Idealized mathematical models for frames.

On this basis, the sum of all k_i of the first-story columns would be equal to the spring constant k_1 in Fig. 3.13b or Fig. 3.13c, and the sum of all k_i of the second-story columns would constitute the spring constant k_2 in the same figure. The dynamic forces $F_1(t)$ and $F_2(t)$ are applied to the two-degree idealized system as shown.

In this manner, the infinite-degree-of-freedom system in Fig. 3.13a has been reduced to an idealized system that has only two degrees of freedom. Such an idealization simplifies a great deal the dynamic analysis of the frame, and it provides accurate information for a design engineer who is concerned with maximum displacements and maximum bending stresses produced from the application of the dynamic forces $F_1(t)$ and $F_2(t)$. The maximum horizontal displacements will occur at the girder levels, and the maximum bending stresses will occur at the ends of the columns. This information can be provided with reasonable accuracy from the solution of the idealized two-degree spring–mass system.

If the frame moves under the influence of viscous damping, then such damping can be incorporated into the two-degree idealized spring–mass system as shown in Fig. 3.13d, where c_1 and c_2 are the viscous damping coefficients. The free undamped frequencies of vibration of the idealized two-degree spring–mass system in Fig. 3.13c may be determined as discussed in Section 2.8, and the damped frequencies of the spring–mass system in Fig. 3.13d may be determined as discussed in Section 3.9. If the forces $F_1(t)$ and $F_2(t)$ are cosine harmonic functions, the dynamic analysis may be carried out as discussed in Section 3.7. The equations developed in this section can be applied directly to determine the horizontal displacements x_1 and x_2 of masses m_1 and m_2, respectively. The acceleration impulse extrapolation method in Section 3.9 may easily be used here to solve the two-degree idealized systems, when $F_1(t)$ and $F_2(t)$ are time functions of some general type, harmonic sine and cosine variations or other possible time function variations.

The total shear force V_1 at the first-story level in Fig. 3.13g, which is the same as the spring force of spring k_1 in Fig. 3.13c, may be determined from the following equation:

$$V_1 = k_1 x_1 \tag{3.167}$$

where x_1 is the displacement of mass m_1. The total shear force V_2 at the top of the second-story columns (Fig. 3.13g), which is the same as the spring force in the k_2 spring in Fig. 3.13c, may be determined from the following expression:

$$V_2 = k_2(x_2 - x_1) \tag{3.168}$$

where x_2 is the displacement of mass m_2.

The total shear force V_1 will be distributed to the first-story columns of the frame in proportion to their stiffnesses. For example, if k_i is the spring stiffness of the ith first-story column, the shear force V_i of this column may be obtained from the equation

$$V_i = \frac{k_i}{k_1} V_1 \tag{3.169}$$

In a similar manner, the second-story shear force V_2 will be distributed to the second-story columns in proportion to their stiffnesses. Therefore, for the ith second-story column of stiffness k_i, the shear force V_i is

$$V_i = \frac{k_i}{k_2} V_2 \tag{3.170}$$

The bending moment M_i at the bottom and at the top of each column of the frame (see Fig. 3.13e) may be computed from the equation

$$M_i = \frac{V_i L_i}{2} \tag{3.171}$$

where L_i is the length of the ith column. If the bottom of the ith column is hinged as in Fig. 3.13f, then

$$M_i = V_i L_i \tag{3.172}$$

The following examples illustrate the application of the above methodologies.

Example 3.9. A steel cantilever beam of length $L = 20$ ft is supporting a weight $W = 200$ lb at its free end as shown in Fig. 3.14a. At its free end, a dynamic force $F(t)$ is applied to the cantilever beam as shown, which has the time variation shown in Fig. 3.14b. The beam has a rectangular uniform cross section of width $b = 4.0$ in. and depth $h = 6.0$ in. and a modulus of elasticity $E = 30 \times 10^3$ ksi. If the beam is under the influence of 20% viscous damping, determine its maximum vertical displacement and its maximum bending stress by using an idealized one-degree spring–mass system. Neglect the weight of the beam.

Solution. The idealized one-degree spring–mass system would be as shown in Fig. 3.14c. The following information is needed to determine the parameters involved and start the dynamic analysis:

$$m = \frac{W}{g} = \frac{200}{386} = 0.5181 \text{ lb} \cdot \text{s}^2/\text{in.}$$

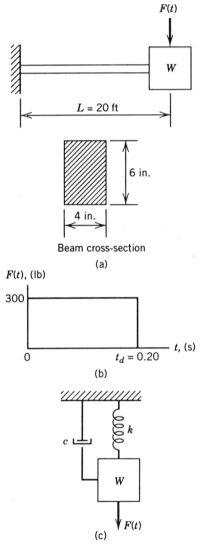

FIGURE 3.14. (a) Cantilever beam supporting a heavy weight W at the free end and acted on by a dynamic load $F(t)$ as shown. (b) Time variation of $F(t)$. (c). Idealized one-degree spring–mass system.

$$I = \frac{bh^3}{12} = \frac{(4)(6)^3}{12} = 72.00 \text{ in.}^4$$

From Eq. (3.163), the stiffness k at the free end of the cantilever beam is

$$k = \frac{3EI}{L^3}$$

$$= \frac{(3)(30) \times (10^6)(72)}{(20 \times 12)^3} = 468.75 \text{ lb/in.}$$

The free vibration ω of the idealized spring–mass system is

$$\omega = \sqrt{\frac{k}{m}} = \sqrt{\frac{468.75}{0.5181}} = 30.079 \text{ rps}$$

$$f = \frac{\omega}{2\pi} = \frac{30.079}{2\pi} = 4.7872 \text{ Hz}$$

$$\tau = \frac{1}{f} = \frac{1}{4.7872} = 0.2089 \text{ s}$$

The critical damping constant c_c is

$$c_c = 2m\omega = (2)(0.5181)(30.079) = 31.1679 \text{ lb·s/in.}$$

and the damping coefficient c is

$$c = \zeta c_c = (0.20)(31.1679) = 6.2336 \text{ lb · s/in.}$$

We select $\Delta t = 0.02$ s, which is approximately equal to one-tenth of the period of vibration τ.

By applying Eq. (3.118), we find

$$\ddot{y}^{(i)} = \frac{2F(t)}{(2)(0.5181) + (6.2336)(0.02)} - \frac{(2)(6.2336)[y^{(i)} - y^{(i-1)}]}{(0.02)[(2)(0.5181) + (6.2336)(0.02)]}$$

$$- \frac{(2)(468.75)y^{(i)}}{(2)(0.5181) + (6.2336)(0.02)} \tag{3.173}$$

$$= 1.7228F(t) - 1344.9424\, y^{(i)} + 537.3793\, y^{(i-1)}$$

By applying Eq. (3.114), the displacement $y^{(1)}$ at the first time station is calculated as follows:

$$y^{(1)} = \frac{\ddot{y}^{(0)}}{2}(\Delta t)^2 = \frac{(1.7228)(300)}{2}(0.02)^2$$

$$= 0.103368 \text{ in.}$$

By applying the acceleration impulse extrapolation method as discussed in Section 3.8, the tabulated values of y are shown in Table 3.5.

Table 3.5 shows that the maximum displacement $y_{max} = 0.97572$ in. occurs at time $t = 0.10$ s. Therefore, the maximum spring force F_{max} of the idealized one-degree spring–mass system is

$$F_{max} = ky_{max} = (468.75)(0.97572)$$

$$= 457.3688 \text{ lb}$$

If F_{max} is applied statically at the free end of the cantilever beam in Fig. 3.14a, the following maximum moment M_{max} is obtained:

$$M_{max} = F_{max}L = (457.3688)(20 \times 12)$$

$$= 109,768.50 \text{ lb} \cdot \text{in.}$$

Thus, the maximum bending stress σ_{max} at the fixed end of the cantilever beam is

$$\sigma_{max} = \frac{M_{max}(h/2)}{I} = \frac{(109,768.50)(3)}{72}$$

$$= 4573.69 \text{ psi}$$

The maximum stress σ_{max} and maximum displacement y_{max}, as calculated above, represent only the effect of the dynamic force $F(t)$. The effect of the dead weight W and the dead weight of the beam may be calculated separately by following the usual static analysis and superimposing the results. For example, a weight $W = 200$ lb will produce the following bending stress σ at the fixed end:

$$\sigma = \frac{(200)(20 \times 12)(3)}{72} = 2000.00 \text{ psi}$$

Therefore, the total bending stress σ_{total} due to $F(t)$ and W is

$$\sigma_{total} = 4573.69 + 2000.00 = 6573.69 \text{ psi}$$

Example 3.10. The two-story steel frame in Fig. 3.15a is part of the framework of a two-story building with 20 ft $\times$ 40 ft floor areas and 20-ft wide walls. The weight per square foot of floor area, which includes the weight of the girders, as well as the weight of the walls, which includes the weight of the columns, are shown in the same figure. All columns are of the same size, with $I = 133.2$ in.[4] and section modulus $S = 26.4$ in.[3]. Determine the maximum horizontal displacements and maximum bending stresses of the frame by using an idealized two-degree spring–mass system. The horizontal

TABLE 3.5. Time Variation of the Displacement y at the Free End of the Cantilever Beam

i	t (s)	$1.7228F(t)$	$-1344.9424y^{(i)}$	$537.3793y^{(i-1)}$	$\ddot{y}^{(i)}$	$\ddot{y}(\Delta t)^2$	$y^{(i)}$ (in.)
0	0	516.84	0	0	516.8400	0.206736	0
1	0.02	516.84	−139.0240	0	377.8160	0.151126	0.103368[a]
2	0.04	516.84	−481.3038	55.5478	91.0840	0.036434	0.357862
3	0.06	516.84	−872.5852	192.3076	−163.4276	−0.065371	0.648790
4	0.08	516.84	−1175.9464	348.6463	−310.4601	−0.124184	0.874347
5	**0.10**	516.84	−1312.2872	469.8560	−325.5912	−0.130236	**0.975720**
6	0.12	516.84	−1273.4681	524.3317	−232.2964	−0.092919	0.946857
7	0.14	516.84	−1109.6784	508.8214	−84.0170	−0.033607	0.825075

[a]Equation (3.114) is used for this displacement.

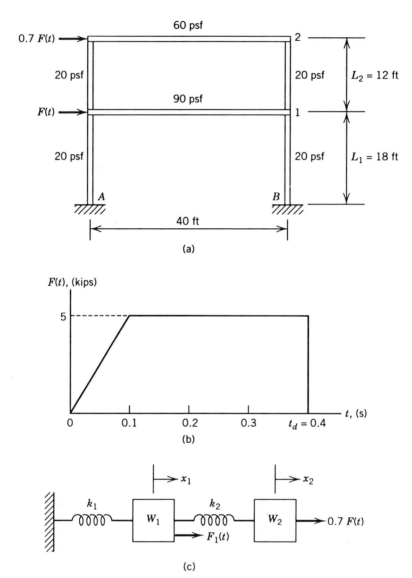

FIGURE 3.15. (a) Two-story frame subjected to the horizontal dynamic loads $F(t)$ and $0.7\ F(t)$ as shown. (b) Time variation of the dyanamic loads. (c) Idealized two-degree spring–mass system.

dynamic forces acting on the frame have the time variation shown in Fig. 3.15b, and the steel modulus of elasticity $E = 30 \times 10^6$ psi. Neglect damping.

Solution. The idealized two-degree spring–mass system would be as shown in Fig. 3.15c. The weights W_1 and W_2 are as follows:

$$W_1 = (90)(20)(40) + (20)\left(\frac{18}{2} + \frac{12}{2}\right)(20)(2) = 84{,}000.0 \text{ lb}$$

$$= 84.0 \text{ kips}$$

$$W_2 = (60)(20)(40) + (20)\left(\frac{12}{2}\right)(20)(2) = 52{,}800.0 \text{ lb}$$

$$= 52.8 \text{ kips}$$

The weight W_1 includes the weight of the first floor plus half the weight of the first- and second-story walls, and the weight W_2 consists of the weight of the second floor plus half the weight of the second-story walls.

By applying Eq. (3.165), the spring constants k_1 and k_2 may be determined as follows:

$$k_1 = (2)\frac{12EI}{L_1^3} = (2)\frac{(12)(30 \times 10^3)(133.2)}{(18 \times 12)^3}$$

$$= 9.52 \text{ kips/in.}$$

$$k_2 = (2)\frac{12EI}{L_2^3} = (2)\frac{(12)(30 \times 10^3)(133.2)}{(12 \times 12)^3}$$

$$= 32.10 \text{ kips/in.}$$

We also have

$$m_1 = \frac{84.0}{386} = 0.2174 \text{ kip} \cdot \text{s}^2/\text{in.}$$

$$m_2 = \frac{52.8}{386} = 0.1366 \text{ kip} \cdot \text{s}^2/\text{in.}$$

The accelerations $\ddot{x}_1$ and $\ddot{x}_2$ of masses m_1 and m_2, respectively, are as shown below:

$$\ddot{x}_1 = \frac{F(t) + k_2(x_2 - x_1) - k_1 x_1}{m_1}$$

$$= \frac{F(t) + 32.1(x_2 - x_1) - 9.52x_1}{0.2174} \tag{3.174}$$

$$= 4.60F(t) + 147.65x_2 - 191.44x_1$$

$$\ddot{x}_2 = \frac{0.7F(t) - k_2(x_2 - x_1)}{m_2}$$

$$= \frac{0.7F(t) - 32.1(x_2 - x_1)}{0.1366} \tag{3.175}$$

$$= 5.12F(t) - 234.99(x_2 - x_1)$$

We also have

$$F(t) = \frac{5}{0.1} t = 50t \qquad 0 \le t \le 0.1 \tag{3.176}$$

$$F(t) = 5 \qquad 0.1 \le t \le 0.4 \tag{3.177}$$

$$F(t) = 0 \qquad t > 0.4 \tag{3.178}$$

By using Eqs. (3.125) and (3.127) and selecting a time increment $\Delta t = 0.03$, the displacements $x_1^{(1)}$ and $x_2^{(1)}$ of masses m_1 and m_2, respectively, at the first time station, are calculated as follows:

$$x_1^{(1)} = \tfrac{1}{6}[2\ddot{x}_1^{(0)} + \ddot{x}_1^{(1)}](\Delta t)^2$$
$$= \tfrac{1}{6}[4.60F(t)^{(1)} + 147.65x_2^{(1)} - 191.44x_1^{(1)}]$$

or

$$1.028716x_1^{(1)} = \tfrac{1}{6}[4.60F(t)^{(1)} + 147.65x_2^{(1)}](0.03)^2 \tag{3.179}$$

$$x_2^{(1)} = \tfrac{1}{6}[2\ddot{x}_2^{(0)} + \ddot{x}_2^{(1)}](\Delta t)^2$$
$$= \tfrac{1}{6}[5.12F(t)^{(1)} + 234.99x_1^{(1)} - 234.99x_2^{(1)}](0.03)^2$$

or

$$x_2^{(1)} = 0.000742F(t)^{(1)} + 0.034048x_1^{(1)} \tag{3.180}$$

Simultaneous solution of Eqs. (3.179) and (3.180) yields

$$x_1^{(1)} = 0.001031 \text{ in.} \tag{3.181}$$

$$x_2^{(1)} = 0.001148 \text{ in.} \tag{3.182}$$

By using Eqs. (3.174), (3.175), (3.129), and (3.130), the displacements $x_1^{(i+1)}$ and $x_2^{(i+1)}$ at any $i+1$ station may be determined as shown in Table 3.6. In this table, we note that the maximum displacements $x_{1\max} = 1.402411$ in. and $x_{2\max} = 1.565492$ in. occur at time $t = 0.54$ s.

By applying Eqs. (3.167) and (3.168), the maximum shear forces $V_{1\max}$ and $V_{2\max}$ are as follows:

$$V_{1\max} = k_1 x_{1\max} = (9.52)(1.402411)$$
$$= 13.35 \text{ kips}$$

$$V_{2\max} = k_2(x_{2\max} - x_{1\max})$$
$$= 32.1(1.565492 - 1.402411)$$
$$= 5.23 \text{ kips}$$

TABLE 3.6. Variation with Time of Displacements x_1 and x_2 of Masses m_1 and m_2, Respectively

i	t (s)	$F(t)$	4.60 $F(t)$	$147.65x_2^{(i)}$	$191.44x_1^{(i)}$	$\ddot{x}_1^{(i)}$	$\ddot{x}_1^{(i)}(\Delta t)^2$
0	0	0	0	0	0	0	0
1	0.03	1.50	6.900	0.169502	0.197375	6.872127	0.006185
2	0.06	3.00	13.800	1.355870	1.578806	13.577064	0.012219
3	0.09	4.50	20.700	4.554117	5.299442	19.954675	0.017959
4	0.12	5.00	23.000	10.715256	12.458149	21.257107	0.019131
5	0.15	5.00	23.000	20.044226	23.279295	19.764931	0.017788
6	0.18	5.00	23.000	32.333135	37.505776	17.827359	0.016045
7	0.21	5.00	23.000	47.303516	54.803912	15.499604	0.013950
8	0.24	5.00	23.000	64.610754	74.772635	12.838119	0.011554
9	0.27	5.00	23.000	83.851764	96.953256	9.898508	0.008909
10	0.30	5.00	23.000	104.575180	120.839417	6.735763	0.006062
11	0.33	5.00	23.000	126.294347	145.886086	3.408261	0.003067
12	0.36	5.00	23.000	148.501350	171.519902	−0.018552	−0.000017
13	0.39	5.00	23.000	170.680890	197.150464	−3.469574	−0.003123
14	0.42	0	0	192.322984	222.183158	−29.860174	−0.026874
15	0.45	0	0	209.531739	242.071094	−32.539355	−0.029285
16	0.48	0	0	221.911749	256.352709	−34.440960	−0.030997
17	0.51	0	0	229.174209	264.700259	−35.526050	−0.031973
18	**0.54**	0	0	231.144894	268.477562	−37.332668	−0.033599
19	0.57	0	0	228.023130	265.822672	−37.799542	−0.034020

[a] See Eqs. (3.181) and (3.182).

$x_1^{(i)}$ (in.)	5.12$F(t)$	$234.99(x_2 - x_1)^{(i)}$	$\ddot{x}_2^{(i)}$	$\ddot{x}_2^{(i)}(\Delta t)^2$	$x_2^{(i)}$ (in.)
0	0	0	0	0	0
0.001031[a]	7.680	0.027494	7.652506	0.006887	0.001148[a]
0.008247	15.360	0.219951	15.140049	0.013626	0.009183
0.027682	23.040	0.743038	22.296962	0.020067	0.030844
0.065076	25.600	1.761485	23.838515	0.021455	0.072572
0.121601	25.600	3.326048	22.273952	0.020047	0.135755
0.195914	25.600	5.421454	20.178546	0.018161	0.218985
0.286272	25.600	8.014099	17.585901	0.015827	0.320376
0.390580	25.600	11.047820	14.552180	0.013097	0.437594
0.506442	25.600	14.444130	11.155870	0.010040	0.567909
0.631213	25.600	18.106214	7.493786	0.006744	0.708264
0.762046	25.600	21.928562	3.671438	0.003304	0.855363
0.895946	25.600	25.806602	−0.206602	−0.000186	1.005766
1.029829	25.600	29.644928	−4.044928	−0.003640	1.155983
1.160589	0	33.361765	−33.361765	−0.030026	1.302560
1.264475	0	36.337914	−36.337914	−0.032704	1.419111
1.339076	0	38.510631	−38.510631	−0.034660	1.502958
1.382680	0	39.822580	−39.822580	−0.035840	1.552145
1.402411	0	38.322404	−38.322404	−0.034490	1.565492
1.388543	0	36.612852	−36.612852	−0.032952	1.544349

Since the stiffnesses of the columns in each floor are equal, the shear V_{1max} will be distributed equally to the two first-story columns, and the shear V_{2max} will be distributed equally to the two second-story columns. Therefore, in accordance with Eq. (3.171), the maximum moment M_{1max} at the top and bottom of each column of the first floor is

$$M_{1max} = \frac{(V_{1max}/2)L_1}{2} = \frac{(13.35/2)(18 \times 12)}{2}$$

$$= 720.9 \text{ kip} \cdot \text{in.}$$

From the same equation, the maximum moment M_{2max} at the top and bottom of each column of the second floor is

$$M_{2max} = \frac{(V_{2max}/2)L_2}{2} = \frac{(5.23/2)(12 \times 12)}{2}$$

$$= 188.30 \text{ kip} \cdot \text{in.}$$

Therefore, the maximum bending stresses σ_{1max} and σ_{2max} at the top and bottom of the first- and second-story columns, respectively, are

$$\sigma_{1max} = \frac{M_{1max}}{S} = \frac{720.9}{26.4}$$

$$= 27.31 \text{ ksi}$$

$$\sigma_{2max} = \frac{M_{2max}}{S} = \frac{188.30}{26.4}$$

$$= 7.13 \text{ ksi}$$

These stresses result strictly from the application of the horizontal dynamic forces $F(t)$ at the first- and second-story levels of the frame.

3.12 DYNAMIC RESPONSE OF ELASTOPLASTIC SYSTEMS

The dynamic analyses in the preceding sections of the chapter were concentrated on elastic systems, where the applied dynamic forces never reached a magnitude that will cause the material to be stressed beyond its elastic limit. In practice, however, there are situations where the material of a system may be permitted to be stressed beyond its elastic limit. For example, if a structural member or a large machine is subjected to a strong blast from a conventional or nuclear explosion or is under the influence of a strong earthquake, then the stresses in such a structure or machine may be permitted to exceed the elastic limit of the material. Such events usually take place only once or twice in the life of a structure, and it would be very uneconomical to design a structure so that the response is all elastic. Therefore, in such situations, the main concern of the designer would be safety. Stresses in the members, as well as displacements, can become inelastic but the design of the structure should be performed in a way that will maintain structural integrity and provide safety for the people involved.

Many ductile materials are characterized by a large yielding range, which permits the material to absorb large amounts of energy before complete failure. The decision that needs to be made by the designer is the amount of energy that a structure should be permitted to absorb for a safe design. A great deal of work on this subject has been performed by various engineers [1, 2, 7, 9], and the results show that a structure can absorb a great deal of inelastic energy before complete collapse. See, in particular, the author's work in refs. 2, 12, 13, and 14.

The analysis in this section concentrates on the inelastic behavior of structures, machines, and their component members that can be idealized and treated as one-degree spring–mass systems. Beams, one-story frames, one-story buildings, and mechanical components are such examples.

The *resistance R* of a structure or a structural component may be defined as the internal force that tends to restore the structure to its unloaded static position. For a spring–mass system such as the one in Fig. 3.16a, the internal restoring force is the force in the spring, and it is equal to ky, where k is the spring constant and y is the spring deflection. The resistance R can have many shapes, as shown in Fig. 3.16b. Curve (2) represents a very ductile material, such as mild steel; curve (1) characterizes the behavior of a brittle material, while curve (3) is that of concrete. For many structural problems, it is often reasonable, and permissible, to use the bilinear resistance form shown in Fig. 3.16c, and this is the one used for the analysis in this section. If more sophisticated methodology is required, the author's work in ref. 2 is very useful.

The analysis in this section is initiated by considering the resistance R and the vertical displacement y of the spring–mass system in Fig. 3.16a. The resistance R is assumed to have the bilinear form shown in Fig. 3.16d. In this figure, the resistance R increases linearly until the elastic limit displacement y_e is attained. At the same point, the resistance R attains its maximum value R_m. For any further increase in y, the resistance is assumed to remain constant and equal to R_m. With increasing y and constant $R = R_m$, two possibilities of structural behavior exist. The one possibility is for y to keep increasing and R to remain constant and equal to R_m until the ductility limit of the material is attained. In such a case, the resistance function may be assumed to be composed of the two lines shown in Fig. 3.16c.

The other possibility is for y to increase and R to remain constant and equal to R_m, until y reaches its maximum value y_m before the ductility limit is attained. In such a case, the structural system is said to rebound. During

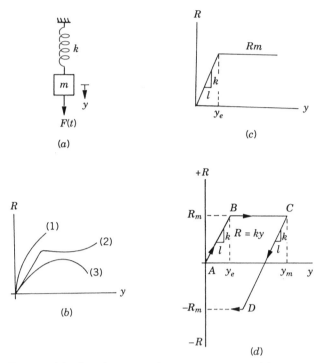

FIGURE 3.16. (a) One-degree spring–mass system. (b) Examples of resistance curves. (c) Bilinear resistance form. (d) Shape parameters of a bilinear resistance when the structure rebounds.

rebound, the resistance R is assumed to decrease linearly along a line parallel to its initial elastic line as shown in Fig. 3.16d, and it will continue to decrease until R becomes equal to $-R_m$.

The differential equation of motion of the one-degree spring–mass system in Fig. 3.16a is

$$m\ddot{y} + R - F(t = 0 \tag{3.183}$$

where $R = ky$. By considering Eq. (3.183) and the resistance form shown in Fig. 3.16d, the following differential equations of motion may be written:

$$m\ddot{y} + ky - F(t) = 0 \qquad 0 \le y \le y_e \tag{3.184}$$

$$m\ddot{y} + R_m - F(t) = 0 \qquad y_e \le y \le y_m \tag{3.185}$$

$$m\ddot{y} + R_m - k(y_m - y) - F(t) = 0 \qquad (2y_m - 2y_e) \le y \le y_m \tag{3.186}$$

For range AB, Eq. (3.184) applies; Eq. (3.185) applies for range BC. When

R_m starts to decrease linearly until it becomes $-R_m$, which is range CD, Eq. (3.186) must be used.

Equations (3.184) through (3.186) may be solved either by a rigorous solution or by a numerical method, such as the acceleration impulse extrapolation method. If a rigorous solution is used, the initial conditions of Eq. (3.184) would be the initial conditions of the motion, such as $y_0 = \dot{y}_0 = 0$ at $t = 0$. The initial conditions of Eq. (3.185) would be the final displacement and final velocity obtained from the solution of Eq. (3.184). Similarly, the initial conditions of Eq. (3.186) are the final displacement and final velocity obtained from the solution of Eq. (3.185).

The above initial conditions would be sufficient for the solution of the problem if the dynamic force $F(t)$ is a continuous function of time that does not become zero before the analysis for y in the above three equations is completed. If $F(t)$ is of finite duration or has discontinuities, additional stages, one for each discontinuity, would have to be included in the solution. For example, if the dynamic load $F(t)$ is a rectangular impulse of time duration t_d, then the following equations would have to be used:

$$m\ddot{y} + ky - F(t) = 0 \qquad t < t_d, \quad y < y_e \qquad (3.187)$$

$$m\ddot{y} + R_m - F(t) = 0 \qquad t < t_d, \quad y > y_e \qquad (3.188)$$

$$m\ddot{y} + ky = 0 \qquad t > t_d, \quad y < y_e \qquad (3.189)$$

$$m\ddot{y} + R_m = 0 \qquad t > t_d, \quad y > y_e \qquad (3.190)$$

$$m\ddot{y} + R_m - k(y_m - y) - F(t) = 0 \qquad t_d > t > t_m \qquad (3.191)$$

$$m\ddot{y} + R_m - k(y_m - y) = 0 \qquad t > t_m, \quad t > t_d \qquad (3.192)$$

Equations (3.187) through (3.190) represent the motion before y_m is attained, and Eqs. (3.191) and (3.192) may be used for the motion when the system is in elastic rebound. The sequence in using these equations depends on the time t_e when the maximum elastic displacement is used and on the time t_m when the maximum displacement y_m is reached.

The following example illustrates the application of the above theory.

Example 3.11. The fixed-fixed uniform steel beam in Fig. 3.17a has a 24W × 84 cross section and supports the weight $W = 40$ kips at its center. The beam is also acted on by the dynamic force $F(t)$, which has the time variation shown in Fig. 3.17b. By following elastoplastic analysis, determine its maximum displacement due to $F(t)$ by using an idealized one-degree spring–mass system. The moment of inertia $I = 2364.3$ in.^{4}, the section modulus $S = 196.3$ in.^{3}, the modulus of elasticity $E = 30 \times 10^6$ psi, and the plastic modulus $Z_p = 224$ in.^{3} [15]. Assume that the beam is weightless and neglect damping.

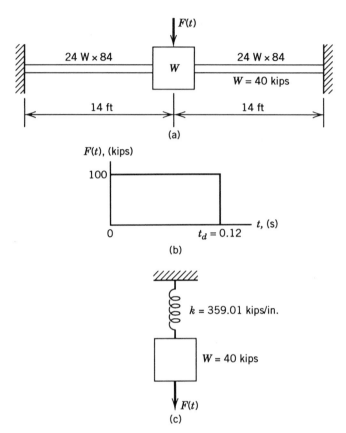

FIGURE 3.17. (a) Fixed–fixed uniform beam supporting a heavy weight W and subjected to a dynamic force $F(t)$. (b) Time variation of $F(t)$. (c) Idealized one-degree spring–mass system for the fixed–fixed beam.

Solution. The stiffness k at the center of the fixed–fixed beam is

$$y = \frac{PL^3}{192EI} = 1$$

or

$$P = k = \frac{192EI}{L^3} = \frac{(192)(30 \times 10^3)(2364.3)}{(28 \times 12)^3}$$

$$= 359.01 \text{ kips/in.}$$

$$m = \frac{W}{g} = \frac{40}{386} = 0.1036 \text{ kips} \cdot \text{s}^2/\text{in.}$$

$$\tau = 2\pi \sqrt{\frac{m}{k}} = 2\pi \sqrt{\frac{0.1036}{359.01}}$$

$$= 0.1067 \text{ s}$$

The plastic moment M_p, based on a yield-point stress σ_y of 30.0 ksi, is

$$M_p = Z_p \sigma_y = (224)(30) = 6720.0 \text{ kip} \cdot \text{in.}$$

This is the ultimate moment that can be carried by the cross section of the member.

The maximum resistance R_m is equal to the load at the center of the beam, which produces the ultimate moment M_p. Since the bending moments at the fixed ends of the beam and at its center are all equal, plastic hinges will develop simultaneously at the ends and at the center of the member. Since $M_p = R_m L/8$, we find that the maximum resistance R_m is

$$R_m = \frac{8M_p}{L} = \frac{(8)(6720)}{(28 \times 12)}$$

$$= 160.0 \text{ kips}$$

The beam also carries the weight $W = 40$ kips, which is located at its center. Therefore, the maximum available resistance R_m is

$$R_m = 160.0 - 40.0 = 120.0 \text{ kips}$$

The resistance function R is assumed to have the form shown in Fig. 3.16d, and the idealized one-degree spring–mass system for the beam in Fig. 3.17a is shown in Fig. 3.17c. On this basis, the elastic deflection y_e is

$$y_e = \frac{R_m}{k} = \frac{120.0}{359.01} = 0.3343 \text{ in.}$$

By applying Eq. (3.183), the general differential equation of motion of the idealized one-degree spring–mass system may be written

$$0.1036\ddot{y} + R - F(t) = 0 \tag{3.193}$$

This equation yields

$$\ddot{y} = 9.6525F(t) - 9.6525R \tag{3.194}$$

Equations (3.187) through (3.192) may be used in conjunction with Eq.

(3.194) to write the expressions for the acceleration $\ddot{y}$ for the various intervals of y and time t. Thus, the following equations for $\ddot{y}$ may be written:

$$\ddot{y} = 965.25 - 3465.344y \qquad t < t_d, \quad y < y_e \tag{3.195}$$

$$\ddot{y} = 965.25 - 1158.30 = -193.05 \qquad t < t_d, \quad y > y_e \tag{3.196}$$

$$\ddot{y} = -3465.344y \qquad t > t_d, \quad y < y_e \tag{3.197}$$

$$\ddot{y} = -1158.30 \qquad t > t_d, \quad y > y_e \tag{3.198}$$

$$\ddot{y} = -1158.30 + 3465.344(y_m - y) + 965.25$$

$$= -193.05 + 3465.344(y_m - y) \qquad t_d > t > t_m \tag{3.199}$$

$$\ddot{y} = -1158.30 + 3465.344(y_m - y) \qquad t > t_m, \quad t > t_d \tag{3.200}$$

The acceleration impulse extrapolation method will be used here to determine the maximum displacement y_m of the member. By using a time interval $\Delta t = 0.01$ s, and applying Eqs. (3.114) and (3.195), the displacement $y^{(1)}$ of the first time interval is as follows:

$$y^{(1)} = \frac{\ddot{y}^{(0)}}{2}(\Delta t)^2 = \frac{965.25}{2}(0.01)^2$$

$$= 0.0483 \text{ in.}$$

For the remaining time intervals, Eq. (3.112) should be used. The results are tabulated in Table 3.7.

Table 3.7 shows that the maximum displacement y_m occurs at time $t = 0.11$ s, and it is equal to 0.9371 in. The results in this table are obtained by using Eq. (3.195) to determine the acceleration $\ddot{y}$ for the time intervals between $t = 0$ and $t = 0.03$ s. Equation (3.196) was used to determine $\ddot{y}$ for $0.04 \leq t \leq 0.12$, and Eq. (3.200) was used for the remaining time intervals. The computations, however, could go on indefinitely until other peaks of the displacement y are obtained.

If we define η as the ratio of y_m and y_e, then we have

$$\eta = \frac{y_m}{y_e} = \frac{0.9371}{0.3343}$$

$$= 2.80$$

The ratio η is often called the *ductility ratio*, and it provides an indication as to how far in the inelastic range a member can be exposed. In practice, ductility ratios of 5 or even higher have been used for structures that have been exposed to blast and earthquake.

TABLE 3.7. Variation of the Displacement y with Time

t (s)	$\ddot{y}$	$\ddot{y}(\Delta t)^2$	y (in.)
0	965.25	0.0965	0
0.01	797.87	0.0798	0.0483[a]
0.02	353.96	0.0354	0.1764
0.03	−212.62	−0.0213	0.3399
0.04	−193.05	−0.0193	0.4821
0.05	−193.05	−0.0193	0.6050
0.06	−193.05	−0.0193	0.7086
0.07	−193.05	−0.0193	0.7929
0.08	−193.05	−0.0193	0.8579
0.09	−193.05	−0.0193	0.9036
0.10	−193.05	−0.0193	0.9300
0.11	−193.05	−0.0193	**0.9371**
0.12	−193.05	−0.0193	0.9249
0.13	−1006.86	−0.1007	0.8934
0.14	−548.75	−0.0549	0.7612
0.15	99.62	0.0100	0.5741

[a]Equation (3.114) is used to determine this displacement.

PROBLEMS

3.1 The one-degree spring–mass systems in Fig. P3.1 are subjected to a harmonic force $F\cos \omega_f t$, where $F = 150.0$ N and $\omega_f = 12.0$ rps. If damping is neglected and we have $k = 10.0$ kN/m and $m = 50.0$ kg, determine the maximum amplitude of the mass m that results from the application of the harmonic force.

3.2 Repeat Problem 3.1 by using various values of the ratio ω_f/ω with ω_f as the variable, and plot the maximum amplitude of mass m versus ω_f/ω by including $\omega_f/\omega = 1.0$. Discuss the results.

3.3 Repeat Problem 3.1 by assuming that the harmonic force is $F\sin \omega_f t$ with $F = 200.0$ N and $\omega_f = 10, 12, 15, 20,$ and 25 rps.

3.4 A weight of 500.0 N is suspended by a spring of stiffness $k = 6000.0$ N/m, and it is forced to vibrate harmonically by a harmonic force of 70.0 N. If the spring–mass system is under the influence of viscous damping with viscous damping constant $c = 200.0$ kg/s, determine (a) the resonant frequency, (b) the amplitude at resonance, and (c) the phase angle at resonance.

3.5 The one-degree spring–mass systems in Fig. P3.5 are subjected to a harmonic force $F\cos \omega_f t$, where $F = 200.0$ N and $\omega_f = 15.0$ rps. The spring–mass system is also under the influence of 10% viscous damping. If the spring constant $k = 12.0$ kN and the mass $m = 70.0$ kg,

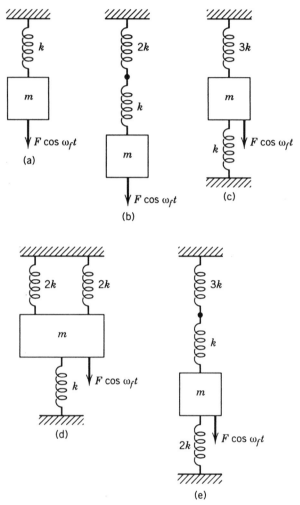

FIGURE P3.1

determine the maximum amplitude of the mass m that is caused by the application of the harmonic force.

3.6 Repeat Problem 3.5 by using various values of the ratio ω_f/ω with ω_f as the variable, and plot the maximum amplitude of the mass m versus ω_f/ω by including $\omega_f/\omega = 1.0$. Discuss the results.

3.7 Repeat Problem 3.5 by assuming that the harmonic force is $F\sin\omega_f t$ with $F = 150.0\,\text{N}$ and $\omega_f = 8, 10, 12, 15, 20$, and 25 rps.

3.8 The weight W of the mass m in Fig. 3.3 is 6000.0 N, and it rubs against a dry surface during the application of a harmonic force $F\cos\omega_f t$,

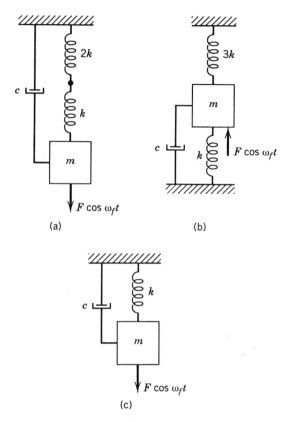

(a)   (b)

(c)

FIGURE P3.5

where $F = 12,000.0$ N and $\omega_f = 25.0$ rps. The spring constant k of the spring–mass system is 550.0 kN/m. Determine the steady-state amplitude of the motion when the coefficient of kinetic friction μ of the dry surface is 0.3.

3.9 Repeat Problem 3.8 for various values of the ratio ω_f/ω with ω_f as the variable, and plot the maximum amplitude of mass m versus ω_f/ω by including $\omega_f/\omega = 1.0$. Discuss the results.

3.10 Repeat problem 3.8 by assuming that friction damping is zero and compare the results.

3.11 The weight W of the spring–mass systems in Fig. P3.11 is 3000.0 N, and it rubs against a dry surface during the application of a harmonic force $F \cos \omega_f t$, where $F = 800.0$ N and $\omega_f = 15.0$ rps. The spring constant $k = 300.0$ kN/m, and the coefficient of kinetic friction μ of the dry surface is 0.25. Determine the steady-state amplitude of the weight.

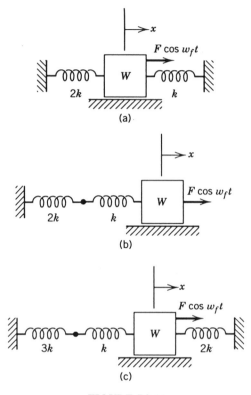

FIGURE P3.11

3.12 Repeat Problem 3.11 for the case where the frequency ω_f of the force is equal to the free undamped frequency of vibration ω of the spring–mass system. Discuss the results.

3.13 The spring–mass systems in Fig. P3.13 are subjected to a suddenly applied constant force of 50.0 kips that is suddenly reduced to zero at time $t = 0.08$ s, as shown in Fig. P3.13b. (a) Determine the expressions for the displacement $y(t)$ of the weight W at times $0 \leq t \leq t_d$ and $t \geq t_d$. (b) Plot the response $y(t)$ versus time t. (c) Find the maximum displacement y_m of the weight W and the time t_{max} at which this displacement occurs. Use the rigorous procedure discussed in Section 3.5, and assume that the spring–mass systems are under the influence of 10% viscous damping.

3.14 Repeat Problem 3.13 by assuming that the time variation of the dynamic force $F(t)$ is as shown in Fig. P3.13c and compare the results.

3.15 Repeat Problem 3.13 by assuming that damping is zero and compare the results.

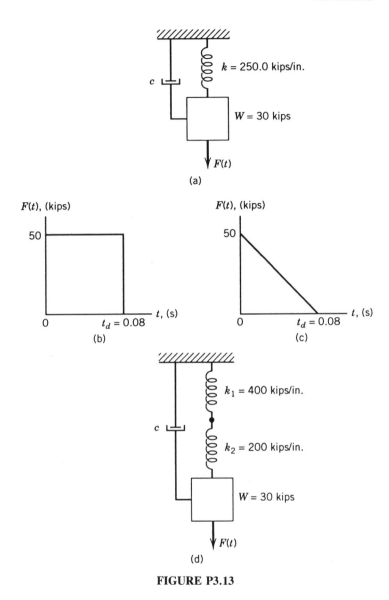

FIGURE P3.13

3.16 Repeat Problem 3.14 by assuming that damping is zero and compare the results.

3.17 The one-degree spring–mass system in Fig. 3.5a is subjected to a dynamic force F that is suddenly applied at time $t = 0$, when the spring–mass system is at rest, and lasts indefinitely as shown in Fig. 3.5b. If the viscous damping constant $c = 0$, determine the expression for the displacement $y(t)$ of the mass m by using the differential

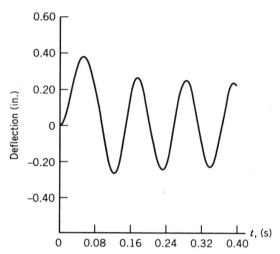

FIGURE A3.13b. Plot of $y(t)$ versus time t for problem in Fig. P3.13a with force function as shown in Fig. P3.13b.

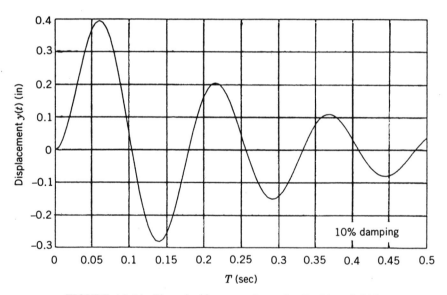

FIGURE A3.14. Plot of $y(t)$ versus time t for Problem 3.14d.

equation $m\ddot{y} + ky = F$ and assuming that $y(t)$ is the sum of the complementary solution $y_c(t)$ and the particular solution $y_p(t)$.

3.18   Repeat Problem 3.17 by assuming that the spring–mass system is moving under the influence of 30% viscous damping. Assume that $F = 800.0$ N, $k = 10.0$ kN/m, and $m = 100.0$ kg.

3.19 Assume that the damping ratio $\zeta = 0.20$, and use Eq. (3.70) to plot the ratio Y_r/y_{s0} versus ω_f/ω for various values of the ratio ω_f/ω by also including the value $\omega_f/\omega = 1.0$.

3.20 Assume that the damping ratio $\zeta = 0.30$, and use Eq. (3.69) to plot the phase angle ϕ for various values of the ratio ω_f/ω by including $\omega_f/\omega = 1.0$.

3.21 Repeat Problem 3.19 for values of $\zeta = 0.1$, 0.2, 0.3, and 0.5 and compare the results.

3.22 Repeat Problem 3.20 for values of $\zeta = 0.1$, 0.2, and 0.3 and compare the results.

3.23 The spring–mass system in Fig. 3.6a has a mass $m = 50.0$ kg, spring stiffness $k = 10.0$ kN/m, and damping ratio $\zeta = 0.10$. The maximum amplitude of the support motion is 3.50 cm, and $\omega_f = 12.0$ rps. Determine the displacement y of the mass m at times $t = 0.1$, 0.2, 0.3, and 0.5 s.

3.24 The two-degree spring–mass systems in Fig. P3.24 are subjected to harmonic forces $F_1 \cos \omega_f t$ and $F_2 \cos \omega_f t$, where $F_1 = 300.0$ lb, $F_2 = 200.0$ lb, and $\omega_f = 17.5$ rps. The spring–mass systems are under the influence of viscous damping with viscous damping coefficients $c_1 = 2.0$ lb · s/in. and $c_2 = 1.0$ lb · s/in. The spring stiffnesses are $k_1 = 500.0$ lb/in., $k_2 = 300.0$ lb/in., and $k_3 = 250.0$ lb/in., and the masses are $m_1 = 0.90$ lb · s^2/in. and $m_2 = 0.30$ lb · s^2/in. By using the procedure discussed in Section 3.7, determine the maximum displacement of masses m_1 and m_2.

3.25 Repeat Problem 3.24 with $c_1 = c_2 = 0$ and compare the results. Make appropriate observations regarding these results.

3.26 Repeat Problem 3.24 with $c_2 = 0$ and compare the results.

3.27 Repeat Problem 3.5 by using the acceleration impulse extrapolation method and compare the results.

3.28 Repeat Problem 3.13 by using the acceleration impulse extrapolation method and compare the results.

3.29 Repeat Problem 3.14 by using the acceleration impulse extrapolation method and compare the results.

3.30 Repeat Problem 3.24 by using the acceleration impulse extrapolation method and compare the results.

3.31 Repeat Problem 3.25 by using the acceleration impulse extrapolation method and compare the results.

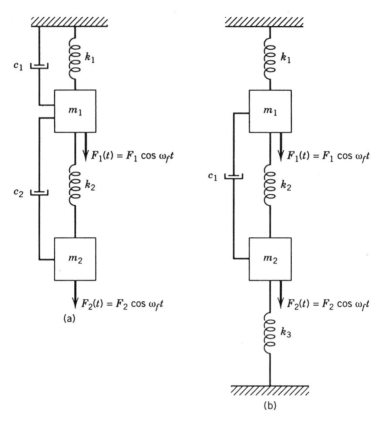

(a)

(b)

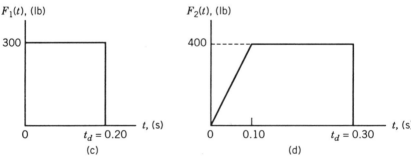

(c)

(d)

FIGURE P3.24

3.32 Repeat Problem 3.26 by using the acceleration impulse extrapolation method and compare the results.

3.33 Repeat Problem 3.24 by assuming that $F_1(t)$ and $F_2(t)$ have the time variation shown in Fig. P3.24c and that $F_2(t) = 0.7 F_1(t)$.

3.34 Repeat Problem 3.24 by assuming that $F_1(t)$ and $F_2(t)$ have the time variation shown in Fig. P3.24d and that $F_2(t) = 0.5\,F_1(t)$.

3.35 The spring–mass system in Fig. 3.11a is acted on by the periodic force $F(t)$ shown in Fig. 3.10a. The maximum magnitude of the force F is 50 lb, $T/2 = 0.10$ s, and the motion is under the influence of 10% viscous damping. Determine the steady-state response of the spring–mass system by using Fourier series analysis.

3.36 Repeat Problem 3.35 by assuming that damping is zero and compare the results.

3.37 Repeat Problem 3.35 by using for $F(t)$ the time variation shown in Fig. 3.10c.

3.38 Repeat Problem 3.35 by using for $F(t)$ the time variation shown in Fig. 3.10d.

3.39 Repeat Problem 3.37 by assuming that damping is zero and compare the results.

3.40 Repeat Problem 3.38 by assuming that damping is zero and compare the results.

3.41 Repeat Problem 3.35 for $T/2 = 0.1$, 0.2, and 0.3 s and compare the results.

3.42 The uniform steel beams in Fig. P3.42 support a weight W that is located as shown in the figure. In addition, the weight W is subjected to a dynamic force $F(t)$ of 10 kips, which is suddenly applied at time $t = 0$ and suddenly removed at $t = 0.2$ s. Determine the maximum displacement and the maximum bending stress in each case by using an appropriate idealized one-degree spring–mass system and a rigorous solution. Neglect the weight of the beam and assume that damping is zero.

3.43 Repeat Problem 3.42 by using the acceleration impulse extrapolation method and compare the results.

3.44 Repeat Problem 3.42 by assuming that $F(t)$ is a 10 kip force applied suddenly at time $t = 0$ and decreased linearly to zero at $t = 0.2$ s. Compare the results.

3.45 Repeat Problem 3.44 by using the acceleration impulse extrapolation method.

3.46 Repeat Problem 3.42 by assuming that the motion is taking place under the influence of 10% viscous damping.

3.47 Repeat Problem 3.44 by assuming that the motion is under the influence of 10% damping.

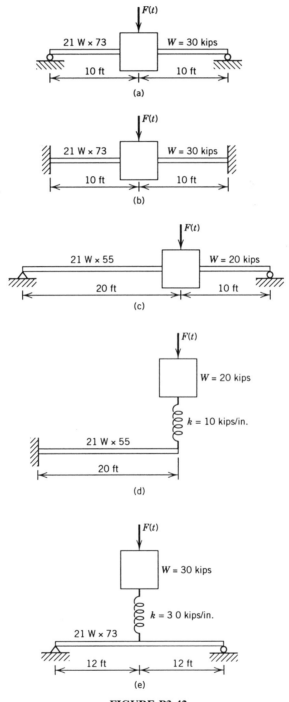

FIGURE P3.42

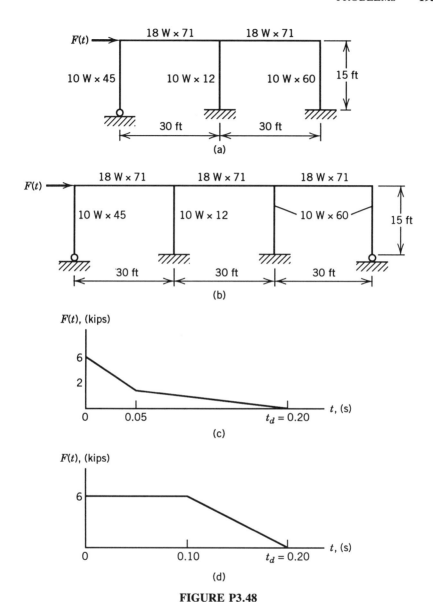

FIGURE P3.48

3.48 The steel frames in Fig. P3.48 are subjected to a concentrated horizontal force $F(t)$ at the girder level with magnitude and time variation as shown in Fig. P3.48c. In each case, the girders are assumed to be of infinite rigidity compared to the columns. By using an idealized one-degree spring–mass system and applying a rigorous solution, determine the maximum horizontal displacement of the frame and the maximum bending stresses in each column. Neglect damping.

3.49 Repeat Problem 3.48 by using the acceleration impulse extrapolation method to solve the idealized one-degree system and compare the results.

3.50 Repeat Problem 3.49 by assuming that the motion is under the influence of 10% viscous damping and compare the results.

3.51 Repeat Problem 3.48 by assuming that the dynamic force $F(t)$ has the time variation shown in Fig. P3.48d.

3.52 Repeat Problem 3.51 by using the acceleration impulse extrapolation method to solve the idealized one-degree system and compare the results.

3.53 Repeat Problem 3.52 by assuming that the motion is under the influence of 10% viscous damping and compare the results.

3.54 Solve the problem in Example 3.10 by assuming that $F(t)$ has the time variation and magnitude shown in Fig. P3.48c.

3.55 Solve the problem in Example 3.10 by assuming that $F(t)$ has the time variation and magnitude shown in Fig. P3.48d.

3.56 Solve Problem 3.54 by assuming that the left end support A of the first story of the frame is a hinge and compare the results.

3.57 Solve Problem 3.55 by assuming that the support of the frame at point B is a hinge and compare the results.

3.58 Solve the problem in Example 3.11 by assuming that $F(t)$ is a 100-kip force applied suddenly at time $t = 0$ and reduced linearly to zero at $t = 0.12$ s. Compare the results.

3.59 Solve the problem in Example 3.11 by assuming that the ends of the beam are simply supported and $F(t)$ has the variation shown in Fig. 3.17b, with $F(t) = 70.0$ kips. Compare the results.

3.60 Design the fixed–fixed beam in Example 3.11 so that the ductility ratio $\eta = 5$. Vary the beam size so that this ductility ratio is obtained.

3.61 Design the simply supported beam in Problem 3.59 for a ductility ratio $\eta = 3$. Find the beam size that will satisfy this condition.

4 Dynamic and Vibration Response of Continuous Systems

4.1 INTRODUCTION

This chapter deals with the dynamic and vibration analysis of various types of elastic systems that possess continuous mass and elasticity throughout their length. Special characteristics of such problems will be investigated by following rigorous solutions that describe their behavior. Axially restrained beam columns on rigid or elastic supports will be investigated in more detail, in order to pinpoint special types of static and flutter instabilities that characterize the behavior of such continuous systems. Other types of problems to be included in the analysis are vibration of continuous beams, beams subjected to dynamic excitations, the effects of shear and rotatory inertia, vibration of strings, stretched membranes, and thin plates. The concept and theory of dynamically equivalent systems are also introduced in this chapter. The concept and theory provide essential information that helps to better understand the vibration and dynamic behavior of complicated structural and mechanical engineering systems. Additional information on this subject is provided in Section 5.16.

4.2 VIBRATION RESPONSE OF CONTINUOUS BEAMS

The differential equation of motion that describes the free undamped vibration of beams was derived in Section 1.9, and in Chapter 2 it was used for the solution of single-span beam problems. The methodology will be extended in this section for the solution of continuous beams, where the stiffness EI can vary from span to span but retains a constant value within the length of each span. Problems where the stiffness EI is permitted to vary in any arbitrary manner along the length of the member are treated in Chapter 8. It should be noted, however, that the evaluation of the free frequencies of vibration of continuous beams and their corresponding mode shapes is a rather involved problem, and the practicing engineer often resorts to approximate numerical solutions that satisfy established design criteria.

Two specific cases of continuous beams will be examined here. In the first case it will be assumed that all spans of the continuous beams have the same stiffness EI and that the end supports are simply supported. This would cover a fairly large number of practical problems. In the second case, the stiffness of the continuous beam will be permitted to vary from span to span, and the end supports may be other than simply supported.

4.2.1   Uniform Stiffness Continuous Beams with Simply Supported Ends

The procedure may be initiated by considering the n-span uniform stiffness beam in Fig. 4.1a. The solution given by Eq. (2.114) may be used for any span of the continuous beam. Therefore, for an intermediate span L_j, such as the one in Fig. 4.1b, we may write Eq. (2.114) as follows:

$$y_j = A_j \cosh \lambda_j x + B_j \sinh \lambda_j x + C_j \cos \lambda_j x + D_j \sin \lambda_j x \qquad (4.1)$$

where A_j, B_j, C_j, and D_j are constants that can be determined by applying

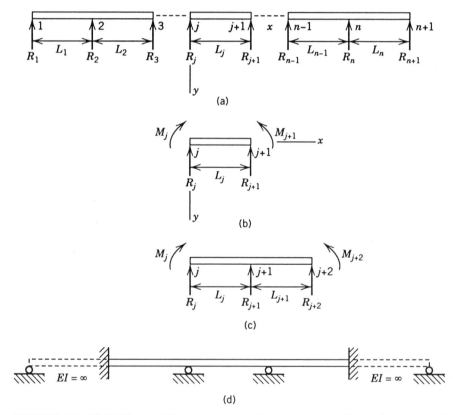

(a)

(b)

(c)

(d)

FIGURE 4.1. (a) Uniform stiffness n-span continuous beam with simply supported ends. (b) Span of length L_j of the continuous beam. (c) Spans of length L_j and L_{j+1} of the continuous beam. (d) Continuous beam with fixed end supports.

the boundary conditions of the member. Thus, by applying the boundary condition $y_j = 0$ at $x = 0$, Eq. (4.1) yields

$$C_j = -A_j \tag{4.2}$$

On this basis, Eq. (4.1) yields

$$y_j = B_j \sinh \lambda_j x + D_j \sin \lambda_j x + C_j(\cos \lambda_j x - \cosh \lambda_j x) \tag{4.3}$$

By following the same procedure, the expression y_{j+1} for the L_{j+1} span may be written

$$y_{j+1} = B_{j+1} \sinh \lambda_{j+1} x + D_{j+1} \sin \lambda_{j+1} x$$
$$+ C_{j+1}(\cos \lambda_{j+1} x - \cosh \lambda_{j+1} x) \tag{4.4}$$

The first and second derivatives of Eq. (4.3) with respect to x are written as follows:

$$\frac{dy_j}{dx} = B_j \lambda_j \cosh \lambda_j x + D_j \lambda_j \cos \lambda_j x - C_j \lambda_j(\sin \lambda_j x + \sinh \lambda_j x) \tag{4.5}$$

$$\frac{d^2 y_j}{dx^2} = B_j \lambda_j^2 \sinh \lambda_j x + D_j \lambda_j^2 \sin \lambda_j x - C_j \lambda_j^2(\cos \lambda_j x + \cosh \lambda_j x) \tag{4.6}$$

At $x = 0$, Eqs. (4.5) and (4.6) yield the following expressions:

$$\frac{dy_j}{dx} = \lambda_j(B_j + D_j) \tag{4.7}$$

$$\frac{d^2 y_j}{dx^2} = -2C_j \lambda_j^2 \tag{4.8}$$

Equation (4.7) shows that the rotation at the end j of span L_j is proportional to $B_j + D_j$, and Eq. (4.8) shows that at the same end the constant C_j is proportional to the end curvature $d^2 y_j/dx^2$. Since the ends of the continuous beam are simply supported, it may be concluded that the constants C_1 and C_{n+1} are both zero. By considering Eq. (4.4) and taking the first and second derivatives with respect to x, we will obtain equations similar to Eqs. (4.5) and (4.6), except that in this case we will have $j + 1$ instead of j.

By applying again the boundary condition that the deflection at support $j+1$ of span L_j is zero, Eq. (4.3) yields

$$B_j \sinh \lambda_j L_j + D_j \sin \lambda_j L_j + C_j(\cos \lambda_j L_j - \cosh \lambda_j L_j) = 0 \tag{4.9}$$

In Fig. 4.1c, the two continuity conditions for any two consecutive spans L_j and L_{j+1} are

$$\left(\frac{dy_j}{dx}\right)_{x=L_j} = \left(\frac{dy_{j+1}}{dx}\right)_{x=0} \tag{4.10}$$

$$\left(\frac{d^2y_j}{dx^2}\right)_{x=L_j} = \left(\frac{d^2y_{j+1}}{dx^2}\right)_{x=0} \tag{4.11}$$

Application of the above two continuity conditions yields

$$B_j \cosh \lambda_j L_j + D_j \cos \lambda_j L_j - C_j(\sin \lambda_j L_j + \sinh \lambda_j L_j) = B_{j+1} + D_{j+1} \tag{4.12}$$

$$B_j \sinh \lambda_j L_j - D_j \sin \lambda_j L_j - C_j(\cos \lambda_j L_j + \cosh \lambda_j L_j) = -2C_{j+1} \tag{4.13}$$

If Eqs. (4.9) and (4.13) are added, we find

$$C_j \cosh \lambda_j L_j - B_j \sinh \lambda_j L_j = C_{j+1} \tag{4.14}$$

If the same equations are subtracted, we obtain

$$C_j \cos \lambda_j L_j + D_j \sin \lambda_j L_j = C_{j+1} \tag{4.15}$$

Simultaneous solution of Eqs. (4.14) and (4.15) yields

$$B_j = \frac{C_j \cosh \lambda_j L_j - C_{j+1}}{\sinh \lambda_j L_j} \tag{4.16}$$

$$D_j = \frac{C_{j+1} - C_j \cos \lambda_j L_j}{\sin \lambda_j L_j} \tag{4.17}$$

Addition of Eqs. (4.16) and (4.17), after some simplifications, yields

$$B_j + D_j = C_j(\coth \lambda_j L_j - \cot \lambda_j L_j) - C_{j+1}(\operatorname{csch} \lambda_j L_j - \csc \lambda_j L_j) \tag{4.18}$$

or, in a more compact form,

$$B_j + D_j = C_j \theta_j - C_{j+1} \psi_j \tag{4.19}$$

where

$$\theta_j = \coth \lambda_j L_j - \cot \lambda_j L_j \tag{4.20}$$

$$\psi_j = \operatorname{csch} \lambda_j L_j - \csc \lambda_j L_j \tag{4.21}$$

By following the same procedure, the following equation may also be written:

$$B_{j+1} + D_{j+1} = C_{j+1}\theta_{j+1} - C_{j+2}\psi_{j+1} \tag{4.22}$$

where

$$\theta_{j+1} = \coth \lambda_{j+1} L_{j+1} - \cot \lambda_{j+1} L_{j+1} \tag{4.23}$$

$$\psi_{j+1} = \operatorname{csch} \lambda_{j+1} L_{j+1} - \csc \lambda_{j+1} L_{j+1} \tag{4.24}$$

By substituting Eqs. (4.16), (4.17), and (4.22) into Eq. (4.12) and simplifying, the following equation is obtained:

$$C_j \psi_j - C_{j+1}(\theta_j + \theta_{j+1}) + C_{j+2}\psi_{j+1} = 0 \tag{4.25}$$

Equation (4.25) applies for any two consecutive spans of the continuous beam in Fig. 4.1a. For the n span of the constant stiffness continuous beam, the following set of equations may be written:

$$-C_2(\theta_1 + \theta_2) + C_3\psi_2 = 0 \tag{4.26}$$

$$C_2\psi_2 - C_3(\theta_2 + \theta_3) + C_4\psi_3 = 0 \tag{4.27}$$

$$C_3\psi_3 - C_4(\theta_3 + \theta_4) + C_5\psi_4 = 0 \tag{4.28}$$

$$\cdot$$
$$\cdot$$
$$\cdot$$

$$C_{n-1}\psi_{n-1} - C_n(\theta_{n-1} + \theta_n) = 0 \tag{4.29}$$

In the above equations the constants C_1 and C_{n+1} are taken as equal to zero because the ends of the continuous beam are simply supported.

If the continuous beam has two spans, Eq. (4.26) yields

$$\theta_1 + \theta_2 = 0 \tag{4.30}$$

where

$$\theta_1 = \coth \lambda_1 L_1 - \cot \lambda_1 L_1 \tag{4.31}$$

$$\theta_2 = \coth \lambda_2 L_2 - \cot \lambda_2 L_2 \tag{4.32}$$

In Eqs. (4.31) and (4.32), the parameters λ_1 and λ_2 are equal because both spans of the beam have the same uniform mass and uniform stiffness. The lengths, however, can be different. Thus, with $\lambda_1 = \lambda_2 = \lambda$, the product

λL_2 in Eq. (4.32) may be expressed in terms of the product λL_1, and both Eqs. (4.31) and (4.32) would be a function of λL_1 only.

If the beam in Fig. 4.1a consists of three spans, Eqs. (4.26) and (4.27) yield

$$-(\theta_1 + \theta_2)C_2 + \psi_2 C_3 = 0 \tag{4.33}$$

$$\psi_2 C_2 - (\theta_2 + \theta_3)C_3 = 0 \tag{4.34}$$

For a solution other than the trivial one, the determinant of the coefficients C_2 and C_3 must be equal to zero. This yields

$$\begin{vmatrix} -(\theta_1 + \theta_2) & \psi_2 \\ \psi_2 & -(\theta_2 + \theta_3) \end{vmatrix} = 0 \tag{4.35}$$

Expansion of the determinant yields the following frequency equation:

$$(\theta_1 + \theta_2)(\theta_2 + \theta_3) - \psi_2^2 = 0 \tag{4.36}$$

In Eq. (4.36), we can express again the product λL_2 and λL_3 in terms of the product λL_1. In this manner, Eq. (4.36) would be a function of λL_1 only. The values of λL_1 that satisfy Eq. (4.36) may be used to determine the free frequencies of vibration of the continuous beam, because λ is a function of ω as shown by Eqs. (2.205) and (2.217); that is,

$$\omega = \lambda^2 \sqrt{\frac{EI}{m}} \tag{4.37}$$

where m is the mass per unit length of the member. The smallest value of λL_1 yields the fundamental frequency of vibration of the beam. The values of λL_1 may be obtained graphically by plotting λL_1 versus $[(\theta_1 + \theta_2)(\theta_2 + \theta_3) - \psi_2^2]$ for various values of λL_1. A similar graphical procedure may be followed to solve Eq. (4.30), where the continuous beam consists of two spans. In this case, λL_1 may be plotted versus $(\theta_1 + \theta_2)$ for various values of λL_1.

4.2.2   A More General Case of Continuous Beam

In this case, each span of the continuous beam can have different stiffness EI, and, consequently, the values of λ will be different. In addition, the end supports of the beam can be other than simply supported, and the constants C_1 and C_{n+1} do not have to be zero. By considering two consecutive spans of length L_j and L_{j+1}, as shown in Fig. 4.1c, and proceeding as in the previous case, an expression analogous to Eq. (4.25) may be obtained. The form of this equation is as follows:

$$
C_j \psi_j - C_{j+1} \left(\frac{\lambda_{j+1}^2}{\lambda_j^2} \cdot \frac{(EI)_{j+1}}{(EI)_j} \theta_j + \frac{\lambda_{j+1}}{\lambda_j} \theta_{j+1} \right)
$$
$$
+ C_{j+2} \frac{\lambda_{j+2}^2}{\lambda_j \lambda_{j+1}} \cdot \frac{(EI)_{j+2}}{(EI)_{j+1}} \psi_{j+1} = 0 \tag{4.38}
$$

Equation (4.38) has the same physical interpretation as Eq. (4.25). The expressions for the θ's and the ψ's are the same as the ones given by Eqs. (4.20), (4.21), (4.23), and (4.24); the λ's are related to the frequencies ω as shown by Eqs. (2.205), (2.217), and (4.37).

By considering each consecutive pair of the n-span continuous beam in Fig. 4.1a and applying Eq. (4.38), the following set of equations is obtained:

$$
C_1 \psi_1 - C_2 \left(\frac{\lambda_2^2}{\lambda_1^2} \cdot \frac{(EI)_2}{(EI)_1} \theta_1 + \frac{\lambda_2}{\lambda_1} \theta_2 \right) + C_3 \frac{\lambda_3^2}{\lambda_1 \lambda_2} \cdot \frac{(EI)_3}{(EI)_2} \psi_2 = 0 \tag{4.39}
$$

$$
C_2 \psi_2 - C_3 \left(\frac{\lambda_3^2}{\lambda_2^2} \cdot \frac{(EI)_3}{(EI)_2} \theta_2 + \frac{\lambda_3}{\lambda_2} \theta_3 \right) + C_4 \frac{\lambda_4^2}{\lambda_2 \lambda_3} \cdot \frac{(EI)_4}{(EI)_3} \psi_3 = 0 \tag{4.40}
$$

$$
C_3 \psi_3 - C_4 \left(\frac{\lambda_4^2}{\lambda_3^2} \cdot \frac{(EI)_4}{(EI)_3} \theta_3 + \frac{\lambda_4}{\lambda_3} \theta_4 \right) + C_5 \frac{\lambda_5^2}{\lambda_3 \lambda_4} \cdot \frac{(EI)_5}{(EI)_4} \psi_4 = 0 \tag{4.41}
$$

$$
\cdot
$$
$$
\cdot
$$
$$
\cdot
$$

$$
C_{n-1} \psi_{n-1} - C_n \left(\frac{\lambda_n^2}{\lambda_{n-1}^2} \cdot \frac{(EI)_n}{(EI)_{n-1}} \theta_{n-1} + \frac{\lambda_n}{\lambda_{n-1}} \theta_n \right)
$$
$$
+ C_{n+1} \frac{\lambda_{n+1}^2}{\lambda_{n-1} \lambda_n} \cdot \frac{(EI)_{n+1}}{(EI)_n} \psi_n = 0 \tag{4.42}
$$

The above set of equations is analogous to Eqs. (4.26) through (4.29). They can be derived from the one above if the EI's and λ's are assumed to be constant for every span of a continuous beam with simply supported ends.

Equation (4.38) can also be expressed in terms of the three moments M_j, M_{j+1}, and M_{j+2} at supports j, $j+1$, and $j+2$, respectively, in Fig. 4.1c.

This can be accomplished by considering Eq. (4.8) and noting that

$$(EI)_j \frac{d^2 y_j}{dx^2} = -2(EI)_j \lambda_j^2 C_j = -M_j \tag{4.43}$$

which yields

$$C_j = \frac{M_j}{2(EI)_j \lambda_j^2} \tag{4.44}$$

In a similar manner, the following equations may be written:

$$C_{j+1} = \frac{M_{j+1}}{2(EI)_{j+1} \lambda_{j+1}^2} \tag{4.45}$$

$$C_{j+2} = \frac{M_{j+2}}{2(EI)_{j+2} \lambda_{j+2}^2} \tag{4.46}$$

By substituting Eqs. (4.44) through (4.46) into Eq. (4.38), the following expression is obtained:

$$\frac{M_j}{2(EI)_j \lambda_j^2} \psi_j - \frac{M_{j+1}}{2} \left(\frac{\theta_j}{(EI)_j \lambda_j^2} + \frac{\theta_j}{(EI)_{j+1} \lambda_j \lambda_{j+1}} \right)$$
$$+ \frac{M_{j+2}}{2(EI)_{j+1} \lambda_j \lambda_{j+1}} \psi_{j+1} = 0 \tag{4.47}$$

Equation (4.47) is known as the *three-moment equation*, because it involves the three moments M_j, M_{j+1}, and M_{j+2} at supports j, $j+1$, and $j+2$, respectively, of the two consecutive spans in Fig. 4.1c. Therefore, the set of expressions given by Eqs. (4.39) through (4.42) may also be written in terms of the moments at the supports of the continuous beam. On this basis, the following set of equations may be written:

$$\frac{M_1}{2(EI)_1 \lambda_1^2} \psi_1 - \frac{M_2}{2} \left(\frac{\theta_1}{(EI)_1 \lambda_1^2} + \frac{\theta_2}{(EI)_2 \lambda_1 \lambda_2} \right) + \frac{M_3}{2(EI)_2 \lambda_1 \lambda_2} \psi_2 = 0 \tag{4.48}$$

$$\frac{M_2}{2(EI)_2 \lambda_2^2} \psi_2 - \frac{M_3}{2} \left(\frac{\theta_2}{(EI)_2 \lambda_2^2} + \frac{\theta_3}{(EI)_3 \lambda_2 \lambda_3} \right) + \frac{M_4}{2(EI)_3 \lambda_2 \lambda_3} \psi_3 = 0 \tag{4.49}$$

$$\frac{M_3}{2(EI)_3 \lambda_3^2} \psi_3 - \frac{M_4}{2} \left(\frac{\theta_3}{(EI)_3 \lambda_3^2} + \frac{\theta_4}{(EI)_4 \lambda_3 \lambda_4} \right) + \frac{M_5}{2(EI)_4 \lambda_3 \lambda_4} \psi_4 = 0 \tag{4.50}$$

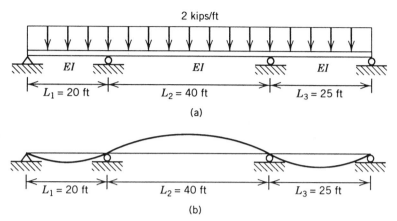

FIGURE 4.2. (a) Uniform three-span continuous beam of constant stiffnes EI. (b) Shape of the fundamental mode of vibration.

$$\frac{M_{n-1}}{2(EI)_{n-1}\lambda_{n-1}^2}\psi_{n-1} - \frac{M_n}{2}\left(\frac{\theta_{n-1}}{(EI)_{n-1}\lambda_{n-1}^2} + \frac{\theta_n}{(EI)_n\lambda_{n-1}\lambda_n}\right)$$
$$+ \frac{M_{n-1}}{2(EI)_n\lambda_{n-1}\lambda_n}\psi_n = 0 \tag{4.51}$$

If the end supports are fixed (e.g., see Fig. 4.1d), an additional three-moment equation for each end support can be written by assuming that there is a beam span extending outside the end support, which has an infinite stiffness EI. In this manner, the number of the unknown support moments is equal to the available number of three-moment equations.

The following example illustrates the application of the method.

Example 4.1. For the three-span continuous beam in Fig. 4.2a, determine its free frequencies of vibration by using the procedure discussed in this section. The stiffness EI is constant throughout the length of the member, and the attached weight per unit length of the beam is 2.0 kips/ft. Neglect damping.

Solution. Since the stiffness of the continuous beam is constant throughout its length, Eq. (4.36) may be used to determine the roots λL_1 and, consequently, its free frequencies of vibration. By carrying out the required mathematical operations, we rewrite Eq. (4.36) as follows:

$$\theta_2^2 + \theta_1\theta_2 + \theta_1\theta_3 + \theta_2\theta_3 - \psi_2^2 = 0 \qquad (4.52)$$

where

$$\theta_1 = \coth \lambda L_1 - \cot \lambda L_1 \qquad (4.53)$$

$$\theta_2 = \coth \lambda L_2 - \cot \lambda L_2 \qquad (4.54)$$

$$\theta_3 = \coth \lambda L_3 - \cot \lambda L_3 \qquad (4.55)$$

$$\psi_2 = \operatorname{csch} \lambda L_2 - \csc \lambda L_2 \qquad (4.56)$$

We note that $\lambda L_2 = 2\lambda L_1$ and $\lambda L_3 = 1.25\lambda L_1$. On this basis, Eqs. (4.53) through (4.56) may be written only in terms of λL_1, and, consequently, Eq. (4.52) would be only in terms of λL_1. The roots of λL_1 that satisfy Eq. (4.52) may then be used to determine the free frequencies of vibration of the continuous beam. The smallest value of λL_1 that satisfies Eq. (4.52) can be used to determine the fundamental free frequency of vibration and so on. This could be done graphically by assuming values of λL_1 and substituting them into Eq. (4.52). If a graph of λL_1 versus Eq. (4.52) is made, the zero intersects are the values of λL_1 that satisfy Eq. (4.52). The smallest value of λL_1 that satisfies Eq. (4.52) is 1.927. Therefore, by using Eq. (4.37), the fundamental frequency of vibration ω_1 may be determined as follows:

$$\omega_1 = \lambda^2 \sqrt{\frac{EI}{m}}$$

$$= \left(\frac{1.927}{L_1}\right)^2 \sqrt{\frac{EI/2}{(12)(386)}}$$

$$= \left(\frac{1.927}{(20)(12)}\right)^2 \sqrt{\frac{(12)(386)EI}{2}}$$

$$= 3.1025 \times 10^{-3} \sqrt{EI} \text{ rps}$$

If the continuous beam is made of a 30 W $\times$ 132 steel cross section with moment of inertia $I = 5770.0$ in.[4] and modulus of elasticity $E = 30 \times 10^3$ ksi, the fundamental frequency of its free vibration is

$$\omega_1 = (3.1025 \times 10^{-3}) \sqrt{(30 \times 10^3)(5770)}$$

$$= 40.82 \text{ rps}$$

In hertz, the fundamental frequency f_1 is

$$f_1 = \frac{\omega_1}{2\pi} = \frac{40.82}{2\pi} = 6.50 \text{ Hz}$$

When the roots of λL_1 that satisfy Eq. (4.52) are determined, the mode shapes corresponding to each root of λL_1 may be obtained by using Eq. (4.1) for each value of λL_1 and applying appropriate boundary conditions to determine the constants involved. Each span of the continuous beam is represented by a mode shape of the form given by Eq. (4.1). It should be noted here that the continuity conditions at the internal supports in terms of deflection and rotation must be satisfied. The mode shape corresponding to the fundamental frequency ω_1 would have the shape shown in Fig. 4.2b.

Additional values of λL_1 that satisfy Eq. (4.52) may be obtained in a similar manner. The procedure may be made very convenient by using the computer program in Appendix B. The first three values of λL_1 that are obtained by using this program are $(\lambda L_1)_1 = 1.92676$, $(\lambda L_1)_2 = 3.82382$, and $(\lambda L_1)_3 = 5.47257$.

4.3   TRANSVERSE VIBRATION OF STRINGS AND CABLES

In this section, the transverse free vibration of strings, cables, or cords is investigated. Consider, for example, the stretched string in Fig. 4.3a, and assume that it vibrates freely in the transverse direction as shown. The

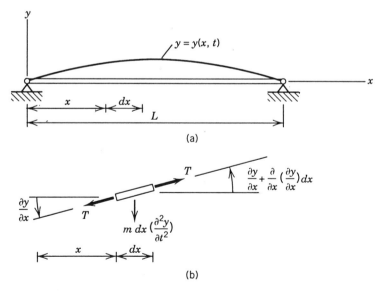

FIGURE 4.3. (a) Vibrating string or cable. (b) Free-body diagram of an element dx of the string.

dynamic elastic line $y(x, t)$ of the string is a function of both x and time t. The tension in the stretched string is T, and the deformations are assumed to be small so that T remains constant along the length of the string.

A free-body diagram of a segment of the string of length dx is shown in Fig. 4.3b. The inertia force is $m\, dx(\partial^2 y/\partial t^2)$, where m is the mass per unit length of the string. The differential equation of equilibrium is derived by using D'Alembert's principle of dynamic equilibrium and setting equal to zero the sum of the forces in the vertical direction. Thus, by assuming that the sines of the small angles $\partial y/\partial x$ and $[\partial y/\partial x + (\partial^2 y/\partial x^2)\, dx]$ are equal to the angles, we have

$$T\frac{\partial y}{\partial x} - T\left(\frac{\partial y}{\partial x} + \frac{\partial^2 y}{\partial x^2}\, dx\right) + m\, dx\left(\frac{\partial^2 y}{\partial t^2}\right) = 0$$

or

$$\frac{\partial^2 y}{\partial x^2} - \frac{m}{T}\frac{\partial^2 y}{\partial t^2} = 0 \tag{4.57}$$

Equation (4.57) is the differential equation of motion of the string. Note that this equation has the same form as Eqs. (1.50) and (1.58). Therefore, the solution $y(x, t)$ can be written directly

$$y(x, t) = Y(x)f(t) \tag{4.58}$$

$$= \left(A \sin\frac{\omega x}{c} + B \cos\frac{\omega x}{c}\right)\sin(\omega t + \phi)$$

where

$$c = \sqrt{\frac{T}{m}} \tag{4.59}$$

ϕ is the phase angle, and

$$Y(x) = A \sin\frac{\omega x}{c} + B \cos\frac{\omega x}{c} \tag{4.60}$$

$$f(t) = C \sin(\omega t + \phi) \tag{4.61}$$

The constants A and B in Eq. (4.58), as well as ω and ϕ, can be evaluated from the two boundary conditions of the string and the two initial conditions of the motion. The following example illustrates the application of the theory.

Example 4.2. Determine the free frequencies of vibration and the corresponding mode shapes for a string of length L that is fixed at both ends. Also, write the general equation of its lateral vibration.

Solution. The boundary conditions of the string suggest that the lateral deflection y is zero for all values of time t at $x = 0$ and $x = L$. The first boundary condition yields $B = 0$. From the second condition, we find

$$A \sin \frac{\omega L}{c} = 0 \tag{4.62}$$

In order to have vibration, the constant A must be different from zero. Therefore, we should have

$$\sin \frac{\omega L}{c} = 0 \tag{4.63}$$

The values of $\omega L/c$ that make $\sin(\omega L/c)$ equal to zero are as follows:

$$\frac{\omega L}{c} = n\pi \qquad n = 1, 2, 3, \ldots \tag{4.64}$$

Thus,

$$\omega = \frac{n\pi c}{L} = \frac{n\pi}{L} \sqrt{\frac{T}{m}} \qquad n = 1, 2, 3, \ldots \tag{4.65}$$

In hertz, the natural frequencies f_n are

$$f_n = \frac{\omega}{2\pi} = \frac{n}{2L} \sqrt{\frac{T}{m}} \qquad n = 1, 2, 3, \ldots \tag{4.66}$$

The corresponding modes, by Eq. (4.60), are

$$Y_n(x) = A_n \sin \frac{n\pi x}{L} \qquad n = 1, 2, 3, \ldots \tag{4.67}$$

From Eq. (4.58), the general expression for the lateral vibration of the string is

$$y(x,t) = \sum_{n=1}^{\infty} A_n \sin \frac{n\pi x}{L} \sin(\omega t + \phi) \tag{4.68}$$

where ω is given by Eq. (4.65) and ϕ is the phase angle. The constants A_n and ϕ can be determined from the initial conditions of the motion, which can be given in terms of a prescribed displacement and velocity at an initial time $t = t_0$.

4.4   VIBRATION OF STRETCHED MEMBRANES

The general differential equation of motion for freely vibrating membranes and its solution are examined in this section. The membrane is assumed to be uniformly stretched, so that its large tension T per unit length is the same in all directions. In its stretched equilibrium position, the plane of the membrane coincides with the xy plane of the x, y, z coordinate system of axes in Fig. 4.4b. The membrane vibrates in the z direction as shown in Fig. 4.4a. Thus, its time-dependent deflection $w(x, y, t)$ depends also on x and y and takes place in the z direction.

The differential equation of motion is derived by considering the free-body diagram of the element $ABCD$ in Fig. 4.4b, of sides dx and dy, and applying Newton's second law of motion. By assuming small deformations, the tension T acting on side AD of the element is inclined at an angle $\theta \approx \partial w / \partial x$. Since w is variable (Fig. 4.4a), the tension T on side BC is inclined

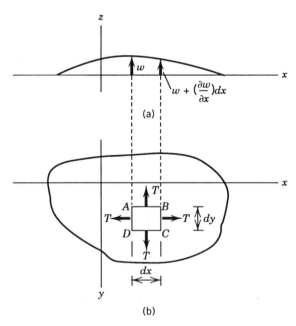

FIGURE 4.4. (a) Deflection configuration in the z direction of a stretched membrane. (b) Top view and element of the stretched membrane.

by an angle $\theta + (\partial \theta / \partial x)\, dx \approx (\partial w / \partial x) + (\partial^2 w / \partial x^2)\, dx$. In a similar manner, the tension T on sides AB and DC is inclined at the angles $\partial w / \partial y$ and $(\partial w / \partial y) + (\partial^2 w / \partial y^2)\, dy$, respectively. The inertia force acting on the element $ABCD$ is $\rho\, dx\, dy(\partial^2 w / \partial t^2)$, where ρ is the mass of the membrane per unit area.

By applying Newton's second law of motion, we have

$$(\rho\, dx\, dy)\frac{\partial^2 w}{\partial t^2} = -(T\, dy)\frac{\partial w}{\partial x} + (T\, dy)\left(\frac{\partial w}{\partial x} + \frac{\partial^2 w}{\partial x^2}\, dx\right)$$

$$-(T\, dx)\frac{\partial w}{\partial y} + (T\, dx)\left(\frac{\partial w}{\partial y} + \frac{\partial^2 w}{\partial y^2}\, dy\right)$$

or

$$\frac{\partial^2 w}{\partial x^2} + \frac{\partial^2 w}{\partial y^2} = \frac{\rho}{T}\frac{\partial^2 w}{\partial t^2} \tag{4.69}$$

Equation (4.69) is the differential equation of motion of the vibrating membrane. Its solution can be obtained as in preceding sections by assuming a solution for w and separating variables. Assume that $w(x, y, t)$ is the product of the functions $X(x)$, $Y(y)$, and $f(t)$; that is,

$$w(x, y, t) = X(x)Y(y)f(t) \tag{4.70}$$

By substituting Eq. (4.70) into Eq. (4.69) and separating variables, we find

$$\frac{\partial^2 X(x)/\partial x^2}{X(x)} + \frac{\partial^2 Y(y)/\partial y^2}{Y(y)} = \frac{\partial^2 f(t)/\partial t^2}{c^2 f(t)} \tag{4.71}$$

where

$$c = \sqrt{\frac{T}{\rho}} \tag{4.72}$$

Equation (4.71) is satisfied for all values if each term of the equation is equal to a constant. This may be accomplished as follows:

$$\frac{\partial^2 X(x)/\partial x^2}{X(x)} = -\beta_1^2 \tag{4.73}$$

$$\frac{\partial^2 Y(y)/\partial y^2}{Y(y)} = -\beta_2^2 \tag{4.74}$$

$$\frac{\partial^2 f(t)/\partial t^2}{c^2 f(t)} = -\beta_1^2 - \beta_2^2 \tag{4.75}$$

The solutions of these three differential equations are

$$X(x) = A \sin \beta_1 x + B \cos \beta_1 x \tag{4.76}$$

$$Y(y) = C \sin \beta_2 y + D \cos \beta_2 y \tag{4.77}$$

$$f(t) = E \sin ct(\beta_1^2 + \beta_2^2)^{1/2} + F \cos ct(\beta_1^2 + \beta_2^2)^{1/2} \tag{4.78}$$

It should be noted here that

$$\beta_1^2 + \beta_2^2 = \left(\frac{\omega}{c}\right)^2 \tag{4.79}$$

In this manner, the solution given by Eq. (4.70) is completely determined. The constants involved can be determined from the boundary conditions of the membrane and the initial conditions of the motion. The following example illustrates the application of the theory.

Example 4.3. Determine the free frequencies of vibration for a rectangular membrane of sides a and b. Also, write the general solution for its free transverse vibration.

Solution. The rectangular membrane is assumed to be tightly held at the ends; an initial displacement defined by a function $g(x, y)$ is applied at $t = 0$. On this basis, the boundary conditions of the membrane, as well as the initial time conditions of the motion, are written as follows:

$$w(0, y, t) = 0 \quad \text{(a)} \qquad w(x, 0, t) = 0 \quad \text{(c)}$$

$$w(a, y, t) = 0 \quad \text{(b)} \qquad w(x, b, t) = 0 \quad \text{(d)}$$

$$w(x, y, 0) = g(x, y) \quad \text{(e)}$$

$$\dot{w}(x, y, 0) = 0 \qquad \text{(f)}$$

The boundary conditions of the membrane are represented by conditions (a), (b), (c), and (d) and the time conditions by (e) and (f). Thus, from boundary condition (a), we find $B = 0$. From boundary condition (b), we have

$$A \sin \beta_1 a = 0 \tag{4.80}$$

Since A should be different from zero, the values of $\beta_1 a$ that make $\sin \beta_1 a = 0$ are as follows:

$$\beta_1 a = m\pi \quad \text{or} \quad \beta_1 = \frac{m\pi}{a} \qquad m = 1, 2, 3, \ldots \tag{4.81}$$

Boundary condition (c) yields $D = 0$, and boundary condition (d) yields

$$C \sin \beta_2 b = 0 \tag{4.82}$$

Thus, for $C \neq 0$, the values $\beta_2 b$ that make $\sin \beta_2 b = 0$ are as follows:

$$\beta_2 b = n\pi \quad \text{or} \quad \beta_2 = \frac{n\pi}{b} \qquad n = 1, 2, 3, \ldots \tag{4.83}$$

Time condition (f) yields $E = 0$.

By using the results obtained from the application of the boundary and time conditions, we find that Eqs. (4.76), (4.77), and (4.78) can be written as follows:

$$X_m(x) = A_m \sin \frac{m\pi x}{a} \tag{4.84}$$

$$Y_n(y) = C_n \sin \frac{n\pi y}{b} \tag{4.85}$$

$$f(t) = F \cos ct \left(\frac{m^2 \pi^2}{a^2} + \frac{n^2 \pi^2}{b^2} \right)^{1/2} \tag{4.86}$$

Thus, by Eqs. (4.79), (4.81), (4.83), and (4.86), it can be concluded that the frequencies of vibration ω of the membrane may be obtained from the equation

$$\omega = c\pi \left(\frac{m^2}{a^2} + \frac{n^2}{b^2} \right)^{1/2} \qquad m = 1, 2, 3, \ldots \quad n = 1, 2, 3, \ldots \tag{4.87}$$

In hertz, the frequencies f are

$$f = \frac{\omega}{2\pi} = \frac{c}{2} \left(\frac{m^2}{a^2} + \frac{n^2}{b^2} \right)^{1/2} \qquad m = 1, 2, 3, \ldots \quad n = 1, 2, 3, \ldots \tag{4.88}$$

The constant c is given by Eq. (4.72). The various combinations of the values of the integers m and n yield the various free frequencies of vibration of the stretched membrane.

By Eq. (4.70) and the results given by Eqs. (4.84), (4.85), and (4.86), the general solution $w(x, y, t)$ may be written as follows:

$$w(x, y, t) = X(x)Y(y)f(t)$$

$$= \sum_{m=1}^{\infty} \sum_{n=1}^{\infty} A_{mn} \sin \frac{m\pi x}{a} \sin \frac{n\pi y}{b} \cos \pi ct \left(\frac{m^2}{a^2} + \frac{n^2}{b^2}\right)^{1/2} \quad (4.89)$$

The constant A_{mn} in this equation can be determined by specifying an initial time displacement distribution $g(x, y)$ for the membrane.

4.5 DIFFERENTIAL EQUATION OF MOTION OF THIN RECTANGULAR PLATES

The differential equation of motion of uniform thin rectangular plates that are subjected to transverse dynamic loads of intensity $q(x, y, t)$ is derived in this section. Such a plate is shown in Fig. 4.5a, and the dynamic load distribution on the surface of the plate is shown in Fig. 4.5b. The sides of the plate are designated as a and b, as shown in Fig. 4.5a, and the thickness of the thin plate is h, which is considered to be small compared to the other dimensions of the plate. It is further assumed that the external dynamic loading $q(x, y, t)$ is applied normal to the surface of the plate, its middle surface is not strained, and the effects of damping, shear, and rotatory inertia are neglected.

The derivation of the differential equation of motion may be initiated by considering an element of the rectangular plate of sides dx and dy and of thickness h. The free-body diagram of this element is shown in Fig. 4.5c, where V_x and V_y are the dynamic shear forces per unit length that are parallel to the y and x axes, respectively, M_x and M_y are the moments, M_{xy} is the twisting moment, and ρ is the volume density of the material of the plate. The positive directions of the x, y, z rectangular coordinate system of axes of the plate is shown in Fig. 4.5d.

By considering the z direction of the thin plate element and equating to zero the sum of all the forces acting in this direction, we obtain the following equation:

$$\frac{\partial V_x}{\partial x} + \frac{\partial V_y}{\partial y} + q(x, y, t) + \rho h \ddot{w} = 0 \quad (4.90)$$

In Eq. (4.90), the term $\rho h \ddot{w}$ is the inertia force, which is assumed to act

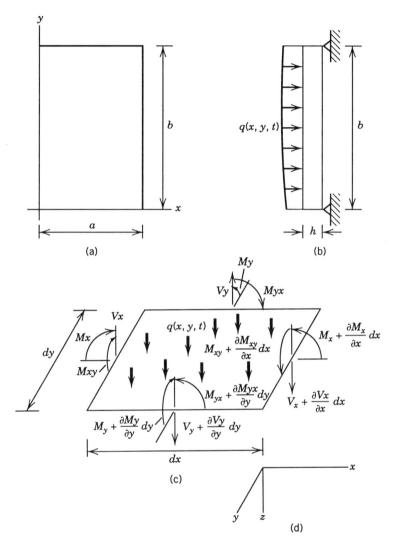

FIGURE 4.5. (a) Thin rectangular plate of sides a and b. (b) Distribution of the dynamic load $g(x, y, t)$. (c) Free-body diagram of an element of the rectangular plate. (d) Positive directions of the x, y, z rectangular coordinate system.

upward in order to conform with D'Alembert's principle of dynamic equilibrium, and w is the deflection of the plate in the z direction.

If we consider the sum of the moments about the x and y axes of the element and equate them individually to zero, we obtain the following two equations:

$$\frac{\partial M_{xy}}{\partial x} - \frac{\partial M_y}{\partial y} + V_y = 0 \tag{4.91}$$

$$\frac{\partial M_{yx}}{\partial y} + \frac{\partial M_x}{\partial x} - V_x = 0 \tag{4.92}$$

In Eqs. (4.91) and (4.92), the higher-order terms were neglected as being small, and M_{xy} may be assumed to be equal to M_{yx}. By using Eqs. (4.90), (4.91), and (4.92), the following equation may be derived:

$$\frac{\partial^2 M_x}{\partial x^2} + \frac{\partial^2 M_y}{\partial y^2} - 2\frac{\partial^2 M_{xy}}{\partial x\,\partial y} = q(x, y, t) - \rho h \ddot{w} \tag{4.93}$$

By referring to the theory of rectangular thin plates [16], the following equations for M_x, M_y, and M_{xy} may be written:

$$M_x = -D\left(\frac{\partial^2 w}{\partial x^2} + \nu\frac{\partial^2 w}{\partial y^2}\right) \tag{4.94}$$

$$M_y = -D\left(\frac{\partial^2 w}{\partial y^2} + \nu\frac{\partial^2 w}{\partial x^2}\right) \tag{4.95}$$

$$M_{xy} = -M_{yx} = D(1 - \nu)\frac{\partial^2 w}{\partial x\,\partial y} \tag{4.96}$$

where

$$D = \frac{Eh^3}{12(1 - \nu^2)} \tag{4.97}$$

In the above equations, the symbol ν is used to denote the Poisson ratio and E is Young's modulus of elasticity.

By substituting the expressions given by Eqs. (4.94), (4.95), and (4.96) into Eq. (4.93) and carrying out the required mathematical manipulations, the following differential equation is obtained:

$$D\left(\frac{\partial^4 w}{\partial x^4} + 2\frac{\partial^4 w}{\partial x^2\,\partial y^2} + \frac{\partial^4 w}{\partial y^4}\right) + \rho h \ddot{w} = q(x, y, t) \tag{4.98}$$

Equation (4.98) is the partial differential equation of motion for rectangular thin plates that are subjected to a dynamic load excitation $q(x, y, t)$.

The differential equation of motion for the free vibration of thin rectangular plates may be obtained from Eq. (4.98) by making $q(x, y, t) = 0$. This yields

$$D\left(\frac{\partial^4 w}{\partial x^4} + 2\frac{\partial^4 w}{\partial x^2 \partial y^2} + \frac{\partial^4 w}{\partial y^4}\right) + \rho h \ddot{w} = 0 \tag{4.99}$$

Equations (4.98) and (4.99) can be applied for the solution of rectangular thin plates of various boundary conditions. The solution of such problems, however, is often very difficult, and the practicing engineer usually resorts to approximate methods of analysis such as finite element, finite difference, and modal analysis. These methods are discussed later in the text.

4.6 VIBRATION ANALYSIS OF SIMPLY SUPPORTED RECTANGULAR THIN PLATES OF UNIFORM THICKNESS

In this section, we consider a thin rectangular plate that is simply supported along its four edges. The free vibration analysis of the plate may be initiated by considering Eq. (4.99) and assuming that the solution $w(x, y, t)$ is of the following form:

$$w(x, y, t) = X(x)Y(y)g(t) \tag{4.100}$$

where $X(x)$ is a function of x only, $Y(y)$ is a function of y only, and $g(t)$ depends only on the time t.

By substituting Eq. (4.100) into Eq. (4.99) and differentiating, we find

$$\begin{aligned} X''''(x)Y(y)g(t) &+ 2X''(x)Y''(y)g(t) + X(x)Y''''(y)g(t) \\ &+ \frac{\rho h}{D}X(x)Y(y)\ddot{g}(t) = 0 \end{aligned} \tag{4.101}$$

If we divide all terms of Eq. (4.101) by the product $X(x)Y(y)g(t)$ and separate variables, we obtain

$$\frac{X''''(x)Y(y) + 2X''(x)Y''(y) + X(x)Y''''(y)}{\rho h X(x)Y(y)/D} = -\frac{\ddot{g}(t)}{g(t)} \tag{4.102}$$

Equation (4.102) may be satisfied for all values if each side of the equation is equal to a constant. If this constant is assumed to be ω^2, the following two differential equations may be written:

$$X''''(x)Y(y) + 2X''(x)Y''(y) + X(x)Y''''(y) - \frac{\rho h \omega^2}{D} X(x)Y(y) = 0 \qquad (4.103)$$

$$\ddot{g}(t) + \omega^2 g(t) = 0 \qquad (4.104)$$

The solution of Eq. (4.104) is the solution of a freely vibrating one-degree spring–mass system, and it has the well-known form given by Eq. (2.5); that is,

$$g(t) = A \sin \omega t + B \cos \omega t \qquad (4.105)$$

For Eq. (4.103), we may assume solutions of the form

$$w(x, y) = X(x)Y(y) = \sum_n \sum_p \sin \frac{n \pi x}{a} \sin \frac{p \pi y}{b} \qquad (4.106)$$

where a and b are the lengths of the rectangular plate in the x and y directions, respectively. On this basis, we find

$$\frac{n^4 \pi^4}{a^4} + 2 \frac{n^2 \pi^2}{a^2} \cdot \frac{p^2 \pi^2}{b^2} + \frac{p^4 \pi^4}{b^4} = \frac{\rho h \omega^2}{D} \qquad (4.107)$$

and

$$\omega_{np} = \pi^2 \left(\frac{n^2}{a^2} + \frac{p^2}{b^2} \right) \sqrt{\frac{D}{\rho h}} \qquad (4.108)$$

The values of ω_{np} in Eq. (4.108) are chosen in a way that will satisfy Eq. (4.107). The number of the free frequencies of vibration ω_{np} of the plate is infinite, because there are infinite combinations of the integer values of n and p. The fundamental frequency of vibration corresponds to the value $n = p = 1$. On this basis, Eq. (4.108) yields

$$\omega_{11} = \pi^2 \left(\frac{1}{a^2} + \frac{1}{b^2} \right) \sqrt{\frac{D}{\rho h}} \qquad (4.109)$$

Every frequency ω_{np} is associated with a mode shape that is given by Eq. (4.106). Note that this equation satisfies the boundary conditions of the plate at the simply supported edges, because the sines are zero at the boundaries of the plate.

4.7 VIBRATION RESPONSE OF BEAM COLUMNS

In this section, we investigate the effect of an axial tensile or compressive force on the free flexural vibrations of a beam. For this purpose, we consider first the simply supported beam in Fig. 4.6a that is loaded statically by a distributed load $q(x)$ and an axial force P, which can be either in tension or in compression. The deflection configuration of this static system and a free-body diagram of an element of the beam are shown in Figs. 4.6b and 4.6c, respectively.

The differential equation representing the static deflection $y(x)$ of this

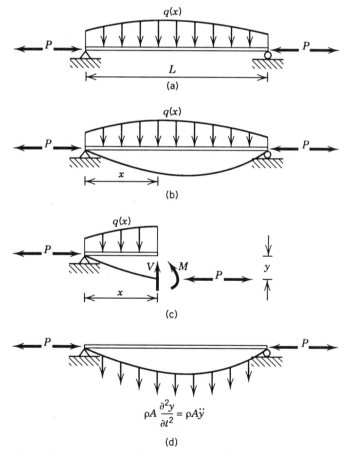

FIGURE 4.6. (a) Beam column subjected to a static load $q(x)$ per unit of length. (b) Deflection configuration of the static beam. (c) Free-body diagram of an element of the static beam. (d) Deformation configuration of a vibrating beam column.

member is given by the following equation:

$$EI\frac{d^2y}{dx^2} = -M \pm Py \tag{4.110}$$

The $+Py$ term in Eq. (4.110) is associated with a tensile axial force, $-Py$ represents the effect of the compressive force P, and M is the bending moment produced by the vertical load $q(x)$.

If we differentiate Eq. (4.110) twice with respect to x, we find the following equation:

$$\frac{d^2}{dx^2}\left(EI\frac{d^2y}{dx^2}\right) = q(x) \pm P\frac{d^2y}{dx^2} \tag{4.111}$$

In order to form the differential equation of motion for the free vibration of the beam column, we examine the deflection configuration of the member shown in Fig. 4.6d. In this figure, the beam is vibrating freely in its transverse direction. The only forces participating in this motion are the inertial forces $\rho A\, \partial^2 y/\partial t^2 = \rho A\ddot{y}$ and the restraining axial force P, where A is the cross-sectional area. If we compare the problem in Fig. 4.6d with the static problem in Fig. 4.6a, we conclude that the differential equation for the free vibration of the beam in Fig. 4.6d may be obtained from Eq. (4.111) by replacing $q(x)$ with the inertia force $(-\rho A\ddot{y}) = -\rho A\, \partial^2 y/\partial t^2$ per unit length. On this basis, we find

$$\frac{\partial^2}{\partial x^2}\left(EI\frac{\partial^2y}{\partial x^2}\right) = -\rho A\frac{\partial^2y}{\partial t^2} \pm P\frac{\partial^2y}{\partial x^2} \tag{4.112}$$

or

$$\frac{\partial^2}{\partial x^2}\left(EI\frac{\partial^2y}{\partial x^2}\right) \mp P\frac{\partial^2y}{\partial x^2} = -\rho A\frac{\partial^2y}{\partial t^2} \tag{4.113}$$

The positive sign in the second term on the right-hand side of Eq. (4.113) corresponds to a compressive axial load P, and the negative sign is used for a tensile axial load P. The quantity $\rho A = m$ is the mass per unit length of the member.

The stiffness EI in Eq. (4.113) may be either variable or constant throughout the length of the member. If it is constant (i.e., prismatic member of isotropic material), Eq. (4.113) yields

$$EI\frac{\partial^4y}{\partial x^4} \mp P\frac{\partial^2y}{\partial x^2} = -\rho A\frac{\partial^2y}{\partial t^2} \tag{4.114}$$

4.7.1 Uniform Beam Column with Axial Tensile Force

The differential equation of motion for this case is

$$EI\frac{\partial^4 y}{\partial x^4} - P\frac{\partial^2 y}{\partial x^2} = -\rho A\frac{\partial^2 A}{\partial t^2} \tag{4.115}$$

The solution $y(x,t)$ of Eq. (4.115) may be assumed to have the following form:

$$y(x,t) = Y(x)(C_1 \sin \omega t + C_2 \cos \omega t) \tag{4.116}$$

By substituting Eq. (4.116) into Eq. (4.115) and using the method of variable separation, as was done in earlier sections, the following equation may be obtained:

$$EI\frac{d^4 Y(x)}{dx^4} - P\frac{d^2 Y(x)}{dx^2} - \rho A\omega^2 Y(x) = 0 \tag{4.117}$$

where ω denotes the frequencies of vibration of the member.

The solution $Y(x)$ of Eq. (4.117) may be assumed to have the form

$$Y(x) = Be^{\lambda x} \tag{4.118}$$

as is customary for such types of differential equations. By substituting Eq. (4.118) into Eq. (4.117) and carrying out the required differentiations, we find

$$EI\lambda^4 - PL^2 - \rho A\omega^2 = 0 \tag{4.119}$$

The roots $\lambda_{1,2}^2$ of Eq. (4.119) are

$$\lambda_{1,2}^2 = \frac{P}{2EI} \pm \sqrt{\frac{P^2}{4(EI)^2} + \frac{\rho A\omega^2}{EI}} \tag{4.119a}$$

Note that the λ's in Eq. (4.119a) are functions of the frequency ω of the axially restrained beam.

With known λ's, Eq. (4.118) can be expressed in the following trigonometric form:

$$Y(x) = B_1 \cosh \lambda_1 x + B_2 \sinh \lambda_1 x + B_3 \cos \lambda_2 x + B_4 \sin \lambda_2 x \tag{4.120}$$

The constants B_1, B_2, B_3, and B_4 may be determined by applying appropriate boundary conditions.

4.7.2   Uniform Beam Column with Axial Compressive Force

In this case, the differential equation of motion is as follows:

$$EI \frac{\partial^4 y}{\partial x^4} + P \frac{\partial^2 y}{\partial x^2} = -\rho A \frac{\partial^2 y}{\partial t^2} \tag{4.121}$$

Since P in Eq. (4.121) may be considered as either positive or negative, the solution given by Eq. (4.120) is also applicable for this case.

The following example illustrates the application of the above methodologies.

Example 4.4. Determine the expression for the flexural free frequencies of vibration ω_n of a uniform simply supported beam of length L that is subjected to axial compressive or tensile forces.

Solution. We use Eq. (4.120) and apply the following boundary conditions:

$$\text{at} \quad x = 0 \qquad Y(0) = 0 \tag{4.122}$$

$$Y''(0) = 0 \tag{4.123}$$

$$\text{at} \quad x = L \qquad Y(L) = 0 \tag{4.124}$$

$$Y''(L) = 0 \tag{4.125}$$

The boundary conditions given by Eqs. (4.122) and (4.123) require that

$$B_1 = B_3 = 0 \tag{4.126}$$

Therefore,

$$Y(x) = B_2 \sinh \lambda_1 x + B_4 \sin \lambda_2 x \tag{4.127}$$

Application of the boundary conditions given by Eqs. (4.124) and (4.125) yields

$$B_2 \sinh \lambda_1 L + B_4 \sin \lambda_2 L = 0 \tag{4.128}$$

$$B_2 \sinh \lambda_1 L - B_4 \sin \lambda_2 L = 0 \tag{4.129}$$

Algebraic addition and subtraction of Eqs. (4.128) and (4.129) yield the following expressions:

$$B_2 \sinh \lambda_1 L = 0 \tag{4.130}$$

$$B_4 \sin \lambda_2 L = 0 \tag{4.131}$$

In Eq. (4.130), the constant B_2 must be zero, because $\sinh \lambda_1 L$ cannot be zero. In Eq. (4.131), the constant B_4 cannot be zero, because if it is zero, the trivial solution of no vibration is obtained. Therefore, we must have

$$\sin \lambda_2 L = 0 \tag{4.132}$$

The roots of $\lambda_2 L$ that satisfy Eq. (4.132) are

$$\lambda_2 L = n\pi \qquad n = 1, 2, 3, \ldots \tag{4.133}$$

Therefore,

$$\lambda_2 = \frac{n\pi}{L} \qquad n = 1, 2, 3, \ldots \tag{4.134}$$

By using the expression for λ_2 given by Eq. (4.119a) and substituting into Eq. (4.134), we obtain

$$\frac{P}{2EI} - \sqrt{\frac{P^2}{4(EI)^2} + \frac{\rho A \omega^2}{EI}} = \frac{n^2 \pi}{L^2} \qquad n = 1, 2, 3, \ldots \tag{4.135}$$

By solving Eq. (4.135) for ω, we find

$$\omega_n = \frac{\pi^2}{L^2} \sqrt{\frac{EI}{\rho A}} \left(\sqrt{n^4 - \frac{n^2 P L^2}{\pi^2 EI}} \right) \qquad n = 1, 2, 3, \ldots \tag{4.136}$$

Equation (4.136) provides the expression for the free frequencies of vibration ω_n of the beam column. Examination of Eq. (4.136) reveals that, if the axial load P is compressive, the value of P in Eq. (4.136) is positive, and the value of P will be negative if the axial load P is tensile. This shows that a compressive axial load P decreases the frequency ω_n with increasing P, and the frequency ω_n increases with an increasing tensile load.

Equation (4.136) may also be written as a function of the ratio P/P_{cr}, where P_{cr} is the *Euler buckling load*. For a simply supported beam, the smaller critical load is given by the expression

$$P_{cr} = \frac{\pi^2 EI}{L^2} \tag{4.137}$$

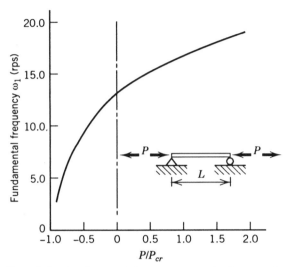

FIGURE 4.7. Plot of the fundamental frequency ω_1 of a beam column showing the influence of the P/P_{cr} ratio.

By substituting Eq. (4.137) in Eq. (4.136) and rearranging, we find

$$\omega_n = \frac{\pi^2}{L^2}\sqrt{\frac{EI}{\rho A}}\left[\sqrt{n^2\left(n^2 - \frac{P}{P_{cr}}\right)}\right] \qquad n = 1, 2, 3, \ldots \qquad (4.138)$$

Equation (4.138) shows that, for $n = 1$, we have the fundamental frequency of vibration ω_1. For a compressive axial load P, this frequency approaches zero as the ratio P/P_{cr} approaches unity. For a tensile axial load P, the sign under the radical in Eq. (4.138) becomes positive, and the frequency ω_1 increases with increasing P/P_{cr}. This is very useful information for practical applications.

The graph in Fig. 4.7 shows the variation of the fundamental frequency ω_1 with increasing or decreasing ratio P/P_{cr}. The beam is assumed to be simply supported with length $L = 200$ in., cross-sectional area $A = 100$ in.^{2}, modulus of elasticity $E = 3 \times 10^6$ psi, moment of inertia $I = 1000$ in.^{4}, and mass per unit length equal to unity.

4.8 ELASTICALLY SUPPORTED BEAMS WITH AXIAL RESTRAINTS

The problem of elastically supported beams with axial and vertical restraints has received particular attention by the author [2, 17–19, 24] and other investigators [20, 21] during recent years, because of its inherent peculiarities associated with their static and vibrational behavior. When axial and vertical

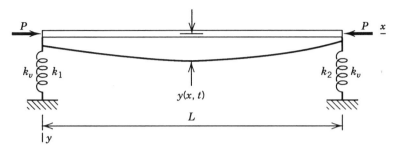

FIGURE 4.8. Uniform elastically supported beam subjected to an axial compressive load P.

restraints are applied to members that are supported by end springs, static (divergent) and flutter instabilities develop that make the member behave in a very unstable manner. In order to pinpoint some of the peculiarities of such types of problems, an elastically supported uniform beam with a spring of stiffness k_v at each end and subjected to an axial compressive load at each end, as shown in Fig. 4.8, is investigated. For additional information on this subject, the reader may consult the author's work in ref. 2.

The differential equation of motion, which includes the effects of shear and rotatory inertia, is derived by the author in Chapter 12 of ref. 2, and it is written as follows:

$$\frac{EI(K'AG)}{K'AG + P} y'''' - \frac{EI\rho}{K'AG + P}\left(A + \frac{K'AG}{E}\right)\ddot{y}'' + m\ddot{y}$$
$$+ Py'' + \frac{I\rho^2 A}{K'AG + P}\ddddot{y} = 0 \tag{4.139}$$

In Eq. (4.139), A is the cross-sectional area of the member, ρ is the material density, P is the axial compressive force, m is the mass per unit length of the member, E is the modulus of elasticity, G is the shear modulus, K' is the shear coefficient, and I is the cross-sectional moment of inertia.

The general solution $y(x, t)$ of Eq. (4.139) may be assumed to be of the form

$$y(x, t) = Y(x)e^{i\omega t} \tag{4.140}$$

where ω in this equation denotes the free frequency of vibration of the member.

By substituting Eq. (4.140) into Eq. (4.139) and carrying out the required mathematical operations, the following differential equation is obtained:

$$\frac{d^4Y(\xi)}{d\xi^4} + \lambda^2 \frac{d^2Y(\xi)}{d\xi^2} - \gamma^2 Y(\xi) = 0 \tag{4.141}$$

In Eq. (4.141), $\xi = x/L$ is a dimensionally independent variable, and

$$\lambda^2 = \frac{\rho\omega^2 L^2}{K'G} + \frac{\rho\omega^2 L^2}{E} + \frac{PL^2(K'AG + P)}{EI(K'AG)} \tag{4.142}$$

$$\gamma^2 = \frac{m\omega^2(K'AG + P)}{EI(K'AG)}\left(1 - \frac{\rho^2\omega^2 IA}{(K'AG + P)m}\right) \tag{4.142a}$$

The general solution of Eq. (4.141) may be written

$$y(\xi) = C_1 \cosh \beta\xi + C_2 \sinh \beta\xi + C_3 \cos \bar\beta\xi + C_4 \sin \bar\beta\xi \tag{4.143}$$

where

$$\beta = \left[-\frac{\lambda^2}{2} + \left(\frac{\lambda^4}{4} + \gamma^2\right)^{1/2}\right]^{1/2} \tag{4.144}$$

$$\bar\beta = \left[\frac{\lambda^2}{2} + \left(\frac{\lambda^4}{4} + \gamma^2\right)^{1/2}\right]^{1/2} \tag{4.145}$$

The constants C_1, C_2, C_3, and C_4 may be determined by applying the boundary conditions of the motion. This will lead to a frequency determinant that can be used to determine the free frequencies of vibration of the member. See details in Chapter 12 of ref. 2.

The following example illustrates some of the important aspects associated with this problem.

Example 4.5. Assume that the member in Fig. 4.8 has a length $L = 100$ in., depth $h = 8$ in., width $b = 6$ in., modulus $E = 30 \times 10^6$ psi, moment of inertia $I = 256$ in.[4], shear coefficient $K' = \frac{2}{3}$, and Poisson ratio $\nu = 0.25$. The mass of the member per cubic inch is 0.0088 lb · s^2/in. Investigate the response of this system for various combinations of the end spring constants k_v and axial compressive force P. Discuss the results as well as other important aspects of the problem. See also ref. 2.

Solution. The solution was carried out by using Eq. (4.143) and applying the following boundary conditions:

$$V(0) + k_v Y(0) = 0 \qquad (4.146)$$

$$V(1) - k_v Y(1) = 0 \qquad (4.147)$$

$$M(0) = 0 \qquad (4.148)$$

$$M(1) = 0 \qquad (4.149)$$

These four boundary conditions yield four equations in terms of the constants C_1, C_2, C_3, and C_4, which in matrix form are written as follows:

$$\begin{bmatrix} V(0) \\ M(0) \\ -V(1) \\ -M(1) \end{bmatrix} = \begin{bmatrix} 0 & \bar{A} & 0 & \bar{B} \\ \bar{C} & 0 & \bar{D} & 0 \\ \bar{A}\mathrm{sh} & \bar{A}\mathrm{ch} & -\bar{B}s & \bar{B}c \\ \bar{C}\mathrm{ch} & \bar{C}\mathrm{sh} & Dc & \bar{D}s \end{bmatrix} \begin{bmatrix} C_1 \\ C_2 \\ C_3 \\ C_4 \end{bmatrix} \qquad (4.150)$$

where $\mathrm{ch} = \cosh \beta$; $\mathrm{sh} = \sinh \beta$; $c = \cos \bar{\beta}$; $s = \sin \bar{\beta}$

$$\bar{A} = \bar{M}(\beta^3 + \lambda^2\beta) \qquad (4.151)$$

$$\bar{B} = \bar{M}(-\bar{\beta}^3 + \lambda^2\bar{\beta}) \qquad (4.152)$$

$$\bar{C} = \bar{E}\left(\beta^2 + \frac{\rho\omega^2 L^2}{K'G}\right) \qquad (4.153)$$

$$\bar{D} = \bar{E}\left(\bar{\beta}^2 + \frac{\rho\omega^2 L^2}{K'G}\right) \qquad (4.154)$$

$$\bar{E} = -\frac{EIK'AG}{K'AG + P} \qquad (4.155)$$

$$\tilde{M} = \frac{EIK'AG/(K'AG + P)}{[1 - \rho\omega^2 I/(K'AG + P)]} \qquad (4.156)$$

Details of the solution may be found in Chapter 12 of ref. 2.

Vibration analysis may be performed by setting equal to zero the square determinant on the right-hand side of Eq. (4.150). This establishes the frequency equation as a function of the horizontal compressive force P when the determinant is expanded. Therefore, by keeping all other parameters constant and varying only the axial force, its influence on the natural flexural frequencies of vibration and rigid-body motion may be determined in each case. The *bisection method* [22] may be used to determine the eigenfrequencies of the lateral vibration and rigid-body motion.

We assumed here that the end springs of the beam have equal spring constants k_v, and the cases examined here are for values of $k_v = 10$, 100, 500, 1000, 2000, and 5000 lb/in. For each case of k_v, the first two frequencies

TABLE 4.1. Natural Frequencies ω_1 and ω_2 for $k_v = 10$, 100, and 500 lb/in.

$k_v = 10$ lb/in.			$k_v = 100$ lb/in.			$k_v = 500$ lb/in.		
P (lb)	ω_1 (rps)	ω_2 (rps)	P (lb)	ω_1 (rps)	ω_2 (rps)	P (lb)	ω_1 (rps)	ω_2 (rps)
0	0.689	1.189	0	2.177	3.758	0	4.877	8.402
100	0.689	1.063	500	2.177	3.565	2.000	4.877	8.059
200	0.689	0.921	1,000	2.177	3.362	4,000	4.877	7.701
300	0.689	0.752	1,500	2.177	3.144	6,000	4.877	7.325
355	0.689	0.689	2,000	2.177	2.911	8,000	4.877	6.929
			2,500	2.177	2.658	10,000	4.877	6.509
			3,000	2.177	2.377	12,000	4.877	6.059
			3,350	2.177	2.177	14,000	4.877	5.574
						16,000	4.877	5.024
						16,550	4.877	4.877

TABLE 4.2. Natural Frequencies ω_1 and ω_2 for $k_v = 1000$, 2000, and 5000 lb/in.

$k_v = 1000$ lb/in.			$k_v = 2000$ lb/in.			$k_v = 5000$ lb/in.		
P (lb)	ω_1 (rps)	ω_2 (rps)	P (lb)	ω_1 (rps)	ω_2 (rps)	P (lb)	ω_1 (rps)	ω_2 (rps)
0	6.876	11.882	0	9.713	16.546	0	11.882	20.572
5,000	6.876	11.272	12,000	9.713	15.761	10,000	11.882	19.876
10,000	6.876	10.628	24,000	9.713	14.647	20,000	11.882	19.153
15,000	6.876	9.941	30,000	9.713	14.058	30,000	11.882	18.402
20,000	6.876	9.294	39,000	9.713	13.122	40,000	11.882	17.620
25,000	6.876	8.402	45,000	9.713	12.461	50,000	11.882	16.800
30,000	6.876	7.516	51,000	9.713	11.762	60,000	11.882	15.939
33,501	6.876	6.876	57,000	9.713	11.019	70,000	11.882	15.028
			63,000	9.713	10.222	80,000	11.882	14.058
			66,065	9.713	9.713	95,564	11.882	11.882

ω_1 and ω_2, in radians per second (rps), are determined by varying the axial compressive load P and keeping all other parameters of the member the same. The results are shown in Tables 4.1 and 4.2.

In the first column of Table 4.1, the axial load P was permitted to vary from 0 to 355 lb. When $P = 355$ lb, we note that the first and second frequencies ω_1 and ω_2, respectively, assume the value of $\omega_1 = \omega_2 = 0.689$ rps. This indicates that, at $P = 355$ lb, the elastically supported beam experiences a state of flutter instability. Since ω_1 and ω_2 are the frequencies corresponding, respectively, to vertical translational and rotational rigid-body type vibrational motions, we can conclude that the system becomes unstable when the rotational and translational motions are of the same frequency (coincide). In the same table, as well as Table 4.2, similar instabilities are observed for other values of the spring constants k_v. We also note that as the vertical

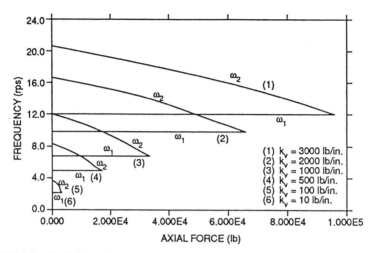

FIGURE 4.9. Variation of ω_1 and ω_2 with increasing axial load P for various values of k_v.

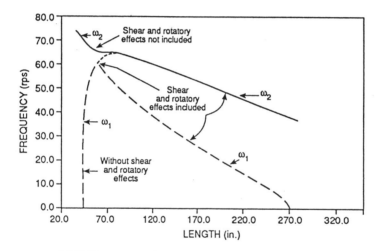

FIGURE 4.10. Effect of shear and rotatory inertia.

spring k_v becomes stiffer, it requires higher values of axial compressive load for ω_1 and ω_2 to coincide.

The results are plotted in Fig. 4.9 for the various cases of k_v. In this graph, the points where $\omega_1 = \omega_2$ are also known as *points of double instability*.

In Fig. 4.10, the effects of shear and rotatory inertia are examined. In this figure, the frequencies ω_1 and ω_2 are plotted by varying the length L of the member up to a value of 275 in. and keeping k_v constant at 50,000 lb/in. The axial force $P = 10^6$ lb, and the mass m per unit length of the member is

$0.4224 \, \text{lb} \cdot \text{s}^2/\text{in}$. By examining this figure, we note that the frequencies ω_1 and ω_2 will coincide only when the effects of shear and rotatory inertia are considered; that is, the existence of flutter instability will be recognized only if the effects of shear and rotatory inertia are considered.

4.9   DYNAMIC RESPONSE OF BEAMS SUBJECTED TO EXTERNAL HARMONIC EXCITATIONS

The dynamic and vibration response and characteristics of uniform single-span beams subjected to harmonic excitations are considered in this section. The general theory will be developed by considering three particular cases of single-span beams and external harmonic excitations. These cases are (a) a uniform simply supported beam, as shown in Fig. 4.11a; (b) a uniform cantilever beam subjected to a harmonic force $F(t) = F \sin \omega_f t$ at the free end, as shown in Fig. 4.11b; and (c) a uniform simply supported beam subjected to a vertical harmonic displacement $y_s(t) = y_0 \sin \omega_f t$ at the end supports, as shown in Fig. 4.11d. In all three cases, ω_f is the frequency of the external excitation. Once the general methodology for these three cases is established, other cases of beams with different boundary conditions and different distributions of harmonic excitations may be treated in a similar manner.

4.9.1   Uniform Simply Supported Beam Subjected to Distributed Harmonic Excitations

We consider here the case of a uniform simply supported beam subjected to a uniformly distributed harmonic excitation $q(t, x)$ of the form

$$q(t, x) = q \sin \omega_f t \tag{4.157}$$

where ω_f is the frequency, in radians per second, of the external harmonic excitation.

The differential equation of motion for such problems is derived in Section 1.9 and is given by Eq. (1.53). For the loading case given by Eq. (4.157), the differential equation of motion is as follows:

$$EI \frac{\partial^4 y}{\partial x^4} + m \frac{\partial^2 y}{\partial t^2} = q \sin \omega_f t \tag{4.158}$$

We are going to obtain the steady-state solution of Eq. (4.158), which is produced by the application of the harmonic force of Eq. (4.157). The transient free vibration response is neglected because it often dies out under the influence of damping, which is usually present in the system. Under these

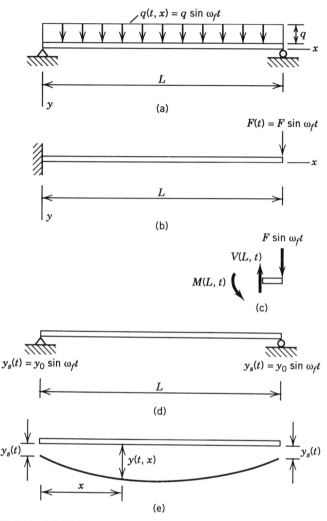

FIGURE 4.11. (a) Uniform simply supported beam subjected to a uniform external harmonic excitation. (b) Uniform cantilever beam subjected to a concentrated harmonic force at the free end. (c) Free-body diagram of an element of the cantilever beam. (d) Uniform simply supported beam subjected to a harmonic vertical support motion. (e) Deformation configuration of the simply supported beam due to harmonic support motion.

conditions, it would be reasonable to assume that the particular solution of Eq. (4.158) would be of the form

$$y = Y \sin \omega_f t \tag{4.159}$$

where Y is a function of x that needs to be determined.

By substituting Eq. (4.159) into Eq. (4.158) and carrying out the required mathematical operations, we obtain the following ordinary differential equation:

$$\frac{d^4Y}{dx^4} - \lambda^4 Y = \frac{q}{EI} \tag{4.160}$$

where

$$\lambda^4 = \frac{m\omega_f^2}{EI} \tag{4.161}$$

The complementary or homogeneous solution $Y_c(x)$ is the same as the solution of Eq. (2.203), and it is given by the equation

$$Y(x) = A_1 \cosh \lambda x + A_2 \sinh \lambda x + A_3 \cos \lambda x + A_4 \sin \lambda x \tag{4.162}$$

The particular solution Y_p of Eq. (4.160), since q/EI is constant along the length of the member, may be assumed to be equal to a constant, C_0; that is,

$$Y_p = C_0 \tag{4.163}$$

By substituting Eq. (4.163) into Eq. (4.160), we find $C_0 = -q/m\omega_f^2$, and, consequently,

$$Y_p = -\frac{q}{m\omega_f^2} \tag{4.164}$$

Thus, the complete solution $Y(x)$ is given by

$$Y(x) = A_1 \cosh \lambda x + A_2 \sinh \lambda x + A_3 \cos \lambda x + A_4 \sin \lambda x - \frac{q}{m\omega_f^2} \tag{4.165}$$

In Eq. (4.165), the constants A_1, A_2, A_3, and A_4 may be determined by using known boundary conditions of the uniform beam.

The end boundary conditions of the simply supported beam are as follows:

$$\text{at } x = 0 \qquad Y(0) = 0 \tag{4.166}$$

$$\frac{d^2Y(0)}{dx^2} = 0 \tag{4.167}$$

$$\text{at } x = L \qquad Y(L) = 0 \tag{4.168}$$

$$\frac{d^2Y(L)}{dx^2} = 0 \qquad (4.169)$$

By using Eq. (4.165) and applying the boundary conditions given by Eqs. (4.166) and (4.167), we find

$$A_1 = A_3 = \frac{q}{2\omega_f^2 m} \qquad (4.170)$$

Application of the end boundary conditions given by Eqs. (4.168) and (4.169) yields

$$A_2 = \frac{q}{2\omega_f^2 m} \tanh \frac{\lambda L}{2} \qquad (4.171)$$

$$A_4 = \frac{q}{2\omega_f^2 m} \tan \frac{\lambda L}{2} \qquad (4.172)$$

By using Eq. (4.162) and substituting for the constants the expressions obtained by Eqs. (4.170), (4.171), and (4.172), the following solution $Y(x)$ is obtained:

$$Y(x) = \frac{q}{\omega_f^2 m} \left(\frac{\cos \lambda(L/2 - x)}{2\cos(\lambda L/2)} + \frac{\cosh \lambda(L/2 - x)}{2\cosh(\lambda L/2)} - 1 \right) \qquad (4.173)$$

On this basis, Eq. (4.159) yields

$$y(t, x) = \frac{1}{\omega_f^2 m} \left(\frac{\cos \lambda(L/2 - x)}{2\cos(\lambda L/2)} + \frac{\cosh \lambda(L/2 - x)}{2\cosh(\lambda L/2)} - 1 \right) \sin \omega_f^2 t \qquad (4.174)$$

Equation (4.174) is the solution $y(t, x)$ of Eq. (4.158), and it provides the time variation of the amplitude of the motion, as well as the variation along the length of the member (with respect to x).

We note that, when λL assumes the values of π, 3π, 5π, . . . , the first term on the right-hand side of Eq. (4.174) approaches infinity. Since λ is a function of the frequency ω_f, as shown by Eq. (4.161), the phenomenon of resonance occurs whenever the forcing frequency ω_f approaches one of the odd-numbered free frequencies of vibration of the member. Equation (4.161) becomes identical to Eq. (2.205) when $\omega_f = \omega$, which means resonance.

With the known expression for $y(t, x)$, the bending moment M and shear force V at any time t and any distance x may be determined by differentiation from Eq. (4.174); that is,

$$M = -EI \frac{\partial^2 y}{\partial x^2} \qquad (4.175)$$

$$V = \frac{\partial M}{\partial x} \qquad (4.176)$$

With known M and V, the dynamic stress may be determined by using formulas from the mechanics of solids.

4.9.2 Uniform Cantilever Beam Subjected to a Vertical Harmonic Force at the Free End of the Beam

Consider now the uniform cantilever beam that is loaded with a concentrated harmonic force $F(t)$ at its free end as shown in Fig. 4.11b; that is,

$$F(t) = F \sin \omega_f t \qquad (4.177)$$

where ω_f is the frequency of the harmonic force. The initial displacement and initial velocity at time $t = 0$ are assumed to be zero.

The governing differential equation of motion for the length L of the member between the fixed support and just to the left of the applied harmonic force $F \sin \omega_f t$ is the same as Eq. (4.158) when the load $q \sin \omega_f t$ is made equal to zero; that is,

$$EI \frac{\partial^4 y}{\partial x^4} + m \frac{\partial^2 y}{\partial t^2} = 0 \qquad (4.178)$$

The steady-state motion produced by the force given by Eq. (4.177) will be investigated here. The solution $y(x, t)$ of this motion must satisfy Eq. (4.178) and the prescribed boundary conditions of the member.

We assume the solution $y(x, t)$ to be of the form

$$y(x, t) = Y(x)g(t) \qquad (4.179)$$

where $Y(x)$ is a function of x only and $g(t)$ is a function of time t only. By substituting Eq. (4.179) into Eq. (4.178) and separating variables, we obtain

$$\frac{EI}{mY(x)} \frac{\partial^4 Y(x)}{\partial x^4} = -\frac{\ddot{g}(t)}{g(t)} \qquad (4.180)$$

The two sides of Eq. (4.180) will be equal to each other only if each one is equal to a constant. This constant is taken equal to ψ^2, and, consequently, Eq. (4.180) yields the following two differential equations:

$$\frac{d^4Y(x)}{dx^4} - \frac{m\psi^2}{EI} Y(x) = 0 \tag{4.181}$$

$$\ddot{g}(t) + \psi^2 g(t) = 0 \tag{4.182}$$

If it is assumed that $g(t) = \sin \omega_f t$, a reasonable assumption since the applied force is sinusoidal, and it is substituted into Eq. (4.182), we obtain

$$\psi^2 = \omega_f^2 \tag{4.183}$$

On this basis, the differential equation given by Eq. (4.181) may be written

$$\frac{d^4Y(x)}{dx^4} - \lambda^4 Y(x) = 0 \tag{4.184}$$

where

$$\lambda^4 = \frac{m\omega_f^2}{EI} \tag{4.185}$$

The solution of Eq. (4.184) is similar to the homogeneous solution of Eq. (4.160), which is given by Eq. (4.162); that is,

$$Y(x) = A_1 \cosh \lambda x + A_2 \sinh \lambda x + A_3 \cos \lambda x + A_4 \sin \lambda x \tag{4.186}$$

Thus, by Eq. (4.179), the solution $y(x, t)$ is

$$y(x, t) = (A_1 \cosh \lambda x + A_2 \sinh \lambda x + A_3 \cos \lambda x + A_4 \sin \lambda x) \sin \omega_f t \tag{4.187}$$

The four constants A_1, A_2, A_3, and A_4 should be determined by using the boundary conditions of the member. These boundary conditions are as follows:

$$\text{at } x = 0 \qquad Y(0) = 0 \tag{4.188}$$

$$\frac{dY(0)}{dx} = 0 \tag{4.189}$$

$$\text{at } x = L \qquad \frac{d^2Y(L)}{dx^2} = 0 \tag{4.190}$$

$$-EI\frac{d^3Y(L)}{dx^3} = F \tag{4.191}$$

The boundary condition given by Eq. (4.191) has been obtained by considering the free-body diagram in Fig. 4.11c and applying vertical equilibrium.

By using Eq. (4.186) and applying the boundary conditions given by Eqs. (4.188) and (4.189), we obtain

$$A_1 = -A_3 \tag{4.192}$$

$$A_2 = -A_4 \tag{4.193}$$

Therefore, Eq. (4.186) yields

$$Y(x) = A_3(-\cosh \lambda x + \cos \lambda x) + A_4(-\sinh \lambda x + \sin \lambda x) \tag{4.194}$$

Using Eq. (4.194) and applying the boundary conditions given by Eqs. (4.190) and (4.191), we determine the constants A_3 and A_4, which are written as follows:

$$A_3 = \frac{F}{2\lambda^3 EI} \frac{\sinh \lambda L + \sin \lambda L}{[1 + (\cosh \lambda L)(\cos \lambda L)/} \tag{4.195}$$

$$A_4 = -\frac{F}{2\lambda^3 EI} \frac{\cosh \lambda L + \cos \lambda L}{[1 + (\cosh \lambda L)(\cos \lambda L)]} \tag{4.196}$$

Therefore, Eq. (4.186) can be written

$$Y(x) = \frac{F}{2\lambda^3 EI[1 + (\cosh \lambda L)(\cos \lambda L)]} [(\sinh \lambda L + \sin \lambda L)(\cosh \lambda x - \cos \lambda x)$$

$$- (\cosh \lambda L + \cos \lambda L)(\sinh \lambda x - \sin \lambda x)] \tag{4.197}$$

The displacement $y(x, t)$ caused by $F(t) = F \sin \omega_f t$ can be determined from the equation

$$y(x, t) = Y(x) \sin \omega_f t \tag{4.198}$$

where $Y(x)$ is given by Eq. (4.197).

At the free end of the member ($x = L$), the deflection $y(L, t)$ is

$$y(L, t) = \frac{F}{\lambda^3 EI} \frac{(\cosh \lambda L)(\sin \lambda L) - (\sinh \lambda L)(\cos \lambda L)}{[1 + (\cosh \lambda L)(\cos \lambda L)]} \sin \omega_f t \tag{4.199}$$

Resonance will occur again for this problem when the forced frequency ω_f coincides with any one of the free frequencies of vibration of the cantilever beam. With known $y(x, t)$, the dynamic moments and the dynamic shears may be obtained by applying Eqs. (4.175) and (4.176), respectively.

4.9.3 Uniform Simply Supported Beam Subjected to Harmonic Excitations at the End Supports

In this case, the uniform simply supported beam in Fig. 4.11d is subjected at its ends to a vertical support harmonic displacement $y_s(t)$ that varies with time as follows:

$$y_s(t) = y_0 \sin \omega_f t \qquad (4.200)$$

where ω_f is the frequency of its excitation, and y_0 is the maximum amplitude of the support motion.

The differential equation of motion for the beam is again given by Eq. (4.178). The steady-state response due to the support motion will be considered here. Therefore, the vertical displacement $y(x, t)$ (see Fig. 4.11e) is assumed to have the form

$$y(x, t) = Y(x)g(t) \qquad (4.201)$$

where $Y(x)$ is a function of x only, and $g(t)$ is a function of time t only. Since the excitation is harmonic, it is reasonable to assume that $g(t)$ is harmonic. Therefore,

$$g(t) = \sin \omega_f t \qquad (4.202)$$

and, consequently,

$$y(t, x) = Y(x) \sin \omega_f t \qquad (4.203)$$

By substituting Eq. (4.203) into Eq. (4.178) and carrying out the required mathematical operations, we find

$$\frac{d^4 Y(x)}{dx^4} - \lambda^4 Y(x) = 0 \qquad (4.204)$$

where

$$\lambda^4 = \frac{m\omega_f^2}{EI} \qquad (4.205)$$

The solution $Y(x)$ of Eq. (4.204) is again the same as the one for Eq. (4.184), and it is written as follows:

$$Y(x) = A_1 \cosh \lambda x + A_2 \sinh \lambda x + A_3 \cos \lambda x + A_4 \sin \lambda x \qquad (4.206)$$

The end boundary conditions of the simply supported beam are as follows:

$$\text{at } x = 0 \qquad Y(0) = y_0 \qquad (4.207)$$

$$\frac{d^2Y(0)}{dx^2} = 0 \qquad (4.208)$$

$$\text{at } x = L \qquad Y(L) = 0 \qquad (4.209)$$

$$\frac{d^2Y(L)}{dx^2} = 0 \qquad (4.210)$$

By using Eq. (4.206) and applying the boundary conditions given by Eqs. (4.207) and (4.208), we find

$$A_1 = \frac{y_0}{2} \qquad (4.211)$$

$$A_3 = \frac{y_0}{2} \qquad (4.212)$$

Thus, from Eq. (4.206), we obtain

$$Y(x) = \frac{y_0}{2}(\cosh \lambda x + \cos \lambda x) + A_2 \sinh \lambda x + A_4 \sin \lambda x \qquad (4.213)$$

$$\frac{d^2Y(x)}{dx^2} = \frac{y_0 \lambda^2}{2}(\cosh \lambda x - \cos \lambda x) + \lambda^2(A_3 \sinh \lambda x - A_4 \sin \lambda x) \qquad (4.214)$$

By using Eqs. (2.213) and (2.214) and applying the boundary conditions given by Eqs. (4.209) and (4.210), the constants A_2 and A_4 may be determined as follows:

$$A_2 = \frac{y_0}{2} \frac{(1 - \cosh \lambda L)}{\sinh \lambda L} \qquad (4.215)$$

$$A_4 = \frac{y_0}{2} \frac{(1 - \cos \lambda L)}{\sin \lambda L} \qquad (4.216)$$

Thus, Eq. (4.213) yields

$$Y(x) = \frac{y_0}{2}\left[(\cosh \lambda x + \cos \lambda x) + \frac{(1 - \cosh \lambda L)}{\sinh \lambda L}\sinh \lambda x\right.$$

$$\left. + \frac{(1 - \cos \lambda L)}{\sin \lambda L}\sin \lambda x\right]$$

(4.217)

On this basis, the solution given by Eq. (4.203) may be written

$$y(t, x) = \frac{y_0}{2}\left[(\cosh \lambda x + \cos \lambda x) + \frac{(1 - \cosh \lambda L)}{\sinh \lambda L}\sinh \lambda x\right.$$

$$\left. + \frac{(1 - \cos \lambda L)}{\sin \lambda L}\sin \lambda x\right]\sin \omega_f t$$

(4.218)

Equation (4.218) yields the total displacement of the beam in Fig. 4.11d at any distance x and time t. The displacement $y_b(t, x)$ of the beam relative to the motion of the end supports (Fig. 4.11e) is given by the equation

$$y_b(t, x) = y(t, x) - y_s(t)$$
$$= [Y(x) - y_0]\sin \omega_f t$$

or

$$y_b(t, x) = \frac{y_0}{2}\left[(\cosh \lambda x + \cos \lambda x) + \frac{(1 - \cosh \lambda L)}{\sinh \lambda L}\sinh \lambda x\right.$$

$$\left. + \frac{(1 - \cos \lambda L)}{\sin \lambda L}\sin \lambda x - 2\right]\sin \omega_f t$$

(4.219)

The following example illustrates the application of the above theory.

Example 4.6. The uniform simply supported beam in Fig. 4.11d is subjected to a vertical harmonic motion $y_s(t) = y_0 \sin \omega_f t$ at its end supports, where $y_0 = 2$ in. and $\omega_f = 4\pi$ rps. The stiffness EI of the beam is 30×10^6 kips · in.^{2}, and its weight w per unit length is 0.6 kip/ft. Determine the maximum deflection $y_{b_{max}}$ of the beam and its maximum bending stress caused by the support displacements. The length L of the members is equal to 30.0 ft.

Solution. From Eq. (4.205), we find

$$\lambda^4 = \frac{m\omega_f^2}{EI} = \frac{(0.6)(4\pi)^2}{(12)(386)(30 \times 10^6)} = 681.84 \times 10^{-12}$$

or

$$\lambda = 0.0051$$

By using the calculated value of λ and Eq. (4.219), we obtain

$$y_b(t, x) = [\cosh(0.0051x) + \cos(0.0051x) - 0.725 \sinh(0.0051x)$$
$$+ 1.308 \sin(0.0051x) - 2] \sin(4\pi t) \tag{4.220}$$

At $x = L/2 = 180.0$ in., Eq. (4.220) yields

$$y_b(t, L/2) = [1.452 + 0.607 - (0.725)(1.052) + 1.308(0.794) - 2] \sin(4\pi t)$$
$$= 0.335 \sin(4\pi t) \tag{4.221}$$

The maximum displacement y_b of the member, relative to the support displacement of $y_0 = 2.0$ in., will occur when $\sin(4\pi t)$ in Eq. (4.221) is equal to 1, that is, when $4\pi t = [(2n - 1)/2]\pi$, or $t = (2n - 1)/8$, and $n = 1, 2, 3, \ldots$. Thus,

$$y_{b_{max}} = \pm 0.335 \text{ in.}$$

The total maximum displacement y_{max} at the center of the member would be

$$y_{max}(L/2) = \pm(y_{b_{max}} + y_0)$$
$$= \pm 2.335 \text{ in.}$$

By using Eq. (4.175), the bending moment at any distance x and time t may be obtained as follows:

$$M(t, x) = -EI\frac{\partial^2 y_b}{\partial x^2} = -EI(0.0051)^2[\cosh(0.0051x) - \cos(0.0051x)$$
$$- 0.725 \sinh(0.0051x) - 1.308 \sin(0.0051x)] \sin(4\pi t)$$

or

$$M(t, x) = (0.0051)^2 EI \sin(4\pi t)[\cos(0.0051x) + 1.308 \sin(0.0051x)$$
$$+ 0.725 \sinh(0.0051x) - \cosh(0.0051x)] \qquad (4.222)$$

The maximum bending moment occurs at $x = L/2 = 180.0$ in. Thus,

$$M(t, L/2) = (0.0051)^2 EI(0.9577) \sin(4\pi t)$$
$$= 747.31 \sin(4\pi t) \qquad (4.223)$$

and

$$M_{max} = \pm 747.31 \text{ kip} \cdot \text{in.}$$

The maximum bending stress σ_{max} will occur at the center section of the member and can be determined from the equation

$$\sigma_{max} = \pm \frac{M_{max}}{S}$$
$$= \pm \frac{747.31}{S} \qquad (4.224)$$

where S is the section modulus of the member. If the moment of inertia I of the member is 1000 in.^{4} and the beam has a square cross section of sides b, then

$$S = \frac{b^3}{6} \qquad (4.225)$$

Also,

$$\frac{b^4}{12} = I = 1000 \text{ in.}^4$$

$$b = 10.4664 \text{ in.}$$

From Eq. (4.225), we find

$$S = \frac{(10.4664)^3}{6} = 191.09 \text{ in.}^3$$

and from Eq. (4.224),

$$\sigma_{max} = \pm \frac{747.31}{191.09} = 3.9108 \text{ ksi}$$

4.10   THE DYNAMIC HINGE CONCEPT

The dynamic hinge concept, as developed by the author and his collaborators [1, 7, 23–25], is based on the idea that when a member is vibrating freely in one of its infinite modes of vibration, all the elements of the member, as well as any portion of the member, vibrate with the same frequency ω. This idea suggests that if a small portion of a member is separated from the complete structure and its end boundary conditions are properly satisfied, then this portion will vibrate with the same frequency ω as the complete structure.

Consider, for example, the continuous beam in Fig. 4.12a, and assume that it vibrates in one of its infinite modes of vibration, say, the one shown in Fig. 4.12b. The stiffness EI of the beam may vary in any arbitrary manner along the length of each span of the member, and it can also be uniform throughout the length of the member. The distribution of the inertia forces, when the member vibrates in the mode shape shown in Fig. 4.12b, is shown in Fig. 4.12c. If the exact magnitudes of the inertia forces in Fig. 4.12c are known, the exact mode shape in Fig. 4.12b may be obtained if these inertia forces are applied statically to the member, as shown in Fig. 4.12c.

When the member vibrates in the mode shape shown in Fig. 4.12c, or in any other mode of vibration of the member, there will be points on the elastic line representing the mode of vibration where the curvature changes sign. Therefore, the bending moment at these points, known as *inflection points*, must be zero. The number of such inflection points is dependent on the shape of the mode under consideration. For the mode shape in Fig. 4.12c, we can easily observe that two such points are located at G and H as shown in the figure. Additional inflection points, however, may be found at other locations.

It was stated earlier that if a portion of a member, say, the one in Fig. 4.12a, is separated from the complete member and its end boundary conditions are appropriately satisfied, then this portion, if excited, will vibrate with the frequency of the complete member. When the member in Fig. 4.12a vibrates in the mode shape shown in Fig. 4.12b and has the inertia force distribution shown in Fig. 4.12c, a convenient portion that would easily satisfy its end boundary condition requirements is portion AG or HE. Note that the inertia forces in Fig. 4.12c are acting in the direction of the mode shape; that is, if the amplitude of the mode is in the upward direction, the inertia force is also acting in the upward direction. In this case, we consider portion HE in Fig. 4.12c and separating it from the complete member. Its free-body diagram is shown in Fig. 4.12d. The mode amplitude at point H is denoted as δ_H, the bending moment M_H at point H is zero, and the shear force at point H is designated as V_H.

If it is now assumed that the initial member is ideally hinged at point H, or at all points where the bending moment is zero, the presence of such hinges will not alter the shape of the bending moment diagram produced by

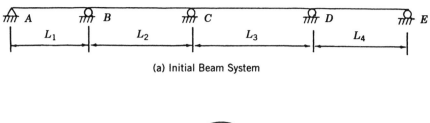

(a) Initial Beam System

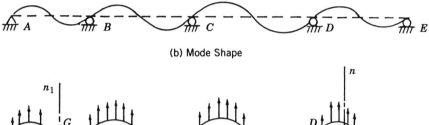

(b) Mode Shape

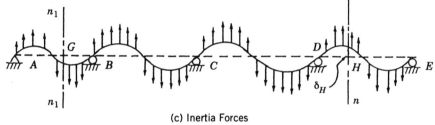

(c) Inertia Forces

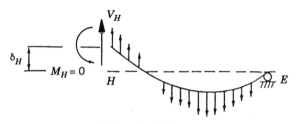

(d) Free Body Diagram

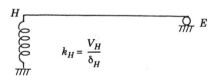

(e) Dynamically Equivalent System

FIGURE 4.12. (a) Initial complete system. (b) Arbitrarily selected mode shape of the beam. (c) Distribution of inertia forces corresponding to the arbitrarily selected mode shape. (d) Free-body diagram of a selected portion of the member. (e) Dynamically equivalent system.

the inertia forces of the mode considered, and, consequently, there will be no changes in the shape of the mode. Therefore, for the mode considered, it would be reasonable to assume that the end boundary conditions of the isolated portion HE can be satisfied as shown in Fig. 4.12e. The shear force V_H shown in Fig. 4.12d, which represents the influence of portion $ABCDH$ on portion HE, is replaced by a linear spring with a spring constant k_H that can be determined from the expression

$$k_H = \frac{V_H}{\delta_H} \qquad (4.226)$$

The system in Fig. 4.12e is termed the *dynamically equivalent system* for the considered mode shape, and point H is the *dynamic hinge point*. The free undamped frequency of vibration of the complete member in Fig. 4.12a, vibrating in the mode shown in Fig. 4.12c, is equal to the fundamental frequency of the dynamically equivalent system in Fig. 4.12e. The fundamental frequency of the dynamically equivalent system in Fig. 4.12e may be determined by using known methods of vibration analysis.

The end supporting conditions of the dynamically equivalent system, in the most general case, can be made to consist always of a simple support on the one end and a spring of constant k at the other end, or both ends to be elastically supported by springs. For example, if we consider a single-span beam fixed at both ends and vibrating in the fundamental mode of vibration, there will be only two points of zero moment along the length of the complete member, and, consequently, there will be only two dynamic hinges. The portion of the member between these two dynamic hinges constitutes a dynamically equivalent system.

It is important to note here that if there is a way to predict the shape of the mode of a structure that vibrates at a frequency ω, then this frequency can be determined directly by using a dynamically equivalent system, as discussed above to determine ω. By using this method, we find that there is no need to know what the values of the preceding frequencies are, as is usually required with other known methods of analysis, in order to determine the nth frequency of vibration.

The following sections of this chapter illustrate the application of the dynamic hinge concept to various types of simple problems by using exact dynamically equivalent systems as well as approximate solutions, which greatly simplify the solution of complicated problems. Note that the concept of the dynamic hinge is general and can be applied for the solution of many complicated structural and mechanical problems, such as beams, shafts, and frames.

4.11   DERIVATION AND SOLUTION OF EXACT DYNAMICALLY EQUIVALENT SYSTEMS

The concept and theory of the dynamic hinge will be used here to derive exact dynamically equivalent systems for simple beam problems. The purpose in this section is (a) to verify the concept and theory regarding the derivation of exact dynamically equivalent systems and (b) to provide the basis for the development of the theory for the derivation of an approximate dynamically equivalent system. The approximate dynamically equivalent system is the one suggested by the author to be used for a reasonable solution of complicated practical engineering problems.

Consider the uniform stiffness beam in Fig. 4.13 of constant mass m per unit length. The free undamped frequencies ω of the beam, in radians per second, and the corresponding mode shapes $Y(x)$ can be determined by solving the differential equation

$$\frac{d^4Y(x)}{dx^4} - \lambda^4 Y(x) = 0 \tag{4.227}$$

as discussed in Section 2.11. In fact, Eq. (4.227) is similar to Eq. (2.203) in Section 2.11. The expressions for the parameter λ in Eq. (4.227) and for the values of the free undamped frequencies ω are as follows:

$$\lambda^4 = \frac{m\omega^2}{EI} \tag{4.228}$$

$$\omega = \lambda^2 \sqrt{\frac{EI}{m}} \tag{4.229}$$

The solution of Eq. (4.227), as discussed in Section 2.11, is given by the expression

$$Y(x) = C_1 \cosh \lambda x + C_2 \sinh \lambda x + C_3 \cos \lambda x + C_4 \sin \lambda x \tag{4.230}$$

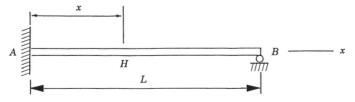

FIGURE 4.13. Beam of constant stiffness EI that is fixed at support A and simply supported at support B.

By applying the boundary conditions of zero deflection and zero rotation at $x = 0$ for the member in Fig. 4.13, we find

$$C_3 = -C_1 \tag{4.231}$$

$$C_4 = -C_2 \tag{4.232}$$

The boundary conditions of zero deflection and zero moment at $x = L$ yield

$$(\cosh \lambda L - \cos \lambda L)C_1 + (\sinh \lambda L - \sin \lambda L)C_2 = 0 \tag{4.233}$$

$$(\cosh \lambda L + \cos \lambda L)C_1 + (\sinh \lambda L + \sin \lambda L)C_2 = 0 \tag{4.234}$$

By setting equal to zero the determinant of the coefficients of C_1 and C_2 in Eqs. (4.233) and (4.234) and then expanding the determinant, we find the following frequency equation:

$$(\cosh \lambda L)(\sin \lambda L) - (\sinh \lambda L)(\cos \lambda L) = 0 \tag{4.235}$$

From Eq. (4.235), we find that the first three roots of λ have the values shown below:

$$\lambda_1 = \frac{3.9266}{L} \tag{4.236}$$

$$\lambda_2 = \frac{7.0686}{L} \tag{4.237}$$

$$\lambda_3 = \frac{10.2102}{L} \tag{4.238}$$

By using Eqs. (4.231) and (4.232) and also expressing C_2 in terms of C_1 by using Eq. (4.234), the solution given by Eq. (4.230) may be written as follows:

$$Y(x) = C_1\left[(\cosh \lambda x - \cos \lambda x) - \left(\frac{\cosh \lambda L + \cos \lambda L}{\sinh \lambda L + \sin \lambda L}\right)(\sinh \lambda x - \sin \lambda x)\right] \tag{4.239}$$

For each value of λ obtained from Eq. (4.235), we can determine the corresponding mode shape from Eq. (4.239) and the value of the free frequency ω from Eq. (4.229). The constant C_1 in Eq. (4.239) can be assumed equal to unity, since ω is independent of the actual amplitudes of the mode, as discussed in earlier sections.

4.11.1 Location of Points of Zero Moment for Each Mode

The points of zero moment for each mode can be determined from Eq. (4.239) by taking the second derivative of $Y(x)$ with respect to x, setting it equal to zero, and then solving for the values of x from the resulting equation. In other words, for each value of λ, it is required in Fig. 4.13 to determine point H where the bending moment is zero, so that portion HB is the one to be used for the derivation of the dynamically equivalent system. Although for higher modes of vibration we may have more than one point of zero moment, we are looking here for the one that is closest to support B, so that HB is the portion to be used to derive the dynamically equivalent system, since this portion would be the simplest one to use.

By using Eq. (4.239) and the values of λ_1, λ_2, and λ_3 given by Eqs. (4.236), (4.237), and (4.238), respectively, the points H of zero moment for the first three modes of vibration are determined, and they are located at the following distances x from the fixed support A in Fig. 4.13:

$$x_1 = 0.264216L \qquad \text{(fundamental mode)} \qquad (4.240)$$

$$x_2 = 0.553559L \qquad \text{(second mode)} \qquad (4.241)$$

$$x_3 = 0.692365L \qquad \text{(third mode)} \qquad (4.242)$$

4.11.2 Dynamically Equivalent Systems for the First Three Modes of Vibration

The derivation of the exact dynamically equivalent system for each mode, as discussed in Section 4.10, requires the computation of the shear force V_H and deflection δ_H at the dynamic hinge point H, in order to determine the spring constant k_H from Eq. (4.226). For the problem at hand, the deflection δ_H for each of the three modes of vibration may be obtained from Eq. (4.239) by using the values of λ given by Eqs. (4.236), (4.237), and (4.238) and the values of x given by Eqs. (4.240), (4.241), and (4.242).

The values of the shear force V_H at the dynamic hinge point H for each mode of vibration may be obtained from the following expression:

$$V = -EI\frac{d^3Y(x)}{dx^3}$$

$$= -C_1EI\lambda^3\left[\sinh \lambda x - \sin \lambda x - \left(\frac{\cosh \lambda L + \cos \lambda L}{\sinh \lambda L + \sin \lambda L}\right)(\cosh \lambda x + \cos \lambda x)\right]$$

$$(4.243)$$

For $\lambda = \lambda_1$ and $x = x_1$, Eq. (4.243) will yield the value of the shear force V_H for the first mode of vibration. In a similar manner, the values of V_H for the remaining modes of vibration may be obtained.

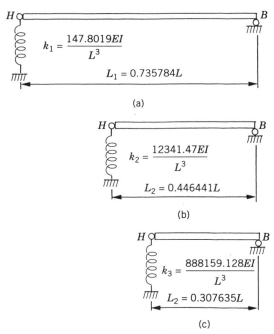

FIGURE 4.14. Dynamically equivalent systems for (a) the fundamental mode of vibration, (b) the second mode of vibration, and (c) the third mode of vibration.

The exact dynamically equivalent systems for the first three modes of vibration of the beam in Fig. 4.13 are shown in Figs. 4.14a, 4.14b, and 4.14c. The stiffness EI and mass m per unit length of the dynamically equivalent systems are the same as the corresponding ones of portion HB of the complete system in Fig. 4.13. The spring constants k_1, k_2, and k_3, in Figs. 4.14a, 4.14b, and 4.14c, respectively, are determined from Eq. (4.226), by using the appropriate values for V_H and δ_H in each mode, as explained above. The values of k_1, k_2, and k_3, for the fundamental, second, and third modes of vibration, are as follows:

$$k_1 = \frac{147.8019EI}{L^3} \qquad \text{(fundamental mode)} \qquad (4.244)$$

$$k_2 = \frac{12{,}341.47EI}{L^3} \qquad \text{(second mode)} \qquad (4.245)$$

$$k_3 = \frac{888{,}159.128EI}{L^3} \qquad \text{(third mode)} \qquad (4.246)$$

where L in the above three equations is the total length of the complete member in Fig. 4.13. The equivalent lengths L_1, L_2, and L_3, in Figs. 4.14a, 4.14b, and 4.14c, respectively, are as follows:

$$L_1 = L - x_1 = L - 0.264216L = 0.735784L \qquad (4.247)$$

$$L_2 = L - x_2 = L - 0.553559L = 0.446441L \qquad (4.248)$$

$$L_3 = L - x_3 = L - 0.692365L = 0.307635L \qquad (4.249)$$

The direct solution of the dynamically equivalent systems in Fig. 4.14 will yield, in each case, the free frequency of vibration of the complete system. For example, if we use the dynamically equivalent system in Fig. 4.14b and solve it directly to determine its fundamental frequency of vibration, this frequency will be equal to the second frequency of vibration of the complete beam in Fig. 4.13.

4.11.3 Solution of the Exact Dynamically Equivalent System

The dynamically equivalent systems shown in Fig. 4.14 may be thought of as having the general form shown in Fig. 4.15, where L_e denotes the equivalent length when the complete member vibrates at a specific mode of vibration, and k_e is the respective spring constant at the dynamic hinge point of the mode, which is used to construct the dynamically equivalent system. The fundamental frequency of the dynamically equivalent system in Fig. 4.15 is equal to the frequency of the complete system corresponding to the mode that is used to derive the dynamically equivalent system.

The differential equation of motion that can be used to solve the dynamically equivalent system is Eq. (4.227); consequently, the solution is represented by Eq.(4.230). The boundary conditions that can be used to solve the dynamically equivalent system in Fig. 4.15 are as follows:

$$\text{at} \quad x = 0 \qquad Y(0) = 0 \qquad (4.250)$$

$$\frac{d^2 Y(0)}{dx^2} = 0 \qquad (4.251)$$

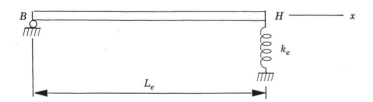

FIGURE 4.15. Dynamically equivalent system corresponding to a mode of vibration of the complete member.

$$\text{at } x = L_e \qquad \frac{d^2Y(L_e)}{dx^2} = 0 \tag{4.252}$$

$$EI\frac{d^3Y(L_e)}{dx^3} = k_e\delta \tag{4.253}$$

where δ is the deflection of the system in Fig. 4.15 at the elastic support point H of spring constant k_e.

By using Eq. (4.230) and applying the above four boundary conditions and then setting the determinant of the coefficients C_1, C_2, C_3, and C_4 equal to zero, the following frequency equation is obtained:

$$\coth \lambda L_e - \cot \lambda L_e = \frac{2k_e}{EI\lambda^2} \tag{4.254}$$

By using the smallest value of λ that satisfies Eq. (4.254) and substituting it into Eq. (4.229), the frequency ω obtained from this equation would be equal to the frequency of the mode of the complete system that is represented by the dynamically equivalent system in Fig. 4.15. For example, the fundamental mode of the complete member in Fig. 4.13 is represented by the dynamically equivalent system in Fig. 4.14a. In this case, we have $L_e = L_1 = 0.735784L$ and $k_e = k_1 = 147.8019EI/L^3$. By substituting these values into Eq. (4.254) and solving for the smallest value of λ, we obtain

$$\lambda = \lambda_1 = \frac{3.9266}{L} \tag{4.255}$$

which is identical to the value of λ given by Eq. (4.236) that is obtained by solving the complete member in Fig. 4.13.

The second and third mode frequencies of the member in Fig. 4.13 are represented by the dynamically equivalent systems in Figs. 4.14b and 4.14c, respectively. For the second mode frequency, we have $L_e = L_2 = 0.44644L$ and $k_e = k_2 = 12{,}341.47EI/L^3$. In this case, the smallest value of λ obtained from Eq. (4.254) is

$$\lambda = \lambda_2 = \frac{7.06856}{L} \tag{4.256}$$

which is identical to the one given by Eq. (4.237) of the complete member. In a similar manner, by using the dynamically equivalent system in Fig. 4.14c, we find that the smallest value of λ obtained from Eq. (4.254) is

$$\lambda = \lambda_3 = \frac{10.2102}{L} \tag{4.257}$$

This value is identical to the one given by Eq. (4.238), which is determined by using the complete member.

By using Eq. (4.239) and applying the boundary conditions given by Eqs. (4.250) and (4.251), we conclude that $C_1 = C_3 = 0$. By applying the boundary condition given by Eq. (4.253), we find

$$C_2 = C_4 \frac{\lambda^3 EI \cos \lambda L_e + k_e \sin \lambda L_e}{\lambda^3 EI \cosh \lambda L_e - k_e \sinh \lambda L_e} \tag{4.258}$$

Therefore, Eq. (4.239) yields

$$Y(x) = C_4 \left[\frac{\lambda^3 EI \cos \lambda L_e + k_e \sin \lambda L_e}{\lambda^3 EI \cosh \lambda L_e - k_e \sinh \lambda L_e} (\sinh \lambda x + \sin \lambda x) \right] \tag{4.259}$$

Equation (4.259) may be used to determine the fundamental mode shapes of the dynamically equivalent systems in Fig. 4.14. The constant C_4 in this equation may be assumed equal to unity.

By using Eq. (4.239) and the values of λ given by Eqs. (4.236), (4.237), and (4.238), the shapes of the fundamental, second, and third modes of the complete member in Fig. 4.13 are determined, and they are plotted in Figs. 4.16a, 4.16b, and 4.16c, respectively. By using Eq. (4.259) and the appropriate values of k_e and L_e, the mode shapes of the dynamically equivalent systems corresponding to fundamental, second, and third modes of vibration are determined, and they are shown in Figs. 4.17a, 4.17b, and 4.17c, respectively, on page 253. In Figs. 4.16 and 4.17, the letter H represents the dynamic hinge point that is used to construct the dynamically equivalent systems.

Comparison of the results in Figs. 4.16 and 4.17 reveals the following:

1. The mode shapes of portion HB in Fig. 4.16 and the mode shapes of portion HB of the corresponding dynamically equivalent systems in Fig. 4.17 are identical.

2. The dynamic hinge point H for the higher modes in Figs. 4.16b and 4.16c is located very near to a node point where vibration amplitude is zero.

3. The amplitude of vibration at the dynamic hinge point for the higher modes is very small compared to the amplitudes at other locations. For example, in Fig. 4.17b, the dynamic hinge amplitude is 0.02821 in. compared to the maximum amplitude of 0.98536 in. In Fig. 4.17c, the amplitude at the dynamic hinge is 0.00003 in., while the maximum

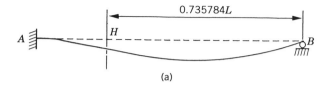

(a)

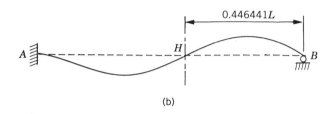

(b)

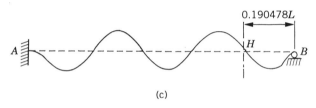

(c)

FIGURE 4.16. (a) Fundamental mode shape of the complete member. (b) Mode shape of the second mode of the complete member. (c) Mode shape of the third mode of the complete member.

amplitude is 0.99692 in. This indicates that the amplitude of the dynamic hinge point H approaches zero for higher modes of vibration.

The importance of the above observations becomes very clear in the next section, where approximate dynamically equivalent systems are derived.

4.12   DERIVATION OF APPROXIMATE DYNAMICALLY EQUIVALENT SYSTEMS

The analysis in the preceding two sections has shown that the exact frequency of a member vibrating at a mode of vibration may be determined by solving a simple dynamically equivalent system, such as the one shown in Fig. 4.12e or Fig. 4.15. The theory and concept suggest that a practical solution of this vibration problem may be obtained if the mode shapes of a member are

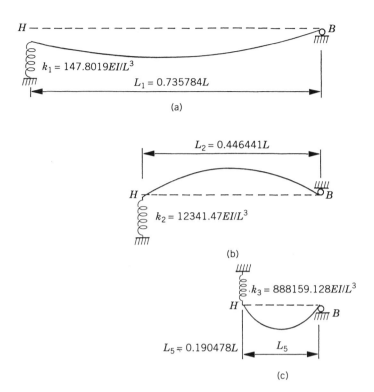

FIGURE 4.17. Fundamental mode shapes of the dynamically equivalent systems corresponding to (a) the fundamental mode of the complete member, (b) the second mode of the complete member, and (c) the fifth mode of the complete member.

known, or if they can be predicted with reasonable accuracy for practical applications. In such a case, if the predicted mode shapes are exact, the exact solution of the problem is obtained. If not, then the accuracy of the solution would depend on the accuracy of the predicted mode shapes. It should be pointed out here, however, that for every predicted mode shape there corresponds a dynamically equivalent system that can be used to determine directly the frequency of this mode without knowing the frequencies of the preceding modes, as is usually required for other approximate methods of vibration analysis.

A reasonable approach regarding the derivation of accurate dynamically equivalent systems would be the use of pseudostatic systems and the assumption that the elastic line of a pseudostatic system is everywhere proportional to the dynamic elastic line corresponding to the mode of vibration under consideration.

When a member vibrates at a mode of vibration, the loads participating in its vibrational motion are usually the dead weight of the member and possibly some additional dead weights that are rigidly attached to the member during motion. For example, a girder of a vibrating bridge would carry its own weight and some part of the bridge deck, since the bridge deck is often attached to the bridge girders by shear connectors. If we wish to calculate the free frequency of vibration of this mode, a pseudostatic system may be used for this purpose, which is constructed by assuming that all loads participating in the vibratory motion are acting in the direction of the dynamic elastic line of the considered mode. Such a system is termed *pseudostatic* and produces an elastic line, called the *pseudostatic elastic line*, which is closely proportional to the exact dynamic elastic line of the mode. This is tantamount to assuming that the inertia loads of the mode under consideration are the dead weight of the member and its rigidly attached dead weights, with their direction arranged to act in the direction of the shape of the mode. For example, if the mode amplitudes are directed upward, then the dead weights of the member are assumed to be directed upward, and vice versa.

Consider, for example, the continuous beam shown in Fig. 4.12a, and assume that the beam vibrates freely in the mode shown in Fig. 4.12b. The actual inertia loads are acting in the direction shown in Fig. 4.12c. If the node points are known (points of zero amplitude), a pseudostatic system corresponding to this mode may be constructed as shown in Fig. 4.18a. The weight w of the member per unit length is the only dead weight that participates in the vibratory motion of the member, and it is assumed to act in the direction of the elastic line of the mode in Fig. 4.12b, which is identical in direction to the inertia forces in Fig. 4.12c. The pseudostatic elastic line produced by the weight distributed as shown in Fig. 4.18a is shown in Fig. 4.18b. The pseudostatic elastic line in Fig. 4.18b, for practical purposes, would be a reasonable approximation of the actual mode shape in Fig. 4.12b. Therefore, point H of zero moment in Fig. 4.18b is the dynamic hinge point, and it can be determined by using static analysis to solve the static problem in Fig. 4.18a and to determine point H of zero moment. There are other points of zero moment along the length of the member in Fig. 4.18a, but we have selected the one closest to support E because it yields a simple, statically determinate, dynamically equivalent system.

The approximate dynamically equivalent system is shown in Fig. 4.18c. Equation (4.226) may be used here to determine the spring constant k_H at the dynamic hinge point H. The shear force V_H and deflection δ_H that are required for the computation of k_H are the shear force and deflection at point H in Fig. 4.18a, and they can be determined by static analysis. The computation of the fundamental frequency of the dynamically equivalent system in Fig. 4.18c may be carried out by using any convenient method of vibration analysis. This frequency would be in close agreement with the exact frequency of the mode considered. The solution of the dynamically equivalent system in Fig. 4.18c for its fundamental mode of vibration may easily be obtained by using the following well-known equation [1, 7]:

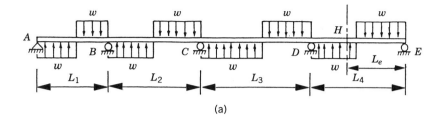

(a)

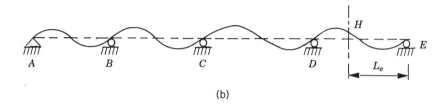

(b)

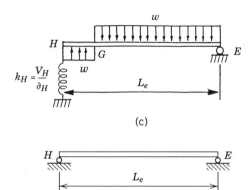

(c)

(d)

FIGURE 4.18. (a) Pseudostatic system. (b) Pseudostatic elastic line. (c) Approximate dynamically equivalent system. (d) Approximate dynamically equivalent system with both ends simply supported.

$$\omega^2 = g \frac{\sum\limits_{i=1}^{n} W_i y_i}{\sum\limits_{i=1}^{n} W_i y_i^2} \tag{4.260}$$

The weights W_i in the above equation represent lumped weights, which are obtained by lumping the weight w in Fig. 4.18c at discrete points along the length of the dynamically equivalent system. For example, the weight w between H and G can be lumped at the middle of the length HG with upward direction. The weight w between G and E could be lumped with three equal concentrated weights acting downward and located at the center of each $GE/3$ length. Usually, three to five lumped weights are sufficient for an accurate solution of the problem. The amplitudes y_i in Eq. (4.260) are the static deflections, which are produced by the weights W_i. They can be determined by using static analysis for deflection of beams, which can be found in standard texts dealing with the mechanics of solids.

Use of the dynamic hinge concept is particularly convenient for beams consisting of many spans. The location of the dynamic hinge point H, as well as the computation of the shear force V_H and deflection δ_H, can be performed accurately by using only two to three spans neighboring the dynamic hinge point H. It is usually unnecessary to use the complete system for this purpose. For example, the dynamic hinge point H, as well as V_H and δ_H in Fig. 4.18c, could be determined accurately by using only spans BC, CD, and DE in Fig. 4.18a. In fact, for many practical situations, just considering spans CD and DE would yield good results. This is reasonable, because the effect of the load of span AB on point H is extremely small, and it could be neglected. Also, the effect of the load of span BC on point H is often small, and it could be neglected.

Another important point to note here is that for the higher modes of vibration the dynamic hinge point gets very close to a node point, and, consequently, k_H becomes increasingly large for higher modes of vibration. See, for example, the exact values of k in Fig. 4.14. For the first mode, we have $k_1 = 147.8019EI/L^3$, while for the third mode, we have $k_3 = 888{,}159.128EI/L^3 = 6009.12k_1$. In such cases, it would be very reasonable and quite appropriate to assume that the spring k_H has infinite rigidity. On this basis, the dynamically equivalent system for the higher modes of vibration would be a beam of length L_e and simply supported on rigid supports as shown in Fig. 4.18d. Since H is located very close to the node point for higher modes of vibration, we may also assume that the node point and the dynamic hinge point H coincide. In this case, the fundamental frequency of vibration of the dynamically equivalent system in Fig. 4.18d may be determined from the following simple equation:

$$\omega = \frac{\pi^2}{L_e^2} \sqrt{\frac{EI}{m}} \tag{4.261}$$

which is the frequency equation, in radians per second, for the fundamental frequency ω of a simply supported beam. The following examples illustrate the application of the theory.

Example 4.7. Determine the free frequencies of vibration of the uniform

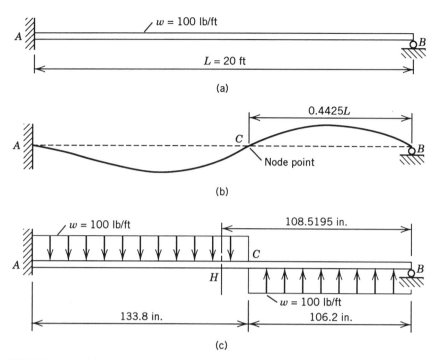

FIGURE 4.19. (a) Original single-span beam. (b) Shape of the second mode of the member. (c) Psuedostatic system corresponding to the second mode of vibration.

beam in Fig. 4.19a, by using approximate dynamically equivalent systems (a) with one end simply supported and the other end elastically supported by spring k_H, and (b) with both ends simply supported on rigid supports. The weight w per unit length of the member is 100 lb/ft. Compare the results with the exact solutions obtained by using the complete member.

Solution. Assume that the member in Fig. 4.19a is vibrating in its second mode of free vibration, as shown in Fig. 4.19b. The node point C is located at a distance of $0.4425L$ from the end support B. The pseudostatic system corresponding to the second mode may be obtained by assuming that the weight w per unit length of the member is applied as a load on the member and in the direction of the second mode, as shown in Fig. 4.19c. Between A and C, the direction of w is downward because the amplitudes of the second mode are directed downward as shown in Fig. 4.19b. The weight w is acting upward between C and B because the vibration amplitudes are directed upward in this range.

The dynamic hinge point H is determined by using the pseudostatic system in Fig. 4.19c and determining the point of zero moment by applying methods from the mechanics of solids. It is found that point H is located at a distance of 108.5195 in. to the left of the end support B. At the dynamic hinge point, the shear force V_H and deflection δ_H are determined by using the pseudostatic system, and they are as follows:

$$V_H = 0.20696wL \tag{4.262a}$$

$$\delta_H = (8.3392)\frac{wL^4}{10^5 EI} \tag{4.262b}$$

where w and L have the values shown in Fig. 4.19a, and $EI = 150 \times 10^6$ lb $\cdot$ in^2. By using Eq. (4.226), we find

$$k_H = \frac{V_H}{\delta_H} = 26{,}928.795 \text{ lb/in.} \tag{4.263}$$

The dynamically equivalent system that corresponds to the second mode of the free vibration of the member is shown in Fig. 4.20a. We note in this figure that the node point C is located close to the dynamic hinge point at a distance of 2.32 in. The distributed weight w of the dynamically equivalent system in Fig. 4.20a is lumped at discrete points along its length, as shown in Fig. 4.20b. The lumped weight W_1 represents the total weight between points H and C, which is lumped at the center of this portion—point 1 in Fig. 4.20b. The lumped weights W_2 are obtained by dividing the length CB into four segments and lumping the total weight of each segment at the center of the segment.

The total static deflections y_1, y_2, y_3, y_4, and y_5 under the load concentration points 1, 2, 3, 4, and 5, respectively, are computed by using well-known methods from the mechanics of solids, and they are as follows:

$$y_1 = 0.018722 \text{ in.}$$

$$y_2 = 0.058642 \text{ in.}$$

$$y_3 = 0.105895 \text{ in.}$$

$$y_4 = 0.100984 \text{ in.}$$

$$y_5 = 0.041964 \text{ in.}$$

By using Eq. (4.260), the fundamental frequency of the approximate dynamically equivalent system in Fig. 4.20b is

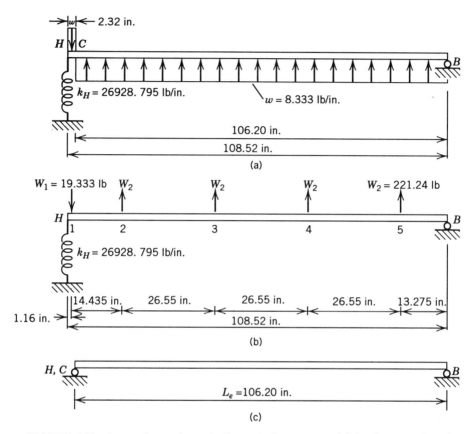

FIGURE 4.20. Approximate dynamically equivalent systems (a) for the second mode, (b) with lumped weights, and (c) with both end supports simply supported.

$$\omega^2 = \frac{\sum\limits_{i=1}^{5} W_i\, y_i}{\sum\limits_{i=1}^{5} W_i\, y_i^2} = 4478.666459$$

or

$$\omega = 66.92 \text{ rps}$$

The exact frequency of the second mode of vibration of the complete system in Fig. 4.19a is 72.31 rps. Therefore, the frequency obtained by using the approximate dynamically equivalent system in Fig. 4.20b is 7.45% lower compared to the exact value of this frequency. However, most of the error here is attributed to mathematical manipulations and rounding errors in the computation of V_H, δ_H, and, consequently, k_H. Since δ_H is very close to

TABLE 4.3. Exact and Approximate Values of the Frequencies of the First, Second, Fifth, and Seventh Modes of Vibration Obtained by Using Approximate Dynamically Equivalent Systems

Mode	Approximate Dynamically Equivalent System ω (rps)	Exact Value ω (rps)	Error (%)
1	21.80	22.30	2.24
2	69.86	72.31	3.39
5	396.29	393.87	0.60
7	752.81	751.12	0.20

the node point C, its value is very small in the higher modes of vibration, and, consequently, small errors during the computation of δ_H influence appreciably the final result.

A more accurate solution for the computation of the second mode frequency is to assume that points H and C in Fig. 4.20a coincide. On this basis, since the deflection δ_H becomes equal to δ_C and, consequently, equal to zero, then k_H is infinitely rigid. A very accurate dynamically equivalent system in this case would be the one shown in Fig. 4.20c, which is a simply supported beam of length $L_e = 106.20$ in. The fundamental frequency of the dynamically equivalent system in Fig. 4.20c may be obtained exactly by using Eq. (4.261); that is,

$$\omega = \frac{\pi^2}{L_e^2} \sqrt{\frac{EI}{m}} = \frac{\pi^2}{(106.20)^2} \sqrt{\frac{(150 \times 10^6)(386 \times 12)}{100}}$$

$$= 72.94 \text{ rps}$$

which is 0.87% higher than the exact value of 72.31 rps. If the length L_e is taken to be equal to the length HB in Fig. 4.20a, that is, $L_e = 108.52$ in., we find from Eq. (4.261) that $\omega = 69.86$ rps, which is 3.39% lower than the exact value.

Table 4.3 compares the values of the frequencies obtained for the first, second, fifth, and seventh modes of vibration by using approximate dynamically equivalent systems. For the first mode, the model shown in Fig. 4.15 is used by locating the dynamic hinge point H and calculating k_H. For the second, fifth, and seventh modes, the model shown in Fig. 4.18d is used with length L_e equal to the actual distance between hinge point H and support E. This table shows that the error is becoming insignificant for the higher modes, because for higher modes the dynamic hinge point H tends to coincide with the position of node point C. Therefore, for the higher modes of vibration, it is satisfactory for practical purposes to use the approximate

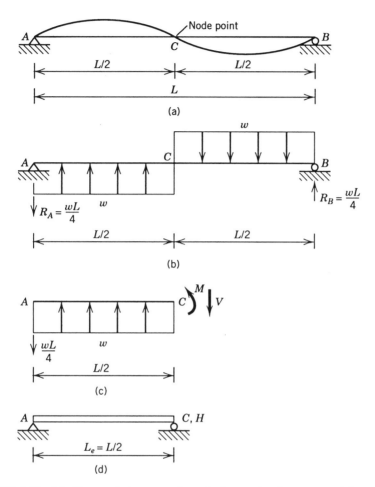

FIGURE 4.21. (a) Shape of the second mode for a simply supported beam. (b) Pseudostatic system corresponding to the second mode. (c) Free-body diagram. (d) Dynamically equivalent system for the second mode.

dynamically equivalent system shown in Fig. 4.20c, where points H and C coincide. Further discussion on this subject is included later in the text.

Example 4.8. Determine the higher free frequencies of vibration of the simply supported beam in Fig. 4.21a by using approximate dynamically equivalent systems. Compare the results with the ones obtained by using the complete member.

Solution. When the member vibrates in the second mode of vibration, the shape of this mode would be as shown in Fig. 2.14c. The shape of this mode is again drawn as shown in Fig. 4.21a, where C in this figure is a node point. The weight of the member per unit length is w, and its length is L. The pseudostatic system for this mode is obtained by assuming that the weight w is acting in the direction shown in Fig. 4.21b. By taking moments about B in Fig. 4.21b and setting the sum of these moments equal to zero, we find that $R_A = wL/4$. By considering the free-body diagram in Fig. 4.21c and setting the sum of the moments about C equal to zero, we find

$$\Sigma M_C = 0 = -\frac{wL}{4} \cdot \frac{L}{2} + \frac{wL}{2} \cdot \frac{L}{4} + M_C$$

or

$$M_C = 0$$

In this case, we find that the point of zero moment H and the node point C coincide, and the dynamically equivalent system is as shown in Fig. 4.21d. The length L_e of this system is equal to $L/2$, and its fundamental frequency of vibration may be determined by using Eq. (4.261); that is,

$$\omega = \frac{\pi^2}{L_e^2} \sqrt{\frac{EI}{m}} = \frac{\pi^2}{(L/2)^2} \sqrt{\frac{EI}{m}}$$

$$= \frac{4\pi^2}{L^2} \sqrt{\frac{EI}{m}}$$

This value is equal to the value given by Eq. (2.231) when $n = 2$. Therefore, the dynamically equivalent system in this case yields the exact value of the second mode of vibration.

If the third mode shape in Fig. 2.14d is considered and the above procedure is followed, we find that node points C and D are also points of zero moment, and any one of portions AC, CD, and DB may be used to determine the exact frequency of the third mode. In this case, $L_e = L/3$. Note that a pattern is being established here, indicating that the length L_e of the dynamically equivalent system may be expressed as

$$L_e = \frac{L}{n} \qquad n = 1, 2, 3, \ldots \tag{4.264}$$

where n indicates the mode number. Thus, by substituting Eq. (4.264) into Eq. (4.261), we obtain

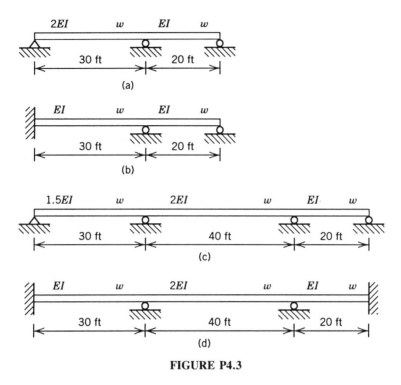

FIGURE P4.3

$$\omega_n = \frac{\pi^2}{(L/n)^2}\sqrt{\frac{EI}{m}} = \frac{n^2\pi^2}{L^2}\sqrt{\frac{EI}{m}} \tag{4.265}$$

which is identical to the exact solution given by Eq. (2.231).

PROBLEMS

4.1 Solve the problem of Example 4.1 by assuming that the three-span continuous beam in Fig. 4.2a is fixed at the left end support. Compare the results.

4.2 Resolve the problem in Example 4.1 by assuming that the stiffness $(EI)_2$ of the second span is twice the stiffness of the first span. Compare the results.

4.3 For the continuous beams shown in Fig. P4.3, determine the first three free frequencies of vibration by using the three-moment equation. The stiffness $EI = 45 \times 10^6 \, \text{kip} \cdot \text{in.}^2$, and the weight w per unit length of the beam is 0.8 kips/ft.

4.4 For the continuous beam problem in Example 4.1, determine and plot the mode shapes corresponding to the first three frequencies of its vibration.

4.5 Repeat Problem 4.3 in order to determine the mode shapes corresponding to the first three frequencies of vibration. Plot these mode shapes.

4.6 A steel wire of length $L = 20$ ft is stretched by a force $T = 500$ lb. The wire is held fixed at the ends, and the diameter d of its circular cross section is 0.25 in. The material weighs 0.284 lb/in.^{3}. Determine the first three free frequencies of its transverse vibration and the corresponding mode shapes.

4.7 A wire is stretched between elastic supports by a force T, and the wire is permitted to vibrate freely in the transverse direction. The vertical springs at the end supports have the same spring constant k. By assuming that the supports are massless, determine the free frequencies of vibration of the wire and the corresponding mode shapes.

4.8 Repeat Problem 4.7 by assuming that the left support is fixed and compare the results.

4.9 Repeat Problem 4.7 by assuming that the left support is hinged and compare the results.

4.10 Repeat Problem 4.7 by assuming that the ends are hinged and compare the results.

4.11 Determine the velocity of the transverse waves of the steel wire in Problem 4.6.

4.12 For a rectangular membrane of sides $a = 30$ in. and $b = 60$ in., determine the first four frequencies of its free transverse vibration. The boundary conditions are the same as the ones given in Example 4.3, and the weight of the membrane per unit area is 0.03 lb/in.^{2}.

4.13 Repeat Problem 4.12 for $a = 30$ in. and $b = 120$ in., and for $a = 30$ in. and $b = 240$ in. Compare the results.

4.14 Consider a rectangular steel plate of sides $a = 90$ in. and $b = 60$ in. and of thickness $h = 2.0$ in. The modulus of elasticity $E = 30 \times 10^6$ psi. The Poisson ratio $v = 0.25$, and the mass m per square inch of plate area is 0.00146 lb · s^2/in. If the ends of the plate are simply supported, determine the first three frequencies of its transverse vibration. Also, write the expressions for the corresponding mode shapes.

4.15 Repeat Problem 4.14 for sides $a = 120$ in. and $b = 60$ in. Compare the results.

4.16 Repeat Problem 4.14 for sides $a = b = 90$ in. Compare the results.

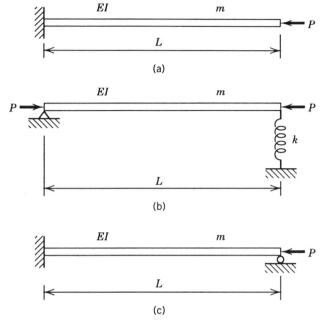

FIGURE P4.17

4.17 Determine the expressions for the flexural free frequencies of vibration and the corresponding mode shapes of the uniform beams supported as shown in Fig. P4.17. The stiffness EI and mass m are constant throughout the length of the members.

4.18 Determine the fundamental frequency of vibration for the uniform beams in Fig. P4.17. Assume that $P/P_{cr} = 0.3$, $EI = 45 \times 10^6$ psi, $L = 35$ ft, and $m = 1.5$ lb · s²/in.

4.19 Repeat Problem 4.18 for $P/P_{cr} = 0, 0.1, 0.2, 0.4, 0.6, 0.8$, and 1.0 and plot the results. Make appropriate comments regarding the results.

4.20 Repeat Problem 4.18 by assuming that P is a tensile axial load.

4.21 Repeat Problem 4.19 by assuming that P is a tensile axial load.

4.22 Derive Eq. (4.139) by using appropriate free-body diagrams.

4.23 Apply the boundary conditions given by Eqs. (4.146) through (4.149) and derive Eq. (4.150).

4.24 Repeat Problem 4.23 and derive an equation similar in format to Eq. (4.150) by using appropriate boundary conditions in terms of vertical deflection and rotation at the ends of the member.

4.25   For the beam problem in Example 4.5, determine the first two frequencies of vibration ω_1 and ω_2 when $k_v = 100$ lb/in. and $P = 3350$ lb. Discuss the results.

4.26   Repeat Problem 4.25 when $k_v = 100$ lb/in. and $P = 0$, 1000, and 2500 lb. Discuss the results.

4.27   Repeat Problem 4.25 when $k_v = 1000$ lb/in. and $P = 0$, 15,000, 25,000, and 33,501 lb. Discuss the results.

4.28   Repeat Problem 4.25 when $k_v = 5000$ lb/in. and $P = 0$, 10,000, 30,000, 50,000, 70,000, and 95,564 lb. Discuss the results.

4.29   For the uniform steel beams loaded as shown in Fig. P4.29, determine in each case the expressions for the deflection and rotation along the length of the dynamically loaded members, as well as the maximum deflection. The weight w per unit length of each beam is 0.8 kip/ft, $E = 30 \times 10^6$ psi, moment of inertia $I = 1500$ in.^{4}, $F = 30$ kips, and $\omega_f = 30$ rps.

4.30   Repeat Problem 4.29 in order to determine the maximum bending stress in each member, which is caused by the application of the harmonic loading. Neglect the stresses due to the weight of the member.

4.31   Repeat Problem 4.30 by varying the magnitude of the dynamic harmonic loads while retaining all other parameters the same. Plot the results as stress versus dynamic load.

4.32   For a uniform bar that is fixed at both ends, determine exact dynamically equivalent systems for the first, second, and third modes of vibration by using the dynamic hinge concept.

4.33   For a uniform simply supported beam, determine an exact dynamically equivalent system for the tenth mode of vibration using the dynamic hinge concept.

4.34   By using the exact dynamically equivalent system for the first mode of vibration obtained in Problem 4.32, solve it exactly in order to obtain the fundamental free frequency of vibration. Use the procedure discussed in Section 4.11.

4.35   Repeat Problem 4.34 in order to obtain the second frequency of the free vibration of the fixed–fixed bar.

4.36   Repeat Problem 4.34 in order to obtain the third frequency of the free vibration of the fixed–fixed bar.

4.37   By using the exact dynamically equivalent system for the tenth mode of a simply supported beam obtained in Problem 4.33, solve it exactly to obtain directly the tenth free frequency of vibration of the bar.

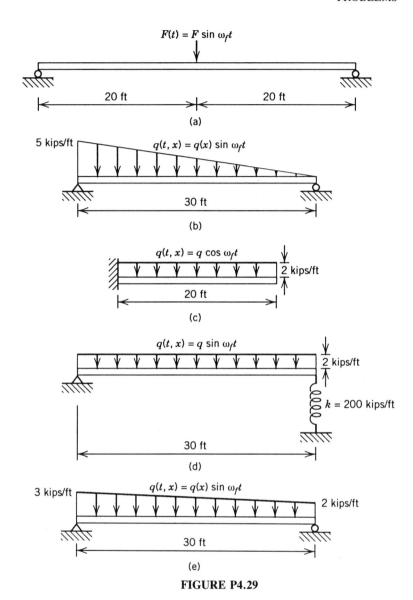

FIGURE P4.29

4.38 For a uniform beam that is fixed at both ends and vibrating in its second mode of vibration, determine the location of the node point and the location of the dynamic hinge point.

4.39 For a uniform simply supported beam vibrating in its fifth mode of vibration, determine the locations of all node points and all dynamic hinge points.

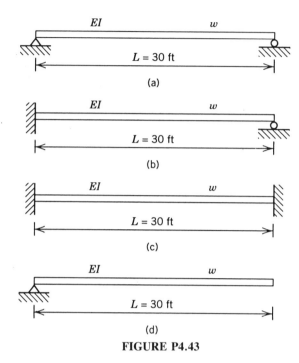

FIGURE P4.43

4.40 For a uniform simply supported beam vibrating in its 15th mode of vibration, determine the locations of all node points and all dynamic hinge points.

4.41 For a uniform beam that is fixed at both ends and vibrating in its fifth mode of vibration, determine the locations of all node points and all dynamic hinge points.

4.42 For a uniform bar with both ends free and vibrating in its third mode of vibration, determine the locations of all node points and all dynamic hinge points.

4.43 For the uniform beam cases supported as shown in Fig. P4.43, determine the approximate locations of all dynamic hinge points for the first five modes of their free vibration by following the procedure discussed in Section 4.12. The stiffness EI is constant and equal to 30×10^6 kips · in.^{2}, and the weight w per unit length of the member is 0.6 kip/in.

4.44 Solve Problem 4.43 by using an exact solution and compare the results.

4.45 For the uniform beam cases of Problem 4.43, determine the locations of all node points by using exact solutions and indicate their relative

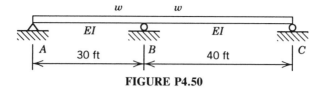

FIGURE P4.50

locations with respect to the dynamic hinge points obtained in Problem 4.43.

4.46 By using the results obtained in Problem 4.43 and following the procedure discussed in Section 4.12, determine for each case the approximate dynamically equivalent systems corresponding to the first five modes of vibration of the member.

4.47 Repeat Problem 4.46 by assuming that the locations of the dynamic hinge points coincide with the locations of the node points.

4.48 By using the approximate dynamically equivalent systems obtained in Problem 4.46 and applying Eq. (4.260), determine in each case the first five frequencies of the free vibration of the beam.

4.49 By using the approximate dynamically equivalent systems obtained in Problem 4.47, determine in each case the first five frequencies of the free vibration of the member. Compare with the results obtained in Problem 4.48.

4.50 For the two-span continuous beam shown in Fig. P4.50, determine the fundamental frequency of its free undamped vibration by using an approximate dynamically equivalent system as discussed in Section 4.12. The uniform stiffness EI of the member is 150×10^6 kips·in.^{2}, and the uniform weight w per unit length is 1.0 kip/ft.

4.51 Repeat Problem 4.3 by using appropriate approximate dynamically equivalent systems and compare the results.

5 Commonly Used Methods of Vibration Analysis

5.1 INTRODUCTION

In this chapter, some of the most popular methods of vibration analysis are included. It is extremely important to comprehend as many methods of vibration as possible, because each method has advantages and disadvantages, and you want to use it where it serves you best. In the preceding chapters, exact rigorous solutions have been obtained for many problems, and many of these solutions can be generalized and put into convenient equations that can be used repeatedly for the solution of many everyday practical situations. There are situations, however, where such solutions are extremely difficult to obtain for the more complicated theoretical or practical situations. For such cases, approximate methods of analysis have been developed, which provide reasonable answers for practical applications. We have to remember here that the purpose of all these methods and methodologies is to provide a way for the practicing engineer to design a structure or mechanical component in a safe and functional manner.

The approximate methods of analysis can represent mathematical models based on the simplification of exact rigorous solutions, or they can be numerical solutions of appropriately derived differential equations of motion. Since, however, all such methods are approximate, extreme caution should be taken in their application to physical situations, and the results should be examined carefully to make sure that the physical situation is appropriately represented in the results. This is usually accomplished by trying to gain as much knowledge as possible about the physics of the problem to be solved.

In this chapter we discuss the method developed by Lord Rayleigh, Stodola's method and the associated iteration procedure, Myklestad's method, utilization of transfer matrices, and the concept of the dynamic hinge. These methods are all approximate, and both advantages and disadvantages will be pointed out during the discussion. The very popular finite element method is treated in Chapter 6, and subsequent chapters are involved with specialized methods of vibration analysis for the solution of complicated vibration and dynamics problems.

5.2 RAYLEIGH'S ENERGY METHOD

An approximate method for the computation of the fundamental frequency of vibration of structural and mechanical problems was developed and published by Lord Rayleigh [26] in 1877. This method is readily applicable to simpler types of structures, and it has been used extensively since its inception in 1877. The required equations for the computation of the free frequencies of vibrating elastic structures are derived by equating their total maximum potential energy to their total maximum kinetic energy at the extreme deflection configurations of the structure.

The application of the method, as it was initially derived, was somewhat limited to the determination of the fundamental frequency of free undamped vibration of relatively simple problems, but with modern developments in science and technology, it was extended to the computation of higher frequencies of vibration of even more complicated vibration problems.

To derive the required frequency equations, we consider the uniform single-span beam in Fig. 5.1a. The fundamental frequency mode shape is shown by the dashed lines in the same figure. The free vibration of the member, as stated in earlier chapters, is considered to be harmonic. Therefore, if we consider an element of the beam at a distance x from the left support, this element will vibrate transversely in a harmonic manner. The weight per unit length of the member is w, the mass dm of the element is

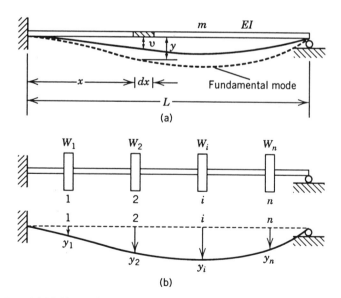

FIGURE 5.1. (a) Uniform single-span beam vibrating harmonically in its fundamental mode of vibration. (b) Uniform single-span beam supporting the concentrated attached weights $W_1, W_2, \ldots, W_i, \ldots, W_n$.

$w\,dx/g$, where g is the acceleration of gravity, and ω is the fundamental frequency of vibration.

Since the member vibrates harmonically, the dynamic displacement v at any time t (see Fig. 5.1a) may be represented by either a sine or a cosine variation; that is,

$$v(t, x) = y(x) \cos \omega t \tag{5.1}$$

or

$$v(t, x) = y(x) \sin \omega t \tag{5.2}$$

where $y(x)$ in the above two expressions is the maximum amplitude of the element and, consequently, the amplitude of the element at the fundamental mode configuration.

In accordance with Eq. (5.1), or Eq (5.2), the maximum velocity $\dot{v}_{\max}$ of the element is ωy, and, consequently, its maximum kinetic energy dT is

$$dT = \tfrac{1}{2}(dm)\dot{v}_{\max}^2$$

$$= \frac{1}{2}\left(\frac{w\,dx}{g}\right)(\omega y)^2 \tag{5.3}$$

$$= \frac{w}{2g}\,\omega^2 y^2\,dx$$

If we consider all the elements of the member, the maximum kinetic energy T of the whole member is as follows:

$$T = \frac{w}{2g}\,\omega^2 \int_0^L y^2\,dx \tag{5.4}$$

The maximum potential energy U of the member is equal to its flexural elastic strain energy in one of its extreme amplitude configurations. For the fundamental mode in Fig. 5.1a, such an extreme configuration occurs when $v = y$. Thus, if the amplitude y and rotation θ are considered to be small, we can write the following expressions:

$$d\theta = \frac{dx}{\rho} = \frac{d^2 y}{dx^2}\,dx \tag{5.5}$$

$$M = \frac{EI}{\rho} = -EI\,\frac{d^2 y}{dx^2} \tag{5.6}$$

where ρ is the radius of curvature and M is the bending moment. Thus, the maximum potential energy dU of the element is

$$dU = \frac{m\,d\theta}{2}$$

$$= \frac{EI}{2}\left(\frac{d^2y}{dx^2}\right)^2 dx \tag{5.7}$$

If we consider all the elements of the member, the following expression for the maximum potential energy of the member is obtained:

$$U = \frac{EI}{2}\int_0^L \left(\frac{d^2y}{dx^2}\right)^2 dx \tag{5.8}$$

Note that the kinetic energy is zero when the potential energy is maximum and vice versa.

In accordance with Rayleigh's method, the frequency equation for the fundamental mode of vibration of the uniform beam may be obtained by equating the expressions given by Eqs. (5.4) and (5.8). By doing so, we obtain

$$\omega^2 = \frac{EIg}{w}\frac{\int_0^L (d^2y/dx^2)^2\,dx}{\int_0^L y^2\,dx} \tag{5.9}$$

Equation (5.9) may be used to determine the fundamental frequency of the free vibration of the member, provided that the expression $y(x)$ of the fundamental mode, as a function of x, is known or can be determined.

If the stiffness EI and weight w are both variable along the length of the member, then EI in the nominator and w in the denominator should be kept within their respective integrals; that is,

$$\omega^2 = g\frac{\int_0^L EI\,(d^2y/dx^2)^2\,dx}{\int_0^L wy^2\,dx} \tag{5.10}$$

The application of Eq. (5.9), or Eq. (5.10), to practical engineering problems requires that $y(x)$ of the fundamental mode be known in order to perform the required integrations. However, such an expression is not usually known, and the first step in applying the method would be to assume a reasonable shape for $y(x)$. If the correct $y(x)$ is assumed, or a $y(x)$ that is everywhere proportional to the true shape, then the exact frequency will be obtained. If the assumed shape for $y(x)$ is not the exact one, the resulting frequency will be higher than the exact value, because an approximate $y(x)$

implies the existence of additional restraints that cause the system to vibrate at a higher frequency. For many structural problems, a reasonable shape to choose for $y(x)$ is the static deflection produced by the dead weight of the structure.

The following examples illustrate the application of the method.

Example 5.1. Determine the free fundamental frequency of vibration of a uniform simply supported beam by assuming that $y(x)$ is the static deflection curve produced by the uniformly distributed weight w of the member. The stiffness EI is constant.

Solution. The deflection $y(x)$ due to the uniformly distributed weight w of the beam is given by the expression

$$y(x) = \frac{w}{2EI}(x^4 - 2Lx^3 + L^3x)$$
$$= C(x^4 - 2Lx^3 + L^3x)$$

$$(5.11)$$

where

$$C = \frac{w}{2EI}$$

$$(5.12)$$

On this basis, we have

$$\int_0^L \left(\frac{d^2y}{dx^2}\right)^2 dx = 144C^2 \int_0^L (x^4 - 2Lx^3 + L^2x^2) \, dx$$
$$= \frac{144C^2L^5}{30}$$

$$(5.13)$$

$$\int_0^L y^2 \, dx = C^2 \int_0^L (x^4 - 2Lx^3 + L^3x)^2 \, dx$$
$$= \frac{31C^2L^9}{630}$$

$$(5.14)$$

By substituting Eqs. (5.13) and (5.14) into Eq. (5.9) and taking into consideration that the mass m per unit length of the member is equal to w/g, we find

$$\omega^2 = \frac{97.7}{L^4} \cdot \frac{EI}{m}$$

or

$$\omega = \frac{9.88}{L^2} \sqrt{\frac{EI}{m}} \tag{5.15}$$

The exact solution is

$$\omega = \frac{\pi^2}{L^2} \sqrt{\frac{EI}{m}} = \frac{9.87}{L^2} \sqrt{\frac{EI}{m}} \tag{5.16}$$

For this problem, Rayleigh's method yields very accurate results. Practically, the exact fundamental frequency is obtained. Therefore, in this case, the static deflection represents very accurately the shape of the fundamental mode of vibration.

Example 5.2. Determine the free fundamental frequency of vibration of the uniform two-span continuous beam shown in Fig. 5.2a by using Rayleigh's energy method and the static deflection curve due to the weight w of the beam.

Solution. Equations (5.9) and (5.10) may also be applied to continuous beams of various boundary conditions. For such types of problems, however, special consideration must be given to the selection of the deflection $y(x)$ to represent the shape of the fundamental mode, as discussed at the end of this problem. The solution here is obtained by assuming that $y(x)$ is the shape produced by the static weight w, applied as shown in Fig. 5.2a.

To determine $y(x)$, we apply first the three-moment equation to determine the internal moment M_B shown in Fig. 5.2b. On this basis, the value of M_B is found to be equal to 162.5 kip · ft. The expression for the three-moment equation for static analysis can be found in many elementary texts [27] dealing with the mechanics of solids. From the free-body diagrams in Fig. 5.2b, by applying statics, we find

$$R_A = 9.58 \text{ kips}$$
$$R_B = 44.48 \text{ kips}$$
$$R_C = 15.94 \text{ kips}$$

By using the free-body diagrams in Fig. 5.2b and applying the double integration method, we find that the deflection curves $y_1(x)$ and $y_2(x)$ for spans L_1 and L_2, respectively, are as follows:

$$EIy_1(x) = -R_A \frac{x^3}{6} + \frac{wx^4}{24} + R_A \frac{L_1^2 x}{6} - \frac{wL_1^3 x}{24} \tag{5.17}$$

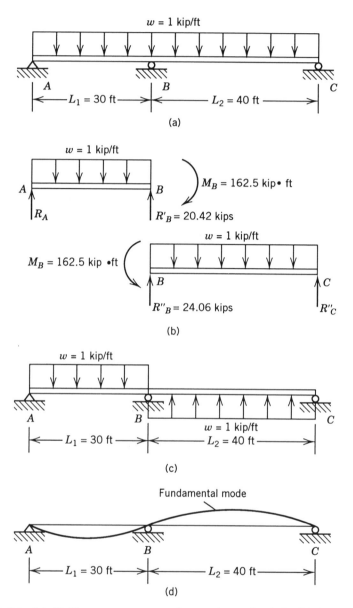

FIGURE 5.2. (a) Uniform two-span continuous beam loaded with its own dead weight w. (b) Free-body diagram for each span of the continuous beam. (c) Pseudo-static system with the weight w assumed to act in the direction of the fundamental mode. (d) Shape of the fundamental mode.

0

$$EIy_2(x) = M_B \frac{x^2}{2} - R''_B \frac{x^3}{6} + \frac{wx^4}{24} - M_B \frac{L_2x}{2} + R''_B \frac{L_2^2x}{6} - \frac{wL_2^3x}{24} \quad (5.18)$$

where $R''_B = 24.06$ kips. Note that $R_B = R'_B + R''_B$.

From Eq. (5.8), the maximum potential energy U of the continuous beam is

$$
\begin{aligned}
U &= \frac{EI}{2} \int_0^L \left(\frac{d^2y}{dx^2}\right)^2 dx \\
&= \frac{1}{2EI}\left[\int_0^{L_1}\left(-R_Ax + \frac{wx^2}{2}\right)^2 dx + \int_0^{L_2}\left(M_B - R''_Bx + \frac{wx^2}{2}\right)^2 dx\right] \\
&= \frac{1}{2EI}\left[R_A^2 \frac{L_1^3}{3} - R_A \frac{wL_1^4}{4} + \frac{w^2L_1^5}{20} + M_B^2L_2 - M_BR''_BL_2^2 \right. \\
&\qquad\qquad \left. + M_B \frac{wL_2^3}{3} + R''^2_B \frac{L_2^3}{3} - R''_B \frac{wL_2^4}{4} + \frac{w^2L_2^5}{20}\right]
\end{aligned}
$$
$$(5.19)$$

By substituting the values of w, L_1, L_2, R_A, R''_B, and M_B in Eq. (5.19), we find

$$U = \frac{371.52(10^6)}{EI} \quad (5.20)$$

By using Eq. (5.4), the kinetic energy T of the continuous beam is

$$
\begin{aligned}
T &= \frac{\omega^2}{2g} \int_0^L wy^2\, dx \\
&= \frac{\omega^2}{2gEI}\left[\int_0^{L_1}\left(-R_A \frac{wx^3}{6} + \frac{w^2x^4}{24} + R_A \frac{wL_1^2x}{6} - \frac{w^2L_1^3x}{24}\right)^2 dx \right. \\
&\qquad + \int_0^{L_2}\left(M_B \frac{wx^2}{2} - R''_B \frac{wx^3}{6} + \frac{w^2x^4}{24} - M_B \frac{wL_2x}{2}\right. \\
&\qquad\qquad \left.\left. + R''_B \frac{wL_2^2x}{6} - \frac{w^2L_2^3x}{24}\right)^2 dx\right] \\
&= \frac{\omega^2 w}{2gEI}\left[R_A^2 \frac{L_1^7}{252} - R_A \frac{wL_1^8}{576} - R_A \frac{313L_1^5}{15} + \frac{w^2L_1^9}{5184}\right.
\end{aligned}
$$

$$+ \frac{313wL_1^6}{72} + \frac{(313)^2L_1^3}{3} + M_B^2\frac{L_2^5}{20} - M_BR_B''\frac{L_2^6}{36}$$

$$+ M_B\frac{wL_2^7}{168} + M_B\frac{503L_2^4}{4} + R_B''^2\frac{L_2^7}{252} - R_B''\frac{wL_2^8}{576} \qquad (5.21)$$

$$\left. - R_B''\frac{503L_2^5}{15} + \frac{w^2L_2^6}{5184} + \frac{503wL_2^6}{72} + \frac{(503)^2L_2^3}{3} \right]$$

By substituting in Eq. (5.21) the values of R_A, R_B'', M_B, w, L_1, and L_2, we find

$$T = \frac{\omega^2}{64.4(EI)^2}[(6.13)(1728)(144)(10^9)]$$

$$= \frac{23,685.41(10^9)\omega^2}{(EI)^2} \qquad (5.22)$$

The fundamental frequency ω of the two-span continuous beam in Fig. 5.2a may now be obtained by equating the expressions given by Eqs. (5.20) and (5.22). This yields

$$\omega = 3.96(10^{-3})\sqrt{EI} \text{ rps} \qquad (5.23)$$

If the two-span continuous beam were solved by using the three-moment equation, as discussed in Section 4.2, the value of this frequency would be

$$\omega = 3.49(10^{-3})\sqrt{EI} \text{ rps} \qquad (5.24)$$

Therefore, in this case, use of the static deflection curve produced by the weight w in Fig. 5.2a would yield a value that is 13.47% higher than the one obtained by using the three-moment equation.

The solution of the two-span continuous beam in Fig. 5.2a by using Rayleigh's method can be improved a great deal by assuming that the shape $y(x)$ of the fundamental mode is the one produced by the pseudostatic system in Fig. 5.2c. For more information on pseudostatic systems, refer to Sections 4.10 through 4.12. In Fig. 5.2c, the distributed weight w of the beam is assumed to act in the direction of the fundamental mode in Fig. 5.2d. The deflection $y(x)$ produced by the pseudostatic system in Fig. 5.2c is a pseudostatic one, and it closely approximates the shape of the fundamental mode. If the solution is repeated by using the pseudostatic curve $y(x)$ and Rayleigh's energy method, as was done above, the value of ω would be almost identical to the one obtained by the three-moment equation. Therefore, for continuous beams, the pseudostatic deflection curve $y(x)$ should be used to obtain an accurate solution to the problem. For single-span beams, the static and pseudostatic deflection curves are identical, and, consequently, the simply supported beam solution in Example 5.1 provided accurate results.

5.3 RAYLEIGH'S METHOD FOR MEMBERS WITH LUMPED AND/OR CONCENTRATED ATTACHED WEIGHTS

Consider again the beam in Fig. 5.1a and assume that, in addition to its dead weight w, the beam also carries the securely attached concentrated weights $W_1, W_2, \ldots, W_i, \ldots, W_n$, as shown in Fig. 5.1b. The dynamic deflections $y_1, y_2, \ldots, y_i, \ldots, y_n$, which are caused by these weights under the weight concentration points, are shown in the same figure. In this case, the total maximum kinetic energy T produced by the dead weight w of the member and the attached W_i weights is as follows:

$$T = \frac{w\omega^2}{2g} \int_0^L y^2 \, dx + \sum_{i=1}^n \frac{W_i}{2g} \omega^2 y_i^2 \qquad (5.25)$$

By considering again the deflection $y(x)$ of the member at an extreme deflection configuration, the expression for its maximum potential energy U has the same form as Eq. (5.8), but y will be replaced by y_d, which takes into consideration the effect of the concentrated weights. By equating U and T, we find the following expression for the frequency equation:

$$\omega^2 = EIg \, \frac{\int_0^L (d^2 y_d / dx^2)^2 \, dx}{w \int_0^L y^2 \, dx + \sum_{i=1}^n W_i y_i^2} \qquad (5.26)$$

Again, we need to know the expression for the y's, or make appropriate assumptions for their shape, in order to obtain a value for ω from Eq. (5.26).

If we select the static, or the pseudostatic, deflection as the shape for the dynamic elastic line, then the work done by gravity on the attached W_i weights in Fig. 5.1b is

$$\frac{1}{2} \sum_{i=1}^n W_i y_i$$

where y_i is the static deflection under the W_i weight. In a similar manner, the work done by gravity on the distributed weight w of the member can be calculated by integrating $wy \, dx/2$ over the whole length of the member. Therefore, the general expression for the fundamental frequency ω of a beam may be written as follows:

$$\omega^2 = g \frac{\int_0^L wy\, dx + \sum_{i=1}^{n} W_i y_i}{\int_0^L wy^2\, dx + \sum_{i=1}^{n} W_i y_i^2} \tag{5.27}$$

Equation (5.27) is general, and it may be used for any variation of the distributed weight w and for any number of concentrated attached weights. In calculating the y's, we can also take into consideration shear and flexural displacements, as well as elastic support displacements. The stiffness EI of the member can also be variable, but the computation of the y's becomes more difficult. The methods discussed in Chapter 8 help a great deal to reduce the mathematical complexity of such problems.

In practice, the vibration analysis of systems with infinite degrees of freedom is carried out by reducing the system to one with finite degrees of freedom. This is accomplished by lumping the weight (or mass) of the member at discrete points along its length. Such an approximation provides reasonable results for the computation of the fundamental frequency of a member and also makes the application of Rayleigh's method more convenient. On this basis, the integrals in Eq. (5.27) become summations, and we can write the following expression for the fundamental frequency ω:

$$\omega^2 = g \frac{\sum_{i=1}^{n} W_i y_i}{\sum_{i=1}^{n} W_i y_i^2} \tag{5.28}$$

where the summations in the above equation include both the lumped weights and the securely attached concentrated weights. The deflection y_i under the concentrated weight W_i is the total static (or pseudostatic) deflection produced at this point when both lumped and attached weights are applied.

5.3.1 Repetitive Approach for Accuracy Improvement

A better approximation of the fundamental frequency and its corresponding mode shape may be obtained by using a repetitive approach. Such a repetitive approach may be initiated by using the results obtained for ω, when Eq. (5.28) is first applied, and the values of y_i and W_i used in this equation. If the value of ω obtained from Eq. (5.28) is designated as ω_1, the inertia forces W_i' corresponding to this frequency may be determined from the following expression:

$$W_i' = \frac{W_i}{g} \omega_1^2 y_i \qquad i = 1, 2, 3, \ldots \tag{5.29}$$

The new W_i' loads may be applied to the member as static loads, in order to calculate the new deflections y_i' produced by these loads. The procedure is the same as the one used to calculate the static displacement y_i due to W_i. The new value ω_2 of the fundamental frequency of the member can be obtained from the following equation:

$$\omega_2^2 = g \, \frac{\displaystyle\sum_{i=1}^{n} W_i' y_i'}{\displaystyle\sum_{i=1}^{n} W_i (y_i')^2} \tag{5.30}$$

The procedure may be repeated until the value of ω obtained from the last trial is about the same as the one obtained from the preceding trial. It only requires a few repetitions for an accurate value of the fundamental frequency.

The following examples illustrate the application of the above methodologies.

Example 5.3. The beam in Fig. 5.3a supports the attached weights W_1, W_2, W_3, and W_4 as shown. Determine the fundamental frequency of vibration of the beam by using Rayleigh's energy method. The stiffness EI is constant throughout the length of the member and is equal to $420 \times 10^6 \, \text{kip} \cdot \text{in.}^2$. The beam is assumed to be weightless, but it maintains its elastic properties between weight concentration points. The magnitude of each weight is $W_1 = 15.47$ kips, $W_2 = 4.18$ kips, $W_3 = 7.86$ kips, and $W_4 = 10.71$ kips.

Solution. The static deflection curve will be used here to determine the fundamental frequency of the member. The deflections y_1, y_2, y_3, and y_4 under the weight concentration points 1, 2, 3, and 4, respectively, are shown in Fig. 5.3b. They can easily be determined statically by using well-known formulas that can be found in texts on elementary mechanics [27] or handbooks. Figure (5.3c) shows that y_1, y_2, y_3, and y_4 can be assumed to consist of two parts. The first part involves the deflections y_1', y_2', y_3', and y_4' that can be obtained from the rigid-body movement of the beam about the left support, point 5 in Fig. 5.3c. The second part involves the flexural deflections y_2'', y_3'', and y_4'' that can be determined by assuming that the member is rigidly supported at points 1 and 5 (simply supported beam). Superposition of the two parts yields the total static deflections y_1, y_2, y_3, and y_4.

On this basis, the total deflections y_1, y_2, y_3, and y_4 are as follows:

$$y_1 = 0.0441 \text{ in.}$$

$$y_2 = 0.0532 \text{ in.}$$

$$y_3 = 0.0529 \text{ in.}$$

$$y_4 = 0.0341 \text{ in.}$$

It should be noted here that use of the static deflection curve is tantamount

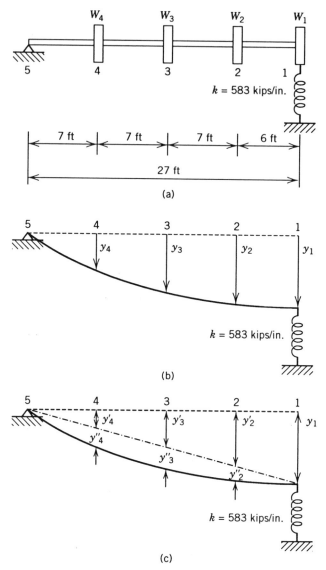

FIGURE 5.3. (a) Uniform single-span beam supporting four attached weights as shown. (b) Static deflection configuration of the member. (c) Static deflection configuration showing the rigid-body and flexural deformations of the member.

to assuming that W_1, W_2, W_3, and W_4 are equal to the inertia loads that produce the fundamental mode of vibration of the member.

We also have

$$\sum_{i=1}^{4} W_i y_i = (15.47)(0.0441) + (4.18)(0.0532) + (7.86)(0.0529)$$

$$+ (10.71)(0.0341)$$

$$= 1.685$$

$$\sum_{i=1}^{4} W_i y_i^2 = (15.47)(0.0441)^2 + (4.18)(0.0532)^2 + (7.86)(0.0529)^2$$

$$+ (10.71)(0.0341)^2$$

$$= 0.0764$$

By applying Eq. (5.28), we find

$$\omega^2 = g\frac{(1.685)}{0.0764)} = \frac{(386)(1.685)}{0.0764}$$

$$= 8520.0 \text{ rad}^2/\text{s}^2$$

or

$$\omega = 92.3 \text{ rps}$$

$$f = \frac{\omega}{2\pi} = \frac{92.3}{2\pi} = 14.69 \text{ Hz}$$

The value obtained here by using the static deflection curve is very accurate, and, for practical purposes, there is no need to repeat the procedure by using Eq. (5.30).

Example 5.4. For the uniform three-span continuous beam in Fig. 5.4a, determine its fundamental frequency of vibration by using Rayleigh's energy method. The beam supports the attached weight $w = 900$ lb/ft, which includes the dead weight of the beam. The constant stiffness $EI = 17.4 \times 10^8$ kip · in.^{2}, and the span lengths are as shown in the figure.

Solution. The three-span continuous beam in Fig. 5.4a is a girder of an existing bridge in the state of Michigan, and the load $w = 900$ lb/ft. acting on the beam includes the dead weight of the girder and the contributory area of the bridge deck. The solution of this problem provides an accurate value for the fundamental frequency of the bridge. The bridge was also tested by the Michigan Department of Transportation in the early 1950s, and experimental values are available for comparison of the results.

By dividing each span of the beam into three equal segments as shown in

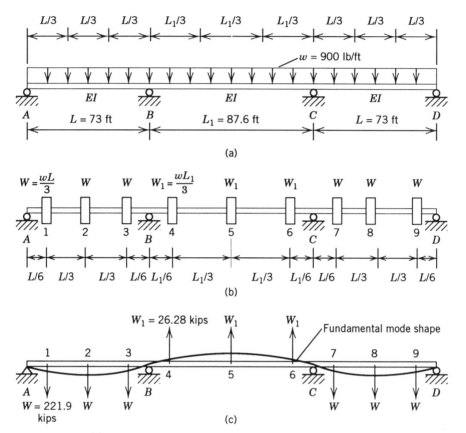

FIGURE 5.4. (a) Uniform three-span continuous beam loaded with its own weight and a uniformly distributed attached weight. (b) Three-span continuous beam with lumped weights. (c) Pseudostatic system with the loads assumed to act in the direction of the fundamental mode as shown.

Fig. 5.4a and then lamping the total weight of each segment at the center of each segment, we obtain the results shown in Fig. 5.4b. On this basis, the values of each W and each W_1 are as follows:

$$W = \frac{wL}{3} = \frac{(0.9)(73)}{3} = 21.90 \text{ kips} \tag{5.31}$$

$$W_1 = \frac{wL_1}{3} = \frac{(0.9)(87.6)}{3} = 26.28 \text{ kips} \tag{5.32}$$

On this basis, we can say that the first step was to lump the weight of the girder and the attached weight, as shown in Fig. 5.4b.

In applying Rayleigh's energy method, we need to assume a shape for

$y(x)$ for the fundamental mode of vibration. For this purpose, we use the pseudostatic system in Fig. 5.4c, where all weights are assumed to act in the direction of the shape of the fundamental mode. In this manner, the pseudostatic deflection $y(x)$ that is produced by the loads in Fig. 5.4c will approximate very closely the shape of the fundamental mode.

The values of the total deflections $y_1, y_2, \ldots, y_9$ under the load concentration points $1, 2, 3, \ldots, 9$, respectively, may be determined by applying elementary methods of mechanics, and they are as follows:

$$y_1 = 0.1912 \text{ in.} \qquad y_6 = -0.2632 \text{ in.}$$
$$y_2 = 0.3558 \text{ in.} \qquad y_7 = 0.1792 \text{ in.}$$
$$y_3 = 0.1792 \text{ in.} \qquad y_8 = 0.3558 \text{ in.}$$
$$y_4 = -0.2632 \text{ in.} \qquad y_9 = 0.1912 \text{ in.}$$
$$y_5 = -0.5365 \text{ in.}$$

By using Eq. (5.28), the fundamental frequency of vibration may be determined as follows:

$$\omega^2 = g \, \frac{\displaystyle\sum_{i=1}^{9} W_i y_i}{\displaystyle\sum_{i=1}^{9} W_i y_i^2}$$

$$= (386) \, \frac{60.1757}{19.7574}$$

$$= 1175.65$$

or

$$\omega = 34.29 \text{ rps.}$$

In hertz, this frequency is

$$f = \frac{\omega}{2\pi} = 5.45 \text{ Hz}$$

If we select the deflection y_3 as a reference value and divide all deflections $y_1, y_2, y_3, \ldots, y_9$ by y_3, the shape of the fundamental mode will be characterized by the following amplitudes β_i $(i = 1, 2, 3, \ldots, 9)$:

$$\beta_1 = \frac{y_1}{y_3} = \frac{0.1912}{0.1792} = 1.067$$

$$\beta_2 = \frac{y_2}{y_3} = \frac{0.3557}{0.1792} = 1.985$$

$$\beta_3 = \frac{y_3}{y_3} = 1.000$$

$$\beta_4 = \frac{y_4}{y_3} = \frac{0.2632}{0.1792} = -1.469$$

$$\beta_5 = \frac{y_5}{y_3} = \frac{0.5365}{0.1792} = -2.994$$

$$\beta_6 = \beta_4 = -1.469$$

$$\beta_7 = \beta_3 = 1.000$$

$$\beta_8 = \beta_2 = 1.985$$

$$\beta_9 = \beta_1 = 1.067$$

This is a normalized shape of the fundamental mode, since all amplitudes y_i $(i = 1, 2, 3, \ldots, 9)$ are divided by the common amplitude y_3.

If better accuracy is desired, Eq. (5.30) may be used to repeat the procedure. For this equation, the new loads W_i' may be determined from Eq. (5.29) as follows:

$$W_i' = \frac{W_i}{g} \omega^2 y_i \qquad (5.33)$$

Since we have expressed y_i as $y_i = y_3 \beta_i$, Eq. (5.33) yields

$$\begin{aligned} W_i' &= \frac{W_i}{g} \omega^2 y_3 \beta_i \\ &= \frac{y_3 \omega^2}{g} W_i \beta_i \end{aligned} \qquad (5.34)$$

In Eq. (5.34), the quantity $y_3 \omega^2/g$ is constant for all weights W_i', and it can be omitted. This is permissible, because the absolute value of the inertia force is indeterminate, and it is of no importance. The frequency ω depends only on the shape of the mode and not on the absolute value of the dynamic amplitudes in that mode. On this basis, Eq. (5.34) yields

$$W_i' = W_i \beta_i \qquad (5.35)$$

By using Eq. (5.35), the new inertia loads W_i' are as follows:

$$W_1' = W_1 \beta_1 = (21.9)(1.067) = 23.36 \text{ kips}$$
$$W_2' = W_2 \beta_2 = (21.9)(1.985) = 43.47 \text{ kips}$$

$$W'_3 = W_3\beta_3 = (21.9)(1.000) = 21.90 \text{ kips}$$
$$W'_4 = W_4\beta_4 = -(26.28)(1.469) = -38.61 \text{ kips}$$
$$W'_5 = W_5\beta_5 = -(26.28)(2.994) = -78.68 \text{ kips}$$
$$W'_6 = W_6\beta_6 = -(26.28)(1.469) = -38.61 \text{ kips}$$
$$W'_7 = W_7\beta_7 = (21.9)(1.000) = 21.90 \text{ kips}$$
$$W'_8 = W_8\beta_8 = (21.9)(1.985) = 43.47 \text{ kips}$$
$$W'_9 = W_9\beta_9 = (21.9)(1.067) = 23.36 \text{ kips}$$

The new approximation of the fundamental mode may be obtained by applying the new W'_i loads in the same way the W_i loads were applied in Fig. 5.4c. The new deflections $y'_1, y'_2, y'_3, \ldots, y'_9$ may be calculated as before, and they are as follows:

$$y'_1 = 0.3234 \text{ in.} \qquad y'_6 = -0.4974 \text{ in.}$$
$$y'_2 = 0.6413 \text{ in.} \qquad y'_7 = 0.3119 \text{ in.}$$
$$y'_3 = 0.3119 \text{ in.} \qquad y'_8 = 0.6413 \text{ in.}$$
$$y'_4 = -0.4974 \text{ in.} \qquad y'_9 = 0.3234 \text{ in.}$$
$$y'_5 = -1.0731 \text{ in.}$$

We also have

$$\sum_{i=1}^{9} W'y'_i = 207.3657$$

$$\sum_{i=1}^{9} W_i(y'_i)^2 = 70.1219$$

By substituting into Eq. (5.30), the value ω_2 of the fundamental frequency from the second trial is as follows:

$$\omega_2^2 = g \frac{\displaystyle\sum_{i=1}^{9} W'_i y'_i}{\displaystyle\sum_{i=1}^{9} W_i(y'_i)^2}$$

$$= (386) \frac{(207.3657)}{(70.1219)}$$

$$= 1141.49$$

or

$$\omega_2 = 33.79 \text{ rps}$$

and

$$f = \frac{\omega_2}{2\pi} = 5.38 \text{ Hz}$$

The value obtained from the second trial differs by an amount of only 1.48% when it is compared with the value obtained from the first trial. This shows that the deflection curve obtained from the pseudostatic system in Fig. 5.4c provides a good approximation of the shape of the fundamental mode of vibration of the continuous beam. If better accuracy is required, the procedure may be repeated in a similar manner.

Based on the results obtained from the second trial, the normalized shape of the fundamental mode is represented by the following amplitudes:

$$\beta_1' = \frac{y_1'}{y_3'} = \frac{0.3234}{0.3119} = 1.037$$

$$\beta_2' = \frac{y_2'}{y_3'} = \frac{0.6413}{0.3119} = 2.056$$

$$\beta_3' = \frac{y_3'}{y_3'} = 1.00$$

$$\beta_4' = \frac{y_4'}{y_3'} = \frac{0.4974}{0.3119} = -1.595$$

$$\beta_5' = \frac{y_5'}{y_3'} = -\frac{1.0731}{0.3119} = -3.441$$

$$\beta_6' = \frac{y_6'}{y_3'} = -\frac{0.4974}{0.3119} = -1.595$$

$$\beta_7' = \frac{y_7'}{y_3'} = 1.00$$

$$\beta_8' = \frac{y_8'}{y_3'} = \frac{0.6413}{0.3119} = 2.056$$

$$\beta_9' = \frac{y_9'}{y_3'} = \frac{0.3234}{0.3119} = 1.037$$

These amplitudes provide an improvement compared to the normalized amplitudes obtained from the first trial. The frequency obtained from the experimental results is only about 1% different compared to the value obtained above.

5.4 STODOLA'S METHOD WITH AN ITERATION PROCEDURE

In the preceding two sections, the method of Lord Rayleigh was used to determine the fundamental frequencies and the corresponding mode shapes of freely vibrating beams and related structural and mechanical components. Another convenient method that can be used for the same purpose is Stodola's method combined with an iteration procedure. This method is general and readily applicable to spring–mass systems and to beam-like structures, where their weight or mass can be lumped at discrete points in a manner similar to the one shown in Fig. 5.4b.

In this section, the application of Stodola's method is limited to structural members, where the stiffness EI can vary from one span to the other but does not vary within a span. It should be pointed out, however, that the method is general and it applies to members of variable stiffness as well. This problem, because of its special nature, is treated by itself in Chapter 8.

Stodola's method is predicated on the idea that, at a principal mode of vibration of constant circular frequency ω, the vibrating system is acted on by inertia forces $m_1 \ddot{y}_i = -m_i \omega^2 y_i$, where y_i is the dynamic amplitude of mass m_i at a position i. Then, for each mass concentration point i, an equation of motion is written that expresses the amplitude y_i in terms of flexibility coefficients a_{ij} and the inertia loads $m_i \omega^2 y_i$ causing the y_i amplitude. A *flexibility* (or *deflection*) *coefficient* a_{ij} is defined as the deflection at point i caused by the application of a unit load at point j. See also Appendix E. Application of the Maxwell–Mohr theorem of reciprocity will expedite the evaluation of these coefficients. This idea suggests that if the inertia loads of the mth normal mode of a vibrating structural system are applied statically to the structure, then the resulting shape of the deflection curve is the mth normal mode of its vibration.

Assume that the continuous beam in Fig. 5.5a vibrates freely at a mode of vibration as shown in Fig. 5.5b. The concentrated masses $m_1, m_2, \ldots, m_n$ in Fig. 5.5a may be assumed to consist of the mass of the member lumped at discrete points along its length, as well as other weights that are securely attached to the beam. The amplitudes $y_1, y_2, \ldots, y_n$ under the mass concentration points $1, 2, \ldots, n$, respectively, are shown in Fig. 5.5b. The inertia forces $m_1 \omega^2 y_1, m_2 \omega^2 y_2, \ldots, m_n \omega^2 y_n$ acting at points $1, 2, \ldots, n$, respectively, are shown in Fig. 5.5c. These are the forces that produce the amplitudes $y_1, y_2, \ldots, y_n$ in Fig. 5.5b.

The influence (or deflection) coefficients a_{ij} may be determined as shown in Figs. 5.5d and 5.5e. In Fig. 5.5d, a unit load is applied statically at point 1. The deflections $a_{11}, a_{21}, \ldots, a_{i1}, \ldots, a_{n1}$ at points $1, 2, \ldots, i, \ldots, n$, respectively, caused by the application of the unit load at point 1, are termed

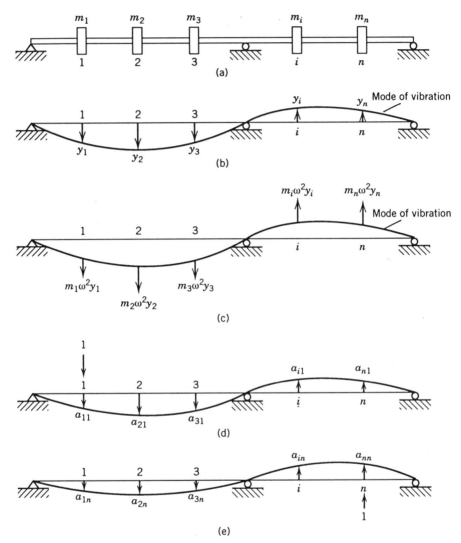

FIGURE 5.5. (a) Two-span continuous beam with n concentrated masses. (b) Mode of vibration showing the amplitudes $y_1, y_2, \ldots, y_n$. (c) Inertia forces corresponding to the mode of vibration of the member. (d) Unit load applied at point 1. (e) Unit load applied at point n.

influence (or *deflection*) *coefficients*. They can be determined by applying known methods from the mechanics of materials dealing with computation of deflections. In Fig. 5.5e, the unit load is applied at point n, and $a_{1n}, a_{2n}, a_{3n}, \ldots, a_{in}, \ldots, a_{nn}$ are the deflection coefficients caused by this unit load. Similar coefficients may be obtained by applying individually unit loads at the remaining points.

The equations of motion for the deflections $y_1, y_2, \ldots, y_i, \ldots, y_n$ at points $1, 2, \ldots, i, \ldots, n$, respectively, which are produced by the inertia loads in Fig. 5.5c, may be written as follows:

$$y_1 = a_{11}(m_1\omega^2 y_1) + a_{12}(m_2\omega^2 y_2) + \cdots + a_{1i}(m_i\omega^2 y_i) + \cdots + a_{1n}(m_n\omega^2 y_n)$$
$$\tag{5.36}$$
$$y_2 = a_{21}(m_1\omega^2 y_1) + a_{22}(m_2\omega^2 y_2) + \cdots + a_{2i}(m_i\omega^2 y_i) + \cdots + a_{2n}(m_n\omega^2 y_n)$$
$$\tag{5.37}$$
$$y_i = a_{i1}(m_1\omega^2 y_1) + a_{i2}(m_2\omega^2 y_2) + \cdots + a_{ii}(m_i\omega^2 y_i) + \cdots + a_{in}(m_n\omega^2 y_n)$$
$$\tag{5.38}$$
$$y_n = a_{n1}(m_1\omega^2 y_1) + a_{n2}(m_2\omega^2 y_2) + \cdots + a_{ni}(m_i\omega^2 y_i) + \cdots + a_{nn}(m_n\omega^2 y_n)$$
$$\tag{5.39}$$

These equations may be written in matrix form as follows:

$$
\begin{bmatrix} y \\ y_2 \\ \vdots \\ y_i \\ \vdots \\ y_n \end{bmatrix} = \omega^2
\begin{bmatrix}
a_{11}m_1 & a_{12}m_2 & \cdots & a_{1i}m_i & \cdots & a_{1n}m_n \\
a_{21}m_1 & a_{22}m_2 & \cdots & a_{2i}m_i & \cdots & a_{2n}m_n \\
\vdots & \vdots & \ddots & \vdots & \ddots & \vdots \\
a_{i1}m_1 & a_{i2}m_2 & \cdots & a_{ii}m_i & \cdots & a_{in}m_n \\
\vdots & \vdots & \ddots & \vdots & \ddots & \vdots \\
a_{n1}m_1 & a_{n2}m_2 & \cdots & a_{ni}m_i & \cdots & a_{nn}m_n
\end{bmatrix}
\begin{bmatrix} y_1 \\ y_2 \\ \vdots \\ y_i \\ \vdots \\ y_n \end{bmatrix}
\tag{5.40}
$$

In a more compact form, Eq. (5.40) may be written

$$\{y\} = \omega^2 [M]\{y\} \tag{5.41}$$

where

$$\{y\} = \begin{bmatrix} y_1 \\ y_2 \\ \vdots \\ y_i \\ \vdots \\ y_n \end{bmatrix} \tag{5.42}$$

$$[M] = \begin{bmatrix} a_{11}m_1 & a_{12}m_2 & \cdots & a_{1i}m_i & \cdots & a_{1n}m_n \\ a_{21}m_1 & a_{22}m_2 & \cdots & a_{2i}m_i & \cdots & a_{2n}m_n \\ \vdots & \vdots & \ddots & \vdots & \vdots & \vdots \\ a_{i1}m_1 & a_{i2}m_2 & \cdots & a_{ii}m_i & \cdots & a_{in}m_n \\ \vdots & \vdots & \ddots & \vdots & \vdots & \vdots \\ a_{n1}m_1 & a_{n2}m_2 & \cdots & a_{ni}m_i & \cdots & a_{nn}m_n \end{bmatrix} \tag{5.43}$$

Equation (5.40), or Eq. (5.41), is known as the *frequency equation*.

Solution of the frequency equation is convenient by using an iteration procedure. Since the exact amplitudes y in Eq. (5.40), or Eq. (5.41), are not known, the iteration procedure may be initiated by assuming arbitrary values for the deflections $y_1, y_2, \ldots, y_n$ in the right-hand column of Eq. (5.40) and carrying out the required matrix multiplication. In fact, we can start by assuming $y_1 = y_2 = \cdots = y_n = 1.00$. After the matrix multiplication is carried out, the resulting column may be normalized by dividing each term in the column by any one of the terms, preferably the smallest one. The resulting column matrix will contain values of the deflections y that represent a better approximation of the fundamental mode shape. These new values of y may now be used in Eq. (5.40) to repeat the procedure.

By using the normalized column each time, the above iteration procedure is rapidly convergent, and it usually requires two to four repetitions to stabilize the amplitudes y in the right-hand column of Eq. (5.40). When the normalized amplitudes of the last trial are about the same as the ones obtained from the preceding trial, we say that the matrix equation (5.40) is converged to the desired accuracy, and the fundamental frequency of vibration may be determined by solving the matrix equation. The examples at the end of this section illustrate the iteration procedure.

Another approach to this problem would be to divide each term in Eqs. (5.36) through (5.39) by ω^2. If we designate $\lambda = 1/\omega^2$, Eqs. (5.36) through (5.39) may be rewritten as follows:

$$(a_{11}m_1 - \lambda)y_1 + a_{12}m_2 y_2 + \cdots + a_{1n}m_n y_n = 0 \tag{5.44}$$

$$a_{21}m_1 y_1 + (a_{22}m_2 - \lambda)y_2 + \cdots + a_{2n}m_n y_n = 0 \tag{5.45}$$

$$\vdots$$

$$a_{n1}m_1 y_1 + a_{n2}m_2 y_2 + \cdots + (a_{nm}m_n - \lambda)y_n = 0 \tag{5.46}$$

If we wish to obtain a nontrivial solution, the determinant of the coefficients of the y's in Eqs. (5.44) through (5.46) must be made equal to zero. In this manner, the following frequency determinant is obtained:

$$\begin{vmatrix} a_{11}m_1 - \lambda & a_{12}m_2 & \cdots & a_{1n}m_n \\ a_{21}m_1 & a_{22}m_2 - \lambda & \cdots & a_{2n}m_n \\ \vdots & \vdots & \ddots & \vdots \\ a_{n1}m_1 & a_{n2}m_2 & \cdots & a_{nn}m_n - \lambda \end{vmatrix} = 0 \qquad (5.47)$$

The expansion of the frequency determinant given by Eq. (5.47) leads to the following frequency equation:

$$\lambda^n + c_1\lambda^{n-1} + c_2\lambda^{n-2} + \cdots + c_{n-1}\lambda + c_n = 0 \qquad (5.48)$$

where $c_1, c_2, \ldots, c_n$ in this equation are known constants. The solution of Eq. (5.48) will provide the n roots $\lambda_1, \lambda_2, \ldots, \lambda_n$ of λ and, consequently, the n values of the frequencies ω, because $\lambda = 1/\omega^2$. For each value of λ, the corresponding mode shape is determined by using Eqs. (5.44) through (5.46). For $n > 3$, the expansion of the determinant and the solution of Eq. (5.48) become tedious, and use of existing computer programs is advisable.

The work in this section deals only with the use of the iteration procedure for the computation of the fundamental frequency of vibration of beam problems. The computation of higher frequencies of vibration using Stodola's method and an iteration procedure is given in the next section. The following example illustrates the application of the method.

Example 5.5. Rework Example 5.3 by using Stodola's method in combination with an iteration procedure.

Solution. The a_{ij} deflection coefficients may be calculated by applying a unit load each time at points 1, 2, 3, and 4 and computing the deflections produced each time at points 1, 2, 3, and 4. For example, if we apply a unit load at point 1 as shown in Fig. 5.6a, the deflection coefficients a_{11}, a_{21}, a_{31}, and a_{41} are the ones shown in Fig. 5.6b, and they can be determined by using known methods from the mechanics of solids. When the unit load is applied at point 2, the a_{12}, a_{22}, a_{32}, and a_{42} deflection coefficients are shown in Fig. 5.6d. In accordance with the Maxwell–Mohr theorem of reciprocity, we should have $a_{12} = a_{21}$, or, in general, $a_{ij} = a_{ji}$. In Fig. 5.6b, the deflections due to a unit load at point 1 involve only rigid-body movement of the member about point 5. In Fig. 5.6d, the deflections produced by a unit load at point 2 are composed in part by the rigid-body rotation of the member about point 5, and a second part due to flexural bending of the member when the ends are assumed to be simply supported. All the deflection coefficients are calculated

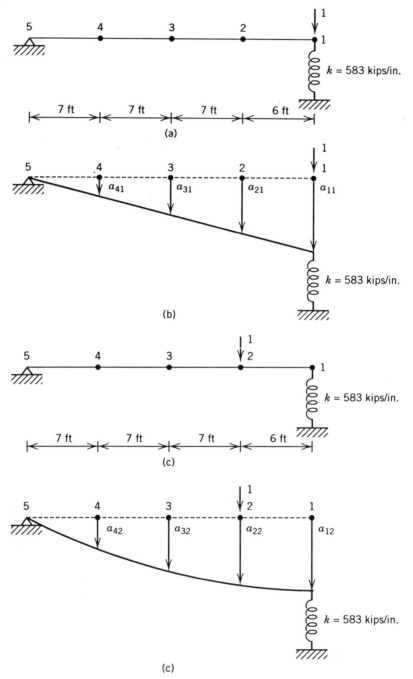

FIGURE 5.6. (a) Unit load applied at point 1. (b) Deflection coefficients produced by the application of a unit load at point 1. (c) Unit load applied at point 2. (d) Deflection coefficients due to the application of a unit load at point 2.

on this basis, and they are as follows:

$$a_{11} = 17.15 \times 10^{-4} \qquad\qquad a_{12} = a_{21} = 13.34 \times 10^{-4}$$
$$a_{21} = 13.34 \times 10^{-4} \qquad\qquad a_{22} = 18.40 \times 10^{-4}$$
$$a_{31} = 8.90 \times 10^{-4} \qquad\qquad a_{32} = 17.50 \times 10^{-4}$$
$$a_{41} = 4.45 \times 10^{-4} \qquad\qquad a_{42} = 10.33 \times 10^{-4}$$

$$a_{13} = a_{31} = 8.90 \times 10^{-4} \qquad a_{14} = a_{41} = 4.45 \times 10^{-4}$$
$$a_{23} = a_{32} = 17.50 \times 10^{-4} \qquad a_{24} = a_{42} = 10.33 \times 10^{-4}$$
$$a_{33} = 21.46 \times 10^{-4} \qquad\qquad a_{34} = a_{43} = 14.11 \times 10^{-4}$$
$$a_{43} = 14.11 \times 10^{-4} \qquad\qquad a_{44} = 11.10 \times 10^{-4}$$

The masses m_1, m_2, m_3, and m_4 at points 1, 2, 3, and 4, respectively, are as follows:

$$m_1 = \frac{W_1}{g} = \frac{15.47}{386} = 40.08 \times 10^{-3} \text{ kip} \cdot \text{s}^2/\text{in.}$$

$$m_2 = \frac{W_2}{g} = \frac{4.18}{386} = 10.83 \times 10^{-3} \text{ kip} \cdot \text{s}^2/\text{in.}$$

$$m_3 = \frac{W_3}{g} = \frac{7.86}{386} = 20.36 \times 10^{-3} \text{ kip} \cdot \text{s}^2/\text{in.}$$

$$m_4 = \frac{W_4}{g} = \frac{10.71}{386} = 27.75 \times 10^{-3} \text{ kip} \cdot \text{s}^2/\text{in.}$$

By substituting into Eq. (5.40), we find

$$\begin{bmatrix} y_1 \\ y_2 \\ y_3 \\ y_4 \end{bmatrix} = (10^{-7})\omega^2 \begin{bmatrix} 687.37 & 144.47 & 181.20 & 123.49 \\ 534.67 & 199.27 & 356.30 & 286.66 \\ 356.71 & 189.53 & 436.93 & 391.55 \\ 178.36 & 111.87 & 287.28 & 308.03 \end{bmatrix} \begin{bmatrix} y_1 \\ y_2 \\ y_3 \\ y_4 \end{bmatrix}$$

If we take the smallest element of the square matrix as a common factor of the elements of the matrix, which is 111.87, the above matrix equation yields

$$\begin{bmatrix} y_1 \\ y_2 \\ y_3 \\ y_4 \end{bmatrix} = (111.87)(10^{-7})\omega^2 \begin{bmatrix} 6.1444 & 1.2914 & 1.6187 & 1.1039 \\ 4.7794 & 1.7813 & 3.1849 & 2.5624 \\ 3.1886 & 1.6942 & 3.9057 & 3.5000 \\ 1.5944 & 1.0000 & 2.5680 & 2.7535 \end{bmatrix} \begin{bmatrix} y_1 \\ y_2 \\ y_3 \\ y_4 \end{bmatrix}$$

$$(5.49)$$

The iteration procedure is initiated by assuming $y_1 = y_2 = y_3 = y_4 = 1.0$ and carrying out the required matrix multiplication in Eq. (5.49). This yields

$$\begin{bmatrix} y_1 \\ y_2 \\ y_3 \\ y_4 \end{bmatrix} = (111.87)(10^{-7})\omega^2 \begin{bmatrix} 6.1444 + 1.2914 + 1.6187 + 1.1039 \\ 4.7794 + 1.7813 + 3.1849 + 2.5624 \\ 3.1886 + 1.6942 + 3.9057 + 3.5000 \\ 1.5944 + 1.0000 + 2.5680 + 2.7535 \end{bmatrix}$$

or

$$\begin{bmatrix} y_1 \\ y_2 \\ y_3 \\ y_4 \end{bmatrix} = (111.87)(10^{-7})\omega^2 \begin{bmatrix} 10.1584 \\ 12.3080 \\ 12.2885 \\ 7.9159 \end{bmatrix}$$

If we take 7.9159 as the common factor in the above column matrix, we find

$$\begin{bmatrix} y_1 \\ y_2 \\ y_3 \\ y_4 \end{bmatrix} = (7.9159)(111.87)(10^{-7})\omega^2 \begin{bmatrix} 1.2833 \\ 1.5548 \\ 1.5524 \\ 1.0000 \end{bmatrix} \qquad (5.50)$$

The new normalized amplitudes obtained in Eq. (5.50) may be used in Eq. (5.49) to repeat the procedure. This yields

$$\begin{bmatrix} y_1 \\ y_2 \\ y_3 \\ y_4 \end{bmatrix} = (111.87)(10^{-7})\omega^2 \begin{bmatrix} 6.1444 & 1.2914 & 1.6187 & 1.1039 \\ 4.7794 & 1.7813 & 3.1849 & 2.5624 \\ 3.1886 & 1.6942 & 3.9057 & 3.5000 \\ 1.5944 & 1.0000 & 2.5680 & 2.7535 \end{bmatrix} \begin{bmatrix} 1.2833 \\ 1.5548 \\ 1.5524 \\ 1.0000 \end{bmatrix}$$

$$= (111.87)(10^{-7})\omega^2 \begin{bmatrix} 13.5098 \\ 16.4096 \\ 16.2892 \\ 10.3410 \end{bmatrix}$$

$$= (10.341)(111.87)(10^{-7})\omega^2 \begin{bmatrix} 1.3064 \\ 1.5868 \\ 1.5752 \\ 1.0000 \end{bmatrix} \tag{5.51}$$

The procedure may now be repeated by using the normalized amplitudes obtained in Eq. (5.51). On this basis, we have

$$\begin{bmatrix} y_1 \\ y_2 \\ y_3 \\ y_4 \end{bmatrix} = (111.87)(10^{-7})\omega^2 \begin{bmatrix} 6.1444 & 1.2914 & 1.6187 & 1.1039 \\ 4.7794 & 1.7813 & 3.1849 & 2.5624 \\ 3.1886 & 1.6942 & 3.9057 & 3.5000 \\ 1.5944 & 1.0000 & 2.5680 & 2.7535 \end{bmatrix} \begin{bmatrix} 1.3064 \\ 1.5868 \\ 1.5752 \\ 1.0000 \end{bmatrix}$$

$$= (111.87)(10^{-7})\omega^2 \begin{bmatrix} 13.7299 \\ 16.6497 \\ 16.5063 \\ 10.4683 \end{bmatrix}$$

or

$$\begin{bmatrix} y_1 \\ y_2 \\ y_3 \\ y_4 \end{bmatrix} = (10.4683)(111.87)(10^{-7})\omega^2 \begin{bmatrix} 1.3116 \\ 1.5905 \\ 1.5768 \\ 1.0000 \end{bmatrix} \tag{5.52}$$

The matrix now is very closely converged to the correct values of the normalized amplitudes y_1, y_2, y_3, and y_4. For practical purposes, we can assume that the matrix is converged and the desired accuracy is obtained.

The fundamental frequency of vibration ω, in radians per second, may be obtained by using any one of the rows in Eq. (5.52). For example, if we use the first row, we have

$$1.3116 = (10.4683)(111.87)(10^{-7})\omega^2(1.3116)$$

$$\omega^2 = 8539.06$$

$$\omega = 92.40 \text{ rps}$$

The value obtained in Example 5.3 by using Rayleigh's method is 92.30 rps. For practical purposes, these two frequencies are considered to be identical.

5.5   COMPUTATION OF HIGHER FREQUENCIES OF FREE VIBRATION

In the preceding two sections, Stodola's method was used to determine the fundamental frequency of free vibration and the corresponding mode shape of beam problems. In this section, Stodola's method in conjunction with the iteration procedure will be used to determine higher frequencies of free vibration and their corresponding mode shapes. The procedure is best explained by using a numerical example.

Assume that the uniform simply supported beam in Fig. 5.7a supports five concentrated, rigidly attached weights at the indicated positions. The values of the weights are $W_1 = 11.5$ kips, $W_2 = 8.19$ kips, $W_3 = 4.33$ kips, $W_4 = 2.57$ kips, and $W_5 = 1.31$ kips. The modulus of elasticity $E = 30 \times 10^6$ psi, and the uniform moment of inertia $I = 3413.33$ in.^{4}. We wish to determine the first two frequencies of the beam's free vibration and the corresponding mode shapes by using Stodola's method and the iteration procedure.

The first step in the procedure is to determine the a_{ij} deflection coefficients by individually applying unit loads at points 1, 2, 3, 4, and 5. This part can be accomplished as discussed in the preceding section. Since the beam is simply supported, handbook formulas are available for this purpose. By applying such formulas, the a_{ij} deflection coefficients are calculated, and they are as follows:

$$
\begin{array}{lll}
a_{11} = 0.0057 & a_{12} = a_{21} = 0.0123 & a_{13} = a_{31} = 0.0130 \\
a_{21} = 0.0123 & a_{22} = 0.0310 & a_{23} = a_{32} = 0.0348 \\
a_{31} = 0.0130 & a_{32} = 0.0348 & a_{33} = 0.0440 \\
a_{41} = 0.0095 & a_{42} = 0.0264 & a_{43} = 0.0348 \\
a_{51} = 0.0034 & a_{52} = 0.0095 & a_{53} = 0.0130
\end{array}
$$

$$
\begin{array}{ll}
a_{14} = a_{41} = 0.0095 & a_{15} = a_{51} = 0.0034 \\
a_{24} = a_{42} = 0.0264 & a_{25} = a_{52} = 0.0095 \\
a_{34} = a_{43} = 0.0348 & a_{35} = a_{53} = 0.0130 \\
a_{44} = 0.0310 & a_{45} = a_{54} = 0.0123 \\
a_{54} = 0.0123 & a_{55} = 0.0057
\end{array}
$$

By substituting in Eq. (5.40), the following matrix equation is obtained:

$$
\begin{bmatrix} y_1 \\ y_2 \\ y_3 \\ y_4 \\ y_5 \end{bmatrix} = \frac{\omega^2}{(10^3)g} \begin{bmatrix} 65.50 & 100.80 & 56.30 & 24.40 & 4.53 \\ 141.60 & 254.10 & 150.80 & 66.70 & 12.50 \\ 149.60 & 285.20 & 190.40 & 89.40 & 17.10 \\ 109.20 & 212.60 & 150.80 & 79.70 & 16.20 \\ 39.70 & 77.80 & 56.30 & 31.60 & 7.50 \end{bmatrix} \begin{bmatrix} y_1 \\ y_2 \\ y_3 \\ y_4 \\ y_5 \end{bmatrix}
$$

By taking as a common factor the smallest element of the matrix, we have

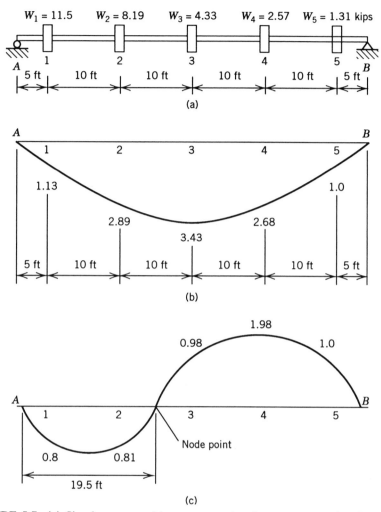

FIGURE 5.7. (a) Simply supported beam supporting five concentrated weights. (b) Shape of the fundamental mode. (c) Shape of the second mode of vibration.

$$\begin{bmatrix} y_1 \\ y_2 \\ y_3 \\ y_4 \\ y_5 \end{bmatrix} = \frac{4.53\omega^2}{(10^3)g} \begin{bmatrix} 14.40 & 22.20 & 12.40 & 5.40 & 1.00 \\ 31.20 & 56.00 & 33.30 & 14.70 & 2.80 \\ 3.30 & 62.90 & 42.00 & 19.70 & 3.80 \\ 24.00 & 46.90 & 33.30 & 17.60 & 3.60 \\ 8.80 & 17.20 & 12.40 & 7.00 & 1.70 \end{bmatrix} \begin{bmatrix} y_1 \\ y_2 \\ y_3 \\ y_4 \\ y_5 \end{bmatrix} \quad (5.53)$$

The fundamental frequency of vibration and its corresponding mode shape are obtained by using Eq. (5.53) and applying the iteration procedure as discussed in the preceding section. On this basis, the matrix equation converges to the following values:

$$\begin{bmatrix} y_1 \\ y_2 \\ y_3 \\ y_4 \\ y_5 \end{bmatrix} = \frac{(4.534)(122.40)\omega^2}{(10^3)g} \begin{bmatrix} 1.13 \\ 2.89 \\ 3.43 \\ 2.68 \\ 1.00 \end{bmatrix} \tag{5.54}$$

By using the last row of Eq. (5.54), the fundamental frequency of vibration is obtained as follows:

$$1.00 = \frac{(4.534)(122.40)\omega^2}{(10^3)g}(1.00)$$

$$\omega = \sqrt{\frac{(10^3)(386)}{(4.534)(122.40)}} = 26.37 \text{ rps}$$

The shape of the first mode is defined by the following amplitudes:

$$y_1 = 1.13$$
$$y_2 = 2.89$$
$$y_3 = 3.43$$
$$y_4 = 2.68$$
$$y_5 = 1.00$$

The shape of the fundamental mode is shown plotted in Fig. 5.7b.

To determine the frequency of vibration of the second mode by using Stodola's method and the iteration procedure, we write the equations of motion for the deflections y_1, y_2, y_3, y_4, and y_5 in the following manner:

$$C^{(n)}y_1^{(n)} = a_{11}W_1y_1^{(n)} + a_{12}W_2y_2^{(n)} + a_{13}W_3y_3^{(n)} + a_{14}W_4y_4^{(n)} + a_{15}W_5y_5^{(n)} \tag{5.55}$$

$$C^{(n)}y_2^{(n)} = a_{21}W_1y_1^{(n)} + a_{22}W_2y_2^{(n)} + a_{23}W_3y_3^{(n)} + a_{24}W_4y_4^{(n)} + a_{25}W_5y_5^{(n)} \tag{5.56}$$

$$C^{(n)}y_3^{(n)} = a_{31}W_1y_1^{(n)} + a_{32}W_2y_2^{(n)} + a_{33}W_3y_3^{(n)} + a_{34}W_4y_4^{(n)} + a_{35}W_5y_5^{(n)} \tag{5.57}$$

$$C^{(n)}y_4^{(n)} = a_{41}W_1y_1^{(n)} + a_{42}W_2y_2^{(n)} + a_{43}W_3y_3^{(n)} + a_{44}W_4y_4^{(n)} + a_{45}W_5y_5^{(n)} \tag{5.58}$$

$$C^{(n)}y_5^{(n)} = a_{51}W_1y_1^{(n)} + a_{52}W_2y_2^{(n)} + a_{53}W_3y_3^{(n)} + a_{54}W_4y_4^{(n)} + a_{55}W_5y_5^{(n)} \tag{5.59}$$

where

$$C = \frac{(10^3)g}{\omega^2} \tag{5.60}$$

and W_1, W_2, W_3, W_4, and W_5 are the attached concentrated weights, which are located as shown in Fig. 5.7a. Equations (5.55) through (5.59) are applicable to all modes of vibration. In these equations, the superscript (n) denotes the mode number, and, in this case, it takes the values $n = 1, 2, 3, 4,$ and 5.

Since the amplitudes of the fundamental mode are known, an orthogonality relation between the first and second mode amplitudes may be written by using Eq. (2.260). Thus, by using Eq. (2.260) for $i = 1$ and $j = 2$, we may write the following equation:

$$(11.50)(1.13)y_1^{(2)} + (8.19)(2.89)y_2^{(2)} + (4.33)(3.43)y_3^{(2)}$$
$$+ (2.57)(2.68)y_4^{(2)} + (1.31)(1.00)y_5^{(2)} = 0$$

or

$$12.00y_1^{(2)} + 23.67y_2^{(2)} + 14.85y_3^{(2)} + 6.89y_4^{(2)} + 1.3y_5^{(2)} = 0 \tag{5.61}$$

Equation (5.61) represents the orthogonality relation between the amplitudes of the first and second modes of vibration.

In order to use an iteration procedure again and have the matrix equation converge to the second frequency and mode of vibration, the order of the matrix in Eq. (5.53) must be reduced by one; that is, the 5×5 matrix must be reduced to a 4×4 matrix. This can be accomplished by using Eq. (5.61) and eliminating one of the amplitudes in Eqs. (5.55) through (5.59). Some care should be taken as to which one of the displacements should be eliminated. It is important to avoid choosing a displacement that is close to a node point. Here, we choose to eliminate the y_5 displacement, since it is far enough away from the location of the node point of the second mode.

By using Eq. (5.61), we may write $y_5^{(2)}$ as follows:

$$y_5^{(2)} = -9.91y_1^{(2)} - 18.00y_2^{(2)} - 11.34y_3^{(2)} - 5.23y_4^{(2)} \tag{5.62}$$

By substituting Eq. (5.62) into Eqs. (5.55) through (5.59) and eliminating $y_5^{(2)}$ from these equations, the following set of four equations is obtained:

$$C^{(2)}y_1^{(2)} = 20.52y_1^{(2)} + 19.14y_2^{(2)} + 5.06y_3^{(2)} + 0.67y_4^{(2)} \tag{5.63}$$

$$C^{(2)}y_2^{(2)} = 18.00y_1^{(2)} + 29.50y_2^{(2)} + 9.81y_3^{(2)} + 1.44y_4^{(2)} \tag{5.64}$$

$$C^{(2)}y_3^{(2)} = -19.77y_1^{(2)} - 22.40y_2^{(2)} - 2.74y_3^{(2)} + 0.074y_4^{(2)} \tag{5.65}$$

$$C^{(2)}y_4^{(2)} = -51.20y_1^{(2)} - 78.70y_2^{(2)} - 32.09y_3^{(2)} - 4.94y_4^{(2)} \tag{5.66}$$

In matrix form, Eqs. (5.63) through (5.66) may be written as shown below:

$$
\begin{bmatrix} y_1 \\ y_2 \\ y_3 \\ y_4 \end{bmatrix} = \frac{0.074\omega^2}{(10^3)g} \begin{bmatrix} 276.70 & 258.10 & 68.30 & 8.99 \\ 242.80 & 397.90 & 132.30 & 19.39 \\ -266.00 & -302.30 & -36.90 & 1.00 \\ -690.00 & -1061.00 & -432.00 & -66.70 \end{bmatrix} \begin{bmatrix} y_1 \\ y_2 \\ y_3 \\ y_4 \end{bmatrix} \qquad (5.67)
$$

The square matrix in Eq. (5.67) is now reduced to a 4 × 4 square matrix. If we apply the iteration procedure, Eq. (5.67) will converge to the second frequency and mode of vibration.

The iteration procedure for Eq. (5.67) may be initiated by assuming $y_1 = y_2 = y_3 = y_4 = 1.00$ and carrying out the required matrix multiplication; that is,

$$
\begin{bmatrix} y_1 \\ y_2 \\ y_3 \\ y_4 \end{bmatrix} = \frac{0.074\omega^2}{(10^3)g} \begin{bmatrix} 276.70 + & 258.10 + & 68.30 + & 8.99 \\ 242.80 + & 397.90 + & 132.30 + & 19.39 \\ -266.00 - & 302.30 - & 36.90 + & 1.00 \\ -690.00 - & 1061.00 - & 432.00 - & 66.70 \end{bmatrix}
$$

$$
= \frac{0.074\omega^2}{(10^3)g} \begin{bmatrix} 612.09 \\ 792.39 \\ -604.20 \\ -2249.70 \end{bmatrix} \qquad (5.68)
$$

$$
= \frac{(0.074)(604.20)\omega^2}{(10^3)g} \begin{bmatrix} 1.01 \\ 1.31 \\ -1.00 \\ -3.72 \end{bmatrix}
$$

The procedure may be repeated by using Eq. (5.67) and the normalized amplitudes obtained in Eq. (5.68). This yields

$$
\begin{bmatrix} y_1 \\ y_2 \\ y_3 \\ y_4 \end{bmatrix} = \frac{(0.074)\omega^2}{(10^3)g} \begin{bmatrix} 515.84 \\ 562.05 \\ -631.49 \\ -1406.69 \end{bmatrix}
$$

$$= \frac{(0.074)(515.84)\omega^2}{(10^3)g} \begin{bmatrix} 1.00 \\ 1.09 \\ -1.22 \\ -2.73 \end{bmatrix}$$

By repeating the procedure about four additional times, the matrix equation converges to the following values:

$$\begin{bmatrix} y_1 \\ y_2 \\ y_3 \\ y_4 \end{bmatrix} = \frac{(0.074)(430.3)\omega^3}{(10^3)g} \begin{bmatrix} 1.00 \\ 1.01 \\ -1.23 \\ -2.47 \end{bmatrix} \tag{5.69}$$

Equation (5.69) provides the amplitudes $y_1^{(2)}$, $y_2^{(2)}$, $y_3^{(2)}$, and $y_4^{(2)}$ under the weights W_1, W_2, W_3, and W_4, respectively. The amplitude $y_5^{(2)}$ may be determined from Eq. (5.62), which yields $y_5^{(2)} = -1.247$. Therefore, the amplitudes under the weight concentration points are

$$y_1^{(2)} = 1.00$$
$$y_2^{(2)} = 1.01$$
$$y_3^{(2)} = -1.23$$
$$y_4^{(2)} = -2.47$$
$$y_5^{(2)} = -1.25$$

Since in the first mode the amplitude $y_5^{(1)}$ was selected to be equal to unity, the normalized shape of the second mode of vibration will be constructed on the same basis. Therefore, by dividing all amplitudes by the absolute value of the amplitude $y_5^{(2)}$, we find

$$y_1^{(2)} = 0.80$$
$$y_2^{(2)} = 0.81$$
$$y_3^{(2)} = -0.98$$
$$y_4^{(2)} = -1.98$$
$$y_5^{(2)} = -1.00$$

The shape of the second mode of vibration is shown plotted in Fig. 5.7c.
The frequency of vibration of the second mode may be determined by

using Eq. (5.69), which yields

$$1.00 = \frac{(0.074)(430.30)\,\omega^2}{(10^3)g}(1.00)$$

or

$$\omega = \sqrt{\frac{(386)(10^3)}{(0.074)(430.30)}} = 110.10 \text{ rps}$$

Additional higher frequencies of vibration may be obtained by writing an orthogonality relation between the amplitudes of the second and third modes of vibration. This equation can be used to reduce the problem into three equations of motion and three amplitudes, yielding a 3×3 matrix. Application of the iteration procedure will converge the matrix equation to the third frequency and mode of vibration. The procedure may be continued to determine as many higher frequencies of vibration as desired. It should be noted, however, that, since the method is approximate, the error increases every time we calculate a higher frequency of vibration. Therefore, only a limited number of higher frequencies can be obtained with reasonable accuracy. For the above problem, the first two or three frequencies will have reasonable accuracy, but the accuracy of the remaining two higher frequencies is questionable. On this basis, use of the results obtained for higher frequencies of vibration by using approximate methods of analysis should be done with caution.

5.6   MYKLESTAD'S METHOD

A general method that can be used for the computation of the free frequencies of vibration of uniform and variable stiffness members was developed independently by N. O. Myklestad [28] and M. A. Prohl [29]. The method was initially developed for uncoupled bending vibration, but later it was extended to both coupled and uncoupled bending vibrations. In this section, Myklestad's method will be developed for the solution of the uncoupled bending vibration of members of either uniform or variable stiffness EI. Coupling torsional effects or the effects of nonlinear deformations and the like are not included in the analysis.

In general, Myklestad's method may be thought of as an extension of the Holzer method that is used extensively for the solution of torsional vibration problems. We should realize, however, that the solution of flexural vibration problems is a great deal more difficult, because we are involved with the computation of four elastic parameters for each length segment of the beam. For torsional vibrations, we need to compute only one elastic parameter. The solution of many problems by using this method can be accomplished

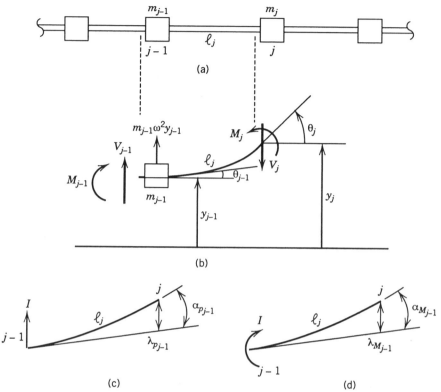

FIGURE 5.8. (a) Beam with lumped masses at the juncture point of each segment. (b) Free-body diagram of a segment of length ℓ_j. (c) Unit load applied at station $j - 1$. (d) Unit moment applied at station $j - 1$.

manually, as was done in Rayleigh's method and Stodola's method, but for more complicated problems the use of computer software will make the solution more convenient.

In using this method, the member should be divided into a sufficiently large number of weightless length segments, and assume that the modulus of elasticity E and moment of inertia I are constant throughout the length of the segment and equal to the mean value of the segment. The total mass of the member is converted into lumped masses at the juncture point of each segment, as shown in Fig. 5.8a, in a manner somewhat similar to the one used in the methods of Rayleigh and Stodola in the preceding sections.

If the member vibrates at a free frequency of vibration ω, the free-body diagram of a segment between stations $j - i$ and j in Fig. 5.8a will be as shown in Fig. 5.8b. Examining the various quantities involved in the segment of length ℓ_j, we note that the inertia force $m_{j-1}\omega^2 y_{j-1}$ of mass m_{j-1} is acting upward, and so we can apply D'Alembert's principle of dynamic equilibrium. Four equations may be written by using the segment in Fig. 5.8b. Two equations will involve the shear force V_j and bending moment M_j, with an additional two equations for the slope θ_j and deflection y_j. These four quantities are shown at the right end of the segment in Fig. 5.8b.

By using the segment in Fig. 5.8b and applying the two equilibrium conditions, one for forces in the vertical direction and one for moments, the following two equations are obtained:

$$V_j = V_{j-1} + m_{j-1}\omega^2 y_{j-1} \tag{5.70}$$

$$M_j = M_{j-1} + V_{j-1}\ell_j + m_{j-1}\omega^2 y_{j-1}\ell_j \tag{5.71}$$

The equations for the slope θ_j and deflection y_j may be written as follows:

$$
\begin{aligned}
\theta_j &= \theta_{j-1} + \frac{\ell_{j-1}^2}{2(EI)_j}(V_{j-1} + m_{j-1}\omega^2 y_{j-1}) + \frac{\ell_j}{(EI)_j}M_{j-1} \\
&= \theta_{j-1} + \frac{\ell_j^2}{2(EI)_j}V_j + \frac{\ell_j}{(EI)_j}M_{j-1}
\end{aligned}
\tag{5.72}
$$

$$
\begin{aligned}
y_j &= y_{j-1} + \theta_{j-1}\ell_j + \frac{\ell_j^3}{6(EI)_j}(V_{j-1} + m_{j-1}\omega^2 y_{j-1}) + \frac{\ell_j^2}{2(EI)_j}M_{j-1} \\
&= y_{j-1} + \theta_{j-1}\ell_j + \frac{\ell_j^3}{6(EI)_j}V_j + \frac{\ell_j^2}{2(EI)_j}M_{j-1}
\end{aligned}
\tag{5.73}
$$

In Eq. (5.72), the coefficient $\alpha_{p_{j-1}}$, where

$$\alpha_{p_{j-1}} = \frac{\ell^2_{\,j}}{2(EI)_j} \tag{5.74}$$

is the slope of the member at station j relative to a tangent drawn at station $j-1$, caused by a unit force applied at station $j-1$. This is shown in Fig. 5.8c. The coefficient $\alpha_{M_{j-1}}$, where

$$\alpha_{M_{j-1}} = \frac{\ell_j}{(EI)_j} \tag{5.75}$$

represents the slope at station j relative to a tangent at station $j-1$, which is caused by the application of a unit moment as shown in Fig. 5.8d.

In Eq. (5.73), the coefficient λ_{pj-1}, where

$$\lambda_{pj-1} = \frac{\ell_j^3}{6(EI)_j} \tag{5.76}$$

is the vertical distance from station j to the tangent at station $j - 1$, produced by a unit load at $j - 1$ as shown in Fig. 5.8c. The coefficient λ_{Mj-1}, where

$$\lambda_{Mj-1} = \frac{\ell_j^2}{2(EI)_j} \tag{5.77}$$

is the vertical distance from station j to the tangent at station $j - 1$ in Fig. 5.8d, caused by a unit moment at $j - 1$. These coefficients were computed by using the moment–area method. They can also be derived from Eqs. (5.70) and (5.71) by using known expressions from the mechanics of solids, which relate y, θ, M, and V.

Assume here that the number of ℓ_j segments is n—that is, $j = 1, 2, 3, \ldots, n$—and that the lumped masses $m_0, m_1, m_2, \ldots, m_n$ are located at stations $0, 1, 2, \ldots, n$, respectively. For each station, Eqs. (5.70) through (5.73) may be used to determine the arguments y, θ, M, and V. The procedure may be initiated at station 0, where the first mass concentration point is located. At station 0, two boundary conditions are known, which may be applied to Eqs. (5.70) through (5.73) to define two out of the four quantities y_0, θ_0, M_0, and V_0. At the same station, if desired, the normalizing value of unity can be assigned to the third quantity. Therefore, at station 0, there will be either two unknown quantities or one unknown.

At this point, assume that there are two unknown quantities at station 0, and assume a value of the frequency ω. On this basis, the quantities θ_1, y_1, V_1, and M_1 at the second station may be determined by using Eqs. (5.70) through (5.73), in terms of numerical values, and the two unknowns at station 0. This procedure may continue from station to station successively, until we reach the last station, which is station n. The four quantities at station n are now expressed in terms of numerical values and the two unknowns from station 0. The two boundary conditions of station n provide the criteria for the computation of the free frequencies of the member. The procedure is illustrated in the following example.

Example 5.6. Determine the frequency determinant and the frequency equation for a uniform simply supported beam by using Myklestad's method.

Solution. We assume here that the left end of the beam is station 0, and the right end support is station n. At station 0 the boundary conditions are $y_0 =$

$M_0 = 0$. At station 1, use of Eqs. (5.70) through (5.73) yields

$$V_1 = V_0 \tag{5.78}$$

$$M_1 = \ell_1 V_0 \tag{5.79}$$

$$\theta_1 = \theta_0 + \frac{\ell_1^2}{2(EI)_1} V_0 \tag{5.80}$$

$$y_1 = \ell_1 \theta_0 + \frac{\ell_1^2}{6(EI)_1} V_0 \tag{5.81}$$

In a similar manner, by using the results obtained in Eqs. (5.78) through (5.81), and again using Eqs. (5.70) through (5.73), we can write the following expressions for the second station:

$$V_2 = m_1 \omega^2 \ell_1 \theta_0 + \left(1 + \frac{m_1 \omega^2 \ell_1^3}{6(EI)_1}\right) V_0 \tag{5.82}$$

$$M_2 = m_1 \omega^2 \ell_1 \ell_2 \theta_0 + \left(\ell_1 + \ell_2 + \frac{m_1 \omega^2 \ell_1^3}{6(EI)_1}\right) V_0 \tag{5.83}$$

$$\theta_2 = \left(1 + \frac{m_1 \omega^2 \ell_2^2 \ell_1}{2(EI)_2}\right) \theta_0$$

$$+ \left(\frac{\ell_1^2}{2(EI)_1} + \frac{\ell_1 \ell_2}{(EI)_2} + \frac{\ell_2^2}{2(EI)_2} + \frac{m_1 \omega^2 \ell_1^3 \ell_2^2}{12(EI)_1(EI)_2}\right) V_0 \tag{5.84}$$

$$y_2 = \left(\ell_1 \ell_2 + \frac{m_1 \omega^2 \ell_1 \ell_2^3}{6(EI)_2}\right) \theta_0$$

$$+ \left(\frac{\ell_1^3}{6(EI)_1} + \frac{\ell_1^2 \ell_2}{2(EI)_1} + \frac{\ell_2^3}{6(EI)_2} + \frac{m_1 \omega^2 \ell_1^3 \ell_2^3}{36(EI)_1(EI)_2}\right) V_0 \tag{5.85}$$

If we proceed in a similar manner from one station to the next, we will reach the nth station. We can easily observe here that the quantities y, θ, M, and V at each station are linear combinations of the unknown quantities θ_0 and V_0 of the first station. On this basis, we conclude that the following expressions for y_n, θ_n, M_n, and V_n at the nth station may be written:

$$y_n = A_1 \theta_0 + A_2 V_0 \tag{5.86}$$

$$\theta_n = B_1 \theta_0 + B_2 V_0 \tag{5.87}$$

$$M_n = C_1 \theta_0 + C_2 V_0 \tag{5.88}$$

$$V_n = D_1 \theta_0 + D_2 V_0 \tag{5.89}$$

The pairs of constants A_1,A_2, B_1,B_2, C_1,C_2, and D_1,D_2 are the coefficients of θ_0 and V_0 in Eqs. (5.86) through (5.89), respectively, and involve quantities similar to the ones shown in Eqs. (5.82) through (5.85). Therefore, if a value for the frequency ω is assumed, the numerical values of these constants may be obtained.

At the nth station of the simply supported beam, the boundary conditions are $y_n = M_n = 0$, and Eqs. (5.86) and (5.88) yield

$$A_1\theta_0 + A_2V_0 = 0 \tag{5.90}$$

$$C_1\theta_0 + C_2V_0 = 0 \tag{5.91}$$

For a solution other than the trivial one, the determinant of the coefficients of θ_0 and V_0 in Eqs. (5.90) and (5.91) must be zero; that is,

$$\begin{vmatrix} A_1 & A_2 \\ C_1 & C_2 \end{vmatrix} = 0 \tag{5.92}$$

Equation (5.92) is known as the frequency determinant. By expanding the determinant in Eq. (5.92), we obtain

$$A_1C_2 - C_1A_2 = 0 \tag{5.93}$$

Equation (5.93) is the frequency equation. Since A_1, A_2, C_1, and C_2 are functions of the frequency ω, we conclude that the values of ω that satisfy Eq. (5.93) are the free frequencies of vibration of the simply supported beam.

For each value of the frequency ω, the corresponding mode shape may be determined from Eqs. (5.70) through (5.73). We have to use these equations for each station j between 0 and n. For example, at station 2, the value of y_2 may be obtained from Eq. (5.85). In these computations, the normalized value of unity may be used for V_0, and since ω is known, the value of θ_0 may be determined by using either Eq. (5.90) or Eq. (5.91). In this manner, numerical values of y are obtained at all stations, and, consequently, the mode shape for each frequency ω is completely defined. It should be noted, however, that Myklestad's method is intended to be used for more complicated problems, where closed-form solutions are very difficult to apply. Since this example involves a uniform simply supported beam and exact closed-form solutions are easily available, it would not be wise to use Myklestad's method. It is, however, a good example to illustrate the mechanics regarding the application of the method.

The expressions given by Eqs. (5.70) through (5.73) may be put into a matrix form, in order to obtain a relationship between the quantities y, θ, M, and V at stations j and $j - 1$. This mathematical formulation introduces the concept of *transfer matrices*, which are used extensively for the solution of structural and mechanical problems. On this basis, we have

$$
\begin{bmatrix} y \\ \theta \\ M \\ V \end{bmatrix}^{j} = \begin{bmatrix} \left(\dfrac{m_{j-1}\omega^2 \ell_j^3}{6(EI)_j} + 1 \right) & \ell_j & \dfrac{\ell_j^2}{2(EI)_j} & \dfrac{\ell_j^3}{6(EI)_j} \\[2mm] \dfrac{m_{j-1}\omega^2 \ell_j^2}{2(EI)_j} & 1 & \dfrac{\ell_j}{(EI)_j} & \dfrac{\ell_j^2}{2(EI)_j} \\[2mm] m_{j-1}\omega^2 \ell_j & 0 & 1 & \ell_j \\[2mm] m_{j-1}\omega^2 & 0 & 0 & 1 \end{bmatrix} \begin{bmatrix} y \\ \theta \\ M \\ V \end{bmatrix}^{j-1}
\tag{5.94}
$$

In a more compact form, Eq. (5.94) may be written

$$
\{p\}_j = [N]_{j-1}\{p\}_{j-1}
\tag{5.95}
$$

where $\{p\}_j$ and $\{p\}_{j-1}$ are the column matrices of the quantities y, θ, M, and V at stations j and $j-1$, respectively, and

$$
[N]_{j-1} = \begin{bmatrix} \left(\dfrac{m_{j-1}\omega^2 \ell_j^3}{6(EI)_j} + 1 \right) & \ell_j & \dfrac{\ell_j^2}{2(EI)_j} & \dfrac{\ell_j^3}{6(EI)_j} \\[2mm] \dfrac{m_{j-1}\omega^2 \ell_j^2}{2(EI)_j} & 1 & \dfrac{\ell_j}{(EI)_j} & \dfrac{\ell_j^2}{2(EI)_j} \\[2mm] m_{j-1}\omega^2 \ell_j & 0 & 1 & \ell_j \\[2mm] m_{j-1}\omega^2 & 0 & 0 & 1 \end{bmatrix}
\tag{5.96}
$$

is a square matrix that is known as the transfer matrix for station $j-1$. With the easy availability of computers to carry out the bulk of the computational work involved, transfer matrices are used rather extensively today for the solution of many structural and mechanical problems.

5.7 TRANSFER MATRICES

In the preceding sections, various methods have been discussed that can be used for the computation of free frequencies of vibration of structural and mechanical systems with multiple and infinite degrees of freedom. The objectives in this section are similar, but a Myklestad–Holzer type analysis, known as transfer matrices, is used. The equations involved are expressed in matrix form—for example, see Eq. (5.96)—and the computations can easily be handled by a digital computer, since they involve basic operations of matrix algebra. See also refs. 30 and 31 for additional information.

In this section and the following sections, we deal with the formulation of transfer matrices for various types of structural and mechanical problems and their application to various types of linearly elastic systems.

5.7.1 Transfer Matrices for Free Vibration of Single-Degree Spring–Mass Systems

Consider the spring–mass system in Fig. 5.9a, which consists of masses m_{j-1}, m_j, and m_{j+1} that are connected by the linear springs of constants k_{j-1}, k_j, k_{j+1}, and k_{j+2}. The masses are permitted to move only in the horizontal direction, and the displacements in the springs are denoted by x_{j-1}, x_j, and x_{j+1}, as shown in the figure. The free-body diagram in Fig. 5.9b depicts the forces and the displacements to the right and to the left of the spring of constant k_j and to the right and left of mass m_j, when the

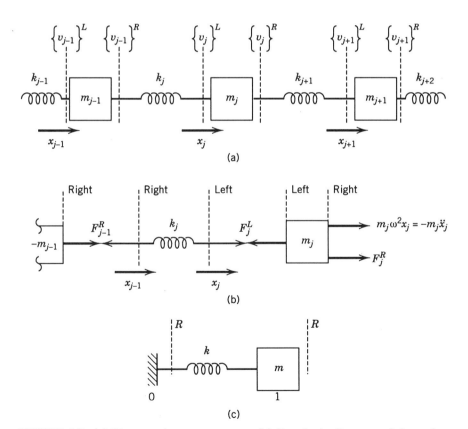

FIGURE 5.9. (a) Linear spring–mass system. (b) Free-body diagrams of the springs and masses when the spring–mass system vibrates at a frequncy ω. (c) Single-degree spring–mass system.

system vibrates freely at a frequency ω. In the work that follows, the letter L is used to indicate left-hand side, and R is used to indicate right-hand side.

By using the free-body diagram of spring k_j in Fig. 5.9b, the following two equations are written:

$$F_{j-1}^R = F_j^L \tag{5.97}$$

which expresses the equilibrium condition of spring k_j, and

$$x_j^L - x_{j-1}^R = \frac{F_{j-1}^R}{k_j}$$

or

$$x_j^L = x_{j-1}^R + \frac{F_{j-1}^R}{k_j} \tag{5.98}$$

which is obtained from the stiffness property of spring k_j.

Equations (5.97) and (5.98) may be written in matrix form:

$$\begin{bmatrix} x \\ F \end{bmatrix}_j^L = \begin{bmatrix} 1 & 1/k_j \\ 0 & 1 \end{bmatrix} \begin{bmatrix} x \\ F \end{bmatrix}_{j-1}^R \tag{5.99}$$

or, in a more compact form,

$$\{v_j\}^L = [F]\{v_{j-1}\}^R \tag{5.100}$$

where

$$\{v_j\}^L = \begin{bmatrix} x \\ F \end{bmatrix}_j^L \qquad \{v_{j-1}\}^R = \begin{bmatrix} x \\ F \end{bmatrix}_{j-1}^R \tag{5.101}$$

$$[F] = \begin{bmatrix} 1 & 1/k_j \\ 0 & 1 \end{bmatrix} \tag{5.102}$$

The column matrices in Eq. (5.101) are known as *state vectors*, which involve the displacements x_j and x_{j-1} at points j and $j - 1$, respectively, and the internal forces F_j and F_{j-1} at the same points. The square matrix in Eq. (5.102) is known as the *field transfer matrix* and relates the state vectors $\{v_j\}^L$ and $\{v_{j-1}\}^R$.

By considering again the free-body diagram of the mass m_j in Fig. 5.9b, another set of two equations is written:

$$x_j^R = x_j^L \tag{5.103}$$

indicating that the displacements x_j to the left and right of the rigid mass m_j are equal, and

$$F_j^R = -m_j\omega^2 x_j + F_j^L \qquad (5.104)$$

which relates the forces acting on the mass m_j. Equations (5.103) and (5.104), in matrix notation, are as follows:

$$\begin{bmatrix} x \\ F \end{bmatrix}_j^R = \begin{bmatrix} 1 & 0 \\ -m_j\omega^2 & 1 \end{bmatrix} \begin{bmatrix} x \\ F \end{bmatrix}_j^L \qquad (5.105)$$

or

$$\{v_j\}^R = [P]\{v_j\}^L \qquad (5.106)$$

In Eq. (5.106),

$$\{v_j\}^R = \begin{bmatrix} x \\ F \end{bmatrix}_j^R \quad \text{and} \quad \{v_j\}^L = \begin{bmatrix} x \\ F \end{bmatrix}_j^L \qquad (5.107)$$

are the state vectors to the right and left of the mass m_j, and

$$[P] = \begin{bmatrix} 1 & 0 \\ -m_j\omega^2 & 1 \end{bmatrix} \qquad (5.108)$$

is the *point transfer matrix* relating the state vectors $\{v_j\}^R$ and $\{v_j\}^L$ to the right and left of the mass m_j.

Equations (5.100) and (5.106) can be combined to relate the state vectors $\{v_{j-1}\}^R$ and $\{v_j\}^R$ in Fig. 5.9a. This is accomplished by substituting Eq. (5.100) into Eq. (5.106). This yields

$$\{v_j\}^R = [P][F]\{v_{j-1}\}^R \qquad (5.109)$$

In expanded form, Eq. (5.109) yields

$$\begin{bmatrix} x \\ F \end{bmatrix}_j^R = \begin{bmatrix} 1 & 0 \\ -m_j\,\omega^2 & 1 \end{bmatrix} \begin{bmatrix} 1 & 1/k_j \\ 0 & 1 \end{bmatrix} \begin{bmatrix} x \\ F \end{bmatrix}_{j-1}^R$$

or, after carrying out the multiplication of the matrices $[P]$ and $[F]$,

$$\begin{bmatrix} x \\ F \end{bmatrix}_j^R = \begin{bmatrix} 1 & 1/k_j \\ -m_j\omega^2 & (1 - m_j\omega^2/k_j) \end{bmatrix} \begin{bmatrix} x \\ F \end{bmatrix}_{j-1}^R \qquad (5.110)$$

Example 5.7. Determine the free frequency of vibration of the single-degree spring–mass system in Fig. 5.9c.

Solution. Equation (5.110) can be applied here to solve the single-degree-of-freedom spring–mass system in Fig. 5.9c. The state vectors at stations 0 and 1 are

$$\{v_0\}^R = \begin{bmatrix} x \\ F \end{bmatrix}_0^R = \begin{bmatrix} 0 \\ F \end{bmatrix}_0^R \quad \text{and} \quad \{v_1\}^R = \begin{bmatrix} x \\ F \end{bmatrix}_1^R = \begin{bmatrix} x \\ 0 \end{bmatrix}_1^R$$

Hence, by applying Eq. (5.110), we find

$$\begin{bmatrix} x \\ 0 \end{bmatrix}_1^R = \begin{bmatrix} 1 & 1/k \\ -m\omega^2 & (1 - m\omega^2/k) \end{bmatrix} \begin{bmatrix} 0 \\ F \end{bmatrix}_0^R \qquad (5.111)$$

Equation (5.111) yields

$$1 - \frac{\omega^2}{k} = 0$$

or

$$\omega = \sqrt{\frac{k}{m}}$$

This is the free undamped frequency of vibration of the spring–mass system in Fig. 5.9c.

5.7.2 Transfer Matrices for Free Vibration of Single-Degree, Torsional, Disk–Shaft Systems

Consider now the freely vibrating system in Fig. 5.10a, consisting of the rigid disks of mass moments of inertia J_{j-2}, J_{j-1}, J_j, and J_{j+1} that are connected to the elastic shafts of torsional stiffnesses $k_{t_{j-1}}$, k_{t_j}, and $k_{t_{j+1}}$. The angular displacements of the shafts and disks are designated by ϕ_{j-2}, ϕ_{j-1}, ϕ_j, and ϕ_{j+1}, as shown in the same figure. The procedure in deriving the transfer matrices is identical to the one used for the spring–mass system in Fig. 5.9a. Thus, from the free-body diagram of the shaft in Fig. 5.10b, the following expressions are written:

$$\phi_j^L - \phi_{j-1}^R = \frac{T_{j-1}^R}{k_{t_j}}$$

or

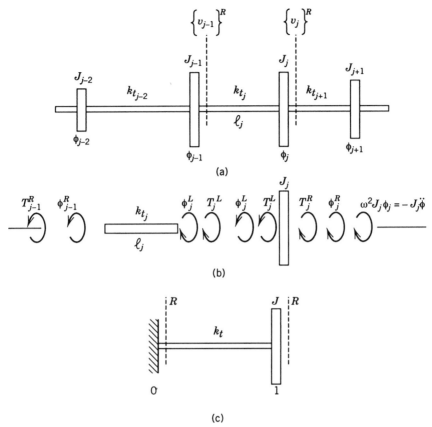

FIGURE 5.10. (a) Torsional disk–shaft system vibrating freely at a frequency of vibration. (b) Free-body diagram of disks and shafts. (c) Single-degree disk–shaft system.

$$\phi_j^L = \phi_{j-1}^R + \frac{T_{j-1}^R}{k_{t_j}} \tag{5.112}$$

and

$$T_j^L = T_{j-1}^R \tag{5.113}$$

In matrix form, Eqs. (5.112) and (5.113) are expressed as

$$\begin{bmatrix} \phi \\ T \end{bmatrix}_j^L = \begin{bmatrix} 1 & 1/k_{t_j} \\ 0 & 1 \end{bmatrix} \begin{bmatrix} \phi \\ T \end{bmatrix}_{j-1}^R \tag{5.114}$$

or

$$\{v_j\}^L = [F]\{v_{j-1}\}^R \qquad (5.115)$$

From the free-body diagram of the disk J_j in Fig. 5.10b, a set of two equations is written as follows:

$$\phi_j^R = \phi_j^L \qquad (5.116)$$

indicating that the angular twist ϕ_j remains unchanged on either side of the disk, and

$$T_j^R = -\omega^2 J_j \phi_j + T_j^L \qquad (5.117)$$

Equations (5.116) and (5.117) are written in matrix form as shown below:

$$\begin{bmatrix} \phi \\ T \end{bmatrix}_j^R = \begin{bmatrix} 1 & 0 \\ -\omega^2 J_j & 1 \end{bmatrix} \begin{bmatrix} \phi \\ T \end{bmatrix}_j^L \qquad (5.118)$$

or

$$\{v_j\}^R = [P]\{v_j\}^L \qquad (5.119)$$

Equations (5.115) and (5.119) can be combined as before to relate the state vectors $\{v_{j-1}\}^R$ and $\{v_j\}^R$ in Fig. 5.10a. By substituting Eq. (5.115) into Eq. (5.119), we find

$$\{v_j\}^R = [P][F]\{v_{j-1}\}^R \qquad (5.120)$$

In expanded form, Eq. (5.120) yields

$$\begin{bmatrix} \phi \\ T \end{bmatrix}_j^R = \begin{bmatrix} 1 & 0 \\ -\omega^2 J_j & 1 \end{bmatrix} \begin{bmatrix} 1 & 1/k_{t_j} \\ 0 & 1 \end{bmatrix} \begin{bmatrix} \phi \\ T \end{bmatrix}_{j-1}^R$$

By carrying out the required multiplication of the two matrices $[P]$ and $[F]$, the above expression yields

$$\begin{bmatrix} \phi \\ T \end{bmatrix}_j^R = \begin{bmatrix} 1 & 1/k_{t_j} \\ -\omega^2 J_j & (1 - \omega^2 J_j/k_{t_j}) \end{bmatrix} \begin{bmatrix} \phi \\ T \end{bmatrix}_{j-1}^R \qquad (5.121)$$

Equation (5.121) can be used to determine the free undamped frequency of vibration of a single-degree-of-freedom torsional system.

Example 5.8. Determine the torsional frequency of vibration of the single-degree disk–shaft system in Fig. 5.10c by using transfer matrices.

Solution. The state vectors at positions 0 and 1 in Fig. 5.10c are as follows:

$$\{v_0\}^R = \begin{bmatrix} \phi \\ T \end{bmatrix}_0^R = \begin{bmatrix} 0 \\ T \end{bmatrix}_0^R \quad \text{and} \quad \{v_1\}^R = \begin{bmatrix} \phi \\ T \end{bmatrix}_j^R = \begin{bmatrix} \phi \\ 0 \end{bmatrix}_1^R$$

Therefore, Eq. (5.121) is written as shown below:

$$\begin{bmatrix} \phi \\ 0 \end{bmatrix}_1^R = \begin{bmatrix} 1 & 1/k_t \\ -\omega^2 J & (1 - \omega^2 J/k_t) \end{bmatrix} \begin{bmatrix} 0 \\ T \end{bmatrix}_0^R \tag{5.122}$$

Equation (5.122) yields

$$1 - \frac{\omega^2 J}{k_t} = 0$$

or

$$\omega = \sqrt{\frac{k_t}{J}}$$

This is the free torsional frequency of vibration, in radians per second, of the disk-shaft system in Fig. 5.10c.

5.8   TRANSFER MATRICES FOR FORCED VIBRATION OF ONE-DEGREE SPRING-MASS SYSTEMS

Consider the spring-mass system in Fig. 5.11a, which is excited harmonically by the harmonic forces shown in the same figure. The undamped steady-state case will be considered here, in which the system vibrates with the frequency ω_f of the exciting force. The amplitude response will be determined by using transfer matrices. Transfer matrices can also be used to solve the more general problem of forced undamped vibration, which includes both the transient and the steady-state conditions.

By considering the free-body diagram of mass m_j in Fig. 5.11b and proceeding in a manner similar to the one discussed in the preceding section, the following two equations are obtained:

$$X_j^R = X_j^L \tag{5.123}$$

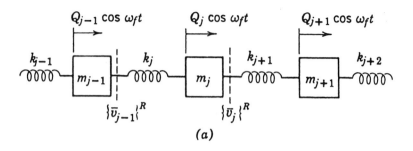

(a)

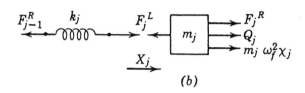

(b)

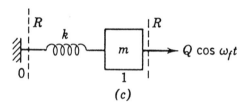

(c)

FIGURE 5.11. (a) Spring–mass system vibrating harmonically under the action of harmonic forces. (b) Free-body diagrams of spring and mass. (c) Single-degree spring–mass system subjected to a harmonic force.

indicating that the displacements on either side of mass m_j are the same, and

$$F_j^R = F_j^L - m_j \omega_f^2 X_j^L - Q_j \tag{5.124}$$

expressing the dynamic equilibrium of the mass m_j.

Equations (5.123) and (5.124) can be expressed in matrix form as follows:

$$
\begin{bmatrix} X \\ F \\ \hline 1 \end{bmatrix}_j^R =
\begin{bmatrix} 1 & 0 & | & 0 \\ -m_j \omega_f^2 & 1 & | & -Q_j \\ \hline 0 & 0 & | & 1 \end{bmatrix}
\begin{bmatrix} X \\ F \\ \hline 1 \end{bmatrix}_j^L \tag{5.125}
$$

or, in short form,

$$\{\bar{v}_j\}^R = [\bar{P}]\{\bar{v}_j\}^L \tag{5.126}$$

In the above equation, the state vectors

$$\{\bar{v}_j\}^R = \begin{bmatrix} X \\ F \\ \hline 1 \end{bmatrix}_j^R \quad \text{and} \quad \{\bar{v}_j\}^L = \begin{bmatrix} X \\ F \\ \hline 1 \end{bmatrix}_j^L \tag{5.127}$$

are the so-called extended state vectors, and

$$[\bar{P}] = \begin{bmatrix} 1 & 0 & 0 \\ -m_j\omega_f^2 & 1 & -Q_j \\ \hline 0 & 0 & 1 \end{bmatrix} \tag{5.128}$$

is the extended transfer point matrix. It may easily be verified, by multiplying $[P]$ by $\{\bar{v}_j\}^L$, that Eq. (5.125) yields Eqs. (5.123) and (5.124) and the identity $1 \equiv 1$.

From the free-body diagram of the spring k_j in Fig. 5.11b, the following two equations can be obtained:

$$F_j^L = F_{j-1}^R \tag{5.129}$$

and

$$X_j^L - X_{j-1}^R = \frac{F_{j-1}^R}{k_j}$$

or

$$X_j^L = X_{j-1}^R + \frac{F_{j-1}^R}{k_j} \tag{5.130}$$

Thus, by using Eqs. (5.129) and (5.130), we can write the following matrix equation:

$$\begin{bmatrix} X \\ F \\ \hline 1 \end{bmatrix}_j^L = \begin{bmatrix} 1 & 1/k_j & 0 \\ 0 & 1 & 0 \\ \hline 0 & 0 & 1 \end{bmatrix} \begin{bmatrix} X \\ F \\ \hline 1 \end{bmatrix}_{j-1}^R \tag{5.131}$$

or

$$\{\bar{v}_j\}^L = [\bar{F}]\{\bar{v}_{j-1}\}^R \tag{5.132}$$

where $\{\bar{v}_j\}^L$ and $\{\bar{v}_{j-1}\}^R$ are the state vectors, and $[\bar{F}]$ is the extended field matrix.

Equations (5.126) and (5.132) may be combined to yield a relationship between the vectors $\{\bar{v}_{j-1}\}^R$ and $\{\bar{v}_j\}^R$ in Fig. 5.11a; that is, by substituting Eq. (5.132) into Eq. (5.126), the result is

$$\{\bar{v}_j\}^R = [\bar{P}][\bar{F}]\{\bar{v}_{j-1}\}^R \tag{5.133}$$

By carrying out the matrix multiplication, Eq. (5.133) yields the following matrix equation:

$$\begin{bmatrix} X \\ F \\ \hline 1 \end{bmatrix}_j^R = \left[\begin{array}{cc|c} 1 & 1/k_j & 0 \\ -m_j\omega_f^2 & (1 - m_j\omega_f^2/k_j) & -Q_j \\ \hline 0 & 0 & 1 \end{array} \right] \begin{bmatrix} X \\ F \\ \hline 1 \end{bmatrix}_{j-1}^R \tag{5.134}$$

Equation (5.134) may be used to determine the dynamic response of a single-degree spring–mass system subjected to a harmonic force.

Example 5.9. Determine the dynamic response of the single-degree spring–mass system in Fig. 5.11c, which is subjected to the harmonic force $Q \cos \omega_f t$, where ω_f is the frequency of the force. Use appropriate transfer matrices.

Solution. The state vectors $\{\bar{v}_0\}^R$ and $\{\bar{v}_1\}^R$ at positions 0 and 1, respectively, in Fig. 5.11c are as follows:

$$\{\bar{v}_0\}^R = \begin{bmatrix} 0 \\ F \\ \hline 1 \end{bmatrix}_0^R \quad \text{and} \quad \{\bar{v}_1\}^R = \begin{bmatrix} X \\ 0 \\ \hline 1 \end{bmatrix}_1^R \tag{5.135}$$

Hence, Eq. (5.134) yields

$$\begin{bmatrix} X \\ 0 \\ \hline 1 \end{bmatrix}_1^R = \left[\begin{array}{cc|c} 1 & 1/k & 0 \\ -m\omega_f^2 & (1 - m\omega_f^2/k) & -Q \\ \hline 0 & 0 & 1 \end{array} \right] \begin{bmatrix} 0 \\ F \\ \hline 1 \end{bmatrix}_0^R \tag{5.136}$$

Equation (5.136), by expansion, yields the following two equations:

$$X_1 = \frac{F_0}{k} \tag{5.137}$$

$$F_0\left(1 - \frac{m\omega_f^2}{k}\right) - Q = 0 \tag{5.138}$$

From Eq. (5.137), we find

$$F_0 = kX_1 \tag{5.139}$$

Thus, by substituting Eq. (5.139) into Eq. (5.138), we find

$$kX_1\left(1 - \frac{m\omega_f^2}{k}\right) - Q = 0$$

or

$$Q = kX_1 - m\omega_f^2 X_1 \tag{5.140}$$

Equation (5.140) yields

$$X_1 = \frac{Q}{k - m\omega_f^2} \tag{5.141}$$

Equation (5.141) is identical to Eq. (3.6), which was obtained in Section 3.2.

5.9 TRANSFER MATRICES FOR FREE VIBRATION OF FLEXURAL ONE-DEGREE SYSTEMS

Transfer matrices can also be used to determine the frequency response of beam-like systems. A member with continuous mass and elasticity and of either uniform or variable stiffness EI can be reduced to one whose mass is concentrated at discrete points along its length, as shown in Fig. 5.12a. The member retains its elastic properties between mass concentration points, but it is assumed to be massless between these points.

At a deflection configuration represented by the solid line in Fig. 5.12a, the state vector $\{v_j\}$ at station j will again represent the displacements and the corresponding internal forces at this station. Relations between the state-vectors at sequential stations and over mass concentration points will be derived as before, by using field and point transfer matrices.

Consider the system in Fig. 5.12, which is assumed to vibrate freely in its vertical direction and from its static equilibrium position. At an extreme amplitude configuration corresponding to a principal mode of vibration, the free-body diagram of the mass m_j is shown in Fig. 5.12b, where $-m_j\ddot{y}_j = m_j\omega^2 y_j$ is the inertia force, V_j is the internal shear force, and M_j is the internal bending moment.

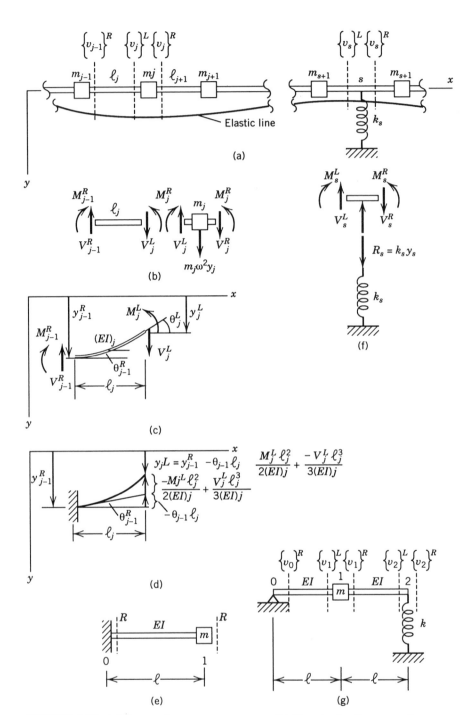

FIGURE 5.12. (a) Lumped parameter system. (b) Free-body diagrams of mass m_j and segment ℓ_j. (c) Free-body diagram of segment ℓ_j in deflection configuration. (d) Free-body diagram of ℓ_j with fixed left end. (e) One degree flexural system. (g) One-degree flexural system on elastic support.

The point transfer matrix relating the state vectors $\{v_j\}^L$ and $\{v_j\}^R$ in Fig. 5.12a may be determined from the relations establishing continuity for moment, deflection, and slope across the lumped mass m_j and from the equilibrium condition of all the forces, including the inertia force $m_j\omega^2 y_j$, in the vertical direction. Hence, from continuity conditions, we have

$$y_j^R = y_j^L \qquad (5.142)$$

$$\theta_j^R = \theta_j^L \qquad (5.143)$$

$$M_j^R = M_j^L \qquad (5.144)$$

and from equilibrium conditions, we have

$$V_j^R = V_j^L - m_j\omega^2 y_j \qquad (5.145)$$

The deflection y_j^L and rotation θ_j^L are shown in the free-body diagram in Fig. 5.12c.

In matrix form, Eqs. (5.142) to (5.145) inclusive are written as follows:

$$
\begin{bmatrix} y \\ \theta \\ M \\ V \end{bmatrix}_j^R =
\begin{bmatrix} 1 & 0 & 0 & 0 \\ 0 & 1 & 0 & 0 \\ 0 & 0 & 1 & 0 \\ -m_j\omega^2 & 0 & 0 & 1 \end{bmatrix}
\begin{bmatrix} y \\ \theta \\ M \\ V \end{bmatrix}_j^L \qquad (5.146)
$$

where

$$
\{v_j\}^R = \begin{bmatrix} y \\ \theta \\ M \\ V \end{bmatrix}_j^R \quad \text{and} \quad \{v_j\}^L = \begin{bmatrix} y \\ \theta \\ M \\ V \end{bmatrix}_j^L \qquad (5.147)
$$

are the state vectors, and

$$
[P] = \begin{bmatrix} 1 & 0 & 0 & 0 \\ 0 & 1 & 0 & 0 \\ 0 & 0 & 1 & 0 \\ -m_j\omega^2 & 0 & 0 & 1 \end{bmatrix} \qquad (5.148)
$$

is the point transfer matrix.

By choosing the first element of the state vector to be $-y$, all the elements of the point matrix may be written as positive; that is,

$$\begin{bmatrix} -y \\ \theta \\ M \\ V \end{bmatrix}_j^R = \begin{bmatrix} 1 & 0 & 0 & 0 \\ 0 & 1 & 0 & 0 \\ 0 & 0 & 1 & 0 \\ m_j\omega^2 & 0 & 0 & 1 \end{bmatrix} \begin{bmatrix} -y \\ \theta \\ M \\ V \end{bmatrix}_j^L \tag{5.149}$$

or

$$\{v_j\}^R = [P]\{v_j\}^L \tag{5.150}$$

Next, the field transfer matrix for the massless element ℓ_j will be derived. The internal-forces acting on the sides of this element are shown in Fig. 5.12b. Its deformed configuration is shown in Fig. 5.12c. The equilibrium condition in the vertical direction requires the sum of all the forces acting in this direction to be equal to zero; that is,

$$V_j^L - V_{j-1}^R = 0 \tag{5.151}$$

The moment equilibrium condition, that is, the summation of the moments about point $j-1$ of all the forces and moments acting on the element ℓ_j, yields

$$M_j^L - M_{j-1}^R - V_j^L \ell_j = 0 \tag{5.152}$$

Thus, we have

$$V_j^L = V_{j-1}^R \tag{5.153}$$

and

$$M_j^L = M_{j-1}^R + V_j^L \ell_j \tag{5.154}$$

Two additional relations may be written by considering the deflection y_j^L and slope θ_j^L. These expressions are as follows:

$$y_j^L = y_{j-1}^R - \theta_{j-1}^R \ell_j + \left[-\frac{M_j^L \ell_j^2}{2(EI)_j} + \frac{V_j^L \ell_j^3}{3(EI)_j} \right] \tag{5.155}$$

$$\theta_j^L = \theta_{j-1}^R + \left[\frac{M_j^L \ell_j}{(EI)_j} - \frac{V_j^L \ell_j^2}{2(EI)_j} \right] \tag{5.156}$$

The terms in the square brackets can be obtained by assuming the element of length ℓ_j as a cantilever beam fixed at the left end—the $j - 1$ end—and acted on by the moment M_j^L and shear force V_j^L at the j end. These expressions can be found in handbooks dealing with mechanics of solids. As an illustration, the vectors representing the terms in Eq. (5.155) are shown in Fig. 5.12d.

By using Eqs. (5.153) and (5.154), Eqs. (5.155) and (5.156) are written as follows:

$$-y_j^L = -y_j^R + \ell_j \theta_{j-1}^R + \frac{\ell_j^2}{2(EI)_j} M_{j-1}^R + \frac{\ell_j^3}{6(EI)_j} V_{j-1}^R \qquad (5.157)$$

$$\theta_j^L = \theta_{j-1}^R + \frac{\ell_j}{(EI)_j} M_{j-1}^R + \frac{\ell_j^2}{2(EI)_j} V_{j-1}^R \qquad (5.158)$$

The above two equations relate the elements of the state vector $\{v_j\}^L$ in terms of the elements of the state vector $\{v_{j-1}\}^R$. The minus sign in y_j is placed so that the elements of the matrix are all positive. This, of course, is not necessary, but it is used for convenience in the computations involved.

In matrix form, Eqs. (5.153), (5.154), (5.157), and (5.158) are written as shown below:

$$\begin{bmatrix} -y \\ \theta \\ M \\ V \end{bmatrix}_j^L = \begin{bmatrix} 1 & \ell_j & \dfrac{\ell_j^2}{2(EI)_j} & \dfrac{\ell_j^3}{6(EI)_j} \\ 0 & 1 & \dfrac{\ell_j}{(EI)_j} & \dfrac{\ell_j^2}{2(EI)_j} \\ 0 & 0 & 1 & \ell_j \\ 0 & 0 & 0 & 1 \end{bmatrix} \begin{bmatrix} -y \\ \theta \\ M \\ V \end{bmatrix}_{j-1}^R \qquad (5.159)$$

or

$$\{v_j\}^L = [F]\{v_{j-1}\}^R \qquad (5.160)$$

where $[F]$ is the field transfer matrix relating the state vectors $\{v_j\}^L$ and $\{v_{j-1}\}^R$ in Fig. 5.12a.

By substituting Eq. (5.159), or (5.160), into Eq. (5.149), or (5.150), respectively, a matrix relating the state vectors $\{v_{j-1}\}^R$ and $\{v_j\}^R$ in Fig. 5.12a is obtained; that is,

$$\{v_j\}^R = [P][F]\{v_{j-1}\}^R \qquad (5.161)$$

By carrying out the required matrix multiplication, the result is

$$
\begin{bmatrix} -y \\ \theta \\ M \\ V \end{bmatrix}_j^L = \begin{bmatrix} 1 & \ell_j & \dfrac{\ell_j^2}{2(EI)_j} & \dfrac{\ell_j^3}{6(EI)_j} \\[2ex] 0 & 1 & \dfrac{\ell_j}{(EI)_j} & \dfrac{\ell_j^2}{2(EI)_j} \\[2ex] 0 & 0 & 1 & \ell_j \\[2ex] m_j\omega^2 & m_j\omega^2\ell_j & \dfrac{m_j\omega^2\ell_j^2}{2(EI)_j} & \left(1 + \dfrac{m_j\omega^2\ell_j^3}{6(EI)_j}\right) \end{bmatrix} \begin{bmatrix} -y \\ \theta \\ M \\ V \end{bmatrix}_{j-1}^R
\tag{5.162}
$$

Equation (5.162) may be used to determine the free frequency of vibration of uniform weightless one-degree flexural systems.

Example 5.10. Determine the free frequency of vibration of the uniform weightless cantilever beam in Fig. 5.12e, carrying a mass m at the free end.

Solution. In this case, station j becomes station 1 in Fig. 5.12e, and station $j - 1$ is station 0 in the figure.

For this problem, it may be noted that $y_0 = \theta_0 = M_1^R = V_1^R = 0$. Thus, the last two rows of the matrix in Eq. (5.162) yield the following expressions:

$$
M_0^R + \ell V_0^R = 0
\tag{5.163}
$$

$$
m\omega^2 \frac{\ell^2}{2EI} M_0^R + \left(1 + \frac{m\omega^2\ell^3}{6EI}\right) V_0^R = 0
\tag{5.164}
$$

For a nontrivial solution, the determinant of the coefficients of M_0^R and V_0^R of the above equations should be equal to zero. Hence, we should have

$$
\begin{vmatrix} 1 & \ell \\[2ex] m\omega^2 \dfrac{\ell^2}{2EI} & \left(1 + \dfrac{m\omega^2\ell^3}{6EI}\right) \end{vmatrix} = 0
\tag{5.165}
$$

Expansion of the determinant in Eq. (5.165) yields

$$
\omega^2 = \frac{3EI}{m\ell^3}
$$

or

$$\omega = \sqrt{\frac{3EI}{m\ell^3}} \tag{5.166}$$

In Eq. (5.166), ω is the free frequency of the cantilever beam in Fig. 5.12e, in radians per second.

5.9.1 Flexural Systems Involving Elastic Supports

If a beam is supported elastically at a point s by a linear spring of constant k_s as shown in Fig. 5.12a, a point transfer matrix relating the state vectors $\{v_s\}^L$ and $\{v_s\}^R$ may easily be obtained. The four equations relating the elements of the state vectors to the left and right sides of the spring are as follows:

$$-y_s^R = -y_s^L \tag{5.167}$$

$$\theta_s^R = \theta_s^L \tag{5.168}$$

$$M_s^R = m_s^L \tag{5.169}$$

$$V_s^R = k_s y_s + V_s^L \tag{5.170}$$

In matrix form, the above four equations are written as follows:

$$\begin{bmatrix} -y \\ \theta \\ M \\ V \end{bmatrix}_s^R = \begin{bmatrix} 1 & 0 & 0 & 0 \\ 0 & 1 & 0 & 0 \\ 0 & 0 & 1 & 0 \\ -k_s & 0 & 0 & 1 \end{bmatrix} \begin{bmatrix} -y \\ \theta \\ M \\ V \end{bmatrix}_s^L \tag{5.171}$$

or

$$\{v_s\}^R = [P]\{v_s\}^L \tag{5.172}$$

Example 5.11. Determine the free frequency of vibration of the elastically supported uniform weightless beam in Fig. 5.12g, which supports the mass m. Use appropriate transfer matrices.

Solution. Consider the weightless uniform beam in Fig. 5.12g, which is simply supported at the left end, elastically supported at the other end by a linear spring of constant k, and carrying a mass m at midspan. Its free frequency of vibration can be determined by using transfer matrices. The state vector $\{v_0\}^R$ is related to the state vector $\{v_2\}^R$ by the following matrix expression:

$$\{v_2\}^R = [P_{\text{spring}}][F_{\text{beam}}][P_{\text{mass}}][F_{\text{beam}}]\{v_0\}^R = [M]\{v_0\}^R \tag{5.173}$$

where

$$[M] = [P_{\text{spring}}][F_{\text{beam}}][P_{\text{mass}}][F_{\text{beam}}] \qquad (5.174)$$

Equation (5.173) was derived by eliminating the intermediate state vectors $\{v_1\}^L$, $\{v_1\}^R$, and $\{v_2\}^L$ in Fig. 5.12g. This is accomplished as follows:

$$\{v_2\}^R = [P_2]\{v_2\}^L \qquad (5.175)$$

and

$$\{v_2\}^L = [F_2]\{v_1\}^R \qquad (5.176)$$

By substituting Eq. (5.176) into Eq. (5.175), we find

$$\{v_2\}^R = [P_2][F_2]\{v_1\}^R \qquad (5.177)$$

Also,

$$\{v_1\}^R = [P_1]\{v_1\}^L \qquad (5.178)$$

Hence, from Eqs. (5.177) and (5.178), we find

$$\{v_2\}^R = [P_2][F_2][P_1]\{v_1\}^L \qquad (5.179)$$

Similarly,

$$\{v_1\}^L = [F_1]\{v_0\}^R \qquad (5.180)$$

Finally, from Eqs. (5.179) and (5.180), we find

$$\{v_2\}^R = [P_2][F_2][P_1][F_1]\{v_0\}^R \qquad (5.181)$$

which is the same as Eq. (5.173), relating the state vectors $\{v_0\}^R$ and $\{v_2\}^R$ in Fig. 5.12g.

The matrix product $[M]$ in Eq. (5.174) can be computed by applying the multiplication scheme discussed in Appendix C, or by using available computer programs. By using the multiplication scheme, we find

$$[F_{\text{beam}}] \rightarrow \begin{bmatrix} 1 & \ell & \dfrac{\ell^2}{2EI} & \dfrac{\ell^3}{6EI} \\ 0 & 1 & \dfrac{\ell}{EI} & \dfrac{\ell^2}{2EI} \\ 0 & 0 & 1 & \ell \\ 0 & 0 & 0 & 1 \end{bmatrix}$$

$$[P_{\text{mass}}] \rightarrow \begin{bmatrix} 1 & 0 & 0 & 0 \\ 0 & 1 & 0 & 0 \\ 0 & 0 & 1 & 0 \\ m\omega^2 & 0 & 0 & 1 \end{bmatrix} \begin{bmatrix} 1 & \ell & \dfrac{\ell^2}{2EI} & \dfrac{\ell^3}{6EI} \\[2ex] 0 & 1 & \dfrac{\ell}{EI} & \dfrac{\ell^2}{2EI} \\[2ex] 0 & 0 & 1 & \ell \\[2ex] m\omega^2 & m\omega^2\ell & \dfrac{m\omega^2\ell^2}{2EI} & \left(1 + \dfrac{m\omega^2\ell^3}{6EI}\right) \end{bmatrix}$$

$$[F_{\text{beam}}] \rightarrow \begin{bmatrix} 1 & \ell & \dfrac{\ell^2}{2EI} & \dfrac{\ell^3}{6EI} \\[2ex] 0 & 1 & \dfrac{\ell}{EI} & \dfrac{\ell^2}{2EI} \\[2ex] 0 & 0 & 1 & \ell \\[2ex] 0 & 0 & 0 & 1 \end{bmatrix} \begin{bmatrix} \ell\left(2 + \dfrac{m\omega^2\ell^3}{6EI}\right) & \dfrac{\ell^3}{EI}\left(\dfrac{4}{3} + \dfrac{m\omega^2\ell^3}{36EI}\right) \\[2ex] \left(1 + \dfrac{m\omega^2\ell^3}{2EI}\right) & \dfrac{\ell^2}{EI}\left(2 + \dfrac{m\omega^2\ell^3}{12EI}\right) \\[2ex] m\omega^2\ell^2 & \ell\left(2 + \dfrac{m\omega^2\ell^3}{6EI}\right) \\[2ex] m\omega^2\ell & \left(1 + \dfrac{m\omega^2\ell^3}{6EI}\right) \end{bmatrix}$$

$$[P_{\text{spring}}] \rightarrow \begin{bmatrix} 1 & 0 & 0 & 0 \\ 0 & 1 & 0 & 0 \\ 0 & 0 & 1 & 0 \\ -k & 0 & 0 & 1 \end{bmatrix} \begin{bmatrix} \ell\left(2 + \dfrac{m\omega^2\ell^3}{6EI}\right) & \dfrac{\ell^3}{EI}\left(\dfrac{4}{3} + \dfrac{m\omega^2\ell^3}{36EI}\right) \\[2ex] \left(1 + \dfrac{m\omega^2\ell^3}{2EI}\right) & \dfrac{\ell^2}{EI}\left(2 + \dfrac{m\omega^2\ell^3}{12EI}\right) \\[2ex] m\omega^2\ell^2 & \ell\left(2 + \dfrac{m\omega^2\ell^3}{6EI}\right) \\[2ex] m\omega^2\ell - k\ell\left(2 + \dfrac{m\omega^2\ell^3}{6EI}\right) & \left[1 + \dfrac{m\omega^2\ell^3}{6EI} - \dfrac{k\ell^3}{EI}\left(\dfrac{4}{3} + \dfrac{m\omega^2\ell^3}{36EI}\right)\right] \end{bmatrix}$$

In the above multiplication, the first and third columns are omitted. These columns will be zero in the final results, because the elements y_0 and M_0 of the state vector $\{v_0\}^R$ are zero. These are the boundary conditions of the problem at station 0. The boundary conditions at station 2, involving the elements M_2^R and V_2^R of the state vector $\{v_2\}^R$, are

$$M_2^R = V_2^R = 0 \tag{5.182}$$

Thus, Eq. (5.173) becomes

$$\begin{bmatrix} -y \\ \theta \\ 0 \\ 0 \end{bmatrix}_2^R = \begin{bmatrix} \ell\left(2 + \dfrac{m\omega^2\ell^3}{6EI}\right) & \dfrac{\ell^3}{EI}\left(\dfrac{4}{3} + \dfrac{m\omega^2\ell^3}{36EI}\right) \\[2ex] \left(1 + \dfrac{m\omega^2\ell^3}{2EI}\right) & \dfrac{\ell^2}{EI}\left(2 + \dfrac{m\omega^2\ell^3}{12EI}\right) \\[2ex] m\omega^2\ell^2 & \ell\left(2 + \dfrac{m\omega^2\ell^3}{6EI}\right) \\[2ex] \left[m\omega^2\ell - k\ell\left(2 + \dfrac{m\omega^2\ell^3}{6EI}\right)\right] & \left[1 + \dfrac{m\omega^2\ell^3}{6EI} - \dfrac{k\ell^3}{EI}\left(\dfrac{4}{3} + \dfrac{m\omega^2\ell^3}{36EI}\right)\right] \end{bmatrix} \begin{bmatrix} 0 \\ \theta \\ 0 \\ V \end{bmatrix}_0^R \tag{5.183}$$

The last two rows in the above expression yield

$$m\omega^2\ell^2\theta_0^R + \ell\left(2 + \frac{m\omega^2\ell^3}{6EI}\right)V_0^R = 0 \tag{5.184}$$

$$\left[m\omega^2\ell - k\ell\left(2 + \frac{m\omega^2\ell^3}{6EI}\right)\right]\theta_0^R$$

$$+ \left[1 + \frac{m\omega^2\ell^3}{6EI} - \frac{k\ell^3}{EI}\left(\frac{4}{3} + \frac{m\omega^2\ell^3}{36EI}\right)\right]V_0^R = 0 \tag{5.185}$$

For other than a trivial solution, the determinant of the coefficients of θ_0^R and V_0^R of the above two homogeneous equations must be equal to zero. Hence,

$$\begin{vmatrix} m\omega^2\ell & \ell\left(2 + \frac{m\omega^2\ell^3}{6EI}\right) \\ m\omega^2\ell - k\ell\left(2 + \frac{m\omega^2\ell^3}{6EI}\right) & 1 + \frac{m\omega^2\ell^3}{6EI} - \frac{k\ell^3}{EI}\left(\frac{4}{3} + \frac{m\omega^2\ell^3}{36EI}\right) \end{vmatrix} = 0 \tag{5.186}$$

This is known as the frequency determinant. The solution of this determinant yields

$$\omega^2 = \frac{4k}{m(1 + 2k\ell^3/3EI)} \tag{5.187}$$

This is the natural frequency, squared, of the beam in Fig. 5.12g, in units of radians per second. Thus,

$$\omega = \sqrt{\frac{4k}{m(1 + 2k\ell^3/3EI)}} \tag{5.188}$$

5.10 TRANSFER MATRICES FOR FREE VIBRATION OF SPRING–MASS SYSTEMS WITH TWO OR MORE DEGREES OF FREEDOM

The basic theory and concepts of transfer matrices have been already discussed in the preceding sections. Their application to systems with two or

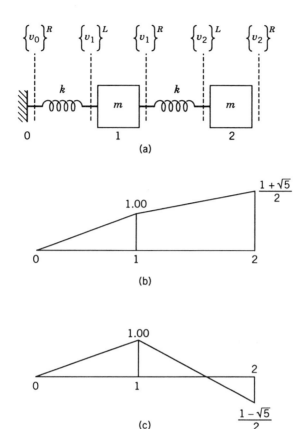

FIGURE 5.13. (a) Two-degree spring–mass system. (b) Shape of the first mode. (c) Shape of the second mode.

more degrees of freedom is rather similar. Consider, for example, the spring–mass system in Fig. 5.13a, constrained to move in the horizontal direction only. Its natural frequencies of vibration may be determined by using transfer matrices. The relation between the state vectors $\{v_0\}^R$ and $\{v_2\}^R$ in Fig. 5.13a is as follows:

$$\{v_2\}^R = [P_2][F_2][P_1][F_1]\{v_0\}^R \qquad (5.189)$$

where $[P_1]$ and $[P_2]$ are the point transfer matrices for stations 1 and 2, respectively, and $[F_1]$ and $[F_2]$ are the field transfer matrices for the springs. The above relation between the state vectors $\{v_2\}^R$ and $\{v_0\}^R$ was obtained by eliminating the intermediate state vectors. These state vectors are

$$\{v_2\}^R = [P_2]\{v_2\}^L \quad \text{and} \quad \{v_2\}^L = [F_2]\{v_1\}^R \qquad (5.190)$$

Thus,

$$\{v_2\}^R = [P_2][F_2]\{v_1\}^R \tag{5.191}$$

Also,

$$\{v_1\}^R = [P_1]\{v_1\}^L \tag{5.192}$$

Thus,

$$\{v_2\}^R = [P_2][F_2][P_1]\{v_1\}^L \tag{5.193}$$

Continuing in a similar manner, the required relation is obtained.
If $n + 1$ stations are involved, Eq. (5.189) takes the form

$$\{v_n\}^R = [P_n][F_n][P_{n-1}][F_{n-1}] \cdots [P_1][F_1]\{v_0\}^R \tag{5.194}$$

The multiplication of the matrices $[P_n][F_n] \cdots [F_1]$ yields the matrix $[M]$, and it can be carried out in the usual way.

Example 5.12. Determine the free frequencies of vibration and the corresponding mode shapes for the two-degree spring–mass system in Fig. 5.13a. Use appropriate transfer matrices.

Solution. The relation between the state vectors $\{v_2\}^R$ and $\{v_0\}^R$ in Fig. 15.13a is given by Eq. (5.189). The matrix multiplications in Eq. (5.189) are carried out as in the preceding section by using the following multiplication scheme (see also Appendix C):

$$[F_1] \rightarrow \begin{bmatrix} 1 & \dfrac{1}{k} \\ 0 & 1 \end{bmatrix}$$

$$[P_1] \rightarrow \begin{bmatrix} 1 & 0 \\ -m\omega^2 & 1 \end{bmatrix} \begin{bmatrix} \dfrac{1}{k} \\ 1 - \dfrac{m\omega^2}{k} \end{bmatrix} \tag{5.195}$$

$$[F_2] \rightarrow \begin{bmatrix} 1 & \dfrac{1}{k} \\ 0 & 1 \end{bmatrix} \begin{bmatrix} \dfrac{1}{k}\left(2 - \dfrac{m\omega^2}{k}\right) \\ 1 - \dfrac{m\omega^2}{k} \end{bmatrix}$$

$$[P_2] \rightarrow \begin{bmatrix} 1 & 0 \\ -m\omega^2 & 1 \end{bmatrix} \begin{bmatrix} & \frac{1}{k}\left(2 - \frac{m\omega^2}{k}\right) \\ & \frac{m^2\omega^4}{k^2} - \frac{3m\omega^2}{k} + 1 \end{bmatrix}$$

The multiplications in the first column are omitted, because they are not required for the end results. This can easily be noted from the beginning of the problem by observing the boundary conditions. The boundary conditions at stations 0 and 2 are $X_0^R = F_2^R = 0$, yielding the state vectors

$$\{v_0\}^R = \begin{bmatrix} x \\ F \end{bmatrix}_0^R = \begin{bmatrix} 0 \\ F \end{bmatrix}_0^R \quad \text{and} \quad \{v_2\}^R = \begin{bmatrix} x \\ F \end{bmatrix}_2^R = \begin{bmatrix} x \\ 0 \end{bmatrix}_2^R \tag{5.196}$$

Hence, Eq. (5.189) yields

$$\begin{bmatrix} x \\ 0 \end{bmatrix}_2^R = \begin{bmatrix} & \frac{1}{k}\left(2 - \frac{m\omega^2}{k}\right) \\ & \frac{m^2\omega^4}{k^2} - \frac{3m\omega^2}{k} + 1 \end{bmatrix} \begin{bmatrix} 0 \\ F \end{bmatrix}_0^R \tag{5.197}$$

The second row of this matrix expression yields the equation

$$\frac{m^2\omega^4}{k^2} - \frac{3m\omega^2}{k} + 1 = 0 \tag{5.198}$$

or

$$\omega^4 - \frac{3k}{m}\omega^2 + \frac{k^2}{m^2} = 0 \tag{5.199}$$

which is known as the frequency equation. This equation yields the roots

$$\omega_1^2 = \frac{k}{m}\left(\frac{3 - \sqrt{5}}{2}\right) \tag{5.200}$$

$$\omega_2^2 = \frac{k}{m}\left(\frac{3 + \sqrt{5}}{2}\right) \tag{5.201}$$

which are the squares of the two free frequencies of the two-degree spring–mass system in Fig. 5.13a.

With ω_1^2 and ω_2^2 calculated, the shapes of the two normal modes corresponding to these frequencies can easily be determined. From the matrix multiplications, Eq. (5.195), the individual state vectors at stations 1 and 2, in terms of the state vector at station 0, are as follows:

$$\{v_1\}^L = [F_1]\{v_0\}^R = \left[\begin{array}{c|c} 1 & \\ \frac{1}{k} & \\ 1 & \end{array}\right] \begin{bmatrix} 0 \\ F \end{bmatrix}_0^R \tag{5.202}$$

$$\{v_1\}^R = [P_1][F_1]\{v_0\}^R = \left[\begin{array}{c|c} & \frac{1}{k} \\ 1 - \dfrac{m\omega^2}{k} & \end{array}\right] \begin{bmatrix} 0 \\ F \end{bmatrix}_0^R \tag{5.203}$$

$$\{v_2\}^L = [F_2][P_1][F_1]\{v_0\}^R = \left[\begin{array}{c|c} \dfrac{1}{k}\left(2 - \dfrac{m\omega^2}{k}\right) & \\ 1 - \dfrac{m\omega^2}{k} & \end{array}\right] \begin{bmatrix} 0 \\ F \end{bmatrix}_0^R \tag{5.204}$$

$$\{v_2\}^R = [P_2][F_2][P_1][F_1]\{v_0\}^R = \left[\begin{array}{c|c} \dfrac{1}{k}\left(2 - \dfrac{m\omega^2}{k}\right) & \\ \dfrac{m^2\omega^4}{k^2} - \dfrac{3m\omega^2}{k} + 1 & \end{array}\right] \begin{bmatrix} 0 \\ F \end{bmatrix}_0^R \tag{5.205}$$

In the above expressions, by substituting for ω^2 the values of ω_1^2 and ω_2^2 from Eqs. (5.200) and (5.201), respectively, and assuming that $F_0 = 1$, the following results are obtained. For $\omega^2 = \omega_1^2$, we find

$$\{v_1\}^L = \begin{bmatrix} 0 & \dfrac{1}{k} \\ 0 & 1 \end{bmatrix} \qquad \{v_1\}^R = \begin{bmatrix} 0 & \dfrac{1}{k} \\ 0 & \left(\dfrac{-1 + \sqrt{5}}{2}\right) \end{bmatrix}$$

$$\{v_2\}^L = \begin{bmatrix} 0 & \dfrac{1 + \sqrt{5}}{2k} \\ 0 & \left(\dfrac{-1 + \sqrt{5}}{2}\right) \end{bmatrix} \qquad \{v_2\}^R = \begin{bmatrix} 0 & \dfrac{1 + \sqrt{5}}{2k} \\ 0 & 0 \end{bmatrix}$$

and for $\omega^2 = \omega_2^2$, we have

$$\{v_1\}^L = \begin{bmatrix} 0 & \dfrac{1}{k} \\ 0 & 1 \end{bmatrix} \qquad \{v_1\}^R = \begin{bmatrix} 0 & \dfrac{1}{k} \\ 0 & -\left(\dfrac{1+\sqrt{5}}{2}\right) \end{bmatrix}$$

$$\{v_2\}^L = \begin{bmatrix} 0 & \left(\dfrac{1-\sqrt{5}}{2k}\right) \\ 0 & -\left(\dfrac{1+\sqrt{5}}{2}\right) \end{bmatrix} \qquad \{v_2\}^R = \begin{bmatrix} 0 & \dfrac{1-\sqrt{5}}{2k} \\ 0 & 0 \end{bmatrix}$$

It should be noted that the first row in the above matrix expressions represents the amplitudes at stations 1 and 2. The amplitude at station 0 is zero. By arbitrarily setting the amplitude x_1 at station 1 equal to one, the normal mode shape corresponding to the free frequency ω_1 is characterized by the amplitudes

$$x_0 = 0 \qquad x_1 = 1 \qquad x_2 = \frac{1+\sqrt{5}}{2} \tag{5.206}$$

The normal mode shape corresponding to the free frequency ω_2 is given by the following amplitudes:

$$x_0 = 0 \qquad x_1 = 1 \qquad x_2 = \frac{1-\sqrt{5}}{2} \tag{5.207}$$

The mode shapes for ω_1 and ω_2 are shown in Figs. 5.13b and 5.13c, respectively.

By using the orthogonality property given by Eq. (2.260), we find

$$\sum_{i-1}^{i=2} m_i x_i^{(1)} x_i^{(2)} = (1)(1)(1) + (1)(\tfrac{1}{2})(1+\sqrt{5})(\tfrac{1}{2})(1-\sqrt{5}) = 1 - 1 = 0$$

indicating that the computed normal modes are in conformity with their orthogonality property.

The free frequencies of vibration and the corresponding normal modes of multiple-degree-of-freedom spring–mass systems can be determined in a similar manner. When these quantities are computed, the dynamic responses of such systems under the action of time-varying forces can be computed by applying the method of modal analysis, as discussed in Chapter 7.

5.11   TRANSFER MATRICES FOR FREE VIBRATION OF TORSIONAL SYSTEMS WITH TWO OR MORE DEGREES OF FREEDOM

The use of transfer matrices for torsional systems with two or more degrees of freedom will be illustrated by using an example. Consider the freely vibrating torsional system in Fig. 5.14. The relation between the state vectors $\{v_0\}^R$ and $\{v_2\}^R$ can be obtained as in the preceding sections by eliminating the intermediate state vectors $\{v_1\}^L$, $\{v_1\}^R$, and $\{v_2\}^L$. By doing so, the relation is written as follows:

$$\{v_2\}^R = [P_2][F_2][P_1][F_1]\{v_0\}^R = [M]\{v_0\}^R \tag{5.208}$$

where

$$[M] = [P_2][F_2][P_1][F_1] \tag{5.209}$$

The matrix product $[M]$ can be determined by applying the usual multiplication scheme. Since $\phi_0^R = 0$, the multiplications required to obtain the first column in the matrix $[M]$ can be neglected. Hence,

$$[F_1] \rightarrow \begin{bmatrix} 1 & \dfrac{1}{k_t} \\ 0 & 1 \end{bmatrix}$$

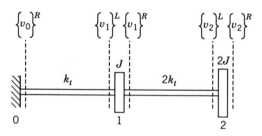

FIGURE 5.14. Two-degree disk–shaft system.

$$[P_1] \rightarrow \begin{bmatrix} 1 & 0 \\ -\omega^2 J & 1 \end{bmatrix} \begin{bmatrix} \dfrac{1}{k_t} \\ 1 - \dfrac{\omega^2 J}{k_t} \end{bmatrix}$$

$$[F_2] \rightarrow \begin{bmatrix} 1 & \dfrac{1}{2k_t} \\ 0 & 1 \end{bmatrix} \begin{bmatrix} \dfrac{3k_t - \omega^2 J}{2k_t^2} \\ \left(1 - \dfrac{\omega^2 J}{k_t}\right) \end{bmatrix} \qquad (5.210)$$

$$[P_2] \rightarrow \begin{bmatrix} 1 & 0 \\ -2\omega^2 J & 1 \end{bmatrix} \begin{bmatrix} \dfrac{3k_t - \omega^2 J}{2k_t^2} \\ \dfrac{\omega^4 J^2}{k_t^2} - \dfrac{4\omega^2 J}{k_t} + 1 \end{bmatrix}$$

Thus, Eq. (5.208) yields

$$\begin{bmatrix} \phi \\ 0 \end{bmatrix}_2^R = \begin{bmatrix} \dfrac{3k_t - \omega^2 J}{2k_t^2} \\ \dfrac{\omega^4 J^2}{k_t^2} - \dfrac{4\omega^2 J}{k_t} + 1 \end{bmatrix} \begin{bmatrix} 0 \\ T \end{bmatrix}_0^R \qquad (5.211)$$

Equation (5.211) yields the following characteristic equation:

$$\frac{\omega^4 J^2}{k_t^2} - \frac{4\omega^2 J}{k_t} + 1 = 0$$

or

$$\omega^4 - \frac{4k_t}{J}\omega^2 + \frac{k_t^2}{J^2} = 0 \qquad (5.212)$$

The solution of the above algebraic equation yields the two free frequencies of the torsional system in Fig. 5.14. They are as follows:

$$\omega_1^2 = \frac{k_t}{J}(2 - \sqrt{3}) \qquad (5.213)$$

$$\omega_2^2 = \frac{k_t}{J}(2 + \sqrt{3}) \qquad (5.214)$$

With ω_1^2 and ω_2^2 evaluated, the shape of the normal modes corresponding to these two frequencies can be obtained in a manner similar to the one used in the preceding section. In other words, the values of the angular displacements at stations 1 and 2 can be obtained from Eq. (5.210) by substituting for ω^2 the values of ω_1^2 and ω_2^2 in exactly the same manner as in the preceding section. Thus, for the first normal mode, we have

$$\phi_0 = 0 \qquad \phi_1 = 1 \qquad \phi_2 = \frac{1 + \sqrt{3}}{2} \tag{5.215}$$

and for the second normal mode, we have

$$\phi_0 = 0 \qquad \phi_1 = 1 \qquad \phi_2 = \frac{1 - \sqrt{3}}{2} \tag{5.216}$$

These results satisfy the orthogonality property given by Eq. (2.260).

The solution of other torsional problems can be obtained in a similar manner.

5.12   TRANSFER MATRICES FOR FREE VIBRATION OF FLEXURAL SYSTEMS WITH TWO DEGREES OF FREEDOM

The natural frequency response of free vibrating flexural systems with two degrees of freedom may be investigated by using transfer matrices. Consider, for example, the weightless beam in Fig. 5.15. At stations 1 and 2, there are two concentrated masses of magnitudes m and $2m$, respectively. The flexural stiffness EI of the beam between mass concentration points is as shown in the figure. It is required to obtain the natural frequency response of this system while it is vibrating in the transverse direction.

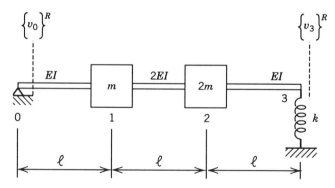

FIGURE 5.15. Two-degree flexural system supporting two masses as shown.

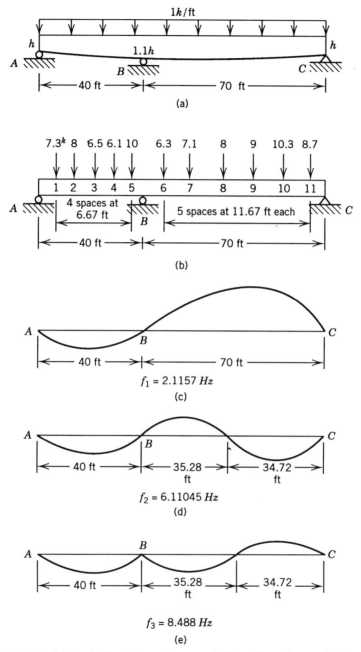

FIGURE A5.15. (a) Statically indeterminate variable stiffness beam. (b) Equivalent system of uniform stiffness. (c) First mode shape. (d) Second mode shape. (e) Third mode shape. (See also Page 779, answer to Problem 5.15)

339

The relation between the state vectors $\{v_0\}^R$ and $\{v_3\}^R$ in Fig. 5.15 can be obtained by eliminating the intermediate state vectors. The elimination procedure is already explained in the preceding sections. The relation between vectors $\{v_0\}^R$ and $\{v_3\}^R$ can be written as follows:

$$\{v_3\}^R = [P_3][F_3][P_2][F_2][P_1][F_1]\{v_0\}^R = [M]\{v_0\}^R \tag{5.217}$$

where

$$[M] = [P_3][F_3][P_2][F_2][P_1][F_1] \tag{5.218}$$

The derivation of the field and point transfer matrices for flexural systems is discussed in Section 5.9. To facilitate the matrix multiplications in Eq. (5.218), the following simplifications regarding the transfer matrices are introduced.

The field transfer matrix relating two state vectors $\{v_j\}^L$ and $\{v_{j-1}\}^R$ of a beam segment ℓ_j between masses m_j and m_{j-1} is given by Eq. (5.159) or (5.160); that is,

$$
\begin{bmatrix} -y \\ \theta \\ M \\ V \end{bmatrix}_j^L =
\begin{bmatrix}
1 & \ell_j & \dfrac{\ell_j^2}{2(EI)_j} & \dfrac{\ell_j^3}{6(EI)_j} \\
0 & 1 & \dfrac{\ell_j}{(EI)_j} & \dfrac{\ell_j^2}{2(EI)_j} \\
0 & 0 & 1 & \ell_j \\
0 & 0 & 0 & 1
\end{bmatrix}
\begin{bmatrix} -y \\ \theta \\ M \\ V \end{bmatrix}_{j-1}^R
\tag{5.219}
$$

In Eq. (5.219), we introduce the following notation:

$$y^* = \frac{y}{\ell} \qquad M^* = \frac{M\ell}{EI} \tag{5.220}$$

$$\theta^* = \theta \qquad V^* = \frac{V\ell^2}{EI}$$

On this basis, the four expressions represented in matrix form by Eq. (5.219)—that is, Eqs. (5.153), (5.154), (5.157), and (5.158)—may be expressed in matrix form as follows:

$$
\begin{bmatrix} -y^* \\ \theta^* \\ M^* \\ V^* \end{bmatrix}_j^L =
\begin{bmatrix}
1 & 1 & \frac{1}{2} & \frac{1}{6} \\
0 & 1 & 1 & \frac{1}{2} \\
0 & 0 & 1 & 1 \\
0 & 0 & 0 & 1
\end{bmatrix}
\begin{bmatrix} -y^* \\ \theta^* \\ M^* \\ V^* \end{bmatrix}_{j-1}^R
\tag{5.221}
$$

This is a much simpler expression to use for matrix multiplications. The only changes that have been made in the original expressions are that Eq. (5.157) was divided through by ℓ_j, Eq. (5.154) was multiplied by $\ell_j/(EI)_j$, and Eq. (5.153) was multiplied by $\ell_j^2/(EI)_j$. Any variations in the stiffness EI and lengths ℓ of the beam segments between mass concentration points can easily be taken into account by factors such as $a(EI)$ and $b\ell$. Thus, the use of transfer matrices is not limited to uniform ℓ_j segments. The stiffness EI between the individual ℓ_j elements can also be different.

Hence, Eq. (5.221) can be written in a more general form as follows:

$$
\begin{bmatrix} -y^* \\ \theta^* \\ M^* \\ V^* \end{bmatrix}_j^L = \begin{bmatrix} 1 & b & b^2/2a & b^3/6a \\ 0 & 1 & b/a & b^2/2a \\ 0 & 0 & 1 & b \\ 0 & 0 & 0 & 1 \end{bmatrix} \begin{bmatrix} -y^* \\ \theta^* \\ M^* \\ V^* \end{bmatrix}_{j-1}^R
\tag{5.222}
$$

or

$$
\{v_j^*\}^L = [F_j^*]\{v_{j-1}^*\}^R
\tag{5.223}
$$

To conform with the new notation given to the elements of the state vectors, the point transfer matrix of a mass m_j relating the state vectors $\{v_j\}^R$ and $\{v_j\}^L$ in Eq. (5.149) is written as shown below:

$$
\begin{bmatrix} -y^* \\ \theta^* \\ M^* \\ V^* \end{bmatrix}_j^R = \begin{bmatrix} 1 & 0 & 0 & 0 \\ 0 & 1 & 0 & 0 \\ 0 & 0 & 1 & 0 \\ \omega^2\ell_j^3 m_j/(EI)_j & 0 & 0 & 1 \end{bmatrix} \begin{bmatrix} -y^* \\ \theta^* \\ M^* \\ V^* \end{bmatrix}_j^L
\tag{5.224}
$$

or

$$
\{v_j^*\}^R = [P_j^*]\{v_j^*\}^L
\tag{5.225}
$$

In this case, Eq. (5.145) was multiplied through by $\ell_j^2/(EI)_j$, and y_j^* was substituted by y_j/ℓ_j. All other changes are obvious.

In a similar manner, the point transfer matrix in Eq. (5.171), relating the state vectors at the left-hand and right-hand sides of a linear spring supporting a beam, is written as shown:

$$
\begin{bmatrix} -y^* \\ \theta^* \\ M^* \\ V^* \end{bmatrix}_s^R = \begin{bmatrix} 1 & 0 & 0 & 0 \\ 0 & 1 & 0 & 0 \\ 0 & 0 & 1 & 0 \\ -k_s\ell_s^3/(EI)_s & 0 & 0 & 1 \end{bmatrix} \begin{bmatrix} -y^* \\ \theta^* \\ M^* \\ V^* \end{bmatrix}_s^L
\tag{5.226}
$$

or

$$
\{v_s^*\}^R = [P_s^*]\{v_s^*\}^L
\tag{5.227}
$$

Under the new arrangement for the field and point transfer matrices, the matrix product in Eq. (5.218) for the problem in Fig. 5.15 can easily be carried out. The boundary conditions at station 0 are $y_0^{*R} = M_0^{*R} = 0$, indicating that the first and third columns of the matrix $[M]$ need not be calculated because they will be zero in the final calculations. The boundary conditions on the right-hand side of station 3 are $M_3^{*R} = V_3^{*R} = 0$.

The matrix product $[M]$ is carried out by using the matrix multiplication scheme, and the final result for Eq. (5.217) is as follows:

$$
\begin{bmatrix} -y^* \\ \theta^* \\ 0 \\ 0 \end{bmatrix}_3^R =
\begin{bmatrix}
\dfrac{\omega^4\ell^6 m^2}{36(EI)^2} + \dfrac{5\omega^2\ell^3 m}{3EI} + 3 \\[2mm]
\dfrac{\omega^4\ell^6 m^2}{12(EI)^2} + \dfrac{15\omega^2\ell^3 m}{4EI} + 1 \\[2mm]
\dfrac{\omega^4\ell^6 m^2}{6(EI)^2} + \dfrac{6\omega^2\ell^3 m}{EI} \\[2mm]
\dfrac{\omega^4\ell^6 m^2}{36(EI)^2}\left(6 - \dfrac{k\ell^3}{EI}\right) + \dfrac{5\omega^2\ell^3 m}{EI}\left(3 - \dfrac{k\ell^3}{EI}\right) - \dfrac{3k\ell^3}{EI}
\end{bmatrix}
\begin{bmatrix}
\dfrac{\omega^4\ell^6 m^2}{216(EI)^2} + \dfrac{\omega^2\ell^3 m}{2EI} + \dfrac{41}{12} \\[2mm]
\dfrac{\omega^4\ell^6 m^2}{72(EI)^2} + \dfrac{31\omega^2\ell^3 m}{24EI} + \dfrac{15}{12} \\[2mm]
\dfrac{\omega^4\ell^6 m^2}{36(EI)^2} + \dfrac{7\omega^2\ell^3 m}{3EI} + 3 \\[2mm]
\dfrac{\omega^4\ell^6 m^2}{216(EI)^2}\left(6 - \dfrac{k\ell^3}{EI}\right) + \dfrac{\omega^4\ell^3 m}{6EI}\left(13 - \dfrac{3k\ell^3}{EI}\right) - \dfrac{41k\ell^3}{12EI} + 1
\end{bmatrix}
\begin{bmatrix} 0 \\ \theta^* \\ 0 \\ V^* \end{bmatrix}
\tag{5.228}
$$

The last two rows of this matrix expression yield

$$\left[\frac{\omega^4\ell^6 m^2}{6(EI)^2} + \frac{6\omega^2\ell^3 m}{EI}\right]\theta_0^{*R} + \left[\frac{\omega^4\ell^6 m^2}{36(EI)^2} + \frac{7\omega^2\ell^3 m}{3EI} + 3\right\}V_0^{*R} = 0 \qquad (5.229)$$

$$\left[\frac{\omega^4\ell^6 m^2}{36(EI)^2}\left(6 - \frac{k\ell^3}{EI}\right) + \frac{5\omega^2\ell^3 m}{3EI}\left(3 - \frac{k\ell^3}{EI}\right) - \frac{3k\ell^3}{EI}\right]\theta_0^{*R}$$

$$+ \left[\frac{\omega^4\ell^6 m^2}{216(EI)^2}\left(6 - \frac{k\ell^3}{EI}\right) + \frac{\omega^2\ell^3 m}{6EI}\left(13 - \frac{3k\ell^3}{EI}\right) - \frac{41k\ell^3}{12EI} + 1\right]V_0^{*R} = 0$$

$$(5.230)$$

or

$$\left[\frac{\omega^4\ell^4 m^2}{6EI} + 6\omega^2\ell m\right]\theta_0^R + \left[\frac{\omega^4\ell^6 m^2}{36(EI)^2} + \frac{7\omega^2\ell^3 m}{3EI} + 3\right]V_0^R = 0 \qquad (5.231)$$

$$\left[\frac{\omega^4\ell^4 m^2}{36EI}\left(6 - \frac{k\ell^3}{EI}\right) + \frac{5}{3}\omega^2\ell m\left(3 - \frac{k\ell^3}{EI}\right) - 3k\ell\right]\theta_0^R$$

$$(5.232)$$

$$+ \left[\frac{\omega^4\ell^6 m^2}{216(EI)^2}\left(6 - \frac{k\ell^3}{EI}\right) + \frac{\omega^2\ell^3 m}{6EI}\left(13 - \frac{3k\ell^3}{EI}\right) - \frac{41k\ell^3}{12EI} + 1\right]V_0^R = 0$$

For a nontrivial solution, the determinant of the coefficients of θ_0^R and V_0^R in the above two equations must be equal to zero; that is,

$$\begin{vmatrix} \left[\frac{\omega^4\ell^4 m^2}{6EI} + 6\omega^2\ell m\right] & \left[\frac{\omega^4\ell^6 m^2}{36(EI)^2} + \frac{7\omega^2\ell^3 m}{3EI} + 3\right] \\[2mm] \left[\frac{\omega^4\ell^4 m^2}{36EI}\left(6 - \frac{k\ell^3}{EI}\right) + \frac{5}{3}\omega^2\ell m\left(3 - \frac{k\ell^3}{EI}\right) - 3k\ell\right] & \left[\frac{\omega^4\ell^6 m^2}{216(EI)^2}\left(6 - \frac{k\ell^3}{EI}\right) + \frac{\omega^2\ell^3 m}{6EI}\left(13 - \frac{3k\ell^3}{EI}\right) - \frac{41k\ell^3}{12EI} + 1\right] \end{vmatrix} = 0$$

$$(5.233)$$

This is the frequency determinant, which by expansion yields the following characteristic equation:

$$\omega^4\left[\frac{35k\ell^6 m^2}{72(EI)^2} + \frac{\ell^3 m^2}{EI}\right] - \omega^2\left[\frac{17k\ell^3 m}{2EI} + 9m\right] + 9k = 0 \qquad (5.234)$$

Equation (5.234) may be solved for ω to yield the free frequencies of vibration of the flexural system in Fig. 5.15.

By assuming that $\ell = 200$ in., $EI = 30 \times 10^6$ kip · in.^{2}, $m = 5.18 \times 10^{-3}$ kip · s^2/in., and $k = 300$ kips/in., Eq. (5.234) yields

$$\omega^4 - 12{,}800\omega^2 + 9.68(10^6) = 0 \tag{5.235}$$

When Eq. (5.235) is solved for ω^2, we find the following two roots:

$$\omega_1^2 = 800 \quad \text{and} \quad \omega_2^2 = 12{,}000$$

The two free frequencies of vibration of the system are

$$\omega_1 = 28.15 \text{ rad/s} \quad \text{and} \quad \omega_2 = 109.50 \text{ rad/s}$$

In hertz, these two frequencies are

$$f_1 = \frac{\omega_1}{2\pi} = 4.51 \text{ Hz}$$

$$f_2 = \frac{\omega_2}{2\pi} = 17.40 \text{ Hz}$$

With the frequencies ω_1 and ω_2 calculated, the corresponding normal modes can be computed in a manner similar to the one used in the preceding two sections. In other words, the state vectors at stations 1, 2, and 3 are expressed in terms of the state vector at station 0 in the usual way. Since $y_0^{*R} = M_0^{*R} = 0$, the unknowns at station 0 are θ_0^{*R} and V_0^{*R}, and their relation can be obtained from Eq. (5.229). When one amplitude is assumed to be unity, the other amplitude can be evaluated. Thus, the state vectors at stations 1, 2, and 3 may be computed.

5.13 TRANSFER MATRICES FOR MULTIPLE-DEGREE-OF-FREEDOM FLEXURAL SYSTEMS

Transfer matrices will be used here to study the free frequency response of flexural multiple-degree-of-freedom systems. In this case, appreciable amounts of computational work will be saved if the field and point transfer matrices are combined into one matrix. This may be accomplished as follows.

The field transfer matrix relating the state vectors $\{v_j\}^L$ and $\{v_{j-1}\}^R$ of the beam segment ℓ_j between masses m_j and m_{j-1} is given by Eq. (5.221); that is,

$$
\begin{bmatrix} -y^* \\ \theta^* \\ M^* \\ V^* \end{bmatrix}_j^L = \begin{bmatrix} 1 & 1 & \frac{1}{2} & \frac{1}{6} \\ 0 & 1 & 1 & \frac{1}{2} \\ 0 & 0 & 1 & 1 \\ 0 & 0 & 0 & 1 \end{bmatrix} \begin{bmatrix} -y^* \\ \theta^* \\ M^* \\ V^* \end{bmatrix}_{j-1}^R
\tag{5.236}
$$

The point matrix of a mass m_j, relating the state vectors $\{v_j\}^R$ and $\{v_j\}^L$, is given by Eq. (5.224); that is,

$$
\begin{bmatrix} -y^* \\ \theta^* \\ M^* \\ V^* \end{bmatrix}_j^R = \begin{bmatrix} 1 & 0 & 0 & 0 \\ 0 & 1 & 0 & 0 \\ 0 & 0 & 1 & 0 \\ \omega^2 \ell_j m_j/(EI)_j & 0 & 0 & 1 \end{bmatrix} \begin{bmatrix} -y^* \\ \theta^* \\ M^* \\ V^* \end{bmatrix}_j^L
\tag{5.237}
$$

By using Eqs. (5.236) and (5.237), the matrix relating the state vectors $\{v_j\}^R$ and $\{v_{j-1}\}^R$ may be obtained from the relation

$$
\{v_j^*\}^R = [P^*][F^*][v_{j-1}^*]^R
\tag{5.238}
$$

where $[P^*]$ and $[F^*]$ are the point and field transfer matrices, respectively; that is,

$$
\begin{bmatrix} -y^* \\ \theta^* \\ M^* \\ V^* \end{bmatrix}_j^R = \begin{bmatrix} 1 & 0 & 0 & 0 \\ 0 & 1 & 0 & 0 \\ 0 & 0 & 1 & 0 \\ \omega^2 \ell_j^3 m_j/(EI)_j & 0 & 0 & 1 \end{bmatrix} \begin{bmatrix} 1 & 1 & \frac{1}{2} & \frac{1}{6} \\ 0 & 1 & 1 & \frac{1}{2} \\ 0 & 0 & 1 & 1 \\ 0 & 0 & 0 & 1 \end{bmatrix} \begin{bmatrix} -y^* \\ \theta^* \\ M^* \\ V^* \end{bmatrix}_{j-1}^R
\tag{5.239}
$$

The $[P^*]$ and $[F^*]$ matrices in Eq. (5.238) or (5.239) may be combined into one matrix $[N^*]$, yielding

$$
\{v_j^*\}^R = [N^*]\{v_{j-1}^*\}^R
\tag{5.239a}
$$

or

$$
\begin{bmatrix} -y^* \\ \theta^* \\ M^* \\ V^* \end{bmatrix}_j^R = \begin{bmatrix} 1 & 1 & \frac{1}{2} & \frac{1}{6} & & 0 \\ 0 & 1 & 1 & \frac{1}{2} & & 0 \\ 0 & 0 & 1 & 1 & & 0 \\ 0 & 0 & 0 & 1 & & \omega^2 \ell_j^3 m_j/(EI)_j \end{bmatrix} \begin{bmatrix} -y^* \\ \theta^* \\ M^* \\ V^* \end{bmatrix}_{j-1}^R
\tag{5.240}
$$

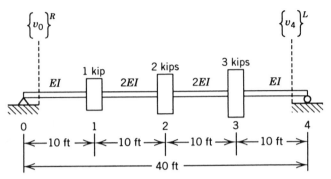

FIGURE 5.16. Weightless, simply supported beam supporting three attached concentrated weights.

To obtain the values for the quantities y_j^{*R}, θ_j^{*R}, and M_j^{*R}, the required matrix multiplication on the right-hand side of Eq. (5.240) may be carried out in the usual way, since they are continuous at station j. The value of V_j^{*R} may be determined from the following product:

$$
\begin{bmatrix} 0 & 0 & 0 & 1 & \vline & \dfrac{\omega^2 \ell_j^3 m_j}{(EI)_j} \end{bmatrix}
\begin{bmatrix} -y^{*R} \\ \theta_{j-1}^{*R} \\ M_{j-1}^{*R} \\ V_{j-1}^{*R} \\ \hline -y_j^{*R} \end{bmatrix}
[V_j^{*R}] \tag{5.241}
$$

Under these conditions of matrix multiplication for Eq. (5.240), Eqs. (5.239) and (5.239a) will yield identical results.

The following example illustrates the application of the above procedure for the calculation of the natural frequencies of vibration of multiple-degree-of-freedom systems. It should be pointed out, however, that the point and field transfer matrices for spring–mass systems may also be combined into one matrix in a similar manner.

Example 5.13. The weightless beam in Fig. 5.16 supports three attached weights at the indicated locations. By using appropriate transfer matrices, determine the free frequencies of vibration of the member and the corresponding mode shapes.

Solution. The simply supported beam in Fig. 5.16 supports three weights that are attached to the member at points 1, 2, and 3 as shown. The member is assumed to be weightless, but the elastic properties throughout its length are retained. The value of the stiffness EI is 30×10^6 kips $\cdot$ in^2.

The relation between the state vectors $\{v_0^*\}^R$ and $\{v_4^*\}^L$ can be found in the usual way, and it is as follows:

$$\{v_4^*\}^L = [F_4^*][P_3^*][F_3^*][P_2^*][F_2^*][P_1^*][F_1^*]\{v_0^*\}^R \qquad (5.242)$$

The point and field transfer matrices for the various stations are combined as in Eq. (5.240), and the multiplication scheme is as shown in Table 5.1. The multiplication of the matrices is carried out as shown in the same table. Since the boundary conditions at station 0 are $y_0^{*R} = M_0^{*R} = 0$, the computations for the first and third columns are omitted. The boundary conditions at station 4 are $y_4^{*L} = M_4^{*L} = 0$. On this basis, Eq. (5.242) may be written

$$
\begin{bmatrix} 0 \\ \theta^* \\ 0 \\ V^* \end{bmatrix}_4^L
=
\begin{bmatrix} A_1 & B_1 \\ A_2 & B_2 \\ A_3 & B_3 \\ A_4 & B_4 \end{bmatrix}
\begin{bmatrix} 0 \\ \theta^* \\ 0 \\ V^* \end{bmatrix}_0^R
\qquad (5.243)
$$

From Eq. (5.243), the frequency determinant is written

$$|M| = \begin{vmatrix} A_1 & B_1 \\ A_3 & B_3 \end{vmatrix} = 0 \qquad (4.244)$$

The elements A_1, A_3, B_1, and B_3 of the above determinant are functions of the frequencies ω of the beam. Upon expansion, the resulting frequency equation will be of the form

$$(\omega^2)^3 + C_1(\omega^2)^2 + C_2\omega^2 + C_3 = 0 \qquad (5.245)$$

The three roots of Eq. (5.245), namely, ω_1^2, ω_2^2, and ω_3^2, are the squares of the three free frequencies of the beam in Fig. 5.16.

It may be noted, however, that the computations in Table 5.1, as well as in Eq. (5.245), become very difficult to handle when the degree of freedom of the system is more than two. For such cases, it is advisable to use computer software or to apply numerical techniques. If numerical techniques are used, various values may be assigned for ω^2 in Table 5.1, and the corresponding values of the frequency determinant may be obtained from Eq. (5.244). The zero intersects of the graph of the determinant $|M|$ versus $0.000149\omega^2$ will yield the required free frequencies of vibration of the beam; that is, the values of ω^2 that make Eq. (5.244) equal to zero are the natural frequencies, squared, of the system.

TABLE 5.1. Multiplication Scheme for the Matrix Product in Eq. (5.242)

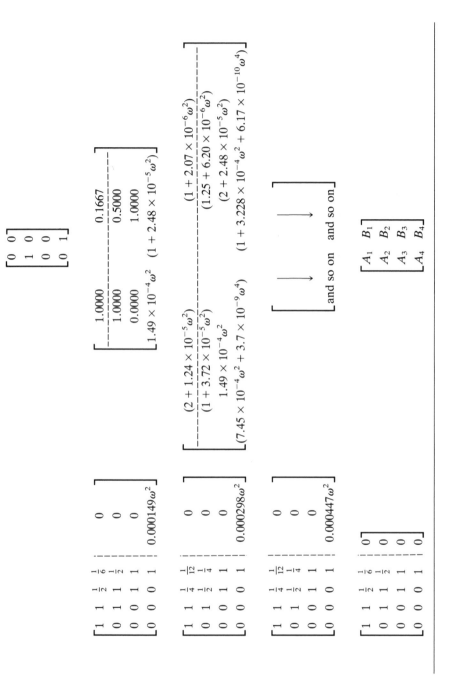

$$
\begin{bmatrix} 1 & 1 & \frac{1}{2} & \frac{1}{6} & & 0 \\ 0 & 1 & 1 & \frac{1}{2} & & 0 \\ 0 & 0 & 1 & 1 & & 0 \\ 0 & 0 & 0 & 1 & & 0.000149\omega^2 \end{bmatrix}
\begin{bmatrix} 1 & 1 & \frac{1}{4} & \frac{1}{12} & & 0 \\ 0 & 1 & 1 & \frac{1}{4} & & 0 \\ 0 & 0 & 1 & 1 & & 0 \\ 0 & 0 & 0 & 1 & & 0.000298\omega^2 \end{bmatrix}
\begin{bmatrix} 1 & 1 & \frac{1}{4} & \frac{1}{12} & & 0 \\ 0 & 1 & 1 & \frac{1}{4} & & 0 \\ 0 & 0 & 1 & 1 & & 0 \\ 0 & 0 & 0 & 1 & & 0.000447\omega^2 \end{bmatrix}
\begin{bmatrix} 1 & 1 & \frac{1}{2} & \frac{1}{6} & & 0 \\ 0 & 1 & 1 & \frac{1}{2} & & 0 \\ 0 & 0 & 1 & 1 & & 0 \\ 0 & 0 & 0 & 1 & & 0 \end{bmatrix}
$$

$$
\begin{bmatrix} 1.0000 & & 0.1667 \\ 1.0000 & & 0.5000 \\ 0.0000 & & 1.0000 \\ 1.49 \times 10^{-4}\omega^2 & & (1 + 2.48 \times 10^{-5}\omega^2) \end{bmatrix}
$$

$$
\begin{bmatrix} (2 + 1.24 \times 10^{-5}\omega^2) & & (1 + 2.07 \times 10^{-6}\omega^2) \\ (1 + 3.72 \times 10^{-5}\omega^2) & & (1.25 + 6.20 \times 10^{-6}\omega^2) \\ 1.49 \times 10^{-4}\omega^2 & & (2 + 2.48 \times 10^{-5}\omega^2) \\ (7.45 \times 10^{-4}\omega^2 + 3.7 \times 10^{-9}\omega^4) & & (1 + 3.228 \times 10^{-4}\omega^2 + 6.17 \times 10^{-10}\omega^4) \end{bmatrix}
$$

$$
\longrightarrow \qquad \longrightarrow
$$

$$
\text{and so on} \qquad \text{and so on} \qquad \text{and so on}
$$

$$
\begin{bmatrix} A_1 & B_1 \\ A_2 & B_2 \\ A_3 & B_3 \\ A_4 & B_4 \end{bmatrix}
$$

TABLE 5.2. Values Obtained from Table 5.1 When $\omega^2 = 400$

$$\begin{bmatrix} 0 & 0 \\ 1 & 0 \\ 0 & 0 \\ 0 & 1 \end{bmatrix}$$

$$\left[\begin{array}{cccc|c} 1 & 1 & \frac{1}{2} & \frac{1}{6} & 0 \\ 0 & 1 & 1 & \frac{1}{2} & 0 \\ 0 & 0 & 1 & 1 & 0 \\ 0 & 0 & 0 & 1 & 0.0596 \end{array}\right] \quad \left[\begin{array}{cc} 1.0000 & 0.1667 \\ \hline 1.0000 & 0.5000 \\ 0.0000 & 1.0000 \\ 0.0596 & 1.0099 \end{array}\right]$$

$$\left[\begin{array}{cccc|c} 1 & 1 & \frac{1}{4} & \frac{1}{12} & 0 \\ 0 & 1 & \frac{1}{2} & \frac{1}{4} & 0 \\ 0 & 0 & 1 & 1 & 0 \\ 0 & 0 & 0 & 1 & 0.1192 \end{array}\right] \quad \left[\begin{array}{cc} 2.0050 & 1.0008 \\ \hline 1.0149 & 1.2525 \\ 0.0596 & 2.0099 \\ 0.0620 & 1.1292 \end{array}\right]$$

$$\left[\begin{array}{cccc|c} 1 & 1 & \frac{1}{4} & \frac{1}{12} & 0 \\ 0 & 1 & \frac{1}{2} & \frac{1}{4} & 0 \\ 0 & 0 & 1 & 1 & 0 \\ 0 & 0 & 0 & 1 & 0.1788 \end{array}\right] \quad \left[\begin{array}{cc} 3.0400 & 2.8498 \\ \hline 1.0602 & 2.5400 \\ 0.1216 & 3.1391 \\ 0.6050 & 1.6392 \end{array}\right]$$

$$\left[\begin{array}{cccc|c} 1 & 1 & \frac{1}{2} & \frac{1}{6} & 0 \\ 0 & 1 & 1 & \frac{1}{2} & 0 \\ 0 & 0 & 1 & 1 & 0 \\ 0 & 0 & 0 & 1 & 0 \end{array}\right] \quad \left[\begin{array}{cc} 4.2618 & 7.2326 \\ 1.4843 & 6.4987 \\ 0.7266 & 4.7783 \\ 0.6050 & 1.6392 \end{array}\right]$$

For example, by assuming $\omega^2 = 400$ in Table 5.1, the computations for the matrix product are shown in Table 5.2. Thus, the frequency determinant in Eq. (5.244) yields

$$\begin{vmatrix} 4.2618 & 7.2326 \\ 0.7266 & 4.7783 \end{vmatrix} = 14.89$$

By assuming $\omega^2 = 1000$ in Table 5.1, the frequency determinant in Eq.

(5.244) yields

$$\begin{vmatrix} 5.2564 & 7.5832 \\ 3.0550 & 5.9564 \end{vmatrix} = 7.08$$

The procedure may continue in a similar manner until a sufficient number of values are accumulated to plot the required graph. The zero intersects in this graph are the values of the free frequencies of vibration of the member. Since three weights are involved and the beam is assumed to be weightless, three free frequencies of vibration will be obtained.

When the values of ω are determined as discussed above, the corresponding normal mode shapes are computed as discussed in Sections 5.10, 5.11, and 5.12. For each value of ω, the relation of the state vectors at stations 1, 2, and 3 with the state vector at station 0 can be obtained from Table 5.1. From here on, the procedure for determining the normal mode amplitudes is the same as the one used in the preceding three sections. At this point, the reader should note the similarity between the procedure used here and the one by Myklestad, as discussed in Section 5.6.

5.14 CONSTRUCTION OF TRANSFER MATRICES OF CONTINUOUS BEAMS OVER RIGID SUPPORTS

If the supports of a continuous beam are elastic springs, the methodology used in the preceding sections for the construction of transfer matrices is directly applicable. If, however, these supports are rigid, then we have to construct transfer matrices that relate the state vectors to the right and left of the rigid support.

To illustrate the procedure, we consider the two-span continuous beam in Fig. 5.17a. From the free-body diagram in Fig. 5.17b, we note that the unknown discontinuity is the reaction Q_j at station j. We also know that the vertical deflection of the member at station j is zero. The procedure for the construction of transfer matrices, which will be used for the computation of the free frequencies of vibration of the member, is similar to the one used in the preceding sections. The member is divided into a desirable number of segments, and the mass of each segment is lumped at the juncture point of the segment. We assume here this is accomplished, and the transfer matrices $[B]$ and $[C]$ in Fig. 5.17, relating the state vectors $\{v_0\}^R$ with $\{v_j\}^L$ and $\{v_j\}^R$ with $\{v_n\}^L$, respectively, are constructed by following the procedures discussed in the preceding sections.

The boundary conditions at station 0 are $y_0 = M_0 = 0$, and, consequently, the unknown quantities at this station are the rotation θ_0 and shear force V_0. If we start from station 0, with unknowns θ_0 and V_0, and move along the

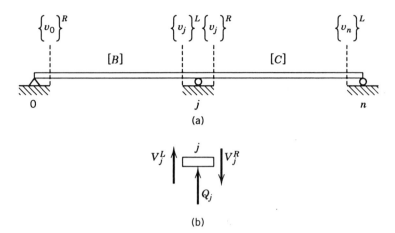

(a)

(b)

FIGURE 5.17. (a) Two-span continuous beam on rigid supports. (b) Free-body diagram of the rigid support j.

length of the member, we will reach the discontinuity Q_j at station j, which must be taken into consideration.

The state vectors $\{v_0\}^R$ and $\{v_j\}^L$ are related by the following equation:

$$\{v_j\}^L = [B]\{v_0\}^R \tag{5.246}$$

Since y_0 and M_0 are zero at station 0, the first and third columns of the matrix $[B]$ in Eq. (5.246) may be omitted. On this basis, by neglecting the elements of the first and third columns of matrix $[B]$, Eq. (5.246) may be written as follows:

$$\{v_j\}^L = \begin{bmatrix} b_{12} & b_{14} \\ b_{22} & b_{24} \\ b_{32} & b_{34} \\ b_{42} & b_{44} \end{bmatrix} \begin{bmatrix} \theta \\ V \end{bmatrix}_0^R \tag{5.247}$$

By using the free-body diagram in Fig. 5.17b, the following expression may be written at station j:

$$V_j^R = V_j^L + Q_j \tag{5.248}$$

The discontinuity Q_j must now be introduced, because the vector containing all the unknowns is $\{\theta_0 \quad V_0 \quad Q_0\}$. With this in mind, a relation between the state vectors $\{v_0\}^R$ and $\{v_j\}^R$ may be written as follows:

$$\begin{bmatrix} b_{12} & b_{14} & 0 \\ b_{22} & b_{24} & 0 \\ b_{32} & b_{34} & 0 \\ b_{42} & b_{44} & 1 \end{bmatrix} \begin{bmatrix} \theta_0 \\ V_0 \\ Q_j \end{bmatrix} = \begin{bmatrix} y \\ \theta \\ M \\ V \end{bmatrix}^R_j \tag{5.249}$$

The reaction Q_j at station j in Fig. 5.17b is taken into consideration in Eq. (5.249) by attaching a quantity equal to unity as the third element of the fourth row of matrix $[B]$. This transfer matrix formulation allows transfer directly from station 0 to the right of station j. The boundary condition at station j is that the deflection y_j^R is zero. On this basis, we can write the following relation by using Eq. (5.249):

$$b_{12}\theta_0 + b_{14}V_0 = 0 \tag{5.250}$$

To transfer just to the left of station n, it is only necessary to multiply the enlarged matrix $[B]$ by the transfer matrix $[C]$; that is,

$$[D] = [C][B] = \begin{bmatrix} c_{11} & c_{12} & c_{13} & c_{14} \\ c_{21} & c_{22} & c_{23} & c_{24} \\ c_{31} & c_{32} & c_{33} & c_{34} \\ c_{41} & c_{42} & c_{43} & c_{44} \end{bmatrix} \begin{bmatrix} b_{12} & b_{14} & 0 \\ b_{22} & b_{24} & 0 \\ b_{32} & b_{34} & 0 \\ b_{42} & b_{44} & 1 \end{bmatrix} \tag{5.251}$$

or

$$[D] = [C][B] = \begin{bmatrix} d_{11} & d_{12} & d_{13} \\ d_{21} & d_{22} & d_{23} \\ d_{31} & d_{32} & d_{33} \\ d_{41} & d_{42} & d_{43} \end{bmatrix} \tag{5.252}$$

This completes the procedure, and, consequently, the relation between the state vectors $\{v_0\}^R$ and $\{v_n\}^L$ in Fig. 5.17a is as follows:

$$\begin{bmatrix} d_{11} & d_{12} & d_{13} \\ d_{21} & d_{22} & d_{23} \\ d_{31} & d_{32} & d_{33} \\ d_{41} & d_{42} & d_{43} \end{bmatrix} \begin{bmatrix} \theta_0 \\ V_0 \\ Q_j \end{bmatrix} = \begin{bmatrix} y \\ \theta \\ M \\ V \end{bmatrix}^L_n \tag{5.253}$$

At station n, the boundary conditions are $y_n^L = M_n^L = 0$, and from Eq. (5.253), the following two expressions can be written:

$$d_{11}\theta_0 + d_{12}V_0 + d_{13}Q_j = 0 \tag{5.254}$$

$$d_{31}\theta_0 + d_{32}V_0 + d_{33}Q_j = 0 \tag{5.255}$$

Note that it is not required to calculate the elements of the second and fourth rows of matrix $[D]$ in Eq. (5.252), since $y_n^L = M_n^L = 0$.

Equations (5.250), (5.254), and (5.255) constitute a system of three homogeneous equations in terms of the unknown quantities θ_0, V_0, and Q_j. For a solution other than the trivial one, we set equal to zero the determinant of the coefficients of θ_0, V_0, and Q_j of these three homogeneous equations. This determinant is as follows:

$$\begin{vmatrix} b_{12} & b_{14} & 0 \\ d_{11} & d_{12} & d_{13} \\ d_{31} & d_{32} & d_{33} \end{vmatrix} = 0 \tag{5.256}$$

The elements of the determinant in Eq. (5.256) are functions of the free frequencies ω of the continuous beam in Fig. 5.17a. Therefore, the values of ω that satisfy Eq. (5.256) are the free frequencies of vibration of the continuous beam in Fig. 5.17a. A computer program is usually available to help in the computation of the values of ω and, consequently, in the determination of the corresponding mode shapes. Note that each discontinuity Q_j will increase the order of the determinant in Eq. (5.256) by one. It would be a 2×2 determinant if we did not have the discontinuity Q_j. The problem becomes much larger when many internal rigid supports are involved, and use of computer software would be the answer. Transfer matrix techniques are, in fact, developed and used extensively today because of the availability of computers to carry out the lengthy computations involved.

5.15 CONTINUOUS BEAMS WITH INTERNAL HINGES AND ON RIGID SUPPORTS

Consider the case where the continuous beam in Fig. 5.18 has an internal frictionless hinge at station i. The discontinuities now are two, the reaction Q_j at station j and the slope change ϕ_i at station i. On this basis, the deflection at station j is zero, and the bending moment at station i is zero. The construction of transfer matrices for the vibrating continuous beam may be carried out in the usual way, except that in this case both Q_i and ϕ_i must be introduced into the analysis. We assume at this point that the transfer

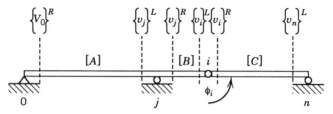

FIGURE 5.18. Continuous beam with internal hinges.

matrices $[A]$, (B), and $[C]$ in Fig. 5.18 between stations 0 and j, j and i, and i and n, respectively, are constructed.

At station 0, the boundary conditions are $y_0 = M_0 = 0$, and the relation between the state vectors $\{v_0\}^R$ and $\{v_j\}^L$ is written as follows:

$$\{v_j\}^L = \begin{bmatrix} a_{12} & a_{14} \\ a_{22} & a_{24} \\ a_{32} & a_{34} \\ a_{42} & a_{44} \end{bmatrix} \begin{bmatrix} \theta \\ V \end{bmatrix}_0^R \tag{5.257}$$

At station j, the reaction Q_j can be introduced as discussed in Section 5.14, and, consequently, a relation between the state vectors $\{v_0\}^R$ and $\{v_j\}^R$ may be written as follows:

$$\begin{bmatrix} a_{12} & a_{14} & 0 & 0 \\ a_{22} & a_{24} & 0 & 0 \\ a_{32} & a_{34} & 0 & 0 \\ a_{42} & a_{44} & 1 & 0 \end{bmatrix} \begin{bmatrix} \theta_0 \\ V_0 \\ Q_j \\ \phi_i \end{bmatrix} = \begin{bmatrix} y \\ \theta \\ M \\ V \end{bmatrix}_j^R \tag{5.258}$$

In the above expression, the vector containing the four unknowns is $\{\theta_0 \ \ V_0 \ \ Q_j \ \ \phi_i\}$. The reaction Q_j in Eq. (5.258) is taken into account by using the quantity of unity in the third element of the fourth row of the transfer matrix $[A]$.

By using Eq. (5.258) and the condition $y_j^R = 0$, the following expression can be written:

$$\alpha_{12}\theta_0 + a_{14}V_0 = 0 \tag{5.259}$$

To transfer just to the left of station i, the enlarged matrix $[A]$ in Eq. (5.258) is multiplied by the transfer matrix $[B]$. In this manner, the following matrix

$[D]$ is obtained:

$$[D] = [B][A] = \begin{bmatrix} b_{11} & b_{12} & b_{13} & b_{14} \\ b_{21} & b_{22} & b_{23} & b_{24} \\ b_{31} & b_{32} & b_{33} & b_{34} \\ b_{41} & b_{42} & b_{43} & b_{44} \end{bmatrix} \begin{bmatrix} a_{12} & a_{14} & 0 & 0 \\ a_{22} & a_{24} & 0 & 0 \\ a_{32} & a_{34} & 0 & 0 \\ a_{42} & a_{44} & 1 & 0 \end{bmatrix} \quad (5.260)$$

or

$$[D] = \begin{bmatrix} d_{11} & d_{12} & d_{13} & 0 \\ d_{21} & d_{22} & d_{23} & 0 \\ d_{31} & d_{32} & d_{33} & 0 \\ d_{41} & d_{42} & d_{43} & 0 \end{bmatrix} \quad (5.261)$$

At station i, because of the discontinuity ϕ_i in the slope, the following equation may be written:

$$\theta_i^R = \theta_i^L + \phi_i \quad (5.262)$$

where ϕ_i is the slope change. The relation between the state vectors $\{v_0\}^R$ and $\{v_i\}^R$ is then written as shown below:

$$\begin{bmatrix} d_{11} & d_{12} & d_{13} & 0 \\ d_{21} & d_{22} & d_{23} & 1 \\ d_{31} & d_{32} & d_{33} & 0 \\ d_{41} & d_{42} & d_{43} & 0 \end{bmatrix} \begin{bmatrix} \theta_0 \\ V_0 \\ Q_j \\ \phi_i \end{bmatrix} = \begin{bmatrix} y \\ \theta \\ M \\ V \end{bmatrix}_j^R \quad (5.263)$$

The slope discontinuity ϕ_i is introduced in Eq. (5.263) by placing a quantity of unity in the fourth element of the second row of the square matrix.

At station i, the bending moment M_i is zero, and on this basis, the following expression may be written:

$$d_{31}\theta_0 + d_{32}V_0 + d_{33}Q_j = 0 \quad (5.264)$$

Equation (5.264) takes into account that the bending moment at station i is zero.

The final step would be to transfer from the right side of station i to the left side of station n. This is accomplished by multiplying the square matrix $[D]$ in Eq. (5.263) by the transfer matrix $[C]$. This yields the combined matrix $[E]$ shown below:

$$[E] = [C][D] = \begin{bmatrix} c_{11} & c_{12} & c_{13} & c_{14} \\ c_{21} & c_{22} & c_{23} & c_{24} \\ c_{31} & c_{32} & c_{33} & c_{34} \\ c_{41} & c_{42} & c_{43} & c_{44} \end{bmatrix} \begin{bmatrix} d_{11} & d_{12} & d_{13} & 0 \\ d_{21} & d_{22} & d_{23} & 1 \\ d_{31} & d_{32} & d_{33} & 0 \\ d_{41} & d_{42} & d_{43} & 0 \end{bmatrix} \tag{5.265}$$

or

$$[E] = \begin{bmatrix} e_{11} & e_{12} & e_{13} & e_{14} \\ e_{21} & e_{22} & e_{23} & e_{24} \\ e_{31} & e_{32} & e_{33} & e_{34} \\ e_{41} & e_{42} & e_{43} & e_{44} \end{bmatrix} \tag{5.266}$$

The relation between the state vectors $\{v_0\}^R$ and $\{v_n\}^L$ may now be written as follows:

$$\begin{bmatrix} e_{11} & e_{12} & e_{13} & e_{14} \\ e_{21} & e_{22} & e_{23} & e_{24} \\ e_{31} & e_{32} & e_{33} & e_{34} \\ e_{41} & e_{42} & e_{43} & e_{44} \end{bmatrix} \begin{bmatrix} \theta_0 \\ V_0 \\ Q_j \\ \phi_i \end{bmatrix} = \begin{bmatrix} y \\ \theta \\ M \\ V \end{bmatrix}^L_n \tag{5.267}$$

At station n, we can use the boundary conditions that the deflection y_n and the bending moment M_n are zero, and from Eq. (5.267), we can write the following two expressions by using these two boundary conditions:

$$e_{11}\theta_0 + e_{12}V_0 + e_{13}Q_j + e_{14}\phi_i = 0 \tag{5.268}$$

$$e_{31}\theta_0 + e_{32}V_0 + e_{33}Q_j + e_{34}\phi_i = 0 \tag{5.269}$$

Equations (5.259), (5.264), (5.268), and (5.269) form a system of four homogeneous equations involving the unknowns θ_0, V_0, Q_j, and ϕ_i. For a nontrivial solution, the determinant of their coefficients should be zero. This frequency determinant is as follows:

$$\begin{vmatrix} a_{12} & a_{14} & 0 & 0 \\ d_{31} & d_{32} & d_{33} & 0 \\ e_{11} & e_{12} & e_{13} & e_{14} \\ e_{31} & e_{32} & e_{33} & e_{34} \end{vmatrix} = 0 \tag{5.270}$$

The elements of the determinant in Eq. (5.270) are functions of the frequencies ω of the continuous beam in Fig. 5.18. The values of ω that satisfy Eq. (5.270) are the free frequencies of vibration of the beam. In this case, a fourth-order frequency determinant is obtained, because we have two discontinuities, namely, Q_j and ϕ_i.

5.16 MODE SHAPE DIAGNOSIS BY USING THE DYNAMIC HINGE CONCEPT

The theory and concept of the dynamic hinge, as well as the derivation of exact and approximate dynamically equivalent systems, are discussed at length in Sections 4.10 through 4.12. It was proved that the free frequencies of vibration of a structure or a machine may be determined by using only a small portion of the complete structure and appropriately satisfying its end boundary conditions.

This concept may also be used to check if the mode shapes obtained by approximate methods of analysis, or by using computer codes, are correct. To illustrate this idea, we consider the two-span continuous beam in Fig. 5.19a. The beam has a rectangular cross-section of width $b = 5.14$ in. and depth $h = 16.0$ in. The modulus of elasticity $E = 30 \times 10^6$ psi, and the span lengths are as shown in the figure. The first four free frequencies of vibration of the beam and the corresponding mode shapes were obtained by using the computer code A.D.I.N.A., and the mode shapes are as shown in Fig. 5.19. The same first four frequencies and the corresponding mode shapes are also determined by using Stodola's method, and they are shown in Fig. 5.20.

We consider first the fourth mode shape in Fig. 5.19d, which was obtained by using the computer code A.D.I.N.A. The free frequency of vibration of this mode was found to be $f_4 = 19.18$ Hz. By using the fourth mode shape, which is drawn again in Fig. 5.21a, and the pseudostatic system in Fig. 5.21b, where the weight of the member is assumed to act in the direction of the amplitudes of the mode shape, the points of zero moment are determined by using static analysis, and they are located at points H, G, and J, as shown in Fig. 5.21c. Since the dynamic hinge points H, G, and J are shown in Fig. 5.21a to be located very close to the node points D, E, and F, we can assume that the spring constants k at the dynamic hinge points are infinite in value. The relative value of the spring constants at the dynamic hinge points of the first four mode shapes are shown in Table 5.3. The dynamically equivalent systems for portions AH, HG, and JC are as shown in Fig. 5.21d.

The fundamental frequency of each of the dynamically equivalent systems in Fig. 5.21d is obtained by using Eq. (4.261) for simply supported beams,

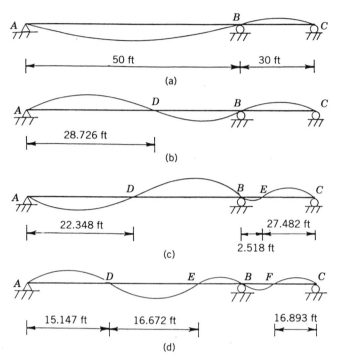

FIGURE 5.19. Mode shapes of a two-span continuous beam by using A.D.I.N.A.: (a) first mode, (b) second mode, (c) third mode, and (d) fourth mode.

and the results are as follows:

Portion *AH*	$f_4 = 26.90$ Hz
Portion *HG*	$f_4 = 16.47$ Hz
Portion *JC*	$f_4 = 17.49$ Hz
Complete system with A.D.I.N.A.	$f_4 = 19.18$ Hz

The above results show that the frequencies of the three portions are not the same, or even closely identical, to the one obtained by using the complete system, and, consequently, the fourth mode shape obtained by using A.D.I.N.A. is incorrect. The node points would have to be adjusted in a way that will make the frequencies of the three portions the same or nearly identical. However, it is interesting to note here that the average frequency of the three portions is 20.29 Hz, which is 5.78% higher than the value obtained by A.D.I.N.A.

Consider now the third mode shape in Fig. 5.22a, which was obtained by using Stodola method with 16 lumped masses and the iteration procedure. The pseudostatic system for this mode is shown in Fig. 5.22b, and its moment diagram is shown in Fig. 5.22c. From this figure, we note that the points of

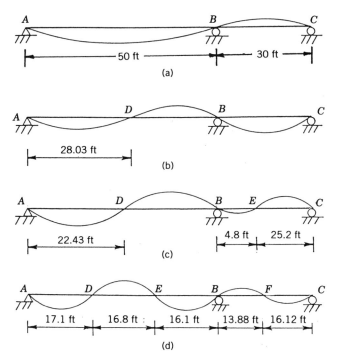

FIGURE 5.20. Mode shapes of a two-span continuous beam by using Stodola's method: (a) first mode, (b) second mode, (c) third mode, and (d) fourth mode.

zero moment (dynamic hinge points) are located at points H and F, which are close to the node points E and D in Fig. 5.22a. The dynamically equivalent systems for portions AH and FC are shown in Fig. 5.22d. We can assume here that the spring stiffnesses k_H and k_F are infinitely large since the dynamic hinge points are close to the node points E and D. This is reasonable; most probably we would introduce more error in calculating k_H and k_F, since the deflections at the dynamic hinge points H and F are very small. The results are as follows:

Portion AH	$f_3 = 10.10\,\text{Hz}$
Portion FC	$f_3 = 10.61\,\text{Hz}$
Complete system by Stodola	$f_3 = 10.28\,\text{Hz}$

In this case, reasonable agreement is obtained between all three values shown above, and the third mode shape obtained by Stodola's method would be fairly accurate. In all cases examined, the average value of the portions closely approximates the value obtained by using the complete system.

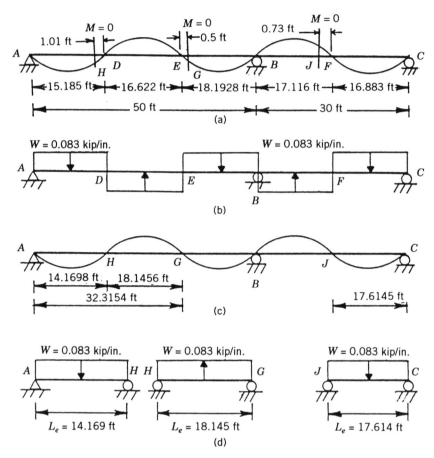

FIGURE 5.21. (a) Fourth mode shape of a two-span continuous beam by using A.D.I.N.A. (b) Pseudostatic system. (c) Moment diagram of the pseudostatic system. (d) Dynamically equivalent systems for portions AH, HG, and JC.

TABLE 5.3. Values of the Spring Constants at the Dynamic Hinge Points of the First Four Mode Shapes

	Spring Stiffness (kips/ft)		
Mode Number	k_1	k_2	k_3
1	441.9	—	—
2	1085.0	2075.0	—
3	2741.0	3886.0	2035.0
4	6028.6	7195.4	4020.8

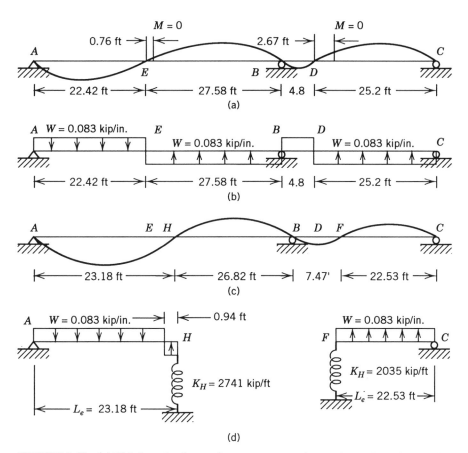

FIGURE 5.22. (a) Third mode shape of a two-span continuous beam by using Stodola's method. (b) Pseudostatic system. (c) Moment diagram of the pseudostatic system. (d) Dynamically equivalent systems for portions *AH* and *FC*.

PROBLEMS

5.1 Rework Example 5.1 by assuming that $y(x) = y_0 \sin(\pi x/L)$, where y_0 is the maximum amplitude of the mode shape, and compare the results.

5.2 For a uniform single-span beam fixed at both ends, determine its fundamental frequency of vibration by using Rayleigh's method. Assume that $y(x)$ is the static deflection curve produced by the uniformly distributed weight w of the member. The stiffness EI is constant.

5.3 For a uniform single-span beam fixed at one end and simply supported at the other, determine its fundamental frequency of vibration by using Rayleigh's method. Assume that $y(x)$ is the static deflection curve produced by the uniformly distributed weight w of the member. The exact value of this frequency is

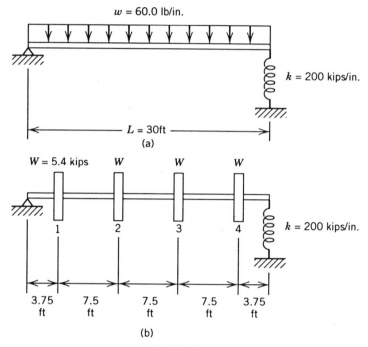

FIGURE P5.5

$$f = 2.45 \sqrt{\frac{gEI}{wL^4}} \text{ Hz}$$

5.4 Repeat Example 5.2 by assuming that the shape $y(x)$ of the fundamental mode of vibration is the one produced by the pseudostatic system in Fig. 5.2c, where the weight w is assumed to act in the direction of the fundamental mode as shown. Compare the results.

5.5 The uniform single-span beam in Fig. P5.5a is hinged at one end and elastically supported at the other by a linear spring of constant $k = 200$ kips/in. Determine the fundamental frequency of its free vibration by using Rayleigh's energy method. Assume that $y(x)$ is the static deflection curve produced by its attached distributed weight $w = 60$ lb/in. The uniform stiffness $EI = 30 \times 10^9$ lb · in.^{2}.

5.6 Rework Problem 5.5 by lumping the distributed weight w of the beam into four concentrated weights as shown in Fig. P5.5b. Each concentrated weight $W = 5.4$ kips is determined by dividing the member into four equal segments and concentrating the weight W of each segment at the center of the segment. Use Rayleigh's method in conjunction with Eqs. (5.28) and (5.30) and compare the results.

5.7 Rework Example 5.3 by taking into consideration the weight $w = 5.0$ lb/in. of the member and compare the results. *Hint*: Lump half of the weight of the member at point 2 and the other half at point 4.

5.8 Repeat Example 5.2 by lumping the weight of the beam in Fig. 5.2c at three points in each span and compare the results. Use Eq. (5.30) to repeat the procedure once. The stiffness EI is constant throughout the length of the member.

5.9 The uniform beams in Fig. P5.9 have constant stiffness $EI = 45 \times 10^6$ kip $\cdot$ in.^{2} and weight $w = 50$ lb/in. By lumping the weight at discrete points along their length and applying Rayleigh's method, determine in each case the fundamental frequency of their free vibration and the corresponding mode shapes. Repeat the procedures as required for an accurate solution.

5.10 Repeat Problem 5.6 by using Stodola's method and the iteration procedure. Compare the results.

5.11 Repeat Problem 5.8 by using Stodola's method and the iteration procedure. Compare the results.

5.12 Repeat Problem 5.9 by applying Stodola's method with an iteration procedure. Compare the results.

5.13 For the beam in Problem 5.6, determine the first two frequencies of vibration and their corresponding mode shapes by applying Stodola's method and an iteration procedure.

5.14 The simply supported beam in Fig. P5.14 is supporting five concentrated weights at the positions shown in the figure. The magnitudes of the concentrated weights are $W_1 = 8.16$ kips, $W_2 = 10.10$ kips, $W_3 = 9.20$ kips, $W_4 = 8.50$ kips, and $W_5 = 5.90$ kips. Determine the first three frequencies of its free vibration and the corresponding mode shapes by applying Stodola's method and the iteration procedure. The modulus of elasticity $E = 30 \times 10^6$ psi, and the constant moment of inertia $I = 3413.33$ in.^{4}.

5.15 The two-span uniform continuous beam in Fig. P5.15 carries 11 attached concentrated weights at the indicated positions. The magnitudes of the weights are $W_1 = 7.3$ kips, $W_2 = 8.0$ kips, $W_3 = 6.5$ kips, $W_4 = 6.1$ kips, $W_5 = 10.0$ kips, $W_6 = 6.3$ kips, $W_7 = 7.1$ kips, $W_8 = 8.0$ kips, $W_9 = 9.0$ kips, $W_{10} = 10.3$ kips, and $W_{11} = 8.7$ kips. Determine the first three frequencies of its free vibration and the corresponding mode shapes by using Stodola's method and the iteration procedure. The modulus of elasticity $E = 30 \times 10^6$ psi, and the constant moment of inertia $I = 3413.33$ in.^{4}.

5.16 For the problem in Fig. P5.9d, determine the first three frequencies

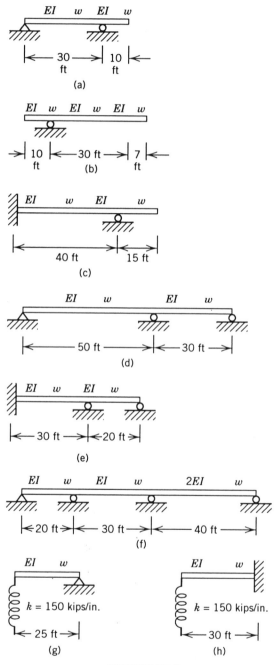

FIGURE P5.9

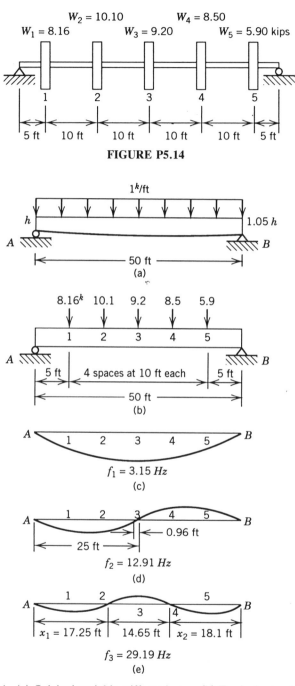

FIGURE P5.14

FIGURE A5.14. (a) Original variable stiffness beam. (b) Equivalent uniform stiffness system. (c) First mode shape. (d) Second mode shape. (e) Third mode shape. (See also Page 779, answer to Problem 5.14)

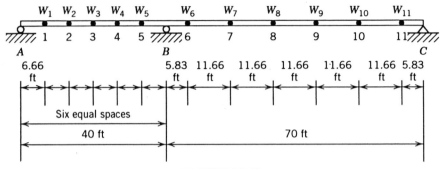

FIGURE P5.15

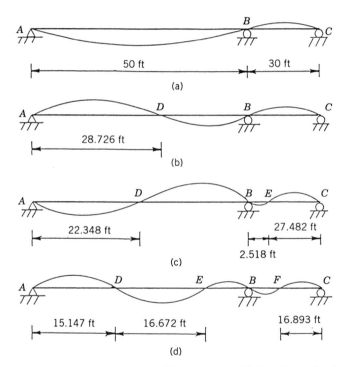

FIGURE A5.16. (a) First, (b) second, (c) third, and (d) fourth mode shapes. (See also Page 779, answer to Problem 5.16)

of its free vibration and the corresponding mode shapes by using Stodola's method and the iteration procedure. The constant stiffness $EI = 52.63 \times 10^9$ psi, and the attached weight w per unit of length of the member is 82.0 lb/in.

5.17 By using Myklestad's method, determine the first two frequencies of

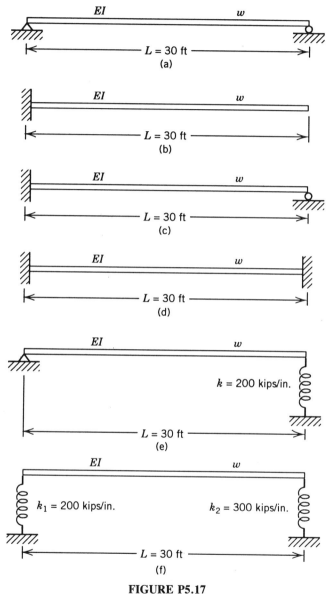

FIGURE P5.17

vibration and the corresponding mode shapes for the beams in Fig.
P5.17. The constant stiffness $EI = 30 \times 10^6$ kip $\cdot$ in.^{2} and the uniform
weight w per unit length is 50 lb/in. Neglect damping.

5.18 Determine the fundamental frequency of vibration of the variable
stiffness cantilever beam, which carries the attached weight $w =$

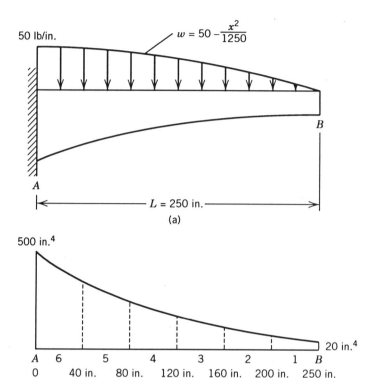

FIGURE P5.18

$[50 - (x^2/1250)]$ as shown in Fig. P5.18a. The moment of inertia variation is as shown in Fig. P5.18b, and it has the following average values at the six intervals between points A and B:

$$I_0 = I_B = 20.0 \text{ in.}^4 \quad I_4 = 197.5 \text{ in.}^4$$

$$I_1 = 32.5 \text{ in.}^4 \quad I_5 = 302.5 \text{ in.}^4$$

$$I_2 = 67.5 \text{ in.}^4 \quad I_6 = 430.0 \text{ in.}^4$$

$$I_3 = 120.0 \text{ in.}^4 \quad I_A = 500.0 \text{ in.}^4$$

The modulus of elasticity $E = 30 \times 10^6$ psi.

5.19 Solve Example 2.1 by using transfer matrices.

5.20 Solve Example 2.2 by using transfer matrices.

5.21 Solve Example 2.8 by using transfer matrices.

5.22 Solve Example 2.10 by using transfer matrices.

5.23 Solve Example 2.11 by using transfer matrices.

5.24 Solve Example 2.13 by using transfer matrices.

5.25 Solve Example 2.14 by using transfer matrices. Use three equal-length segments and lump the mass of each segment at the juncture points.

5.26 Solve Example 3.1 by using transfer matrices.

5.27 Solve Problem 3.1 by using transfer matrices.

5.28 Solve Problem 3.11 by using transfer matrices.

5.29 The beams shown in Fig. P5.29 are assumed to be of negligible weight, and they are supporting the attached weights as shown. By using transfer matrices for flexural one-degree systems, determine for each member the frequency and mode shape of its free undamped vibration.

5.30 By using appropriate transfer matrices, determine the free frequencies of vibration and the corresponding mode shapes of the free undamped spring–mass systems shown in Fig. P5.30.

5.31 The beams shown in Fig. P5.31 are assumed to be of negligible weight, and they are supporting the attached weights as shown. By using appropriate transfer matrices, determine for each member the free undamped frequencies of vibration and the corresponding mode shapes.

5.32 Solve Problem 5.14 by using transfer matrices.

5.33 Solve Problem 5.15 by using transfer matrices.

5.34 Solve Problem 5.16 by using transfer matrices.

5.35 Solve Problem 5.18 by using transfer matrices.

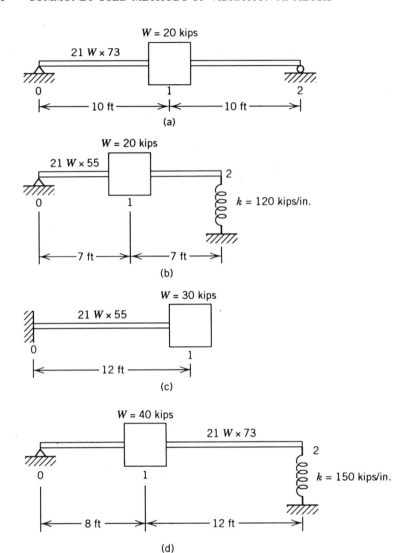

FIGURE P5.29

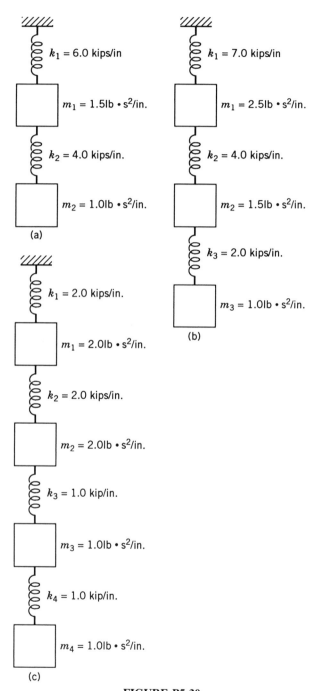

(a)

$k_1 = 6.0$ kips/in

$m_1 = 1.5$lb $\cdot$ s^2/in.

$k_2 = 4.0$ kips/in.

$m_2 = 1.0$lb $\cdot$ s^2/in.

(b)

$k_1 = 7.0$ kips/in

$m_1 = 2.5$lb $\cdot$ s^2/in.

$k_2 = 4.0$ kips/in.

$m_2 = 1.5$lb $\cdot$ s^2/in.

$k_3 = 2.0$ kips/in.

$m_3 = 1.0$lb $\cdot$ s^2/in.

(c)

$k_1 = 2.0$ kips/in.

$m_1 = 2.0$lb $\cdot$ s^2/in.

$k_2 = 2.0$ kips/in.

$m_2 = 2.0$lb $\cdot$ s^2/in.

$k_3 = 1.0$ kip/in.

$m_3 = 1.0$lb $\cdot$ s^2/in.

$k_4 = 1.0$ kip/in.

$m_4 = 1.0$lb $\cdot$ s^2/in.

FIGURE P5.30

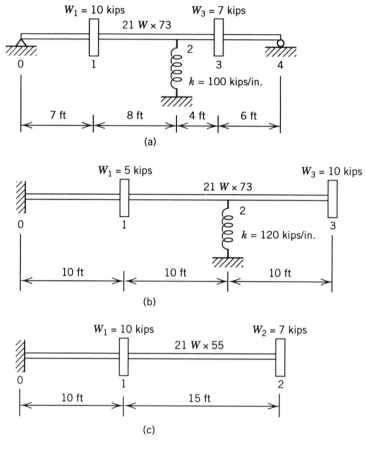

(a)

(b)

(c)

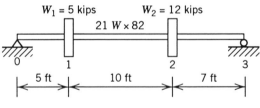

(d)

FIGURE P5.31

6 Finite Element Method*

6.1 INTRODUCTION

The origin of the finite element method can be traced back to a paper [32] published in 1943 by the mathematician Courant, who suggested the division of a continuum problem into discrete triangles, with approximation of the solution across the triangles. The development of the method, however, can be attributed to the intense interest by the aircraft industry. The method became a reality in a paper by Turner, Clough, Martin, and Topp [33], which was published in 1956. Clough [34, 91], the lone member of that group who was not in the aerospace area, later coined the term *finite element method* in his 1960 paper.

Another major contemporary in the development of the finite element method was Argyris [35]. Argyris' work was published in the *Journal of Aircraft Engineering* in 1954 and 1955, but he did not use the term finite element. A conference on matrix methods in structural mechanics [36] was held in 1966 at Wright–Patterson Air Force Base in Ohio, and at this time implementation of the finite element method onto computers had begun. Under a NASA project, the computer program NASTRAN (NASA Structural Analysis Program) was developed.

The basic concept of the finite element approach is to subdivide a large complex structure into a finite number of simple elements, such as beam elements, quadrilaterals, or triangles, and the complex differential equations are then solved for the simple elements. Assemblage of the elements into a global matrix transforms the problem from a differential equations formulation over a continuum to a linear algebra problem that is readily solvable by using computers. The method is similar to the stiffness method used to analyze trusses and frames, except that the truss and frame members are now subdivided into very small simple elements that are later assembled into global matrices.

In static analysis, the finite element method solves equations of the form

$$[K]\{U\} = \{R\} \tag{6.1}$$

---

*This chapter was contributed by Dr. Paul A. Bosela in collaboration with the author.

where $[K]$ is the global stiffness matrix, $\{U\}$ is a vector of nodal displacements, and $\{R\}$ is a vector of nodal forces.

The dynamic analysis usually starts with an equation of the form

$$([K] - \omega^2[M])\{U\} = \{R\} \tag{6.2}$$

where ω in this equation denotes the frequencies of vibration and $[M]$ is the mass matrix. For a free vibration, we have $\{R\} = 0$, and Eq. (6.2) reduces to the following expression:

$$([K] - \omega^2[M])\{U\} = \{0\} \tag{6.3}$$

For a nontrivial solution, we have

$$[K] = \omega^2[M] \tag{6.4}$$

which represents the *generalized eigenvalue problem*. Various algorithms may be used to solve for the frequencies (*eigenvalues*) and the corresponding mode shapes (*eigenvectors*) of a given structure or machine.

6.2   ELEMENT STIFFNESS AND MASS MATRICES

Numerous elements have been developed using direct stiffness extremization of a functional, weighted residuals, and energy balance approaches. Many of these more advanced formulations may be found in the works by Bathe [37], Huebner and Thornton [38], Cook [39], Martin [40], and Martin and Carey [41]. A brief discussion of some of the simple element formulations is included in this section.

6.2.1   Axial Elements

Consider the truss element shown in Fig. 6.1, which has the degrees of freedom u_1 and u_2 at points 1 and 2, respectively. Since the members of a truss are loaded axially, the element would be subject to axial forces f_1 and f_2 as shown in the figure, and, consequently, u_1 and u_2 would represent axial displacements. We can assume here that the displacement function $u(x)$ may

FIGURE 6.1. Truss element.

be represented by the following linear equation:

$$u(x) = b_1 + b_2 x \qquad (6.5)$$

where b_1 and b_2 are constants. In matrix form, Eq. (6.5) may be written as follows:

$$u(x) = \{1 \quad x\} \begin{Bmatrix} b_1 \\ b_2 \end{Bmatrix} \qquad (6.6)$$

By using Eq. (6.5) and applying the boundary conditions $u(0) = u_1$ and $u(L) = u_2$, where L is the length of the elements, we find

$$u(0) = b_1 + (0)b_2 \qquad (6.7)$$

$$u(L) = b_2 + Lb_2 \qquad (6.8)$$

Equations (6.7) and (6.8) are written in matrix form as follows:

$$\begin{Bmatrix} u_1 \\ u_2 \end{Bmatrix} = \begin{bmatrix} 1 & 0 \\ 1 & L \end{bmatrix} \begin{Bmatrix} b_1 \\ b_2 \end{Bmatrix} \qquad (6.9)$$

The same two equations may also be written in a matrix form that relates b_1 and b_2 and u_1 and u_2 in the following way:

$$\begin{Bmatrix} b_1 \\ b_2 \end{Bmatrix} = \begin{bmatrix} L & 0 \\ -1 & 1 \end{bmatrix} \begin{Bmatrix} u_1 \\ u_2 \end{Bmatrix} \qquad (6.10)$$

By substituting Eq. (6.10) into Eq. (6.6), we find

$$u(x) = \left\{ 1 - \frac{x}{L} \quad \frac{x}{L} \right\} \begin{Bmatrix} u_1 \\ u_2 \end{Bmatrix} \qquad (6.11)$$

or, in general,

$$u(x) = \{H_1 \quad H_2\} \begin{Bmatrix} u_1 \\ u_2 \end{Bmatrix} \qquad (6.12)$$

where H_i in Eq. (6.12) are referred to as *shape functions* or *interpolation functions*.

The axial strain ε of the element in Fig. 6.1 is the rate of change of $u(x)$ with respect to x. Therefore, by differentiating Eq. (6.11) with respect to x, we find

$$\varepsilon = \left\{ -\frac{1}{L} \quad \frac{1}{L} \right\} \left\{ \begin{matrix} u_1 \\ u_2 \end{matrix} \right\} \tag{6.13}$$

By applying *Hooke's law*, the normal stress σ at cross sections along the length of the element is given by the expression

$$\sigma = E\varepsilon \tag{6.14}$$

where E is Young's modulus of elasticity and ε is given by Eq. (6.13).

The total energy Π stored in the element is defined by the expression

$$\Pi = \frac{1}{2} \int_0^L EA \left(\frac{du}{dx} \right)^T \left(\frac{du}{dx} \right) dx - f_1 u_1 - f_2 u_2 \tag{6.15}$$

where A is the cross-sectional area of the member. By using Eq. (6.12), the expression given by Eq. (6.15) may be written as follows:

$$\Pi = \frac{1}{2} \int_0^L EA \{ u_1 \quad u_2 \} \left\{ \begin{matrix} -\dfrac{1}{L} \\ \dfrac{1}{L} \end{matrix} \right\} \left\{ -\frac{1}{L} \quad \frac{1}{L} \right\} \left\{ \begin{matrix} u_1 \\ u_2 \end{matrix} \right\} dx - \{ f_1 \quad f_2 \} \left\{ \begin{matrix} u_1 \\ u_2 \end{matrix} \right\} \tag{6.16}$$

When EA is constant, we have

$$\int_0^L dx = L \tag{6.17}$$

and Eq. (6.16) yields

$$\Pi = \frac{EAL}{2} \{ u_1 \quad u_2 \} \begin{bmatrix} \dfrac{1}{L^2} & -\dfrac{1}{L^2} \\ -\dfrac{1}{L^2} & \dfrac{1}{L^2} \end{bmatrix} \begin{bmatrix} u_1 \\ u_2 \end{bmatrix} - \{ f_1 \quad f_2 \} \left\{ \begin{matrix} u_1 \\ u_2 \end{matrix} \right\}$$

$$= \frac{1}{2} \{ u_1 \quad u_2 \} \frac{AE}{L} \begin{bmatrix} 1 & -1 \\ -1 & 1 \end{bmatrix} \begin{bmatrix} u_1 \\ u_2 \end{bmatrix} - \{ f_1 \quad f_2 \} \left\{ \begin{matrix} u_1 \\ u_2 \end{matrix} \right\} \tag{6.18}$$

For a stationary condition, we have

$$\frac{\partial \Pi}{\partial u_i} = 0 \tag{6.19}$$

On this basis, use of Eq. (6.19) yields

$$\frac{AE}{L}\begin{bmatrix} 1 & -1 \\ -1 & 1 \end{bmatrix}\begin{Bmatrix} u_1 \\ u_2 \end{Bmatrix} = \begin{Bmatrix} f_1 \\ f_2 \end{Bmatrix} \tag{6.20}$$

By comparing Eq. (6.20) with Eq. (6.1), we conclude that the stiffness matrix $[K]$ of the element is

$$[K] = \frac{AE}{L}\begin{bmatrix} 1 & -1 \\ -1 & 1 \end{bmatrix} \tag{6.21}$$

In a similar manner, various other element stiffness matrices may be obtained.

6.2.2 Bernoulli Beam

Consider now the two-node Bernoulli beam shown in Fig. 6.2. The beam has three degrees of freedom at each node, namely, horizontal translation, vertical translation, and rotation about an axis perpendicular to the plane of the paper. The six degrees of freedom of the Bernoulli beam are numbered from 1 to 6 as shown in the figure.

The total potential energy Π of the beam equals the strain energy U less the potential V of the external loads; that is, $\Pi = U - V$. The strain energy U can be written

$$\begin{aligned} U &= \int_0^L \left[\frac{EA}{2}\left[\frac{du_0}{dx} + \frac{1}{2}\left[\frac{dW}{dx}\right]^2\right]^2 + \frac{EI}{2}\left[\frac{d^2W}{dx^2}\right]^2 \right] dx \\ &= \int_0^L \left[\frac{EA}{2}\left[\underbrace{\left[\frac{du_0}{dx}\right]^2}_{\{1\}} + \underbrace{\frac{du_0}{dx}\left[\frac{dW}{dx}\right]}_{\{2\}} + \frac{1}{4}\left[\frac{dW}{dx}\right]^4\right] + \frac{EI}{2}\underbrace{\left[\frac{d^2W}{dx^2}\right]^2}_{\{3\}} \right] dx \end{aligned} \tag{6.22}$$

FIGURE 6.2. Bernoulli beam two-node element.

One can then approximate the displacements using the Hermitian interpolation polynomials as shape function, such that

$$u_0(x) = [H_1 \quad H_4]\begin{bmatrix} u_1 \\ u_4 \end{bmatrix} \tag{6.23}$$

$$W(x) = [H_2 \quad H_3 \quad H_5 \quad H_6]\begin{bmatrix} u_2 \\ u_3 \\ u_5 \\ u_6 \end{bmatrix} \tag{6.24}$$

The shape functions are

$$\begin{aligned}
H_1 &= 1 - \frac{x}{L} & H_4 &= \frac{X}{L} \\
H_2 &= 1 - 3\left(\frac{x}{L}\right)^2 + 2\left(\frac{x}{L}\right)^3 & H_5 &= 3\left(\frac{x}{L}\right)^2 - 2\left(\frac{x}{L}\right)^3 \\
H_3 &= x - \frac{2x^2}{L} + \frac{x^3}{L^2} & H_6 &= -\frac{x^2}{L} + \frac{x^3}{L^2}
\end{aligned} \tag{6.25}$$

By substituting the expressions given by Eq. (6.25) into Eqs. (6.23) and (6.24) and performing the required multiplication and taking the derivatives, we find

$$u = H_1 u_1 + H_4 u_4$$

$$\frac{du}{dx} = H_1' u_1 + H_4' u_4$$

$$W = H_2 u_2 + H_3 u_3 + H_5 u_5 + H_6 u_6$$

$$\frac{dW}{dx} = H_2' u_2 + H_3' u_3 + H_5' u_5 + H_6' u_6$$

$$\frac{d^2 W}{dx^2} = H_2'' u_2 + H_3'' u_3 + H_5'' u_5 + H_6'' u_6$$

Now consider term {1} of Eq. (6.22):

Term {1}

$$U = \int_0^L \frac{EA}{2}\left[\frac{du_0}{dx}\right]^2 dx = \int_0^L \frac{EA}{2}\left[\frac{du_0}{dx}\right]^T \left[\frac{du_0}{dx}\right] dx$$

$$= \tfrac{1}{2}[u_1 \ \ u_4]^T \int_0^L EA \begin{bmatrix} -\dfrac{1}{L} \\[2mm] \dfrac{1}{L} \end{bmatrix} \begin{bmatrix} -\dfrac{1}{L} & \dfrac{1}{L} \end{bmatrix} dx \begin{bmatrix} u_1 \\ u_4 \end{bmatrix}$$

(6.26)

For constant EA, Eq. (6.26) yields

$$\tfrac{1}{2}[u_1 \ \ u_4]^T \frac{EA}{L}\begin{bmatrix} 1 & -1 \\ -1 & 1 \end{bmatrix}\begin{bmatrix} u_1 \\ u_4 \end{bmatrix}$$

(6.27)

Next, we consider term {3} of Eq. (6.22):

Term {3}

$$U = \int_0^L \frac{EI}{2}\left[\frac{d^2W}{dx^2}\right]^2 dx = \int_0^L \frac{EI}{2}\left[\frac{\partial^2W}{\partial x^2}\right]^T \left[\frac{d^2W}{dx^2}\right] dx$$

(6.28)

$$= \tfrac{1}{2}[u_2 \ \ u_3 \ \ u_5 \ \ u_6] \int_0^L EI \begin{bmatrix} H_2'' \\ H_3'' \\ H_5'' \\ H_6'' \end{bmatrix} [H_2'' \ \ H_3'' \ \ H_5'' \ \ H_6''] \, dx \begin{bmatrix} u_2 \\ u_3 \\ u_5 \\ u_6 \end{bmatrix}$$

(6.29)

By taking the second derivatives of the shape functions, we find

$$H_2'' = -\frac{6}{L^2} + \frac{12x}{L^3}$$

$$H_3'' = -\frac{4}{L} + \frac{6x}{L^2}$$

$$H_5'' = \frac{6}{L^2} - \frac{12x}{L^3}$$

(6.30)

$$H_6'' = -\frac{2}{L} + \frac{6x}{L^2}$$

Substituting the expressions given by Eq. (6.30) into Eq. (6.29) and keeping EI as constant, we find

$$\frac{1}{2}[u_2 \quad u_3 \quad u_5 \quad u_6]\frac{EI}{L^3}\begin{bmatrix} 12 & 6L & -12 & 6L \\ 6L & 4L^2 & -6L & 2L^2 \\ -12 & -6L & 12 & -6L \\ 6L & 2L^2 & -6L & 4L^2 \end{bmatrix}\begin{bmatrix} u_2 \\ u_3 \\ u_5 \\ u_6 \end{bmatrix} \qquad (6.31)$$

From Eqs. (6.27) and (6.31), the elastic stiffness matrix of the beam is

$$[K_e] = \frac{EI}{L^3}\begin{bmatrix} AL^2/I & 0 & 0 & -AL^2/I & 0 & 0 \\ 0 & 12 & 6L & 0 & -12 & 6L \\ 0 & 6L & 4L^2 & 0 & -6L & 2L^2 \\ -AL^2/I & 0 & 0 & AL^2/I & 0 & 0 \\ 0 & -12 & -6L & 0 & 12 & -6L \\ 0 & 6L & 2L^2 & 0 & -6L & 4L^2 \end{bmatrix} \qquad (6.32)$$

Note that the stiffness matrix in Eq. (6.32) is a function of Young's modulus E, moment of inertia I, and length L. It does not consider the effects of the preload. Term {2} of Eq. (6.22) will be used later to develop the geometric stiffness matrix.

The Bernoulli beam element can effectively be used to model most beams, where bending stresses far exceed the effects of shear, and where the lower frequencies are of prime concern. For short, stubby beams with high shear loading, the Bernoulli beam model may not provide satisfactory results.

6.2.3 Timoshenko Beam

In dynamic analysis, transverse shear and rotary inertia can have a significant effect on higher modes. Since the Bernoulli beam does not include transverse shear, the Timoshenko beam was developed. A detailed derivation is provided by Bellini [42].

The Bernoulli formulation assumed that the slope of a beam at a point is $\partial w/\partial x$, where w is due to bending only. The total displacement w, however, is due to both bending and shear. Thus, we have

$$\frac{\partial w}{\partial x} = \frac{\partial w_b}{\partial x} + \frac{\partial w_s}{\partial x}$$

$$= \beta + \gamma \qquad (a)$$

where β is the slope produced by bending, and γ is the slope caused by shear deformation. The normal strain e_{xx} is

$$e_{xx} = z \frac{\partial \beta}{\partial x}$$

where z is the distance from the neutral axis.

For a beam with a distributed load $q(x)$, the total potential Π can be written

$$\Pi = \frac{1}{2} \int_0^L \left[EI \left(\frac{\partial \beta}{\partial x} \right)^2 + GAK \left(\frac{\partial w}{\partial x} - \beta \right)^2 \right] \partial x - \int_0^L q(x)\, w\, \partial x \qquad \text{(b)}$$

Setting $\delta \Pi = 0$ and integrating by parts, we find the following equations:

$$\frac{EI \partial^2 \beta}{\partial x^2} + GAK \left(\frac{\partial w}{\partial x} - \beta \right) = 0 \qquad \text{(c)}$$

$$GAK \left(\frac{\partial^2 w}{\partial x^2} - \frac{\partial \beta}{\partial x} \right) + q(x) = 0 \qquad \text{(d)}$$

At this point, we let $q(x) = 0$ and assume the following interpolating polynomials:

$$w(x) = a_0 + a_1 x + a_2 x^2 + a_3 x^3 \qquad \text{(e)}$$
$$\beta(x) = b_0 + b_1 x + b_2 x^2 \qquad \text{(f)}$$

Substituting (e) and (f) into (c) and (d) and equating coefficients, we find

$$\begin{Bmatrix} w \\ \beta \end{Bmatrix} = \begin{bmatrix} 1 & x & x^2 & x^3 \\ 0 & 1 & 2x & \left(3x^2 + \dfrac{6EI}{GAK} \right) \end{bmatrix} \begin{Bmatrix} a_0 \\ a_1 \\ a_2 \\ a_3 \end{Bmatrix}$$

By applying the boundary conditions

$$w(0) = w_1, \quad \beta(0) = \theta_1, \quad w(L) = w_2, \quad \text{and} \quad \beta(L) = \theta_2$$

and letting $\alpha = 6EI/GAK$, we find

$$
\begin{bmatrix}
1 & 0 & 0 & 0 \\
0 & 1 & 0 & \dfrac{6EI}{GAK} \\
1 & L & L^2 & L^3 \\
0 & 1 & 2L & \left(\dfrac{6EI}{GAK} + 3L^2\right)
\end{bmatrix}
\begin{bmatrix}
a_0 \\
a_1 \\
a_2 \\
a_3
\end{bmatrix}
=
\begin{bmatrix}
w_1 \\
\theta_1 \\
w_2 \\
\theta_2
\end{bmatrix}
$$

By inverting the matrix and performing additional mathematical manipulations, we obtain

$$
\begin{Bmatrix} w \\ \beta \end{Bmatrix} = \frac{1}{L(2\alpha + L^2)}
\left[
\begin{array}{cc|c}
(L^3 - 3Lx^2 + 2x^3) + 2\alpha(L - x) & (L^3x - 2L^2x^2 + Lx^3) + \alpha(Lx - x^2) \\
-6(Lx - x^2) & (L^3 - 4L^2x + 3Lx^2) + 2\alpha(L - x)
\end{array}
\right.
$$

$$
\left.
\begin{array}{c|c}
(3Lx^2 - 2x^3) + 2\alpha x & (-L^2x^2 + Lx^3) + \alpha(-Lx + x^2) \\
6(Lx - x^2) & (-2L^2x + 3Lx^2) + 2\alpha x
\end{array}
\right]
\begin{Bmatrix}
w_1 \\
\theta_1 \\
w_2 \\
\theta_2
\end{Bmatrix}
\tag{g}
$$

At this point, we recall the Hermitian polynomials used for the Bernoulli beam, which are as follows:

$$
{}^H h_1 = 1 - 3\left(\frac{x}{L}\right)^2 + 2\left(\frac{x}{L}\right)^3
$$

$$
{}^H h_2 = \frac{x}{L} - 2\left(\frac{x}{L}\right)^2 + \left(\frac{x}{L}\right)^3
$$

$$
{}^H h_3 = 3\left(\frac{x}{L}\right)^2 - 2\left(\frac{x}{L}\right)^3
$$

$$
{}^H h_4 = -\left(\frac{x}{L}\right)^2 + \left(\frac{x}{L}\right)^3
$$

We also add the following interpolating polynomials for shear

$$
{}^g h_1 = 1 - \frac{x}{L}
$$

$$
{}^g h_2 = \frac{x}{L}
$$

$$^sh_3 = -\frac{x}{L} + \left(\frac{x}{L}\right)^2 = -\,^sh_1\,^sh_2$$

By substituting the above shape functions into (g), we find

$$\begin{Bmatrix} w \\ \beta \end{Bmatrix} = \frac{1}{L(2\alpha + L^2)} \begin{bmatrix} L^{3H}h_1 + 2\alpha L\,^sh_1 & L^{4H}h_2 + \alpha L^{2s}h_3 \\ L^{3H}h_1' & L^{3H}h_2' + 2\alpha L\,^sh_1 \end{bmatrix}$$

$$\begin{bmatrix} L^{3H}h_3 + 2\alpha L\,^sh_2 & L^{4H}h_4 + \alpha L^{2s}h_3 \\ L^{3H}h_3' & L^{3H}h_4' + 2\alpha L\,^sh_2 \end{bmatrix} \begin{Bmatrix} w_1 \\ \theta_1 \\ w_2 \\ \theta_2 \end{Bmatrix}$$

For convenience, we let

$$\phi = \frac{2\alpha}{L^2} = \frac{12EI}{L^2 GAK}$$

and, consequently, we have

$$\begin{Bmatrix} w \\ \beta \end{Bmatrix} = \frac{1}{(1+\phi)} \begin{bmatrix} ^Hh_1 + \phi\,^sh_1 & L\left(^Hh_2 - \frac{\phi}{2}\,^sh_3\right) & ^Hh_3 + \phi\,^sh_2 \\ ^Hh_1' & L^Hh_2' + \phi\,^sh_1 & ^Hh_3' \end{bmatrix}$$

$$\begin{bmatrix} L\left(^Hh_4 + \frac{\phi}{2}\,^sh_3\right) \\ ^Hh_4' + \phi\,^sh_2 \end{bmatrix} \begin{Bmatrix} w_1 \\ \theta_1 \\ w_2 \\ \theta_2 \end{Bmatrix}$$

The strain can be represented in matrix form as follows:

$$\{e\} = [B]\{\hat{u}\}$$

where

$$e_{xx} = \frac{\partial u}{\partial x}, \quad u = z\beta, \quad \text{and} \quad \frac{\partial w}{\partial x} = \beta + \gamma$$

or

$$\{e\} = \left\{ \begin{matrix} e_{xx} \\ \gamma_{xy} \end{matrix} \right\} = \left\{ \begin{matrix} z\beta' \\ w'(-\beta) \end{matrix} \right\}$$

$$= \frac{1}{(1+\phi)} \begin{bmatrix} z\,^{H}h_1'' & z(L\,^{H}h_2'' + \phi^{s}h_1') & z\,^{H}h_3'' & z(L\,^{H}h_4'' + \phi^{s}h_2') \\ -\dfrac{\phi}{L} & -\dfrac{\phi}{2} & \dfrac{\phi}{L} & -\dfrac{\phi}{2} \end{bmatrix} \{\hat{u}\}$$

Hence,

$$[K] = \int_0^L \int_A [B]^T [c][B] \, \partial A \, \partial z$$

where

$$[c] = \begin{bmatrix} E & \\ & G \end{bmatrix}$$

Since we have

$$\int \partial A = KA, \quad \int z^2 \partial A = I, \quad \text{and} \quad \int z \, \partial A = 0$$

the following matrix $[K]$ may be obtained with additional mathematical manipulations:

$$[K] = \frac{EI}{L^3} \frac{1}{(1+\phi)} \begin{bmatrix} 12 & 6L & -12 & 6L \\ 6L & (4+\phi)L^2 & -6L & (2-\phi)L^2 \\ -12 & -6L & 12 & -6L \\ 6L & (2-\phi)L^2 & -6L & (4+\phi)L^2 \end{bmatrix} \tag{h}$$

If we include axial deformations, it becomes

$$[K_e] = \frac{EI\left(\dfrac{1}{1+\phi}\right)}{L^3}\begin{bmatrix} AL^2(1+\varphi)/I & 0 & 0 & -AL^2(1+\varphi)/I & 0 & 0 \\ 0 & 12 & 6L & 0 & -12 & 6L \\ 0 & 6L & (4+\varphi)L^2 & 0 & -6L & (2-\varphi)L^2 \\ -AL^2(1+\varphi)/I & 0 & 0 & AL^2(1+\varphi)/I & 0 & 0 \\ 0 & -12 & -6L & 0 & 12 & -6L \\ 0 & 6L & (2-\varphi)L^2 & 0 & -6L & (4+\varphi)L^2 \end{bmatrix}$$

$$(6.33)$$

where

$$\varphi = \frac{12EI}{L^2 K'AG} \tag{6.34}$$

In Eq. (6.34), the symbol G is used to denote the shear modulus, and K' is the shear factor that depends on the shape of the cross section of the member. Values of K' are usually available in the literature. For rectangular cross sections, for example, K' would be equal to 5/6. As the product $K'AG$ becomes very large, φ approaches zero, and, consequently, we have

$$[K_e]_{\text{Timoshenko}} = [K_e]_{\text{Bernoulli}} \tag{6.35}$$

6.2.4 Plate Element

Numerous plate elements have been developed by various researchers, although most are used implicitly in the finite element programs. One that is stated explicitly by Yang [43] is the four-node rectangular element in Fig. 6.3, which has 16 degrees of freedom. The degrees of freedom numbered as 13, 14, 15, and 16 correspond to a second-order twist term such as $\partial^2 w/\partial x\, \partial y$. This element is obtained by combining the shape functions for the elementary beam element in both the x and y directions; that is,

$$w(x, y) = f_1(x)f_1(y)w_1 + f_2(x)f_1(y)w_2 + f_2(x)f_2(y)w_3$$

$$+ f_1(x)f_2(y)w_4 + f_3(x)f_1(y)\left(\frac{\partial w}{\partial x}\right)_1 + f_4(x)f_1(y)\left(\frac{\partial w}{\partial x}\right)_2$$

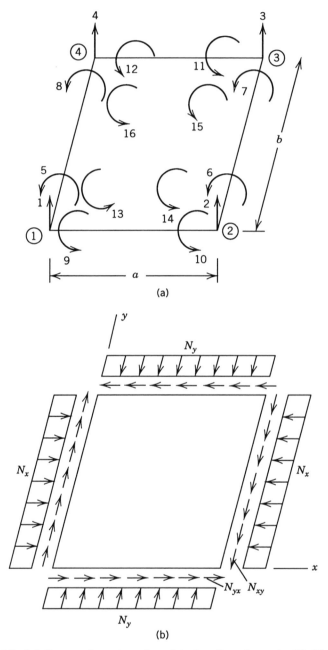

FIGURE 6.3. (a) Four-node rectangular plate bending element with 16 degrees of freedom. (b) In-plane normal and shear forces on quadrilateral membrane element.

$$+ f_4(x) f_2(y) \left(\frac{\partial w}{\partial x}\right)_3 + f_3(x) f_2(y) \left(\frac{\partial w}{\partial x}\right)_4 + f_1(x) f_3(y) \left(\frac{\partial w}{\partial x}\right)_1$$

$$+ f_2(x) f_3(y) \left(\frac{\partial w}{\partial y}\right)_2 + f_2(x) f_4(y) \left(\frac{\partial w}{\partial y}\right)_3$$

$$+ f_1(x) f_4(y) \left(\frac{\partial w}{\partial y}\right)_4 \tag{a}$$

where

$$f_1(x) = 1 - 3\left(\frac{x}{a}\right) + 2\left(\frac{x}{a}\right)^3 \tag{b}$$

$$f_2(x) = 3\left(\frac{x}{a}\right)^2 - 2\left(\frac{x}{a}\right)^3 \tag{c}$$

$$f_3(x) = x\left(\frac{x}{a} - 1\right)^2 \tag{d}$$

$$f_4(x) = x\left[\left(\frac{x}{a}\right)^2 - \frac{x}{a}\right] \tag{e}$$

$$f_1(y) = 1 - 3\left(\frac{y}{a}\right) + 2\left(\frac{y}{a}\right)^3 \tag{f}$$

$$f_2(y) = 3\left(\frac{y}{a}\right)^2 - 2\left(\frac{y}{a}\right)^3 \tag{g}$$

$$f_3(y) = y\left(\frac{y}{a} - 1\right)^2 \tag{h}$$

$$f_4(y) = y\left[\left(\frac{y}{a}\right)^2 - \frac{y}{a}\right] \tag{i}$$

This yields a rectangle with 12 degrees of freedom. In addition, the second-order twist derivative terms $\partial^2 w/\partial x\,\partial y$ are added at each corner node, to permit constant shear strain. These are degrees of freedom 13, 14, 15, and 16. In terms of the physical degrees of freedom, we obtain

$$w(x, y) = \sum_{i=1}^{16} f_i(x, y) q_i$$

$$
= \frac{1}{a^3 b^3} \Big[(a^3 + 2x^3 - 3ax^2)(b^3 + 2y^3 - 3by^2) w_1
$$

$$
+ (3ax^2 - 2x^3)(b^3 + 2y^3 - 3by^2) w_2
$$

$$
+ (3ax^2 - 2x^3)(3by^2 - 2y^3) w_3
$$

$$
+ (a^3 + 2x^2 - 3ax^2)(3by^2 - 2y^3) w_4
$$

$$
+ ax(x - a)^2 (b^3 + 2y^3 - 3by^2) \left(\frac{\partial w}{\partial x} \right)_1
$$

$$
+ a(x^3 - ax^2)(b^3 + 2y^3 - 3by^2) \left(\frac{\partial w}{\partial x} \right)_2
$$

$$
+ a(x^3 - ax^2)(3by^2 - 2y^3) \left(\frac{\partial w}{\partial x} \right)_3
$$

$$
+ ax(x - a)^2 (3by^2 - 2y^3) \left(\frac{\partial w}{\partial x} \right)_4
$$

$$
+ b(a^3 + 2x^3 - 3ax^2) y(y - b)^2 \left(\frac{\partial w}{\partial y} \right)_1
$$

$$
+ b(3ax^2 - 2x^3) y(y - b)^2 \left(\frac{\partial w}{\partial y} \right)_2
$$

$$
+ b(3ax^2 - 2x^3)(y^3 - by^2) \left(\frac{\partial w}{\partial y} \right)_3
$$

$$
+ b(a^3 + 2x^3 - 3ax^2)(y^3 - by^2) \left(\frac{\partial w}{\partial y} \right)_4
$$

$$
+ abxy(x - a)^2 (y - b)^2 \left(\frac{\partial^2 w}{\partial x \, \partial y} \right)_1
$$

$$
+ abxy(x^2 - ax)(y - b)^2 \left(\frac{\partial^2 y}{\partial x \, \partial y} \right)_2
$$

$$
+ abxy(x^2 - ax)(y^2 - by) \left(\frac{\partial^2 w}{\partial x \, \partial y} \right)_3 \tag{j}
$$

$$
+ abxy(x - a)^2 (y^2 - by) \left(\frac{\partial^2 w}{\partial x \, \partial y} \right)_4 \Big]
$$

On this basis, the elastic stiffness matrix $[K_e]$ for this plate element is as follows:

$$[K_e] = \begin{bmatrix}
A_1 & A_2 & A_3 & A_4 & A_5 & A_6 & A_7 & A_8 & A_9 & A_{10} & A_{11} & A_{12} & A_{13} & A_{14} & A_{15} & A_{16} \\
B_1 & B_2 & B_3 & B_4 & B_5 & B_6 & B_7 & B_8 & B_9 & B_{10} & B_{11} & B_{12} & B_{13} & B_{14} & B_{15} & B_{16} \\
C_1 & C_2 & C_3 & C_4 & C_5 & C_6 & C_7 & C_8 & C_9 & C_{10} & C_{11} & C_{12} & C_{13} & C_{14} & C_{15} & C_{16} \\
D_1 & D_2 & D_3 & D_4 & D_5 & D_6 & D_7 & D_8 & D_9 & D_{10} & D_{11} & D_{12} & D_{13} & D_{14} & D_{15} & D_{16} \\
E_1 & E_2 & E_3 & E_4 & E_5 & E_6 & E_7 & E_8 & E_9 & E_{10} & E_{11} & E_{12} & E_{13} & E_{14} & E_{15} & E_{16} \\
F_1 & F_2 & F_3 & F_4 & F_5 & F_6 & F_7 & F_8 & F_9 & F_{10} & F_{11} & F_{12} & F_{13} & F_{14} & F_{15} & F_{16} \\
G_1 & G_2 & G_3 & G_4 & G_5 & G_6 & G_7 & G_8 & G_9 & G_{10} & G_{11} & G_{12} & G_{13} & G_{14} & G_{15} & G_{16} \\
H_1 & H_2 & H_3 & H_4 & H_5 & H_6 & H_7 & H_8 & H_9 & H_{10} & H_{11} & H_{12} & H_{13} & H_{14} & H_{15} & H_{16} \\
I_1 & I_2 & I_3 & I_4 & I_5 & I_6 & I_7 & I_8 & I_9 & I_{10} & I_{11} & I_{12} & I_{13} & I_{14} & I_{15} & I_{16} \\
J_1 & J_2 & J_3 & J_4 & J_5 & J_6 & J_7 & J_8 & J_9 & J_{10} & J_{11} & J_{12} & J_{13} & J_{14} & J_{15} & J_{16} \\
K_1 & K_2 & K_3 & K_4 & K_5 & K_6 & K_7 & K_8 & K_9 & K_{10} & K_{11} & K_{12} & K_{13} & K_{14} & K_{15} & K_{16} \\
L_1 & L_2 & L_3 & L_4 & L_5 & L_6 & L_7 & L_8 & L_9 & L_{10} & L_{11} & L_{12} & L_{13} & L_{14} & L_{15} & L_{16} \\
M_1 & M_2 & M_3 & M_4 & M_5 & M_6 & M_7 & M_8 & M_9 & M_{10} & M_{11} & M_{12} & M_{13} & M_{14} & M_{15} & M_{16} \\
N_1 & N_2 & N_3 & N_4 & N_5 & N_6 & N_7 & N_8 & N_9 & N_{10} & N_{11} & N_{12} & N_{13} & N_{14} & N_{15} & N_{16} \\
P_1 & P_2 & P_3 & P_4 & P_5 & P_6 & P_7 & P_8 & P_9 & P_{10} & P_{11} & P_{12} & P_{13} & P_{14} & P_{15} & P_{16} \\
Q_1 & Q_2 & Q_3 & Q_4 & Q_5 & Q_6 & Q_7 & Q_8 & Q_9 & Q_{10} & Q_{11} & Q_{12} & Q_{13} & Q_{14} & Q_{15} & Q_{16}
\end{bmatrix}$$

$$(6.36)$$

where

$$A_1 = B_2 = C_3 = D_4 = \frac{12d(65a^4 + 42a^2b^2 + 65b^4)}{175a^3b^3}$$

$$A_2 = B_1 = C_4 = D_3 = \frac{6d(45a^4 - 84a^2b^2 - 130b^4)}{175a^3b^3}$$

$$A_3 = B_4 = C_1 = D_2 = -\frac{18d(15a^4 - 28a^2b^2 + 15b^4)}{175a^3b^3}$$

$$A_4 = B_3 = C_2 = D_1 = -\frac{6d(130a^4 + 84a^2b^2 - 45b^4)}{175a^3b^3}$$

$$A_5 = D_8 = E_1 = \frac{2d[55a^4 + 21a^2b^2(5\nu + 1) + 195b^4]}{175a^2b^3}$$

$$B_5 = C_8 = E_2 = H_3 = \frac{d(65a^4 - 42a^2b^2 - 390b^4)}{175a^2b^3}$$

$$A_7 = D_6 = F_4 = G_1 = \frac{d(65a^4 - 42a^2b^2 + 135b^4)}{175a^2b^3}$$

$$A_8 = D_5 = E_4 = H_1 = -\frac{d[110a^4 + 42a^2b^2(5\nu + 1) - 135b^4]}{175a^2b^3}$$

$$A_9 = B_{10} = I_1 = J_2 = \frac{2d[195a^4 + 21a^2b^2(5\nu + 1) + 55b^4]}{175a^3b^2}$$

$$A_{10} = B_9 = I_2 = J_1 = \frac{d[135a^4 - 42a^2b^2(5\nu + 1) - 110b^4]}{175a^3b^2}$$

$$A_{11} = B_{12} = K_1 = L_2 = \frac{d(135a^4 - 42a^2b^2 + 65b^4)}{175a^3b^2}$$

$$A_{12} = B_{11} = K_2 = L_1 = \frac{d(390a^4 + 42a^2b^2 - 65b^4)}{175a^3b^2}$$

$$A_{13} = C_{15} = M_1 = P_3 = \frac{d[110a^4 + 7a^2b^2(10\nu + 1) + 110b^4]}{350a^2b^2}$$

$$A_{14} = C_{16} = F_9 = H_{11} = I_6 = K_8 = N_1 = Q_3$$
$$= -\frac{d[65a^4 - 7a^2b^2(5\nu + 1) - 110b^4]}{350a^2b^2}$$

$$A_{15} = C_{13} = F_{12} = H_{10} = J_8 = L_6 = M_3 = P_1 = -\frac{d(65a^4 - 7a^2b^2 + 65b^4)}{350a^2b^2}$$

$$B_6 = C_7 = F_2 = G_3 = -\frac{2d[55a^4 + 21a^2b^2(5\nu + 1) + 195b^4]}{175a^2b^3}$$

$$B_8 = C_5 = E_3 = H_2 = -\frac{d(65a^4 - 42a^2b^2 + 135b^4)}{175a^2b^3}$$

$$B_{14} = D_{16} = N_2 = Q_4 = -\frac{d[110a^4 + 7a^2b^2(10\nu + 1) + 110b^4]}{350a^2b^2}$$

$$A_6 = D_7 = F_1 = G_4 = -\frac{d(65a^4 - 42a^2b^2 - 390b^4)}{175a^2b^3}$$

$$C_{10} = D_9 = I_4 = J_3 = -\frac{d(390a^4 + 42a^2b^2 - 65b^4)}{175a^3b^2}$$

$$C_{11} = D_{12} = L_4 = K_3 = -\frac{2d[195a^4 + 21a^2b^2(5\nu + 1) + 55b^4]}{175a^3b^2}$$

$$C_{12} = D_{11} = K_4 = L_3 = -\frac{d[135a^4 - 42a^2b^2(5\nu + 1) - 110b^4]}{175a^3b^2}$$

$$C_9 = D_{10} = I_3 = J_4 = -\frac{d(135a^4 - 42a^2b^2 + 65b^4)}{175a^3b^2}$$

$$B_{15} = D_{13} = F_{11} = H_9 = I_8 = K_6 = M_4 = P_2$$
$$= -\frac{d[110a^4 + 7a^2b^2(5\nu + 1) - 65b^4]}{350a^2b^2}$$

$$B_7 = C_6 = F_3 = G_2 = \frac{d[110a^4 + 42a^2b^2(5\nu + 1) - 135b^4]}{175a^2b^3}$$

$$E_5 = F_6 = G_7 = H_8 = \frac{4d(5a^4 + 14a^2b^2 + 65b^4)}{175ab^3}$$

$$E_6 = F_5 = G_8 = H_7 = -\frac{d(15a^4 + 14a^2b^2 - 130b^4)}{175ab^3}$$

$$E_7 = F_8 = G_5 = H_6 = \frac{d(15a^4 + 14a^2b^2 + 45b^4)}{175ab^3}$$

$$E_8 = F_7 = G_6 = H_5 = -\frac{2d(10a^4 + 28a^2b^2 - 45b^4)}{175ab^3}$$

$$E_9 = I_5 = G_{11} = K_7 = \frac{d[110a^4 + 7a^2b^2(60\nu + 1) + 110b^4]}{350a^2b^2}$$

$$B_{13} = D_{15} = E_{10} = G_{12} = J_5 = L_7 = M_2 = P_4$$
$$= \frac{d[65a^4 - 7a^2b^2(5\nu + 1) - 110b^4]}{350a^2b^2}$$

$$B_{16} = D_{14} = E_{11} = G_9 = I_7 = K_5 = N_4 = Q_2$$
$$= \frac{d(65a^4 - 7a^2b^2 + 65b^4)}{350a^2b^2}$$

$$A_{16} = C_{14} = E_{12} = G_{10} = J_7 = L_5 = N_3 = Q_1$$
$$= \frac{d[110a^4 + 7a^2b^2(5\nu + 1) - 65b^4]}{350a^2b^2}$$

$$E_{13} = F_{14} = M_5 = N_6 = \frac{2d[15a^4 + 7a^2b^2(5\nu + 1) + 55b^4]}{525ab^2}$$

$$E_{14} = F_{13} = M_6 = N_5 = -\frac{d[45a^4 + 7a^2b^2(5\nu + 1) - 110b^4]}{1050ab^2}$$

$$E_{15} = F_{16} = P_5 = Q_6 = -\frac{d(45a^4 + 7a^2b^2 + 65b^4)}{1050ab^2}$$

$$E_{16} = F_{15} = P_6 = Q_5 = \frac{d(30a^4 + 14a^2b^2 - 65b^4)}{525ab^2}$$

$$F_{10} = J_6 = H_{12} = L_8 = -\frac{d[110a^4 + 7a^2b^2(60\nu + 1) + 110b^4]}{350a^2b^2}$$

$$G_{13} = H_{14} = M_7 = N_8 = \frac{d(45a^4 + 7a^2b^2 + 65b^4)}{1050ab^2}$$

$$G_{14} = M_8 = N_7 = H_{13} = -\frac{d(30a^4 + 14a^2b^2 - 65b^4)}{525ab^2}$$

$$G_{15} = H_{16} = P_7 = Q_8 = -\frac{2d[15a^4 + 7a^2b^2(5\nu + 1) + 55b^4]}{525ab^2}$$

$$H_4 = \frac{2d[55a^4 + 21a^2b^2(5\nu + 1) + 195b^4]}{175a^2b^3}$$

$$I_{10} = J_9 = K_{12} = L_{11} = \frac{2d(45a^4 - 28a^2b^2 - 10b^4)}{175a^3b}$$

$$I_{11} = J_{12} = K_9 = L_{10} = \frac{d(45a^4 + 14a^2b^2 + 15b^4)}{175a^3b}$$

$$I_{12} = J_{11} = K_{10} = L_9 = \frac{d(130a^4 - 14a^2b^2 - 15b^4)}{175a^3b}$$

$$I_{13} = L_{16} = M_9 = Q_{12} = \frac{2d[55a^4 + 7a^2b^2(5\nu + 1) + 15b^4]}{525a^2b}$$

$$I_{14} = L_{15} = N_9 = P_{12} = -\frac{d(65a^4 - 14a^2b^2 - 30b^4)}{525a^2b}$$

$$I_{15} = L_{14} = N_{12} = P_9 = -\frac{d(65a^4 + 7a^2b^2 + 45b^4)}{1050a^2b}$$

$$I_{16} = M_{12} = L_{13} = Q_9 = \frac{d[110a^4 - 7a^2b^2(5\nu + 1) - 45b^4]}{1050a^2b}$$

$$J_{14} = K_{15} = N_{10} = P_{11} = -\frac{2d[55a^4 + 7a^2b^2(5\nu + 1) + 15b^4]}{525a^2b}$$

$$J_{15} = K_{14} = N_{11} = -\frac{d[110a^4 - 7a^2b^2(5\nu + 1) - 45b^4]}{1050a^2b}$$

$$I_9 = J_{10} = K_{11} = L_{12} = \frac{4d(65a^4 + 14a^2b^2 + 5b^4)}{175a^3b}$$

$$M_{13} = N_{14} = P_{15} = Q_{16} = \frac{4d(15a^4 + 14a^2b^2 + 15b^4)}{1575ab}$$

$$M_{14} = N_{13} = P_{16} = Q_{15} = -\frac{d(45a^4 + 14a^2b^2 - 30b^4)}{1575ab}$$

$$M_{15} = N_{16} = P_{13} = Q_{14} = -\frac{d(45a^4 - 7a^2b^2 + 45b^4)}{3150ab}$$

$$M_{16} = N_{15} = P_{14} = Q_{13} = \frac{d(30a^4 - 14a^2b^2 - 45b^4)}{1575ab}$$

$$H_{15} = G_{16} = P_8 = Q_7 = \frac{d[45a^4 + 7a^2b^2(5\nu + 1) - 110b^4]}{1050ab^2}$$

$$J_{16} = K_{13} = M_{11} = Q_{10} = \frac{d(65a^4 + 7a^2b^2 + 45b^4)}{1050a^2b}$$

$$P_{10} = -\frac{d[110a^4 - 7a^2b^2(5\nu + 1) - 45b^4]}{1050a^2b}$$

$$J_{13} = K_{16} = M_{10} = Q_{11} = \frac{d(65a^4 - 14a^2b^2 - 30b^4)}{525a^2b}$$

where

$$d = \frac{Eh^3}{12(1 - \nu^2)} \tag{6.37}$$

h is the thickness of the plate, w is the vertical displacement of the plate, and ν is the Poisson ratio.

6.2.5 Mass Matrices

To express the inertia forces acting on the finite element, it is necessary to define a mass matrix for this element. The simplest way to do this is to assign or lump the mass at the nodes, where translational degrees of freedom are designated. This procedure results in a diagonal mass matrix, in which there would be no mass coefficients corresponding to diagonal degrees of freedom.

By using virtual work, the M_{ij} coefficients can also be determined by using the following equation:

$$M_{ij} = \int_0^L \rho(x)A(x)\Phi_i(x)\Phi_j(x)\,dx \tag{6.38}$$

where $\Phi_i(x)$ and $\Phi_j(x)$ are interpolating functions. If the Hermitian polynomials are used as the interpolating functions $\Phi_i(x)$ and $\Phi_j(x)$, as well as for determining the stiffness matrix, then the resultant matrix is referred to as the *consistent mass matrix*.

For a uniform beam, the consistent mass matrix $[M]$ is as follows:

$$[M] = \frac{mL}{420} \begin{bmatrix} 156 & 54 & 22L & -13L \\ 54 & 156 & 13L & -22L \\ 22L & 13L & 4L^2 & -3L^2 \\ -13L & -22L & -3L^2 & 4L^2 \end{bmatrix} \tag{6.39}$$

where $m = \rho A$, ρ is the mass per unit volume, and A is the cross-sectional area. Note that the mass matrix here is symmetric.

Numerous plate mass matrices have also been developed by various investigators. We note here the one developed by Yang [43], which is expressed explicitly and involves a rectangular mass with four nodes and 16 degrees of freedom. See also Fig. 6.3. This mass matrix is shown on page 395, where h is the thickness of the rectangular mass, and a and b are the sides.

The coefficients M_{ij} of the mass matrix have been determined by using the following equation:

$$M_{ij} = \rho h \int_0^b \int_0^a f_i(x, y) f_j(x, y) \, \partial x \, \partial y$$

where $f_i(x, y)$ and $f_j(x, y)$ are given by Eq. (j) in Section 6.2.4.

6.2.6 Effective Loads

The *load vector* simply specifies the forces acting at the respective degrees of freedom. A generalized load vector $\{R\}$ must be determined when the load is distributed, rather than concentrated at a certain degree of freedom. For example, consider a Bernoulli beam element with a uniformly distributed load $R(x, t) = w$, acting in a positive direction. The equation for the generalized load vector is

$$R_i(t) = \int_0^L R(x, t) H_i(x) \, dx \qquad \text{when} \quad i = 2, 3, 5, 6 \tag{a}$$

$$R_i(t) = 0 \qquad \text{when} \quad i = 1, 4 \tag{b}$$

If we substitute the shape vectors used for the Bernoulli beam, Eq. (6.25),

$$[M] = \rho abh \times$$

$$
\begin{bmatrix}
\frac{169}{1225} & & & & & & & & & & & & & & & \\[4pt]
\frac{117}{2450} & \frac{129}{1225} & & & & & & & & & & & & & & \\[4pt]
\frac{81}{4900} & \frac{117}{2450} & \frac{169}{1225} & & & & & & & & & & & & & \\[4pt]
\frac{117}{2450} & \frac{81}{4900} & \frac{117}{2450} & \frac{169}{1225} & & & & & & & & & & & & \\[4pt]
\frac{143a}{7350} & \frac{169a}{14700} & \frac{143a}{7350} & \frac{33a}{4900} & \frac{13a^2}{3675} & & & & & & & & & & & \\[4pt]
\frac{-169a}{14700} & \frac{-143a}{7350} & \frac{-169a}{14700} & \frac{-39a}{9800} & \frac{-13a^2}{4900} & \frac{13a^2}{3675} & & & & & & & & & & \\[4pt]
\frac{-39a}{9800} & \frac{-33a}{4900} & \frac{-39a}{9800} & \frac{-169a}{14700} & \frac{-9a^2}{9800} & \frac{3a^2}{2450} & \frac{13a^2}{3675} & & & & & & & & & \\[4pt]
\frac{33a}{4900} & \frac{39a}{9800} & \frac{33a}{4900} & \frac{143a}{7350} & \frac{3a^2}{2450} & \frac{-9a^2}{9800} & \frac{-13a^2}{4900} & \frac{13a^2}{3675} & & & & & & & & \\[4pt]
\frac{143b}{7350} & \frac{33b}{4900} & \frac{39b}{9800} & \frac{143b}{7350} & \frac{121ab}{44100} & \frac{143ab}{88200} & \frac{169ab}{176400} & \frac{143ab}{88200} & \frac{13b^2}{3675} & & & & & & & \\[4pt]
\frac{39b}{9800} & \frac{143b}{9800} & \frac{169b}{14700} & \frac{169b}{14700} & \frac{-143ab}{88200} & \frac{-121ab}{44100} & \frac{-143ab}{88200} & \frac{-169ab}{176400} & \frac{-13b^2}{4900} & \frac{13b^2}{3675} & & & & & & \\[4pt]
\frac{-39b}{9800} & \frac{169b}{9800} & \frac{-33b}{4900}? & & \frac{-169ab}{176400} & \frac{-143ab}{88200} & \frac{-121ab}{44100} & \frac{-143ab}{88200} & \frac{-9b^2}{9800} & \frac{3b^2}{2450} & \frac{13b^2}{3675} & & & & & \\[4pt]
\frac{-169b}{14700} & \frac{-39b}{9800} & & & \frac{143ab}{88200} & \frac{169ab}{176400} & \frac{143ab}{88200} & \frac{121ab}{44100} & \frac{3b^2}{2450} & \frac{-9b^2}{9800} & \frac{-13b^2}{4900} & \frac{13b^2}{3675} & & & & \\[4pt]
\frac{121ab}{44100} & \frac{143ab}{88200} & \frac{169ab}{176400} & \frac{143ab}{88200} & \frac{13a^2b}{22050} & \frac{-11a^2b}{29400} & \frac{-13a^2b}{58800} & \frac{13a^2b}{44100} & \frac{13ab^2}{22050} & \frac{-11ab^2}{29400} & \frac{-13ab^2}{58800} & \frac{13ab^2}{44100} & \frac{a^2b^2}{11025} & & & \\[4pt]
\frac{-143ab}{88200} & \frac{-121ab}{44100} & \frac{-143ab}{88200} & \frac{-169ab}{176400} & \frac{-13a^2b}{44100} & \frac{11a^2b}{22050} & \frac{13a^2b}{29400} & \frac{-11a^2b}{22050} & \frac{11ab^2}{22050} & \frac{-13ab^2}{58800} & \frac{-11ab^2}{22050} & \frac{13ab^2}{29400} & \frac{-a^2b^2}{14700} & \frac{a^2b^2}{11025} & & \\[4pt]
\frac{169ab}{176400} & \frac{143ab}{88200} & \frac{121ab}{44100} & \frac{143ab}{88200} & \frac{-13a^2b}{58800} & \frac{11a^2b}{29400} & \frac{13a^2b}{22050} & \frac{-11a^2b}{22050} & \frac{13ab^2}{58800} & \frac{-11ab^2}{29400} & \frac{-13ab^2}{22050} & \frac{11ab^2}{22050} & \frac{a^2b^2}{19600} & \frac{-a^2b^2}{14700} & \frac{a^2b^2}{11025} & \\[4pt]
\frac{-143ab}{88200} & \frac{-169ab}{176400} & \frac{-143ab}{88200} & \frac{-121ab}{44100} & \frac{13a^2b}{44100} & \frac{-11a^2b}{22050} & \frac{-13a^2b}{22050} & \frac{11a^2b}{22050}? & \frac{11ab^2}{22050} & \frac{13ab^2}{44100} & \frac{11ab^2}{22050} & \frac{-13ab^2}{58800} & \frac{-a^2b^2}{14700} & \frac{a^2b^2}{19600} & \frac{-a^2b^2}{14700} & \frac{a^2b^2}{11025}
\end{bmatrix}
$$

S Y M M E T R I C

(6.40)

we obtain

$$R = \begin{bmatrix} 0 \\ wL/2 \\ wL^2/12 \\ 0 \\ wL/2 \\ -wL^2/12 \end{bmatrix} \tag{c}$$

6.3. ELEMENT COMPLETENESS AND COMPATIBILITY

To ensure monotonic convergence, the elements must be complete and compatible. *Monotonic convergence* means that as the finite element mesh is refined, the accuracy of the finite element solution increases. *Compatibility* means that the element has the capability to represent the rigid body modes and constant strain states.

The element stiffness matrix should possess a complete set of rigid-body modes. In other words, the element should be able both to translate and to rotate as a rigid body (no deformation) without developing internal stresses due to that motion. The need for this capability can be demonstrated by applying a transverse load to the center of a cantilever beam. We note then that the tip of the beam rotates, although it is in a stress-free state.

For example, let us consider the Bernoulli beam element in Fig. 6.2. The two translational and one rotational rigid-body modes may be expressed in matrix form $\{U\}$ as follows:

$$\{U\} = \begin{bmatrix} 1 & 0 & 0 \\ 0 & 1 & -L\theta/2 \\ 0 & 0 & \theta \\ 1 & 0 & 0 \\ 0 & 1 & L\theta/2 \\ 0 & 0 & \theta \end{bmatrix} \tag{6.41}$$

where θ is the angle of rotation. If we apply Eq. (6.1), the result is

$$\{R\} = \begin{bmatrix} 0 & 0 & 0 \\ 0 & 0 & 0 \\ 0 & 0 & 0 \\ 0 & 0 & 0 \\ 0 & 0 & 0 \\ 0 & 0 & 0 \end{bmatrix} \tag{6.42}$$

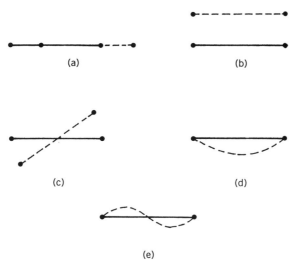

FIGURE 6.4. (a) Horizontal translational rigid-body mode. (b) Vertical translational rigid-body mode. (c) Rotational rigid-body mode. (d, e) Bending modes.

which implies that no internal forces, or corresponding internal stresses, are developed during the rigid-body translations or the rigid-body rotation.

Similarly, if we solve the vibration problem represented by Eq. (6. 2), we will obtain three zero eigenvalues and the corresponding eigenvectors, which are rigid-body mode shapes.

The lack of a complete set of rigid-body modes causes *grounding*, or an erroneously stiff element that is missing one or more zero eigenvalues. This grounding phenomenon typically occurs when a member with free–free boundary conditions is subjected to a preload, such as the pretensioned photovoltaic arrays that are used in aerospace applications (i.e., International Space Station). It does not normally cause problems for constrained structures that are subjected to static loading, such as the ones for which we want to predict critical buckling loads [44–47].

Finite element stiffness matrices should also possess the ability to represent constant strain states. This can intuitively be seen by refining a mesh in such a way that each element is sufficiently small, so that the difference in strain within each element is negligible. On this basis, the strain in each element may be considered constant.

Figure 6.4 illustrates the rigid-body modes and some of the bending modes of the Bernoulli beam element. Figures 6.4a and 6.4b represent horizontal and vertical translational rigid-body modes, respectively, while Fig. 6.4c shows the rotational rigid-body mode. Figures 6.4d and 6.4e illustrate bending modes of vibration.

Element compatibility means that displacements across element boundaries and within the elements are continuous. Thus, no gaps occur during loading. It has been proved, however, that satisfactory results may be obtained by incompatible elements, provided that the assemblage passes the *patch test*, which verifies completeness of the assemblage. This test consists of the application of specific nodal displacements corresponding to a constant strain state. If the resultant strains do represent a constant strain state, then the assemblage passes the patch test.

6.4   HIGHER ORDER EFFECTS

The elastic stiffness matrices developed in the preceding section include bending stresses. The effects of shear are included in the case of the Timoshenko beam. The presence of an axial force introduces additional stiffness terms, resulting in the *geometric stiffness matrix*, or *initial stress stiffness matrix*.

For the truss element in Fig. 6.1, which has two degrees of freedom, the geometric stiffness matrix $[K_g]$ is as follows:

$$[K_g] = \frac{P}{L}\begin{bmatrix} 1 & -1 \\ -1 & 1 \end{bmatrix} \tag{6.43}$$

where P is the axial preload, which may be tensile or compressive.

Consider the Bernoulli beam in Fig. 6.2, and the second term of Eq. (6.22), which we neglected in the derivation of Bernoulli's $[K_e]$.

Term {2}

$$\int_0^L \frac{EA}{2}\left[\frac{du_0}{dx}\right]\left[\frac{dW}{dx}\right]^2 dx = \int_0^L \frac{EA}{2}\frac{du_0}{dx}\left[\frac{dW}{dx}\right]^T \frac{dW}{dx} dx$$

$$= \tfrac{1}{2}[u_2 \quad u_3 \quad u_5 \quad u_6]\int_0^L EA\frac{du_0}{dx}\begin{bmatrix} H_2' \\ H_3' \\ H_5' \\ H_6' \end{bmatrix} \tag{6.44}$$

$$\times [H_2' \quad H_3' \quad H_5' \quad H_6']\, dx \begin{bmatrix} u_2 \\ u_3 \\ u_5 \\ u_6 \end{bmatrix}$$

By taking the first derivatives of the shape functions, we find

$$H_2' = -\frac{6x}{L^2} + \frac{6x^2}{L^3} \qquad \text{(a)}$$

$$H_3' = 1 - 4\left(\frac{x}{L}\right) + 3\left(\frac{x}{L}\right)^2 \qquad \text{(b)}$$

$$H_5' = \frac{6x}{L^2} - \frac{6x^2}{L^3} \qquad \text{(c)}$$

$$H_6' = -2\left(\frac{x}{L}\right) + 3\left(\frac{x}{L}\right)^2 \qquad \text{(d)}$$

$$(6.45)$$

We let

$$EA\frac{\partial u_0}{\partial x} = P \tag{6.46}$$

where P is the constant axial force in the member and is considered positive when the member is in tension. On this basis, Eq. (6.44) can be written as follows:

$$\tfrac{1}{2}\begin{bmatrix} u_2 & u_3 & u_5 & u_6 \end{bmatrix}\frac{P}{30L}\begin{bmatrix} 36 & 3L & -36 & 3L \\ 3L & 4L^2 & -3L & -L^2 \\ -36 & -3L & 36 & -3L \\ 3L & -L^2 & -3L & 4L^2 \end{bmatrix}\begin{bmatrix} u_2 \\ u_3 \\ u_5 \\ u_6 \end{bmatrix} = [Kg] \tag{6.47}$$

Hence, the Bernoulli consistent geometric stiffness matrix K_g is

$$[K_g] = \frac{P}{30L}\begin{bmatrix} 0 & 0 & 0 & 0 & 0 & 0 \\ 0 & 36 & 3L & 0 & -36 & 3L \\ 0 & 3L & 4L^2 & 0 & -3L & -L^2 \\ 0 & 0 & 0 & 0 & 0 & 0 \\ 0 & -36 & -3L & 0 & 36 & -3L \\ 0 & 3L & -L^2 & 0 & -3L & 4L^2 \end{bmatrix} \tag{6.48}$$

$[K_g]$ is necessary when modeling a beam with an axial preload, including static buckling and dynamic analysis of preloaded beams. $[K_g]$ is also referred to as the initial stress stiffness matrix. Suppose that we consider rigid-body motion only. If we apply Eq. (6.1) and use Eqs. (6.41) and (6.48), we find

$$
[K_g]\{U\} = \frac{P}{30L}
\begin{bmatrix}
0 & 0 & 0 & 0 & 0 & 0 \\
0 & 36 & 3L & 0 & -36 & 3L \\
0 & 3L & 4L^2 & 0 & -3L & -L^2 \\
0 & 0 & 0 & 0 & 0 & 0 \\
0 & -36 & -3L & 0 & 36 & -3L \\
0 & 3L & -L^2 & 0 & -3L & 4L^2
\end{bmatrix}
\begin{bmatrix}
1 & 0 & 0 \\
0 & 1 & -L\theta/2 \\
0 & 0 & \theta \\
1 & 0 & 0 \\
0 & 1 & L\theta/2 \\
0 & 0 & \theta
\end{bmatrix}
$$

$$
=
\begin{bmatrix}
0 & 0 & 0 \\
0 & 0 & -P\theta \\
0 & 0 & 0 \\
0 & 0 & 0 \\
0 & 0 & P\theta \\
0 & 0 & 0
\end{bmatrix}
\tag{6.49}
$$

The terms $\pm P\theta$ in Eq. (6.49) are fictitious forces generated during rigid-body rotation. Hence, the geometric stiffness matrix given by Eq. (6.48) is incomplete. It lacks rigid-body rotational capability, although it has been shown to accurately predict buckling loads when sufficient elements are used [48].

The source of the fictitious forces (grounding) has been investigated in detail [46] and can be eliminated by development of a directed force correction matrix at the global level [47]. This can be done by using traditional development of $[K_g]$ from the horizontal component of force and applying Argyris' load correction method [49] on the vertical component to generate the directed force correction factors.

For example, consider the two-beam element model with directed force P as shown in Fig. 6.5b, and let

$$
\phi \approx \frac{u_8 - u_2}{L}
\tag{6.50}
$$

If one neglects the change in the axial component of P that occurs during rotation, as is customarily done ($P \cos \phi \approx P$), we obtain the consistent geometric stiffness matrix.

Suppose we retain the vertical component, $P \sin \phi$, and use Argyris' approach to develop a load correction matrix. The load vector for this force becomes

$$
R^{DFC} = [P \sin(u_8 - u_2)/L, 0, 0, 0, -P \sin(u_8 - u_2)/L, 0]
\tag{6.51}
$$

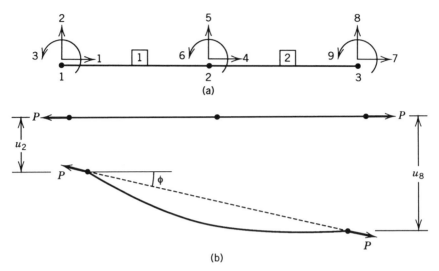

FIGURE 6.5. (a) Two-element model. (b) Directed force two-element representation.

The load correction matrix is generated using the following equation:

$$[K^{DFC}] = \left[\frac{dR_i^{DF}}{du_i}\right] \tag{6.52}$$

Equation (6.52) yields

$$[\mathrm{K}^{DFC}] = \begin{bmatrix} -\dfrac{P}{L}\cos\left(\dfrac{u_8 - u_2}{L}\right) & 0 & 0 & 0 & \dfrac{P}{L}\cos\left(\dfrac{u_8 - u_2}{L}\right) & 0 \\ 0 & 0 & 0 & 0 & 0 & 0 \\ 0 & 0 & 0 & 0 & 0 & 0 \\ 0 & 0 & 0 & 0 & 0 & 0 \\ \dfrac{P}{L}\cos\left(\dfrac{u_8 - u_2}{L}\right) & 0 & 0 & 0 & -\dfrac{P}{L}\cos\left(\dfrac{u_8 - u_2}{L}\right) & 0 \\ 0 & 0 & 0 & 0 & 0 & 0 \end{bmatrix} \tag{6.53}$$

For small rotation, $\cos(u_8 - u_2)/L \approx 1$, and $[K^{DFC}]$ reduces to the following form:

$$
[K^{DFC}] =
\begin{bmatrix}
-P/L & 0 & 0 & 0 & P/L & 0 \\
0 & 0 & 0 & 0 & 0 & 0 \\
0 & 0 & 0 & 0 & 0 & 0 \\
0 & 0 & 0 & 0 & 0 & 0 \\
P/L & 0 & 0 & 0 & -P/L & 0 \\
0 & 0 & 0 & 0 & 0 & 0
\end{bmatrix}
\tag{6.54}
$$

At this point, the global stiffness matrix is obtained by assembling two consistent geometric stiffness matrices and adding $[K^{DFC}]$. Multiplying this global stiffness matrix by a matrix of rigid-body modes does not result in any fictitious forces.

For our two-element model, suppose we let

$$
\begin{aligned}
A &= 48 \text{ in.}^2 & m &= 0.03525 \text{ s}^2/\text{in.}^2 \\
E &= 30 \times 10^6 \text{ psi} & L &= 100 \text{ in.} \\
I &= 1000 \text{ in.}^4 & P &= 60{,}000 \text{ kips}
\end{aligned}
$$

By assembling $[K]$ and multiplying by the rigid-body modes, we find

$$
[K] \times [\text{rigid-body modes}] =
\begin{bmatrix}
0 & 0 & 0 \\
0 & 0 & -6 \times 10^7 \\
0 & 0 & 0 \\
0 & 0 & 0 \\
0 & 0 & 0 \\
0 & 0 & 0 \\
0 & 0 & 0 \\
0 & 0 & 6 \times 10^7 \\
0 & 0 & 0
\end{bmatrix}
\tag{6.55}
$$

As expected, large pseudoforces occurred during rigid-body rotation.

Suppose we now add the directed force correction matrix to $[K]$. Then the global stiffness matrix becomes

$$
\begin{bmatrix}
0.288\text{E}8 & 0 & 0 & -0.288\text{E}8 & 0 & 0 & 0 & 0 & 0 \\
0 & 0.372\text{E}7 & 0.78\text{E}8 & 0 & -0.432\text{E}7 & 0.78\text{E}8 & 0 & 0 & 0 \\
0 & 0.78\text{E}8 & 0.28\text{E}10 & 0 & -0.78\text{E}8 & 0.11\text{E}10 & 0 & 0 & 0 \\
-0.288\text{E}8 & 0 & 0 & 0.576\text{E}8 & 0 & 0 & -0.288\text{E}8 & 0 & 0 \\
0 & -0.432\text{E}7 & -0.78\text{E}8 & 0 & 0.864\text{E}7 & 0 & 0 & -0.432\text{E}7 & 0.78\text{E}8 \\
0 & 0.78\text{E}8 & 0.11\text{E}10 & 0 & 0 & 0.56\text{E}10 & 0 & -0.78\text{E}8 & 0.11\text{E}10 \\
0 & 0 & 0 & -0.288\text{E}8 & 0 & 0 & 0.288\text{E}8 & 0 & 0 \\
0 & 0.6\text{E}6 & 0 & 0 & -0.432\text{E}7 & -0.78\text{E}8 & 0 & 0.372\text{E}7 & -0.78\text{E}8 \\
0 & 0 & 0 & 0 & 0.78\text{E}8 & 0.11\text{E}10 & 0 & -0.78\text{E}8 & 0.28\text{E}1
\end{bmatrix}
$$

The matrix of rigid-body modes is

$$
\begin{bmatrix}
1 & 0 & 0 \\
0 & 1 & -50 \\
0 & 0 & 1 \\
1 & 0 & 0 \\
0 & 1 & 0 \\
0 & 0 & 1 \\
1 & 0 & 0 \\
0 & 1 & 50 \\
0 & 0 & 1
\end{bmatrix}
$$

where columns 1 and 2 represent unit translation in the x and y directions, respectively, and column 3 represents a unit rotation about the z axis. See also Figs. 6.4a, 6.4b, and 6.4c. Thus,

$$
[K_T]
\begin{bmatrix}
1 & 0 & 0 \\
0 & 1 & -50 \\
0 & 0 & 1 \\
1 & 0 & 0 \\
0 & 1 & 0 \\
0 & 0 & 1 \\
1 & 0 & 0 \\
0 & 1 & 50 \\
0 & 0 & 1
\end{bmatrix}
=
\begin{bmatrix}
0 & 0 & 0 \\
0 & 0 & 0 \\
0 & 0 & 0 \\
0 & 0 & 0 \\
0 & 0 & 0 \\
0 & 0 & 0 \\
0 & 0 & 0 \\
0 & 0 & 0 \\
0 & 0 & 0
\end{bmatrix}
$$

Hence, the pseudo forces have been eliminated.

Numerous other higher order beam stiffness and mass matrices have been developed by various investigators. The matrices include rotary inertia and shear effects. Paz and Dung [50, 51] investigated the power series expansion of the dynamic stiffness matrix, from both Bernoulli–Euler and a Timoshenko beam approach, and further investigations were also performed by Bosela, Fertis, and Shaker [44–47].

Various geometric stiffness matrices have also been developed for plate elements. Yang's [43] quadrilateral geometric stiffness matrix for a plate element with 16 degrees of freedom, Fig. 6.3b, is generated as follows:

$$
K_{g_{ij}} = \int_0^b \int_0^a \left\{ N_x \left(\frac{\partial f_i}{\partial x} \right) \frac{\partial f_j}{\partial x} + Ny \left(\frac{\partial f_i}{\partial y} \right) \frac{\partial f_j}{\partial y} \right.
$$
$$
\left. + N_{xy} \left[\left(\frac{\partial f_i}{\partial x} \right) \frac{\partial f_j}{\partial y} + \left(\frac{\partial f_i}{\partial y} \right) \frac{\partial f_j}{\partial x} \right] \right\} \partial x \, \partial y
$$

where $f_i(x, y)$ are given by Eq. (j) in Section 6.2.4. For an element with a distributed load in the x direction only, we have

$$[K_g] = \begin{bmatrix}
A_1 & A_2 & A_3 & A_4 & A_5 & A_6 & A_7 & A_8 & A_9 & A_{10} & A_{11} & A_{12} & A_{13} & A_{14} & A_{15} & A_{16} \\
B_1 & B_2 & B_3 & B_4 & B_5 & B_6 & B_7 & B_8 & B_9 & B_{10} & B_{11} & B_{12} & B_{13} & B_{14} & B_{15} & B_{16} \\
C_1 & C_2 & C_3 & C_4 & C_5 & C_6 & C_7 & C_8 & C_9 & C_{10} & C_{11} & C_{12} & C_{13} & C_{14} & C_{15} & C_{16} \\
D_1 & D_2 & D_3 & D_4 & D_5 & D_6 & D_7 & D_8 & D_9 & D_{10} & D_{11} & D_{12} & D_{13} & D_{14} & D_{15} & D_{16} \\
E_1 & E_2 & E_3 & E_4 & E_5 & E_6 & E_7 & E_8 & E_9 & E_{10} & E_{11} & E_{12} & E_{13} & E_{14} & E_{15} & E_{16} \\
F_1 & F_2 & F_3 & F_4 & F_5 & F_6 & F_7 & F_8 & F_9 & F_{10} & F_{11} & F_{12} & F_{13} & F_{14} & F_{15} & F_{16} \\
G_1 & G_2 & G_3 & G_4 & G_5 & G_6 & G_7 & G_8 & G_9 & G_{10} & G_{11} & G_{12} & G_{13} & G_{14} & G_{15} & G_{16} \\
H_1 & H_2 & H_3 & H_4 & H_5 & H_6 & H_7 & H_8 & H_9 & H_{10} & H_{11} & H_{12} & H_{13} & H_{14} & H_{15} & H_{16} \\
I_1 & I_2 & I_3 & I_4 & I_5 & I_6 & I_7 & I_8 & I_9 & I_{10} & I_{11} & I_{12} & I_{13} & I_{14} & I_{15} & I_{16} \\
J_1 & J_2 & J_3 & J_4 & J_5 & J_6 & J_7 & J_8 & J_9 & J_{10} & J_{11} & J_{12} & J_{13} & J_{14} & J_{15} & J_{16} \\
K_1 & K_2 & K_3 & K_4 & K_5 & K_6 & K_7 & K_8 & K_9 & K_{10} & K_{11} & K_{12} & K_{13} & K_{14} & K_{15} & K_{16} \\
L_1 & L_2 & L_3 & L_4 & L_5 & L_6 & L_7 & L_8 & L_9 & L_{10} & L_{11} & L_{12} & L_{13} & L_{14} & L_{15} & L_{16} \\
M_1 & M_2 & M_3 & M_4 & M_5 & M_6 & M_7 & M_8 & M_9 & M_{10} & M_{11} & M_{12} & M_{13} & M_{14} & M_{15} & M_{16} \\
N_1 & N_2 & N_3 & N_4 & N_5 & N_6 & N_7 & N_8 & N_9 & N_{10} & N_{11} & N_{12} & N_{13} & N_{14} & N_{15} & N_{16} \\
P_1 & P_2 & P_3 & P_4 & P_5 & P_6 & P_7 & P_8 & P_9 & P_{10} & P_{11} & P_{12} & P_{13} & P_{14} & P_{15} & P_{16} \\
Q_1 & Q_2 & Q_3 & Q_4 & Q_5 & Q_6 & Q_7 & Q_8 & Q_9 & Q_{10} & Q_{11} & Q_{12} & Q_{13} & Q_{14} & Q_{15} & Q_{16}
\end{bmatrix}$$

$$(6.56)$$

where

$$A_1 = B_2 = C_3 = D_4 = \frac{78N_x b}{175a} \qquad A_2 = B_1 = C_4 = D_3 = -\frac{78N_x b}{175a}$$

$$A_4 = B_3 = C_2 = D_1 = \frac{27N_x b}{175a} \qquad A_3 = B_4 = C_1 = D_2 = -\frac{27N_x b}{175a}$$

$$A_5 = A_6 = D_7 = D_8 = \frac{13N_x b}{350} \qquad B_5 = B_6 = C_7 = C_8 = -\frac{13N_x b}{350}$$

$$A_7 = A_8 = D_5 = D_6 = \frac{9N_x b}{700} \qquad B_7 = B_8 = C_5 = C_6 = -\frac{9N_x b}{700}$$

$$A_9 = B_{10} = C_{12} = D_{11} = \frac{11N_x b^2}{175a} \qquad A_{10} = B_9 = C_{11} = D_{12} = -\frac{11N_x b^2}{175a}$$

$$A_{11} = B_{12} = C_{10} = D_9 = \frac{13N_x b^2}{350a} \qquad A_{12} = B_{11} = C_9 = D_{10} = -\frac{13N_x b^2}{350a}$$

$$A_{13} = A_{14} = C_{15} = C_{16} = \frac{11N_x b^2}{2100} \qquad B_{13} = B_{14} = D_{15} = D_{16} = -\frac{11N_x b^2}{2100}$$

$$B_{15} = B_{16} = D_{13} = D_{14} = \frac{13N_x b^2}{4200} \qquad A_{15} = A_{16} = C_{13} = C_{14} = -\frac{13N_x b^2}{4200}$$

$$E_1 = F_1 = G_4 = H_4 = \frac{13N_x b}{350} \qquad E_2 = F_2 = G_3 = H_3 = -\frac{13N_x b}{350}$$

$$E_4 = F_4 = G_1 = H_1 = \frac{9N_x b}{700}$$

$$E_3 = F_3 = G_2 = H_2 = -\frac{9N_x b}{700}$$

$$E_5 = F_6 = G_7 = H_8 = \frac{26N_x ab}{525}$$

$$E_6 = F_5 = G_8 = H_7 = -\frac{13N_x ab}{1050}$$

$$E_7 = F_8 = G_5 = H_6 = -\frac{3N_x ab}{700}$$

$$E_8 = F_7 = G_6 = H_5 = \frac{3N_x ab}{175}$$

$$E_9 = F_9 = G_{11} = H_{11} = \frac{11N_x b^2}{2100}$$

$$E_{10} = F_{10} = G_{12} = H_{12} = -\frac{11N_x b^2}{2100}$$

$$E_{11} = F_{11} = G_9 = H_9 = \frac{13N_x b^2}{4200}$$

$$E_{12} = F_{12} = G_{10} = H_{10} = -\frac{13N_x b^2}{4200}$$

$$E_{13} = F_{14} = \frac{11N_x ab^2}{1575}$$

$$G_{15} = H_{16} = -\frac{11N_x ab^2}{1575}$$

$$G_{16} = H_{15} = \frac{11N_x ab^2}{6300}$$

$$E_{14} = F_{13} = -\frac{11N_x ab^2}{6300}$$

$$E_{15} = F_{16} = \frac{13N_x ab^2}{12600}$$

$$G_{13} = H_{14} = -\frac{13N_x ab^2}{12600}$$

$$G_{14} = H_{13} = \frac{13N_x ab^2}{3150}$$

$$E_{16} = F_{15} = -\frac{13N_x ab^2}{3150}$$

$$I_1 = J_2 = K_4 = L_3 = \frac{11N_x b^2}{175a}$$

$$I_2 = J_1 = K_3 = L_4 = -\frac{11N_x b^2}{175a}$$

$$I_4 = J_3 = K_1 = L_2 = \frac{13N_x b^2}{350a}$$

$$I_3 = J_4 = K_2 = L_1 = -\frac{13N_x b^2}{350a}$$

$$I_5 = I_6 = K_7 = K_8 = \frac{11N_x b^2}{2100}$$

$$J_5 = J_6 = L_7 = J_8 = -\frac{11N_x b^2}{2100}$$

$$I_7 = I_8 = K_5 = K_6 = \frac{13N_x b^2}{4200}$$

$$J_7 = J_8 = L_5 = L_6 = -\frac{13N_x b^2}{4200}$$

$$I_9 = J_{10} = K_{11} = L_{12} = \frac{2N_x b^3}{175a}$$

$$I_{10} = K_{12} = L_{11} = J_9 = -\frac{2N_x b^3}{175a}$$

$$I_{11} = J_{12} = K_9 = L_{10} = \frac{3N_x b^3}{350a}$$

$$I_{12} = J_{11} = K_{10} = L_9 = -\frac{3N_x b^3}{350a}$$

$$I_{13} = I_{14} = L_{15} = L_{16} = M_9 = N_9 = Q_{12} = P_{12} = \frac{N_x b^3}{1050}$$

$$J_{13} = J_{14} = K_{15} = K_{16} = M_{10} = P_{11} = Q_{11} = -\frac{N_x b^3}{1050}$$

$$J_{15} = J_{16} = K_{13} = K_{14} = M_{11} = N_{11} = P_{10} = Q_{10} = \frac{N_x b^3}{1400}$$

$$I_{15} = I_{16} = L_{13} = L_{14} = M_{12} = N_{12} = P_9 = Q_9 = -\frac{N_x b^3}{1400}$$

$$M_1 = N_1 = P_3 = Q_3 = \frac{11 N_x b^2}{2100} \qquad M_2 = N_2 = P_4 = Q_4 = -\frac{11 N_x b^2}{2100}$$

$$M_4 = N_4 = P_2 = Q_2 = \frac{13 N_x b^2}{4200} \qquad M_3 = N_3 = P_1 = Q_1 = -\frac{13 N_x b^2}{4200}$$

$$M_5 = N_6 = \frac{11 N_x ab^2}{1575} \qquad M_6 = N_5 = -\frac{11 N_x ab^2}{6300}$$

$$M_8 = N_7 = \frac{13 N_x ab^2}{3150} \qquad M_7 = N_8 = -\frac{13 N_x ab^2}{12600}$$

$$M_{13} = N_{14} = P_{15} = Q_{16} = \frac{2 N_x ab^3}{1575} \qquad M_{14} = N_{13} = P_{16} = Q_{15} = -\frac{N_x ab^3}{3150}$$

$$M_{15} = N_{16} = P_{13} = Q_{14} = \frac{N_x ab^3}{4200} \qquad M_{16} = N_{15} = P_{14} = Q_{13} = -\frac{N_x ab^3}{1050}$$

$$P_5 = Q_6 = \frac{13 N_x ab^2}{12600} \qquad N_{10} = -\frac{N_x b^3}{1050} \qquad P_6 = Q_5 = -\frac{13 N_x ab^2}{3150}$$

$$P_8 = Q_7 = \frac{11 N_x ab^2}{6300} \qquad P_7 = Q_8 = -\frac{11 N_x ab^2}{1575}$$

6.5   ASSEMBLAGE

Global stiffness and mass matrices are assembled in the following way:

$$[K]_{\text{Global}} = \sum [K_i] \tag{6.57}$$

$$[M]_{\text{Global}} = \sum [M_i] \tag{6.58}$$

Consider, for example, the beam shown in Fig. 6.2 with the addition of a midpoint node as shown in Fig. 6.5a.

The stiffness matrices $[K_1]$ and $[K_2]$ for the member elements 1 and 2 of lengths L_1 and L_2, respectively, are as follows:

$$[K_1] = \frac{E_1 I_1}{L_1^3} \begin{bmatrix} \dfrac{A_1 L_1^2}{I_1} & 0 & 0 & \dfrac{-A_1 L_1^2}{I_1} & 0 & 0 \\ 0 & 12 & 6L_1 & 0 & -12 & 6L_1 \\ 0 & 6L_1 & 4L_1^2 & 0 & -6L_1 & 2L_1^2 \\ \dfrac{-A_1 L_1^2}{I_1} & 0 & 0 & \dfrac{A_1 L_1^2}{I_1} & 0 & 0 \\ 0 & -12 & -6L_1 & 0 & 12 & -6L_1 \\ 0 & 6L_1 & 2L_1^2 & 0 & -6L_1 & 4L_1^2 \end{bmatrix} \tag{6.59}$$

$$[K_2] = \frac{E_2 I_2}{L_2^3} \begin{bmatrix} \dfrac{A_2 L_2^2}{I_2} & 0 & 0 & \dfrac{-A_2 L_2^2}{I_2} & 0 & 0 \\ 0 & 12 & 6L_2 & 0 & -12 & 6L_2 \\ 0 & 6L_2 & 4L_2^2 & 0 & -6L_2 & 2L_2^2 \\ \dfrac{-A_2 L_2^2}{I_2} & 0 & 0 & \dfrac{A_2 L_2^2}{I_2} & 0 & 0 \\ 0 & -12 & -6L_2 & 0 & 12 & -6L_2 \\ 0 & 6L_2 & 2L_2^2 & 0 & -6L_2 & 4L_2^2 \end{bmatrix} \tag{6.60}$$

If we consider the case where node 2 in Fig. 6.5a is at midspan and the beam is uniformly homogeneous, we have $A_1 = A_2 = A$, $I_1 = I_2 = I$, and $L_1 = L_2 = L$. For this case and in accordance with Eq. (6.57), the global stiffness matrix is as follows:

$$[K]_{\text{Global}} = \frac{EI}{L^3} \begin{bmatrix} \dfrac{AL^2}{I} & 0 & 0 & \dfrac{-AL^2}{I} & 0 & 0 & 0 & 0 & 0 \\ 0 & 12 & 6L & 0 & -12 & 6L & 0 & 0 & 0 \\ 0 & 6L & 4L^2 & 0 & -6L & 2L^2 & 0 & 0 & 0 \\ \dfrac{-AL^2}{I} & 0 & 0 & \dfrac{2AL^2}{I} & 0 & 0 & \dfrac{-AL^2}{I} & 0 & 0 \\ 0 & -12 & -6L & 0 & 24 & 0 & 0 & -12 & 6L \\ 0 & 6L & 2L^2 & 0 & 0 & 8L^2 & 0 & -6L & 2L^2 \\ 0 & 0 & 0 & \dfrac{-AL^2}{I} & 0 & 0 & \dfrac{AL^2}{I} & 0 & 0 \\ 0 & 0 & 0 & 0 & -12 & -6L & 0 & 12 & -6L \\ 0 & 0 & 0 & 0 & 6L & 2L^2 & 0 & -6L & 4L^2 \end{bmatrix} \tag{6.61}$$

In a similar manner, for the same beam and element case as above, the global mass matrix $[M]_{\text{Global}}$ may be obtained. For the two elements in Fig. 6.5a, where $L_1 = L_2 = L$, $I_1 = I_2 = I$, and $A_1 = A_2 = A$, we have

$$[M_1] = \frac{mL}{420} \begin{bmatrix} 140 & 0 & 0 & 70 & 0 & 0 \\ 0 & 156 & 22L & 0 & 54 & -13L \\ 0 & 22L & 4L^2 & 0 & 13L & -3L^2 \\ 70 & 0 & 0 & 140 & 0 & 0 \\ 0 & 54 & 13L & 0 & 156 & -22L \\ 0 & -13L & -3L & 0 & -22L & 4L^2 \end{bmatrix} \tag{6.62}$$

$$[M_2] = \frac{mL}{420} \begin{bmatrix} 140 & 0 & 0 & 70 & 0 & 0 \\ 0 & 156 & 22L & 0 & 54 & -13L \\ 0 & 22L & 4L^2 & 0 & 13L & -3L^2 \\ 70 & 0 & 0 & 140 & 0 & 0 \\ 0 & 54 & 13L & 0 & 156 & -22L \\ 0 & -13L & -3L & 0 & -22L & 4L^2 \end{bmatrix} \tag{6.63}$$

The global mass matrix may be obtained in accordance with Eq. (6.58), and it is as follows:

$$[M]_{\text{Global}} = \frac{mL}{420} \begin{bmatrix} 140 & 0 & 0 & 70 & 0 & 0 & 0 & 0 & 0 \\ 0 & 156 & 22L & 0 & 54 & -13L & 0 & 0 & 0 \\ 0 & 22L & 4L^2 & 0 & 13L & -3L^2 & 0 & 0 & 0 \\ 70 & 0 & 0 & 280 & 0 & 0 & 70 & 0 & 0 \\ 0 & 54 & 13L & 0 & 312 & 0 & 0 & 54 & -13L \\ 0 & -13L & -3L^2 & 0 & 0 & 8L^2 & 0 & 13L & -3L^2 \\ 0 & 0 & 0 & 70 & 0 & 0 & 140 & 0 & 0 \\ 0 & 0 & 0 & 0 & 54 & 13L & 0 & 156 & -22L \\ 0 & 0 & 0 & 0 & -13L & -3L & 0 & -22L & 4L^2 \end{bmatrix} \tag{6.64}$$

In the above example, assembling was very straightforward, because the element coordinate directions and the global coordinate system are the same. In the general case, where the element and global coordinate systems are not the same, the finite element program must convert the element stiffness and mass coefficients to the global coordinate system by using a transformation algorithm.

Assume that beam segment 2 in Fig. 6.5a is now oriented as shown in Fig. 6.6. For beam element 1, we note that the element degrees of freedom and the coordinate directions coincide with the global degrees of freedom and directions; that is, $u_1 = \bar{u}_1$, $u_2 = \bar{u}_2$, and so on. For beam element 2, however, the member axes x', y' are rotated by an angle α with respect to the x, y global system of axes. Consequently, the local and global degrees of freedom of element 2 are related as follows:

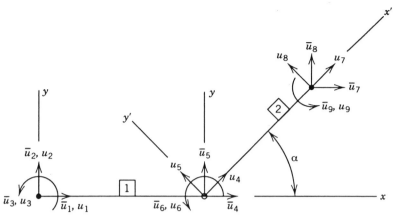

FIGURE 6.6. Two-element mesh with beam axes oriented in different directions.

$$u_4 = \bar{u}_4 \cos \alpha + \bar{u}_5 \sin \alpha \tag{6.65}$$

$$u_5 = -\bar{u}_4 \sin \alpha + \bar{u}_5 \cos \alpha \tag{6.66}$$

$$u_6 = \bar{u}_6 \tag{6.67}$$

$$u_7 = \bar{u}_7 \cos \alpha + \bar{u}_8 \sin \alpha \tag{6.68}$$

$$u_8 = \bar{u}_7 \sin \alpha + \bar{u}_8 \cos \alpha \tag{6.69}$$

$$u_9 = \bar{u}_9 \tag{6.70}$$

On this basis, we can construct a *transformation matrix* $[C]$, which relates the element to the global degrees of freedom. The transformation matrix $[C]$ for element 2 is as follows:

$$[C] = \begin{bmatrix} \cos \alpha & \sin \alpha & 0 & 0 & 0 & 0 \\ -\sin \alpha & \cos \alpha & 0 & 0 & 0 & 0 \\ 0 & 0 & 1 & 0 & 0 & 0 \\ 0 & 0 & 0 & \cos \alpha & \sin \alpha & 0 \\ 0 & 0 & 0 & -\sin \alpha & \cos \alpha & 0 \\ 0 & 0 & 0 & 0 & 0 & 1 \end{bmatrix} \tag{6.71}$$

However, we have to verify that the transformation matrix given by Eq. (6.71) is an *orthogonal matrix*. This condition is checked by the following identity:

$$[C]^T = [C]^{-1} \tag{6.72}$$

The basic equation of motion for an undamped system in terms of the local coordinate system may be written as follows:

$$[M_e]\{\ddot{u}\}_e + [K_e]\{u\}_e = \{R_e\}_e \tag{6.73}$$

where $[M_e]$ is the local mass matrix, $[K_e]$ is the local stiffness matrix, $[R_e]$ is the matrix of the local forces and moments on a joint, and $\{u\}_e$ are the displacements in the direction of the local degrees of freedom.

We can transform to the global coordinate system by premultiplying and postmultiplying each of the two terms on the left-hand side of Eq. (6.73) by $[C]^T$ and $[C]$, respectively. This mathematical manipulation yields

$$[\bar{M}]_e\{\ddot{\bar{u}}_i\}_e + [\bar{K}]_e\{\bar{u}_i\}_e = \{\bar{R}_i\}_e \tag{6.74}$$

where

$$[\bar{M}]_e = [C]^T[M_e][C] \tag{6.75}$$

$$[\tilde{K}]_e = [C]^T[K_e][C] \tag{6.76}$$

Also, $\{\bar{u}_i\}_e$ are the nodal displacements in the direction of the global degrees of freedom, and $\{\bar{R}_i\}_e$ is the matrix of nodal forces and moments in the direction of the global degrees of freedom.

As an application, consider, for example, the beam in Fig. 6.6, and assume that $L_1 = L_2 = 240$ in., Young's modulus of elasticity $E = 30 \times 10^6$ psi, moment of inertia $I = 288$ in.^{4}, $A = 24$ in.^{2}, $\rho = 0.28$ lb/in.^{3}, $\alpha_1 = 0°$, and $\alpha_2 = 30°$. On this basis, from Eq. (6.59), the element stiffness matrix for beam element 1 is

$$[K]_1 = \begin{bmatrix} 3 \times 10^6 & 0 & 0 & -3 \times 10^6 & 0 & 0 \\ 0 & 7500 & 9 \times 10^5 & 0 & -7500 & 9 \times 10^5 \\ 0 & 9 \times 10^5 & 1.44 \times 10^8 & 0 & -9 \times 10^5 & 7.2 \times 10^7 \\ -3 \times 10^6 & 0 & 0 & 3 \times 10^6 & 0 & 0 \\ 0 & -7500 & -9 \times 10^5 & 0 & 7500 & -9 \times 10^5 \\ 0 & 9 \times 10^5 & 7.2 \times 10^7 & 0 & -9 \times 10^5 & 1.44 \times 10^8 \end{bmatrix} \tag{6.77}$$

Since the element degrees of freedom correspond to the global degrees of freedom, it is not required to pre- and postmultiply $[K]_1$ by the transformation matrix $[C]$. If we were to substitute $\alpha = 0°$ into the transformation matrix, it would become the identity matrix, and we would have $[I]^T[K][I] = [K]$, and, hence, $[\bar{K}]_1 = [K]_1$.

For beam element 2, the stiffness matrix with respect to the member coordinate system is

$$[K]_2 = \begin{bmatrix} 3 \times 10^6 & 0 & 0 & -3 \times 10^6 & 0 & 0 \\ 0 & 7500 & 9 \times 10^5 & 0 & -7500 & 9 \times 10^5 \\ 0 & 9 \times 10^5 & 1.44 \times 10^8 & 0 & -9 \times 10^5 & 7.2 \times 10^7 \\ -3 \times 10^6 & 0 & 0 & 3 \times 10^6 & 0 & 0 \\ 0 & -7500 & -9 \times 10^5 & 0 & 7500 & -9 \times 10^5 \\ 0 & 9 \times 10^5 & 7.2 \times 10^7 & 0 & -9 \times 10^5 & 1.44 \times 10^8 \end{bmatrix}$$

$$(6.78)$$

To obtain the required transformation matrix, we must substitute $\alpha = 30°$ into Eq. (6.71). This yields

$$[C] = \begin{bmatrix} 0.866025 & \frac{1}{2} & 0 & 0 & 0 & 0 \\ -\frac{1}{2} & 0.866025 & 0 & 0 & 0 & 0 \\ 0 & 0 & 1 & 0 & 0 & 0 \\ 0 & 0 & 0 & 0.866025 & \frac{1}{2} & 0 \\ 0 & 0 & 0 & -\frac{1}{2} & 0.866025 & 0 \\ 0 & 0 & 0 & 0 & 0 & 1 \end{bmatrix} \quad (6.79)$$

By applying the equation

$$[\bar{K}]_2 = [C]^T [K]_2 [C] \tag{6.80}$$

we obtain

$$[\bar{K}]_2 = \begin{bmatrix} 2.25187 \times 10^6 & 1.29579 \times 10^6 & -4.5 \times 10^5 \\ 1.29579 \times 10^6 & 7.55625 \times 10^5 & 7.79422 \times 10^5 \\ -4.5 \times 10^5 & 7.94422 \times 10^5 & 1.44 \times 10^8 \\ -2.25187 \times 10^6 & -1.29579 \times 10^6 & 4.5 \times 10^5 \\ -1.29579 \times 10^6 & -7.55625 \times 10^5 & -7.79422 \times 10^5 \\ -4.5 \times 10^5 & 7.79442 \times 10^5 & 7.2 \times 10^7 \end{bmatrix}$$

$$\begin{matrix} -2.25187 \times 10^6 & -1.29579 \times 10^6 & -4.5 \times 10^5 \\ -1.29579 \times 10^6 & -7.55625 \times 10^5 & 7.79422 \times 10^5 \\ 4.5 \times 10^5 & -7.79422 \times 10^5 & 7.2 \times 10^7 \\ 2.25187 \times 10^6 & 1.29579 \times 10^6 & 4.5 \times 10^5 \\ 1.29579 \times 10^6 & 7.55625 \times 10^5 & -7.79422 \times 10^5 \\ 4.5 \times 10^5 & -7.79422 \times 10^5 & 1.44 \times 10^8 \end{matrix} \Bigg] \qquad (6.81)$$

By using Eq. (6.57) in combination with Eqs. (6.77) and (6.81), the global stiffness matrix $[\bar{K}]$ is given by

$$
[\bar{K}] = \begin{bmatrix}
3 \times 10^5 & 0 & 0 & -3 \times 10^6 & 0 & 0 \\
0 & 7500 & 9 \times 10^5 & 0 & 9 \times 10^5 & 9 \times 10^5 \\
0 & 9 \times 10^5 & 1.44 \times 10^8 & 0 & 7.2 \times 10^7 & 7.2 \times 10^7 \\
-3 \times 10^6 & 0 & 0 & 3 \times 10^6 & 0 & 0 \\
0 & -7500 & -9 \times 10^5 & 0 & -9 \times 10^5 & -9 \times 10^5 \\
0 & 9 \times 10^5 & 7.2 \times 10^7 & 0 & 1.44 \times 10^8 & 1.44 \times 10^8
\end{bmatrix}
$$

$$
+ \begin{bmatrix}
2.25187 \times 10^6 & 1.29579 \times 10^6 & -4.5 \times 10^5 \\
1.29579 \times 10^6 & 7.55625 \times 10^5 & 7.79422 \times 10^5 \\
-4.5 \times 10^5 & 7.94422 \times 10^5 & 1.44 \times 10^8 \\
-2.25187 \times 10^6 & -1.29579 \times 10^6 & 4.5 \times 10^5 \\
-1.29579 \times 10^6 & -7.55625 \times 10^5 & -7.79422 \times 10^5 \\
-4.5 \times 10^5 & 7.79422 \times 10^5 & 7.2 \times 10^7
\end{bmatrix}
$$

$$
\begin{bmatrix}
-2.25187 \times 10^6 & -1.29579 \times 10^6 & -4.5 \times 10^5 \\
-1.29579 \times 10^6 & -7.55625 \times 10^5 & 7.79422 \times 10^5 \\
4.5 \times 10^5 & -7.79422 \times 10^5 & 7.2 \times 10^7 \\
2.25187 \times 10^6 & 1.29579 \times 10^6 & 4.5 \times 10^5 \\
1.29579 \times 10^6 & 7.55625 \times 10^5 & -7.79422 \times 10^5 \\
4.5 \times 10^5 & -7.79422 \times 10^5 & 1.44 \times 10^8
\end{bmatrix}
$$

or

$$
[\bar{K}] = \begin{bmatrix}
3 \times 10^6 & 0 & 0 & -3 \times 10^6 & 0 \\
0 & 7500 & 9 \times 10^5 & 0 & -700 \\
0 & 9 \times 10^5 & 1.44 \times 10^8 & 0 & -9 \times 10^5 \\
-3 \times 10^6 & 0 & 0 & 2.25187 \times 10^6 & 1.29579 \times 10^6 \\
0 & -7500 & -9 \times 10^5 & 1.29579 \times 10^6 & 7.55625 \times 10^5 \\
0 & 9 \times 10^5 & 7.2 \times 10^7 & -4.5 \times 10^5 & 7.94422 \times 10^5 \\
0 & 0 & 0 & -2.25187 \times 10^6 & -1.29579 \times 10^6 \\
0 & 0 & 0 & -1.29579 \times 10^6 & -7.55625 \times 10^5 \\
0 & 0 & 0 & -4.5 \times 10^5 & 7.79422 \times 10^5
\end{bmatrix}
$$

$$
\begin{bmatrix}
0 & 0 & 0 & 0 \\
9 \times 10^5 & 0 & 0 & 0 \\
7.2 \times 10^7 & 0 & 0 & 0 \\
-4.5 \times 10^5 & -2.25187 \times 10^6 & -1.29579 \times 10^6 & -4.5 \times 10^5 \\
7.79422 \times 10^5 & -1.29579 \times 10^6 & -7.55625 \times 10^5 & 7.79422 \times 10^5 \\
1.44 \times 10^6 & 4.5 \times 10^5 & -7.79422 \times 10^5 & 7.2 \times 10^7 \\
4.5 \times 10^5 & 2.25187 \times 10^6 & 1.29579 \times 10^6 & 4.5 \times 10^5 \\
-7.79422 \times 10^5 & 1.29579 \times 10^6 & 7.55625 \times 10^5 & -7.79422 \times 10^5 \\
7.2 \times 10^7 & 4.5 \times 10^5 & -7.79422 \times 10^5 & 1.44 \times 10^8
\end{bmatrix}
$$

$$\tag{6.82}$$

6.6 BOUNDARY CONDITIONS

Consider a uniform simply supported beam that is modeled as shown in Fig. 6.5a. If the axial deformation involving degrees of freedom 1, 4, and 7 is neglected, the stiffness and mass matrices may be obtained from Eqs. (6.61) and (6.64), respectively, and they are as follows:

$$[K] = \frac{EI}{L^3} \begin{bmatrix} 12 & 6L & -12 & 6L & 0 & 0 \\ 6L & 4L^2 & -6L & 2L^2 & 0 & 0 \\ -12 & -6L & 24 & 0 & -12 & 6L \\ 6L & 2L^2 & 0 & 8L^2 & -6L & 2L^2 \\ 0 & 0 & -12 & -6L & 12 & -6L \\ 0 & 0 & 6L & 2L^2 & -6L & 4L^2 \end{bmatrix} \tag{6.83}$$

$$[M] = \frac{mL}{420} \begin{bmatrix} 156 & 22L & 54 & -13L & 0 & 0 \\ 22L & 4L^2 & 13L & -3L^2 & 0 & 0 \\ 54 & 13L & 312 & 0 & 54 & -13L \\ -13L & -3L^2 & 0 & 8L^2 & 13L & -3L^2 \\ 0 & 0 & 54 & 13L & 156 & -22L \\ 0 & 0 & -13L & -3L^2 & -22L & 4L^2 \end{bmatrix} \tag{6.84}$$

Since the beam is simply supported, nodes 1 and 3 are restrained from moving in the transverse direction. Therefore, the transverse displacements u_2 and u_8 are zero. The equation for the free undamped vibration of the beam may be written as follows:

$$[M]\{\ddot{u}\} + [K]\{u\} = \{0\} \tag{6.85}$$

With the help of Eqs. (6.2), (6.83), and (6.84), we can rewrite Eq. (6.85) in the following manner:

$$\begin{bmatrix} \frac{EI}{L^3} \begin{bmatrix} 12 & 6L & -12 & 6L & 0 & 0 \\ 6L & 4L^2 & -6L & 2L^2 & 0 & 0 \\ -12 & -6L & 24 & 0 & -12 & 6L \\ 6L & 2L^2 & 0 & 8L^2 & -6L & 2L^2 \\ 0 & 0 & -12 & -6L & 12 & -6L \\ 0 & 0 & 6L & 2L^2 & -6L & 4L^2 \end{bmatrix} \\ -\omega^2 \frac{mL}{420} \begin{bmatrix} 156 & 22L & 54 & -13L & 0 & 0 \\ 22L & 4L^2 & 13L & -3L^2 & 0 & 0 \\ 54 & 13L & 312 & 0 & 54 & -13L \\ -13L & -3L^2 & 0 & 8L^2 & 13L & -3L^2 \\ 0 & 0 & 54 & 13L & 156 & -22L \\ 0 & 0 & -13L & -3L^2 & -22L & 4L^2 \end{bmatrix} \end{bmatrix} \begin{bmatrix} u_2 \\ u_3 \\ u_5 \\ u_6 \\ u_8 \\ u_9 \end{bmatrix} = \begin{bmatrix} R_2 \\ R_3 \\ R_5 \\ R_6 \\ R_8 \\ R_9 \end{bmatrix}$$

$$\tag{6.86}$$

Since $u_2 = u_8 = 0$, we can delete the corresponding rows and columns in Eq. (6.86). On this basis, and also noting that the Rs are zero in this case, Eq. (6.86) yields

$$\left[\frac{EI}{L^3} \begin{bmatrix} 4L^2 & -6L & 2L^2 & 0 \\ -6L & 24 & 0 & 6L \\ 2L^2 & 0 & 8L^2 & 2L^2 \\ 0 & 6L & 2L^2 & 4L^2 \end{bmatrix} \right.$$

$$\left. - \omega^2 \frac{\rho A L}{420} \begin{bmatrix} 4L^2 & 13L & -3L^2 & 0 \\ 13L & 312 & 0 & -13L \\ -3L & 0 & 8L^2 & -3L^2 \\ 0 & -13L & -3L^2 & 4L^2 \end{bmatrix} \right] \begin{bmatrix} u_3 \\ u_5 \\ u_6 \\ u_9 \end{bmatrix} = \begin{bmatrix} 0 \\ 0 \\ 0 \\ 0 \end{bmatrix} \qquad (6.87)$$

Equation (6.87) may be solved by setting the following determinant equal to zero:

$$|[K] - \omega^2 [M]| = 0 \qquad (6.88)$$

Equation (6.88) represents the generalized eigenvalue problem.

6.7 NATURAL FREQUENCIES OF VIBRATION AND THEIR CORRESPONDING MODE SHAPES

Various methods are usually available for the solution of the generalized eigenvalue problem that is encountered in free vibration problems. The one presented in this section is the *Jacobi method*. Other methods of analysis may be found in the specialized books dealing with finite element and other numerical methods [37, 39,52–54].

The basic Jacobi method is used here to solve for the eigenvalues of the standard form

$$[A]\{x\} = \lambda\{x\} \qquad (6.89)$$

Equation (6.89) may also be written in the following form:

$$[[A] - \lambda[I]]\{x\} = \{0\} \qquad (6.90)$$

Our free vibration problem, however, is of the form

$$[[K] - \omega^2[M]]\{u\} = \{0\} \qquad (6.91)$$

The λ values (eigenvalues) in Eq. (6.90) correspond to the values of the free frequencies ω^2 in Eq. (6.91). We note, however, that the basic Jacobi method does not apply directly here, because the mass matrix $[M]$ is not the identity matrix $[I]$. If, however, the mass matrix $[M]$ is positive definite, a Cholesky decomposition can be performed on $[M]$ to convert the generalized eigenvalue to the standard form. The Cholesky decomposition consists of representing $[M]$ as the product of two matrices $[L]$ and $[L]^T$, where $[L]$ is a lower triangular matrix, and, subsequently, $[L]^T$ is its upper triangular transpose matrix.

Consider, for example, the reduced mass matrix given by Eq. (6.87). If $\rho A = 1$ and $L = 420$, the mass matrix yields

$$[M] = \begin{bmatrix} 705600 & 5460 & -529200 & 0 \\ 5460 & 312 & 0 & -5460 \\ -529200 & 0 & 1411200 & -529200 \\ 0 & -5460 & -529200 & 705600 \end{bmatrix} \qquad (6.92)$$

The matrix $[L]$ and the L_{ij} elements of this matrix may be determined from the elements of m_{ij} of the mass matrix as follows:

$$L_{11} = \sqrt{m_{11}} \qquad (6.93)$$

$$L_{i1} = \frac{m_{i1}}{L_{11}} \qquad i = 2, 3, \ldots, n \qquad (6.94)$$

$$L_{ii} = \sqrt{m_{ii} - \sum_{k=1}^{i-1} L_{ik}L_{ik}} \qquad i = 2, 3, \ldots, n \qquad (6.95)$$

$$L_{ij} = \frac{m_{ij} - \sum_{k=1}^{j-1} L_{ik}L_{jk}}{L_{jj}} \qquad i = j+1, j+2, \ldots, n \qquad (6.96)$$

By following the procedure given by Eqs. (6.93) to (6.96), we find

$$L_{11} = \sqrt{m_{11}} = \sqrt{705600} = 840$$

$$L_{21} = \frac{m_{21}}{L_{11}} = \frac{5460}{840} = 6.5$$

$$L_{22} = \sqrt{m_{22} - L_{21}L_{21}} = \sqrt{312 - 6.5^2}$$

$$= 16.424$$

$$L_{31} = \frac{m_{31}}{L_{11}} = -\frac{529200}{840} = -630$$

$$L_{32} = \frac{m_{32} - L_{31}L_{21}}{L_{22}} = \frac{0 - (-630)(6.5)}{16.424}$$

$$= 249.33$$

$$L_{33} = \sqrt{m_{33} - L_{31}L_{31} - L_{32}L_{32}}$$

$$= \sqrt{1411200 - (630)^2 - (249.33)^2}$$

$$= 975.774$$

$$L_{41} = \frac{m_{41}}{L_{11}} = 0$$

$$L_{42} = \frac{m_{42} - L_{41}L_{31}}{L_{22}} = \frac{-5460}{16.424}$$

$$= -332.44$$

$$L_{43} = \frac{m_{43} - L_{41}L_{31} - L_{42}L_{32}}{L_{33}}$$

$$= \frac{-529200 - (-332.44)(249.33)}{975.774}$$

$$= 457.394$$

$$L_{44} = \sqrt{m_{44} - L_{41}^2 - L_{42}^2 - L_{43}^2}$$

$$= \sqrt{705600 - (322.44)^2 - (457.394)^2}$$

$$= 621.188$$

On this basis, the matrix $[L]$, its transpose $[L]^T$, and the product $[L][L]^T$ are as follows:

$$[L] = \begin{bmatrix} 840.0 & 0 & 0 & 0 \\ 6.5 & 16.424 & 0 & 0 \\ -630.0 & 249.330 & 975.774 & 0 \\ 0 & -332.440 & -457.394 & 621.188 \end{bmatrix} \tag{6.97}$$

$$[L]^T = \begin{bmatrix} 840 & 6.500 & -630.000 & 0 \\ 0 & 16.424 & 249.330 & -332.440 \\ 0 & 0 & 975.774 & -457.394 \\ 0 & 0 & 0 & 621.188 \end{bmatrix} \quad (6.98)$$

$$[L][L]^T = \begin{bmatrix} 7.056 \times 10^5 & 5460.000 & -5.292 \times 10^5 & 0 \\ 5460 & 311.997 & -0.00408163 & -5459.990 \\ -5.292 \times 10^5 & -0.00408163 & 1.4112 \times 10^6 & -5.292 \times 10^5 \\ 0 & -5459.990 & -5.2920 \times 10^5 & 7.056 \times 10^5 \end{bmatrix}$$
$$(6.99)$$

We note here that the elements of $[L][L]^T$ given by Eq. (6.99) are closely identical to the elements of the mass matrix $[M]$ given by Eq. (6.92).

Hence, if we have the identity

$$[K]\{u\} = \omega^2[M]\{u\} \quad (6.100)$$

and substitute for $[M]$ the expression

$$[M] = [L][L]^T \quad (6.101)$$

we find

$$[K]\{u\} = \omega^2[L][L]^T\{u\} \quad (6.102)$$

If we premultiply both sides of Eq. (6.102) by $[L]^{-1}$, we find

$$[L]^{-1}[K]\{u\} = \omega^2[L]^{-1}[L][L]^T\{u\} \quad (6.103)$$

or

$$[L]^{-1}[K][L^{-1}]^T[L]^T\{u\} = \omega^2[L]^{-1}[L][L]^T\{u\} \quad (6.104)$$

or

$$[L]^{-1}[K][L^{-1}]^T[L]^T\{u\} = \omega^2[L]^T\{u\} \quad (6.105)$$

Equation (6.105) may be written in the standard eigenvalue form

$$[A]\{x\} = \lambda\{x\} \qquad (6.106)$$

by making the following substitutions:

$$[A] = [L]^{-1}[K][L^{-1}]^T \qquad (6.107)$$

$$\lambda = \omega^2 \qquad (6.108)$$

$$\{x\} = [L]^T\{u\} \qquad (6.109)$$

If we use the example problem in Section 6.6 and assume that $E = 30 \times 10^6$, $I = 100$, $m = 1$, and $L = 420$, we find

$$[K] = \begin{bmatrix} 2.85714 \times 10^7 & -1.02040 \times 10^5 & 1.42857 \times 10^7 & 0 \\ -1.02040 \times 10^5 & 971.81700 & 0 & 1.02040 \times 10^5 \\ 1.42857 \times 10^7 & 0 & 5.71428 \times 10^7 & 1.42857 \times 10^7 \\ 0 & 1.02040 \times 10^5 & 1.42857 \times 10^7 & 2.85714 \times 10^7 \end{bmatrix}$$

$$(6.110)$$

$$[L]^{-1} = \begin{bmatrix} 0.00119047 & 0 & 0 & 0 \\ -4.71142 \times 10^{-4} & 0.0608863 & 1.03869 \times 10^{-8} & 0 \\ 8.89006 \times 10^{-4} & -0.0155576 & 0.00102482 & 0 \\ 4.02456 \times 10^{-4} & 0.0211289 & 7.54605 \times 10^{-4} & 0.00160981 \end{bmatrix}$$

$$(6.111)$$

Thus, by using Eq. (6.107), we find

$$[A] = [L]^{-1}[K][L^{-1}]^T \qquad (6.112)$$

$$= \begin{bmatrix} 40.4923 & -23.421300 & 49.5572 & 23.955700 \\ -23.4213 & 15.798900 & -26.0558 & -0.728542 \\ 49.5572 & -26.055800 & 111.6840 & 89.304300 \\ 23.9557 & -0.728541 & 89.3043 & 160.234000 \end{bmatrix}$$

The Jacobi method provides a solution for the standard eigenvalue problem, by performing a series of transformations on the symmetric matrix $[A]$. These transformations are used to zero out the off-diagonal terms one at a time. One such sequence of zeroing out all the off-diagonal terms is called a *sweep*. During the process, previously zeroed terms will become nonzero.

However, with additional sweeps, the off-diagonal terms will all approach zero.

The rotational matrix $[P]$ has the following form:

$$
[P_n] = \begin{bmatrix} 1 & & & \\ & \cos\theta & -\sin\theta & \\ & \sin\theta & \cos\theta & \\ & & & 1 \end{bmatrix} \begin{matrix} \\ i \\ j \\ \\ \end{matrix}
\qquad (6.113)
$$

The matrix $[P_n]$ is an orthogonal matrix, which means that

$$[P]^T[P] = [I] \qquad (6.114)$$

It follows that the transformed matrix $[P]^T[A][P]$ has the same eigenvalues and eigenvectors as the original matrix $[A]$. For a diagonal matrix, the eigenvalues are the diagonal terms.

The angle θ is selected such that

$$\tan 2\theta = \frac{2A_{ij}}{A_{ii} - A_{jj}} \qquad \text{for} \quad A_{ii} \neq A_{jj} \qquad (6.115)$$

$$\theta = \pi/4 \qquad \text{for} \quad A_{ii} = A_{jj} \qquad (6.116)$$

Note that the angle θ is based on a new matrix $[A]$ for each rotation.

The transformation procedure is as follows:

1. Let $[A_n] = [A]$.
2. Select an off-diagonal term i, j, to eliminate, that is, $1, 2$.
3. Use Eq. (6.115) or Eq. (6.116) to determine θ.
4. Generate matrix $[P_n]$ based on the current i, j and the current θ.
5. $[A_{n+1}] = [P_n]^T[A_n][P_n]$ (6.117)
6. Select the next successive off-diagonal term, say, $1, 3$.
7. Use equations to determine a new θ, using the $[A_{n+1}]$ terms in the equation.
8. Generate $[P_{n+1}]$ based on the new i, j and new θ.
9. $[A_{n+2}] = [P_{n+1}][A_{n+1}][P_n]$ (6.118)
10. Continue this process by repeating steps 6 through 9, until all the off-diagonal terms have been zeroed. This concludes one sweep.
11. At this point, some of the previously zeroed terms will no longer be

zero. Thus, additional sweeps may be performed until all off-diagonal terms have converged to zero.

Consider again the following matrix $[A]$, which is formed by using the matrix product $[L]^{-1}[K][L^{-1}]^T$; that is,

$$[A_1] = \begin{bmatrix} 40.4923 & -23.421300 & 49.5572 & 23.955700 \\ -23.4213 & 15.798900 & -26.0558 & -0.728542 \\ 49.5572 & -26.055800 & 111.6840 & 89.304300 \\ 23.9557 & -0.728541 & 89.3043 & 160.234000 \end{bmatrix} \quad (6.119)$$

For $i = 1$, $j = 2$, Eq. (6.115) yields

$$\tan 2\theta = \frac{2(-23.4213)}{40.4923 - 15.7989}$$

$$\theta = -9.9676 \text{ rad}$$

Therefore,

$$[P_1] = \begin{bmatrix} \cos(-9.9676) & -\sin(-9.9676) & 0 & 0 \\ \sin(-9.9676) & \cos(-9.9676) & 0 & 0 \\ 0 & 0 & 1 & 0 \\ 0 & 0 & 0 & 1 \end{bmatrix} \quad (6.120)$$

and

$$[A_2] = [P_1]^T [A_1][P_1]$$

$$= \begin{bmatrix} 54.62190 & -3.21724 \times 10^{-4} & -55.89270 & -20.8884 \\ -3.63535 \times 10^{-4} & 1.67922 & -3.28866 & -11.7506 \\ 55.89270 & -3.28869 & 111.6840 & 89.3042 \\ -20.88840 & -11.75060 & 89.30420 & 160.2340 \end{bmatrix}$$

$$(6.121)$$

where

$$[P_1]^T = \begin{bmatrix} -0.856252 & 0.516554 & 0 & 0 \\ -0.516554 & -0.856252 & 0 & 0 \\ 0 & 0 & 1 & 0 \\ 0 & 0 & 0 & 1 \end{bmatrix} \tag{6.122}$$

Note that $A_{21,2} = A_{22,1} \approx 0$.

For $i = 1$ and $j = 3$, we have

$$\tan 2\theta = \frac{2(-55.8927)}{54.6219 - 111.6840}$$

$$\theta = 9.97418 \text{ rad}$$

Thus,

$$[P_2] = \begin{bmatrix} \cos(9.97418) & 0 & -\sin(9.97418) & 0 \\ 0 & 1 & 0 & 0 \\ \sin(9.97418) & 0 & \cos(9.97418) & 0 \\ 0 & 0 & 0 & 1 \end{bmatrix}$$

$$= \begin{bmatrix} -0.852837 & 0 & 0.522176 & 0 \\ 0 & 1 & 0 & 0 \\ -0.522176 & 0 & -0.852837 & 0 \\ 0 & 0 & 0 & 1 \end{bmatrix}$$

$$[A_3] = [P_2]^T [A_2][P_2]$$

$$= \begin{bmatrix} 20.39930 & 1.71755 & -8.00721 \times 10^{-4} & -28.8181 \\ 1.71757 & 1.66921 & 2.80450 & -11.7506 \\ -7.95559 \times 10^{-4} & 2.80454 & 145.90600 & -87.0693 \\ -28.81810 & -11.75060 & -87.06930 & 160.2340 \end{bmatrix} \tag{6.123}$$

Note that $[A_3]_{1,3} \approx [A_3]_{3,1} \approx 0$. However, the elements $[A_3]_{1,2} = [A_3]_{2,1}$ are now nonzero.

For $i = 1$ and $j = 4$, we have

$$\tan 2\theta = \frac{2(-28.8181)}{20.3993 - 160.234}$$

$$\theta = 3.33707 \text{ rad}$$

$$[P_3] = \begin{bmatrix} \cos(3.33707) & 0 & 0 & -\sin(3.33707) \\ 0 & 1 & 0 & 0 \\ 0 & 0 & 1 & 0 \\ \cos(3.33707) & 0 & 0 & \cos(3.33707) \end{bmatrix} \tag{6.124}$$

$$[A_4] = [P_3]^T [A_3][P_3]$$

$$= \begin{bmatrix} 14.69310 & 0.59753 & 16.9126 & -2.90575 \times 10^{-4} \\ 0.59751 & 1.66921 & 2.8045 & 11.86040 \\ 16.91260 & 2.80454 & 145.9060 & 85.41100 \\ -2.95117 \times 10^{-4} & 11.86040 & 85.4110 & 165.94000 \end{bmatrix}$$

For $i = 2$ and $j = 3$, we have

$$\tan 2\theta = \frac{2(2.8045)}{1.6621 - 145.906}$$

$$\theta = -3.16102 \text{ rad}$$

$$[P_4] = \begin{bmatrix} 1 & 0 & 0 & 0 \\ 0 & \cos(-3.16102) & -\sin(-3.16102) & 0 \\ 0 & \sin(-3.16102) & \cos(-3.16102) & 0 \\ 0 & 0 & 0 & 1 \end{bmatrix} \tag{6.125}$$

$$[A_5] = [P_4]^T [A_4][P_4]$$

$$= \begin{bmatrix} 14.69310 & -0.268872 & -16.92100 & -2.90575 \times 10^{-4} \\ -0.268853 & 1.614700 & 9.64153 \times 10^{-4} & -10.19890 \\ -16.921000 & 10.117600 \times 10^{-4} & 145.96100 & -85.62520 \\ -2.951170 \times 10^{-4} & -10.198900 & -85.62520 & 165.94000 \end{bmatrix}$$

$$\tag{6.126}$$

For $i = 2$ and $j = 4$, we have

$$\tan 2\theta = \frac{2(-10.1989)}{1.6147 - 165.94}$$

$$\theta = 3.20334 \text{ rad}$$

$$[P_5] = \begin{bmatrix} 1 & 0 & 0 & 0 \\ 0 & \cos(3.20334) & 0 & -\sin(3.20334) \\ 0 & 0 & 1 & 0 \\ 0 & \sin(3.20334) & 0 & \cos(3.20334) \end{bmatrix} \tag{6.127}$$

$$[A_6] = [P_5]^T[A_5][P_5]$$

$$= \begin{bmatrix} 14.693100 & 0.268378 & -16.92100 & -163.01700 \times 10^{-4} \\ 0.268359 & 0.984120 & 5.28286 & -3.37418 \times 10^{-4} \\ -16.921000 & 5.282810 & 145.96100 & 85.46210 \\ -0.016296 & -3.336800 \times 10^{-4} & 85.46210 & 166.57000 \end{bmatrix}$$

$$\tag{6.128}$$

Finally, for $i = 3$ and $j = 4$, we have

$$\tan 2\theta = \frac{2(85.4621)}{145.961 - 166.570}$$

$$\theta = 2.41619 \text{ rad}$$

$$[P_6] = \begin{bmatrix} 1 & 0 & 0 & 0 \\ 0 & 1 & 0 & 0 \\ 0 & 0 & \cos(2.41619) & -\sin(2.41619) \\ 0 & 0 & \sin(2.41619) & \cos(2.41619) \end{bmatrix} \tag{6.129}$$

$$[A_7] = [P_6]^T[A_6][P_6]$$

$$= \begin{bmatrix} 14.693100 & 0.268378 & 12.65000 & 11.23820 \\ 0.268359 & 0.984120 & -3.95303 & -3.50459 \\ 12.650000 & -3.952990 & 70.18470 & -6.43607 \times 10^{-4} \\ 11.238200 & -3.504560 & -6.39594 \times 10^{-4} & 242.34700 \end{bmatrix}$$

$$\tag{6.130}$$

This concludes the first sweep, with only the 3, 4 and 4, 3 terms being close to zero. Additional sweeps are required until all the off-diagonal terms become sufficiently close to zero. Note that if a particular off-diagonal term is sufficiently close to zero, say, within a specified tolerance, rotation with respect to that term may be skipped within the sweep.

By continuing the procedure and using five sweeps, the final result is

$$
\begin{bmatrix}
11.5692 & 0 & 0 & 0 \\
0 & 0.5916 & 0 & 0 \\
0 & 0 & 73.0955 & 0 \\
0 & 0 & 0 & 242.9541
\end{bmatrix}
\tag{6.131}
$$

Note that the diagonal terms of the matrix in Eq. (6.131) are the squares of the free frequencies of vibration of the simply supported beam in units of radians per second.

The corresponding eigenvectors (mode shapes) may be determined by substituting into Eq. (6.91) the values of ω_i^2 obtained in Eq. (6.131). It could be determined more easily, however, by simply multiplying the transposes of the plane rotation matrices. On this basis, the eigenvectors would be represented by the columns of the following square matrix:

$$
[P_1^T P_2^T P_3^T \cdots P_{\text{final}}^T]
\tag{6.132}
$$

For the example problem discussed in this section, we have the following results:

$$
\omega_1^2 = 0.5916
$$

$$
\omega_1 = 0.7692 \text{ rps}
\tag{6.133}
$$

$$
\{\phi_1\} =
\begin{bmatrix}
1.0 \\
267.4 \\
0 \\
-1.0
\end{bmatrix}
\tag{6.134}
$$

$$
\omega_2^2 = 11.5692
$$

$$
\omega_2 = 3.4014 \text{ rps}
\tag{6.135}
$$

$$
\{\phi_2\} =
\begin{bmatrix}
1 \\
0 \\
-1 \\
1
\end{bmatrix}
\tag{6.136}
$$

$$
\omega_3^2 = 73.0955
$$

$$\omega_3 = 8.5496 \text{ rps} \tag{6.137}$$

$$\{\phi_3\} = \begin{bmatrix} 1.0 \\ -45.9 \\ 0 \\ -1.0 \end{bmatrix} \tag{6.138}$$

$$\omega_4^2 = 242.9541$$
$$\omega_4 = 15.5870 \text{ rps} \tag{6.139}$$

$$\{\phi_4\} = \begin{bmatrix} 1 \\ 0 \\ 1 \\ 1 \end{bmatrix} \tag{6.140}$$

Note that these eigenvectors are unique with an arbitrary magnification constant. They only represent the shape of the vibration curve and not actual displacements. Also note that the eigenvectors are orthogonal with respect to the mass and stiffness matrices; that is,

$$\{\phi_i\}^T [M]\{\phi_j\} = 0 \qquad i \neq j \tag{6.141}$$
$$\{\phi_i\}^T [K]\{\phi_j\} = 0$$

Distinct eigenvalues produce unique eigenvectors. Multiple eigenvalues produce eigenvectors which are not unique. But we may still choose orthogonal eigenvectors. It would be advantageous for us to mass orthonormalize the eigenvectors so that

$$[\Phi]^T [M][\Phi] = I \tag{6.142}$$

This procedure will facilitate the solution of the forced vibration problem when modal superposition is used.

6.8 DYNAMIC RESPONSE

When a structure or a mechanical system is subjected to a dynamic excitation, the general matrix equation representing the resulting motion is as follows:

$$[M]\ddot{u} + [c]\dot{u} + [K]u = \{R\} \tag{6.143}$$

where $[M]$ is the global mass matrix, $[c]$ is the damping matrix, $[K]$ is the global stiffness matrix, and $\{R\}$ is the excitation force.

The matrix equation may be solved either by direct integration or by using modal superposition. Due to the presence of off-diagonal terms in the matrices $[M]$ and $[K]$, the equations of motion represented by Eq. (6.143) are coupled. Direct integration of Eq. (6.143) requires the use of step-by-step numerical integration, which can be carried out by various procedures, such as Newmark–Beta, central difference, Houbolt, Wilson θ, and Runge–Kutta.

If mode superposition is used, the matrix equation is first transformed into the realm of the eigenvectors by pre- and post-multiplying by the mode shape matrix obtained from the solution of the free vibration problem. The equations of motion are uncoupled by the transformation, and the resultant uncoupled equations are then solved by using numerical integration procedures, such as Newmark–Beta. An "exact" solution of the uncoupled differential equations may also be possible, although the finite element program customarily uses a numerical integration algorithm.

Both numerical integration and mode superposition introduce approximations, regardless of the numerical integration technique used. For large meshes, mode superposition may be more efficient, particularly in the cases where the inclusion of all the modes may not be required for a satisfactory solution of the problem. Various modal selection strategies have been developed to determine which modes should be included; for example, see refs. 55 to 62.

Some of the numerical integration techniques, such as the Runge–Kutta, may erroneously converge to infinity in certain problems if the time step increment is comparatively large. Therefore, only the Newmark–Beta technique is presented here. Other numerical integration techniques commonly used in finite element dynamic response may be found in the literature [37, 52, 54].

6.8.1 Direct Integration

A step-by-step procedure regarding the solution for the dynamic response of a structural or a mechanical system by using the Newmark–Beta method may be stated as follows:

Step 1. We assemble the mass $[M]$, stiffness $[K]$, damping $[c]$, and load $[R]$ matrices at the initial time. For the Newmark–Beta (constant acceleration) method, we let $\alpha = 0.25$ and $\delta = 0.5$. We also choose a time step Δt

Step 2. Initialize the displacement, ^0u, velocity $^0\dot{u}$, and acceleration $^0\ddot{u}$ vectors.

Step 3. Determine the following integration constants:

$$a_0 = \frac{1}{\alpha(\Delta t)^2} \tag{6.144}$$

$$a_1 = \frac{\delta}{\alpha(\Delta t)} \tag{6.145}$$

$$a_2 = \frac{1}{\alpha(\Delta t)} \tag{6.146}$$

$$a_3 = \frac{1}{2\alpha} - 1 \tag{6.147}$$

$$a_4 = \frac{\delta}{\alpha} - 1 \tag{6.148}$$

$$a_5 = \frac{(\Delta t)}{2}\left(\frac{\delta}{\alpha} - 2\right) \tag{6.149}$$

$$a_6 = (\Delta t)(1 - \delta) \tag{6.150}$$

$$a_7 = \delta(\Delta t) \tag{6.151}$$

Step 4. Form $\hat{K}$, where

$$\hat{K} = K + a_0 M + a_1 c \tag{6.152}$$

Step 5. Triangularize $\hat{K}$, where

$$\hat{K} = LDL^T \tag{6.153}$$

This is done to solve for $^{t+\Delta t}u$ in step 7, although other methods may be used. For a detailed description of the LDL^T and back substitution algorithm, see Bathe [37].

Step 6. Calculate the effective loads; that is,

$$^{t+\Delta t}\hat{R} = {}^{t+\Delta t}R + M(a_0{}^t u + a_2{}^t \dot{u} + a_3{}^t \ddot{u}) + c(a_1{}^t u - a_4{}^t \dot{u} + a_5{}^t \ddot{u}) \tag{6.154}$$

Step 7. Solve for the displacements $^{t+\Delta t}u$, using LDL^T and back substitution:

$$\hat{K}^{t+\Delta t}u = {}^{t+\Delta t}\hat{R} \tag{6.155}$$

Step 8. Calculate $^{t+\Delta t}u$ and $^{t+\Delta t}\ddot{u}$:

$$^{t+\Delta t}\ddot{u} = a_0(^{(t+\Delta t}u - {}^t u) - a_2{}^t\dot{u} - a_3{}^t\ddot{u} \qquad (6.156)$$

$$^{t+\Delta t}\dot{u} = {}^t\dot{u} + a_6{}^t\ddot{u} + a_7{}^{t+\Delta t}\ddot{u} \qquad (6.157)$$

Step 9. Proceed with the next time step $t = t + \Delta t$.

Step 10. Repeat steps 6 to 9 until T_{max} is exceeded.

The following example illustrates the application of the above procedure.

Example 6.1. For a system initially at rest, determine the displacement at times $t = 0.1$ and $t = 0.2$ by using direct integration and the Newmark–Beta method. The given data are as follows:

$$[M] = \begin{bmatrix} 2 & 0 \\ 0 & 1 \end{bmatrix} \qquad (6.158)$$

$$[K] = \begin{bmatrix} 6 & -2 \\ -2 & 2 \end{bmatrix} \qquad (6.159)$$

$$\{R\} = \begin{bmatrix} 0 \\ 10 \end{bmatrix} \qquad (6.160)$$

$$[c] = 0 \qquad (6.161)$$

Solution. We apply step 1 and assemble the $[M]$, $[K]$, $\{R\}$, and $[c]$ matrices, which are given by Eqs. (6.158), (6.159), (6.160), and (6.161), respectively. We also let $\alpha = 0.25$, $\delta = 0.5$, and $\Delta t = 0.1$.

By following step 2, we initialize the displacement, velocity, and acceleration vectors. This yields the following results:

$$^0u = \begin{bmatrix} 0 \\ 0 \end{bmatrix} \qquad ^0\dot{u} = \begin{bmatrix} 0 \\ 0 \end{bmatrix}$$

$$M\ddot{u} + Ku = R$$

$$\begin{bmatrix} 2 & 0 \\ 0 & 1 \end{bmatrix} {}^0\ddot{u} + \begin{bmatrix} 6 & -2 \\ -2 & 2 \end{bmatrix} \begin{bmatrix} 0 \\ 0 \end{bmatrix} = \begin{bmatrix} 0 \\ 10 \end{bmatrix}$$

$$2\,{}^0\ddot{u}_1 = 0 \qquad {}^0\ddot{u}_1 = 0$$

$$^0\ddot{u}_2 = 10$$

$$^0\ddot{u} = \begin{bmatrix} 0 \\ 10 \end{bmatrix}$$

From step 3, the integration constants are

$$a_0 = \frac{1}{\alpha(\Delta t)^2} = \frac{1}{(0.25)(0.1)^2} = 400$$

$$a_1 = \frac{\delta}{\alpha(\Delta t)} = \frac{0.5}{(0.25)(0.1)} = 20$$

$$a_2 = \frac{1}{\alpha(\Delta t)} = \frac{1}{(0.25)(0.1)} = 40$$

$$a_3 = \frac{1}{2\alpha} - 1 = \frac{1}{2(0.25)} - 1 = 1$$

$$a_4 = \frac{\delta}{\alpha} - 1 = \frac{0.5}{0.25} - 1 = 1$$

$$a_5 = \frac{\Delta t}{2}\left(\frac{\delta}{\alpha} - 2\right) = \frac{0.1}{2}\left(\frac{0.5}{0.25} - 2\right) = 0$$

$$a_6 = \Delta t(1 - \delta) = (0.1)(1 - 0.5) = 0.05$$

$$a_7 = \delta(\Delta t) = (0.5)(0.1) = 0.05$$

Now we use step 4 to form $\hat{K}$. Thus, by applying Eq. (6.152), we have

$$\hat{K} = \begin{bmatrix} 6 & -2 \\ -2 & 2 \end{bmatrix} + 400\begin{bmatrix} 2 & 0 \\ 0 & 1 \end{bmatrix} + 0$$

$$= \begin{bmatrix} 806 & -2 \\ -2 & 402 \end{bmatrix}$$

By using step 5, we triangularize $\hat{K}$ by using Eq. (6.153). This yields

$$\begin{bmatrix} 806 & -2 \\ -2 & 402 \end{bmatrix} \rightarrow \begin{bmatrix} 806 & -2 \\ 0 & 401.995 \end{bmatrix} \rightarrow \begin{bmatrix} 1 & -0.002481 \\ 0 & 1 \end{bmatrix}$$

$$\underset{\hat{K}}{\begin{bmatrix} 806 & -2 \\ -2 & 402 \end{bmatrix}} = \underset{L}{\begin{bmatrix} 1 & 0 \\ -0.002481 & 1 \end{bmatrix}} \underset{D}{\begin{bmatrix} 806 & 0 \\ 0 & 401.995 \end{bmatrix}} \underset{L^T}{\begin{bmatrix} 1 & -0.002481 \\ 0 & 1 \end{bmatrix}}$$

By using step 6, we calculate the effective loads. Thus, by using Eq.

(6.154), we find

$$
{}^{0.1}\hat{R} = \begin{bmatrix} 0 \\ 10 \end{bmatrix} + \begin{bmatrix} 2 & 0 \\ 0 & 1 \end{bmatrix} (400\,{}^{t}u + 40\,{}^{t}\dot{u} + {}^{t}\ddot{u})
$$

$$
= \begin{bmatrix} 0 \\ 10 \end{bmatrix} + \begin{bmatrix} 2 & 0 \\ 0 & 1 \end{bmatrix} \left(\begin{bmatrix} 0 \\ 0 \end{bmatrix} + \begin{bmatrix} 0 \\ 0 \end{bmatrix} + \begin{bmatrix} 0 \\ 10 \end{bmatrix} \right)
$$

$$
= \begin{bmatrix} 0 \\ 20 \end{bmatrix}
$$

We use step 7 here to solve for the displacements u. Thus, by using Eq. (6.155), we write

$$
\hat{K}\,{}^{0.1}u = {}^{0.1}\hat{R}
$$

or, by substitution, we find

$$
\begin{bmatrix} 806 & -2 \\ -2 & 402 \end{bmatrix} \begin{bmatrix} {}^{0.1}u_1 \\ {}^{0.1}u_2 \end{bmatrix} = \begin{bmatrix} 0 \\ 20 \end{bmatrix}
$$

$$
\begin{bmatrix} 806 & -2 \\ 0 & 401.995 \end{bmatrix} \begin{bmatrix} {}^{0.1}u_1 \\ {}^{0.1}u_2 \end{bmatrix} = \begin{bmatrix} 0 \\ 20 \end{bmatrix}
$$

$$
{}^{0.1}u_2 = \frac{20}{401.995} = 0.049752
$$

$$
806\,{}^{0.1}u_1 - 2(0.049752) = 0
$$

$$
{}^{0.1}u_1 = 0.00012345
$$

$$
{}^{0.1}u = \begin{bmatrix} 0.00012345 \\ 0.04975200 \end{bmatrix}
$$

This result could also be determined by using LDL^T; that is,

$$
\hat{R} = \begin{bmatrix} 0 \\ 20 \end{bmatrix}
$$

$$
v_1 = R_1 = 0
$$

$$
v_2 = R_2 - L_{12}^T v_1
$$

$$
= 20 - (-0.002481)(0)
$$

$$
= 20
$$

$$
\bar{v} = D^{-1} v
$$

$$= \begin{bmatrix} 806 & 0 \\ 0 & 401.995 \end{bmatrix}^{-1} \begin{bmatrix} 0 \\ 20 \end{bmatrix}$$

$$= \begin{bmatrix} 0 \\ 0.0497518 \end{bmatrix}$$

$$u_2 = \bar{v}_2 = 0.0497518$$

$$u_1 = \bar{v}_1^{(1)} = \bar{v}_1^{(2)} - L_{12}u_2$$

$$= 0 - (-0.002481)(0.0497518)$$

$$= 0.00012345$$

$$^{0.1}u = \begin{bmatrix} 0.00012345 \\ 0.0497180 \end{bmatrix}$$

We use step 8 to determine $^{t+\Delta t}\dot{u}$ and $^{t+\Delta t}\ddot{u}$. Thus, by using Eq. (6.156), we find

$$^{0.1}\ddot{u} = 400\left(\begin{bmatrix} 0.00012345 \\ 0.04975200 \end{bmatrix} - \begin{bmatrix} 0 \\ 0 \end{bmatrix}\right) - 40\begin{bmatrix} 0 \\ 0 \end{bmatrix} - \begin{bmatrix} 0 \\ 10 \end{bmatrix}$$

$$^{0.1}\ddot{u} = \begin{bmatrix} 0.04938 \\ 9.90080 \end{bmatrix}$$

By using Eq. (6.157), we find

$$^{0.1}\dot{u} = \begin{bmatrix} 0 \\ 0 \end{bmatrix} + 0.05\begin{bmatrix} 0 \\ 10 \end{bmatrix} + 0.05\begin{bmatrix} 0.04938 \\ 9.90080 \end{bmatrix}$$

$$^{0.1}\dot{u} = \begin{bmatrix} 0.002469 \\ 0.995040 \end{bmatrix}$$

We use step 9 to proceed with the next time step $t = 0.1 + 0.1 = 0.2$, and we follow step 10 by repeating steps 6 to 9. By repeating step 6, we find

$$^{0.2}\hat{R} = \begin{bmatrix} 0 \\ 10 \end{bmatrix} + \begin{bmatrix} 2 & 0 \\ 0 & 1 \end{bmatrix}\left(400\begin{bmatrix} 0.00012345 \\ 0.04975200 \end{bmatrix} + 40\begin{bmatrix} 0.002469 \\ 0.995040 \end{bmatrix} + \begin{bmatrix} 0.04938 \\ 9.90080 \end{bmatrix}\right)$$

$$^{0.2}\hat{R} = \begin{bmatrix} 0.39504 \\ 69.60320 \end{bmatrix}$$

By repeating step 7, we find

$$\hat{K}^{0.2}u = {}^{0.2}\hat{R}$$

$$\hat{K} = \begin{bmatrix} 806 & -2 \\ -2 & 402 \end{bmatrix}$$

$${}^{0.2}\hat{R} = \begin{bmatrix} 0.39504 \\ 69.60320 \end{bmatrix} = v$$

$$L_{12} = -0.002481$$

$$\bar{v} = D^{-1}v = \begin{bmatrix} 806 & 0 \\ 0 & 401.995 \end{bmatrix} \begin{bmatrix} 0.39504 \\ 69.60320 \end{bmatrix}$$

$$= \begin{bmatrix} 0.000490124 \\ 0.173144000 \end{bmatrix}$$

$$u_2 = \bar{v}_2 = 0.173144$$

$$u_1 = \bar{v}_1 = \bar{v}_1^{(2)} - L_{12}u_2$$

$$= 0.000490124 - (-0.002481)(0.173144)$$

$$= 0.000919695$$

$${}^{0.2}u = \begin{bmatrix} 0.000919695 \\ 0.173144000 \end{bmatrix}$$

6.8.2 Modal Analysis

We use here the same example as the one used in Section 6.8.1, and we apply modal analysis. The equation of motion is

$$\begin{bmatrix} 2 & 0 \\ 0 & 1 \end{bmatrix} \begin{bmatrix} \ddot{u}_1 \\ \ddot{u}_2 \end{bmatrix} + \begin{bmatrix} 6 & -2 \\ -2 & 2 \end{bmatrix} \begin{bmatrix} u_1 \\ u_2 \end{bmatrix} = \begin{bmatrix} 0 \\ 10 \end{bmatrix} \qquad (6.162)$$

The equations of motion are coupled due to the presence of the off-diagonal terms in $[K]$.

We assume temporarily that $[R] = 0$, and we are solving for the free frequencies of vibration by using the following equation:

$$\left| \begin{bmatrix} 6 & -2 \\ -2 & 2 \end{bmatrix} - \omega^2 \begin{bmatrix} 2 & 0 \\ 0 & 1 \end{bmatrix} \right| = 0 \qquad (6.163)$$

or

$$\left| \left[\begin{matrix} (6 - 2\omega^2) & -2 \\ -2 & (2 - \omega^2) \end{matrix} \right] \right| = 0 \tag{6.164}$$

Equation (6.164) yields

$$\omega_1 = 1 \tag{6.165}$$

$$\omega_2 = 2 \tag{6.166}$$

For $\omega_1 = 1$, we have

$$\left[\left[\begin{matrix} 6 & -2 \\ -2 & 2 \end{matrix} \right] - (1)^2 \left[\begin{matrix} 2 & 0 \\ 0 & 1 \end{matrix} \right] \right] \left[\begin{matrix} u_1 \\ u_2 \end{matrix} \right] = 0$$

or

$$\left[\begin{matrix} 4u_1 & -2u_2 \\ u_2 & -2u_1 \end{matrix} \right] = 0 \tag{6.167}$$

If we assume that $u_1 = 0.5$, Eq. (6.145) yields $u_2 = 1$. Hence, the shape Φ_1 of the first mode is

$$\Phi_1 = \left[\begin{matrix} 0.5 \\ 1.0 \end{matrix} \right] \tag{6.168}$$

For $\omega_2 = 2$, we have

$$\left[\left[\begin{matrix} 6 & -2 \\ -2 & 2 \end{matrix} \right] - (2)^2 \left[\begin{matrix} 2 & 0 \\ 0 & 1 \end{matrix} \right] \right] \left[\begin{matrix} u_1 \\ u_2 \end{matrix} \right] = 0$$

or

$$\left[\begin{matrix} -2u_1 & -2u_2 \\ -2u_1 & -2u_2 \end{matrix} \right] = 0 \tag{6.169}$$

If we assume that $u_1 = 1$, then Eq. (6.169) yields $u_2 = -1$, and the shape Φ_2 of the second mode is

$$\Phi_2 = \left[\begin{matrix} 1 \\ -1 \end{matrix} \right] \tag{6.170}$$

Since only the relative size of the components of Φ_1 and Φ_2 is important, we can mass orthonormalize the eigenvectors as follows:

$$\Phi_1 = \begin{bmatrix} 0.408248 \\ 0.816496 \end{bmatrix} \tag{6.171}$$

$$\Phi_2 = \begin{bmatrix} 0.577350 \\ -0.577350 \end{bmatrix} \tag{6.172}$$

On this basis, we can write the following:

$$\underbrace{\begin{bmatrix} 0.408248 & 0.816496 \\ 0.577350 & -0.577350 \end{bmatrix}}_{\Phi^T} \cdot \underbrace{\begin{bmatrix} 2 & 0 \\ 0 & 1 \end{bmatrix}}_{M} \cdot \underbrace{\begin{bmatrix} 0.408248 & 0.816496 \\ 0.577350 & -0.577350 \end{bmatrix}}_{\Phi} = \begin{bmatrix} 1 & 0 \\ 0 & 1 \end{bmatrix} \tag{6.173}$$

$$\underbrace{\begin{bmatrix} 0.408248 & 0.816496 \\ 0.577350 & -0.577350 \end{bmatrix}}_{\Phi^T} \cdot \underbrace{\begin{bmatrix} 6 & -2 \\ -2 & 2 \end{bmatrix}}_{K} \cdot \underbrace{\begin{bmatrix} 0.408248 & 0.816496 \\ 0.577350 & -0.577350 \end{bmatrix}}_{\Phi} = \underbrace{\begin{bmatrix} 1 & 0 \\ 0 & 4 \end{bmatrix}}_{\omega^2} \tag{6.174}$$

$$\Phi^T R = \begin{bmatrix} 8.16496 \\ -5.77349 \end{bmatrix} \tag{6.175}$$

The new differential equation can be written as follows:

$$\underbrace{\Phi^T M \Phi \ddot{x}}_{I} + \underbrace{\Phi^T K \Phi x}_{\omega^2} = \Phi^T R$$

or

$$\ddot{x} + \omega^2 x = \Phi^T R \tag{6.176}$$

By using Eqs. (6.174) and (6.175), Eq. (6.176) yields

$$\ddot{x} + \begin{bmatrix} 1 & 0 \\ 0 & 4 \end{bmatrix} x = \begin{bmatrix} 8.16496 \\ -5.77349 \end{bmatrix} \tag{6.177}$$

Thus, for $x = x_1$ and $x = x_2$, Eq. (6.177) yields the following two differential equations:

$$\ddot{x}_1 + x_1 = 8.16496 \tag{6.178}$$

$$\ddot{x}_2 + 4x_2 = -5.77349 \tag{6.179}$$

The values of x_1 and x_2 may be obtained by either using an exact solution of Eqs. (6.178) and (6.179) or by using numerical integration, such as the Newmark–Beta method. Once $[x]$ has been determined, the actual displacements can be obtained from the following transformation:

$$[u] = [\Phi][x] \tag{6.180}$$

6.8.3 Exact Solution

Consider Eq. (6.178). The exact solution of this equation is of the following form:

$$x_1 = C_1 \cos t + C_2 \sin t + 8.16496 \tag{6.181}$$

By using Eq. (6.181) and applying the initial conditions $x_1 = \dot{x}_1 = 0$ at $t = 0$, we find that $C_1 = -8.16496$ and $C_2 = 0$. By substituting into Eq. (6.181), the exact solution is

$$x_1 = -8.16496 \cos t + 8.16496 \tag{6.182}$$

We consider now Eq. (6.179), and we write its general solution x_2 as

$$x_2 = A_1 \cos(2t) + A_2 \sin(2t) - 1.44337 \tag{6.183}$$

By applying the initial conditions $x_2 = \dot{x}_2 = 0$ at $t = 0$, we find $A_1 = 1.44337$ and $A_2 = 0$. Therefore, the solution x_2 is

$$x_2 = 1.44337 \cos(2t) - 1.44337 \tag{6.184}$$

Thus,

$$[x] = \begin{bmatrix} -8.16496 \cos t \\ 1.44337 \cos(2t) - 1.44337 \end{bmatrix} \tag{6.185}$$

Table 6.1 shows the variations of x_1, x_2, u_1, and u_2 with respect to time t, which are the computer results obtained from the exact solution of the differential equation.

TABLE 6.1. Results Obtained from the Exact Solution of the Differential Equation

t	x_1	x_2	u_1	u_2
0.2	0.162756E + 00	−0.113947E + 00	0.657104E − 03	0.198677E + 00
0.4	0.644534E + 00	−0.437771E + 00	0.103823E − 01	0.779007E + 00
0.6	0.142613E + 01	−0.920357E + 00	0.508455E − 01	0.169580E + 01
0.8	0.247638E + 01	−0.148552E + 01	0.153314E + 00	0.287961E + 01
1.0	0.375341E + 01	−0.204402E + 01	0.352209E + 00	0.424476E + 01
1.2	0.520632E + 01	−0.250769E + 01	0.667654E + 00	0.569876E + 01
1.4	0.677719E + 01	−0.280334E + 01	0.114827E + 01	0.715205E + 01
1.6	0.840337E + 01	−0.288427E + 01	0.176543E + 01	0.852655E + 01
1.8	0.100201E + 02	−0.273772E + 01	0.251005E + 01	0.976196E + 01
2.0	0.115628E + 02	−0.238681E + 01	0.334246E + 01	0.108190E + 02
2.2	0.129700E + 02	−0.188696E + 01	0.420556E + 01	0.116794E + 02
2.4	0.141858E + 02	−0.131708E + 01	0.503089E + 01	0.123430E + 02
2.6	0.151614E + 02	−0.767132E + 00	0.574672E + 01	0.128221E + 02
2.8	0.158582E + 02	−0.323949E + 00	0.628703E + 01	0.131352E + 02
3.0	0.162482E + 02	−0.574986E − 01	0.660010E + 01	0.132998E + 02
3.2	0.163160E + 02	−0.984665E − 02	0.665529E + 01	0.133276E + 02
3.4	0.160588E + 02	−0.188517E + 00	0.644714E + 01	0.132208E + 02
3.6	0.154870E + 02	−0.565300E + 00	0.599614E + 01	0.129714E + 02
3.8	0.146232E + 02	−0.108071E + 01	0.534593E + 01	0.125637E + 02
4.0	0.135019E + 02	−0.165338E + 01	0.455756E + 01	0.119789E + 02

6.8.4 Newmark–Beta Method

If the Newmark–Beta method is used, the equation of motion is written

$$\ddot{x} + \begin{bmatrix} 1 & 0 \\ 0 & 4 \end{bmatrix} x = \begin{bmatrix} 8.16496 \\ -5.77349 \end{bmatrix} \tag{6.186}$$

where

$$[\tilde{M}] = [I] = \begin{bmatrix} 1 & 0 \\ 0 & 1 \end{bmatrix} \tag{6.187}$$

$$[\tilde{K}] = \begin{bmatrix} 1 & 0 \\ 0 & 4 \end{bmatrix} \tag{6.188}$$

$$[\tilde{R}] = \begin{bmatrix} 8.16496 \\ -5.77349 \end{bmatrix} \tag{6.189}$$

The initial conditions are

$$^0x = \Phi^T M \, ^0u = \begin{bmatrix} 0 \\ 0 \end{bmatrix} \tag{6.190}$$

$$^0\dot{x} = \Phi^T M \, ^0\dot{u} = \begin{bmatrix} 0 \\ 0 \end{bmatrix} \tag{6.191}$$

Thus, by using Eq. (6.186), we have

$$^0\ddot{x} + \begin{bmatrix} 1 & 0 \\ 0 & 4 \end{bmatrix} \begin{bmatrix} 0 \\ 0 \end{bmatrix} = \begin{bmatrix} 8.16496 \\ -5.77349 \end{bmatrix}$$

or

$$^0\ddot{x} = \begin{bmatrix} 8.16496 \\ -5.77349 \end{bmatrix} \tag{6.192}$$

We let $\alpha = 0.25$, $\delta = 0.5$, and $\Delta t = 0.1$. The integration constants can be determined as discussed earlier, and they are as follows:

$$a_0 = 400$$
$$a_1 = 20$$
$$a_2 = 40$$
$$a_3 = 1$$
$$a_4 = 1$$
$$a_5 = 0$$
$$a_6 = 0.05$$
$$a_7 = 0.05$$

We form $^0\hat{K}$ as follows:

$$^0\hat{K} = \tilde{K} + a_0 \tilde{M} + a_1 \tilde{c}$$
$$= \begin{bmatrix} 1 & 0 \\ 0 & 4 \end{bmatrix} + 400 \begin{bmatrix} 1 & 0 \\ 0 & 1 \end{bmatrix} + 0 \tag{6.193}$$
$$= \begin{bmatrix} 401 & 0 \\ 0 & 404 \end{bmatrix}$$

We calculate the effective loads $^{0.1}\hat{R}$ from the following equation:

$$^{0.1}\hat{R} = {}^{0.1}\tilde{R} + \tilde{M}(a_0\,{}^0x + a_2\,{}^0\dot{x} + a_3\,{}^0\ddot{x}) + c(a_1\,{}^0x + a_4\,{}^0\dot{x} + a_5\,{}^0\ddot{x})$$

$$= \begin{bmatrix} 8.16496 \\ -5.77349 \end{bmatrix} + \begin{bmatrix} 1 & 0 \\ 0 & 1 \end{bmatrix}\left(400\begin{bmatrix} 0 \\ 0 \end{bmatrix} + 40\begin{bmatrix} 0 \\ 0 \end{bmatrix} + \begin{bmatrix} 8.16496 \\ -5.77349 \end{bmatrix}\right) + 0$$

$$= \begin{bmatrix} 16.32992 \\ -11.54698 \end{bmatrix} \tag{6.194}$$

We determine the displacements x from the equation

$$\hat{K}^{t+\Delta t}x = {}^{t+\Delta t}\hat{R} \tag{6.195}$$

or

$$\begin{bmatrix} 401 & 0 \\ 0 & 404 \end{bmatrix}\begin{bmatrix} {}^{0.1}x_1 \\ {}^{0.1}x_2 \end{bmatrix} = \begin{bmatrix} 16.32992 \\ -11.54698 \end{bmatrix} \tag{6.196}$$

Equation (6.196) yields

$$^{0.1}x = \begin{bmatrix} 0.0407230 \\ -0.0285816 \end{bmatrix} \tag{6.197}$$

We determine now the actual displacements u by using the equation $u = \Phi x$; that is,

$$^{0.1}u = \begin{bmatrix} 0.408248 & 0.577350 \\ 0.816496 & -0.577350 \end{bmatrix}\begin{bmatrix} 0.0407230 \\ -0.0285816 \end{bmatrix}$$

$$= \begin{bmatrix} 0.000123497 \\ 0.049751800 \end{bmatrix} \tag{6.198}$$

We update acceleration and velocity as follows:

$$^{0.1}\ddot{x} = a_0({}^{0.1}x - {}^0x) - a_2\,{}^0\dot{x} - {}^0\ddot{x}$$

$$= 400\left(\begin{bmatrix} 0.0407230 \\ -0.0285816 \end{bmatrix} - \begin{bmatrix} 0 \\ 0 \end{bmatrix}\right) - 40\begin{bmatrix} 0 \\ 0 \end{bmatrix} - \begin{bmatrix} 8.16496 \\ -5.77349 \end{bmatrix} \tag{6.199}$$

$$= \begin{bmatrix} 8.12424 \\ -5.65915 \end{bmatrix}$$

$$^{0.1}\dot{x} = {}^{0}\dot{x} + a_6\,{}^{0}\ddot{x} + a_7\,{}^{0.1}\ddot{x}$$

$$= \begin{bmatrix} 0 \\ 0 \end{bmatrix} + 0.05 \begin{bmatrix} 8.16496 \\ -5.77349 \end{bmatrix} + 0.05 \begin{bmatrix} 8.12424 \\ -5.65915 \end{bmatrix} \qquad (6.200)$$

$$= \begin{bmatrix} 0.814460 \\ -0.571632 \end{bmatrix}$$

We calculate $^{t+\Delta t}\hat{R}$ from the equation

$$^{0.2}\hat{R} = {}^{0.2}\tilde{R} + \tilde{M}(a_0\,{}^{0.1}x + a_2\,{}^{0.1}\dot{x} + a_3\,{}^{0.1}\ddot{x}) + c(a_1\,{}^{0.2}x + a_4\,{}^{0.1}\dot{x} + a_5\,{}^{0.1}\ddot{x})$$

$$= \begin{bmatrix} 8.16496 \\ -5.77349 \end{bmatrix} + \begin{bmatrix} 1 & 0 \\ 0 & 1 \end{bmatrix}\left(400\begin{bmatrix} 0.407230 \\ -0.0285816 \end{bmatrix} + 40\begin{bmatrix} 0.814460 \\ -0.571632 \end{bmatrix}\right.$$

$$\left. + \begin{bmatrix} 8.12424 \\ -5.65915 \end{bmatrix}\right) + 0 \qquad (6.201)$$

$$= \begin{bmatrix} 65.1568 \\ -45.73056 \end{bmatrix}$$

We solve now for the displacement x as follows:

$$\hat{K}\,{}^{0.2}x = {}^{0.2}\hat{R}$$

or

$$\begin{bmatrix} 401 & 0 \\ 0 & 404 \end{bmatrix}\begin{bmatrix} {}^{0.2}x_1 \\ {}^{0.2}x_2 \end{bmatrix} = \begin{bmatrix} 65.15680 \\ -45.73056 \end{bmatrix} \qquad (6.202)$$

Equation (6.202) yields

$$^{0.2}x = \begin{bmatrix} 0.162486 \\ -0.113194 \end{bmatrix} \qquad (6.203)$$

Thus, the displacements u are

$$^{0.2}u = \begin{bmatrix} 0.408248 & 0.57735 \\ 0.816496 & -0.57735 \end{bmatrix}\begin{bmatrix} 0.162486 \\ -0.113194 \end{bmatrix}$$

$$= \begin{bmatrix} 0.000981678 \\ 0.198022000 \end{bmatrix} \qquad (6.204)$$

TABLE 6.2. Computer Results for u_1 and u_2 by Using the Newmark–Beta Method

t	u_1	u_2
0.2000	0.981678E − 03	0.198022E + 00
0.4000	0.113766E − 01	0.776557E + 00
0.6000	0.526393E − 01	0.169083E + 01
0.8000	0.155523E + 00	0.287202E + 01
1.0000	0.353969E + 00	0.423507E + 01
1.2000	0.667749E + 00	0.568807E + 01
1.4000	0.114537E + 01	0.714179E + 01
1.6000	0.175843E + 01	0.851816E + 01
1.8000	0.249835E + 01	0.975659E + 01
2.0000	0.332623E + 01	0.108172E + 02
2.2000	0.418587E + 01	0.116811E + 02
2.4000	0.500966E + 01	0.123472E + 02
2.6000	0.572649E + 01	0.128274E + 02
2.8000	0.627062E + 01	0.131398E + 02
3.0000	0.659011E + 01	0.133022E + 02
3.2000	0.665363E + 01	0.133268E + 02
3.4000	0.645466E + 01	0.132165E + 02
3.6000	0.601234E + 01	0.129646E + 02
3.8000	0.536898E + 01	0.125563E + 02
4.0000	0.458448E + 01	0.119737E + 02

The computer results for u_1 and u_2 for various times t by using this methodology are given in Table 6.2.

6.9   CAUTIONS ON UTILIZATION OF COMPUTER SOFTWARE

Modern computer software is an invaluable tool for the solution of dynamics problems, yet there are also inherent and possible catastrophic risks. Before using the software, one should ask oneself the following, in order to reduce the risk of a "computer-aided failure."

1. Do you have sufficient knowledge of structural dynamics and vibration to understand the problem? Having computer software does not make you an expert. You must be sure that you understand the problem and the software and are able to determine whether the results are reasonable.
2. Can the problem be solved rigorously? In many cases, a complex structure cannot be solved rigorously, but the structure can be idealized into a simpler structure and solved rigorously. If one can bound the

solution in this manner, then one can ascertain the reasonableness of the computer solution of the actual structure.

3. Does the computer program accurately predict results from physically testing a model? If not, why? Do not neglect testing wherever possible.

4. Does increasing refinement of the finite element mesh indicate proper convergence?

5. Does the finite element model pass the patch test and possess rigid-body capability? Plot the mode shapes and look at them.

6. Do the program results ensure that the structure will not fail? Remember, the software will only solve for what it has been programmed to do. It will not look for other modes of failure that you have not envisioned. The quantity of data generated may give you a false sense of security that you really understand the structure. This may not be the case. Plot the mode shapes and deflections to provide yourself with a more physical interpretation of the behavior of the structure or system.

7. Do you really understand the software? What are its limitations? Once again, can you obtain reasonable solutions for various known problems solved rigorously?

In addition, there are some general guidelines for meshing the model.

1. Start with a primitive mesh, progressively refine it, and then compare results.

2. Number nodal points in such a way as to minimize the bandwidth of the global matrices. For any element, minimizing the difference between node numbers minimizes the bandwidth of the matrices.

3. Avoid excessive skewness of elements.

PROBLEMS

6.1 Using Eq. (j) in Section 6.2.4, determine the 3,5 term of $[K_e]$ for the 16-degree-of-freedom plate element.

6.2 Using Eq. (6.38) and the Hermitian polynomials, derive the consistent mass matrix.

6.3 Prove that the consistent mass matrix for a beam element with $L = 120$ in. and $m = 0.5$ lb $\cdot$ s^2/in. is positive definite.

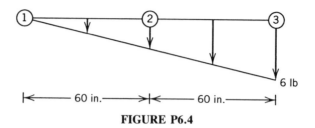

FIGURE P6.4

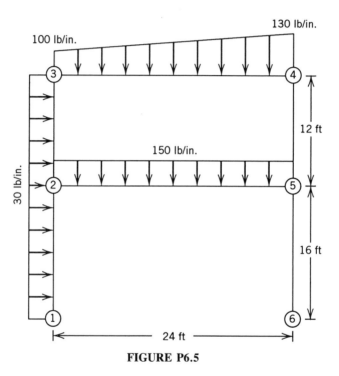

FIGURE P6.5

6.4 Determine the generalized load vector for the Bernoulli beam in Fig. P6.4.

6.5 Determine the effective loads for the frame in Fig. P6.5.

6.6 Assemble the active stiffness matrix for the beam in Fig. P6.6. The cross-sectional area $A = 94.0$ in.^{2}, $E = 30 \times 10^6$ psi, $I = 4140$ in.^{4}, and $m = 0.83$ lb $\cdot$ s^2/in.

6.7 Assemble the global stiffness matrix for the two-element model in Fig. P6.7, and test it for complete rigid-body capability. The cross-sectional area $A = 48.0$ in.^{2}, $E = 30 \times 10^6$ psi, $I = 1000$ in.^{4}, $m = 0.03525$ lb $\cdot$ s^2/in., and $P = 1.2 \times 10^8$ lb.

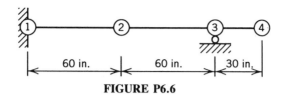

FIGURE P6.6

FIGURE P6.7

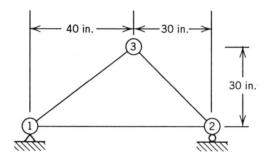

FIGURE P6.8

6.8 For the truss in Fig. P6.8, determine the active global stiffness matrix. The cross-sectional area $A = 48.0$ in.^{2}, $E = 30 \times 10^6$ psi, $I = 1000$ in.^{4}, and $m = 0.05$ lb · s^2/in.

6.9 Repeat Problem 6.8 by assuming that all joints are fixed, and determine the active global stiffness matrix for the rigid frame. Use Bernoulli beam elements.

6.10 Perform a Cholesky decomposition on the following matrix:

$$\begin{bmatrix} 156 & 1320 & 54 & -780 \\ 1320 & 14400 & 780 & -10800 \\ 54 & 780 & 156 & -1320 \\ -780 & -10800 & -1320 & 14400 \end{bmatrix}$$

6.11 Solve for free frequencies of vibration and the corresponding mode

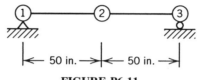

FIGURE P6.11

shapes of the two-element beam model in Fig. P6.11. The cross-sectional area $A = 48.0$ in.^{2}, $E = 30 \times 10^6$ psi, $I = 1000$ in.^{4}, and $m = 0.05$ lb · s^2/in. Neglect shear effects.

6.12 Repeat Problem 6.11 by assuming that the shape factor $K' = 0.666$.

6.13 Solve for the free frequencies of vibration and the corresponding mode shapes of Problem 6.7.

6.14 Solve Problem 6.11 by using the four-element model shown in Fig. P6.14, and compare the results with the ones obtained in Problem 6.11.

6.15 Solve Problem 6.11 by replacing the pin and roller supports by linear springs, as shown in Fig. P6.15.

6.16 Solve for the first three free frequencies of vibration and the corre-

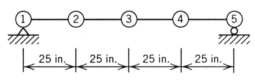

FIGURE P6.14

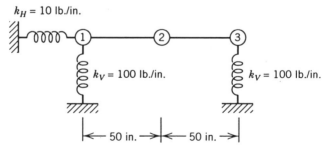

FIGURE P6.15

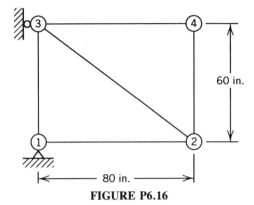

FIGURE P6.16

sponding mode shapes of the truss shown in Fig. P6.16. The cross-sectional area $A = 10.0$ in.^{2}, $E = 30 \times 10^6$ psi, $I = 280$ in.^{4}, and $m = 0.10$ lb · s^2/in.

6.17 Modify the global stiffness matrix of Problem 6.7 to include directed force correction factors, and solve for the free frequencies of vibration and the corresponding mode shapes. Compare the results with the ones obtained in Problem 6.13.

6.18 For the stiffness and mass matrices given below, determine the free frequencies of vibration and the corresponding mode shapes using the Jacobi method.

$$[K] = \begin{bmatrix} 6 & -2 & 0 \\ -2 & 10 & -2 \\ 0 & -2 & 4 \end{bmatrix} \qquad [M] = \begin{bmatrix} 2 & 0 & 0 \\ 0 & 3 & 0 \\ 0 & 0 & 1 \end{bmatrix}$$

6.19 By using the results of Problem 6.18, mass orthonormalize the eigenvectors, and show that

$$\phi^T m \phi = I$$
$$\phi^T K \phi = [\omega^2]$$

6.20 Determine the free frequencies of vibration and the corresponding mode shapes of the beam with a directed force P as shown in Fig. P6.20. The cross-sectional area $A = 40.0$ in.^{2}, $E = 30 \times 10^6$ psi, $I = 568$ in.^{4}, $m = 0.35$ lb · s^2/in., and $P = 100$ kips.

6.21 Perform a LDL^T decomposition on the matrix of Problem 6.10.

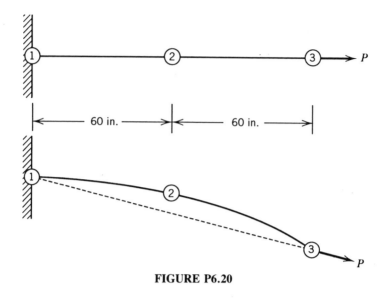

FIGURE P6.20

6.22 Consider the simple system represented by the following equation of motion:

$$\begin{bmatrix} 4 & 0 & 0 \\ 0 & 6 & 0 \\ 0 & 0 & 2 \end{bmatrix} \begin{bmatrix} \ddot{u}_1 \\ \ddot{u}_2 \\ \ddot{u}_3 \end{bmatrix} + \begin{bmatrix} 12 & -4 & 0 \\ -4 & 20 & -4 \\ 0 & -4 & 8 \end{bmatrix} \begin{bmatrix} u_1 \\ u_2 \\ u_3 \end{bmatrix} = \begin{bmatrix} 0 \\ 0 \\ 10 \end{bmatrix}$$

where $^0u = {}^0\dot{u} = 0$. Determine its dynamic response at times $t = 0.2$, 0.4, and 0.6 s by using the Newmark–Beta method with $\delta = 0.5$, $\alpha = 0.25$, and time step $\Delta t = 0.2$ s.

6.23 The solution of the free vibration equation for the system in Problem 6.22 yields the following free frequencies and mass orthonormalized mode shapes:

$$\omega_1 = \sqrt{2} \qquad \phi_1 = \begin{bmatrix} 0.2887 \\ 0.2887 \\ 0.2887 \end{bmatrix}$$

$$\omega_2 = \sqrt{\frac{10}{3}} \qquad \phi_2 = \begin{bmatrix} 0.3873 \\ -0.1291 \\ -0.3873 \end{bmatrix}$$

$$\omega_3 = \sqrt{5} \qquad \phi_3 = \begin{bmatrix} 0.1291 \\ -0.2582 \\ 0.5164 \end{bmatrix}$$

Determine the dynamic response at times $t = 0.2$, 0.4, and $0.6\,\text{s}$ by using modal analysis and direct integration.

6.24 Repeat Problem 6.23 by using modal analysis and the Newmark–Beta method, and compare the results.

7 Utilization of Modal Analysis for Dynamic Response Due to External Excitations

7.1 INTRODUCTION

In this chapter, the method of modal analysis, also known as the modal superposition method, will be discussed rather extensively, because it provides a convenient way to determine the dynamic stresses and displacements of complicated structural and mechanical systems subjected to external dynamic excitations. The modal equations required for the application of this method are derived by using Lagrange's equation, and, consequently, the early part of this chapter deals with the derivation of Lagrange's equation. Since the derivation of Lagrange's equation makes use of energy relations for the systems, as well as generalized coordinates, the modal analysis method may also be thought of as an energy method.

Both modal analysis and Lagrange's equation are extensively used by practicing engineers for the design of complicated structural and mechanical systems that are acted on by various types of external dynamic loads, such as blast overpressures, earthquake acceleration, and gusty wind forces. Theoretically, modal analysis is supposed to be used for linearly elastic systems and for cases where the dynamic forces acting on a structure or on a mechanical system have the same time variation. If, however, these restrictions cannot be removed, then numerical methods such as the acceleration impulse extrapolation method may be used to solve the modal equations.

In this chapter, the modal equations for spring–mass systems and for systems with continuous mass and elasticity are derived, and they are used for the solution of various types of mechanical and structural problems. The modal equations for earthquake structural response are also derived, and they are applied for the solution of structural problems.

7.2 DERIVATION OF LAGRANGE'S EQUATION

Lagrange's equation, which was developed by Lagrange and published in his book *Mechanics Analytique* in 1788, gained wide popularity in the solution of many classical and modern dynamics problems. The main purpose regarding the use of this equation is to determine the differential equation (or equations) that needs to be solved when a structural or mechanical system is subjected to external dynamic excitations or to various types of vibratory motions. Its derivation uses energy expressions for the system under consideration and generalized coordinates. It may be used for systems with any number of degrees of freedom, by including systems with continuous mass and elasticity.

The number of generalized coordinates that define the system under consideration is equal to the number of degrees of freedom of the system. For example, the position of mass m of the vibrating pendulum in Fig. 7.1a of length ℓ is completely defined by the angle ϕ, and, consequently, ϕ may be taken as the generalized coordinate. Therefore, this system has one degree of freedom and one generalized coordinate. The positions of masses m_1 and m_2 in Fig. 7.1b are completely defined by the angles ϕ and ψ, and, thus, the two-degree system is completely defined by the generalized coordinates ϕ and ψ. For the one-degree spring–mass system in Fig. 7.1c, the position of mass m, moving in the vertical direction only, is completely defined by the vertical displacement y, and, consequently, y may be considered as the generalized coordinate.

We begin the derivation of Lagrange's equation by considering the beam in Fig. 7.2a, which supports the weights $W_1, W_2, \ldots, W_j, \ldots, W_n$ at points $1, 2, \ldots, j, \ldots, n$, respectively. The member is also subjected to the concentrated dynamic forces $F_1, F_2, \ldots, F_k, \ldots, F_m$. The beam is assumed to be weightless between the weight concentration points, but it retains its elastic properties throughout its length. On this basis, since we have n concentrated weights on the beam, the member has n degrees of freedom, and, consequently, we need n generalized coordinates to define the position of the weights during motion. Let us say that the required n generalized coordinates are designated as $q_1, q_2, \ldots, q_j, \ldots, q_n$, and they are located at the positions shown in Fig. 7.2b. Since these locations were chosen arbitrarily, other positions could have been selected, because the only requirement is that they should be equal in number to the number of degrees of freedom of the member. In Fig. 7.2b, the deflections $y_1, y_2, \ldots, y_j, \ldots, y_n$, under the weight concentration points $1, 2, \ldots, j, \ldots, n$, respectively, are also shown.

The deflections $y_1, y_2, \ldots, y_n$ are functions of the generalized displacements $q_1, q_2, \ldots, q_n$, and they can be expressed as follows:

$$y_1 = \lambda_1(q_1, q_2, \ldots, q_n) \qquad (7.1)$$

$$y_2 = \lambda_2(q_1, q_2, \ldots, q_n) \qquad (7.2)$$

$$\vdots$$

$$y_n = \lambda_n(q_1, q_2, \ldots, q_n) \qquad (7.3)$$

By considering the generalized displacement q_i and using a virtual change

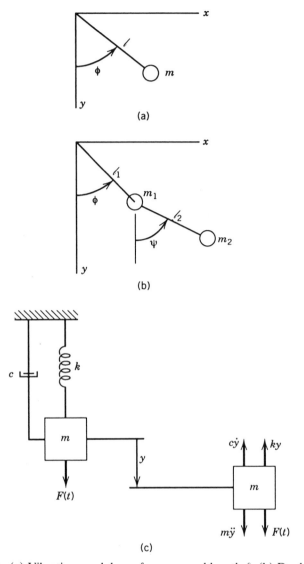

FIGURE 7.1. (a) Vibrating pendulum of mass m and length ℓ. (b) Double pendulum with masses m_1 and m_2 (c) One-degree spring–mass system moving in the vertical direction only.

δq_i, the work δW performed by the external forces moving through the virtual displacement δq_i is equal to the change δU in the internal strain energy of the member. Mathematically, we write

$$\delta U = \delta W \qquad (7.4)$$

or

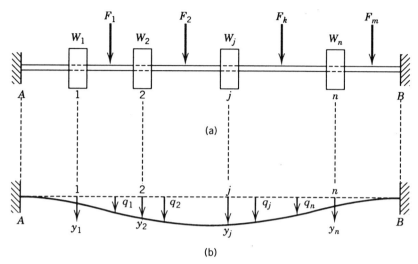

FIGURE 7.2. (a) Single-span beam supporting the weights $W_1, W_2, \ldots, W_n$ and subjected to the dynamic forces $F_1, F_2, \ldots, F_m$. (b) Deflection configuration of the member, showing the displacements $y_1, y_2, \ldots, y_n$ and the generalized coordinates $q_1, q_2, \ldots, q_n$.

$$\frac{\partial U}{\partial q_i} \delta q_i = \frac{\partial W}{\partial q_i} \delta q_i \qquad (7.5)$$

The forces that produce work can be classified in a somewhat general manner as consisting of the external dynamic forces $F_1, F_2, \ldots, F_m$, the inertia forces $W_1 \ddot{y}_1/g$, $W_2 \ddot{y}_2/g, \ldots, W_n \ddot{y}_n/g$, and the forces produced by damping. We now define δW_e, δW_{in}, and δW_d as the virtual work produced by the external forces, the forces of inertia, and the damping forces, respectively. On this basis, we have

$$\frac{\partial W}{\partial q_i} \delta q_i = \frac{\partial W_e}{\partial q_i} \delta q_i + \frac{\partial W_d}{\partial q_i} \delta q_i - \sum_{j=1}^{n} (m_j \ddot{y}_j) \frac{\partial y_j}{\partial q_i} \delta q_i \qquad (7.6)$$

In Eq. (7.6), the symbol y_j denotes the total displacement y_j of mass $m_j = W_j/g$, where g is the acceleration of gravity.

On this basis, by substitution, Eq. (7.5) yields

$$\frac{\partial U}{\partial q_i} \delta q_i = \frac{\partial W_e}{\partial q_i} \delta q_i + \frac{\partial W_d}{\partial q_i} \delta q_i - \sum_{j=1}^{n} (m_j \ddot{y}_j) \frac{\partial y_j}{\partial q_i} \delta q_i \qquad (7.7)$$

By considering the kinetic energy T of the member, we can prove [1, 2] that the following expression applies:

$$- \sum_{j=1}^{n} (m_j \ddot{y}_j) \frac{\partial y_j}{\partial q_i} \delta q_i = - \frac{d}{dt} \frac{\partial T}{\partial \dot{q}_i} \delta q_i + \frac{\partial T}{\partial q_i} \delta q_i \qquad (7.8)$$

By substituting Eq. (7.8) into Eq. (7.7) and simplifying, we obtain the following expression:

$$\frac{d}{dt} \frac{\partial T}{\partial \dot{q}_i} - \frac{\partial T}{\partial q_i} + \frac{\partial U}{\partial q_i} - \frac{\partial W_d}{\partial q_i} = \frac{\partial W_e}{\partial q_i}$$

$$i = 1, 2, \ldots, n \qquad (7.9)$$

Equation (7.9) is known as Lagrange's equation, and the number of such equations is equal to the number of degrees of freedom of the structural or mechanical system under consideration.

The following examples illustrate the application of Lagrange's equation.

Example 7.1. Determine the differential equation of motion for the one-degree spring–mass system in Fig. 7.1c by using Lagrange's equation. The free-body diagram of mass m is shown in the same figure.

Solution. Since the spring–mass system in Fig. 7.1c is moving in the vertical direction only, the degree of freedom is one, and, consequently, the vertical displacement y of mass m is taken as the generalized coordinate. The expressions for the energies T, U, W_d, and W_e are as follows:

$$T = \tfrac{1}{2} m \dot{y}^2 \qquad (7.10)$$

$$U = \tfrac{1}{2} k y^2 \qquad (7.11)$$

$$W_d = (-c\dot{y})y \qquad (7.12)$$

$$W_e = F(t)y \qquad (7.13)$$

By substituting Eqs. (7.10) through (7.13) into Eq. (7.9), with $i = 1$ and $q_i = y$, and simplifying, we find

$$m\ddot{y} + ky + c\dot{y} = F(t) \qquad (7.14)$$

Equation (7.14) is the familiar differential equation of motion for the one-degree spring–mass system in Fig. 7.1c. This is identical to Eq. (1.21) obtained in Section 1.7, and its solution for various types of force functions $F(t)$ is given in Chapter 3.

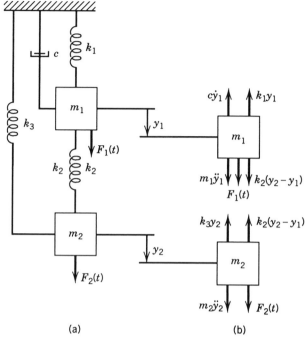

(a)   (b)

FIGURE 7.3. (a) Two-degree spring–mass system moving in the vertical direction only. (b) Free-body diagrams of masses m_1 and m_2.

Example 7.2. Determine the differential equations of motion for the two-degree spring–mass system in Fig. 7.3a by using Lagrange's equation. The free-body diagrams of masses m_1 and m_2 are shown in Fig. 7.3b.

Solution. Since the spring–mass system in Fig. 7.3a is moving in the vertical direction only, it has two degrees of freedom, and the displacements y_1 and y_2 of masses m_1 and m_2, respectively, may be taken as the generalized coordinates; that is, $q_1 = y_1$ and $q_2 = y_2$. The expressions for the energies T, U, W_d, and W_e are as follows:

$$T = \tfrac{1}{2}m_1\dot{y}_1^2 + \tfrac{1}{2}m_2\dot{y}_2^2 \tag{7.15}$$

$$U = \tfrac{1}{2}k_1 y_1^2 + \tfrac{1}{2}k_2(y_2 - y_1)^2 + \tfrac{1}{2}k_3 y_2^2 \tag{7.16}$$

$$W_d = (-c\dot{y}_1)y_1 \tag{7.17}$$

$$W_e = F_1(t)y_1 + F_2(t)y_2 \tag{7.18}$$

We also have

$$\frac{\partial T}{\partial \dot{y}_1} = m_1 \dot{y}_1 \tag{7.19}$$

$$\frac{d}{dt} \frac{\partial T}{\partial \dot{y}_1} = m_1 \ddot{y}_1 \tag{7.20}$$

$$\frac{\partial T}{\partial y_1} = \frac{\partial T}{\partial y_2} = 0 \tag{7.21}$$

$$\frac{\partial U}{\partial y_1} = k_1 y_1 - k_2(y_2 - y_1) \tag{7.22}$$

$$\frac{\partial W_e}{\partial y_1} = F_1(t) \tag{7.23}$$

$$\frac{\partial W_d}{\partial y_1} = - c\dot{y}_1 \tag{7.24}$$

$$\frac{\partial T}{\partial \dot{y}_2} = m_2 \dot{y}_2 \tag{7.25}$$

$$\frac{d}{dt} \frac{\partial T}{\partial \dot{y}_2} = m_2 \ddot{y}_2 \tag{7.26}$$

$$\frac{\partial U}{\partial y_2} = k_2(y_2 - y_1) + k_3 y_2 \tag{7.27}$$

$$\frac{\partial W_e}{\partial y_2} = F_2(t) \tag{7.28}$$

For $i = 1$, we use Eq. (7.9) and Eqs. (7.20) through (7.24) to obtain the first differential equation of motion. The second differential equation of motion is obtained by using Eq. (7.9), for $i = 2$, and Eqs. (7.26) through (7.28). The results are as follows:

$$m_1 \ddot{y}_1 + k_1 y_1 - k_2(y_2 - y_1) + c\dot{y}_1 = F_1(t) \tag{7.29}$$

$$m_2 \ddot{y}_2 + k_3 y_2 + k_2(y_2 - y_1) = F_2(t) \tag{7.30}$$

Equations (7.29) and (7.30) are the two differential equations of motion of the two-degree spring–mass system in Fig. 7.3a.

7.3   DERIVATION OF MODAL EQUATIONS FOR VARIOUS SYSTEMS BY USING LAGRANGE'S EQUATION

In the preceding section, Lagrange's equation has been derived, and it was also applied to determine the differential equations of motion for spring–mass systems. In this section, Lagrange's equation will be used to develop the modal equations for various structural and mechanical systems, which are used to determine the response of the system in terms of both stress and displacement caused by external dynamic excitations. In fact, a modal equation is derived for each mode of vibration, and the response contribution of each mode in terms of stresses and displacements is obtained by using the modal equation. The total response is determined by considering the contribution of all modes of the system. The contribution of the first mode usually predominates very heavily, and most often a numerical addition of the contributions of the first few modes will provide reasonable accuracy for practical applications. This topic will be discussed in greater detail as the work in the chapter progresses.

The modal equations, as they are derived in this section, permit each mode of a multiple or infinite degree-of-freedom system to be represented by an equivalent one-degree-of-freedom system, which can be solved independently by methods discussed in the early chapters of the text.

7.3.1   Modal Equations for Spring–Mass Systems or for Continuous Systems That Can Be Idealized as Spring–Mass Systems

We start the development of modal equations for such spring–mass systems by considering a general case that involves r masses, j springs, and M modes of vibration. We designate as $\dot{a}_{ip}$ the velocity component of mass m_i in the p mode. The total kinetic energy T of the spring–mass system at a time t is

$$T = \sum_{i=1}^{r} \tfrac{1}{2} m_i \left(\sum_{p=1}^{M} \dot{a}_{ip} \right)^2 \qquad (7.31)$$

The total strain energy U stored in the springs of the spring–mass system is

$$U = \sum_{n=1}^{j} \tfrac{1}{2} k_n \left(\sum_{p=1}^{M} \delta_{np} \right)^2 \qquad (7.32)$$

In Eq. (7.32), the quantity δ_{np} is the relative displacement of the ends of spring n in the p mode, and k_n is the constant of the nth spring.

In accordance with the orthogonality properties of normal modes, the cross products in the squared series in Eqs. (7.31) and (7.32) are zero. On this basis, Eqs. (7.31) and (7.32) may be written as follows:

$$T = \sum_{i=1}^{r} \tfrac{1}{2} m_i \sum_{p=1}^{M} \dot{a}_{ip}^2 \tag{7.33}$$

$$U = \sum_{n=1}^{j} \tfrac{1}{2} k_n \sum_{p=1}^{M} \delta_{np}^2 \tag{7.34}$$

The work W_e performed by an external force $F_i(t)$ acting on a mass m_i is as follows:

$$W_e = \sum_{i=1}^{r} F_i(t) \sum_{p=1}^{M} a_{ip} \tag{7.35}$$

If we select arbitrarily a modal displacement Y_p in the p mode, preferably the displacement of one of the masses of the spring–mass system, then all the a_{ip} displacements of the masses may be expressed as a proportion of the modal displacement Y_p. On this basis, we have

$$a_{ip} = Y_p \beta_{ip} \tag{7.36}$$

$$\dot{a}_{ip} = \dot{Y}_p \beta_{ip} \tag{7.37}$$

$$\delta_{np} = Y_p \beta_{\delta np} \tag{7.38}$$

where β_{ip} and $\beta_{\delta np}$ define the characteristic shape of the mode, and they are given by the following expressions:

$$\beta_{ip} = \frac{a_{ip}}{Y_p} = \frac{\dot{a}_{ip}}{\dot{Y}_p} \tag{7.39}$$

$$\beta_{\delta np} = \frac{\delta_{np}}{Y_p} \tag{7.40}$$

With this in mind, Eqs. (7.33), (7.34), and (7.35) may be written as follows:

$$T = \sum_{i=1}^{r} \tfrac{1}{2} m_i \sum_{p=1}^{M} \dot{Y}_p \beta_{ip}^2 \tag{7.41}$$

$$U = \sum_{n=1}^{j} \tfrac{1}{2} k_n \sum_{p=1}^{M} Y_p^2 \beta_{\delta np}^2 \tag{7.42}$$

$$W_e = \sum_{i=1}^{r} F_i(t) \sum_{p=1}^{M} Y_p \beta_{ip} \tag{7.43}$$

We now select the modal displacement Y_p as the generalized coordinate q_i, and we write the following equations:

$$\frac{d}{dt}\left(\frac{\partial T}{\partial \dot{Y}_p}\right) = \ddot{Y}_p \sum_{i=1}^{r} m_i \beta_{ip}^2 \tag{7.44}$$

$$\frac{\partial U}{\partial Y_p} = Y_p \sum_{n=1}^{j} k_n \beta_{\delta np}^2 \tag{7.45}$$

$$\frac{\partial W_e}{\partial Y_p} = \sum_{i=1}^{r} F_i(t) \beta_{ip} \tag{7.46}$$

By using Eq. (7.9), which is Lagrange's equation, and substituting in the equation the values obtained by Eqs. (7.44) through (7.46), the following equation is obtained:

$$m_e \ddot{Y}_p + k_e Y_p = F_e(t) \tag{7.47}$$

where

$$m_e = \sum_{i=1}^{r} m_i \beta_{ip}^2 \tag{7.48}$$

$$k_e = \sum_{n=1}^{j} k_n \beta_{\delta np}^2 \tag{7.49}$$

$$F_e(t) = \sum_{i=1}^{r} F_i(t) \beta_{ip} \tag{7.50}$$

Equation (7.47) is known as the modal equation for the p mode of vibration. It represents an equivalent one-degree-of-freedom spring–mass system, where m_e is the equivalent mass, k_e is the equivalent spring constant, and $F_e(t)$ is the equivalent force. For each system, there are as many equivalent one-degree systems as there are modes of vibration. The kinetic energy, internal strain energy, and work done by all external forces on the equivalent one-degree system are all equal to the same quantities of the complete system while vibrating in a p mode. Thus, each mode of the spring–mass system can be analyzed independently as a one-degree spring–mass system by using the methods discussed in earlier chapters, and the dynamic response contained in each mode can be obtained.

If we divide all the terms of Eq. (7.47) by m_e and use Eqs. (7.48) and (7.50), the resulting expression is as follows:

$$\ddot{Y}_p + \omega_p^2 Y_p = g(t) \frac{\displaystyle\sum_{i=1}^{r} F_i \beta_{ip}}{\displaystyle\sum_{i=1}^{r} m_i \beta_{ip}^2} \tag{7.51}$$

In Eq. (7.51), $g(t)$ is the time function of $F_i(t) = g(t)F_i$, and $\omega_p^2 = k_e/m_e$. The two summations on the right-hand side of Eq. (7.51) can only be computed if the mode amplitudes β_{ip} of the p mode are known. This can be accomplished by solving the initial multidegree spring–mass system by using known methods of vibration analysis. Such methods are discussed in the preceding two chapters. When this is accomplished, the dynamic response of the spring–mass system in terms of stresses and displacements may be determined in each p mode by using Eq. (7.51).

In the development here, we assume that the time function $g(t)$ is the same for all $F_i(t)$ dynamic forces. If this is not true, then the modal equations—Eq. (7.51)—may be solved by using numerical methods, such as the acceleration impulse extrapolation method.

From Eq. (7.47), we note that the modal static displacement Y_{pst} is given by the expression

$$Y_{pst} = \frac{F_e}{k_e} = \frac{F_e}{\omega_p^2 m_e} \tag{7.52}$$

By substituting Eqs. (7.48) and (7.50) into Eq. (7.52), we obtain

$$Y_{pst} = \frac{\sum\limits_{i=1}^{r} F_i \beta_{ip}}{\omega_p^2 \sum\limits_{i=1}^{r} m_i \beta_{ip}^2} \tag{7.53}$$

On this basis, we have

$$Y_p(t) = Y_{pst}\Gamma_p \tag{7.54}$$

$$Y_{pmax} = Y_{pst}\Gamma_{pmax} \tag{7.55}$$

where Γ_p or Γ_{pmax} is the magnification factor in the p mode, and it depends only on the time function $g(t)$ and the frequency ω_p.

The total response may be obtained by superimposing the responses of all modes of the system; that is, the response $y_i(t)$ of mass m_i is

$$y_i(t) = \sum_{p=1}^{M} Y_{pst}\beta_{ip}\Gamma_p \qquad i = 1, 2, 3, \ldots \tag{7.56}$$

In many practical problems, the response of the first mode predominates, and it can be as high as 98% of the total response. In such cases, numerical superposition of the results of the first few modes yields very accurate results and a conservative answer.

If the contributions of the higher modes are appreciable, a more reasonable approach is to take the sum of the fundamental mode response plus the square root of the sum of the squares of the higher modes. If the modal

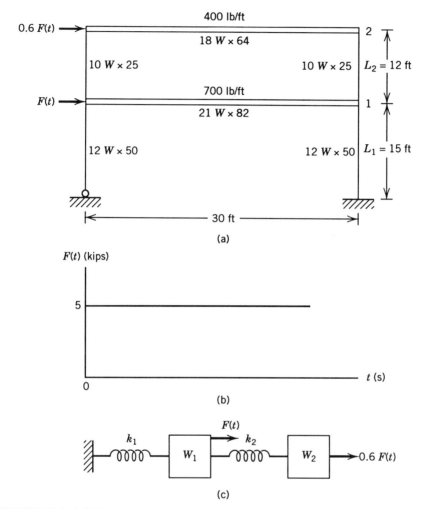

FIGURE 7.4. (a) Two-story frame subjected to horizontal dynamic forces as shown. (b) Time variation of the dynamic forces. (c) Idealized two-degree spring–mass system.

components are random variables, a total maximum response may be obtained by taking the root mean square, that is, the square root of the sum of the squares of the modal maximum responses. This assumption is reasonable for random inputs.

Example 7.3. The two-story frame in Fig. 7.4a is subjected to the horizontal dynamic forces $F(t)$ and $0.6F(t)$ at points 1 and 2, respectively, as shown in the figure. Force $F(t)$ is a suddenly applied force of magnitude 5.0 kips, lasting indefinitely as shown in Fig. 7.4b. The sizes of the columns and girders, the dimensions of the frame, and the weight per unit length at the floor and roof levels are shown in Fig. 7.4a. By idealizing the frame as a two-degree spring–mass system and applying modal analysis, determine the maximum horizontal displacements at points 1 and 2 and the maximum bending stresses in the columns. Assume that the girders are infinitely stiff compared to the stiffness of the columns, and neglect the weight of the columns.

Solution. The procedure discussed in Section 3.11 may be used to determine the idealized two-degree spring–mass system of the frame, which is shown in Fig. 7.4c. On this basis, we have

$$W_1 = (700)(30) = 21,000 \, \text{lb}$$
$$= 21.0 \, \text{kips}$$
$$W_2 = (400)(30) = 12,000 \, \text{lb}$$
$$= 12.0 \, \text{kips}$$
$$k_1 = \frac{3EI}{L_1^3} + \frac{12EI}{L_1^3}$$
$$= \frac{(3)(30 \times 10^3)(394)}{(15 \times 12)^3} + \frac{(12)(30 \times 10^3)(394)}{(15 \times 12)^3} = 6.08 + 24.32$$
$$= 30.4 \, \text{kips/in.}$$
$$k_2 = (2)\frac{12EI}{L_2^3} = (2)\frac{(2)(30 \times 10^3)(133.2)}{(12 \times 12)^3}$$
$$= 32.1 \, \text{kips/in.}$$

Thus,

$$m_1 = \frac{W_1}{g} = \frac{21.0}{386.4} = 0.0543 \, \frac{\text{kip·s}^2}{\text{in.}}$$

$$m_2 = \frac{W_2}{g} = \frac{12.0}{386.4} = 0.0311 \, \frac{\text{kip·s}^2}{\text{in.}}$$

The two free frequencies of vibration and tbe corresponding mode shapes of the idealized two-degree spring–mass system in Fig. 7.4c may be determined by using the procedure discussed in Section 2.8. By following this procedure, we find

$$\omega_1 = 17.56 \text{ rps}$$
$$\omega_2 = 43.30 \text{ rps}$$

The normalized amplitudes of the two mode shapes are as follows:

Mode 1: $x_1 = 1.000$ $x_2 = 1.426$

Mode 2: $x_1 = 1.000$ $x_2 = -1.225$

Now we are ready to use Eq. (7.53) to determine the modal static amplitudes $Y_{1_{st}}$ and $Y_{2_{st}}$ of the first and second modes, respectively. On this basis, we find

$$Y_{1_{st}} = \frac{\sum\limits_{i=1}^{2} F_i \beta_{i1}}{\omega_1^2 \sum\limits_{i=1}^{2} m_i \beta_{i1}^2}$$

$$= \frac{(5)(1.000) + (0.6)(1.426)(5)}{(17.56)^2[(0.0543)(1.000)^2 + (0.0311)(1.426)^2]}$$

$$= 0.256 \text{ in.}$$

$$Y_{2_{st}} = \frac{\sum\limits_{i=1}^{2} F_i \beta_{i2}}{\omega_2^2 \sum\limits_{i=1}^{2} m_i \beta_{i2}^2}$$

$$= \frac{(5)(1.000) + (0.6)(5)(-1.225)}{(43.30)^2[(0.0543)(1.000)^2 + (0.0311)(-1.225)^2]}$$

$$= 0.0070 \text{ in.}$$

From Appendix A, we note that the maximum magnification factor Γ_{max} for the force function in Fig. 7.4b is

$$\Gamma_{max} = 1 - \cos \pi = 1 - (-1) = 2$$

By using Eq. (7.56), the maximum horizontal displacements $x_{1_{max}}$ and $x_{2_{max}}$ at points 1 and 2, respectively, of the frame in Fig. 7.4a are as follows:

$$x_{1_{max}} = \sum_{p=1}^{2} Y_{p_{st}} \beta_{ip} \Gamma_{p_{max}}$$

$$= Y_{1_{st}} \beta_{11} \Gamma_{1_{max}} + Y_{2_{st}} \beta_{12} \Gamma_{2_{max}}$$

$$= (0.256)(1.000)(2.0) + (0.0070)(1.000)(2.0)$$

$$= 0.512 + 0.014$$

$$= 0.526 \text{ in.}$$

$$x_{2\text{max}} = \sum_{p=1}^{2} Y_{p\text{st}}\beta_{ip}\Gamma_{p\text{max}}$$

$$= Y_{1\text{st}}\beta_{21}\Gamma_{1\text{max}} + Y_{2\text{st}}\beta_{22}\Gamma_{2\text{max}}$$

$$= (0.256)(1.426)(2.0) + (0.0070)(-1.225)(2.0)$$

$$= 0.730 + |-0.017|$$

$$= 0.747 \text{ in.}$$

The numerical sums of the response contributions of the first and second modes are used to determine $x_{1\text{max}}$ and $x_{2\text{max}}$. Note that the contributions of the second mode are very small compared to the contributions of the first mode. For $x_{1\text{max}}$, the second mode contributes 0.014 in., and for $x_{2\text{max}}$ the second mode contributes 0.017 in. In practical situations, the response contribution of the second mode could be neglected.

The maximum total shear force $V_{p\text{max}}$ at the first-story level in Fig. 7.4a may be obtained from Eq. (3.167). This value is

$$V_{1\text{max}} = k_1 x_{1\text{max}} = (30.4)(0.526)$$

$$= 15.99 \text{ kips}$$

This shear force will be distributed among the two columns in proportion to their stiffness. The hinged column has a stiffness $k_H = 6.08$ kips/in. and the fixed column has a stiffness $k_F = 24.32$ kips/in. Therefore, the corresponding shear forces V_H and V_F are as follows:

$$V_H = \frac{k_H}{k_1} V_{1\text{max}}$$

$$= \frac{6.08}{30.4}(15.99) = 3.20 \text{ kips}$$

$$V_F = \frac{24.32}{30.4}(15.99) = 12.79 \text{ kips}$$

The bending moments M_H and M_F for the hinged and fixed columns, respectively, can be determined by using the corresponding Eqs. (3.172) and (3.171). On this basis, we have

$$M_H = V_H L_1 = (3.20)(15) = 48.0 \text{ kip·ft}$$

$$= 576 \text{ kip·in.}$$

$$M_F = \frac{V_F L_1}{2} = \frac{(12.79)(15)}{2} = 95.93 \text{ kip·ft}$$

$$= 1151.1 \text{ kip·in.}$$

The section modulus S of the first-story columns is 64.7 in^3. Therefore, the maximum stresses σ_H and σ_F in the hinged and fixed columns, respectively, are as follows:

$$\sigma_H = \frac{M_H}{S} = \frac{576}{64.7} = 8.90 \text{ ksi}$$

$$\sigma_F = \frac{M_F}{S} = \frac{1151.10}{64.7} = 17.79 \text{ ksi}$$

The maximum total shear force $V_{2\text{max}}$, at the roof level of the frame, may be obtained by using Eq. (3.168). This equation yields

$$V_{2\text{max}} = k_2(x_{2\text{max}} - x_{1\text{max}})$$

$$= 32.10(0.747 - 0.526)$$

$$= 7.094 \text{ kips}$$

Since both columns of the second story of the frame have the same stiffness, the shear force $V_{2\text{max}}$ will be divided equally between the two columns. Therefore, each column will be subjected to a shear force $V = 7.094/2 = 3.55$ kips. From Eq. (3.171), the moment M at the top and bottom of each column will be as follows:

$$M = \frac{VL_2}{2} = \frac{(3.55)(12)}{2} = 21.30 \text{ kip·ft}$$

$$= 255.60 \text{ kip·in.}$$

The section modulus S for the second-story columns is 26.4 in.^{3}. Therefore, the maximum bending stress σ_{max} in the column is

$$\sigma_{\text{max}} = \frac{M}{S} = \frac{255.60}{26.4} = 9.68 \text{ ksi}$$

7.3.2 Systems with Continuous Mass and Elasticity

The modal equation for systems with continuous mass and elasticity can also be derived by using Lagrange's equation. The procedure is very similar to the one used for spring–mass systems, except that, in this case, the expressions for the analogous energies will involve integrals and not summations. Consequently, the derived modal equations, in this case, will be very similar in form, but the right-hand side of the equation will involve integrals and not summations. On this basis, if we consider a member of mass m per unit length and loaded with a distributed dynamic load $q(x, t) = q_0(x)g(t)$, the modal equation for a mode p may be written as follows:

$$\ddot{Y}_p + \omega_p^2 Y_p = g(t) \frac{\int_0^L q_0(x)\beta_p(x)\, dx}{m \int_0^L \beta_p^2(x)\, dx} \tag{7.57}$$

The term $\beta_p(x)$ in Eq. (7.57) represents the shape of the p mode of vibration of the member, ω_p is the frequency of the mode, and Y_p is the modal amplitude. Note the similarity of this equation with Eq. (7.51). In the above equation, the mass m of the member is assumed to be uniform throughout its length L.

In this case, the static modal displacement $Y_{p_{st}}$ for a mode p may be determined from the following equation:

$$Y_{p_{st}} = \frac{\int_0^L q(x)\beta_p(x)\, dx}{\omega_p^2 m \int_0^L \beta_p^2(x)\, dx} \tag{7.58}$$

The dynamic modal displacement $Y_p(t)$ of mode p is

$$Y_p(t) = Y_{p_{st}}\Gamma_p \tag{7.59}$$

where Γ_p is the magnification factor for the p mode. The total dynamic displacement $y(t, x)$, at any x and time t, may be determined from the equation

$$y(t, x) = \sum_{p=1}^{\infty} Y_p(t)\beta_p(x) \tag{7.60}$$

Equation (7.60) contains the displacement contributions of all modes of the member. The superposition of these responses may be done in the same way as was done for the spring–mass systems. The same reasoning is involved here.

If the applied dynamic loading consists of concentrated time-varying load functions $F_1(t)$, $F_2(t)$, ..., $F_r(t)$, then the numerator on the right-hand side of Eq. (7.57) becomes a summation that involves one term for each concentrated dynamic load. If the time function $g(t)$ is the same for all concentrated dynamic loads, the modal equation for the p mode may be written as follows:

$$\ddot{Y}_p + \omega_p^2 Y_p = g(t) \frac{\sum\limits_{i=1}^{r} F_i \beta_p(x_i)}{m \int_0^L \beta_p^2(x)\, dx} \tag{7.61}$$

In Eq. (7.61), the term $\beta_p(x_i)$ represents the amplitude of mode p under the force F_i, where $F_i = F_i(t)/g(t)$.

The static modal displacement $Y_{p\mathrm{st}}$ of the p mode may be written as follows:

$$Y_{p\mathrm{st}} = \frac{\sum\limits_{i=1}^{r} F_i \beta_p(x_i)}{\omega_p^2 m \int_0^L \beta_p^2(x)\, dx} \tag{7.62}$$

The dynamic modal displacement of the p mode is

$$Y_p(t) = Y_{p\mathrm{st}} \Gamma_p \tag{7.63}$$

where Γ_p is the magnification factor for the p mode. The total dynamic deflection $y(t, x)$ at any x and time t is as follows:

$$y(t, x) = \sum_{p=1}^{\infty} Y_p(t) \beta_p(x) \tag{7.64}$$

The following examples illustrate the application of the above equation.

Example 7.4. A uniform simply supported beam of length L is loaded by a uniformly distributed dynamic load $q(t, x) = q \sin \omega_f t$ over the whole length of the member, where ω_f is the frequency of the force. Determine the expression for the dynamic displacement $y(t, x)$ of the member by using modal analysis.

Solution. The mode shapes of the uniform simply supported beam are determined in Example 2.14, and they are given by Eq. (2.236). For any p mode, the expression for the mode shape is

$$\beta_p(x) = \sin\frac{p\pi x}{L} \qquad p = 1, 2, 3, \ldots \tag{7.65}$$

On this basis, Eq. (7.58) yields

$$Y_{p\mathrm{st}} = \frac{\int_0^L q \sin\left(\dfrac{p\pi x}{L}\right) dx}{\omega_p^2 m \int_0^L \sin^2\left(\dfrac{p\pi x}{L}\right) dx} \tag{7.66}$$

$$= -\frac{2q}{p\pi\omega_p^2 m}(\cos p\pi - 1)$$

Equation (7.66) shows that when $p = 2, 4, 6, \ldots$, the modal static displacement $Y_{p_{st}}$ is zero. On this basis, the only modes that contribute to the dynamic response are the odd ones. Therefore, we have

$$Y_{p_{st}} = \frac{4q}{p \pi \omega_p^2 m} \qquad p = 1, 3, 5, \ldots \tag{7.67}$$

From Appendix A, the magnification factor Γ_p for any p mode is

$$\Gamma_p = \frac{1}{1 - (\omega_f/\omega_p)^2} \left(\sin \omega_f t - \frac{\omega_f}{\omega_p} \sin \omega_p t \right) \qquad p = 1, 3, 5, \ldots \tag{7.68}$$

In Eq. (7.68), ω_f is the frequency of the force and ω_p is the free frequency of vibration of the p mode of the member. Thus, by Eq. (7.63), we have

$$Y_p(t) = Y_{p_{st}} \Gamma_p$$

$$= \frac{4q}{p \pi \omega_p^2 m [1 - (\omega_f/\omega_p)^2]} \left(\sin \omega_f t - \frac{\omega_f}{\omega_p} \sin \omega_f t \right) \tag{7.69}$$

$$p = 1, 3, 5, \ldots$$

The expression for the total response $y(t, x)$ of the member may be obtained by using Eq. (7.64). This equation yields

$$y(t, x) = \frac{4q}{m \pi} \sum_{p = 1, 3, 5, \ldots} \frac{1}{[1 - (\omega_f/\omega_p)^2] p \omega_p^2} \left(\sin \omega_f t - \frac{\omega_f}{\omega_p} \sin \omega_p t \right) \sin \frac{p \pi x}{L} \tag{7.70}$$

The denominator of Eq. (7.70) indicates that when $\omega_f = \omega_p$, the amplitude $y(t, x)$ becomes infinitely large with time. We also note that ω_p^2 is in the denominator of Eq. (7.70), which indicates that the contribution of the higher modes becomes increasingly smaller. The significance of these contributions depends on how close the magnitudes of ω_p are with respect to each other. If these frequencies are spaced well apart, as they are for this problem, then the contributions of the higher modes are small, and, in many practical situations, they could be neglected.

With known $y(t, x)$, the expressions for the dynamic shear force and bending moment may be obtained by differentiation. See Eqs. (4.175) and (4.176).

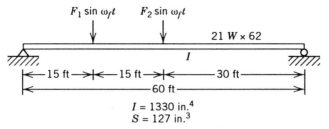

FIGURE 7.5. Simply supported beam loaded with two harmonic forces.

Example 7.5. The uniform simply supported beam in Fig. 7.5 is loaded by two harmonic forces as shown in the same figure. The beam has a 21W × 62 cross-section, with moment of inertia $I = 1330$ in.^{4} and section modulus $S = 127$ in.^{3}. The force frequency $\omega_f = 3\pi$, $F_1 = 10$ kips, and $F_2 = 5$ kips. Determine the expression for the total maximum deflection response of the member produced by the two forces, by considering the contributions of its first three modes. The modulus of elasticity $E = 30 \times 10^6$ psi.

Solution. Since the weight of the member is 62.0 lb/ft, or 5.17 lb/in., the mass m of the beam is

$$m = \frac{5.17}{386.4} = 0.0134 \text{ lb} \cdot \text{s}^2/\text{in.}$$

The free frequencies of vibration ω_p of the simply supported beam may be determined from the following expression:

$$\omega_p = \frac{p^2 \pi^2}{L^2} \sqrt{\frac{EI}{m}} \qquad (7.71)$$

Therefore,

$$\omega_1 = \frac{(1)^2 \pi^2}{(60 \times 12)^2} \sqrt{\frac{(30 \times 10^6)(1330)}{(0.0134)}}$$

$$= 32.85 \text{ rps}$$

$$\omega_2 = 131.40 \text{ rps}$$

$$\omega_3 = 295.70 \text{ rps}$$

The static modal displacement $Y_{p_{st}}$ for a p mode may be determined by using Eq. (7.62). This equation yields

$$Y_{p_{st}} = \frac{2[F_1 \sin(p\pi/4) + F_2 \sin(p\pi/2)]}{\omega_p^2 mL} \tag{7.72}$$

For the first three modes, Eq. (7.72) yields

$$Y_{1_{st}} = \frac{(2)[10\sin(\pi/4) + 5\sin(\pi/2)](10^3)}{(32.85)^2(0.0134)(60)(12)}$$

$$= 2.319 \text{ in.}$$

$$Y_{2_{st}} = \frac{(2)[10\sin(2\pi/4) + 5\sin(2\pi/2)](10^3)}{(131.40)^2(0.0134)(60)(12)}$$

$$= 0.120 \text{ in.}$$

$$Y_{3_{st}} = \frac{(2)[10\sin(3\pi/4) + 5\sin(3\pi/2)](10^3)}{(295.70)^2(0.0134)(60)(12)}$$

$$= 0.00491 \text{ in.}$$

The magification factor Γ_1 for the first mode can be obtained from Eq. (7.68) when $p = 1$. This yields

$$\Gamma_1 = \frac{1}{1 - (3\pi/32.85)^2}\left(\sin(3\pi t) - \frac{3\pi}{32.85}\sin(32.85t)\right)$$

$$= 1.090\left(\sin(3\pi t) - \frac{3\pi}{32.85}\sin(32.85t)\right) \tag{7.73}$$

The time t_m required for Γ_1 to reach its maximum value Γ_{1max} can be determined from Eq. (7.73) by setting equal to zero the derivative of Γ_1 with respect to time. This yields

$$\frac{d\Gamma_1}{dt} = 3\pi(1.090)[\cos(3\pi t) - \cos(32.85t)] = 0 \tag{7.74}$$

By using a trial and error procedure, Eq. (7.74) yields $t_m = 0.30$ s. Therefore, by Eq. (7.73),

$$\Gamma_{1_{st}} = 1.090\left(\sin[3\pi(0.30)] - \frac{3\pi}{32.85}\sin[32.85(0.30)]\right)$$

$$= 0.47$$

In a similar manner, by using Eq. (7.68) and $t = 0.30$ s, we find

$$\Gamma_2 = 0.24$$
$$\Gamma_3 = 0.29$$

By using Eq. (7.63), the modal response $Y_p(t)$ for any mode p can be written as follows:

$$Y_p(t) = Y_{pst}\Gamma_p \qquad (7.75)$$

Thus,

$$Y_1(t) = (2.319)(1.090)\left(\sin(3\pi t) - \frac{3\pi}{32.85}\sin(32.85t)\right) \qquad (7.76)$$

$$= 2.528\sin(3\pi t) - 0.725\sin(32.85t)$$

$$Y_1(t = 0.30) = (2.319)(0.47) = 1.090 \text{ in.}$$

$$Y_2(t) = (0.120)\left[\frac{1}{1 - (3\pi/131.4)^2}\left(\sin(3\pi t) - \frac{3\pi}{131.4}\sin(131.4t)\right)\right]$$

$$= 0.121\sin(3\pi t) - 0.00865\sin(131.4t) \qquad (7.77)$$

$$Y_2(t = 0.30) = 0.0288 \text{ in.}$$

$$Y_3(t) = (0.00491)\left[\frac{1}{1 - (3\pi/295.7)^2}\left(\sin(3\pi t) - \frac{3\pi}{295.7}\sin(295.7t)\right)\right]$$

$$= 0.00491\sin(3\pi t) - 0.000157\sin(295.7t) \qquad (7.78)$$

$$Y_3(t = 0.30) = 0.00142 \text{ in.}$$

From Eq. (7.64), the total response of the member is

$$y(t, x) = \sum_{p=1}^{3} Y_p(t)\beta_p(x) \qquad (7.79)$$

$$= Y_1(t)\sin\frac{\pi x}{L} + Y_2(t)\sin\frac{2\pi x}{L} + Y_3(t)\sin\frac{3\pi x}{L}$$

where $Y_1(t)$, $Y_2(t)$, and $Y_3(t)$ are given by Eqs. (7.76), (7.77), and (7.78), respectively, and $\beta_p(x) = \sin(p\pi x/L)$. At the center of the beam ($x = 30$ ft) and at $t = 0.30$ s, the displacement y_c of the member is

$$y_c = 1.090 + 0.0288 + |-0.00142|$$

$$= 1.120 \text{ in.}$$

We note here that the response of the first mode predominates. It is 97.32%

of the total response of the first three modes. For practical applications, the contribution of the higher modes may be neglected in this case. Note that numerical addition is used to calculate y_c.

7.4 MODAL EQUATIONS FOR BEAMS WITH MOVING LOADS

Consider the uniform simply supported beam in Fig. 7.6, which is acted on by a constant force F moving along the length of the member with constant velocity v. At time $t = 0$, the force F is assumed to be at the left support and moving to the right with constant velocity v. The mode shape of the uniform simply supported beam is given by Eq. (7.65), and the modal equation may be written as follows:

$$\ddot{Y}_p + \omega_p^2 Y_p = \frac{F \sin p \pi v t}{m \int_0^L \sin^2 \dfrac{p \pi x}{L} \, dx} \tag{7.80}$$

By carrying out the integration and simplifying, Eq. (7.80) is written as shown below:

$$\ddot{Y}_p + \omega_p^2 Y_p = \frac{2F}{mL} \sin \frac{p \pi v t}{L} \tag{7.81}$$

Equation (7.81) shows that the time function is $\sin \Omega_p t$, where

$$\Omega_p = \frac{p \pi v}{L} \tag{7.82}$$

In the above equations, m is the mass per unit length of the member.

Since the time function in Eq. (7.81) is sinusoidal, the magnification factor Γ_p for a p mode is given by Eq. (7.68) by replacing ω_p with Ω_p; that is,

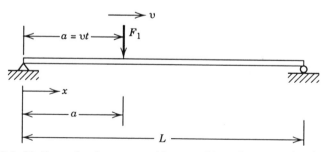

FIGURE 7.6. Uniform simply supported beam subjected to a moving load.

$$\Gamma_p = \frac{1}{1 - (\Omega_p/\omega_p)^2}\left(\sin \Omega_p t - \frac{\Omega_p}{\omega_p}\sin \omega_p t\right) \qquad (7.83)$$

The modal static displacement $Y_{p_{st}}$ is

$$Y_{p_{st}} = \frac{2F}{\omega_p^2 mL} \qquad (7.84)$$

Thus, the modal displacement $Y_p(t)$ is

$$Y_p(t) = \Gamma_p Y_{p_{st}}$$

$$= \frac{2F}{mL} \cdot \frac{1}{\omega_p^2[1 - (\Omega_p/\omega_p)^2]}\left(\sin \Omega_p t - \frac{\Omega_p}{\omega_p}\sin \omega_p t\right) \qquad (7.85)$$

The total response $y(t, x)$ at any t and x is given by the following equation:

$$y(t, x) = \sum_{p=1}^{\infty} Y_p(t)\beta_p(x) \qquad (7.86)$$

where

$$\beta_p(x) = \sin \frac{p\pi x}{L} \qquad (7.87)$$

By using Eqs. (7.85) and (7.87) and substituting into Eq. (7.86), we find

$$y(t, x) = \frac{2F}{mL}\sum_{p=1}^{\infty} \frac{1}{\omega_p^2[1 - (\Omega_p/\omega_p)^2]}\left(\sin \Omega_p t - \frac{\Omega_p}{\omega_p}\sin \omega_p t\right)\sin \frac{p\pi x}{L} \qquad (7.88)$$

Here again the contribution of the higher modes is small, and the displacement $y(t, x)$ becomes infinitely large when Ω_p becomes equal to ω_p.

Example 7.6. The uniform simply supported beam in Fig. 7.6 is subjected to a moving load $F = 30.0$ kips. The load enters the beam at the left support and moves along the beam's length with constant velocity $v = 30$ miles per hour. The length L of the beam is 30.0 ft, its mass $m = 0.00155$ kip $\cdot$ s^2/in., and its uniform stiffness $EI = 30 \times 10^6$ kip $\cdot$ in.^{2}. By considering the first mode response only, determine the expression for the deflection $y(t, x)$ of the member and the maximum deflection at its center.

Solution. The constant velocity v (in units of in./s) is

$$v = \frac{(30)(5280)}{3600} = 44.0 \text{ ft/s}$$

$$= 528.0 \text{ in./s}$$

For $p = 1$, Eq. (7.82) yields

$$\Omega_1 = \frac{\pi(528)}{(30)(12)} = 4.61 \text{ rps}$$

The fundamental free frequency of vibration ω_1 of the simply supported beam is

$$\omega_1 = \frac{\pi^2}{L^2}\sqrt{\frac{EI}{m}} = \frac{\pi^2}{(360)^2}\sqrt{\frac{30 \times 10^6}{0.00155}}$$

$$= 10.59 \text{ rps}$$

By using the above data and Eq. (7.88), we find

$$y(t, x) = \frac{(2)(30)}{(0.00155)(360)}\left[\frac{1}{(10.59)^2[1 - (4.61/10.59)^2]}\right.$$

$$\left. \times \left(\sin 4.61t - \frac{4.61}{10.59}\sin 10.59t\right)\right]\sin\frac{\pi x}{L} \tag{7.89}$$

$$= 1.1837(\sin 4.61t - 0.4353 \sin 10.59t)\sin\frac{\pi x}{L}$$

At the center of the member ($x = L/2$), we have

$$y(t, L/2) = 1.1837(\sin 4.61t - 0.4353 \sin 10.59t) \tag{7.90}$$

The time of maximum response may be obtained by taking the derivative with respect to time t of Eq. (7.90) and setting it equal to zero. It yields the following expression:

$$\cos 4.61t - \cos 10.59t = 0 \tag{7.91}$$

A value of t that satisfies Eq. (7.91) is $t = 0.4134$ s. On this basis, Eq. (7.90) yields

$$y(0.4134, L/2) = 1.1837[\sin(4.61)(0.4134) - (0.4353)\sin(10.59)(0.4134)]$$
$$= 1.6046 \text{ in.}$$

If the load F is applied statically at the center of the member, the deflection y produced by this load at its center is

$$y_{x=L/2} = \frac{FL^3}{48EI}$$

$$= \frac{(30)(360)^3}{(48)(30 \times 10^6)} = 0.972 \text{ in.}$$

This deflection is 39.42% smaller than the deflection obtained by using dynamic analysis.

The position x of the load F at $t = 0.4134$ s is

$$x = vt = (528)(0.4134) = 218.28 \text{ in.}$$

If the load is applied statically at $x = 218.28$ in., the deflection y at the center of the beam is

$$y_{x=L/2} = \frac{(30)(141.72)(180)}{(6)(360)(30 \times 10^6)}[(360)^2 - (141.72)^2 - (180)^2]$$

$$= 0.9107 \text{ in.}$$

This value is 43.24% smaller than the value obtained by dynamic analysis.

7.5 MODAL ANALYSIS FOR UNIFORM SIMPLY SUPPORTED THIN PLATES

In this section, we consider the application of the method of modal analysis for uniform thin rectangular plates that are simply supported on all four edges. The dynamic load $q(t, x, y)$ acting on the plate is assumed to be uniformly distributed throughout the xy area of the plate. It may be expressed as

$$q(t) = qg(t) \tag{7.92}$$

where $g(t)$ represents the time variation of the uniformly distributed dynamic load.

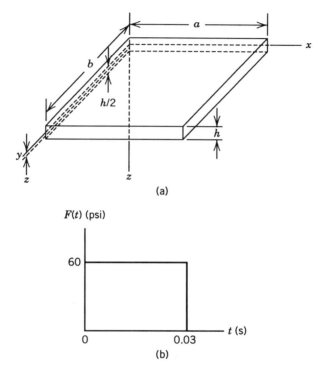

FIGURE 7.7. (a) Uniform simply supported thin plate loaded by a uniformly distributed dynamic load. (b) Time variation of the distributed dynamic load.

We now consider the uniform thin rectangular plate in Fig. 7.7a, of sides a and b as shown. The xy axes are at the neutral surface of the plate, where the bending stresses are assumed to be negligibly small, z is the vertical axis, h is the uniform thickness of the plate, and m is its uniform mass per unit area. From the theory of plates [16], the dynamic displacement w of the plate in the vertical z direction may be assumed to have the following form:

$$W = \sum_{n=1}^{\infty} \sum_{p=1}^{\infty} A_{np} \sin \frac{n\pi x}{a} \sin \frac{p\pi y}{b} \qquad (7.93)$$

In Eq. (7.93), the quantity A_{np} represents the modal displacement at the center of the plate. In addition, Eq. (7.93) is expressed in a way that satisfies the boundary conditions of zero moment and zero displacement of all four edges.

Lagrange's equation, which was derived in Section 7.2, will be used here to derive the appropriate modal equation for any mode of vibration of the above rectangular plate. By neglecting damping and noting that $\partial T/\partial A_{np} =$

0, Lagrange's equation, in this case, may be written as follows:

$$\frac{d}{dt}\left(\frac{\partial T}{\partial \dot{A}_{np}}\right) + \frac{\partial U}{\partial A_{np}} = \frac{\partial W_e}{\partial A_{np}} \tag{7.94}$$

Note the similarity between Eqs. (7.94) and (7.9).

In Eq. (7.93), any possible combination of the integers n and p constitutes a mode shape. We consider here the following general shape for a mode of the plate:

$$w = A_{np} \sin \frac{n\pi x}{a} \sin \frac{p\pi y}{b} \tag{7.95}$$

If we consider an element of the plate of sides dx and dy, the kinetic energy dT of the element may be written as follows:

$$dT = \tfrac{1}{2}m \, dx \, dy \, \dot{w}^2 \tag{7.96}$$

Therefore, the total kinetic energy T may be obtained by using Eq. (7.96) as follows:

$$T = \frac{m}{2} \int_0^a \int_0^b \left(\dot{A}_{np} \sin \frac{n\pi x}{a} \sin \frac{p\pi y}{b}\right)^2 dx \, dy \tag{7.97}$$

By performing the required integrations in Eq. (7.97), we find

$$T = \frac{mab}{8} \dot{A}_{np} \tag{7.98}$$

Therefore,

$$\frac{d}{dt}\left(\frac{\partial T}{\partial \dot{A}_{np}}\right) = \frac{mab}{4} \ddot{A}_{np} \tag{7.99}$$

Equation (7.99) provides the expression for the first term on the left-hand side of Eq. (7.94).

The expression for the second term on the left-hand side of Eq. (7.94) may be obtained by using the equation for the total strain energy U from the theory of thin rectangular plates [16]:

$$U = \frac{Eh^3}{24(1 - \nu^2)} \int_0^a \int_0^b \left[\left(\frac{\partial^2 w}{\partial x^2} \right)^2 + \left(\frac{\partial^2 w}{\partial y^2} \right)^2 + 2\nu \frac{\partial^2 w}{\partial x^2} \frac{\partial^2 w}{\partial y^2} \right.$$
$$\left. + 2(1 - \nu) \left(\frac{\partial^2 w}{\partial x \, \partial y} \right)^2 \right] dx \, dy \qquad (7.100)$$

In Eq. (7.100), ν is the Poisson ratio and E is the modulus of elasticity.

If we substitute for w in Eq. (7.100) the expression given by Eq. (7.95) and carry out the fairly lengthy mathematical manipulations and operations, we find

$$U = \frac{Eh^3 \pi^4 ab A_{np}^2}{96(1 - \nu^2)} \left(\frac{n^2}{a^2} + \frac{p^2}{b^2} \right) \qquad (7.101)$$

Consequently, the second term on the left-hand side of Eq. (7.94) may be written as follows:

$$\frac{\partial U}{\partial A_{np}} = \frac{Eh^3 \pi^4 ab}{48(1 - \nu^2)} A_{np} \left(\frac{n^2}{a^2} + \frac{p^2}{b^2} \right) \qquad (7.102)$$

The expression for the term on the right-hand side of Eq. (7.94) may be determined by using the work W_e done by the externally applied uniformly distributed dynamic load $q(t)$. This is given by the following equation:

$$W_e = q(t) \int_0^a \int_0^b A_{np} \sin \frac{n\pi x}{a} \sin \frac{p\pi y}{b} \, dx \, dy$$
$$= q(t) \frac{ab}{np\pi^2} A_{np}(\cos n\pi - 1)(\cos p\pi - 1) \qquad (7.103)$$

On this basis, the term on the right-hand side of Eq. (7.94) yields

$$\frac{\partial W_e}{\partial A_{np}} = q(t) \frac{ab}{np\pi^2} (\cos n\pi - 1)(\cos p\pi - 1) \qquad (7.104)$$

By substituting Eqs. (7.99), (7.102), and (7.104) into Eq. (7.94) and simplifying, we find

$$\ddot{A}_{np} + \frac{Eh^3 \pi^4}{12(1 - \nu^2)m} \left(\frac{n^2}{a^2} + \frac{p^2}{b^2} \right) A_{np} = q(t) \frac{4}{npm\pi^2} (\cos n\pi - 1)(\cos p\pi - 1)$$
$$(7.105)$$

Equation (7.105) is the modal equation for any np mode of the thin rectangular plate. The format of this equation is similar to the ones derived in preceding sections of the chapter. The coefficient of A_{np} on the right-hand side of Eq. (7.105) is the free frequency squared, ω_i^2, of the np mode of vibration of the plate. On this basis, the frequency ω_i for any np mode may be obtained from the following expression:

$$\omega_i = \pi^2 \left(\frac{Eh^3}{12(1 - \nu^2)m} \right)^{1/2} \left(\frac{n^2}{a^2} + \frac{p^2}{b^2} \right) \tag{7.106}$$

or

$$\omega_i = \pi^2 \left(\frac{D}{m} \right)^{1/2} \left(\frac{n^2}{a^2} + \frac{p^2}{b^2} \right) \tag{7.107}$$

where

$$D = \frac{Eh^3}{12(1 - \nu^2)} \tag{7.108}$$

By considering the various combinations of the integers n and p in Eq. (7.107), the analogous expressions for the frequencies ω_i may be obtained. For example, if we select the values $n = p = 1$, the fundamental frequency of vibration ω_1 of the plate may be obtained. On this basis, Eq. (7.107) yields

$$\omega_1 = \pi^2 \left(\frac{D}{m} \right)^{1/2} \left(\frac{1}{a^2} + \frac{1}{b^2} \right) \tag{7.109}$$

The modal equation corresponding to the fundamental mode of the plate may be obtained from Eq. (7.105) with $n = p = 1$. By combining Eqs. (7.105) and (7.109), the modal equation for the fundamental mode of the plate is written as follows:

$$\ddot{A}_{11} + \omega_1^2 A_{11} = q(t) \frac{16}{m\pi^2} \tag{7.110}$$

Other combinations of the integers n and p lead to analogous expressions for the modal equations of other modes of vibration of the thin plate.

By referring to Eqs. (7.92) and (7.105), we can conclude that the modal static displacement $(A_{np})_{st}$ for an np mode may be obtained from the following expression:

$$(A_{np})_{st} = \frac{(4q/npm\pi^2)(\cos n\pi - 1)(\cos p\pi - 1)}{[Eh^3\pi^4/12m(1 - \nu^2)](n^2/a^2 + p^2/b^2)^2} \tag{7.111}$$

or

$$(A_{np})_{st} = \frac{48q(1 - \nu^2)(\cos n\pi - 1)(\cos p\pi - 1)}{Eh^3np\pi^6(n^2/a^2 + p^2/b^2)^2} \tag{7.112}$$

The dynamic modal displacement A_{np} may be written as follows:

$$A_{np} = (A_{np})_{st}\Gamma_{np} \tag{7.113}$$

where Γ_{np} is the magnification factor associated with the time function $g(t)$ of the np mode. It has the same interpretation and significance as the magnification factors used in the preceding sections of the chapter.

If we consider the fundamental mode of the plate, which corresponds to $n = p = 1$, Eq. (7.112) yields

$$(A_{11})_{st} = \frac{192q(1 - \nu^2)}{Eh^3\pi^6(1/a^2 + 1/b^2)^2} \tag{7.114}$$

The dynamic modal displacement A_{11} of the same mode may be obtained from Eq. (7.113) as follows:

$$A_{11} = (A_{11})_{st}\Gamma_{11} \tag{7.115}$$

With known modal displacement A_{np}, the expression for the dynamic displacement w of the plate may be obtained from Eq. (7.95). By substituting Eq. (7.113) into Eq. (7.95), we find

$$w = (A_{np})_{st}\Gamma_{np} \sin\frac{n\pi x}{a} \sin\frac{p\pi y}{b} \tag{7.116}$$

Equation (7.116) gives the displacement response of the plate corresponding to the np mode. The total response can be obtained by superimposing the responses of all modes. For practical purposes, it is only necessary to consider the response of the first few modes, because the contribution of the higher modes becomes increasingly smaller.

The bending stresses σ_x and σ_y in the x and y directions, respectively, may be determined by using well-known formulas of the theory of plates [16]. These expressions are as follows:

$$\sigma_x = - \frac{Ez}{1 - v^2} \left(\frac{\partial^2 w}{\partial x^2} + v \frac{\partial^2 w}{\partial y^2} \right) \qquad (7.117)$$

$$\sigma_y = - \frac{Ez}{1 - v^2} \left(\frac{\partial^2 w}{\partial y^2} + v \frac{\partial^2 w}{\partial x^2} \right) \qquad (7.118)$$

In Eqs. (7.117) and (7.118), z is the distance from the median plane of the rectangular plate in Fig. 7.7a, and the displacement w is given by Eq. (7.116). The maximum stress occurs at $z = h/2$.

The following example illustrates the application of the above methodology.

Example 7.7. The steel rectangular plate in Fig. 7.7a is subjected to a uniform dynamic pressure $q(t)$ of 60.0 psi, which is applied suddenly at time $t = 0$ and removed suddenly at $t = t_d = 0.03$ s, as shown in Fig 7.7b. The sides of the plate are $a = 120$ in. and $b = 60$ in., the thickness $h = 2$ in., the mass m per unit area of the plate is 0.00146 lb · s^2/in.^{3}, the Poisson ratio $v = 0.25$, and the modulus of elasticity $E = 30 \times 10^6$ psi. Determine the maximum response of the plate by considering only the first mode.

Solution. The static modal displacement $(A_{11})_{st}$ for the first mode may be obtained from Eq. (7.114); that is,

$$(A_{11})_{st} = \frac{(192)(60)[1 - (0.25)^2]}{(30 \times 10^6)(2)^3 \pi^6 [1/(120)^2 + 1/(60)^2]^2}$$

$$= 0.388 \text{ in.}$$

The frequency of vibration ω_1 of the first mode may be determined by using Eq. (7.109), and the plate rigidity D is given by Eq. (7.108). On this basis, we have

$$D = \frac{(30 \times 10^6)(2)^3}{12[1 - (0.25)^2]} = 21.3333 \times 10^6$$

$$\omega_1 = \pi^2 \sqrt{\frac{21.3333 \times 10^6}{0.00146} \left(\frac{1}{(120)^2} + \frac{1}{(60)^2} \right)}$$

$$= 41.97 \pi^2 \text{ rps}$$

The period of vibration τ_1 is

$$\tau_1 = \frac{2\pi}{\omega_1} = \frac{2\pi}{41.97\pi^2} = 0.0152 \text{ s}$$

Therefore,

$$\frac{t_d}{\tau_1} = \frac{0.03}{0.0152} = 1.97$$

By using the equations for Γ in Appendix A, or the graphs in Appendix H, we find that the maximum magnification factor $(\Gamma_{11})_{max} = 2$.

The maximum modal dynamic displacement $(A_{11})_{max}$ may be obtained from Eq. (7.115); that is,

$$(A_{11})_{max} = (A_{11})_{st}(\Gamma_{11})_{max}$$
$$= (0.388)(2) = 0.776 \text{ in.}$$

The maximum dynamic displacement w_{max} occurs at the center of the plate, and it can be determined from Eq. (7.116). From this equation, with $x = a/2$ and $y = b/2$, we find

$$w_{max} = (A_{11})_{st}(\Gamma_{11})_{max} \sin\frac{\pi}{2} \sin\frac{\pi}{2}$$
$$= (0.388)(2)(1)^2$$
$$= 0.776 \text{ in.}$$

The maximum stresses σ_x and σ_y in the x and y directions, respectively, may be determined by using Eqs. (7.117) and (7.118). The maximum stresses occur at the position defined by $z = h/2$, $x = a/2$, and $y = b/2$. On this basis, we have

$$\frac{\partial^2 w}{\partial x^2} = -(A_{11})_{max}\frac{\pi^2}{a^2}$$

$$\frac{\partial^2 w}{\partial y^2} = -(A_{11})_{max}\frac{\pi^2}{b^2}$$

Therefore,

$$(\sigma_x)_{max} = (A_{11})_{max} \frac{Eh\pi^2}{2(1 - \nu^2)} \left(\frac{1}{a^2} + \frac{\nu}{b^2} \right)$$

$$= (0.776) \frac{(30 \times 10^6)(2)\pi^2}{2[1 - (0.25)^2]} \left(\frac{1}{(120)^2} + \frac{0.25}{(60)^2} \right)$$

$$= 34,037 \text{ psi}$$

$$(\sigma_y)_{max} = (A_{11})_{max} \frac{Eh\pi^2}{2(1 - \nu^2)} \left(\frac{1}{b^2} + \frac{\nu}{a^2} \right)$$

$$= (0.776) \frac{(30 \times 10^6)(2)\pi^2}{2[1 - (0.25)^2]} \left(\frac{1}{(60)^2} + \frac{0.25}{(120)^2} \right)$$

$$= 72,334 \text{ psi}$$

7.6 MODAL ANALYSIS FOR EARTHQUAKE RESPONSE

The following analysis provides a brief treatment of the earthquake problem, which is important in the analysis of structures and machines to withstand the catastrophic effects of earthquakes.

7.6.1 Introductory Aspects Regarding Earthquakes

Although the purpose of this text is not to provide an extensive treatment of the earthquake subject, it is, however, appropriate at this point to provide some introductory discussion in order to stress its importance and prepare the student for more advanced work on this subject. The earthquake problem is an old one, maybe as old as the earth itself, and probably there is no part of the world that has not experienced, at least to some degree, the effects of an earthquake shock. Many parts of the United States and the rest of the world have experienced minor, moderate, and severe earthquake shocks. From the frequency of such earthquake activity, we come to the conclusion that there is no part of the world that is immune from the possibility of earthquake damage.

The outer part of the earth, known as the crust, is divided into large irregular blocks called *fault blocks*, and the trace of a *fault plane* is known as the *rift*. The fault blocks are normally in equilibrium, but, at times, relative motion known as *slip* may occur. If the slip is gradual, similar gradual adjustments take place in the continuous strata, and the consequences are not usually observable at the time such adjustments are taking place. If, on the other hand, the slip is abrupt, a series of local movements begin to propagate through the earth, and the surface vibration caused by these movements is what is known today as *tectonic earthquakes*.

After the event of an earthquake, the equilibrium of the adjacent fault blocks will be restored and the movement will cease. However, there is always a possibility for such movement to be repeated if the forces that produced the previous slip continue to exist. The intensities of such earthquakes are obtained by measuring their respective accelerations as a function of time, and records of such accelerations in terms of both vertical and horizontal components are recorded by governmental and private agencies as they occur. From the accelerations, the ground velocities and ground displacements, as a function of time, may be obtained by integration. Figure 7.8 illustrates the ground accelerations, as well as the corresponding ground velocities and displacements, of a strong earthquake that took place in Kern County in California. Such records provide a useful history of past earthquake activity, which can be used to establish design criteria for the engineering problem at hand.

The design of a structure or a machine to resist the effects of an earthquake is a difficult task, because it is practically impossible to predict the intensity and character of the earthquake activity that a mechanical or a structural system will experience in its lifetime. In practice, three methods are commonly used for this purpose. The first one uses a standard earthquake of specific acceleration amplitude and time variation. Since the response of an engineering system depends on both the acceleration amplitude and time variation, the use of such a standard earthquake would be a disadvantage. A second approach, which is considered to be more practical, is to use earthquake response spectra from real earthquake records, but similar disadvantages are also encountered here. The third approach is to treat earthquake ground motions as random variables. This is a more logical approach, since the input is nondeterministic, but it requires a sufficiently large number of regional earthquake records in order to design the system for the most critical earthquake.

The engineer who wishes to specialize in this subject should become familiar with all three methodologies and use what is most appropriate in a physical situation. In addition, when possible, other design considerations that could minimize or isolate the engineering system from the effects of such accelerations should be undertaken.

In this section, it will be shown how the method of modal analysis can be used to determine the earthquake response of simple systems.

7.6.2 Modal Equations for Earthquake Response

The method of modal analysis may be used to determine the dynamic response of systems with many degrees of freedom, when they are subjected to earthquake accelerations. The development of the required modal equations may be initiated by considering the modal equation for spring–mass systems, Eq. (7.51), which was derived in Section 7.3. We write this equation

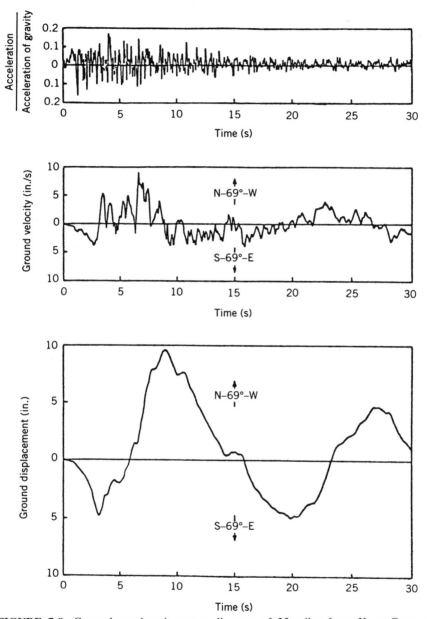

FIGURE 7.8. Ground acceleration at a distance of 35 miles from Kern County, California, and the corresponding velocities and displacements. (Adapted from ref. 63. Accelerations record courtesy of United States Coast and Geodetic Survey.)

again below:

$$\ddot{Y}_p + \omega_p^2 Y_p = g(t) \frac{\sum\limits_{i=1}^{r} F_i \beta_{ip}}{\sum\limits_{i=1}^{r} m_i \beta_{ip}^2} \qquad (7.119)$$

In Eq. (7.119), $g(t)$ is the time function of the applied dynamic forces $F_i(t) = F_i g(t)$.

If desired, the effects of viscous damping may also be incorporated as follows:

$$\ddot{Y}_p + \omega_p^2 Y_p + 2\mu \dot{Y}_p = g(t) \frac{\sum\limits_{i=1}^{r} F_i \beta_{ip}}{\sum\limits_{i=1}^{r} m_i \beta_{ip}^2} \qquad (7.120)$$

where $\mu = \zeta\omega$ and ζ is the damping ratio.

Consider now the one-story frame in Fig. 7.9a that is subjected to an earthquake ground motion, which is represented by the support displacement u_s shown in the figure. The displacement of the top of the frame relative to the ground is denoted as u. On this basis, the total horizontal displacement x of the top of the frame is

$$x = u_s + u \qquad (7.121)$$

If we assume that the girder of the frame is infinitely stiff compared to the columns, the one-degree idealized spring–mass system would be as shown in Fig. 7.9b. Note that the motion is assumed to be under the influence of viscous damping. From the free-body diagram of mass m in Fig. 7.9c, we can write the following differential equation of motion:

$$m\ddot{x} + ku + c\dot{u} = 0 \qquad (7.122)$$

By substituting Eq. (7.121) into Eq. (7.122), we find

$$m\ddot{u} + ku + c\dot{u} = - m\ddot{u}_s \qquad (7.123)$$

Since $k = m\omega^2$, Eq. (7.123) is written as follows:

$$\ddot{u} + \omega^2 u + \frac{c}{m}\dot{u} = - \ddot{u}_s \qquad (7.124)$$

or, since $c/m = 2\zeta\omega = 2\mu$, where $\mu = \zeta\omega$, Eq. (7.124) yields

$$\ddot{u} + 2\mu\dot{u} + \omega^2 u = - \ddot{u}_s \qquad (7.125)$$

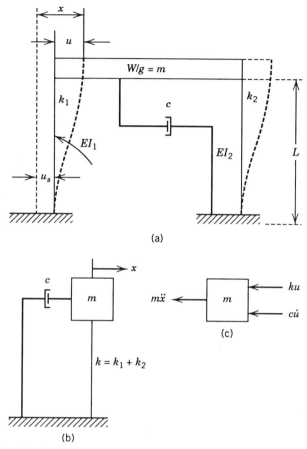

FIGURE 7.9. (a) One-story frame subjected to an earthquake ground motion. (b) Idealized one-degree spring–mass system. (c) Free-body diagram of mass m.

The solution of Eq. (7.125) is

$$u(t) = \frac{1}{\omega} \int_0^t (-\ddot{u}_s) e^{-\mu(t-T)} \sin \omega(t-T)\, dT \qquad (7.126)$$

where $\ddot{u}_s$ is the applied ground or support acceleration, and $u(t)$ is the displacement of the system relative to the ground. See also Section 3.5.2 regarding the solution of Eq. (7.125).

The support acceleration $\ddot{u}_s$ may also be written

$$\ddot{u}_s = \ddot{u}_{s0} f_s(t) \qquad (7.127)$$

where $\ddot{u}_{s0}$ is the maximum support acceleration, and $f_s(t)$ is the time variation.

On this basis, Eq. (7.126) is written as follows:

$$u(t) = - \frac{\ddot{u}_{s0}}{\omega} \int_0^t f_s(T) e^{-\mu(t-T)} \sin \omega(t-T) \, dT \qquad (7.128)$$

If there is no damping, we have $\mu = 0$, and Eq. (7.128) yields

$$u(t) = - \frac{\ddot{u}_{s0}}{\omega} \int_0^t f_s(T) \sin \omega(t-T) \, dT \qquad (7.129)$$

Now we consider Eq. (7.128). We multiply and divide its right-hand side by $m\omega$, and we substitute k for $m\omega^2$. The result is

$$u(t) = - \frac{m\ddot{u}_{s0}\omega}{k} \int_0^t f_s(T) e^{-\mu(t-T)} \sin \omega(t-T) \, dT \qquad (7.130)$$

Equation (7.130) may also be written as follows:

$$u(t) = -u_{st}\Gamma = - \frac{\ddot{u}_{s0}}{\omega^2} \Gamma \qquad (7.131)$$

where

$$u_{st} = \frac{m\ddot{u}_{s0}}{k} = \frac{\ddot{u}_{s0}}{\omega^2} \qquad (7.132)$$

and

$$\Gamma = \omega \int_0^t f_s(T) e^{-\mu(t-T)} \sin \omega(t-T) \, dT \qquad (7.133)$$

The static deflection u_{st} and magnification factor Γ in Eq. (7.131) serve the same purpose as u_{st} and Γ in preceding sections of the chapter. If damping is neglected, Eq. (7.133) yields

$$\Gamma = \omega \int_0^t f_s(t) \sin \omega(t-T) \, dT \qquad (7.134)$$

The above derivations show that support motions produce responses that are equivalent to the ones produced for the case where the applied forces are equal to $-m\ddot{u}_s$, where $\ddot{u}_s$ may be expressed as shown by Eq. (7.127). For spring–mass systems with many degrees of freedom, or for systems that can be idealized as spring–mass systems, the applied force on mass m_i would be $-m_i\ddot{u}_s$. On this basis, the modal equation can readily be obtained from Eq. (7.119) or Eq. (7.120), by replacing $F_{ig}(t)$ with $-m_i\ddot{u}_s$ and also using Eq. (7.127) for $\ddot{u}_s$. On this basis, Eq. (7.119) yields

$$\ddot{Y}_p + \omega_p^2 Y_p = -f_s(t)\ddot{u}_{s0}\frac{\sum\limits_{i=1}^{r} m_i\beta_{ip}}{\sum\limits_{i=1}^{r} m_i\beta_{ip}^2} \tag{7.135}$$

and Eq. (7.120) yields the following expression:

$$\ddot{Y}_p + \omega_p^2 Y_p + 2\mu\dot{Y}_p = -f_s(t)\ddot{u}_{s0}\frac{\sum\limits_{i=1}^{r} m_i\beta_{ip}}{\sum\limits_{i=1}^{r} m_i\beta_{ip}^2} \tag{7.136}$$

Equations (7.135) and (7.136) are the modal equations corresponding to a p mode of vibration of the spring–mass system when it is subjected to a support motion. Note that Eq. (7.136) takes into consideration the effects of viscous damping.

In Eqs. (7.135) and (7.136), the summations on the right-hand side of the equations are constants. For practical purposes, we introduce here the *modal participation factor* Λ_p, which is defined as follows:

$$\Lambda_p = \frac{\sum\limits_{i=1}^{r} m_i\beta_{ip}}{\sum\limits_{i=1}^{r} m_i\beta_{ip}^2} \tag{7.137}$$

On this basis, the modal displacement $Y_p(t)$ relative to the support may be determined from the following expression:

$$Y_p(t) = \Lambda_p u_p(t) \tag{7.138}$$

$$= -\Lambda_p \frac{\ddot{u}_{s0}}{\omega_p^2}\Gamma_p$$

where, for the p mode, $u_p(t)$ is expressed as shown by Eq. (7.131). At a mode of vibration with frequency ω_p, the magnification factor Γ_p may be determined from Eq. (7.133), when viscous damping is considered, and from Eq. (7.134), when damping is neglected.

For the mass m_i, its displacement $u_{ip}(t)$ may be obtained from the following expression:

$$u_{ip}(t) = -\Lambda_p\Gamma_p\beta_{ip}\frac{\ddot{u}_{s0}}{\omega_p^2} \tag{7.139}$$

If the responses of all modes are superimposed, the displacement $u_i(t)$ of mass m_i is

$$u_i(t) = -\sum_{p=1}^{N} \Lambda_p \Gamma_p \beta_{ip} \frac{\ddot{u}_{s0}}{\omega_p^2} \qquad (7.140)$$

where N on the summation sign represents the total number of modes.

7.6.3 Applications for Earthquake Response

The application of the above methodologies is illustrated here by using a numerical example.

Example 7.8. The two-story steel frame in Fig. 7.10a is subjected to an artificial earthquake of ground acceleration $\ddot{u}_s = 0.05g \sin 4\pi t$, where g is the acceleration of gravity and t is time. The duration of the acceleration $\ddot{u}_s$ is 1 s. By applying the method of modal analysis, determine the maximum displacements $u_{1\text{max}}$ and $u_{2\text{max}}$ at the first- and second-story levels, respectively, and the maximum bending stresses in the columns. Assume that the

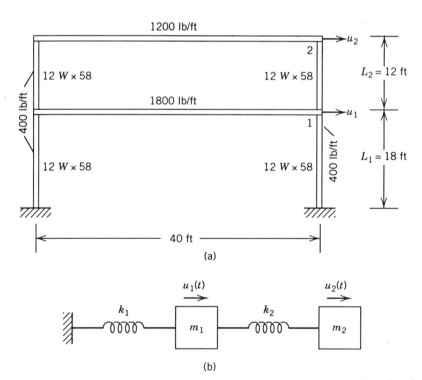

FIGURE 7.10. (a) Two-story frame subjected to an earthquake ground acceleration. (b) Idealized two-degree spring–mass system.

girders of the frame are infinitely stiff compared to the columns, and neglect damping. The required dimensions of the frame and other required quantities are shown in the figure. The modulus of elasticity $E = 30 \times 10^6$ psi.

Solution. The two-story frame will be idealized as a two-degree spring–mass system as shown in Fig. 7.10b. The values of m_1, m_2, k_1, and k_2 are as follows:

$$m_1 = \frac{(1800)(40) + (2)(400)(9 + 6)}{386.4}$$

$$= 217.391 \text{ lb} \cdot \text{s}^2/\text{in.}$$

$$m_2 = \frac{(1200)(40) + (2)(400)(6)}{386.4}$$

$$= 136.646 \text{ lb} \cdot \text{s}^2/\text{in.}$$

$$k_1 = 2\frac{12EI}{L_1^3}$$

$$= \frac{(24)(30 \times 10^6)(475)}{(18 \times 12)^3} = 33,938.328 \text{ lb/in.}$$

$$k_2 = 2\frac{12EI}{L_2^3}$$

$$= \frac{(24)(30 \times 10^6)(475)}{(12 \times 12)^3} = 114,535.108 \text{ lb/in.}$$

The two free frequencies of vibration ω_1 and ω_2 may be obtained from the following equation:

$$\omega_{1,2}^2 = \frac{1}{2}\left(\frac{k_1 + k_2}{m_1} + \frac{k_2}{m_2}\right) \pm \sqrt{\left(\frac{k_1 + k_2}{m_1} + \frac{k_2}{m_2}\right)^2 - 4\frac{k_1 k_2}{m_1 m_2}}$$

$$= 760.573 \pm 667.054$$

Therefore,

$$\omega_1 = 9.567 \text{ rps}$$

$$\omega_2 = 37.810 \text{ rps}$$

$$f_1 = \frac{\omega_1}{2\pi} = 1.523 \text{ Hz}$$

$$f_2 = \frac{\omega_2}{2\pi} = 6.018 \text{ Hz}$$

$$\tau_1 = \frac{1}{f_1} = 0.657 \text{ s}$$

$$\tau_2 = \frac{1}{f_2} = 0.166 \text{ s}$$

For the first mode,

$$C_1 = \frac{k_2}{k_2 - m_2\omega_1^2} = 1.123$$

$$x_2^{(1)} = C_1 x_1^{(1)} = 1.123 x_1^{(1)}$$

Thus, with $x_1^{(1)} = 1.00$, we have the following mode amplitudes:

$$x_1^{(1)} = \beta_{11} = 1.000$$

$$x_2^{(1)} = \beta_{21} = 1.123$$

For the second mode,

$$C_2 = \frac{k_2}{k_2 - m_2\omega_2^2} = -1.417$$

$$x_2^{(2)} = C_2 x_1^{(2)} = -1.417 x_1^{(2)}$$

Thus, with $x_1^{(2)} = 1.00$, we have the following mode amplitudes:

$$x_1^{(2)} = \beta_{12} = 1.000$$

$$x_2^{(2)} = \beta_{22} = -1.417$$

The modal participation factors Λ_1 and Λ_2 may be determined from Eq. (7.137), and they are as follows:

$$\Lambda_1 = \frac{m_1\beta_{11} + m_2\beta_{21}}{m_1\beta_{11}^2 + m_2\beta_{21}^2}$$

$$= 0.952$$

$$\Lambda_2 = \frac{m_1\beta_{12} + m_2\beta_{22}}{m_1\beta_{12}^2 + m_2\beta_{22}^2}$$

$$= 0.0483$$

Since the earthquake ground motion is sinusoidal, the magnification factors Γ_1 and Γ_2 may be obtained from the following general equation (see also Appendix A):

$$\Gamma_p = \frac{1}{[1 - (\omega_f/\omega_p)^2]}\left(\sin \omega_f t - \frac{\omega_f}{\omega_p}\sin \omega t\right) \tag{7.141}$$

For $p = 1$ and 2 and $\omega_f = 4\pi$, Eq. (7.141) yields

$$\Gamma_1 = -1.379 \sin 4\pi t + 1.811 \sin 9.567t \tag{7.142}$$

$$\Gamma_2 = 1.124 \sin 4\pi t - 0.374 \sin 37.810t \tag{7.143}$$

The relative horizontal displacements $u_1(t)$ and $u_2(t)$ may be obtained from Eq. (7.140), and they are as follows:

$$u_1(t) = -\ddot{u}_{s0}\left[\left(\frac{\Lambda_1\Gamma_1}{\omega_1^2}\right)\beta_{11} + \left(\frac{\Lambda_2\Gamma_2}{\omega_2^2}\right)\beta_{12}\right]$$

$$= -(0.05g)\left[\left(\frac{-1.312808 \sin 4\pi t + 1.724074 \sin 9.567t}{(9.567)^2}\right)(1.00)\right.$$

$$\left. + \left(\frac{0.0542892 \sin 4\pi t - 0.0180642 \sin 37.810t}{(37.810)^2}\right)(1.00)\right] \tag{7.144}$$

$$= 0.276 \sin 4\pi t - 0.364 \sin 9.567t + 2.441 \times 10^{-4}\sin 37.810t$$

$$u_2(t) = -\ddot{u}_{s0}\left[\left(\frac{\Lambda_1\Gamma_1}{\omega_1^2}\right)\beta_{21} + \left(\frac{\Lambda_2\Gamma_2}{\omega_2^2}\right)\beta_{22}\right]$$

$$= -(0.05g)\left[\left(\frac{-1.312808 \sin 4\pi t + 1.724072 \sin 9.567t}{(9.567)^2}\right)(1.123)\right.$$

$$\left. + \left(\frac{0.0542892 \sin 4\pi t - 0.0180642 \sin 37.810t}{(37.810)^2}\right)(-1.417)\right] \tag{7.145}$$

$$= 0.312 \sin 4\pi t - 0.409 \sin 9.567t - 3.460 \times 10^{-4}\sin 37.810t$$

The horizontal relative displacement $u_{r2}(t)$ between first-story and second-story levels is

$$u_{r2}(t) = u_2(t) - u_1(t)$$

$$= -0.036 \sin 4\pi t + 0.045 \sin 9.567t \tag{7.146}$$

$$+ 5.901 \times 10^{-4}\sin 37.810t$$

The time of maximum response for $u_1(t)$ and $u_{r2}(t)$ may be obtained by setting equal to zero the derivatives with respect to time t of Eqs. (7.144) and (7.146). This yields $t_m = 0.85$ s. On this basis, we have the following values for $u_1(t)$ and $u_{r2}(t)$:

$$u_1(t = 0.85 \text{ s}) = -0.61236 \text{ in.}$$

$$u_{r2}(t = 0.85 \text{ s}) = -0.07790 \text{ in.}$$

Since the stiffness values of the first-story columns are all the same, the total shear force at the top of the first-story columns will be distributed equally between the two columns. Therefore, the maximum shear force $V_{1\text{max}}$ acting on each column is

$$V_{1\text{max}} = \frac{k_1 u_1(t = 0.85 \text{ s})}{2}$$

$$= \frac{(33{,}936.328)(0.61236)}{2}$$

$$= 10{,}390.62 \text{ lb}$$

In a similar manner, the maximum shear force $V_{2\text{max}}$ acting in each of the second-story columns is found to be as follows:

$$V_{2\text{max}} = \frac{k_2 u_{r2}(t = 0.85 \text{ s})}{2}$$

$$= \frac{(114{,}535.108)(0.07790)}{2}$$

$$= 4461.14 \text{ lb}$$

The maximum bending moments $M_{1\text{max}}$ and $M_{2\text{max}}$ acting on each of the first-story and second-story columns, respectively, are as follows:

$$M_{1\text{max}} = \frac{V_{1\text{max}} L_1}{2}$$

$$= \frac{(10{,}390.62)(18)(12)}{2}$$

$$= 1{,}122{,}186.96 \text{ in.} \cdot \text{lb}$$

$$M_{2\text{max}} = \frac{V_{2\text{max}}L_2}{2}$$

$$= \frac{(4461.14)(12)(12)}{2}$$

$$= 321{,}202.08 \text{ in.} \cdot \text{lb}$$

Consequently, the maximum bending stresses $\sigma_{1\text{max}}$ and $\sigma_{2\text{max}}$ at the top and bottom of each of the first-story and second-story columns, respectively, are as shown below:

$$\sigma_{1\text{max}} = \frac{M_{1\text{max}}}{S}$$

$$= \frac{1{,}122{,}186.96}{78}$$

$$= 14{,}387.01 \text{ psi}$$

$$\sigma_{2\text{max}} = \frac{M_{2\text{max}}}{S}$$

$$= \frac{321{,}202.08}{78}$$

$$= 4117.98 \text{ psi}$$

Since the intensity of the earthquake in the above problem is moderate, reasonable bending stresses have been obtained from the above elastic analysis. For stronger earthquakes, the bending stresses would be much higher, and elastic analysis may not be a reasonable thing to do. Since earthquakes may occur only a few times, if any, during the life of the structural or mechanical system, it would be more appropriate to design for inelastic response and make a decision as to how far in the inelastic range the system should be permitted to be stressed. More information regarding this specialized topic may be found in ref. 1, 2, and 7.

PROBLEMS

7.1   Determine the differential equations of motion for the spring–mass systems in Fig. P1.12 by using Lagrange's equation.

7.2   Determine the differential equations of motion for the spring–mass systems in Fig. P1.15 by using Lagrange's equation.

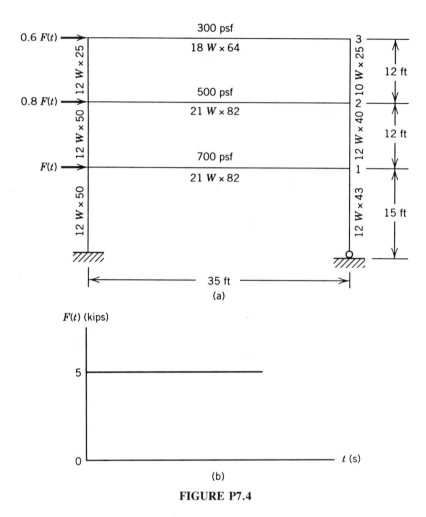

FIGURE P7.4

7.3 Determine the differential equations of motion for the spring–mass systems in Fig. P1.17 by using Lagrange's equation.

7.4 The three-story frame in Fig. P7.4a is subjected to the horizontal dynamic forces $F(t)$, $0.8F(t)$, and $0.6F(t)$ at the first-, second-, and third-story levels, as shown in the figure. The time variation of $F(t)$ is shown in Fig. 7.4b. The girder and column sizes, as well as the dimensions of the frame, are given in Fig. P7.4a. By idealizing the frame as a three-degree spring–mass system and applying modal analysis, determine the maximum horizontal displacements at the floor levels 1, 2, and 3, and the maximum bending stresses in the columns. Assume that the girders are rigid and neglect the weight of the columns. The floor weight distributions are shown in Fig. P7.4a.

7.5   The two-story frame in Fig. P7.5a is subjected to the horizontal dynamic forces $F(t)$ and $0.7F(t)$ at levels 1 and 2, respectively, as shown in the figure. The force $F(t) = 5$ kips is applied suddenly at $t = 0$ and is released suddenly at $t = 0.3$ s, as shown in Fig. P7.5b. The sizes of the columns and girders, the dimensions of the frame, and the weight per unit length at the walls, floor, and roof levels are as shown in Fig. 7.5a. By idealizing the frame as a two-degree spring–mass system and applying modal analysis, determine the maximum horizontal displacements at the floor and roof levels and the maximum bending stresses in the columns. The girders are assumed to be rigid.

7.6   Repeat Problem 7.5 by using the time variation of $F(t)$ shown in Fig. P7.5c.

7.7   For the steel frames shown in Fig. P7.7, determine the maximum horizontal displacements at the girder levels, and the maximum bending stresses in the columns by using the method of modal analysis. The constant dynamic force $F_1(t) = 6$ kips is applied suddently at $t = 0$ and is removed suddenly at $t = 0.2$ s. Assume that the girders are rigid.

7.8   Repeat Problem 7.7 when the force $F_1(t) = 10$ kips is applied suddenly at $t = 0$ and decreases linearly to zero at $t = 0.2$ s.

7.9   If the entire length of each of the uniform beams in Fig. P7.9 is acted on by a uniformly distributed dynamic force of 5 kips applied suddenly at $t = 0$ and of infinite duration, determine the maximum vertical displacement at the center of each span by using modal analysis. The stiffness $EI = 45 \times 10^6$ kip · in.^{2} and the beam weight $w = 0.6$ kips/in. Use the first mode response only.

7.10   Repeat Problem 7.9 by superimposing the dynamic responses of the first three modes.

7.11   The uniform simply supported single-span beam in Fig. P7.11a is loaded by the concentrated dynamic forces $F_1(t)$, $F_2(t)$, and $F_3(t)$ located at $L/4$, $L/2$, and $3L/4$, respectively. If $F_1(t) = F_1 = 10$ kips, $F_2(t) = F_2 = 20$ kips, and $F_3(t) = F_3 = 15$ kips, and all three forces are applied suddenly at $t = 0$ and removed suddenly at $t = 0.3$ s as shown in Fig. P7.11b, determine the maximum vertical displacement at the center of the beam by applying modal analysis. The stiffness $EI = 30 \times 10^6$ kip · in.^{2}, and the weight w of the beam is 0.6 kip/ft.

7.12   Repeat Example 7.6 by considering the response of the first four modes, and compare the results.

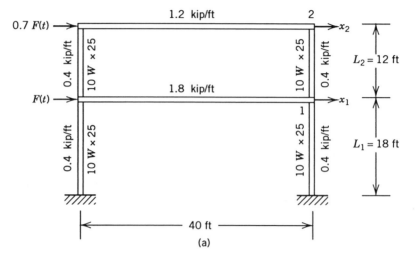

(a)

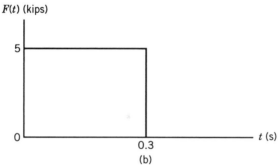

(b)

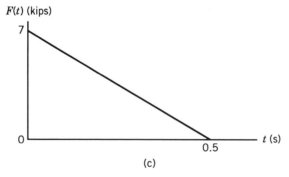

(c)

FIGURE P7.5

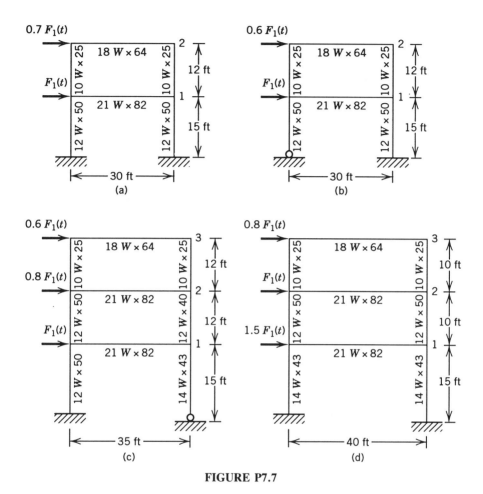

FIGURE P7.7

7.13 Repeat Example 7.6 for velocities $v = 20$, 40, 50, and 60 miles per hour, and compare the results.

7.14 Repeat Example 7.6 for lengths $L = 20$, 40, 50, 60, 80, and 100 ft, and compare the results.

7.15 Repeat Problem 7.13 for lengths $L = 20$ and 50 ft, and compare the results.

7.16 Repeat Example 7.7 by assuming that the dynamic pressure $q(t) = 60$ psi is applied suddenly at $t = 0$ and lasts indefinitely. Compare the results.

7.17 The steel rectangular plate in Fig. 7.7a is subjected to a uniform dynamic pressure of 40 psi applied suddenly at $t = 0$ and lasting indefinitely. The sides of the plate are $a = 120$ in. and $b = 70$ in., the thick-

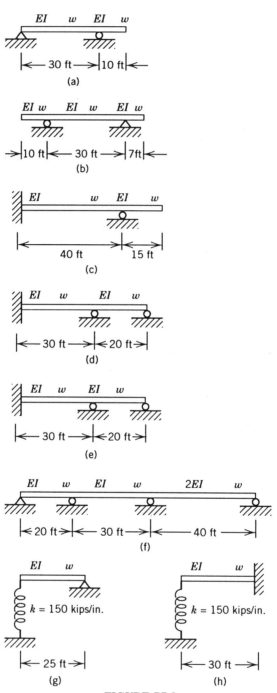

FIGURE P7.9

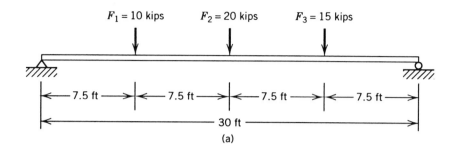

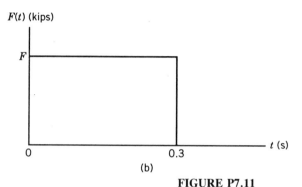

FIGURE P7.11

ness $h = 2$ in., the mass $m = 0.00146$ lb · s²/in.³, the Poisson ratio $\nu =$ 0.25, and the modulus of elasticity $E = 30 \times 10^6$ psi. Determine the maximum displacement of the plate and the maximum bending stresses by considering only the first mode response.

7.18 Repeat Example 7.7 by considering the response of the first two modes, and compare the results.

7.19 Repeat Problem 7.16 by considering the response of the first two modes, and compare the results.

7.20 Repeat Problem 7.17 by considering the response of the first two modes, and compare the results.

7.21 Repeat Example 7.7 with $a = b = 60$ in., and compare the results.

7.22 Repeat Problem 7.16 with $a = b = 60$ in., and compare the results.

7.23 Repeat Example 7.7 by assuming that the pressure $q(t)$ of 60 psi is applied suddenly at $t = 0$ and decreased linearly to zero at $t = t_d = 0.03$ s. Compare the results.

7.24 Repeat Example 7.8 by assuming that $\ddot{u}_s = 0.10g \sin 4\pi t$, and compare the results.

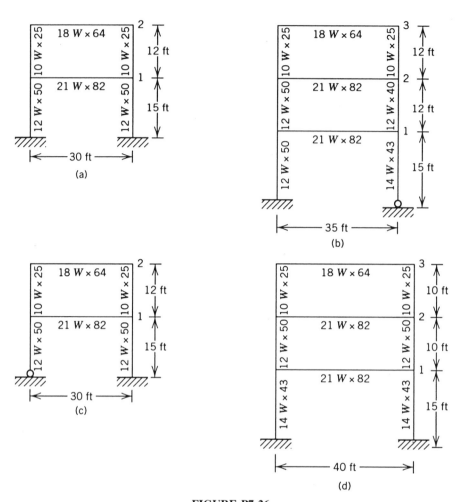

FIGURE P7.26

7.25 Repeat Example 7.8 by assuming that $\ddot{u}_s = 0.15g \sin 4\pi t$, and compare the results.

7.26 The steel frames in Fig. P7.26 are subjected to an artificial earthquake of ground acceleration $\ddot{u}_s = 0.15g \sin 4\pi t$, where g is the acceleration of gravity, and t is time. The duration of $\ddot{u}_s$ is 1 s. By applying the method of modal analysis, determine the maximum bending stresses in the columns of each frame. Assume that the girders are rigid and neglect damping. The modulus of elasticity $E = 30 \times 10^6$ psi.

8 Vibration Response of Variable Stiffness Members

8.1 INTRODUCTION

Most of the research regarding the computation of the free vibration and the corresponding mode shapes of beams and frames is ordinarily limited to constant stiffness members with small static and vibrational amplitudes. Not much work, however, was done in the past for the solution of problems that involve component members with continuously varying stiffness along their lengths. This topic has received more attention today, because more structures and machines are composed of structural elements with continuously varying stiffness along their lengths. For example, the moment of inertia of many concrete girders of highway bridges is variable. Similar observations can be made by examining the wings of an airplane, the component elements of many buildings and machines, tall stacks and poles, and the many futuristic space structures that are expected to overwhelm the minds of our 21st century design engineers.

In addition, we have to remember that our structural and mechanical problems are becoming more nonlinear, and reliable methods to deal with geometric and material nonlinearities must be developed. The solution of such problems is very complex, and it requires methods of analysis that will help one to understand better the physical behavior of nonlinear systems in both static and dynamic senses.

The purpose of this chapter is to provide methods and concepts that can be used to derive accurate and convenient solutions to complicated linear and nonlinear beam problems. Most of the work will be concentrated in illustrating the advantages and disadvantages of such methodologies, and references will be provided regarding the extensive work done by the author and his collaborators [1, 2, 7, 12, 14, 23–25, 64] for further study of these important subjects. Extensive use will be made of equivalent linear, nonlinear, and pseudolinear systems as developed by the author and his collaborators, because they provide accurate and convenient solutions to extremely complicated linear and nonlinear problems. Several methods, such as the finite element, Stodola's, and Galerkin's consistent finite element methods, are used to solve the same problem and compare the results. The level of treatment is prepared in a way that will be appreciated by students at the senior and first-year graduate level, as well as by professional engineers in industry and government.

8.2 DERIVATION OF DIFFERENTIAL EQUATIONS FOR LINEAR AND NONLINEAR SYSTEMS OF UNIFORM AND VARIABLE STIFFNESS

If we consider a member that is subjected to large deformations, say, the one in Fig. 8.1a, its general expression regarding the large deformation of the member is represented by the Euler–Bernoulli equation as shown below:

$$\frac{y''}{[1 + (y')^2]^{3/2}} = -\frac{M_x}{E_x I_x} \tag{8.1}$$

Equation (8.1) is a nonlinear differential equation and represents the exact

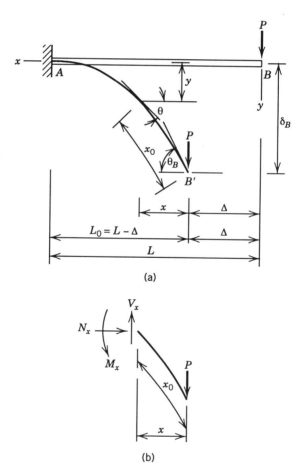

(a)

(b)

FIGURE 8.1. (a) Large deformation of a loaded uniform cantilever beam. (b) Free-body diagram of an element of the beam.

shape of the deflection curve of a flexible member, which is called the *elastica*. The problem of the elastica was first investigated by Bernoulli, Lagrange, Euler, and Plana [65–67], and mathematical solutions of some simple problems have been obtained. The most popular elastica is the solution of a flexible uniform cantilever beam loaded by a concentrated load P at the free end. Solutions were obtained later by Frisch–Fay [68] and other investigators, which are either numerical or closed-form type solutions of rather simple problems. In recent years, the problem of large deformations for many complicated problems was investigated in detail by the author and his collaborators [2, 14,64]. Elastic and inelastic responses, as well as vibration responses are investigated in detail.

In Eq. (8.1), the moment M_x, as well as the modulus of elasticity E_x and moment of inertia I_x, may vary in any arbitrary manner. In Fig. 8.1a, we note that both horizontal and vertical displacements are large. The vertical displacement is denoted by $y(x)$ as a function of x, and the horizontal displacement is denoted by $\Delta(x)$, which is also a function of x. At the free end of the beam, the horizontal displacement is denoted by Δ and the vertical displacement by δ_B. With the free end as the origin, the horizontal displacement $\Delta(x)$ and vertical displacement $y(x)$, at any horizontal distance x, may be obtained by solving Eq. (8.1). Note in the same figure that the coordinate x_0 defines points along the deformed arc length of the member. From ref. 2, we find that x and x_0 are interrelated by the expression

$$x_0(x) = x + \Delta(x) \qquad (8.2)$$

We also know that

$$x_0(x) = \int_0^x [1 + (y')^2]^{1/2}\, dx \qquad (8.3)$$

and, consequently, we have

$$x + \Delta(x) = \int_0^x [1 + (y')^2]^{1/2}\, dx \qquad (8.4)$$

The total length L of the member is

$$L = L_0 + \Delta \qquad (8.5)$$

where Δ is the horizontal movement of the free end of the cantilever beam, and it may be determined from the following expression:

$$L = \int_0^{L_0} [1 + (y')^2]^{1/2}\, dx \qquad (8.6)$$

If the deformations are small, then y' in the denominator of Eq. (8.1) is very small compared to one, and it can be neglected. On this basis, Eq. (8.1) may be written

$$y'' = -\frac{M_x}{E_x I_x} \qquad (8.7)$$

Equation (8.7) is a linear differential equation, since we assumed that the deformations are small. However, both E_x and I_x may vary in any arbitrary manner in a given problem. Such variations complicate the solution of the problem, as will be shown later in the chapter.

For small deformations, we can also assume that $x = x_0$, and, consequently, the horizontal displacements $\Delta(x)$ would be zero. On this basis, the undeformed length L of the member will remain unchanged during deformation and will be equal to the deformed straight length L_0. The use of Eqs. (8.1) through (8.7) will be discussed later in the chapter.

Consider now the flexible cantilever beam in Fig. 8.2a, and assume that w_0 is the securely attached weight, which includes the weight of the member and other securely attached weights. Its large deformed configuration is shown in Fig. 8.2b. In this figure, $y_s(x)$ is the static equilibrium position caused by w_0, and $y_d(x, t)$ is the dynamic amplitude when the beam vibrates about its static equilibrium position $y_s(x)$. The total displacement from the undeformed horizontal configuration is $y(x, t)$, and, consequently,

$$y(x, t) = y_s(x) + y_d(x, t) \qquad (8.8)$$

We are going to derive here the differential equation of motion for the freely vibrating beam. We are not including any other restrictions, except that there is no damping; therefore, the differential equation of motion would be applicable for any beam span.

We begin the derivation by considering the Euler–Bernoulli equation given by Eq. (8.1). We multiply both sides of this equation by $E_x I_x$, and we obtain

$$E_x I_x \frac{y''}{[1 + (y')^2]^{3/2}} = -M_x \qquad (8.9)$$

If we differentiate both sides of Eq. (8.9) with respect to x, we obtain

$$\frac{d}{dx}\left(E_x I_x \frac{y''}{[1 + (y')^2]^{3/2}}\right) = -V(x) \qquad (8.10)$$

where the shear force $V(x) = dM_x/dx$, at any cross section, must be the same

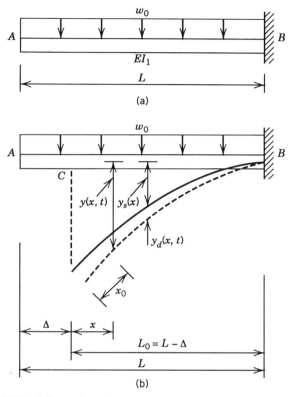

FIGURE 8.2. (a) Undeformed configuration of a uniform cantilever beam. (b) Deformed configuration of the cantilever beam.

whether it is defined in the deformed or undeformed configuration; that is, $V(x_0) = V(x(x_0))$. On this basis, Eq. (8.10) yields

$$\frac{d}{dx}\left[E_xI_x\frac{y''}{[1 + (y')^2]^{3/2}}\right] = -\frac{V(x_0)}{\cos\theta} \tag{8.11}$$

Differentiation of Eq. (8.11) with respect to x yields

$$\frac{d^2}{dx^2}\left(E_xI_x\frac{y''}{[1 + (y')^2]^{3/2}}\right) = -\frac{d}{dx}\left(\frac{V(x_0)}{\cos\theta}\right) \tag{8.12}$$

or

$$\frac{d^2}{dx^2}\left(E_xI_x\frac{y''}{[1 + (y')^2]^{3/2}}\right) = -\frac{1}{\cos\theta}\frac{d}{dx_0}\left(\frac{V(x_0)}{\cos\theta}\right) \tag{8.13}$$

because the expression for the transverse weight is well defined in the unde-formed configuration; that is,

$$V(x_0) = -x_0 w(x_0) \cos \theta \tag{8.14}$$

By performing the required differentiation on the right-hand side of Eq. (8.13), we obtain

$$\frac{d^2}{dx^2}\left(E_x I_x \frac{y''}{[1 + (y')^2]^{3/2}}\right) = -\frac{1}{\cos \theta} w(x_0) \tag{8.15}$$

where $x_0(x)$ is given by Eq. (8.3), and

$$\cos \theta = \frac{1}{[1 + (y')^2]^{1/2}} \tag{8.16}$$

By using Eq. (8.16) and assuming a uniformly distributed weight, that is, $w(x_0) = w_0$, Eq. (8.15) yields

$$\frac{d^2}{dx^2}\left(E_x I_x \frac{y''}{[1 + (y')^2]^{3/2}}\right) = -[1 + (y')^2]^{1/2} w_0 \tag{8.17}$$

For transverse free vibration, the weight w_0 in Eq. (8.17) may be replaced by the inertia force $w_{in} = m\ddot{y}$, where m is the uniform mass density of the member, and $\ddot{y}$ is the relative acceleration. On this basis, the transverse free vibration of a flexible member with uniformly distributed mass or weight is written as follows:

$$\frac{d^2}{dx^2}\left(E_x I_x \frac{y''}{[1 + (y')^2]^{3/2}}\right) + [1 + (y')^2]^{1/2} m\ddot{y} = 0 \tag{8.18}$$

If the mass m is of some arbitrary distribution $m(x_0)$ along the arc length of the member, Eq. (8.18) may be written

$$\frac{d^2}{dx^2}\left(E_x I_x \frac{y''}{[1 + (y')^2]^{3/2}}\right) + [1 + (y')^2]^{1/2} m(x_0)\ddot{y} = 0 \tag{8.19}$$

The solution of Eq. (8.19) is extremely difficult because we cannot use separation of variables, and the frequencies of vibration are amplitude depen-dent. The solution of Eq. (8.19) by using equivalent systems is discussed in greater detail in the following sections of this chapter. See also ref. 2.

If the deformations are small, the rotation y' in Eq. (8.19) is small compared to one, and it could be neglected for practical applications. On this basis, Eq. (8.19) may be written

$$\frac{d^2}{dx^2} (E_x I_x y'') + m\ddot{y} = 0 \tag{8.20}$$

Equation (8.20) is a linear differential equation, where the stiffness $E_x I_x$ and mass m per unit length may vary in any arbitrary manner. Since the static displacement $y_s(x)$ in Eq. (8.8) is small, we can assume that the static deformed and undeformed configurations of the member are the same, and, consequently, $y_s(x)$ is zero. This is a commonly used assumption in the theory of vibrations. Although vibrations are taking place with respect to the static equilibrium position, the above assumption suggests that the static equilibrium position of the member and its undeformed configuration are identical, and, consequently, for Eq. (8.20), we can assume that $y(x, t) = y_d(x, t)$. If both E and I in Eq. (8.20) are constant, or EI is constant, we have

$$EI \frac{\partial^4 y}{\partial x^4} + m\ddot{y} = 0 \tag{8.21}$$

Equation (8.21) is identical to Eq. (2.201), which was derived in Section 2.11.

8.3 EQUIVALENT SYSTEMS FOR LINEAR AND NONLINEAR PROBLEMS OF UNIFORM AND VARIABLE STIFFNESS

In this section, the initial linear or nonlinear system, which is subjected to large or small deformations, with modulus E_x and moment of inertia I_x that can vary in any arbitrary manner, may be transformed into an equivalent pseudolinear system of constant stiffness $E_1 I_1$ that has the same deflection as the initial system with variable stiffness $E_x I_x$. The constant stiffness equivalent system may be solved by applying well-known linear analysis. This concept and methodology may be used to simplify the solution of complex linear and nonlinear problems with complicated loading conditions and stiffness variations.

8.3.1 Constant Stiffness Equivalent Pseudolinear Systems for Flexible Members

Flexible members are associated with large deformations, and their elastic line is represented by the nonlinear Euler–Bernoulli equation, which is

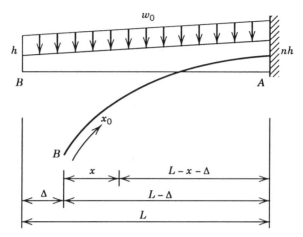

FIGURE 8.3. Tapered cantilever beam loaded with a uniformly distributed loading.

written as shown by Eq. (8.1). In this equation, the moment M_x, modulus of elasticity E_x, and moment of inertia I_x are assumed to vary in any arbitrary manner.

The curvature on the left-hand side of Eq. (8.1) is geometrical in nature, and one must relate the quantities M_x, E_x, and I_x on the right-hand side of the equation with the deformed configuration of the member. When the cross-sectional moment of inertia is variable along the length of the flexible member and the applied loading is distributed, the expressions for these parameters are, in general, nonlinear integral equations of the deformation and contain functions of horizontal displacement. This means that M_x, I_x, and depth h_x of the member are all functions of x and x_0. This is easily observed by examining the variable stiffness cantilever beam in Fig. 8.3.

The derivation of pseudolinear equivalent systems of constant stiffness $E_1 I_1$ may be initiated by assuming that $E_x I_x$ may be expressed as follows:

$$E_x I_x = E_1 I_1 g(x) f(x) \qquad (8.22)$$

where the function $g(x)$ represents the variation of E_x with respect to an arbitrary reference value E_1, and the function $f(x)$ represents the variation of I_x with respect to an arbitrary reference value I_1. For constant E and I, we have $g(x) = f(x) = 1.00$, and $E_x I_x = E_1 I_1 = EI$.

By substituting Eq. (8.22) into Eq. (8.1), we find

$$\frac{y''}{[1 + (y')^2]^{3/2}} = -\frac{M_x}{g(x)f(x)} \cdot \frac{1}{E_1 I_1} \qquad (8.23)$$

By integrating Eq. (8.23) twice, we have

$$y(x) = \frac{1}{E_1 I_1} \int \left(- \int [1 + (y')^2]^{3/2} \frac{M_x}{g(x)f(x)} \, dx \right) dx + C_1 \int dx + C_2 \qquad (8.24)$$

where C_1 and C_2 are the constants of integration, which can be determined by using the boundary conditions of the member.

If we consider now a flexible member of constant stiffness $E_1 I_1$, which has the same length and coordinate reference system of axes as the one used for Eq. (8.24), the expression for its large displacement y_e may be written as follows:

$$y_e(x) = \frac{1}{E_1 I_1} \int \left(- \int [1 + (y')^2]^{3/2} M_e \, dx \right) dx + C_1' \int dx + C_2' \qquad (8.25)$$

where C_1' and C_2' are the constants of integration, and M_e is the bending moment at any cross section x.

The deflections $y(x)$ and $y_e(x)$ are identical if the following conditions are satisfied:

$$C_1 = C_1' \qquad C_2 = C_2' \qquad (8.26)$$

$$\int \left(- \int [1 + (y')^2]^{3/2} \frac{M_x}{g(x)f(x)} \, dx \right) dx = \int \left(- \int [1 + (y_e')^2]^{3/2} M_e \, dx \right) dx \qquad (8.27)$$

The conditions given by Eq. (8.26) may be satisfied if the two members have the same length and boundary conditions. Equation (8.27) may be satisfied by the following two identities:

$$y' = y_e' \qquad (8.28)$$

$$M_e = \frac{M_x}{f(x)g(x)} \qquad (8.29)$$

On this basis, Eq. (8.27) may be reduced to the following identity:

$$[1 + (y')^2]^{3/2} M_e = [1 + (y')^2]^{3/2} \frac{M_x}{f(x)g(x)} \qquad (8.30)$$

Equations (8.23) and (8.30) suggest that the moment M_e' at any cross section x of the pseudolinear system of constant stiffness $E_1 I_1$ may be obtained from the following expression:

$$M'_e = M_e[1 + (y')^2]^{3/2} = \frac{M_x z_e}{f(x)g(x)} \tag{8.31}$$

where

$$z_e = [1 + (y')^2]^{3/2} \tag{8.32}$$

and $\theta = \tan^{-1}(y')$ is the slope of the initial system at any x. With this in mind, Eq. (8.23) takes the following form:

$$y'' = - \frac{M'_e}{E_1 I_1} \tag{8.33}$$

where M'_e is given by Eq. (8.31).

Equation (8.33) is the differential equation of a pseudolinear system of constant stiffness $E_1 I_1$. If M'_e is known, or if it can be determined, then linear analysis may be used to solved the pseudolinear system. The shear force V'_e and loading w'_e of the pseudolinear system may be determined as follows:

$$V'_e = \frac{d}{dx}(M'_e) = \frac{d}{dx}\left(\frac{z_e}{f(x)g(x)}\right)M_x \tag{8.34}$$

$$w'_e = - \frac{d}{dx}(V'_e)\cos\theta = - \frac{d^2}{dx^2}\left(\frac{z_e}{f(x)g(x)}\right)M_x\cos\theta \tag{8.35}$$

8.3.2 Simplified Nonlinear Equivalent Systems of Constant Stiffness $E_1 I_1$

There are many problem situations in nonlinear analysis where the loading conditions involve load combinations of various types. For example, we may have a partially distributed loading combined with concentrated loads along the length of the member. In addition, the moment of inertia along the length of the member may vary in a very arbitrary manner. Since the method of superposition is not applicable for such nonlinear problems, a reasonable solution would be the derivation of simpler equivalent nonlinear systems of constant stiffness $E_1 I_1$, which represent very accurately (or exactly) the original complex nonlinear problem. The simplified equivalent nonlinear system of constant stiffness may then be solved by deriving a pseudolinear system, as discussed in Section 8.3.1, and applying linear analysis.

The derivation of such equivalent systems is initiated by considering Eq. (8.23). We rewrite this equation as follows:

$$\frac{y''}{[1 + (y')^2]^{3/2}} = -\frac{M_e}{E_1 I_1}$$ (8.36)

where

$$M_e = \frac{M_x}{f(x)g(x)}$$ (8.37)

Equation (8.36) is the nonlinear differential equation of the simplified non-linear equivalent system of constant stiffness $E_1 I_1$. The bending moment M_e at any cross section of this equivalent nonlinear system may be obtained from Eq. (8.37). The shear force V_e and the loading w_e may be determined from Eq. (8.37) by differentiation; that is,

$$V_e = \frac{d}{dx}(M_e) = \frac{d}{dx}\left(\frac{M_x}{f(x)g(x)}\right)$$ (8.38)

$$w_e = -\frac{d}{dx}(V_e)\cos\theta = -\frac{d^2}{dx^2}\left(\frac{M_x}{f(x)g(x)}\right)\cos\theta$$ (8.39)

The solution of the initial nonlinear problem may now be obtained from Eq. (8.36). This equation, since it represents a simpler nonlinear problem, may conveniently be solved by using pseudolinear systems, as discussed in Section 8.3.1. To make the solution easier, the shape of the M_e diagram given by Eq. (8.37) may be approximated with a few straight-line segments. On this basis, the equivalent nonlinear system will always be loaded with a few concentrated loads. This approximation is very accurate, and it greatly simplifies the solution of the nonlinear problem. Similar reasoning can be applied for the M_e' diagram of the pseudolinear equivalent system. This is explained in greater detail later in this chapter. See also refs. 1 and 2.

8.3.3 Equivalent Systems for Beams of Variable Stiffness Subjected to Small Deformations

We recall here Eq. (8.36). If the deformations are small, that is, small deflection theory, the rotation y' on the left-hand side of Eq. (8.36) is small compared to one, and it may be neglected. This yields

$$y'' = -\frac{M_e}{E_1 I_1}$$ (8.40)

where M_e is given by Eq. (8.37). Equation (8.40) is a linear differential equation, and it represents an equivalent system of constant stiffness $E_1 I_1$. The bending moment M_e at any cross section along the length of the member may be obtained from Eq. (8.37), and the shear force V_e and loading w_e may be determined by differentiation; that is,

$$V_e = \frac{d}{dx}(M_e) = \frac{d}{dx}\left(\frac{M_x}{f(x)g(x)}\right) \tag{8.41}$$

$$w_e = -\frac{d}{dx}(V_e) = -\frac{d^2}{dx^2}\left(\frac{M_e}{f(x)g(x)}\right)\cos\theta \tag{8.42}$$

Since the rotations are small, $\cos\theta$ in Eq. (8.42) may be assumed to be equal to one.

The theories developed in these last three subsections will be applied in the following sections for the solution of linear and nonlinear problems.

8.4 TAPERED FLEXIBLE CANTILEVER BEAM LOADED WITH A CONCENTRATED LOAD AT THE FREE END

In this section, the theory developed in the preceding section will be used to derive pseudolinear equivalent systems of constant stiffness for flexible tapered cantilever beams. The procedure and methodology will be illustrated by considering the tapered cantilever beam in Fig. 8.4a, which is loaded by a concentrated load P at its free end. The modulus of elasticity E is assumed to be constant, and the moment of inertia I_x at any $0 \leqslant x \leqslant L_0$, where $L_0 = L - \Delta$ and Δ is the horizontal displacement of the free end B, is given by the following equation:

$$I_x = I_B f(x) \tag{8.43}$$

where $I_B = bh^3/12$ is the moment of inertia at the free end of the member, and b is the constant width. The function $f(x)$, which expresses the variation of I_x with respect to the reference value I_B, is as follows:

$$f(x) = \left(1 + \frac{(n-1)}{L}x_0\right)^3 \tag{8.44}$$

where

$$x_0 = \int_0^x [1 + (y')^2]^{1/2}\, dx \tag{8.45}$$

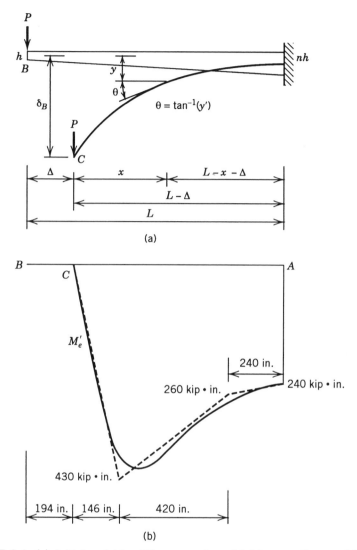

FIGURE 8.4. (a) Initial variable stiffness member. (b) Moment diagram M'_e of the pseudolinear system with its shape approximated with three straight-line segments.

The bending moment M_x at any $0 \leq x \leq L_0$ (Fig. 8.4a) is as follows:

$$M_x = Px \tag{8.46}$$

By substituting Eqs. (8.43) and (8.46) into Eq. (8.23) and observing that

$g(x) = 1$, we obtain

$$\frac{y''}{[1 + (y')^2]^{3/2}} = -\frac{P}{EI_B} \frac{x}{\left(1 + \dfrac{(n-1)}{L} \displaystyle\int_0^x [1 + (y')^2]^{1/2}\, dx\right)^3} \tag{8.47}$$

Equation (8.47) is an integral nonlinear differential equation that is very difficult to solve.

It was stated earlier that I_x is also a nonlinear function of the large deformation, as can be observed from Eqs. (8.43) through (8.45). For uniform and tapered members that are loaded with concentrated loads only, it was shown by the author [2] that the variation of the depth $h(x)$ of the member may be approximated by the following equation:

$$h(x) = (n-1)\left(\frac{1}{n-1} + \frac{x}{L - \Delta}\right)h \tag{8.48}$$

The error in the analysis, when Eq. (8.48) is used, is less than 3%, and it is considered acceptable for practical applications.

By using Eq. (8.48), we can write the variation of I_x as follows:

$$I_x = \frac{bh^3}{12}\left(1 + \frac{(n-1)x}{L - \Delta}\right)^3$$

$$= I_B f(x) \tag{8.49}$$

where

$$I_B = \frac{bh^3}{12} \tag{8.50}$$

$$f(x) = \left(1 + \frac{(n-1)x}{L - \Delta}\right)^3 \tag{8.51}$$

By substituting Eqs. (8.46) and (8.49) into Eq. (8.23), we obtain

$$\frac{y''}{[1 + (y')^2]^{3/2}} = -\frac{P}{EI_B} \cdot \frac{x(L - \Delta)^3}{[(x-1)x + (L - \Delta)]^3} \tag{8.52}$$

On this basis, a closed-form solution may now be obtained rather conveniently by using Eq. (8.52) in place of Eq. (8.47).

By integrating Eq. (8.52) once and determining the constant of integration by applying the boundary condition of zero rotation at $x = L - \Delta$, we obtain

$$y'(x) = \frac{Q(x)}{\{1 - [Q(x)]^2\}^{1/2}} \tag{8.53}$$

where

$$Q(x) = \frac{P(L - \Delta)^3}{EI_B} \left(\frac{2(n - 1)x + (L - \Delta)}{2(n - 1)^2[(x - 1)x + (L - \Delta)]^2} - \frac{2n - 1}{2(n - 1)^2 n^2 (L - \Delta)} \right)$$

$$\tag{8.54}$$

Equations (8.53) and (8.54) are functions of the horizontal displacement Δ of the free end of the beam, and the value of Δ may be determined from Eq. (8.6) by using a trial-and-error procedure. We rewrite this equation below:

$$L = \int_0^{L_0} [1 + (y')^2]^{1/2} \, dx \tag{8.55}$$

where $L_0 = L - \Delta$. A trial-and-error procedure may be initiated by assuming a value of Δ in Eq. (8.53). This yields $y'(x)$ as a function of x only. Then use $y'(x)$ in Eq. (8.55) to obtain a value for the length L of the member. The procedure may be repeated for various values of Δ until the correct length L is obtained. Computer programs make this procedure convenient.

With known Δ, the values of $y'(x)$ at any $0 \leq x \leq L_0$ may be obtained from Eq. (8.53), and the values of z_e may be obtained from Eq. (8.32). With known z_e, the values at any $0 \leq x \leq L_0$ of the bending moment M'_e of the pseudolinear system of constant stiffness EI_B may be determined from Eq. (8.31). Equation (8.35) may be used to determine the loading w'_e acting on the pseudolinear equivalent system. Once the pseudolinear equivalent system is obtained, it can be solved by using well-known methods of linear analysis to determine the deflection $y(x)$ at any $0 \leq x \leq L_0$. The procedure is greatly simplified if the shape of M'_e is approximated with straight-line segments. On this basis, the constant stiffness equivalent pseudolinear systems will always be loaded with concentrated loads. A small number of segments, usually two to four segments, provide excellent accuracy for practical applications. The following numerical example illustrates the procedure.

Example 8.1. The tapered cantilever beam in Fig. 8.4a is loaded with a load $P = 1.0$ kip at its free end B. The length $L = 1000$ in., the stiffness $EI_B = 180 \times 10^3$ kip $\cdot$ in.^{2}, and the taper $n = 1.5$. Determine a pseudolinear equivalent system of constant stiffness EI_B that is loaded with equivalent concentrated loads. Then, by following linear analysis, solve the pseudolinear system to determine the deflection and rotation at the free end of the member.

Solution. By using the given data and Eq. (8.54), we find

$$
\begin{aligned}
Q(x) = {} & \frac{1.0(1000 - \Delta)^3}{180 \times 10^3} \left(\frac{2(1.5 - 1)x + (1000 - \Delta)}{2(1.5 - 1)^2[(1.5 - 1)x + (1000 - \Delta)]^2} \right. \\
& \left. - \frac{(2)(1.5) - 1}{2(1.5 - 1)^2(1.5)^2(1000 - \Delta)} \right) \\
= {} & \frac{1.0(1000 - \Delta)^3}{180 \times 10^3} \frac{x + 1000 - \Delta}{(0.5)(0.5x + 1000 - \Delta)^2} \\
& - \frac{2}{(1.25)(1000 - \Delta)} \Big)
\end{aligned}
$$

The iteration for the computation of the horizontal displacement Δ of the free end may be initiated by assuming a value of $\Delta = 194.0$ in. Thus,

$$
\begin{aligned}
Q(x) = {} & \frac{1.0(1000 - 194)^3}{180 \times 10^3} \left(\frac{x + 1000 - 194}{(0.5)(0.5x + 1000 - 194)^2} \right. \\
& \left. - \frac{2}{(1.25)(1000 - 194)} \right)
\end{aligned}
$$

or

$$
Q(x) = \frac{-1.6040x^2 + 646.4280x + 521,020.9015}{0.25x^2 + 806x + 649,636}
$$

By substituting into Eq. (8.55) and integrating from $x = 0$ to $x = L_0 = 806$ in., we find

$$
L = \int_0^{806} \left[1 + \left(\frac{Q(x)}{1 - [Q(x)]^2]^{1/2}} \right)^2 \right]^{1/2} dx
$$

$$
= 1000 \text{ in.}
$$

In this case, the assumed value of $\Delta = 194$ in. is the correct value for the horizontal displacement of the free end of the member. Simpson's rule may be used to carry out the required integrations.

With $\Delta = 194$ in., we may use Eqs. (8.53), (8.32), (8.46), and (8.31) to determine the required values of y', z_e, M_x, and M'_e, respectively. The results

TABLE 8.1. Values of M'_e of the Pseudolinear System of Constant Stiffness EI_B at Intervals of 100 in.

x (in.)	$f(x)$	y' (rad)	z_e	M_x (kips · in.)	M'_e (kips · in.)
0	1.0000	1.3428	4.9631	0.00	0.00
6	1.0112	1.3423	4.6897	6.00	27.23
106	1.2105	1.2245	3.9514	106.00	346.01
206	1.4345	1.0064	2.8557	206.00	410.09
306	1.6894	0.7827	2.0598	306.00	374.20
406	1.9619	0.5927	1.5708	406.00	325.06
506	2.2682	0.4236	1.2809	506.00	285.75
606	2.6049	0.2728	1.1137	606.00	259.09
706	2.9733	0.1348	1.0274	706.00	243.95
806	3.3750	0.0000	1.0000	806.00	238.81

are listed in Table 8.1. These values are given at intervals of 100 in. with the exception of the first interval. In the computation of these values, we took into consideration that the modulus E is constant, and, consequently, $g(x) = 1.0$. Cases where the modulus E is not constant along the length of a member are examined in detail in ref. 2 and also in the later parts of this chapter.

The moment diagram M'_e of the pseudolinear system is shown plotted by the solid line in Fig. 8.4b. The approximation of its shape with three straight-line segments is shown by the dashed lines in the same figure. Note that the straight-line juncture points can be above or below the solid-line points, so that the areas added or subtracted from the M'_e diagram are approximately balanced. No particular care should be taken to do this, because the rotations and deflections are not very sensitive to such approximation. Even larger errors in these approximations will produce reasonable values for rotations and deflections, because, in theory, integrations are used to obtain these values, which tend to minimize the error involved.

By using statics, the shear force diagram corresponding to the approximated M'_e diagram is shown in Fig. 8.5b, and, consequently, the pseudolinear equivalent system of constant stiffness EI_B and loaded with concentrated loads is shown in Fig. 8.5c. The large deflection at any $0 \le x \le L_0$ of the original system in Fig. 8.4a can be determined accurately by using the pseudolinear system in Fig. 8.5c and applying elementary linear analysis. For example, the large deflection δ_C at the free end B in Fig. 8.4a is equal to the deflection δ_C at point C of the pseudolinear system in Fig. 8.5c. By using the pseudolinear system and applying the moment–area method, we find $\delta_C = 517.8$ in. The value obtained by solving Eq. (8.52) directly by double integration and using Simpson's rule to carry out the required integration is 519.85 in. The difference is only 0.4%. In a similar manner, by using the pseudolinear system and the moment–area method, we find that $y'_C = 1.6772$,

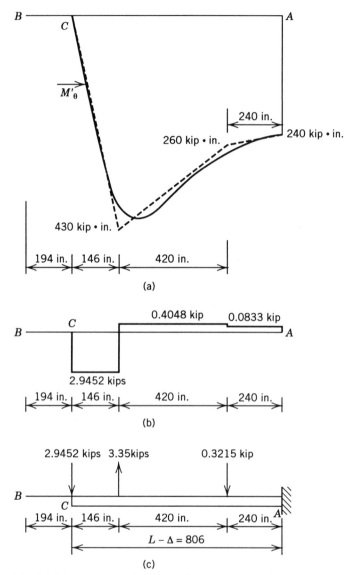

FIGURE 8.5. (a) M'_e diagram of the pseudolinear system and its shape approximation with straight-line segments. (b) Shear force diagram corresponding to the approximated M'_e diagram. (c) Equivalent pseudolinear system of constant stiffness EI_B.

and, consequently, the rotation θ_C at C is $\theta_C = \tan^{-1} y'_C = 52.7°$. The direct solution of Eq. (8.52) yielded $\theta_B = \theta_C = 53.31°$, a difference of only 0.11%.

8.5 ANALYSIS OF TAPERED CANTILEVER BEAMS OF VARIABLE STIFFNESS BY USING SIMPLIFIED NONLINEAR EQUIVALENT SYSTEMS

In this section, the idea of simplified nonlinear equivalent systems of constant stiffness will be used for the analysis of tapered flexible cantilever beams that are subjected to combined loading conditions. The methodology is illustrated by considering the tapered cantilever beam in Fig. 8.6a, which is loaded by a uniformly distributed load $w_0 = 0.005$ kip/in. over half its span and by a concentrated load $P = 1.0$ kip at the free end A. The length L of the member is 1000 in., and the stiffness EI_A at the free end A is 180,000 kip · in.^{2}.

The moment of inertia I_x at any $0 \le x \le L$ is given by the following expression:

$$I_x = I_A \left(\frac{L + 0.5x}{L} \right)^3 \tag{8.56}$$

where

$$I_x = \frac{bh^3}{12} \tag{8.57}$$

$$f(x) = \left(\frac{L + 0.5x}{L} \right)^3 \tag{8.58}$$

and b is the width of the member, which is assumed to be constant. The width, however, could also vary in any arbitrary manner without complicating the procedure.

The bending moment M_x at any $0 \le x \le L$ may be obtained by using statics. The moment M_e of the simplified nonlinear equivalent system of constant stiffness EI_A may be obtained by using Eq. (8.37) with $g(x) = 1.0$. The values of M_e at intervals of 100 in. along the length of the member, as well as M_x and $f(x)$, are shown in Table 8.2. The M_e moment diagram is shown plotted by the solid line in Fig. 8.6b. Its approximation with one straight line AD, as shown by the dashed line in Fig. 8.6b, leads to the simplified nonlinear equivalent system of constant stiffness EI_A shown in Fig. 8.6c. The simplified system is loaded only with a concentrated equivalent load $P_e = 0.52$ kip applied at the free end of the member.

The solution of the simplified nonlinear system in Fig. 8.6c is much simpler than the solution of the initial system in Fig. 8.6a, and it can be obtained

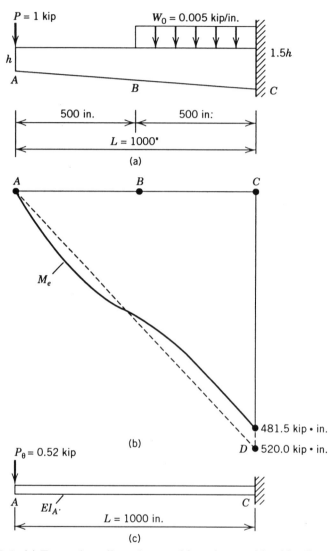

FIGURE 8.6. (a) Tapered cantilever beam subjected to combined loading conditions. (b) Moment diagram M_e of a simplified equivalent system of constant stiffness EI_A. (c) Simplified nonlinear equivalent system of constant stiffness EI_A.

by following pseudolinear analysis as discussed in Section 8.4. In fact, an approximate solution of this particular problem by using elliptic integrals is given by Gere and Timoshenko [27]. The solution of the simplified problem is obtained here by using pseudolinear analysis as discussed in Section 8.4, and the results are as follows:

TABLE 8.2. Values of $f(x)$, M_x, and M_e at Intervals of 100 in.

x (in.)	M_x (kips · in.)	$f(x)$	$M_e = M_x/f(x)$ (kips · in)
0	0	1.0000	0
100	100	1.1576	86.39
200	200	1.3331	150.03
300	300	1.5269	197.25
400	400	1.7280	231.48
500	500	1.9531	256.00
600	625	2.1970	284.48
700	800	2.4604	325.15
800	1025	2.7440	373.54
900	1300	3.0486	426.43
1000	1625	3.3750	481.48

$$L_0 = 755.03 \text{ in.}$$

$$\Delta = 244.97 \text{ in.}$$

$$\delta_A = 593.73 \text{ in.}$$

$$\theta_A = 55.43°$$

The direct solution of the original system in Fig. 8.6a by following pseudolinear analysis was obtained by the author [2], and the results are as shown below:

$$L_0 = 744.19 \text{ in.}$$

$$\Delta = 255.81 \text{ in.}$$

$$\delta_A = 600.10 \text{ in.}$$

$$\theta_A = 58.65°$$

The two results, for practical applications, can be considered identical. The difference is 1.46% for L_0, −4.24% for Δ, 1.06% for δ_A, and −5.49% for θ_A.

8.6 UTILIZATION OF EQUIVALENT SYSTEMS FOR VARIABLE STIFFNESS BEAMS SUBJECTED TO SMALL DEFORMATIONS

In this section, we assume that the deformation of the member is small, and we apply the theory explained in Section 8.3.3 to carry out the analysis of variable stiffness members. For this purpose, we consider the cantilever beam

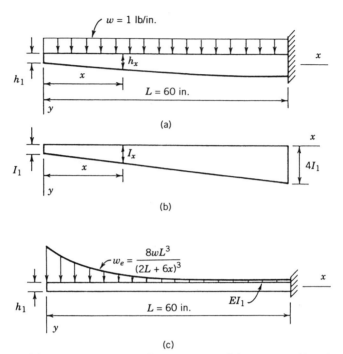

FIGURE 8.7. (a) Original variable stiffness member. (b) Moment of inertia variation. (c) Equivalent system of constant stiffness EI_1.

in Fig. 8.7a, which is loaded with a uniformly distributed load w as shown. The depth $h(x)$ at any distance x from the free end varies in a way that makes the variation of the moment of inertia I_x linear along the length of the member, that is,

$$I_x = \frac{(L + 3x)}{L} I_1 = I_1 f(x) \tag{8.59}$$

where

$$f(x) = \frac{L + 3x}{L} \tag{8.60}$$

By using statics, the bending moment M_x at any distance x is

$$M_x = -\frac{wx^2}{2} \tag{8.61}$$

We assume here that the modulus E is constant, that is, $g(x) = 1$, and we determine the moment M_e of the equivalent system of constant stiffness EI_1 by using Eq. (8.37). This yields

$$M_e = \frac{M_x}{f(x)} = -\frac{wLx^2}{2L - 6x} \tag{8.62}$$

By using Eqs.(8.41) and (8.42), the shear force V_e and loading w_e, respectively, of the equivalent system are determined, and they are as follows:

$$V_e = \frac{d}{dx}(M_e) = -\frac{4wL^2x + 6wLx^2}{(2L + 6x)^2} \tag{8.63}$$

$$w_e = -\frac{d}{dx}(V_e) = \frac{8wL^3}{(2L + 6x)^3} \tag{8.64}$$

The exact equivalent system of constant stiffness EI_1 and its loading w_e are shown in Fig. 8.7c.

We note here that the exact solution of the equivalent system in Fig. 8.7c is still fairly complicated, because w_e is variable. Since we have made the stiffness constant, the initial loading w should be changed to w_e to compensate for the change in stiffness. This mathematical difficulty, however, may be eliminated here by approximating the shape of the M_e diagram with straight-line segments, as was done for the nonlinear system in the preceding sections. For example, if we assume that $L = 60$ in. and $w = 1$ lb/in., Eq. (8.62) yields the M_e diagram shown plotted by the solid line in Fig. 8.8a. If its shape is approximated by the three straight-line segments shown by the dashed lines in the same figure, then the equivalent system of uniform stiffness EI_1 would be as shown in Fig. 8.8b. This system is loaded with concentrated loads only, and it is much easier to solve by linear analysis. By using the equivalent system in Fig. 8.8b and applying the moment–area method, we find that the deflection y_A at the free end is

$$y_A = \frac{491.9 \times 10^3}{EI_1}$$

The exact solution using equivalent systems yields

$$y_A = \frac{489 \times 10^3}{EI_1}$$

$$\text{Error} = \frac{2900}{489 \times 10^3} = \frac{1}{169} < 1\%$$

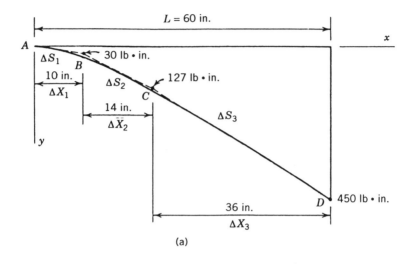

(a)

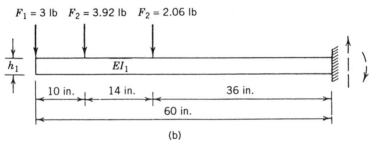

(b)

FIGURE 8.8. (a) M_e diagram of the equivalent system of constant stiffness EI_1, approximated with three straight lines. (b) Equivalent system of constant stiffness EI_1 loaded with three concentrated loads.

The slope θ_A at the same end is $11.63 \times 10^3/EI_1$, and it is practically identical to the exact value.

The method of equivalent systems is general, and it can be applied to both statically determinate and statically indeterminate problems. Extensive work by the author on this subject, for both linear and nonlinear analysis, may be found in refs 1, 2, and 7. The method is particularly convenient for cases with very complicated loading conditions and moments of inertia. It can be applied to beams, plates, and so on, as indicated in the stated references. Some of these problems are treated later in the chapter.

If the problem is statically indeterminate, the M_x in Eq. (8.37) is not readily available. The redundant reactions would have to be determined first in the usual way. Consider, for example, the variable stiffness member in

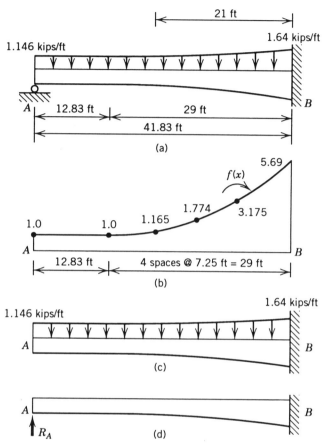

FIGURE 8.9. (a) Original statically indeterminate variable stiffness member. (b) Moment of inertia variation. (c) Cantilever beam with original loading. (d) Cantilever beam loaded with the reaction R_A.

Fig. 8.9a, with stiffness variation $f(x)$ as shown in Fig. 8.9b. By selecting reaction R_A at end A as redundant, the two cantilever beam problems that need to be solved are shown in Figs. 8.9c and 8.9d.

Since the problems in Figs. 8.9c and 8.9d are now statically determinate, M_x for each case can be determined. Note in Fig. 8.9d that M_x would be a function of R_A. By applying the method of equivalent systems as discussed previously, the equivalent system of constant stiffness EI_A for the problem in Fig. 8.9c would be as shown in Fig. 8.10a, and the one for the problem in Fig. 8.9d would be as shown in Fig. 8.10b.

By using the equivalent system in Fig. 8.10a and applying handbook formulas, the deflection δ_A' at the free end A is found to be equal to $-168{,}868/EI_A$. In a similar manner, by using the equivalent system in Fig.

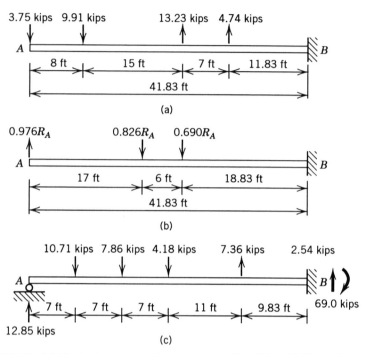

FIGURE 8.10. (a) Equivalent system for member in Fig. 8.9c. (b) Equivalent system for member in Fig. 8.9d. (c) Equivalent system for the original member in Fig. 8.9a.

8.10b, we find that the deflection δ_A'' at the free end is $10,923R_A/EI_A$. The problem In Fig. 8.9a suggests that the deflection δ_A at end A should be zero. Therefore, we must have

$$\delta_A = \delta_A' + \delta_A'' = 0 \qquad (8.65)$$

By substituting the values of δ_A' and δ_A'' in Eq. (8.65) and solving for R_A, we find

$$R_A = 15.46 \text{ kips}$$

With known R_A, the rotations and vertical deflections along the length of the member in Fig. 8.9a may be determined by using the constant stiffness equivalent systems in Figs. 8.10a and 8.10b and superimposing the results at the corresponding points.

Since R_A is known, the moment M_x at any x of the original variable stiffness member in Fig. 8.9a may be determined, and, consequently, by Eq. (8.37), the moment diagram M_e of the equivalent system of constant stiffness EI_A may be determined. The approximation of its shape with straight-line segments, as discussed earlier, leads to the constant stiffness EI_A equivalent system in Fig. 8.10c. Its elastic line is closely identical to that of the original system in Fig. 8.9a. Thus, deflections and rotations at any x may also be determined by using the equivalent system in Fig. 8.10c.

The construction of equivalent systems of constant stiffness is particularly useful for vibration and dynamic analysis of variable stiffness members, as shown in the following sections of this chapter.

Other beam problems may be solved in a similar manner.

8.7 VIBRATION ANALYSIS OF FLEXIBLE BARS OF UNIFORM AND VARIABLE STIFFNESS

Use of equivalent systems will be made here in order to determine the vibration response of flexible members that are subjected to small vibrational amplitudes. The general differential equation of motion for flexible bars is derived in Section 8.2, and it is given by Eq. (8.19). The difficulties associated with the solution of this equation may be reduced by remembering that the vibration of a flexible bar is taking place from the static equilibrium position, which is defined as $y_s(x)$. See Fig. 8.2b. The vibration amplitude $y_d(x, t)$ measured from the static equilibrium position $y_s(x)$ will be assumed to be small. See also refs. 2 and 69.

The considerations made in this analysis are that the slope of the dynamic amplitude $y_d(x, t)$ is small, and that the static equilibrium position $y_s(x)$ is determined by static analysis, as discussed in earlier sections of this chapter. The static amplitudes $y_s(x)$ are assumed to be large, and, consequently, large deformation theory must be used for their computation. The total weight acting on a deformed segment of the member is approximated by a function of the horizontal displacement. This suggests that the arc length position of the deformed segment may be expressed by Eq. (8.2). We also take into consideration that the law of variation of the depth $h(x)$ of the member in the deformed configuration is an integral equation that depends on the large deformation, and this integral will be replaced with a function that contains the horizontal displacement $\Delta(x)$. The time dependence is assumed to be harmonic; that is, $\ddot{y}_d = -\omega^2 y_d$, where y_d is the dynamic displacement measured from the static equilibrium position $y_s(x)$.

8.7.1 Differential Equation of Motion for Flexible Bars with Small Amplitude Vibrations

To derive the differential equation of motion for flexible bars with small vibrational amplitudes, we consider the uniform flexible cantilever beam in Fig. 8.2a. However, the differential equation of motion derived here is also applicable to flexible bars of variable stiffness. The uniformly distributed load w_0 on the beam is assumed to be its uniform weight, which may include other weights that are securely attached to the member, for example, the bridge deck that is securely attached to the bridge girders by shear connectors.

The large static deformation configuration $y_s(x)$ of the member is shown in Fig. 8.2b. From the static equilibrium position $y_s(x)$, the dynamic amplitude $y_d(x, t)$ is also shown in the same figure. Therefore, the total amplitude $y(x, t)$ from the undeformed straight configuration of the member is

$$y(x, t) = y_s(x) \pm y_d(x, t) \tag{8.66}$$

By differentiating Eq. (8.66) with respect to x, we obtain

$$y'(x, t) = y_s'(x) \tag{8.67}$$

In this differentiation, we have assumed that the slope $y_d'(x, t)$ of the dynamic amplitude curve is small compared to the large static slope $y_s'(x)$, and it was neglected.

By differentiating Eq. (8.66) with respect to x once again, we find

$$y''(x, t) = y_s''(x) \pm y_d''(x, t) \tag{8.68}$$

If we differentiate Eq. (8.66) twice with respect to time t, the result is

$$\frac{d^2 y(x, t)}{dt^2} = \frac{d^2 y_d(x, t)}{dt^2} \tag{8.69}$$

By substituting Eqs. (8.68) and (8.69) into Eq. (8.19), we find

$$\frac{d^2}{dx^2}\left(E_x I_x \frac{y_s''(x) \pm y_d''(x, t)}{[1 + (y_s')^2]^{3/2}}\right) + [1 + (y_s')^2]^{1/2} m(x_0)\frac{d^2 y_d}{dt^2} = 0 \tag{8.70}$$

or

$$\frac{d^2}{dx^2}\left(E_x I_x \frac{y_s''(x)}{[1 + (y_s')^2]^{3/2}}\right) \pm \frac{d^2}{dx^2}\left(E_x I_x \frac{y_d''(x)}{[1 + (y_s')^2]^{3/2}}\right)$$
$$+ [1 + (y_s')^2]^{1/2} m(x_0)\frac{d^2 y_d}{dt^2} = 0 \tag{8.71}$$

Equation (8.71) indicates that, from the static equilibrium configuration, we have

$$\frac{d^2}{dx^2}\left(E_x I_x \frac{y_s''(x)}{[1 + (y_s')^2]^{3/2}}\right) = -\frac{w(x_0)}{\cos\theta} \tag{8.72}$$

From the dynamic equilibrium configuration, we have

$$\frac{d^2}{dx^2}\left(E_x I_x \frac{y_d''(x, t)}{[1 + (y_s')^2]^{3/2}}\right) = -[1 + (y_s')^2]^{1/2} m(x_0)\frac{d^2 y_d(x, t)}{dt^2} \tag{8.73}$$

We stated earlier that we will assume that the time function is harmonic; that is,

$$\frac{d^2 y_d(x, t)}{dt^2} = -\omega^2 y_d(x) \tag{8.74}$$

On this basis, Eq. (8.73) yields

$$\frac{d^2}{dx^2}\left(E_x I_x \frac{y_d''(x, t)}{[1 + (y_s')^2]^{3/2}}\right) - \{[1 + (y_s')^2]^{1/2} m(x_0)\}\omega^2 y_d(x) = 0 \tag{8.75}$$

Equation (8.75) may also be expressed in terms of an equivalent variable moment of inertia $I_e(x)$ and an equivalent variable mass density $m_e(x)$, as follows:

$$\frac{d^2}{dx^2}\{E_x I_e(x) y_d''(x)\} - m_e(x)\omega^2 y_d(x) = 0 \tag{8.76}$$

$$I_e(x) = \frac{I_x}{[1 + (y_s')^2]^{3/2}} \tag{8.77}$$

$$m_e(x) = [1 + (y_s')^2]^{1/2} m(x_0) \tag{8.78}$$

where I_x is the moment of inertia at any $0 \le x \le L_0$ of the original member.

Equation (8.76) represents a straight beam of length $L_0 = L - \Delta$, which vibrates with the same frequencies as the initial member does from its static equilibrium position $y_s(x)$. The variation of the equivalent moment of inertia $I_e(x)$ and mass $m_e(x)$ along the length L_0 of the equivalent system may be obtained from Eqs. (8.77) and (8.78), respectively. The variations in $I_e(x)$ and $m_e(x)$ take into account the changes in mass and moment of inertia due to the large static deformations.

In summary, the two differential equations shown below define completely the bending transverse vibration of a flexible member that is experiencing large static amplitudes $y_s(x)$ combined with vibration of small amplitudes:

$$\frac{d^2}{dx^2}\left(E_xI_x\frac{y_s''(x)}{[1 + (y_s')^2]^{3/2}}\right) = -[1 + (y_s')^2]^{1/2}w_0(x_0) \tag{8.79}$$

$$\frac{d^2}{dx^2}[E_xI_e(x)y_d''(x)] - m_e(x)\omega^2 y_d(x) = 0 \tag{8.80}$$

Equations (8.79) and (8.80) must be solved simultaneously for the computation of the frequencies of vibration. Equation (8.79) defines the large static equilibrium configuration, and it is equivalent to the Euler–Bernoulli equation given by Eq. (8.1). Since Eq. (8.79) is very difficult to solve, we can replace this equation by the Euler–Bernoulli equation and solve it by using equivalent systems as discussed in earlier sections. The considerations discussed in the beginning of the section will also be taken into account. In this manner, the static equilibrium position $y_s(x)$ and the rotations $y_s'(x)$ may be determined.

Now that $y_s(x)$ and $y_s'(x)$ are known, Eq. (8.80) may be used to determine the free frequencies of vibration and their corresponding mode shapes. This equation represents the transverse bending vibration of a pseudovariable stiffness member with equivalent mass density m_e and equivalent moment of inertia I_e. The quantities $I_e(x)$ and $m_e(x)$ define the geometry and mass, respectively, of an equivalent straight beam of length L_0. The depth $h_e(x)$ of the equivalent straight member may be obtained from the following equation:

$$h_e(x) = \frac{h(x)}{[1 + (y_s')^2]^{1/2}} \tag{8.81}$$

where $h(x)$ represents the depth variation of the initial system, which may incorporate any arbitrary thickness variation.

The solution of Eq. (8.80) may be obtained by using existing methods of analysis for the free vibration of straight beams. The finite difference method, or Galerkin's finite element method, should yield excellent results. Galerkin's finite element method with equivalent uniform stiffness and mass will be discussed here. See also ref. 2 for more information.

8.7.2   Galerkin's Finite Element Method (GFEM)

In the application of this method to solve Eq. (8.76), we develop an equivalent uniform stiffness and mass approach so that uniform shape functions can be used. On this basis, an equivalent uniform stiffness and an equivalent uniform mass are defined for each element. By using the differential equation as a basis for the kth element, Galerkin's method with uniform shape functions is used to define the equivalent uniform stiffness and mass. In this manner, we can replace the complex variable stiffness and mass with uniform stiffness and mass density. We prefer to solve Eq. (8.76) because it represents the solution of a straight beam, while Eq. (8.75) represents the solution of the original member as a curved beam.

The beam is divided into M elements with a total of $M + 1$ node points. Each element has two degrees of freedom per node, namely, rotation and vertical translation. For the kth element, the differential equation may be expressed as

$$E\tilde{J}_k y_d'''' - \beta_k \omega^2 y_d = 0 \qquad k = 1, 2, 3, \ldots, M \qquad (8.82)$$

where $\tilde{J}_k$ is the equivalent uniform stiffness of the kth element, and β_k is its equivalent uniform mass. On this basis, we can write the following expressions for $\tilde{J}_k$ and β_k:

$$\tilde{J}_k = \frac{1}{2I_1} \left(\frac{f(x_i)}{\{1 + [y_s'(x_i)]^2\}^{3/2}} + \frac{f(x_j)}{\{1 + [y_s'(x_j)]^2\}^{3/2}} \right) \qquad (8.83)$$

$$\beta_k = \tfrac{1}{2} m(x_0)\{\sqrt{1 + [y_s'(x_i)]^2} + \sqrt{1 + [y_s'(x_j)]^2}\} \qquad (8.84)$$

Note that, if $\beta_k = m(x_0)$, the effect of the change due to the large static curvature is ignored.

We require here that the error between the approximate and true solutions be orthogonal to the function used in the approximation. If we start with the differential equation

$$E\tilde{J}_k y_d'''' - \beta_k \omega^2 y_d = 0 \qquad (8.85)$$

and approximate the solution by the equation

$$y_d(x) = \Sigma \mathcal{L}_i U_i \qquad (8.86)$$

then the equation yields

$$E\tilde{J}_k [\mathcal{L}_i U_i]'''' - \beta_k \omega^2 [\mathcal{L}_i U_i] = \varepsilon \qquad (8.87)$$

where ε is the *residual*, or error. The purpose is to make the error as small as possible, and we can do this by requiring that

$$\int \mathcal{L}_i \varepsilon \, dR = 0 \qquad (8.88)$$

for each *basis function* $\mathcal{L}_i$. The integral states that the basis function must be orthogonal to the error.

By substituting for ε in Eq. (8.88), we find

$$\int_0^{L_0} \mathcal{L}_i [E\tilde{J}_k y_d'''' - \beta_k \omega^2 y_d]\, dx = 0 \tag{8.89}$$

Since the interpolation function for y_d is defined over a single element, we rewrite Eq. (8.89) as follows:

$$\sum_{k=1}^{M} \int_0^{L_0} \mathcal{L}_i [E\tilde{J}_k y_d'''' - \beta_k \omega^2 y_d]\, dx = 0 \tag{8.90}$$

The integral, however, must be reduced to one that contains first and second derivatives, in order to define the stiffness and mass matrices. Thus, integrating by parts element by element, we find

$$E\tilde{J}_k \left\{ \left[\mathcal{L}_i^k \frac{d}{dx}(y_d'') \right]_0^{L_k} - \left[\frac{d}{dx} \mathcal{L}_i^k [y_d'] \right]_0^{L_k} + \int_0^{L_k} \frac{d^2}{dx^2} \mathcal{L}_i^k [y_d'']\, dx \right\}$$
$$- \omega^2 \beta_k \int_0^{L_k} \frac{d^2}{dx^2} \mathcal{L}_i^k [y_d]\, dx = 0 \tag{8.91}$$

When the summation over the elements is completed, the first two terms of Eq. (8.91) will drop out, and we obtain the following expression:

$$E\tilde{J}_k \int_0^{L_k} \frac{d^2}{dx^2} [\mathcal{L}_k]\{y_d''\}\, dx - \omega^2 \beta_k \int_0^{L_k} [\mathcal{L}_k]\{y_d\}\, dx = 0 \tag{8.92}$$

If $y_d(x) = [\mathcal{L}]\{U\}$, Eq. (8.92) yields the expression

$$E\tilde{J}_k \int_0^{L_k} [\mathcal{L}_k'']'[\mathcal{L}_k'']\{U\}\, dx - \omega^2 \beta_k \int_0^{L_k} [\mathcal{L}_k]'[\mathcal{L}_k]\{U\}\, dx = 0 \tag{8.93}$$

We rewrite Eq. (8.93) as follows:

$$[k]\{U\} - \omega^2[M]\{U\} = 0 \tag{8.94}$$

and we define the stiffness and mass matrices as shown below:

$$[K^k] = E\tilde{J}_k \int_0^{L_k} [\mathcal{L}_k'']'[\mathcal{L}_k'']\, dx \tag{8.95}$$

$$[M^k] = \beta_k \int_0^{L_k} [\mathcal{L}_k]'[\mathcal{L}_k]\, dx \tag{8.96}$$

The derivation of the element stiffness and mass matrices may be performed as shown in the next section.

8.7.3 Derivation of Element Stiffness and Mass Matrices and the Eigenvalue Problem

Consider an element k that is subject to bending vibration, where $\tilde{J}_k$ and β_k are its equivalent uniform bending stiffness and mass density, respectively. The interpolation function, or shape function, may be obtained by solving the following equation:

$$E\tilde{J}_k \frac{d^4 y_d(x)}{dx^4} = 0 \tag{8.97}$$

By integrating Eq. (8.97) four times, we find

$$y_d(x) = C_1 x^3 + C_2 x^2 + C_3 x + C_4 \tag{8.98}$$

where the constants C_1, C_2, C_3, and C_4 may be determined from the geometric boundary conditions at each node. Since the two degrees of freedom per node are rotation and vertical translation, the boundary conditions may be written as follows:

$$y_d(0) = y_1 \tag{8.99}$$

$$\frac{dy_d(0)}{dx} = \theta_1 \tag{8.100}$$

$$y_d(L_k) = y_2 \tag{8.101}$$

$$\frac{dy_d(L_k)}{dx} = \theta_2 \tag{8.102}$$

By using Eq. (8.98) and the above four boundary conditions, the constants of integration are determined, and the expression for $y_d(x)$ is as follows:

$$y_d(x) = \mathcal{L}_1(x)y_1 + \mathcal{L}_2(x)L_k\theta_1 + \mathcal{L}_3(x)y_2 + \mathcal{L}_4(x)L_k\theta_2 \tag{8.103}$$

where

$$\mathcal{L}_1(x) = 1 - 3\left(\frac{x}{L_k}\right)^2 + 2\left(\frac{x}{L_k}\right)^3 \tag{8.104}$$

$$\mathcal{L}_2(x) = \frac{x}{L_k} - 2\left(\frac{x}{L_k}\right)^2 + \left(\frac{x}{L_k}\right)^3 \tag{8.105}$$

$$\mathcal{L}_3(x) = 3\left(\frac{x}{L_k}\right)^2 - 2\left(\frac{x}{L_k}\right)^3 \tag{8.106}$$

$$\mathcal{L}_4(x) = -\left(\frac{x}{L_k}\right)^2 + \left(\frac{x}{L_k}\right)^3 \tag{8.107}$$

which are known as Hermitian cubics.

By using Eqs. (8.95) and (8.96), the stiffness and mass matrices may be derived, and they are as follows:

$$[K_k] = \frac{E\tilde{J}_k}{L_k^3}\begin{bmatrix} 12 & 6 & -12 & 6 \\ & 4 & -6 & 2 \\ & & 12 & -6 \\ \text{symm} & & & 4 \end{bmatrix} \tag{8.108}$$

$$[M_k] = \frac{\beta_k}{420L_k}\begin{bmatrix} 156 & 22 & 54 & -13 \\ & 4 & 13 & -3 \\ & & 156 & -22 \\ \text{symm} & & & 4 \end{bmatrix} \tag{8.109}$$

In Eq. (8.108), the quantity $E\tilde{J}_k$ is the equivalent uniform bending stiffness for the kth element, and β_k in Eq. (8.109) is the equivalent uniformly distributed mass density of the same element. Note that both stiffness and mass matrices for each element have the same form.

The individual stiffness and mass matrices are then assembled by adding the contributions of all the elements. This yields

$$[K_s] = \sum_{k=1}^{M} [K^k] \tag{8.110}$$

$$[M_s] = \sum_{k=1}^{M} [M^k] \tag{8.111}$$

On this basis, the equation of motion may be written

$$[K_s]\{U\} - \omega^2[M_s]\{U\} = 0 \tag{8.112}$$

The frequencies of vibration and the corresponding mode shapes may be determined by solving the following general eigenvalue problem:

$$[K]\{U\} = \omega^2[M]\{U\} \tag{8.113}$$

where $[K]$ is the stiffness matrix, $[M]$ is the mass matrix, and $\{U\}$ incorporates the nodal coordinates. If we premultiply both sides of Eq. (8.113) by $[K]^{-1}$, the resulting equation is

$$\{U\} = \omega^2[K]^{-1}[M]\{U\} \qquad (8.114)$$

In a more standard form, Eq. (8.114) may be written

$$[A - \lambda I]\{U\} = 0 \qquad (8.115)$$

where

$$\lambda = \frac{1}{\omega^2} \qquad (8.116)$$

A canned eigensolver may be used to determine the eigenvalues and the corresponding eigenvectors, and Eq. (8.116) provides the required free frequencies of vibration for flexible members. The following section provides an application of the theory in this section for the computation of free frequencies of vibration of flexible members. More extensive work on this subject may be found in ref. 2.

8.8   FREE VIBRATION OF FLEXIBLE CANTILEVER BEAMS

The methodologies developed in the preceding section will be applied here, in order to determine the free vibration of flexible cantilever beams. The procedure is illustrated by using a simpler flexible cantilever beam problem. The solution of many challenging flexible beam problems may be found in ref. 2. The main purpose of the discussion here is to illustrate the application of the methodologies involved.

Consider the uniform flexible cantilever beam in Fig. 8.2a, which has a constant stiffness $EI_1 = EI = 180 \times 10^3$ kip $\cdot$ in.^{2} and a length $L = 1000$ in. The uniform loading w_0 is assumed to be equal to 1.5 lb/in., and it consists of the weight of the beam and additional weights that are securely attached to the member and participate in its vibrational motion. The methodology, however, is general, and it can be applied to cases where both $E_x I_x$ and $w_0(x)$ are variable.

In accordance with the discussion in Section 8.7, the static analysis should be carried out first, in order to establish the static equilibrium position $y_s(x)$ of the cantilever beam. This will be accomplished by using pseudolinear analysis as discussed in Section 8.4.

The bending moment M_x at any $0 \leq x \leq L_0$ may be obtained by using statics, and it is as follows:

$$M_x = -\frac{w_0 x}{2} x_0(x) \tag{8.117}$$

The expression for the deformed arc length $x_0(x)$ is given by Eq. (8.3), and it is as follows:

$$x_0(x) = \int_0^x [1 + (y_s')^2]^{1/2} \, dx \tag{8.118}$$

The coordinate $x_0(x)$, as stated earlier in the chapter, is a function of the horizontal displacement $\Delta(x)$, and, by Eq. (8.2), it may be written

$$x_0(x) = x + \Delta(x) \tag{8.119}$$

When one of the end supports of a beam is permitted to move horizontally, approximate expressions for the variation of $\Delta(x)$ may be used that facilitate the solution of the problem without appreciable loss of accuracy. See also Chapter 1 of ref. 2. The function $\Delta(x)$ may be assumed as $\Delta(x) = $ constant $= \Delta$, where Δ is the horizontal displacement of the movable end, or it may have the variations $\Delta(x) = \Delta x/L_0$, $\Delta(x) = \Delta(x/L_0)^{1/2}$, and $\Delta(x) = \Delta \sin(\pi x/2L_0)$. All these assumed variations of $\Delta(x)$ are proved in ref. 2 to provide accurate results.

For the solution of the problem in Fig. 8.2a, we assume that $\Delta(x) = $ constant $= \Delta$, and Eq. (8.119) yields

$$x_0(x) = x + \Delta \tag{8.120}$$

By substituting Eq. (8.120) into Eq. (8.117), we obtain

$$M_x = -\frac{w_0 x}{2}(x + \Delta) \tag{8.121}$$

It should be noted here that the accuracy of the method is not very sensitive to the expression used for $\Delta(x)$, and Eq. (8.120) should lead to accurate results.

By substituting Eq. (8.121) into the Euler–Bernoulli equation, which is Eq. (8.1), we obtain

$$\frac{y_s''}{[1 + (y_s')^2]^{3/2}} = \frac{w_0}{2EI} x(x + \Delta) \tag{8.122}$$

By integrating Eq. (8.122) once and satisfying the boundary condition of zero rotation at $x = L_0$ for the computation of the constant of integration, we find

$$\frac{y_s'}{[1 + (y_s')^2]^{1/2}} = \frac{w_0}{12EI}[2x^3 + 3\Delta x^2 - 2(L - \Delta)^3 - 3\Delta(L \doteq \Delta)^2] \quad (8.123)$$

Solving Eq. (8.123) for $y_s'(x)$, we find

$$y_s'(x) = \frac{Q(x)}{\{1 - [Q(x)]^2\}^{1/2}} \quad (8.124)$$

where

$$Q(x) = \frac{w_0}{12EI}[2x^3 + 3\Delta x^2 - 2(L - \Delta)^3 - 3\Delta(L - \Delta)^2] \quad (8.125)$$

The quantity $y_s'(x)$ in Eq. (8.124) is a function of the horizontal displacement Δ of the free end of the member, and, consequently, Δ needs to be determined. This may be accomplished by using a trial-and-error procedure as discussed in Section 8.4, which involves the following equation:

$$L = \int_0^{L_0} \{1 + [y_s'(x)]^2\}^{1/2} dx \quad (8.126)$$

This equation is similar to Eq. (8.6).

The solution is easier if we introduce the dummy variable ξ in the following way:

$$x = \xi L_0 \quad (8.127)$$

$$dx = L_0 d\xi \quad (8.128)$$

On this basis, Eq. (8.126) yields

$$L = L_0 \int_0^1 \{1 + [y_s'(\xi)]^2\}^{1/2} d\xi \quad (8.129)$$

The trial-and-error procedure is performed by assuming a value of Δ and using Eq. (8.124) to find the expression for $y_s'(x)$. Then, with known $y_s'(x)$, we perform the integration in Eq. (8.126) to determine the length L of the member. We repeat the procedure until the correct length L is obtained. For this problem, the trial-and-error procedure yields $\Delta = 277.25$ in. and $L_0 = L - \Delta = 1000 - 277.25 = 722.75$ in. At the free end, the static rotation θ_s may be obtained from Eq. (8.124), and it has the value $\theta_s = \tan^{-1}(y_s') = 55.69°$.

With known Δ, the expression for $y_s'(x)$ is completely established from Eq. (8.124). The static equilibrium position $y_s(x)$ can be determined in one of two ways. One way would be to integrate Eq. (8.124) once and determine the constant of integration by applying the boundary condition of zero deflection at the fixed end. The second way would be to use pseudolinear equivalent systems, as was done in Section 8.4. The second way is the most convenient to use, so pseudolinear analysis is used here.

At the free end, the vertical deflection δ_A is found to be equal to 638.0 in., which indicates that it is a very flexible member. With known $y_s(x)$, the next step would be to use the differential equation of motion given by Eq. (8.76) to determine the free frequencies of vibration of the flexible member and its corresponding mode shapes. Equation (8.76) represents a pseudovariable stiffness equivalent straight beam of length L_0, as shown in Fig. 8.11a. Its equivalent depth $h_e(x)$, equivalent moment of inertia $I_e(x)$, and equivalent mass $m_e(x)$ may be determined from the following equations:

$$h_e(x) = h_1\{1 - [Q(x)]^2\}^{1/2} \tag{8.130}$$

$$I_e(x) = I_1\{1 - [Q(x)]^2\}^{3/2} \tag{8.131}$$

$$m_e(x) = \frac{m(x)}{\{1 - [Q(x)]^2\}^{1/2}} \tag{8.132}$$

where $Q(x)$ may be determined from Eq. (8.125). Note that the above three equations are obtained from Eqs. (8.81), (8.77), and (8.78), respectively, by substituting for $y_s'(x)$ the expression given by Eq. (8.124).

Galerkin's finite element method in conjunction with equivalent uniform stiffness and equivalent uniform mass, as shown in Fig. 8.11b, may be used to solve the pseudovariable stiffness member in Fig. 8.11a. To apply GFEM with equivalent uniform stiffness $E\tilde{J}$ and equivalent mass β, we subdivide the member into M elements, and Eqs. (8.83) and (8.84) may be used to determine the stiffness $E\tilde{J}$ and mass β, respectively.

By substituting Eq. (8.124) into Eqs. (8.83) and (8.84), the following equations for $\tilde{J}_k$ and β_k are obtained:

$$\tilde{J}_k = \frac{1}{2I_1}\{f(x_i)\{1 - [Q(x_i)]^2\}^{3/2} + f(x_j)\{1 - [Q(x_j)]^2\}^{3/2}\} \tag{8.133}$$

$$\beta_k = \frac{1}{2}\left\{\frac{m(x_i)}{\{1 - [Q(x_i)]^2\}^{1/2}} + \frac{m(x_j)}{\{1 - [Q(x_j)]^2\}^{1/2}}\right\} \tag{8.134}$$

On this basis, the stiffness and mass matrices may be determined from Eqs. (8.108) and (8.109), respectively, and assemblage of the element stiffness

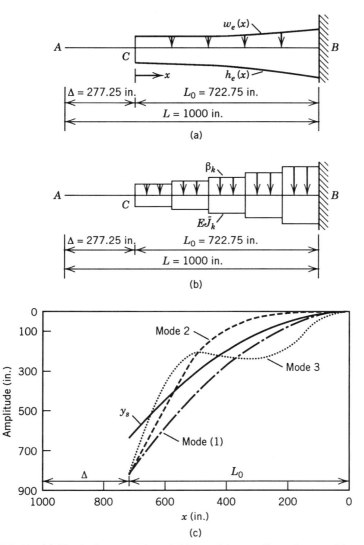

FIGURE 8.11. (a) Equivalent pseudovariable straight cantilever beam of length L_0. (b) Equivalent piecewise uniform straight cantilever beam with equivalent piecewise uniform mass. (c) First three mode shapes of the beam.

and mass matrices may be performed by using Eqs. (8.110) and (8.111), respectively. The boundary conditions of zero vertical displacement and zero rotation at the fixed end of the member may be used in this procedure.

The eigenvalue problem represented by Eq. (8.113) may be solved as discussed earlier in Section 8.7.3, in order to determine the free frequencies of vibration and the corresponding mode shapes of the member. On this

TABLE 8.3. Free Frequencies of Vibration of a Uniform Flexible Cantilever Member Using the Approximation $x_0 = x + \Delta$

(1)	(2)	(3)	(4)	(5)	(6)
		GFEM with Equivalent Uniform Stiffness and Mass		FEM	Difference
ω (rps)	$M = 10$	$M = 20$	$M = 40$	81 Elements	(%)
ω_1	0.9429	0.9453	0.9469	0.9465	0.042
ω_2	4.9058	4.9192	4.9262	4.9217	0.091
ω_3	13.4414	13.4237	13.4254	13.4041	0.159
ω_4	26.4260	26.1823	26.1596	26.0851	0.286
ω_5	43.7239	43.1522	43.1515	42.9672	0.429
ω_6	63.9176	64.4496	64.3845	64.0142	0.578

TABLE 8.4. Free Frequencies of Vibration of a Uniform Flexible Cantilever Beam Using FDM with 81 Elements and Various Approximations of x_0

ω (rps)	Various Cases of x_0			
	$x_0 = x + \Delta$	$x_0 = x + \Delta x/L_0$	$x_0 = x + \Delta (x/L_0)^{1/2}$	$x_0 = x + \Delta \sin(\pi x/2L_0)$
ω_1	0.9465	0.9179	0.9291	0.9119
ω_2	4.9217	4.8487	4.8703	4.9045
ω_3	13.4041	13.3545	13.3678	13.4106
ω_4	26.0851	26.0494	26.0580	26.1030
ω_5	42.9672	42.9428	42.9478	42.9933
ω_6	64.0142	64.0025	64.0034	64.0488

basis, the first six free frequencies of vibration are determined by using GFEM with 10, 20, and 40 elements, that is, $M = 10$, 20, and 40, and the results are shown in Table 8.3. The pseudovariable system in Fig 8.11a is also solved by using the finite difference method (FDM) with 81 elements, and the results are shown in column (5) of Table 8.3. The last column in this table compares the results obtained by FDM with the results obtained by GFEM with $M = 40$. As you can see, the percentage difference between these two methods is small, and, for practical applications, the results may be considered identical. We also note in Table 8.3 that GFEM with $M = 10$ provides very good results when it is compared with $M = 40$. The largest difference is only 1.328%, and it is associated with the frequency ω_5. For practical applications, the accuracy of the results obtained by using $M = 10$ is usually sufficient.

The schematic representation of the first three modes of vibration of the flexible member is shown in Fig. 8.11c. Note in this figure that the free vibration is taking place from the static equilibrium position y_s. In Table 8.4, the first six free frequencies of vibration are determined by using the pseudovariable equivalent system in Fig. 8.11a and applying the FDM with 81 elements. Four cases of the function x_0 have been used, as shown in the table, and the results are compared. The approximation $x_0 = x + \Delta$ is a rather crude approximation of x_0, and it represents an upper limit. All remaining three approximations of x_0 are very reasonable, and they provide a good representation of the actual variation of x_0. However, by observing the results in Table 8.4, we note that all approximations of x_0—even the crude one—yield good results for practical applications. This is very useful information for the design engineer.

8.9 VIBRATION ANALYSIS OF VARIABLE STIFFNESS MEMBERS SUBJECTED TO SMALL DEFORMATIONS

In this section, the method of equivalent systems will be used for the solution of members of any arbitrary stiffness and mass along their lengths. The member is assumed to be subjected to small amplitude vibrations from its static equilibrium position. In this analysis, we do not consider damping, and the effects of shear and rotatory inertia are neglected. The differential equation of motion for such members is given by Eq. (8.20), and it is written here as follows:

$$\frac{d^2}{dx^2}(E_x I_x y'') = -m(x)\ddot{y} \tag{8.135}$$

Equation (8.135) is a linear differential equation of motion, where the stiffness $E_x I_x$ and mass $m(x)$ per unit length of the member can vary in any arbitrary manner. In reality, this is a partial differential equation, where the left-hand side involves derivatives with respect to the variable x and the right-hand side involves time derivatives. In fact, the right-hand side is the inertia force $m\ddot{y}$ of the vibrating member. By applying the method of separation of variables, we can obtain a differential equation that is a function of x only, and an additional equation that is a function of time t only. The solution of the first equation yields the free frequencies of vibration and their corresponding mode shapes as a function x. The second solution provides the time variation of the displacement of every point on the beam, which, in this case, is harmonic. We need to remember here that when a member vibrates at a free frequency ω, all points of the member vibrate with the same frequency.

For small deformations, we usually assume that the initial undeformed straight configuration of the member and the static curved equilibrium position y_s are the same, since accurate solutions may be obtained in this case by using small deflection theory. However, we do not neglect the fact that all vibrations are taking place from the static equilibrium position. The small deformation theory helps us only to assume that the member is straight since y_s is small.

To simplify the solution of the initial complicated variable stiffness problem, we want to replace the member with one that has constant stiffness throughout its length and to compensate for the change in stiffness by using an equivalent mass $m_e(x)$ throughout the length of the member. This will simplify and facilitate the solution of the complicated variable stiffness problem. In this case, the equivalent differential equation that needs to be solved is

$$\frac{d^2}{dx^2}\{E_1I_1y_e''\} = -m_e(x)\ddot{y}_e \tag{8.136}$$

where E_1I_1 is the constant stiffness of the member, and $m_e(x)$ is its equivalent mass.

We remember here that the frequency of a freely vibrating member is a function of both its stiffness and mass. If we change the stiffness, the mass must be changed to compensate for the change in stiffness. We also must remember that if we take the inertia forces $m\ddot{y} = -m\omega^2 y$ of a mode of vibration and apply them statically to the member, the static deflection caused by these loads would be the mode shape of the vibrating member. These three observations constitute the basis for the construction of equivalent systems of constant stiffness that can be used to determine the free vibrations of the initial variable stiffness member.

By taking into consideration the above discussion and the associated observations, we begin the construction of such an equivalent system that obeys Eq. (8.136), by using the method of equivalent systems discussed in Sections 8.3.3 and 8.6. To do this, we consider the initial member of variable stiffness E_xI_x and we load it with its own weight and any additional weights that are attached to the member and participate in its vibrational motion. This loading system produces the static equilibrium position $y_s(x)$, and the member vibrates harmonically about this position. To determine $y_s(x)$, we apply the method discussed in Sections 8.3.3 and 8.6, which uses equivalent systems of constant stiffness E_1I_1 that are loaded with the load w_e, which is different from the loading w of the original system, in order to compensate for the change in the stiffness.

In this case, since the loading in the initial variable stiffness member is weight, the loading on the equivalent system of constant stiffness E_1I_1 would be equivalent weight. Consequently, since these two systems are equivalent, the static equilibrium position $y_s(x)$ may be established by using either one of these two systems. Here, we choose the equivalent system of constant stiffness, since it is the simplest one to use. We also assume here that the constant stiffness equivalent system would be a reasonably accurate system to use for vibration analysis. In other words, we assume that the equivalent mass $m_e(x)$ in Eq. (8.136), which uses constant stiffness E_1I_1, is the same as the $m_e(x)$ in the above equivalent system, which provides the static equilibrium position $y_s(x)$.

This is not unreasonable, because E_xI_xy'' in Eq. (8.135) is the moment M_x produced by the static application of the inertia loads $-m\omega^2y$, where $m = w/g$, and

$$w = \frac{d^2}{dx^2}(E_xI_xy'') = \frac{d^2}{dx^2}(M_x)$$

In Eq. (8.136), we have $m_e = w_e/g$, where, from Eq. (8.42),

$$w_e = \frac{d^2}{dx^2}\left(\frac{M_x}{f(x)g(x)}\right) \quad \text{and} \quad M_e = E_1I_1y_e'' = \frac{E_xI_x}{g(x)f(x)}$$

where $g(x)$ is the variation of E_x with reference value E_1, and $f(x)$ is the variation of I_x with reference value I_1. This rule states that to change E_xI_x to E_1I_1, we should also change $m(x)$ to $m_e(x)$. This procedure provides very accurate and reliable results, and it is illustrated in the following examples. The results are compared with the results obtained by using other methods of analysis.

Example 8.2. The exact values for the first three free frequencies of vibration of the doubly tapered, simply supported beam in Fig. 8.12a were obtained by Heidebrecht (see ref. 70), and they are $f_1 = 188.45$ Hz, $f_2 = 758.08$ Hz, and $f_3 = 1704.90$ Hz. Determine the same three frequencies by using (a) an equivalent system of constant stiffness as discussed earlier in the section, (b) the conventional Stodola's method and iteration procedure, and (c) the finite element method, and compare the results. The mass density of the material is 0.00073386 lb $\cdot$ s^2/in.^{4}, the modulus of elasticity $E = 30 \times 10^6$ psi, and the required dimensions are shown in the figure.

Solution.

A. Use of an Equivalent Constant Stiffness System. The weight that participates in the vibrational motion is only the weight of the variable stiffness member. For convenience, the beam is subdivided into six segments, and the total weight of each segment is lumped at the center of the segment, as shown in Fig. 8.12b. Since the mass density of the material is known, the gravitational weight for each segment can easily be determined, and it is as shown in the figure.

We assume now that the initial system is as shown in Fig. 8.12b, and we apply the method of equivalent systems as discussed in Sections 8.3.3 and 8.6, in order to determine an equivalent system of constant stiffness EI_A that

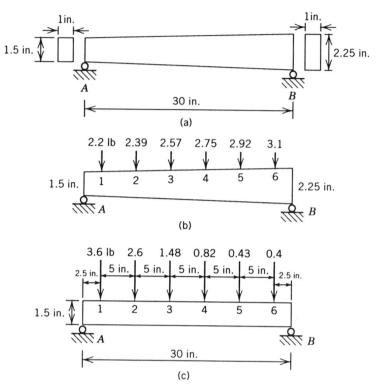

FIGURE 8.12. (a) Initial variable stiffness tapered beam. (b) Initial variable stiffness member with its weight lumped at six points. (c) Equivalent system with constant stiffness EI_A.

is loaded with concentrated equivalent weights, where I_A is the moment of inertia at support A of the original system. This procedure yields the equivalent system shown in Fig. 8.12c. The depth of this equivalent member is 1.5 in. throughout its length, and the constant width is 1.0 in. By using the equivalent system in Fig. 8.12c and applying Stodola's method and the iteration procedure as discussed in Sections 5.4 and 5.5, the first three free frequencies of vibration are determined: $f_1 = 192.25$ Hz, $f_2 = 817.85$ Hz, and $f_3 = 1813.20$ Hz. The mode shapes corresponding to these frequencies are shown in Figs. 8.13b, 8.13c, and 8.13d. Columns (3), (5), and (7) in Table 8.5 show the error percentages of the frequencies when they are compared with the exact values. The accuracy, however, will be improved if we use more lumped masses in Fig. 8.12b and more equivalent weights in Fig. 8.12c.

B. Use of Stodola's Method to Solve the Original System. Stodola's method and the iteration procedure, as discussed in Sections 5.4 and 5.5, are applied

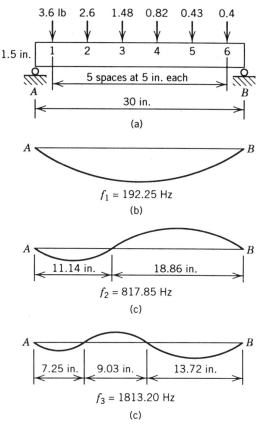

FIGURE 8.13. (a) Equivalent system of constant stiffness EI_A. (b) First mode shape. (c) Second mode shape and node point locations. (d) Third mode shape and node point locations.

here to solve directly the original system in Fig. 8.12b. The required a_{ij} deflection coefficients are determined by applying a unit load each time at points $1, 2, \ldots, 6$ and determining the deflections at the same points. The method of equivalent systems as discussed in Sections 8.3.3 and 8.6 is used for this purpose. For example, if we apply a unit load at point 1, as shown in Fig. 8.14b, the deflection coefficients a_{11}, a_{21}, a_{31}, a_{41}, a_{51}, and a_{61}, at points 1, 2, 3, 4, 5, and 6, respectively, may be determined by using the constant stiffness equivalent system in Fig. 18.14c and applying known methods or formulas. This may be repeated individually for all six points in Fig. 8.14b.

By using the calculated a_{ij} deflection coefficients, we formulate the required matrix equations, as discussed in Sections 5.4 and 5.5, and proceed with the iteration to compute the first three free frequencies of vibration. The

TABLE 8.5. Results Obtained by Various Methods of Vibration Analysis Compared to the Exact Results

(1) Method	(2) f_1 (Hz)	(3) Error (%)	(4) f_2 (Hz)	(5) Error (%)	(6) f_3 (Hz)	(7) Error (%)
Exact	188.45	0	758.08	0	1704.90	0
Equivalent system	192.25	+2.02	817.85	+7.87	1813.20	+6.35
Original member in Fig. 8.12b by Stodola's method	188.05	−0.2	770. 68	+1.66	2725.17	+59.84
Original member in Fig. 8.12a by FEM	187.50	−1.0	747.40	−1.40	1652.00	−3.10
Equivalent system in Fig. 8.12c by FEM	191.70	+1.7	812.80	7.20	1805.00	+5.87

results obtained are $f_1 = 188.05$ Hz, $f_2 = 770.68$ Hz, and $f_3 = 2725.17$ Hz. Columns (3), (5), and (7) of Table 8.5 give the error percentages when they are compared to the exact values. We note here that the first two frequencies are considered accurate for practical purposes, but a very large error is associated with the third frequency. The corresponding mode shapes and the locations of the associated node points are shown in Fig. 8.15. The accuracy of the third frequency will be improved by using more lumped weights in Fig. 8.12b.

C. Use of the Finite Element Method to Solve the Original System. The finite element method is also used here to determine the first three frequencies of vibration and the corresponding mode shapes of the original system in Fig. 8.16a. This is accomplished by using the finite element code A.D.I.N.A. with exact integration and consistent lumped masses. The original member is subdivided into six elements as shown in Fig. 8.16b, which yields a total of 33 nodes. On this basis, the computed values for the first three free frequencies of vibration are $f_1 = 187.50$ Hz, $f_2 = 747.40$ Hz, and $f_3 = 1652.00$ Hz. Columns (3), (5), and (7) in Table 8.5 show the error percentages when compared to the exact values. For practical purposes, the results are considered accurate. The corresponding mode shapes with the associated node points are shown in Fig. 8.17.

D. Use of the Finite Element Method to Solve the Equivalent System. In this case, the constant stiffness equivalent system in Fig. 8.12c, which is shown

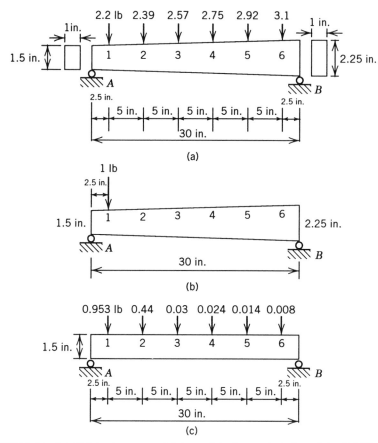

FIGURE 8.14. (a) Original system with its weight lumped at six points. (b) Unit load applied at point 1. (c) Equivalent system of constant stiffness EI_A corresponding to the unit load at point 1.

again in Fig. 8.18a, is also solved by using the finite element method in conjunction with the finite element code A.D.I.N.A. The equivalent system is subdivided into six elements, as shown in Fig. 8.18b, thus providing a total of 33 nodes. The calculated values of the first three free frequencies of vibration are $f_1 = 191.70\ \text{Hz}$, $f_2 = 812.80\ \text{Hz}$, and $f_3 = 1805.00\ \text{Hz}$. Again, columns (3), (5), and (7) of Table 8.5 provide the error percentages when compared to the exact values. The values obtained here are very close to the values obtained by solving the equivalent system in Fig. 8.12c with Stodola's method and the iteration procedure, as one may note by examining the results in Table 8.5. The corresponding three mode shapes are shown in Fig. 8.19.

(*text continues on p. 550*)

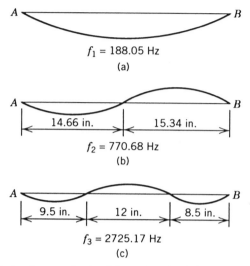

$f_1 = 188.05$ Hz

(a)

14.66 in.   15.34 in.

$f_2 = 770.68$ Hz

(b)

9.5 in.   12 in.   8.5 in.

$f_3 = 2725.17$ Hz

(c)

FIGURE 8.15. Mode shapes obtained by using Stodola's method. (a) First mode shape. (b) Second mode shape. (c) Third mode shape.

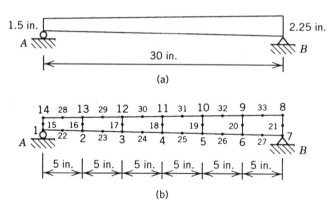

1.5 in.   2.25 in.

30 in.

(a)

14 28 13 29 12 30 11 31 10 32 9 33 8

15 16   17   18   19   20   21

22 2 23 3 24 4 25 5 26 6 27

5 in. 5 in. 5 in. 5 in. 5 in. 5 in.

(b)

FIGURE 8.16. (a) Original variable stiffness member. (b) Original member subdivided into six finite elements.

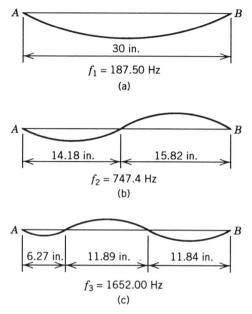

FIGURE 8.17. Mode shapes obtained by using the finite element method to solve the original system. (a) First mode shape. (b) Second mode shape. (c) Third mode shape.

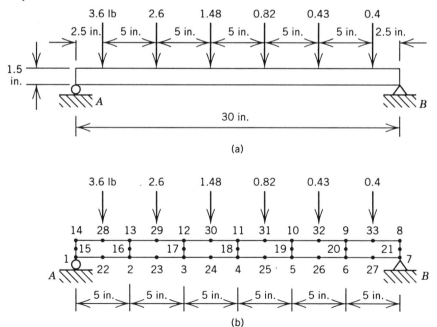

FIGURE 8.18. (a) Equivalent system of constant stiffness EI_A. (b) Equivalent system subdivided into six finite elements.

Example 8.3. A tapered two-span continuous beam is shown in Fig. 8.20, where the loading of 1.0 kip/ft acting on the beam represents its own weight and other securely attached weights that participate in the vibrational motion of the member. The constant width $b = 10.0$ in., depth $h = 16.0$ in., modulus of elasticity $E = 30 \times 10^6$ psi, and n is the taper parameter. Determine the first three free frequencies of vibration and corresponding mode shapes by using (a) a constant stiffness equivalent system, (b) Stodola's method and the iteration procedure, and (c) the finite element method; compare the results.

Solution. The procedure followed here for the three methods of solution is briefly discussed.

A. Use of a Constant Stiffness Equivalent System. The problem in Fig. 8.20 is statically indeterminate, and the redundant reaction should be determined first in order to obtain the moment diagram M_e required for the construction of the equivalent constant stiffness system. For this problem, the vertical reaction R_B at the intermediate support B is taken as redundant. On this basis, we now have a simply supported beam AC loaded with the uniformly distributed load of 1.0 kip/ft, and the same beam AC is loaded with reaction R_B at point B. Here, we can use the method of equivalent systems, discussed in Sections 8.3.3 and 8.6, to determine the vertical deflection at B due to the 1 kip distributed weight and reaction R_B. Since the sum of these two vertical deflections should be zero, R_B can be determined from this boundary condition. The values of R_B for taper parameters $n = 1.1, 1.2, 1.3, 1.4, 1.5$, and 2.0 are shown in Table 8.6, and are determined by using 11 equal lumped weights and 22 equal lumped weights; that is, we lumped the weight of 1 kip/ft at 11 and 22 equally spaced points along the length of the member.

With known R_B, the moment M_x at any section along the length of the member in Fig. 8.20 may be obtained by using statics, and the moment M_e of the equivalent system of constant stiffness EI_A may be determined from the expression $M_e = M_x/f(x)$. The approximation of the shape of M_e with straight-line segments leads to an equivalent system of constant stiffness EI_A and loaded with equivalent concentrated weights. This system is the one to be used for the computation of the free frequencies of vibration and the corresponding mode shapes. Stodola's method and the iteration procedure discussed in Sections 5.4 and 5.5 may be used for this purpose. For the results obtained in this problem, the shape of the M_e diagram is approximated in a way that yields 11 concentrated weights on the equivalent system. The values of the first free frequencies for taper $n = 1.1, 1.3$, and 2.0 are shown in the third column of Table 8.7. The corresponding mode shapes for $n = 2$ are shown in Fig. 8.21.

B. Use of Stodola's Method to Solve the Original System. To compare results, the original member in Fig. 8.20 with taper $n = 1.3$ is solved by using Stodola's method and the iteration procedure. The weight of 1.0 kip/ft is lumped at 11 points along the length of the member. Equal segments of 10 ft

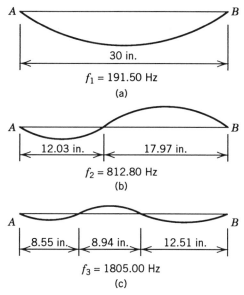

$f_1 = 191.50$ Hz
(a)

$f_2 = 812.80$ Hz
(b)

$f_3 = 1805.00$ Hz
(c)

FIGURE 8.19. Mode shapes obtained by using the finite element method to solve the equivalent system. (a) First mode shape. (b) Second mode shape. (c) Third mode shape.

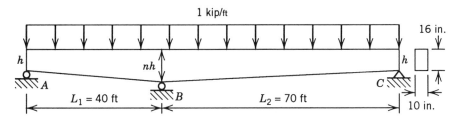

FIGURE 8.20. Tapered two-span continuous beam.

TABLE 8.6. Values of the Redundant Reaction R_B for 11 and 22 Lumped Weights and for Various Values of the Taper n

	R_B (kips)	R_B (kips)
n	11 Lumped Weights	22 Lumped Weights
1.1	73.68	73.80
1.2	74.34	74.46
1.3	74.90	75.02
1.4	75.39	75.50
1.5	75.82	75.94
2.0	77.24	77.28

TABLE 8.7. Values of the First Three Free Frequencies of Vibration Using an Equivalent System, Stodola's Method, and the Finite Element Method

(1)	(2)	(3)	(4)	(5)
nh	Frequency (Hz)	Equivalent System	Stodola's Method	Finite Element Method
1.1h	f_1	2.04		2.01
	f_2	6.01		5.79
	f_3	8.38		8.19
1.3h	f_1	2.43	2.49	2.35
	f_2	6.87	6.91	6.55
	f_3	9.69	9.49	9.29
2h	f_1	4.04		3.20
	f_2	11.09		8.87
	f_3	16.59		12.63

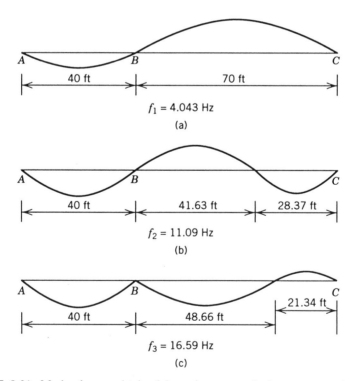

FIGURE 8.21. Mode shapes obtained by using an equivalent system of constant stiffness. (a) First node shape. (b) Second mode shape. (c) Third mode shape.

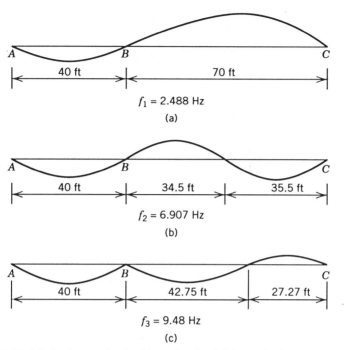

FIGURE 8.22. Mode shapes obtained by using Stodola's method to solve the original member. (a) First mode shape. (b) Second mode shape. (c) Third mode shape.

each are used, and the weight of the segment is lumped at the center of the segment. On this basis, we have 11 weights of 10 kips each. The a_{ij} deflection coefficients are determined by applying a unit load at a load concentration point of the variable stiffness member, and then the method of equivalent systems is used to determine the deflection caused by the unit load at all 11 points. In this manner, 11 problems must be solved since we have 11 cases of unit load applications.

When the computation of all a_{ij} coefficients is completed, Stodola's method and the iteration procedure, as discussed in Sections 5.4 and 5.5, are used to determine the first three frequencies of vibration and the corresponding mode shapes. The results for $n = 1.3$ are shown in Column (4) of Table 8.7, and they are in reasonable agreement with the results obtained by using an equivalent system of constant stiffness. The corresponding three mode shapes and the locations of the node points are shown plotted in Fig. 8.22.

C. Use of the Finite Element Method to Solve the Original Member. The finite element method is also used here to solve the original variable stiffness member shown again in Fig. 8.23a. The member is subdivided into 22 elements as shown in Fig. 8.23b, with a total of 113 nodes. The results obtained

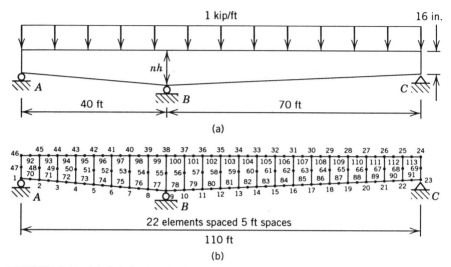

FIGURE 8.23. (a) Original variable stiffness member. (b) Finite element representation of the two-span beam.

for the first three frequencies of vibration for $n = 1.1$, 1.3, and 2.0 are shown in column (5) of Table 8.7.

Examination of the results in Table 8.7 indicates that all three methods are in reasonable agreement for taper $n = 1.3$, and reasonable agreement also is obtained between the equivalent systems method and the finite element method for taper $n = 1.1$ and 1.3. Much larger differences, however, are obtained when $n = 2.0$. The values obtained by the finite element method are much lower than expected, and the consistent approach would have to be used to improve its accuracy for frequencies higher than the second one.

8.10   VIBRATION OF INELASTIC MEMBERS

Since prismatic and nonprismatic members are commonly used as components of many engineering structures and machines (i.e., highway bridges, machine elements, buildings, and space and aircraft structures), it is important to know how these members react when their material is stressed well beyond their elastic limit. Such analysis is usually complicated, particularly when both the modulus of elasticity E_x and moment of inertia I_x vary along the length of the member. The method of equivalent systems, as discussed in the preceding sections of the chapter, provides an accurate and efficient method of analysis for such problems, and it greatly simplifies the computational work required for the solution.

In this section, we derive equivalent systems of uniform stiffness $E_1 I_1$ that can be used for both static and vibration analyses of members when E_x and

I_x can vary in any arbitrary manner along the length of the member. The variation of E_x considered in this section results from the inelastic behavior of the member. For example, when the material of a member is stressed beyond its elastic limit, the modulus of elasticity E_x will vary at cross sections along the length of the member, and this variation must be taken into consideration in the analysis, if an accurate solution of the problem is required. The analysis in this section takes into consideration the variation of both E_x and I_x along the length of the member. We assume, however, that inelastic deformations are small, and small deflection theory may be used in the analysis. If the deformations are large, the theory is still applicable, but the effect of the nonlinearity caused by the large deformation of the member must be taken into consideration. Extensive work on this subject may be found in ref. 2.

8.10.1   Derivation of Equivalent Systems for Inelastic Analysis

When the material of a member is stressed beyond its elastic limit and its moment of inertia at cross sections along its length is variable, then both functions $f(x)$ and $g(x)$ in Eq. (8.37) must be known, or evaluated, in order to determine the moment diagram M_e of the constant stiffness equivalent system and to determine the loading on the equivalent system as discussed in Sections 8.3.3 and 8.6. The method used here to evaluate the function $g(x)$, which represents the variation of E_x with respect to a reference value E_1, is based on the determination of a reduced modulus E_r by using Timoshenko's [71] method. The derivation of equivalent systems for the inelastic analysis of members is illustrated by using the following example.

Consider the tapered cantilever beam in Fig. 8.24a, which is loaded by a uniformly distributed load w as shown in the figure. The material of the member is Monel, which has the stress–strain curve shown in Fig. 8.24b. In practice, the yield stress of Monel is considered to be 50,000 psi at 70°F, with a modulus of elasticity $E = 26 \times 10^6$ psi. We consider the case where the uniform load $w = 1600$ lb/in., the depth h at the free end B is 8 in., the width b of the member is constant and equal to 6 in., and the taper parameter $n = 1.5$. Under this loading condition, the member is stressed well beyond its elastic limit, and, consequently, the modulus of elasticity E is no longer constant along its length.

The moment of inertia I_x at any x in Fig. 8.24a is variable, and its variation is given by the following expression:

$$I_x = I_B f(x) \tag{8.137}$$

where the reference value I_B at the free end B of the member and function

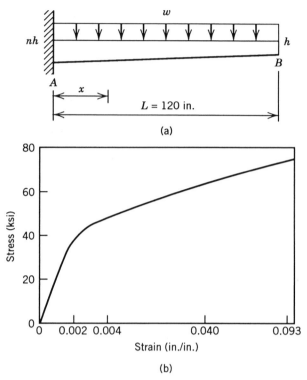

FIGURE 8.24. (a) Tapered cantilever beam loaded with a uniformly distributed load w. (b) Stress–strain curve of Monel.

$f(x)$ are as follows:

$$I_B = \frac{bh^3}{12} \tag{8.138}$$

$$f(x) = \left(\frac{(n-1)(L-x)+L}{L}\right)^3 \tag{8.139}$$

The evaluation of the function $g(x)$ is carried out here by using Timoshenko's method for the evaluation of a reduced modulus E_r. To determine E_r, we need to know the stress–strain curve of the material, which for Monel is as shown in Fig. 8.24b. The shape of the stress–strain curve may be approximated with two, three, or more straight-line segments, depending on practical design requirements. Very accurate results are obtained if two to four straight-line segments are used. This is easily determined by examining the

TABLE 8.8. Values of E for a Two-Line, Three-Line, and Six-Line Approximation of the Stress–Strain Curve of Monel

(1) Modulus E (psi)	(2) Two-Line Approximation	(3) Three-Line Approximation	(4) Six-Line Approximation
E_1	26×10^6	22×10^6	30×10^6
E_2	53×10^3	504×10^3	15×10^6
E_3	–	125×10^3	364×10^3
E_4	–	–	400×10^3
E_5	–	–	244×10^3
E_6	–	–	220×10^3

TABLE 8.9. Values of σ for a Two-Line, Three-Line, and Six-Line Approximation of the Stress–Strain Curve of Monel

(1) Stress σ (psi)	(2) Two-Line Approximation	(3) Three-Line Approximation	(4) Six-Line Approximation
σ_1	50×10^3	48×10^3	30×10^3
σ_2	–	59×10^3	42×10^3
σ_3	–	–	50×10^3
σ_4	–	–	58×10^3
σ_5	–	–	60.2×10^3

shape of the stress–strain curve. Tables 8.8 and 8.9 provide the values of the modulus E and stresses σ, respectively, when the shape of the stress–strain curve of Monel is approximated with two, three, and six straight-line segments.

In order to be somewhat general in the derivation of the appropriate equations, let us use the more general shape of a stress–strain curve shown in Fig. 8.25a. The member is assumed to be free of any axial restraints, and, consequently, the neutral and centroidal axes coincide. Consideration of axial restraints may be found in ref. 2. A section of a beam with rectangular cross section is shown in Fig. 8.25b, where r is the radius of curvature of the neutral surface produced by the bending moment M, and h_1 and h_2 are the distances from the neutral axis to the lower and upper surfaces of the member, respectively. If ε_1 and ε_2 in Fig. 8.25a are the unit elongations of the extreme fibers, and ε is the unit elongation of a fiber at a distance y from

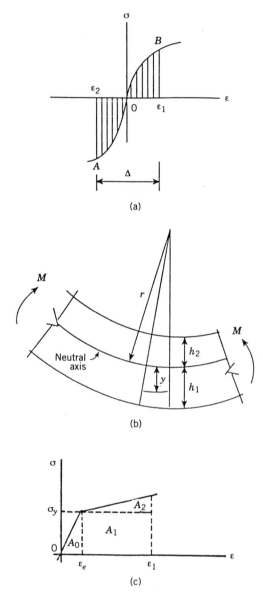

FIGURE 8.25. (a) General shape of a stress–strain curve. (b) Portion of a member with rectangular cross section. (c) Bilinear approximation of a stress–strain curve and areas under the stress–strain curve.

the neutral axis, the following expressions may be written:

$$\varepsilon_1 = \frac{h_1}{r} \tag{8.140}$$

$$\varepsilon_2 = \frac{h_2}{r} \tag{8.141}$$

$$\varepsilon = \frac{y}{r} \tag{8.142}$$

From statics, the following two equations may be used to determine the position of the neutral axis and the radius of curvature r:

$$b \int_{-h_2}^{h_1} \sigma \, dy = 0 \tag{8.143}$$

$$b \int_{-h_2}^{h_1} \sigma y \, dy = M \tag{8.144}$$

where b is the width of the member.
From Eq. (8.142), we have

$$y = \varepsilon r \tag{8.145}$$
$$dy = r \, d\varepsilon \tag{8.146}$$

By substituting Eq. (8.146) into Eq. (8.143), we obtain

$$r \int_{\varepsilon_2}^{\varepsilon_1} \sigma \, d\varepsilon = 0 \tag{8.147}$$

Since $r \neq 0$, Eq. (8.147) suggests that the position of the neutral axis should satisfy the following identity:

$$\int_{\varepsilon_2}^{\varepsilon_1} \sigma \, d\varepsilon = 0 \tag{8.148}$$

We determine now the position of the neutral axis by considering the curve AOB in Fig. 8.25a, where the symbol Δ is used to denote the sum of the absolute values of ε_1 and ε_2. To satisfy Eq. (8.148), we define Δ in a way that will make the two shaded areas in Fig. 8.25a equal. This procedure yields the values of the strains ε_1 and ε_2. Then, from Eqs. (8.140) and

(8.141), we obtain

$$\frac{h_1}{h_2} = \left| \frac{\varepsilon_1}{\varepsilon_2} \right| \tag{8.149}$$

Equation (8.149) defines the position of the neutral axis. From Eq. (8.142), we note that the strain ε is proportional to the distance y from the neutral axis. Therefore, we may also conclude that the curve AOB in Fig. 8.25a represents the bending stress distribution along the depth h of the member, if h is substituted for Δ.

The radius of curvature r may be determined by substituting Eqs. (8.145) and (8.146) into Eq. (8.144), which yields the following expression:

$$br^2 \int_{\varepsilon_2}^{\varepsilon_1} \sigma\varepsilon \, d\varepsilon = M \tag{8.150}$$

We also know that

$$\Delta = \frac{h_1}{r} + \frac{h_2}{r} = \frac{h}{r} \tag{8.151}$$

which can be used to rearrange Eq. (8.150) as follows:

$$\frac{bh^3}{12} \cdot \frac{1}{r} \cdot \frac{12}{\Delta^3} \int_{\varepsilon_2}^{\varepsilon_1} \sigma\varepsilon \, d\varepsilon = M \tag{8.152}$$

For the elastic range, we have the following well-known equation:

$$M = \frac{EI}{r} \tag{8.153}$$

If the proportional limit of the material is exceeded, the curvature produced by the bending moment M can be determined from the expression

$$M = \frac{E_r I}{r} \tag{8.154}$$

where E_r is the reduced modulus, which is defined by the following equation:

$$E_r = \frac{12}{\Delta^3} \int_{\varepsilon_2}^{\varepsilon_1} \sigma\varepsilon \, d\varepsilon \tag{8.155}$$

The integral in Eq. (8.155) represents the moment of the shaded area in Fig.

8.25a with respect to the vertical axis through the origin O. The units of E_r are force per unit area, because the ordinate values of the curve in Fig. 8.25a represent stresses and the abscissa values represent strain. The units are identical to the units of E.

Since the stress and strain levels at cross sections along the length of a member can vary, the reduced modulus E_r will also vary along the length of the member. If the stresses at a cross section are elastic, E_r will be equal to the elastic modulus E. At a given cross section of the member of depth h and moment of inertia I, the reduced modulus E_r may be determined by using a trial-and-error procedure. This procedure may be initiated by assuming values of Δ and using the curve in Fig. 8.25a to determine the extreme elongations ε_1 and ε_2 for each value of Δ. Equation (8.155) may be used to determine E_r. This procedure is explained in detail in Appendix I by using a numerical example.

With known E_r, the bending moment M may be determined from Eq. (8.154), where r can be determined from Eq. (8.151), since Δ is known. Since E_r can vary along the length of the member, the function $g(x)$ representing the variation of E_r with respect to a reference value E may be written as follows:

$$g(x) = \frac{E_r}{E} \tag{8.156}$$

The elastic modulus E is considered here as the reference value.

With known $g(x)$ and $f(x)$, the method of equivalent systems may be applied in the usual way to determine the static or vibration response.

We return now to the problem in Fig. 8.24a in order to apply the above methodology. The bilinear approximation of the stress–strain curve of Monel will be used here to illustrate the procedure. The shape of the stress–strain curve of Monel is the same for both tension and compression, and, consequently, $\varepsilon_1 = \varepsilon_2 = \varepsilon$ and $\Delta = 2\varepsilon$. From Tables 8.8 and 8.9, we note that the bilinear approximation of the stress–strain curve of Monel yields $E_1 = 26 \times 10^6$ psi, $E_2 = 53 \times 10^3$ psi, and yield stress $\sigma_y = 50,000$ psi. Figure 8.25c shows schematically the tension part of the bilinear approximation of the stress–strain curve of Monel and areas A_0, A_1, and A_2 under this curve. Note that A_0 is the elastic part of the curve, while A_1 and A_2 are in the inelastic region. The sum of the first moments of A_0, A_1, and A_2 about the origin O, multiplied by a factor of 2 since the compression part of the curve is identical, provides the value of the integral in Eq. (8.155).

A trial-and-error procedure, such as the one stated above, is used to calculate E_r at sections located at intervals of 3 in. along the length of the member. By assuming values of Δ and using the curve in Fig. 8.25c, we determine for each value of Δ the strain $\varepsilon_1 = \varepsilon_2 = \varepsilon$, and from Eq. (8.155), we determine the corresponding values of E_r. Equation (8.154) is then used

TABLE 8.10. Summary of Reduced Modulus of Elasticity and Required Moment at Cross Sections Along the x Axis of the Member

(1) x (in.)	(2) h_x (in.)	(3) I_x (in.^{4})	(4) Δ (in./in.)	(5) Strain ε (in./in.)	(6) E_r (10^6 psi)	(7) $M_{req} = M_x$ (10^6 in. · lb)
0	12.0	864.00	19.4790×10^{-2}	$\Delta/2$	0.8214	11.52
3	11.9	842.58	9.5345×10^{-2}	$\Delta/2$	1.6222	10.95
6	11.8	821.52	2.1996×10^{-2}	$\Delta/2$	6.7892	10.40
9	11.7	800.81	1.0846×10^{-2}	$\Delta/2$	13.2760	9.86
12	11.6	780.45	7.9961×10^{-3}	$\Delta/2$	17.3300	9.32
15	11.5	760.44	6.6462×10^{-3}	$\Delta/2$	20.0620	8.82
18	11.4	740.77	5.7961×10^{-3}	$\Delta/2$	22.0890	8.32
21	11.3	721.45	5.1962×10^{-3}	$\Delta/2$	23.6000	7.83
24	11.2	702.46	4.7462×10^{-3}	$\Delta/2$	24.6890	7.35
27	11.1	683.82	4.3962×10^{-3}	$\Delta/2$	25.4160	6.88
30	11.0	665.50	4.1462×10^{-3}	$\Delta/2$	25.8010	6.47
33	10.9	647.51	3.8962×10^{-3}	$\Delta/2$	25.9940	6.02
36	10.8	629.86	3.7227×10^{-3}	$\Delta/2$	26.0000	5.64

to determine the required moment M_{req}. At each section, the required moment M_{req} should match the actual moment M_x that is obtained at each section by applying the equations of statics. This is correct as long as the member under consideration is statically determinate. For statically indeterminate members, you may consult ref. 2.

Table 8.10 provides a summary of the computations of E_r at intervals of 3 in. along the length of the member. The values of E_r are given in column (6) of the table, and the required moments M_{req} that match the applied moments M_x are shown in column (7). In the same table, the values of Δ and strain $\varepsilon = \Delta/2$ are also tabulated. Note that the stress at locations beyond 36 in. is below the yield strength of the material, and, consequently, $E_r = E$ in these locations.

Table 8.11 gives the values of $f(x)$, E_r, $g(x)$, $M_x = M_{req}$, and $M_e = M_x/f(x)g(x)$, at intervals of 6 in. along the length of the member. The moment diagram M_e of the equivalent system of constant stiffness $E_1 I_B$ is shown plotted by a solid line in Fig. 8.26a. The approximation of its shape by three straight-line segments, as shown by the dashed line in the same figure, leads to the constant stiffness equivalent system shown in Fig. 8.26b. The deflection at any point along the length of the equivalent system and its rotation are identical to the corresponding ones of the original member in Fig. 8.24a, and they can be determined by using the equivalent system and applying handbook formulas or using elementary methods of linear mechanics. For example, by using the moment–area method, the vertical deflection y_B at the free end B is found to be 8.76 in.

For the bilinear approximation of the stress–strain curve of Monel, the

TABLE 8.11. Values of $f(x)$, $g(x)$, E_r, $M_x = M_{req}$, and M_e for Inelastic Analysis

(1) x (in.)	(2) $f(x)$	(3) E_r (10^6 psi)	(4) $g(x)$	(5) $M_x = M_{req}$ (10^6 in.·lb)	(6) $M_e = M_x/f(x)g(x)$ (10^6 in. · lb)
0	3.38	0.821	0.0316	11.520	108.040
6	3.21	6.789	0.2611	10.397	12.407
12	3.05	17.330	0.6665	9.323	4.588
18	2.89	22.089	0.8496	8.319	3.384
24	2.74	24.689	0.9496	7.349	2.821
30	2.60	25.801	0.9924	6.472	2.509
36	2.46	26.000	1.0000	5.645	2.294
42	2.33	26.000	1.0000	4.867	2.092
48	2.20	26.000	1.0000	4.147	1.888
54	2.07	26.000	1.0000	3.485	1.681
60	1.95	26.000	1.0000	2.880	1.475
66	1.84	26.000	1.0000	2.333	1.269
72	1.73	26.000	1.0000	1.843	1.067
78	1.62	26.000	1.0000	1.411	0.870
84	1.52	26.000	1.0000	1.037	0.682
90	1.42	26.000	1.0000	0.720	0.506
96	1.33	26.000	1.0000	0.461	0.346
102	1.24	26.000	1.0000	0.259	0.209
108	1.16	26.000	1.0000	0.115	0.100
114	1.08	26.000	1.0000	0.029	0.027
120	1.00	26.000	1.0000	0	0

analysis of the nonprismatic beam in Fig. 8.24a was repeated by varying the distributed load w and using values of the taper parameter $n = 1.5$, 1.75, and 2. The results for these three cases of n are shown in Fig. 8.27 on page 565. The starting point P in these curves represents the transition from elastic to inelastic behavior of the member. Note that as n increases, additional load w is required to reach the transition stage. When a critical value of the load w is reached, the deflection increases very rapidly with small changes in w, indicating that the ultimate capacity of the member to resist load and deformation is reached.

The results obtained here by using the bilinear approximation of the stress–strain curve of Monel are reasonably accurate for practical applications. The three-line approximation of the stress–strain curve would improve the accuracy of the results in the critical region of the curves in Fig. 8.27. The curves are somewhat less steep in this region, and the three-line approximation of the stress–strain curve provides all the accuracy we need to define this region.

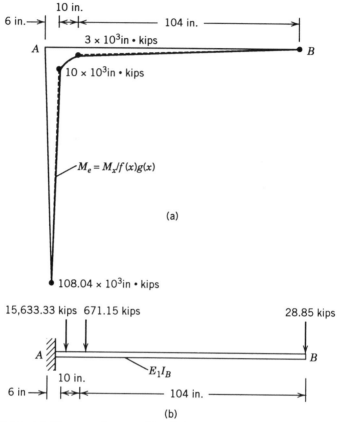

FIGURE 8.26. (a) Moment diagram M_e of the constant stiffness equivalent system with its shape approximated with three straight lines. (b) Equivalent system of constant stiffness $E_1 I_B$.

8.10.2 Vibration Analysis of Inelastic Tapered Cantilever Beams

The fundamental frequency of vibration of the inelastic tapered cantilever beam in Fig. 8.24a will be determined here. We assume that the uniform loading w on the beam represents the weight of the beam and other additional weight that is securely attached to the member. The procedure used in Section 8.9 may also be used here to determine the fundamental frequency of the inelastic member. If we assume that $w = 1600$ lb/in., a portion of the member will be stressed well beyond its elastic limit, and the member will become inelastic. Since vibrations are taking place about the static equilibrium position, the vibration will take place from this position with the member being inelastic.

With respect to the static equilibrium position, an equivalent system of constant stiffness $E_1 I_B$, where E_1 and I_B are the reference values for E_x and

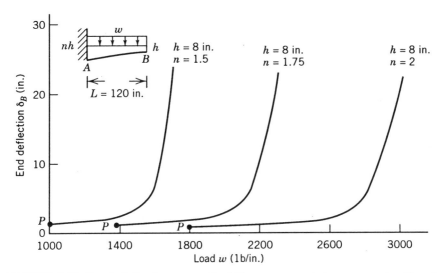

FIGURE 8.27. Load–deflection curves for bilinear stress–strain curve and uniformly distributed loading.

I_x, respectively, may be derived as discussed in the preceding section. Such an equivalent system is already derived for the member in Fig. 8.24a, and it is shown in Fig. 8.26b. The static equilibrium positions of the equivalent system and of the initial system are practically identical. The equivalent system, however, has a constant stiffness $E_1 I_B$, and it is loaded with equivalent concentrated weights. The equivalent system will be used here to determine the fundamental frequency of free vibration of the initial member in Fig. 8.24a. Stodola's method and the iteration procedure will be used for this purpose.

The original system and the equivalent system are again shown in Figs. 8.28a and 8.28b, respectively. By applying unit loads at points 1, 2, and 3 of the equivalent system, the a_{ij} deflection coefficients are determined by using formulas, and they are as follows:

$$a_{11} = \frac{72}{E_1 I_B} \qquad a_{12} = a_{21} = \frac{252}{E_1 I_B} \qquad a_{13} = a_{31} = \frac{2124}{E_1 I_B}$$

$$a_{21} = \frac{252}{E_1 I_B} \qquad a_{22} = \frac{1365.33}{E_1 I_B} \qquad a_{23} = a_{32} = \frac{14677.33}{E_1 I_B}$$

$$a_{31} = \frac{2124}{E_1 I_B} \qquad a_{32} = \frac{14677.33}{E_1 I_B} \qquad a_{33} = \frac{576 \times 10^3}{E_1 I_B}$$

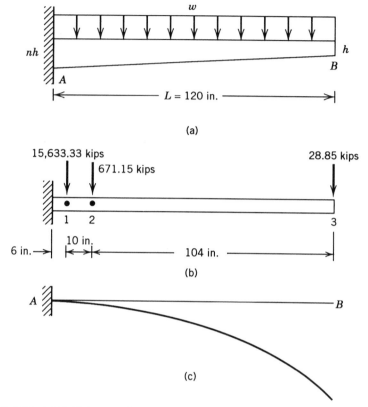

FIGURE 8.28. (a) Initial tapered cantilever beam. (b) Equivalent system of uniform stiffness $E_1 I_B$. (c) Shape of the fundamental mode of vibration.

By using Eq. (5.40) of Chapter 5, the matrix equation is written as follows:

$$\begin{bmatrix} y_1 \\ y_2 \\ y_3 \end{bmatrix} = \frac{(61277 \times 10^3)\omega^2}{E_1 I_B g} \begin{bmatrix} 18.369 & 2.760 & 1.000 \\ 64.292 & 14.954 & 6.910 \\ 567.399 & 160.757 & 271.188 \end{bmatrix} \begin{bmatrix} y_1 \\ y_2 \\ y_3 \end{bmatrix} \quad (8.157)$$

The iteration procedure of Eq. (8.157) may be initiated by assuming that $y_1 = y_2 = y_3 = 1.000$. This yields

$$\begin{bmatrix} y_1 \\ y_2 \\ y_3 \end{bmatrix} = \frac{(61277 \times 10^3)(22.129)\omega^2}{E_1 I_B g} \begin{bmatrix} 1.000 \\ 3.893 \\ 45.160 \end{bmatrix}$$

The new amplitudes may be used to repeat the procedure. After six iterations (or repetitions), the matrix converges as follows:

$$\begin{bmatrix} y_1 \\ y_2 \\ y_3 \end{bmatrix} = \frac{(61277 \times 10^3)(278)\omega^2}{E_1 I_B g} \begin{bmatrix} 1.000 \\ 6.587 \\ 241.382 \end{bmatrix} \qquad (8.158)$$

By using the first row of the above matrix, we find

$$\frac{(61277 \times 10^3)(278)\omega^2}{E_1 I_B g} = 1$$

or

$$\omega^2 = \frac{E_1 I_B g}{(61277 \times 10^3)(278)} \qquad (8.159)$$

The reference value E_1 that is used to derive the equivalent system in Fig. 8.28b is equal to 26×10^6 psi, because the bilinear stress–strain curve is used to derive E_r and, consequently, $g(x)$. The moment of inertia I_B at the free end B is 256 in.[4], and the acceleration of gravity $g = 386.4$ in./s^2 On this basis, Eq. (8.159) yields

$$\omega^2 = \frac{(26 \times 10^6)(256)(386.4)}{(61277 \times 10^3)(278)} = 150.976 \ (\text{rps})^2$$

$$\omega = 12.287 \ \text{rps}$$

$$f = \frac{\omega}{2\pi} = 1.956 \ \text{Hz}$$

If elastic analysis had been used, the fundamental frequency of vibration would be equal to 2.397 Hz, which is 22.55% higher than the one obtained by using inelastic analysis.

The shape of the fundamental mode is characterized by the amplitudes $y_1 = 1.000$, $y_2 = 6.587$, and $y_3 = 241.382$, which are shown in the column on the right-hand side of Eq. (8.158). The mode shape is plotted in Fig. 8.28c.

8.10.3 Vibration Analysis of a Uniform Simply Supported Inelastic Beam

Inelastic analysis will be performed here in order to determine the fundamental frequency of vibration and the corresponding mode shape of the uniform

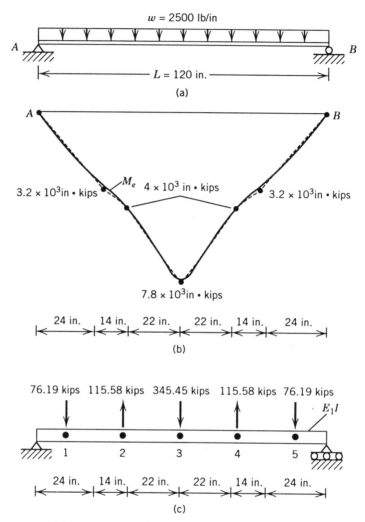

FIGURE 8.29. (a) Initial uniform simply supported beam. (b) Moment diagram M_e of the equivalent system with its shape approximated with six straight-line segments. (c) Equivalent system of constant stiffness $E_1 I$.

simply supported beam shown in Fig. 8.29a. The weight $w = 2500$ lb/in. acting on the beam is assumed to represent its weight and other possible weights that are securely attached to the member. The width $b = 6$ in. and the depth $h = 8$ in. The weight on the member is large enough to cause the member to behave inelastically at the static equilibrium position. Again, the material of the member is Monel, and Tables 8.8 and 8.9 provide the appropriate data for a two-line, three-line, and six-line approximation of the stress–strain curve of Monel.

The function $g(x)$ and, consequently, the constant stiffness equivalent system are derived by using the two-line approximation of the stress–strain curve of Monel. By proceeding as in the preceding two sections of this chapter, the values of Δ, E_r, $g(x)$, $M_{req} = M_x$, and $M_e = M_x/f(x)g(x)$ are determined, and they are shown in Table 8.12. The moment diagram M_e of the equivalent system is plotted as the solid line in Fig. 8.29b. Its shape approximation with six straight-line segments, as shown by the dashed lines in Fig. 8.29b, leads to the constant stiffness equivalent system shown in Fig. 8.29c.

The equivalent system in Fig. 8.29c is loaded with equivalent concentrated weights acting at points 1, 2, 3, 4, and 5. The static deflections y_1, y_2, y_3, y_4, and y_5 at points 1, 2, 3, 4, and 5, respectively, caused by the equivalent weights, establish the static equilibrium position of the member, and they can be determined by using known handbook formulas. They are as follows:

$$y_1 = 0.7390 \text{ in.} \qquad y_1^2 = 0.5461 \text{ in.}^2$$

$$y_2 = 1.0701 \text{ in.} \qquad y_2^2 = 1.1451 \text{ in.}^2$$

$$y_3 = 1.3076 \text{ in.} \qquad y_3^2 = 1.7098 \text{ in.}^2$$

$$y_4 = y_2 = 1.0701 \text{ in.} \qquad y_4^2 = 1.1451 \text{ in.}^2$$

$$y_5 = y_1 = 0.7390 \text{ in.} \qquad y_5^2 = 0.5461 \text{ in.}^2$$

The method of Lord Rayleigh, represented by Eqs. (5.28) and (5.30) in Chapter 5, will be used here to determine the fundamental frequency of vibration. For the first approximation, we use Eq. (5.28), which is as follows:

$$\omega^2 = g \frac{\sum\limits_{i=1}^{5} W_i y_i}{\sum\limits_{i=1}^{5} W_i y_i^2} \tag{8.160}$$

On this basis, we have

$$\sum_{i=1}^{5} W_i y_i = (76.19)(0.7390) + (115.58)(1.0701) + (345.45)(1.3076)$$

$$+ (115.58)(1.0701) + (76.19)(0.7390)$$

$$= 811.6836$$

$$\sum_{i=1}^{5} W_i y_i^2 = (76.19)(0.5461) + (115.58)(1.1451) + (345.45)(1.7098)$$

$$+ (115.58)(1.1451) + (76.19)(0.5461)$$

$$= 938.5667$$

TABLE 8.12. Values of $f(x)$, E_r, $g(x)$, M_{req}, and M_e for Inelastic Analysis of the Simply Supported Beam

(1) x (in.)	(2) h_x (in.)	(3) $f(x)$	(4) Δ (10^{-3} in./in.)	(5) E_r (10^6 psi)	(6) $g(x)$	(7) $M_{req} = M_x$ (10^6 in.·lb)	(8) $M_e = M_x/f(x)g(x)$ (10^6 in.·lb)
0	8	1	0	26.000	1.0000	0	0
6	8	1	1.0276	26.000	1.0000	0.855	0.855
12	8	1	1.9421	26.000	1.0000	1.620	1.620
18	8	1	2.7584	26.000	1.0000	2.295	2.295
24	8	1	3.4615	26.000	1.0000	2.880	2.880
30	8	1	4.0962	25.858	0.9945	3.389	3.408
36	8	1	4.7962	24.571	0.9451	3.771	3.990
42	8	1	5.7961	22.082	0.8493	4.096	4.822
48	8	1	6.9962	19.282	0.7416	4.317	5.821
54	8	1	8.2961	16.787	0.6457	4.457	6.902
60	8	1	8.8961	15.813	0.6082	4.502	7.402
66	8	1	8.2961	16.787	0.6457	4.457	6.902
72	8	1	6.9962	19.282	0.7416	4.317	5.821
78	8	1	5.7961	22.082	0.8493	4.096	4.822
84	8	1	4.7962	24.571	0.9451	3.771	3.990
90	8	1	4.0962	25.858	0.9945	3.389	3.408
96	8	1	3.4615	26.000	1.0000	2.880	2.880
102	8	1	2.7584	26.000	1.0000	2.295	2.295
108	8	1	1.9471	26.000	1.0000	1.620	1.620
114	8	1	1.0276	26.000	1.0000	0.855	0.855
120	8	1	0	26.000	1.0000	0	0

Thus, Eq. (8.160) yields

$$\omega^2 = (386.4)\frac{811.6836}{938.5667} = 334.1633$$

or

$$\omega = 18.28 \text{ rps}$$

The procedure may be repeated for better accuracy by using Eq. (5.30), which is written again below:

$$\omega^2 = g\frac{\displaystyle\sum_{i=1}^{5} W_i' y_i'}{\displaystyle\sum_{i=1}^{r} W_i(y_i')^2} \tag{8.161}$$

In order to use Eq. (8.161), we need to know the new loads W_i' and the new deflections y_i' produced by these loads. We have

$$W_1' = W_1 y_1 = (76.19)(0.739) = 56.30 \text{ kips}$$
$$W_2' = W_2 y_2 = -(115.58)(1.0701) = -123.68 \text{ kips}$$
$$W_3' = W_3 y_3 = (345.45)(1.3076) = 451.71 \text{ kips}$$
$$W_4' = W_2' = -123.68 \text{ kips}$$
$$W_5' = W_i' = 56.30 \text{ kips}$$

By applying statically the new W_i' loads at points 1, 2, 3, 4, and 5 of the equivalent system in Fig. 8.29c, we obtain

$$y_1' = 0.9501 \text{ in.} \qquad (y_1')^2 = 0.9027 \text{ in.}^2$$
$$y_2' = 1.3773 \text{ in.} \qquad (y_2')^2 = 1.8970 \text{ in.}^2$$
$$y_3' = 1.6880 \text{ in.} \qquad (y_3')^2 = 2.8493 \text{ in.}^2$$
$$y_4' = 1.3773 \text{ in.} \qquad (y_4')^2 = 1.8970 \text{ in.}^2$$
$$y_5' = 0.9501 \text{ in.} \qquad (y_5')^2 = 0.9027 \text{ in.}^2$$

On this basis, we have

$$\sum_{i=1}^{5} W_i' y_i' = 1210.1567$$

$$\sum_{i=1}^{5} W_i(y_i')^2 = 1560.3547$$

Equation (8.161) yields

$$\omega^2 = (386.4) \frac{1210.1567}{1560.3547} = 299.6784$$

$$\omega = 17.3112 \text{ rps}$$

$$f = \frac{\omega}{2\pi} = 2.7552 \text{ Hz}$$

No further repetition is required because the next repetition yields about the same value.

If elastic analysis is used to solve the original member in Fig. 8.29a, we have

$$\omega = \frac{\pi^2}{L^2} \sqrt{\frac{EI}{m}}$$

$$= \frac{\pi^2}{(120)^2} \sqrt{\frac{(26 \times 10^6)(256)(386.4)}{2500}}$$

$$= 21.9833 \text{ rps}$$

which is 26.99% higher than the one obtained by using inelastic analysis.

If we assume that $y_1 = 1.000$, the normalized shape of the fundamental mode is characterized by the amplitudes

$$y_1 = 1.0000 \qquad y_4 = 1.4496$$

$$y_2 = 1.4496 \qquad y_5 = 1.0000$$

$$y_3 = 1.7767$$

PROBLEMS

8.1 Solve the problem in Example 8.1 by assuming that $P = 2.0$ kips, and compare the results.

8.2 Solve the problem in Example 8.1 by assuming that the thickness parameter $n = 1$ and 2, and compare the results.

8.3 Solve the problem in Example 8.1 by assuming that $P = 3.0$ kips, and compare the results.

8.4 Repeat Problem 8.1 by using a simplified nonlinear equivalent system

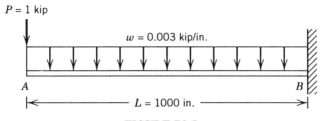

FIGURE P8.5

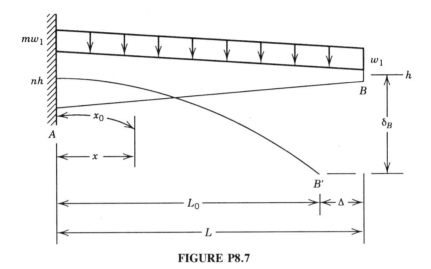

FIGURE P8.7

of constant stiffness EI_B that is loaded with an equivalent concentrated load P_e at the free end. Compare the results.

8.5 The uniform cantilever beam in Fig. P8.5 is loaded by a uniformly distributed load $w = 0.003$ kip/in. and a concentrated load $P = 1.0$ kip at the free end A. The length L of the member is 1000 in. and the constant stiffness $EI = 609 \times 10^3$ kip · in.^{2} By using a simplified non-linear equivalent system of constant stiffness EI that is loaded with an equivalent concentrated load P_e at the free end, determine the rotation θ_A, the vertical displacement δ_A, and the horizontal displacement Δ_A at the free end A.

8.6 Repeat Problem 8.5 with $P = 5.0$ kips, $w = 0.0015$ kip/in., and $EI = 600 \times 10^3$ kip · in.^{2}, and compare the results.

8.7 The doubly tapered flexible cantilever beam in Fig. P8.7 is loaded with a distributed trapezoidal load as shown in the figure. The load $w_1 = 0.005$ kip/in., $EI_B = 180 \times 10^3$ kip · in.^{2}, $L = 1000$ in., and $n =$

$m = 2$. By using a simplified nonlinear equivalent system of constant stiffness EI_A that is loaded with an equivalent concentrated load P_e at the free end, determine the rotation θ_B, the horizontal displacement Δ_B, and the vertical displacement δ_B at the free end B.

8.8 Repeat Problem 8.7 with $m = 2$ and $n = 1$ and 3, and compare the results.

8.9 Assume that the variable stiffness members in Fig. P8.9 are made of steel with width $b = 6$ in. and depth $h_1 = 12$ in. For each case, determine an accurate equivalent system of constant stiffness EI_1 that is loaded with concentrated equivalent loads. The modulus of elasticity $E = 30 \times 10^6$ psi.

8.10 The steel fixed–fixed beam in Fig. P8.10a is loaded by a distributed load as shown. The thickness variation is parabolic, and the values of the stiffness function $f(x)$ at intervals of 7.25 ft are shown in Fig. P8.10b. By applying the method of the equivalent systems as discussed in Section 8.6, determine the reaction R_A and bending moment M_A at the fixed end A. The stiffness $EI_1 = 460 \times 10^6$ kip · in.^{2}, where I_1 is the moment of inertia at the center of the beam.

8.11 By using the method of equivalent systems, determine the reaction R_B at support B of the two-span continuous beam in Fig. P8.11. The constant width $b = 10$ in., and depth $h = 16$ in. Between supports A and B and supports C and B, the depth h_x is arranged so that the moment of inertia is linear between these points. The modulus of elasticity $E = 30 \times 10^6$ psi, and the loading on the beam is as shown.

8.12 A tapered simply supported beam is shown in Fig. P8.12, where the uniformly distributed load of 1.0 kip/ft represents the weight of the beam and any other weights that are securely attached to the member. The constant width b of the member is 10.0 in., $h = 16.0$ in., the taper $n = 1.3$, and the modulus of elasticity $E = 30 \times 10^6$ psi. By using an equivalent system of constant stiffness EI_A, where I_A is the moment of inertia at support A, determine the first three free frequencies of vibration and their corresponding mode shapes.

8.13 Repeat Problem 8.12 by using Stodola's method and the iteration procedure to solve the initial system, and compare the results. Calculate the required a_{ij} deflection coefficients by using the method of equivalent systems.

8.14 Repeat Problem 8.12 by assuming that the taper $n = 2$, and compare the results.

8.15 Repeat Problem 8.14 by using Stodola's method and the iteration procedure to solve the initial system, and compare the results. Calcu-

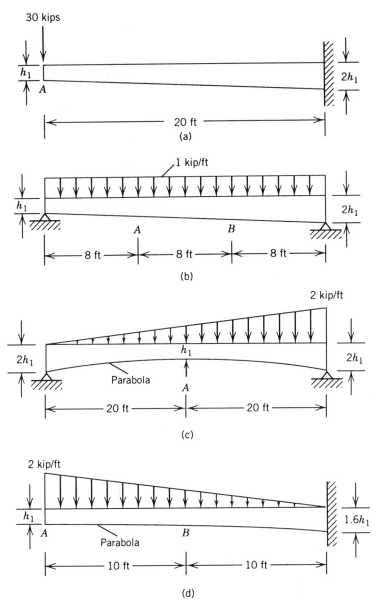

FIGURE P8.9

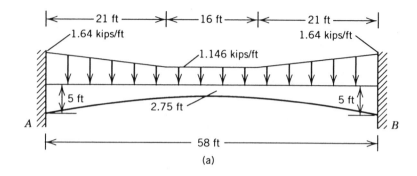

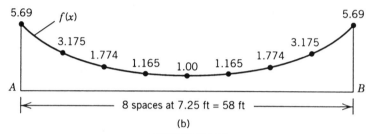

(b)

FIGURE P8.10

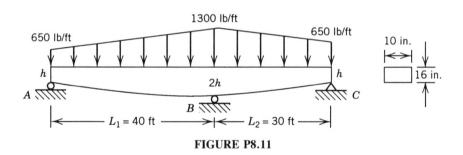

FIGURE P8.11

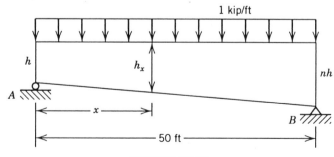

FIGURE P8.12

late the required a_{ij} deflection coefficients by using the method of equivalent systems.

8.16 Solve Problem 8.13 by using the finite element method, and compare the results.

8.17 Solve Problem 8.12 by using the finite element method to solve the constant stiffness equivalent system, and compare the results.

8.18 Solve Problem 8.15 by using the finite element method, and compare the results.

8.19 Solve Problem 8.14 by using the finite element method to solve the constant stiffness equivalent system, and compare the results.

8.20 Consider the variable stiffness fixed–fixed beam in Fig. P8.10a and assume that the distributed loading on the member is the weight of the beam and additional weight that is securely attached to the member. The depth of the member has a parabolic variation, and the variation $f(x)$ of its moment of inertia is shown in Fig. P8.10b. The reference stiffness EI_1 is at the center of the member, and it is equal to $460 \times 10^6 \text{ kip} \cdot \text{in.}^2$. By using an equivalent system of constant stiffness EI_1, determine the fundamental free frequency of vibration of the member and the corresponding mode shape.

8.21 Repeat Problem 8.20 by using Stodola's method and the iteration procedure to solve the original system in Fig. P8.10a, and compare the results. Use the method of equivalent systems to determine the required a_{ij} deflection coefficients.

8.22 Repeat Problem 8.20 by using the finite element method to solve (a) the original variable stiffness member and (b) the constant stiffness equivalent system. Compare the results.

8.23 Consider the tapered two-span continuous beam in Fig. P8.23, and assume that the loading of 1.0 kip/ft represents the weight of the member and other weights that are securely attached to the member. The constant width $b = 10$ in., the smallest depth $h = 6$ in., the modulus of elasticity $E = 30 \times 10^6$ psi, and the span lengths are as shown in

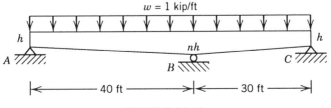

FIGURE P8.23

the figure. By assuming that the taper $n = 1.5$ and by using an equivalent system of constant stiffness EI_1, where I_1 is the moment of inertia at the ends A and C of the member, determine the first three free frequencies of vibration of the member and the corresponding mode shapes.

8.24 Repeat Problem 8.23 for taper $n = 1.2$, and compare the results.

8.25 Repeat Problem 8.23 for taper $n = 2$, and compare the results.

8.26 Rework Problem 8.23 by using Stodola's method and the iteration procedure to solve the original variable stiffness member, and compare the results. Use the method of equivalent systems to determine the required a_{ij} deflection coefficients.

8.27 Rework Problem 8.23 by using the finite element method to solve (a) the original variable stiffness member and (b) the constant stiffness equivalent system. Compare the results.

8.28 Repeat Problem 8.24 by using Stodola's method to solve the original variable stiffness members, and compare the results.

8.29 Consider the tapered cantilever beam in Fig. P8.29 that is loaded with a concentrated load Q at the free end B. Assume that $Q = 125$ kips, $h = 8$ in., width $b = 6$ in., and the taper $n = 1.75$. The material of the member is Monel. By using the three-line approximation of the stress–strain curve of Monel and following inelastic analysis, determine (a) the moment diagram M_e of the equivalent system of constant stiffness E_1I_B and (b) an equivalent system of constant stiffness E_1I_B loaded with concentrated loads. (c) Use the equivalent system to determine the inelastic deflection δ_B at the free end B of the member.

8.30 Repeat Problem 8.29 by using the six-line approximation of the stress–strain curve of Monel, and compare the results.

8.31 Repeat Problem 8.29 by using the bilinear stress–strain curve of Monel, and compare the results.

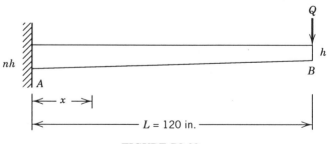

FIGURE P8.29

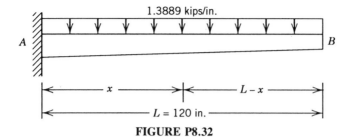

1.3889 kips/in.

A

B

x

$L - x$

$L = 120$ in.

FIGURE P8.32

8.32 Consider the cantilever beam in Fig. P8.32, which is loaded with a uniformly distributed load of 1.3889 kips/in. The moment of inertia I_B at the free end B is 250 in.^{4} and 500 in.^{4} at the fixed end A, and it varies linearly between A and B. The material of the member is Monel. By following inelastic analysis and using the six-line approximation of the stress–strain curve of Monel, determine (a) the moment diagram M_e of the equivalent system of constant stiffness $E_1 I_B$ and (b) an equivalent system of constant stiffness $E_1 I_B$ loaded with concentrated loads. (c) Use the equivalent system to determine the inelastic deflection δ_B at the free end B.

8.33 Repeat Problem 8.32 by using the three-line approximation of the stress–strain curve of Monel, and compare the results.

8.34 Repeat Problem 8.32 by using the two-line approximation of the stress–strain curve of Monel, and compare the results.

8.35 By following inelastic analysis, determine the fundamental frequency of vibration and the corresponding mode shape of the cantilever beam in Problem 8.32. Assume that the uniform load of 1.3889 kips/in. is the weight of the member and additional weight that is securely attached to the member.

8.36 Repeat Problem 8.35 by using the two-line approximation of the stress–strain curve of Monel, and compare the results.

8.37 Repeat Problem 8.35 by assuming that the weight $w = 1200$ lb/in., and compare the results.

8.38 The tapered simply supported beam in Fig. P8.38 is made out of mild steel. The weight distribution w and the dimensions of the beam are as shown in the figure. The width b is constant and equal to 6 in. The four-line approximation of the stress–strain curve of the steel yields $E_1 = 94.366 \times 10^5$ psi, $E_2 = 28.289 \times 10^4$ psi, $E_3 = 3.785 \times 10^4$ psi, $E_4 = -7.971 \times 10^4$ psi, $\sigma_1 = 33.5 \times 10^3$ psi, $\sigma_2 = 55.0 \times 10^3$ psi, and

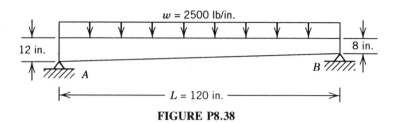

FIGURE P8.38

$\sigma_3 = 61.0 \times 10^3$ psi. By applying inelastic analysis, determine the fundamental frequency of vibration of the beam and its corresponding mode shape.

8.39 Repeat Problem 8.38 by assuming that the weight $w = 2800$ lb/in., and compare the results.

9 Dynamic Response of Idealized Systems by Using Fourier and Laplace Transforms

9.1 INTRODUCTION

In Section 3.10, Fourier series were used to express periodic force functions in terms of sine and cosine functions in order to determine the total dynamic response of a spring–mass system by a superposition of harmonic excitation responses as discussed in Sections 3.2 and 3.3. In this chapter, methods of transformed calculus, which involve Fourier and Laplace transforms, will be used to determine the dynamic response of structural and mechanical systems that are subjected to periodic and nonperiodic excitations. The discussion, however, is geared to spring–mass systems and structural and mechanical systems that can be idealized as spring–mass systems, as discussed in earlier chapters of the text. This method of approach, together with the convolution theorem as treated in this chapter, provides some of the basic groundwork needed for the analysis of systems subjected to random excitations, as discussed in the following chapter. The last part of this chapter deals with the dynamic response of machine foundations that are subjected to various excitations, by including the effects of viscous damping.

9.2 PERIODIC AND NONPERIODIC EXCITATIONS AND DISCRETE SPECTRA

The general theory of periodic and nonperiodic excitation and the associated discrete spectra are discussed in this section.

9.2.1 Periodic Excitations

A force function $f(t)$ acting on an engineering system and defined in the time interval between zero and τ can be represented in the frequency domain by an infinite number of cosine and sine components, provided that the following three conditions are satisfied: (a) $f(t)$ is periodic; (b) $f(t)$ has a finite number of discontinuities; and (c) the following integral exists:

$$\int_{-\tau/2}^{\tau/2} |f(t)| \, dt \tag{9.1}$$

The harmonic components of $f(t)$ have frequencies that are harmonically related to the fundamental one. The amplitude and relative phase of each of these frequency components can be expressed by a Fourier series expansion of $f(t)$, as shown in Section 3.10. The trigonometric form of the Fourier series is written again here as follows:

$$f(t) = \frac{a_0}{\tau} + \frac{2}{\tau} \sum_{n=1}^{\infty} (a_n \cos n\omega_0 t + b_n \sin n\omega_0 t) \tag{9.2}$$

where

$$a_0 = \int_{-\tau/2}^{\tau/2} f(t) \, dt \tag{9.3}$$

$$a_n = \int_{-\tau/2}^{\tau/2} f(t) \cos n\omega_0 t \, dt \tag{9.4}$$

$$b_n = \int_{-\tau/2}^{\tau/2} f(t) \sin n\omega_0 t \, dt \tag{9.5}$$

Also, ω_0 in the above equations is the fundamental frequency, τ is the period, and

$$\omega_0 = \frac{2\pi}{\tau} \tag{9.6}$$

The Fourier series expansion of $f(t)$ may also be written in exponential form as follows:

$$f(t) = \frac{1}{\tau} \sum_{n=-\infty}^{\infty} c_n e^{jn\omega_0 t} \tag{9.7}$$

where

$$c_n = \int_{-\tau/2}^{\tau/2} f(t) e^{-jn\omega_0 t} \, dt \tag{9.8}$$

In Eqs. (9.7) and (9.8), the symbol j is used to denote the imaginary part of a complex number.

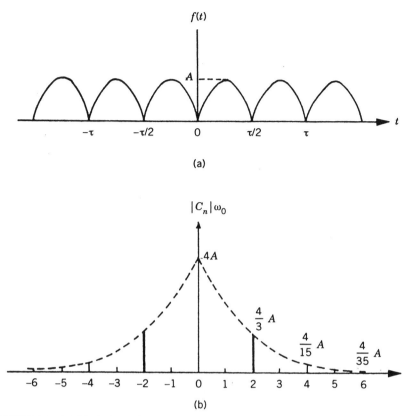

FIGURE 9.1. (a) Variation of the periodic exciting force $f(t)$. (b) Discrete frequency spectrum of $f(t)$.

Equation (9.7) shows that a periodic function contains all frequencies, both negative and positive, which are related harmonically to the fundamental frequency. The magnitude of each frequency component is $|c_n/\tau|$, where $n = 1, 2, 3, \ldots$, and they constitute the discrete spectrum of the waveform.

Example 9.1. If the exciting force $f(t)$ has the waveform shown in Fig. 9.1a, determine its Fourier series representation and the frequency spectrum of the waveform.

Solution. The function $f(t)$ defining the waveform in Fig. 9.1a is as follows:

$$f(t) = A|\sin \omega_0 t| \tag{9.9}$$

where A is the maximum value of $f(t)$, and ω_0 is given by Eq. (9.6).

By substituting Eq. (9.9) into Eq. (9.8), we find

$$
\begin{aligned}
c_n &= \int_{-\tau/2}^{\tau/2} A|\sin \omega_0 t| e^{-jn\omega_0 t}\, dt \\
&= A \int_{-\tau/2}^{\tau/2} |\sin \omega_0 t|(\cos n\omega_0 t - j\sin n\omega_0 t)\, dt \\
&= A \int_{-\tau/2}^{\tau/2} |\sin \omega_0 t| \cos n\omega_0 t\, dt \\
&\quad - jA \int_{-\tau/2}^{\tau/2} |\sin \omega_0 t| \sin n\omega_0 t\, dt
\end{aligned}
\tag{9.10}
$$

Since $f(t)$ in Eq. (9.10) is symmetrical with respect to the vertical axis, the second integral on the right-hand side of the equation can be proved to be zero, and Eq. (9.10) yields

$$
\begin{aligned}
c_n &= 2A \int_{0}^{\tau/2} (\sin \omega_0 t)(\cos n\omega_0 t)\, dt \\
&= A \int_{0}^{\tau/2} [\sin(n+1)\omega_0 t - \sin(n-1)\omega_0 t]\, dt
\end{aligned}
\tag{9.11}
$$

By integrating Eq. (9.11), we find

$$
c_n = -A\left(\frac{\cos(n+1)\pi}{(n+1)\omega_0} - \frac{\cos(n-1)\pi}{(n-1)\omega_0} - \frac{1}{(n+1)\omega_0} - \frac{1}{(n-1)\omega_0}\right)
\tag{9.12}
$$

In Eq. (9.12), we note that, for odd values of n, the quantities $n+1$ and $n-1$ are even, and, consequently, c_n is zero. For even values of n, we have

$$
c_n = \frac{A}{\omega_0} \cdot \frac{4}{(n^2 - 1)}
\tag{9.13}
$$

By substituting Eq. (9.13) into Eq. (9.7), the Fourier series expansion of $f(t)$ is as follows:

$$
f(t) = \sum_{-\infty}^{\infty} -\frac{4A}{\omega_0 \tau}\left(\frac{1}{n^2 - 1}\right)e^{jn\omega_0 t}
\tag{9.14}
$$

It should be noted that even values of n should be used to evaluate the summation in Eq. (9.14).

Figure 9.1b illustrates the discrete frequency spectrum of $f(t)$, where the horizontal axis represents the values of n, and the vertical axis gives the values of $|c_n|\omega_0$. In the same figure, the dashed line represents the envelope of the frequency spectrum, which is shown to decrease as $1/(n^2 - 1)$.

9.2.2 Fourier Transforms for Nonperiodic Excitations

If the force functions are not periodic, a different approach should be followed to examine their frequency domain. This approach is known as the *Fourier transform method*, and it can be initiated by using the Fourier series given by Eq. (9.7). We initiate the procedure by letting τ in Eq. (9.8) approach infinity. In this manner, the function $f(t)$ may be considered to be nonperiodic, and as τ approaches infinity, all Fourier coefficients c_n go to zero, provided that the following condition is satisfied:

$$\int_{t=t_0}^{t=\infty} |f(t)|\,dt < \infty \tag{9.15}$$

On this basis, the fundamental frequency $\omega_0 = 2\pi/\tau$ also goes to zero, and, in this manner, each component of the series represented by Eq. (9.8) is practically equal to the adjacent ones; that is, the smallness of the difference

$$(n + 1)\omega_0 - n\omega_0 = \frac{2\pi}{\tau} \tag{9.16}$$

is characterized by the size of the parameter τ.

We can remove some of the difficulties associated with judging the smallness of differences, by introducing minor adjustments in Eqs. (9.7) and (9.8). we define $\Delta\omega$ as follows:

$$\Delta\omega \equiv \omega_0 = \frac{2\pi}{\tau} \tag{9.17}$$

where the symbol $\equiv$ in Eq. (9.17) is used to denote *equal by definition*. We also have

$$F(n\omega_0) \equiv \tau c_n \tag{9.18}$$

Thus, by Eqs. (9.17) and (9.18), we have

$$\Delta\omega F(n\omega_0) = 2\pi c_n \tag{9.19}$$

On this basis, Eq. (9.8) may be written in the following manner:

$$F(n\omega_0) = \lim_{\tau \to \infty} \int_{-\tau/2}^{\tau/2} f(t) e^{-jn\omega_0 t} \, dt$$

or

$$F(\omega) = \int_{-\infty}^{\infty} f(t) e^{-j\omega t} \, dt \qquad (9.20)$$

where, in Eq. (9.20), ω is substituted for $n\omega_0$. In a similar manner, Eq. (9.7) yields

$$f(t) = \frac{1}{2\pi} \int_{-\infty}^{\infty} F(\omega) e^{j\omega t} \, d\omega \qquad (9.21)$$

Equation (9.20) is known as the *Fourier transform*, and Eq. (9.21) is the *inverse Fourier transform*. Equation (9.20) represents the synthesis of a function of time $f(t)$ that is composed of an infinite number of sines and cosines. In this equation, $F(\omega)$ represents the complex spectrum density of $f(t)$. Equations (9.20) and (9.21) are referred to as the *Fourier transform pair*.

Example 9.2. Determine the frequency spectrum for the waveform shown in Fig. 9.2a.

Solution. From Eq. (9.20), we obtain

$$F(\omega) = \int_{-\infty}^{\infty} f(t) e^{-j\omega t} \, dt$$

$$= \int_{-\infty}^{0} e^{3t} e^{-j\omega t} \, dt + \int_{0}^{\infty} e^{-3t} e^{-j\omega t} \, dt$$

$$= \frac{1}{3 - j\omega} + \frac{1}{3 + j\omega}$$

or

$$F(\omega) = \frac{6}{9 + \omega^2} \qquad (9.22)$$

The plot of Eq (9.22) yields the continuous frequency spectrum shown in Fig. 9.2b.

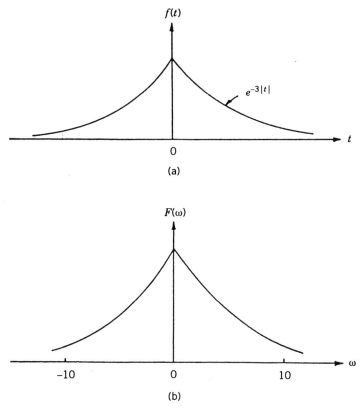

FIGURE 9.2. (a) Variation of the nonperiodic force $f(t)$. (b) Continuous frequency spectrum of $f(t)$.

9.2.3   Laplace Transforms for Nonperiodic Excitations

The force functions that are usually of interest to the engineer are the ones with zero values at $t < 0$ and not converging for $t > 0$. For such cases, it is often possible to multiply the function under consideration by a converging factor $e^{-\sigma t}$, where σ is a constant, and form a new function $\Phi(t)$ that converges. This new function is expressed as follows:

$$\Phi(t) = f(t)e^{-\sigma t} \tag{9.23}$$

where σ is considered to be sufficiently large to make the following integral finite:

$$\int_{-\infty}^{\infty} \Phi(t)\, dt \quad \text{(to be finite)} \tag{9.24}$$

By using Eq. (9.20), the Fourier transform of $\Phi(t)$ is

$$F(\omega) = \int_{-\infty}^{\infty} \Phi(t)\, e^{-j\omega t}\, dt \tag{9.25}$$

Since $f(t)$ is zero for $t < 0$, Eq. (9.25) yields

$$
\begin{aligned}
F(\omega) &= \int_{-\infty}^{0} (0)\, e^{-j\omega t}\, dt + \int_{0}^{\infty} \Phi(t)\, e^{-j\omega t}\, dt \\
&= \int_{0}^{\infty} \Phi(t)\, e^{-j\omega t}\, dt
\end{aligned}
\tag{9.26}
$$

or, by using Eq. (9.23),

$$F(\omega, \sigma) = \int_{0}^{\infty} [f(t)\, e^{-\sigma t}]\, e^{-j\omega t}\, dt \tag{9.27}$$

From Eq. (9.27), we note that the transform now becomes a function of both ω and σ.

Equation (9.27) may also be written as follows:

$$F(\sigma + j\omega) = \int_{0}^{\infty} f(t)\, e^{-(\sigma + j\omega)t}\, dt \tag{9.28}$$

Equation (9.28) is known as the *direct Fourier integral transform* of $f(t)$. We can denote now by s the complex quantity $\sigma + j\omega$ and rewrite Eq. (9.28) in the following manner:

$$F(s) = \int_{0}^{\infty} f(t)\, e^{-st}\, dt \tag{9.29}$$

Equation (9.29) is known as the *Laplace transform* of $f(t)$, which shows that when σ is zero, Eq. (9.29) is identical to Eq. (9.20).

The Laplace transform of a time function $f(t)$ transforms the function to the s plane. With known $F(s)$, the function $f(t)$ can be determined from the following expression:

$$f(t) = \frac{1}{2\pi j} \int_{\sigma - j\omega}^{\sigma + j\omega} F(s)e^{st} dt \qquad (9.30)$$

Equations (9.29) and (9.30) constitute the *Laplace transform pair* that can be used for the solution of many practical engineering problems, as will be shown in the following sections. For convenience in applications, the Laplace transforms of certain functions and derivatives are shown in Appendix J.

9.3 ONE-DEGREE SPRING–MASS SYSTEMS SUBJECTED TO EXTERNAL EXCITATIONS AND VISCOUS DAMPING

We start the analysis by considering the single-degree spring–mass system in Fig. 9.3, which is subjected to a dynamic force $F(t)$ and moves under the influence of viscous damping. The maximum value of the force is F_0 and $f(t)$ is its time variation. The differential equation of motion is as follows:

$$m\ddot{y} + c\dot{y} + ky = F_0 f(t) \qquad (9.31)$$

or, by dividing through by m,

$$\ddot{y} + \frac{c}{m}\dot{y} + \frac{k}{m}y = \frac{F_0}{m}f(t) \qquad (9.32)$$

Since $\omega^2 = k/m$ and $c/m = 2\omega\zeta$, where ζ is the damping ratio, Eq. (9.32)

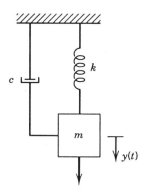

$F(t) = F_0 f(t)$ **FIGURE 9.3.** Single-degree spring–mass system.

yields

$$\ddot{y} + 2\omega\zeta\dot{y} + \omega^2 y = \frac{F_0}{m}f(t) \tag{9.33}$$

By taking the Laplace transform of both sides of Eq. (9.33), we find

$$s^2 Y(s) - sY(0) - \dot{Y}(0) + 2\omega\zeta[sY(s) - Y(0)] + \omega^2 Y(s) = \frac{F_0}{m}F(s) \tag{9.34}$$

where $Y(0)$ and $\dot{Y}(0)$ are the displacement and velocity of the mass, respectively, at time $t = 0$. If $Y(0) = \dot{Y}(0) = 0$, Eq. (9.34) yields

$$Y(s)(s^2 + 2\omega\zeta s + \omega^2) = \frac{F_0}{m}F(s) \tag{9.35}$$

or

$$Y(s) = \frac{[F_0 F(s)]/m}{s^2 + 2\omega\zeta s + \omega^2} \tag{9.36}$$

Equation (9.36) provides a relationship between the input $F_0 F(s)$ and the output $Y(s)$ of the spring–mass system, through the characteristic impedance $Z(s)$, where

$$Z(s) = s^2 + 2\omega\zeta s + \omega^2 \tag{9.37}$$

Therefore, Eq. (9.36) may be written as follows:

$$Y(s) = \frac{[F_0 F(s)]/m}{Z(s)} \tag{9.38}$$

We now consider the case where the time function $f(t)$ in Eq. (9.31) is sinusoidal; that is,

$$f(t) = \sin \omega_f t \tag{9.39}$$

where ω_f is the frequency of the force $F(t)$. The Laplace transform of Eq. (9.39) is

$$F(s) = \frac{\omega_f}{s^2 + \omega_f^2} \tag{9.40}$$

Thus, by substituting Eq. (9.40) into Eq. (9.36), we obtain

$$Y(s) = \frac{(F_0/k)\omega^2 \omega_f}{(s^2 + 2\omega\zeta s + \omega^2)(s^2 + \omega_f^2)} \tag{9.41}$$

If we set the denominator of Eq. (9.41) equal to zero, we can obtain four roots from the resulting equation. If we denote these roots as s_1, s_2, s_3, and s_4, we can write Eq. (9.41) in the following manner:

$$Y(s) = \frac{(F_0/k)\omega^2 \omega_f}{(s - s_1)(s - s_2)(s - s_3)(s - s_4)} \tag{9.42}$$

Now we can find the actual values, or expressions, for s_1, s_2, s_3, and s_4 as follows:

$$(s^2 + 2\omega\zeta s + \omega^2)(s^2 + \omega_f^2) = 0 \tag{9.43}$$

The solution of the algebraic equation given by Eq. (9.43) yields

$$s_1 = -\zeta\omega + j\omega(1 - \zeta^2)^{1/2} \tag{9.44}$$

$$s_2 = -\zeta\omega - j\omega(1 - \zeta^2)^{1/2} \tag{9.45}$$

$$s_3 = -j\omega_f \tag{9.46}$$

$$s_4 = j\omega_f \tag{9.47}$$

By using partial fractions expansion, we can write Eq. (9.42) as follows:

$$Y(s) = \frac{(F_0/k)\omega^2 \omega_f}{(s - s_1)(s - s_2)(s - s_3)(s - s_4)} = \frac{A}{s - s_1} + \frac{B}{s - s_2} + \frac{C}{s - s_3} + \frac{D}{s - s_4} \tag{9.48}$$

where A, B, C, and D are constants that need to be determined. For example, the constant A can be determined by multiplying both sides of Eq. (9.48) by $(s - s_1)$ and evaluating the equation for $s = s_1$. This procedure yields

$$A = \frac{(F_0/k)\omega^2 \omega_f}{(s - s_2)(s - s_3)(s - s_4)} \tag{9.49}$$

where s_2, s_3, and s_4 are given by Eqs. (9.45), (9.46), and (9.47), respectively. At $s = s_1 = -\omega\zeta + j\omega(1 - \zeta^2)^{1/2}$, Eq. (9.49) yields

$$A = \frac{(F_0/k)\omega^2 \omega_f}{2j\omega_d[(\omega\zeta - j\omega_d)^2 + \omega_f^2]} \tag{9.50}$$

where ω_d is the damped free frequency of the spring–mass system, and it is given by the expression

$$\omega_d = \omega(1 - \zeta^2)^{1/2} \tag{9.51}$$

In a similar manner, the constant B in Eq. (9.48) may be obtained by multiplying both sides of the equation by $(s - s_2)$ and evaluating the resulting equation for $s = s_2$. This yields the following expression for the constant B:

$$B = \frac{(F_0/k)\omega^2 \omega_f}{- 2j\omega_d[(\omega\zeta + j\omega_d)^2 + \omega_f^2]} \tag{9.52}$$

The same procedure may be used to determine the constants C and D of Eq. (9.48). On this basis, the following results are obtained:

$$C = \frac{(F_0/k)\omega^2}{- 2j[(\zeta\omega - j\omega_f)^2 + \omega_d^2]} \tag{9.53}$$

$$D = \frac{(F_0/k)\omega^2}{2j[(\zeta\omega + j\omega_f)^2 + \omega_d^2]} \tag{9.54}$$

Note that the constants C and D are complex conjugate numbers; that is, $D = C^*$.

The displacement $y(t)$ of the mass m in Fig. 9.3 may be determined by taking the inverse Laplace transform of Eq. (9.48). The procedure yields

$$y(t) = Ae^{-(\zeta\omega - j\omega_d)t} + Be^{-(\zeta\omega + j\omega_d)t}$$
$$+ Ce^{-j\omega_f t} + D^{j\omega_f t} \tag{9.55}$$

The constants A, B, C, and D are given by Eqs. (9.50), (9.52), (9.53), and (9.54), respectively. It can easily be shown that Eq. (9.55) may also be obtained by following the procedure discussed in Section 3.3. The first two terms in Eq. (9.55) yield the transient part of the response, and the remaining two terms yield the steady-state response, which is produced by the applied force $F(t)$.

Example 9.3. For the one-degree spring–mass system in Fig. 9.3, determine the expression $y(t)$ of the vertical displacement of the mass m. Assume that $m = 3.0 \, \text{lb} \cdot \text{s}^2/\text{in.}$, and $k = 5000 \, \text{lb/in.}$ The spring–mass system is under the influence of viscous damping with damping ratio $\zeta = 0.10$, and it is subjected to a force $F(t) = F_0 \sin \omega_f t$, where $F_0 = 3000 \, \text{lb}$ and $\omega_f = 35 \, \text{rps}$.

Solution. The values of ω and ω_d are as follows:

$$\omega = \sqrt{\frac{k}{m}} = \sqrt{\frac{5000}{3}} = 40.82 \text{ rps}$$

$$\omega_d = \omega\sqrt{1 - \zeta^2} = 40.82\sqrt{1 - (0.1)^2}$$

$$= 40.62 \text{ rps}$$

The constants A, B, C, and D may be determined from Eqs. (9.50), (9.52), (9.53), and (9.54), respectively, and they are as follows:

$$A = \frac{(3000/5000)(40.82)^2(35)}{2j(40.62)[(4.082 - 40.62j)^2 + (35)^2]}$$

$$= \frac{430.72}{331.6217 - 408.3217j}$$

$$B = \frac{430.72}{331.6217 + 408.3217j}$$

$$C = \frac{(3000/5000)(40.82)^2}{-2j[(4.082 - 35.00j)^2 + (40.62)^2]}$$

$$= -\frac{499.8817}{285.74 + 441.6471j}$$

$$D = -\frac{499.8817}{285.74 - 441.6471j}$$

From Eq. (9.55), by substitution, we find

$$y(t) = \frac{430.72}{331.6217 - 408.3217j} e^{-(4.082 - 40.62j)t}$$

$$+ \frac{430.72}{331.6217 + 408.3217j} e^{-(4.082 + 40.62j)t}$$

$$- \frac{499.8817}{285.74 + 441.6471j} e^{-(35j)t}$$

$$- \frac{499.8817}{285.74 - 441.6471j} e^{(35j)t}$$

If desired, further simplifications for the expression of $y(t)$ may be made by using trigonometric identities.

9.4   RESPONSE DUE TO AN IMPULSE AND A UNIT IMPULSE

We now consider the case where the force $F(t)$ acting on the mass m in Fig. 9.3 is an impulse; that is,

$$F(t) = F_0 \delta(t) \tag{9.56}$$

where F_0 is the magnitude of the impulse at the time it occurred, and $\delta(t)$ is the delta Dirac function, which is defined as follows:

$$\delta(t) = 0 \quad \text{for } t \neq 0 \tag{9.57}$$

$$\int \delta(t) \, dt = 1 \tag{9.58}$$

In this case, the function $F(s)$ in Eq. (9.36) would be the Laplace transform of $\delta(t)$. From Appendix J, we note that the Laplace transform of $\delta(t)$ is equal to one, and, consequently, Eq. (9.36) gives

$$Y(s) = \frac{F_0/m}{s^2 + 2\omega\zeta s + \omega^2} \tag{9.59}$$

or

$$Y(s) = \frac{F_0/m}{Z(s)} \tag{9.60}$$

where $Z(s)$ is given by Eq. (9.37). The quantity

$$H(s) = \frac{1}{Z(s)} \tag{9.61}$$

is defined as the *transfer function* of the system.

The method of partial fractions may again be used here to determine the dynamic response of the spring–mass system. We write Eq. (9.59) as follows:

$$Y(s) = \frac{F_0/m}{s^2 + 2\omega\zeta s + \omega^2} = \frac{A}{s - s_1} + \frac{B}{s - s_2} \tag{9.62}$$

where s_1 and s_2 are given by Eqs. (9.44) and (9.45), respectively. The constants A and B in Eq. (9.62) may be evaluated as in the preceding section, and they are as follows:

$$A = \frac{(F_0/k)\,\omega^2}{j2\omega\sqrt{1 - \zeta^2}} \tag{9.63}$$

$$B = \frac{(F_0/k)\,\omega^2}{-j2\omega\sqrt{1 - \zeta^2}} \tag{9.64}$$

The response $y(t)$ may be obtained by taking the inverse Laplace transform of Eq. (9.62), which yields

$$y(t) = Ae^{-(\omega\zeta - j\,\omega_d)t} + Be^{-(\omega\zeta + j\,\omega_d)t} \tag{9.65}$$

where the constants A and B are given by Eqs. (9.63) and (9.64), respectively, and $\omega_d = \omega(1 - \zeta^2)^{1/2}$.

We now consider Eq. (9.59) in conjunction with Eq. (9.61) and assume that $s = j\omega$. This yields

$$Y(j\omega) = \frac{F_0}{m} H(j\omega) \tag{9.66}$$

By using Eq. (9.21), the inverse Fourier transform of Eq. (9.66) is

$$y(t) = \frac{F_0}{2\pi m} \int_{-\infty}^{\infty} H(j\omega)e^{j\omega t}\, d\omega \tag{9.67}$$

This shows that the displacement $y(t)$ of the mass m of the spring–mass system, due to an impulse, is the inverse Fourier transform of $H(j\omega)$.

If the impulse is a unit impulse, implying that $F_0 = 1$, Eq. (9.67) yields

$$y(t) = h(t) = \frac{1}{2\pi m} \int_{-\infty}^{\infty} H(j\omega)e^{j\omega t}\, d\omega \tag{9.68}$$

where the symbol $h(t)$ is used to denote the inverse Fourier transform of $H(j\omega)$. With known $h(t)$, the transfer function $H(j\omega)$ is given by the following equation:

$$H(j\omega) = m \int_{-\infty}^{\infty} h(t)e^{-j\omega t}\, dt \tag{9.69}$$

Equations (9.68) and (9.69) constitute a Fourier transform pair.

9.5   VISCOUSLY DAMPED TWO-DEGREE SPRING–MASS SYSTEMS SUBJECTED TO EXTERNAL EXCITATIONS

Consider the two-degree spring–mass system in Fig. 9.4, which is moving under the influence of viscous damping with damplng constant c. The forces F_1 and F_2 are applied suddenly at $t = 0$ to masses m_1 and m_2, respectively, and are of infinite duration. The differential equations of motion for the two masses are as follows:

$$m_1\ddot{y}_1 + k_1 y_1 - k_2(y_2 - y_1) + c\dot{y} - c(\dot{y}_2 - \dot{y}_1) = F_1 \qquad (9.70)$$

$$m_2\ddot{y}_2 + k_2(y_2 - y_1) + c(\dot{y}_2 - \dot{y}_1) = F_2 \qquad (9.71)$$

The Laplace transforms of Eqs. (9.70) and (9.71), with some rearrangements, are as follows:

$$Y_1(s)(m_1 s^2 + 2cs + k_1 + k_2) - Y_2(s)(cs + k_2) = \frac{F_1}{s} \qquad (9.72)$$

$$-Y_1(s)(cs + k_2) + Y_2(s)(m_2 s^2 + cs + k_2) = \frac{F_2}{s} \qquad (9.73)$$

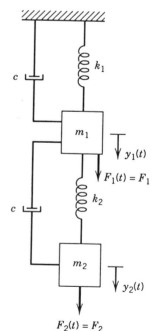

FIGURE 9.4. Two-degree spring–mass system under the influence of viscous damping.

By applying Cramer's rule, the simultaneous solution of Eqs. (9.72) and (9.73) yields

$$Y_1(s) = \frac{\Delta_1(s)}{Z(s)} \tag{9.74}$$

$$Y_2(s) = \frac{\Delta_2(s)}{Z(s)} \tag{9.75}$$

where

$$\Delta_1(s) = \begin{vmatrix} F_1/s & -(cs + k_2) \\ F_2/s & (m_2 s^2 + cs + k_2) \end{vmatrix} \tag{9.76}$$

$$\Delta_2(s) = \begin{vmatrix} (m_1 s^2 + 2cs + k_1 + k_2) & F_1/s \\ -(cs + k_2) & F_2/s \end{vmatrix} \tag{9.77}$$

$$Z(s) = \begin{vmatrix} (m_1 s^2 + 2cs + k_1 + k_2) & -(cs + k_2) \\ -(cs + k_2) & (m_2 s^2 + cs + k_2) \end{vmatrix} \tag{9.78}$$

In Eqs. (9.74) and (9.75), the quantity $1/Z(s)$ is the transfer function $H(s)$ of the spring–mass system.

By expanding the associated determinants, Eqs. (9.74) and (9.75) may also be written as follows:

$$Y_1(s) =$$

$$\frac{F_1 m_2 s^2 + F_1 k_2 + F_1 cs + F_2 cs + F_2 k_2}{s[m_1 m_2 s^4 + 2m_2 cs^3 + (c^2 + m_1 k_2 + m_2 k_1 + m_2 k_2)s^2 + (m_1 c + k_1 c + k_2 c)s + k_1 k_2]} \tag{9.79}$$

$$Y_2(s) =$$

$$\frac{F_2 m_1 s^2 + F_2 k_1 + F_2 k_2 + 2F_2 cs + F_1 cs + F_1 k_2}{s[m_1 m_2 s^4 + 2m_2 cs^3 + (c^2 + m_1 k_2 + m_2 k_1 + m_2 k_2)s^2 + (m_1 c + k_1 c + k_2 c)s + k_1 k_2]} \tag{9.80}$$

By using partial fractions, Eqs. (9.79) and (9.80) may be written as follows:

$$Y_1(s) = \frac{A_0}{s} + \frac{A_1}{s - s_1} + \frac{A_2}{s - s_2} + \frac{A_3}{s - s_3} + \frac{A_4}{s - s_4} \tag{9.81}$$

$$Y_2(s) = \frac{B_0}{s} + \frac{B_1}{s - s_1} + \frac{B_2}{s - s_2} + \frac{B_3}{s - s_3} + \frac{B_4}{s - s_4} \tag{9.82}$$

where $Y_1(s)$ and $Y_2(s)$ may also be written

$$Y_1(s) = \frac{F_1(m_2s^2 + cs + k_2) + F_2(cs + k_2)}{s(s - s_1)(s - s_2)(s - s_3)(s - s_4)} \tag{9.83}$$

$$Y_2(s) = \frac{F_1(cs + k_2) + F_2(m_1s^2 + 2cs + k_1 + k_2)}{s(s - s_1)(s - s_2)(s - s_3)(s - s_4)} \tag{9.84}$$

In the above four equations, the roots s_1, s_2, s_3, and s_4 are the roots of the quartic equation in the denominator of Eqs. (9.79) and (9.80). These roots can be determined by solving the quartic equation, as discussed in Section 2.9. The following identity is used to write Eqs. (9.83) and (9.84):

$$[m_1m_2s^4 + 2m_2cs^3 + (c^2 + m_1k_2 + m_2k_1 + m_2k_2)s^2$$
$$+ (m_1c + k_1c + k_2c)s + k_1k_2] = (s - s_1)(s - s_2)(s - s_3)(s - s_4) \tag{9.85}$$

With known s_1, s_2, s_3, and s_4, the values of the constants A_0, A_1, A_2, A_3, A_4, B_0, B_1, B_2, B_3, and B_4 may be determined by applying the procedure discussed in the preceding section of this chapter.

The time displacements $y_1(t)$ and $y_2(t)$ of masses m_1 and m_2, respectively, may be determined by taking the inverse Laplace transforms of Eqs. (9.81) and (9.82). The procedure yields the following expressions for $y_1(t)$ and $y_2(t)$.

$$y_1(t) = A_0 + A_1e^{s_1t} + A_2e^{s_2t} + A_3e^{s_3t} + A_4e^{s_4t} \tag{9.86}$$

$$y_2(t) = B_0 + B_1e^{s_1t} + B_2e^{s_2t} + B_3e^{s_3t} + B_4e^{s_4t} \tag{9.87}$$

The procedure in determining s_1, s_2, s_3, and s_4, A_0, A_1, A_2, A_3, A_4, B_0, B_1, B_2, B_3, and B_4 is rather lengthy, but a correct closed-form solution is obtained in this manner. Computer programs are usually available to expedite the solution.

Similar procedure may also be used to solve spring–mass systems with two or more degrees of freedom that are under the influence of viscous damping and subjected to various types of external excitations.

9.6 CONVOLUTION AND CONVOLUTION INTEGRALS

An important mathematical concept that relates input and output in the time or frequency domain is *convolution*. If we consider two functions $x(t)$ and $h(t)$ that are well behaved within their corresponding time domain, the convolution for $x(t)$ and $h(t)$ is defined by $y(t)$, which mathematically is written as follows:

$$y(t) = \int_{-\infty}^{\infty} x(t - \rho)h(\rho) \, d\rho \tag{9.88}$$

where ρ is a dummy variable of integration. The operation of convolution may also be written in the following shorthand form:

$$y(t) = x(t) * h(t) \tag{9.89}$$

Now we can determine the Fourier transform pair of $x(t)$ and $h(t)$ by using Eqs. (9.20) and (9.21). The Fourier transform of Eq. (9.88) is denoted by $Y(\omega)$, and it is as follows:

$$Y(\omega) = \int_{-\infty}^{\infty} h(\rho) \, d\rho \int_{-\infty}^{\infty} e^{-j\omega t} x(t - \rho) \, dt \tag{9.90}$$

If the variables in Eq. (9.90) are changed so that $u = t - \rho$, $du = dt$, $t = \infty$, $t = -\infty$, $u = \infty$, and $u = -\infty$, the result is

$$Y(\omega) = \int_{-\infty}^{\infty} h(\rho)e^{-j\omega\rho} \, d\rho \int_{-\infty}^{\infty} x(u)e^{-j\omega u} \, du \tag{9.91}$$

Equation (9.91) may also be written in the following form:

$$Y(\omega) = H(\omega)X(\omega) \tag{9.92}$$

where

$$H(\omega) = \int_{-\infty}^{\infty} h(\rho)e^{-j\omega\rho} \, d\rho \tag{9.93}$$

$$X(\omega) = \int_{-\infty}^{\infty} x(u)e^{-j\omega u} \, du \tag{9.94}$$

The result given by Eq. (9.92) shows that convolution in the time domain is equivalent to multiplication in the frequency domain. We can also prove the reverse; that is, convolution in the frequency domain is equivalent to multiplication in the time domain. Note that the frequency domain of $x(t)$ is represented by $X(\omega)$ in Eq. (9.92).

If we know the frequency response $H(\omega)$ and the input function $x(t)$, we can determine the output $y(t)$ by taking the inverse transform of Eq. (9.92), which yields

$$y(t) = \frac{1}{2\pi} \int_{-\infty}^{\infty} H(\omega) X(\omega) e^{j\omega t} \, d\omega \qquad (9.95)$$

We can also express Eq. (9.95) in terms of the dummy variable ρ as follows:

$$y(t) = \int_{-\infty}^{\infty} h(\rho) x(t - \rho) \, d\rho \qquad (9.96)$$

or in the form

$$y(t) = \int_{-\infty}^{\infty} x(\rho) h(t - \rho) \, d\rho \qquad (9.97)$$

The integrals in Eqs. (9.96) and (9.97) are the convolution integrals, and they provide a direct way to compute the output $y(t)$ resulting from a given input. An example of output $y(t)$ is the displacement $y(t)$ of the mass m in Fig. 9.3, which is produced by a unit impulse, and is represented here by $x(t) = F(t) = F_0 \delta(t)$, with F_0 being equal to one.

The convolution theorem has found many applications in engineering, and it is also used in Chapter 11 of the text to formulate the theory of stochastic (random) analysis. It was also used by Biot [72] in 1943 for earthquake analysis and design purposes. Convolution is often called by a variety of names, such as superposition theorem, Faltungsintegral, Green's theorem, Duhamel's theorem, Borel's theorem, and the Boltzmann–Hopkinson theorem.

9.7 ANALYSIS OF MACHINE FOUNDATIONS SUBJECTED TO EXTERNAL DYNAMIC EXCITATIONS AND VISCOUS DAMPING

In this section, we consider the problems arising from mechanical systems that are supported directly on soil or other similar types of elastic foundations. Such mechanical systems usually involve rotating parts that produce critical dynamic excitations that are transmitted to the immediate or nearby environment. They can also produce very damaging effects to the mechanical system itself, if not appropriately controlled. Considerable studies on soil vibrations and on methods of computing elastic damping constants of soils have been done in the past by Richart, Woods, and Hall [73], D'Appolonia [74], Bycroft [75], Hsieh [76], Toriumi [77], Barkan [78], Fertis [7], and others. The efforts of the author are concentrated on the dynamic coupled motions of foundations with damping when they are subjected to various types of dynamic excitations.

9.7.1 Derivation of the Differential Equations of Motion

We derive here the differential equations of motion of circular footings on an elastic half-space, which are subjected to coupled rocking and sliding motions produced by the various excitations transmitted to the foundation by the mechanical system they support. The case of rectangular bases is also examined here by converting it to an equivalent circular one of radius r_0.

We consider now the circular foundation shown in Fig. 9.5a, which rests on the surface of an elastic half-space. In the figure, h_0 is used to denote the distance of the center of gravity G of the foundation to the surface of the elastic half-space, x_g is the horizontal translation of G, and ψ is the rotation of the rocking motion as shown in Fig. 9.5b. In other words, the horizontal translational motion of G is represented by x_g, and the angle ψ is used for

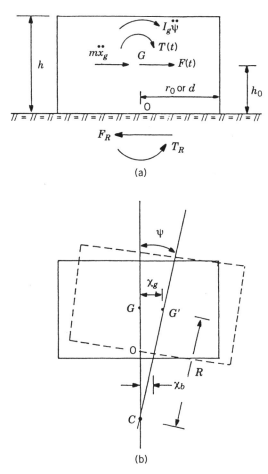

(a)

(b)

FIGURE 9.5. (a) Circular foundation subjected to coupled horizontal and rocking motions. (b) Rotational motion of the foundation.

the coupled rocking motion. For the rocking motion, the center of rotation C in Fig. 9.5b may lie above or below G. If the translatory motion is to the right with clockwise rotational motion, the center C is located below G; it will be located above G if the rotation is counterclockwise. The reactions F_R and T_R produced by the resistance of the soil, as well as the applied horizontal force $F(t)$ and torque $T(t)$ at G, are shown in Fig. 9.5a.

The horizontal translation x_b of the base of the foundation shown in Fig. 9.5b is

$$x_b = x_g - h_0\psi \tag{9.98}$$

The resistances F_R and T_R in Fig. 9.5a are as follows:

$$F_R = c_x\dot{x}_b + k_x x_b \tag{9.99}$$

$$T_R = c_\psi\dot{\psi} + k_\psi\psi \tag{9.100}$$

where c_x and c_ψ are the translational and rotational damping coefficients, respectively, and k_x and k_ψ are the respective translational and rotational spring reactions. These four coefficients may be obtained from the elastic half-space theory that is published in specialized books on soil dynamics or from expressions derived by using a mass–spring–dashpot analog. The ones obtained from Hall's analog are as follows:

$$k_x = \frac{32(1 - \nu)}{7 - 8\nu}Gr_0 \tag{9.101}$$

$$c_x = \frac{18.4(1 - \nu)}{7 - 8\nu}r_0^2\sqrt{G\rho} \tag{9.102}$$

$$k_\psi = \frac{8Gr_0^3}{3(1 - \nu)} \tag{9.103}$$

$$c_\psi = \frac{0.80r_0^4\sqrt{G\rho}}{(1 - \nu)(1 + B_\psi)} \tag{9.104}$$

For uniformly distributed mass, I_ψ, which is the mass moment of inertia about the center of rotation O (Fig. 9.5a), may be obtained from the following equation:

$$I_\psi = \frac{\pi r_0^2 h\gamma}{g}\left(\frac{r_0^2}{r} + \frac{h^2}{3}\right) \tag{9.105}$$

The value of B_ψ in Eq. (9.104) may be determined from the expression

$$B_\psi = \frac{3(1-\nu)}{8} \cdot \frac{I_\psi}{\rho r_0^5}$$ (9.106)

In the above equations, γ is the unit weight of the soil (lb/ft^3), $\rho = \gamma/g$ is the mass density (lb $\cdot$ s^2/ft^4), ν is the Poisson ratio, G is the shear modulus of the soil, g is the acceleration of gravity, r_0 is the radius of the circular foundation, h is its height, m is its mass, and I_g is its moment of inertia about the center of gravity G.

If the foundation has a rectangular shape of width $2c$ along the axis of rotation for rocking, and of length $2d$ in the plane of rotation for rocking, we can convert it into an equivalent circular base of radius r_0 by using the following expressions:

$$r_0 = \left(\frac{4cd}{\pi}\right)^{1/2} \quad \text{(for translation)}$$ (9.107)

$$r_0 = \left(\frac{16cd^3}{3\pi}\right)^{1/4} \quad \text{(for rocking)}$$ (9.108)

The equations of motion for the foundation in Fig. 9.5a may be obtained by applying Newton's second law of motion. They are as follows:

$$m\ddot{x}_g = -F_R + F(t) = -c_x\ddot{x}_b - k_x x_b + F(t)$$ (9.109)

$$I_g\ddot{\psi} = -T_R + T(t) + h_0 F_R$$ (9.110)

By using Eqs. (9.98), (9.99), and (9.100), we can write Eqs. (9.109) and (9.110) as follows:

$$m\ddot{x}_g + c_x\dot{x}_g - c_x h_0\dot{\psi} + k_x x_g - k_x h_0\psi = F(t)$$ (9.111)

$$I_g\ddot{\psi} + (c_\psi + c_x h_0^2)\dot{\psi} + (k_\psi + k_x h_0^2)\psi - h_0 c_x\dot{x}_g - h_0 k_x x_g = T(t)$$ (9.112)

Equations (9.111) and (9.112) are the differential equations of motion for the coupled horizontal translational and rocking motions of the foundation. Translational and rotational displacements of the foundation may be obtained from the solution of the above two equations.

9.7.2 Solution of the Differential Equations of Motion of the Machine Foundation

The solution of Eqs. (9.111) and (9.112) may be obtained by using Laplace transforms, and it is similar to the one used in Section 9.5 for a two-degree spring–mass system. We assume here that $F(t)$ and $T(t)$ are applied suddenly to the foundation and that the time duration is long enough to be assumed of infinite duration. This is usually the case for such types of machine foundation problems. We now rewrite Eqs. (9.111) and (9.112) as follows:

$$\ddot{x}_g + c_1\dot{x}_g - c_2\dot{\psi} + c_3 x_g - c_4\psi = F_0 \tag{9.113}$$

$$\ddot{\psi} + c_5\dot{\psi} + c_6\psi - c_7\dot{x}_g - c_8 x_g = T_0 \tag{9.114}$$

where

$$c_1 = \frac{c_x}{m} \tag{9.115}$$

$$c_2 = \frac{c_x h_0}{m} \tag{9.116}$$

$$c_3 = \frac{k_x}{m} \tag{9.117}$$

$$c_4 = \frac{k_x h_0}{m} \tag{9.118}$$

$$c_5 = \frac{c_\psi + c_x h_0^2}{I_g} \tag{9.119}$$

$$c_6 = \frac{k_\psi + k_x h_0^2}{I_g} \tag{9.120}$$

$$c_7 = \frac{c_x h_0}{I_g} \tag{9.121}$$

$$c_8 = \frac{k_x h_0}{I_g} \tag{9.122}$$

$$F_0 = \frac{F}{m} \tag{9.123}$$

$$T_0 = \frac{T}{I_g} \tag{9.124}$$

where F and T are the suddenly applied horizontal force and torque, respectively.

The Laplace transforms of Eqs. (9.113) and (9.114) are as follows:

$$(s^2 + c_1 s + c_3)X_g(s) - (c_2 s + c_4)\Psi(s) = \frac{F_0}{s} \tag{9.125}$$

$$-(c_7 s + c_8)X_g(s) + (s^2 + c_5 s + c_6)\Psi(s) = \frac{T_0}{s} \tag{9.126}$$

By applying Cramer's rule as in Section 9.5 and carrying out the required expansions for the associated determinants, we find

$$X_g(s) = \frac{F_0(s^2 + c_5 s + c_6) + T_0(c_2 s + c_4)}{s(s^4 + \alpha s^3 + \beta s^2 + \delta s + \gamma)} \tag{9.127}$$

$$\Psi(s) = \frac{T_0(s^2 + c_1 s + c_3) + F_0(c_7 s + c_8)}{s(s^4 + \alpha s^3 + \beta s^2 + \delta s + \gamma)} \tag{9.128}$$

where

$$\alpha = c_1 + c_5 \tag{9.129}$$

$$\beta = c_3 + c_6 + c_1 c_5 - c_2 c_7 \tag{9.130}$$

$$\delta = c_1 c_6 + c_3 c_5 - c_2 c_8 - c_4 c_7 \tag{9.131}$$

$$\gamma = c_3 c_6 - c_4 c_8 \tag{9.132}$$

The common denominator in Eqs. (9.127) and (9.128) involves the fourth degree polynomial $(s^4 + \alpha s^3 + \beta s^2 + \delta s + \gamma)$, which must be factored by using the solution of the following quartic equation:

$$s^4 + \alpha s^3 + \beta s^2 + \delta s + \gamma = 0 \tag{9.133}$$

The four roots s_1, s_2, s_3, and s_4 of Eq. (9.133) may be determined as discussed in Sections 2.9 and 9.5, and the polynomial in Eq. (9.133) may be written as follows:

$$s^4 + \alpha s^3 + \beta s^2 + \delta s + \gamma = (s - s_1)(s - s_2)(s - s_3)(s - s_4) \tag{9.134}$$

By using partial fractions expansion, we can write Eqs. (9.127) and (9.128) as follows:

$$X_g(s) = \frac{F_0(s^2 + c_5 s + c_6) + T_0(c_2 s + c_4)}{s(s - s_1)(s - s_2)(s - s_3)(s - s_4)}$$
$$= \frac{A_1}{s} + \frac{B_1}{s - s_1} + \frac{C_1}{s - s_2} + \frac{D_1}{s - s_3} + \frac{E_1}{s - s_4} \tag{9.135}$$

$$\Psi(s) = \frac{T_0(s^2 + c_1 s + c_3) + F_0(c_7 s + c_8)}{s(s - s_1)(s - s_2)(s - s_3)(s - s_4)}$$
$$= \frac{A_2}{s} + \frac{B_2}{s - s_1} + \frac{C_2}{s - s_2} + \frac{D_2}{s - s_3} + \frac{E_2}{s - s_4} \tag{9.136}$$

The constants A_1, B_1, C_1, D_1, E_1, A_2, B_2, C_2, D_2, and E_2 in Eqs. (9.135) and (9.136) may be determined by using the method of partial fractions, as discussed in Sections 9.3 and 9.5.

The expressions for the time responses $x_g(t)$ and $\psi(t)$ of the horizontal translational motion and rocking, respectively, may be determined by taking the inverse Laplace transforms of Eqs. (9.135) and (9.136). This procedure yields

$$x_g(t) = A_1 + B_1 e^{s_1 t} + C_1 e^{s_2 t} + D_1 e^{s_3 t} + E_1 e^{s_4 t} \tag{9.137}$$

$$\psi(t) = A_2 + B_2 e^{s_1 t} + C_2 e^{s_2 t} + D_2 e^{s_3 t} + E_2 e^{s_4 t} \tag{9.138}$$

The following practical numerical example illustrates the procedure. The trigonometric form of Eqs. (9.137) and (9.138) may also be obtained in the example by using appropriate trigonometric expressions.

Example 9.4. The concrete foundation shown in Fig. 9.6, which supports three shredders, is idealized as a circular one with an equivalent radius $r_0 = 21.60$ ft for translation and $r_0 = 15.14$ ft for rocking. The foundation is resting on soil of shear modulus $G = 21.60$ ksi. If one hammer of a shredder brakes durlng operation, it produces a suddenly applied force $F(t) = F = 72.00$ kips. We assume that the duration of F is long enough to be considered of infinite duration. The center of gravity G of the foundation and shredder is determined, and it is located as shown in Fig. 9.7. The force $F(t)$ produces a torque $T = 8467.20$ in. $\cdot$ kips about G. The mass m of the foundation and shredder is 8.1188 kip $\cdot$ s^2/in., and the Poisson ratio $\nu = 0.3$. Determine the expressions for $x_g(t)$ and $\psi(t)$ and the shaft displacement of the shredder, which rotates at 900 RPM.

Solution. By using the procedure discussed above, the following data are obtained:

$$k_x = \frac{32(0.70)}{7 - 2.4}(21.60)(20.48)(12)$$

$$= 25,849.68 \text{ kips/in.}$$

$$k_\psi = \frac{(8)(21.6)(15.14)^3(12)^3}{(3)(0.7)} = 493,453.015 \times 10^3 \text{ kips} \cdot \text{in.}$$

$$I_g = 103,761.6802 \text{ kip} \cdot \text{in.} \cdot \text{s}^2$$

$$I_\psi = 263,801.4905 \text{ kip} \cdot \text{in.} \cdot \text{s}^2$$

$$\rho = \frac{\gamma}{g} = \frac{130}{32.2} = 4.04 \text{ lb} \cdot \text{s}^2/\text{ft}^4$$

$$= 0.1948 \times 10^{-6} \text{ kip} \cdot \text{s}^2/\text{in.}^4$$

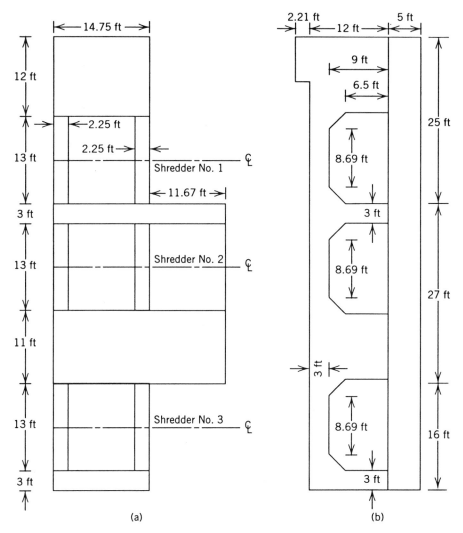

FIGURE 9.6. (a) Top view of the shredder foundation. (b) Side view of the shredder foundation.

$$c_x = \frac{18.4(0.7)}{7-2.4}(20.48)^2(144)\sqrt{(0.1948 \times 10^{-6})(21.60)}$$

$$= 346.8535 \text{ kip} \cdot \text{s/in.}$$

$$B_\psi = \frac{(3)(0.7)}{8} \cdot \frac{263{,}801.4905}{(0.1948 \times 10^{-6})(15.15)^5(12)^3(144)}$$

$$= 1.7959$$

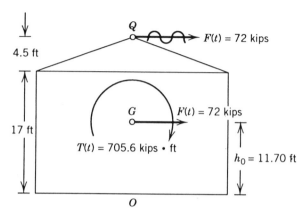

FIGURE 9.7. Schematic representation showing locations of $F(t)$, $T(t)$, and center of gravity G.

$$c_\psi = \frac{(0.80)(15.14)^4(12)^4(0.002051)}{(0.70)(1 + 1.7959)}$$

$$= 913{,}407.26 \text{ kips} \cdot \text{in.} \cdot \text{s}$$

$$c_1 = \frac{(346.8535)}{8.1188} = 42.7223$$

$$c_2 = \frac{(346.8535)(11.70)(12)}{8.1188} = 5998.2056$$

$$c_3 = \frac{25{,}849.68}{8.1188} = 3183.9287$$

$$c_4 = \frac{(25{,}849.68)(11.70)(12)}{8.1188} = 447{,}023.5838$$

$$c_5 = \frac{913{,}407.26 + 346.8535(11.70 \times 12)^2}{103{,}761.8802}$$

$$= 74.6964$$

$$c_6 = \frac{(493{,}453.015 \times 10^3) + 25{,}849.68(11.70 \times 12)^2}{103{,}761.8802}$$

$$= 9666.4212$$

$$c_7 = \frac{(346.8535)(11.70)(12)}{103{,}761.8802} = 0.4693$$

$$c_8 = \frac{(25,849.68)(11.70)(12)}{103,761.8802} = 34.9772$$

$$F_0 = \frac{72}{8.1188} = 8.8683$$

$$T_0 = \frac{8467.2}{103,761.8802} = 0.0816$$

By substituting into Eqs. (9.113) and (9.114), we obtain the following two differential equations of motion:

$$\ddot{x}_g + 42.7223\dot{x}_g - 5998.2056\dot{\psi}$$
$$+ 3183.9287x_g - 447,023.5838\psi = 8.8683 \tag{6.139}$$

$$\ddot{\psi} + 74.6964\dot{\psi} + 9666.4212\psi - 0.4693\dot{x}_g - 34.9772x_g = 0.0816 \tag{9.140}$$

By applying Eqs. (9.129) through (9.132), we find

$$\alpha = 117.4187$$
$$\beta = 13,226.594$$
$$\delta = 231,211.1533$$
$$\gamma = 15,141,562.58$$

Thus, by Eq. (9.133), the quartic equation is as follows:

$$s^4 + 117.4187s^3 + 13,226.594s^2 + 231,211.1533s$$
$$+ 15,141,562.58 = 0 \tag{9.141}$$

The four complex roots s_1, s_2, s_3, and s_4 of Eq. (9.141) are obtained by using the procedure discussed in Section 2.9. Computer programs are also readily available for this purpose. These roots are as follows:

$$s_1 = -0.6593 + 43.1710i \tag{9.142}$$

$$s_2 = -0.6593 - 43.1710i \tag{9.143}$$

$$s_3 = -58.0501 + 88.5411i \tag{9.144}$$

$$s_4 = -58.0501 - 88.5411i \tag{9.145}$$

where $i = \sqrt{-1}$. The above four roots indicate that the damped frequencies of vibration ω_1 and ω_2 are

$$\omega_1 = 43.1710 \text{ rps} \tag{9.146}$$

$$\omega_2 = 88.5411 \text{ rps} \tag{9.147}$$

By applying Eqs. (9.135) and (9.136), we find

$$
\begin{aligned}
X_g(s) &= \frac{8.8683s^2 + 1151.8837s + 122,201.8475}{s(s - s_1)(s - s_2)(s - s_3)(s - s_4)} \\
&= \frac{A_1}{s} + \frac{B_1}{s - s_1} + \frac{C_1}{s - s_2} + \frac{D_1}{s - s_3} + \frac{E_1}{s - s_4}
\end{aligned} \tag{9.148}
$$

$$
\begin{aligned}
\Psi(s) &= \frac{0.0816s^2 + 7.6485s + 569.9969}{s(s - s_1)(s - s_2)(s - s_3)(s - s_4)} \\
&= \frac{A_2}{s} + \frac{B_2}{s - s_1} + \frac{C_2}{s - s_2} + \frac{D_2}{s - s_3} + \frac{E_2}{s - s_4}
\end{aligned} \tag{9.149}
$$

where s_1, s_2, s_3, and s_4 are given by Eqs. (9.142), (9.143), (9.144), and (9.145), respectively.

We can determine the constants A_1, B_1, C_1, D_1, E_1, A_2, B_2, C_2, D_2, and E_2 in Eqs. (9.148) and (9.149) by applying the method of partial fractions, as discussed in Sections 9.3 and 9.5. For example, to evaluate the constant A_1, we multiply both sides of Eq. (9.148) by s and evaluate the equation for $s = 0$. This yields

$$
\begin{aligned}
A_1 &= \frac{122,201.8475}{(0.6593 - 43.1710i)(0.6593 + 43.1710i)(58.0501 - 88.5411i)(58.0501 + 88.5411i)} \\
&= 5.8481 \times 10^{-3}
\end{aligned}
$$

To evaluate the constant B_1, we multiply both sides of Eq. (9.148) by $(s - s_1)$ and evaluate the equation for $s = s_1$. This yields

$$B_1 = -2.9510 \times 10^{-3} + 0.1991 \times 10^{-3}i$$

In a similar manner, the remaining constants are evaluated, and they are as follows:

$$C_1 = -2.9510 \times 10^{-3} - 0.1991 \times 10^{-3}i$$

$$D_1 = 0.02690 \times 10^{-3} - 0.09295 \times 10^{-3}i$$

$$E_1 = 0.02690 \times 10^{-3} + 0.09295 \times 10^{-3}i$$

$$A_2 = 2.7278 \times 10^{-5}$$

$$B_2 = -1.3245 \times 10^{-5} - 0.2157 \times 10^{-5}i$$
$$C_2 = -1.3245 \times 10^{-5} + 0.2157 \times 10^{-5}i$$
$$D_2 = -0.0395 \times 10^{-5} + 0.1409 \times 10^{-5}i$$
$$E_2 = -0.0395 \times 10^{-5} - 0.1409 \times 10^{-5}i$$

By applying Eqs. (9.137) and (9.138) and performing the required mathematical operations, we find

$$
\begin{aligned}
x_g(t) &= 5.8481 \times 10^{-3} + e^{-0.6593t}(10^{-3})[-5.902 \cos(43.171t) \\
&\quad -0.3982 \sin(43.171t)] + e^{-58.0501t}(10^{-3})[0.0538 \cos(88.5411t) \\
&\quad + 0.1859 \sin(88.5411t)]
\end{aligned}
\tag{9.150}
$$

$$
\begin{aligned}
\psi(t) &= 2.7278 \times 10^{-5} + e^{-0.6593t}(10^{-5})[-2.649 \cos(43.171t) \\
&\quad +0.4314 \sin(43.171t)] + e^{-58.0501t}(10^{-5})[-0.079 \cos(88.5411t) \\
&\quad - 0.2818 \sin(88.5411t)]
\end{aligned}
\tag{9.151}
$$

The maximum values of x_g and ψ occur at time $t = 07277$ s, and they are as follows:

$$x_g(t = 0.07277) = 11.4751 \times 10^{-3} \text{ in.} \tag{9.152}$$
$$\psi(t = 0.07277) = 5.251 \times 10^{-5} \text{ radians} \tag{9.153}$$

The distance R to the center of rotation C in Fig. 9.5b is

$$R = \frac{x_g}{\psi} = \frac{11.4751 \times 10^{-3}}{5.251 \times 10^{-5}} = 218.53 \text{ in.}$$

The shaft of the shredder is located at point Q in Fig. 9.7, and the shaft displacement x_Q is as follows:

$$
\begin{aligned}
x_Q &= [R + (21.50 - 11.70)(12)]\psi \\
&= (218.53 + 117.60)(5.251 \times 10^{-5}) \\
&= 0.01765 \text{ in.} \\
&= 17.65 \text{ mils}
\end{aligned}
$$

Allowable levels of maximum deflection are usually associated with such foundations. The above value of 17.65 mils would be very high, since the machine operates at 900 RPM, and the machine may not be able to function appropriately. Damage may occur after a period of time, and it would also be

TABLE 9.1. Response of the Shredder–Foundation System for Three Different Values of the Shear Modulus G of the Soil

(1) Shear Modulus G (psi)	(2) Shaft Displacement x_Q (mils)	(3) ω_1 (rps)	(4) ω_2 (rps)	(5) x_g (mils)
21,600	17.65	43.17	88.44	11.48
100,000	4.66	79.29	189.30	3.03
1,000,000	0.46	250.64	590.62	0.30

difficult to control undesirable effects that are transmitted to environmental structures and related engineering elements. For the above problem, the deflection x_Q should not exceed 4.0 mils for smooth operation. It could go as high as 10.0 mils if the machine operates at 100 RPM or less.

If we decide to neglect damping, the free undamped frequencies of vibration for rocking and sliding may be obtained from the following equation:

$$mI_g\omega^4 - (mk_\psi + mk_xh_0^2 + k_xI_g)\omega^2 + k_xk_\psi = 0 \qquad (9.154)$$

The solution of Eq. (9.154) yields

$$\omega_{1,2} = \frac{(mk_\psi + mk_xh_0^2 + k_xI_g) \mp \sqrt{(mk_\psi + mk_xh_0^2 + k_xI_g)^2 - 4mI_gk_xk_\psi}}{2mI_g}$$

$$(9.155)$$

in units of radians per second.

By substituting in Eq. (9.155) the appropriate values for m, k_ψ, and so on, we find the following values for the undamped frequencies ω_1 and ω_2:

$$\omega_1 = 36.226 \text{ rps} \qquad \text{(undamped)} \qquad (9.156)$$

$$\omega_2 = 107.415 \text{ rps} \qquad \text{(undamped)} \qquad (9.157)$$

By comparing the results with the values obtained in Eqs. (9.146) and (9.147), we note that the undamped frequency ω_1 is 16.09% lower, and the undamped frequency ω_2 is 21.32% higher, which indicates that damping should be considered for an accurate solution of the problem.

By considering again the same shredder–foundation problem and including the effects of soil damping, the response was determined for values of the shear modulus $G = 10^5$ psi and 10^6 psi. The results are shown in Table 9.1. The second column of the table provides the values of the shaft displacement x_Q, columns (3) and (4) show the damped frequencies ω_1 and ω_2, and the fifth column gives the displacements x_g of the center of gravity G of the system. This exemplifies the tremendous influence of the soil rigidity on the response of the shredder–foundation system.

The response to other types of dynamic excitations may be determined in a similar manner. Additional information on the subject may be found in refs. 7 and 73.

PROBLEMS

9.1 Determine the Fourier series expansion of the waveforms shown in Fig. P9.1, and plot in each case the discrete frequency spectrum.

9.2 Determine the Fourier series expansion of the waveform function $f(t) = e^{-t}$, and plot the discrete frequency spectrum.

9.3 Determine the frequency spectrum for the function

$$f(t) = \begin{cases} 0 & t < 0 \\ e^{-t/\tau} \sin \omega_0 t & t > 0 \end{cases}$$

9.4 For the coupled pendulum shown in Fig. P9.4, determine the displacements x_1 and x_2 when the initial displacements at $t = 0$ are $x_1(0) = x_2(0) = 0$, and the initial velocities at $t = 0$ are $\dot{x}_1(0) = b$ and $\dot{x}_2(0) = 0$.

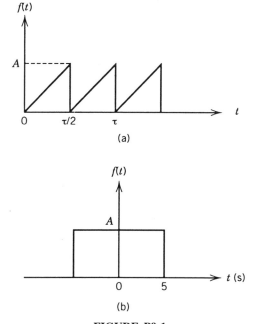

(a)

(b)

FIGURE P9.1.

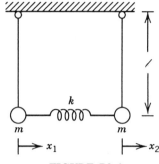

FIGURE P9.4.

9.5 The response of the one-degree spring–mass system in Fig. 9.3, when $F(t) = F_0 \sin \omega_f t$, is given by Eq. (9.55). If the initial conditions at $t = 0$ are $y(0) = \dot{y}(0) = 0$, show that Eq. (9.55) may also be obtained by using the procedure discussed in Section 3.3.

9.6 Consider the one-degree spring–mass system in Fig. 9.3, and assume that the initial conditions at $t = 0$ are $y(0) = y_0$ and $\dot{y}(0) = \dot{y}_0$. Determine the transfer function, the impedance, and the response $y(t)$ of the spring–mass system. Assume that $F(t) = F_0 \sin \omega_f t$ and separate $y(t)$ into transient and steady-state response terms.

9.7 Rework the problem in Example 9.3 by assuming that $\omega_f = 30$ rps, 38 rps, 40.0 rps, and 40.82 rps, and compare the results.

9.8 Repeat Problem 9.6 when the initial conditions at $t = 0$ are $y(0) = y_0$, $\dot{y}(0) = \dot{y}_0$, and the dynamic force $F(t) = F_0 \cos \omega_f t$, and compare the results.

9.9 The response of the two-degree spring–mass system in Fig. 9.4 is given by Eqs. (9.86) and (9.87). Determine the constants A_0, A_1, A_2, A_3, A_4, B_0, B_1, B_2, B_3, and B_4. Assume that $k_1 = 76.46$ lb/in., $k_2 = 11.77$ lb/in., $m_1 = m_2 = 0.10$ lb · s²/in., $F_1 = 40$ lb, $F_2 = 30$ lb, and $c = 1.0$ lb · s/in.

9.10 Determine the response of the two-degree spring–mass system in Fig. 9.4 when $F_1(t) = F_1 \sin \omega_f t$ and $F_2(t) = F_2 \sin \omega_f t$. The initial displacements and velocities at $t = 0$ are zero.

9.11 Repeat Problem 9.10 when $F_1(t) = F_1 \cos \omega_f t$ and $F_2(t) = F_2 \cos \omega_f t$, and compare the results.

9.12 Determine the expressions $y_1(t)$ and $y_2(t)$ for the displacements of the

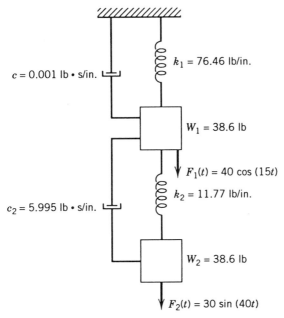

$c = 0.001$ lb · s/in.

$k_1 = 76.46$ lb/in.

$W_1 = 38.6$ lb

$F_1(t) = 40 \cos (15t)$

$k_2 = 11.77$ lb/in.

$c_2 = 5.995$ lb · s/in.

$W_2 = 38.6$ lb

$F_2(t) = 30 \sin (40t)$

FIGURE P9.12.

weights W_1 and W_2, respectively, for the two-degree spring–mass system in Fig. P9.12. All required data are shown in the figure.

9.13 Repeat Problem 9.12 for $c_1 = 1.0$ lb · s/in., $k_2 = 25.0$ lb/in., and $W_1 = 60.0$ lb, and compare the results.

6.14 Rework Example 9.4 for shear modulus $G = 50 \times 10^3$ psi, 100×10^3 psi, 500×10^3 psi, and 10^6 psi, and compare the results with the ones listed in Table 9.1.

9.15 Rework Example 9.4 by assuming that the foundation of the machine is a concrete slab 30.0 ft long, 20.0 ft wide, and 7.0 ft thick. The shaft of the shredder is located at point Q, which is 3.0 ft above the top of the slab, and the shredder weighs 100,000 lb.

10 Variational and Stochastic Approaches

10.1 INTRODUCTION

Two types of methodologies are discussed in this chapter. The first one is the *variational approach*, which involves methodologies that can be used to determine maxima or minima of functionals. *Functionals* are variable values that depend on a variable, or variables, running through a set of functions. The methods that are used to solve a given variational problem are similar to those used to determine maxima and minima of ordinary functions.

Variational methods are extensively used in many engineering applications today. An application of variational calculus may also be attributed to the ancient Greeks, in solving the problem of determining a closed curve having a length ℓ and encircling an area A that is maximum. This curve can be proved to be the circumference of a circle.

The development of the field, however, was influenced by three basic problems. The first one is the brachistochrone problem, which was investigated by Johann Bernoulli in 1696 and requires one to determine the path of quickest descent. The second one is the geodesic problem, which deals with the determination of the line of minimum length lying on a given surface $\Phi(x, y, z) = 0$ and joining two given points of the surface. The third one is the isoperimetric problem, which, as stated earlier, requires one to determine a closed curve of length ℓ encircling an area A that is maximal. According to the ancient Greeks, such a curve should be the circumference of a circle. The second problem was solved by Johann Bernoulli in 1697, and a general solution was provided by J. Lagrange and L. Euler. The third problem was also further investigated by Euler.

The second methodology is the *stochastic* one, which involves the element of prediction. Excitations that involve the element of prediction are called *nondeterministic*, and there is no guarantee that such physical situations will actually occur. Random excitations that are produced by an earthquake, blast, or wind gust are nondeterministic. The response to nondeterministic or random excitations requires some knowledge of transform calculus and statistical concepts. The subject of transform calculus is treated in Chapter 9, and basic statistical concepts such as probability and random variables are included in this chapter.

The purpose in this chapter is to provide an introductory treatment to both variational and stochastic methodologies and apply the methodologies involved to various types of vibration problems. For additional information the reader may consult available references [79–82].

10.2 FUNCTIONALS AND THEIR VARIATIONAL PROPERTIES

Variational calculus provides methodologies for the determination of *maximal* and *minimal* values of *functionals*. Consequently, problems that deal with the determination of maxima and minima of a functional are known as variational problems. The Greek letter Ω is used here to represent a functional, which in mathematical terms is written as follows:

$$\Omega = \Omega(y(x)) \tag{10.1}$$

Functionals are variable values depending on a variable running through a set of functions, or a finite number of such variables, which are determined completely by a definite choice of these variable functions.

If we select points A and B in Fig. 10.1, the length L of the curve that passes through these two points is a functional, because it is fully determined by choosing a definite function $y = y(x)$. When the equation of $y(x)$ is given, the value L may be calculated; that is,

$$L(y(x)) = \Omega(y(x)) = \int_{x_0}^{x_1} [1 + (y')^2]^{1/2} \, dx \tag{10.2}$$

In other words, for a given functional $\Omega = \Omega(y(x))$, there corresponds a unique number Ω to each function $y = y(x)$, as it does for an ordinary function $z = g(x)$, where to each number x there corresponds a unique number z. Therefore, the purpose of the calculus of variation is to provide

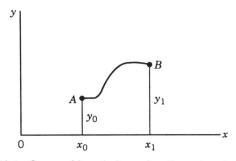

FIGURE 10.1. Curve of length L passing through points A and B.

methods that can be used to determine maximal and minimal values of functionals.

Returning to Eq. (10.1), we note that we have defined a functional Ω as $\Omega(y(x))$, where $y(x)$ is the independent function. This means that for a given function $y(x)$, from a certain class of functions, there corresponds a certain value of Ω. The difference $[y(x) - y_1(x)]$ of two functions $y(x)$ and $y_1(x)$ from a certain class of functions is denoted as δy, and it represents the *variation* δy of the argument $y(x)$ of the functional. A functional would be a *continuous functional* if small variations of $y(x)$ always lead to small variations of $\Omega(y(x))$.

The last definition leads to the question of which variations of the function $y(x)$ are called small, or which curves $y = y(x)$ and $y = y_1(x)$ are considered close to each other. One possibility is to assume that $y(x)$ and $y_1(x)$ are coordinatewise close, which means that the absolute value of their difference $[y(x) - y_1(x)]$ is small for all values of x that define $y(x)$ and $y_1(x)$. We may go further and assume also that the absolute value of the difference $[y'(x) - y_1'(x)]$ of their first derivatives is small, or the difference of all their derivatives is small.

Therefore, we may define *closeness* of two functions $y = y(x)$ and $y = y_1(x)$ as follows: They are close or neighboring in the sense of *closeness of order zero*, if the absolute value of the difference $[y(x) - y_1(x)]$ is small. If the differences $[y(x) - y_1(x)]$ and $[y'(x) - y_1'(x)]$ are small, then they are close in the sense of *closeness of order one*. They are close in the sense of *closeness of order n*, if the absolute values of the differences $[y(x) - y_1(x)]$, $[y'(x) - y_1'(x)], \ldots , [y''(x) - y_1''(x)]$ are small. This also means that if the curves are close in the sense of closeness of order n, they are also close in the sense of closeness of any other order that is less than n. The functional $\Omega(y(x))$ is *linear* if it satisfies the conditions $\Omega(cy(x)) = c\Omega(y(x))$, where c is an arbitrary constant, and $\Omega(y_1(x) + y_2(x)) = \Omega(y_1(x)) + \Omega(y_2(x))$. The variation of a functional is linear in the δy part of its increment, and it plays the same role in the theory of functionals as does the differential in ordinary functions.

If we consider a functional of the form $\Omega(y(x))$, we can define the variation as the derivative of the functional $\Omega(y(x) + \alpha\,\delta y)$ with respect to α, at $\alpha = 0$, where α is a variable, provided that the variation in the sense of the main linear increment of the functional exists. Mathematically, we write the variation as follows:

$$\frac{\partial}{\partial \alpha} \Omega(y(x) + \alpha\,\delta y)\Big|_{\alpha = 0} \tag{10.3}$$

The functional $\Omega(y(x))$ takes on a maximum value along the curve $y = y_0(x)$ if

$$\Delta\Omega = \Omega(y(x)) - \Omega(y_0(x)) \leq 0 \qquad (10.4)$$

which means that all the values of $\Omega(y(x))$ taken along arbitrary curves neighboring $y = y_0(x)$ are not greater than $\Omega(y_0(x))$. If $\Delta\Omega \leq 0$ and $\Delta\Omega = 0$ only when $y(x) = y_0(x)$, then $\Omega(y(x))$ takes on an absolute maximum along the curve $y = y(x)$. When the functional has a minimum value along $y = y_0(x)$, then $\Delta\Omega \geq 0$ for all curves sufficiently close to $y = y_0(x)$.

If the variation of a functional $\Omega(y(x))$ exists and Ω takes on a maximum or a minimum along $y = y_0(x)$, then $\Delta\Omega = 0$ along $y = y_0(x)$. This theorem suggests that the variation of a functional vanishes along the curves, which makes $\Omega(y(x))$ have an extremum. We should remember, however, that the order of closeness should be pointed out distinctly when we are investigating maxima or minima. In the sense of closeness of order zero, maxima or minima are called *strong*; it is called a *weak* maximum or minimum if the sense of closeness is of order one. This distinction is essential when we are considering sufficiency conditions for extrema.

The characteristics, definitions, and theorem, as discussed above, apply as well to functionals involving several independent variables, such as

$$\Omega(y_1(x), y_2(x), \ldots, y_n(x)) \qquad (10.5)$$

$$\Omega(y(x_1, x_2, \ldots, x_n)) \qquad (10.6)$$

$$\Omega(y_1(x_1, x_2, \ldots, x_n), y_2(x_1, x_2, \ldots, x_n), \ldots, y_n(x_1, x_2, \ldots, x_n)) \quad (10.7)$$

$$\Omega(y(x), z(x)) \qquad (10.8)$$

10.3   NECESSARY CONDITIONS FOR A FUNCTIONAL TO HAVE AN EXTREMUM

In this section, we wish to discuss the *necessary conditions* for a functional to have an extremum. We will examine two cases of functionals. The first one involves functionals with *fixed boundaries*, while functionals with *movable boundaries* are examined in the second case. The necessary conditions for a functional to have an extremum are represented by the *Euler equation*, which was derived by Euler in 1744 and carries his name.

10.3.1   Euler Equation for Functionals with Fixed Boundaries

We consider again Fig. 10.1, where the boundaries A and B of all admissible curves are fixed. At the boundaries, we have $y(x_0) = y_0$ and $y(x_1) = y_1$. The functional has the form

$$\Omega(y(x)) = \int_{x_o}^{x_1} F(x, y(x), y'(x)) \, dx \tag{10.9}$$

and we assume that the third derivative of the function $F(x, y, y')$ exists.

In general, the necessary condition for a functional to have an extremum is that its variation should vanish. We assume now that an extremum occurs along a curve $y = y(x)$, which possesses a second-order derivative, and we take an arbitrary admissible curve $y = y^*(x)$ neighboring to $y = y(x)$. In this manner, we formulate a one-parameter family of curves, namely, $y(x, \alpha) = y(x) + \alpha(y^*(x) - y(x))$, which contains the curves $y = y(x)$ and $y = y^*(x)$. The parameter α is chosen so that, for $\alpha = 0$, we have $y = y(x)$, and, for $\alpha = 1$, we have $y = y^*(x)$. The difference $y^*(x) - y(x)$ is the variation δy of the function $y(x)$, which is also a function of x that can be differentiated once or more times to yield $(\delta y)'$, $(\delta y)''$, and so on.

If we consider that the value of the functional given by Eq. (10.9) is taken along the family of curves $y = y(x, \alpha)$ only, we have a function of the variable α; that is,

$$\Omega(y(x, \alpha)) = \Phi(\alpha) \tag{10.10}$$

because, in this instance, the value $\Omega(y(x, \alpha))$ depends only on the choice of α. For $\alpha = 0$, we have $y = y(x)$, and, consequently, for $\alpha = 0$, this function takes on an extremum with respect to any neighboring admissible curve, which includes any neighboring curve of the family $y = y(x, \alpha)$.

The necessary condition for the function $\Phi(\alpha)$ to have an extremum for $\alpha = 0$ is its derivative $\Phi'(0)$ to vanish; that is,

$$\Phi'(0) = 0 \tag{10.11}$$

By carrying out the required mathematics, we find

$$\Phi'(0) = \int_{x_0}^{x_1} [F_y(x, y(x), y'(x))\delta y + F_{y'}(x, y(x), y'(x))\delta y'] \, dx \tag{10.12}$$

We know that $\Phi'(0)$ is called the variation of the functional and is designated as $\Delta\Omega$. Consequently, the necessary condition for a functional Ω to have an extremum is $\Delta\Omega = 0$. This yields

$$\Delta\Omega = \int_{x_0}^{x_1} \left(F_y - \frac{d}{dx} F_{y'} \right) \delta y \, dx = 0 \tag{10.13}$$

Since $(F_y - dF_{y'}/dx)$ taken along the curve $y = y(x)$, which gives an extremum, is a continuous function, and the variation δy is an arbitrary function, we must have

$$F_y - \frac{d}{dx} F_{y'} = 0 \qquad (10.14)$$

and $y = y(x)$ is a solution of the second-order differential equation given by Eq. (10.14). Explicitly, Eq. (10.14) may be written

$$F_y - F_{xy'} - F_{yy'} y' - F_{y'y'} y'' = 0 \qquad (10.15)$$

Equation (10.14), or Eq. (10.15), is known as the *Euler equation*, and it was first obtained by Euler in 1744. The integral curves $y = y(x, C_1, C_2)$ of Eq. (10.14), or Eq. (10.15), are called *extremals*, and only extremals can make the functional given by Eq. (10.9) have an extremum. These curves can be determined by solving the Euler equation and using the boundary conditions $y(x_0) = y_0$ and $y(x_1) = y_1$ to determine the constants C_1 and C_2.

It should be pointed out, however, that the Euler equation provides only the necessary condition for a functional to have an extremum, and there is no guarantee that such an extremum really constitutes an extremum. It cannot even tell if the extremum is a maximum or a minimum. The answers to these questions may be obtained only by examining the *sufficiency conditions*. For many engineering problems, these answers can be obtained from the physical interpretation of the given problem, and there is no need to refer to sufficiency conditions. Sufficiency conditions could become very valuable tools in research work.

The Euler equation for other types of functionals with fixed boundaries may be obtained in a similar manner. Appendix L provides the necessary conditions for an extremum of various types of functionals with fixed boundaries.

The solution of variational problems may also be given in parametric representation, where the functions and their derivatives will involve time. A well-known variational principle involving time is *Hamilton's principle*, which is written as follows:

$$\Omega = \int_{t_0}^{t_1} (T - U) \, dt \qquad (10.16)$$

where T and U are the kinetic and potential energies, respectively, of the system.

Example 10.1. Determine the curves that make the following functional have an extremum:

$$\Omega(y(x)) = \int_0^{\pi/2} [(y')^2 - y^2] \, dx$$
$$y(0) = 0 \qquad y(\pi/2) = 1$$

Solution. We have

$$F_y = \frac{\partial}{\partial y} F((y')^2 - y^2)$$

$$= \frac{\partial}{\partial y} ((y')^2 - y^2) = -2y$$

$$F_{y'} = \frac{\partial}{\partial y'} F((y')^2 - y^2)$$

$$= \frac{\partial}{\partial y'} ((y')^2 - y^2) = 2y'$$

By using the Euler equation given by Eq. (10.14), we find

$$y'' + y = 0 \tag{10.17}$$

The solution of Eq. (10.17) is

$$y = C_1 \cos x + C_2 \sin x \tag{10.17a}$$

By using Eq. (10.17a) and applying the boundary conditions $y(0) = 0$ and $y(\pi/2) = 1$, we find $C_1 = 0$ and $C_2 = 1$. Therefore, the only curve that makes the functional $\Omega(y(x))$ have an extremum is the curve $y = \sin x$.

10.3.2   Functionals with Movable Boundaries

In this section, one or both points A and B of a functional are assumed to be movable. Therefore, the family of curves that form such a functional have common end points, but their end points are permitted to expand since they are movable. To start the discussion, we consider again the following functional of the simplest form:

$$\Omega = \int_{x_0}^{x_1} F(x, y, y') \, dx \tag{10.18}$$

If a curve $y = y(x)$ yields an extremum when the boundary points are

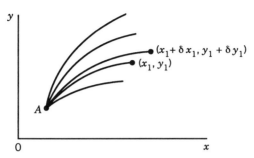

FIGURE 10.2. Schematic representation of functionals with moving boundaries.

movable, then the same curve gives an extremum with respect to a more restricted class of curves that have the same boundary points as $y = y(x)$. Thus, in order to have an extremum, the function $y(x)$ should be a solution of the Euler equation given by Eq. (10.14), or Eq. (10.15); that is, we should have

$$F_y - \frac{d}{dx} F_{y'} = 0 \tag{10.19}$$

As stated earlier, Eq. (10.19) may be satisfied by a family of curves of the form

$$y = y(x, C_1, C_2) \tag{10.20}$$

where C_1 and C_2 are constants that can be determined from the boundary conditions.

When the boundaries are fixed, the constants C_1 and C_2 may be determined from the boundary conditions $y(x_0) = y_0$ and $y(x_1) = y_1$. When one or both boundaries are movable, then one or both of these boundary conditions do not hold true. Thus, additional boundary conditions must be developed that take into consideration the boundary movements. We can accomplish this by incorporating the movement of the boundaries in the variation $\delta\Omega$ of the functional and making $\delta\Omega = 0$. In this manner, we can derive the necessary conditions by selecting curves that make the functional have an extremum.

We consider now the functional given by Eq. (10.18), and we assume that end point A is fixed, while end point B is movable. The extremals that pass through point $A(x_0, y_0)$ from the pencil of extremals $y = (x, C_1)$ are as shown in Fig. 10.2. If the boundary end B moves from position (x_1, y_1) to position $(x_1 + \delta x_1, y_1 + \delta y_1)$, as shown in Fig. 10.2, then the variation $\delta\Omega$ of the functional $\Omega(y(x, C_1))$ taken along extremals of the pencil $y = y(x, C_1)$ can be determined. However, it should be pointed out that the functional, which is taken only along the curves of the pencil, turns into a function of x_1 and y_1, and, consequently, its variation turns into the differential of this function. Therefore, by making $\delta\Omega = 0$, the necessary condition for the functional to have an extremum is found to be as follows:

$$(F - y'F_{y'})|_{x = x_1} \, \delta x_1 + F_{y'}|_{x = x_1} \, dy_1 = 0 \tag{10.21}$$

If the variations δx_1 and δy_1 are assumed to be independent—a case that is often required—then we have

$$(F - y'F_{y'})|_{x = x_1} = 0 \tag{10.22}$$

$$F_{y'}|_{x = x_1} = 0 \tag{10.23}$$

For example, if the end point $B(x_1, y_1)$ is permitted to move along the curve $y_1 = \Phi(x_1)$, then $\delta y_1 \approx \Phi'(x_1)\delta x_1$, and Eq. (10.21) yields

$$[F + (\Phi' - y')F_{y'}]\delta x_1 = 0 \tag{10.24}$$

Since δx_1 varies arbitrarily, we have

$$(F + (\Phi' - y')F_{y'})|_{x = x_1} = 0 \tag{10.25}$$

Equation (10.25) is known as the *transversality condition*, which establishes a relation between directional coefficients Φ' and y' at the end point.

The transversality condition, together with $y_1 = \Phi(x_1)$, provides a way to distinguish one or more extremals from the pencil $y = y(x, C)$ that can give an extremum. Therefore, if the end point $A(x_0, y_0)$ can move along a curve $y = \Psi(x)$, then, by the same reasoning, we can write the transversality condition at the point $A(x_0, y_0)$ as follows:

$$(F + (\Psi' - y')F_{y'})_{x = x_0} = 0 \tag{10.26}$$

In a similar manner, the transversality conditions for other forms of functionals may be obtained. If we consider, for example, the functional

$$\Omega = \int_{x_0}^{x_1} F(x, y, z, y', z') \, dx \tag{10.27}$$

and assume that point $B(x_1, y_1, z_1)$ is movable and point $A(x_0, y_0, z_0)$ is fixed, an extremum can be obtained only along integral curves that satisfy the following Euler equations:

$$F_y - \frac{d}{dx} F_{y'} = 0 \tag{10.28}$$

$$F_z - \frac{d}{dx} F_{z'} = 0 \tag{10.29}$$

The transversality condition with δx_1 being arbitrary is found to be as follows:

$$[F + (\Phi' - y')F_{y'} + (\Psi' - z')F_{z'}]_{x = x_1} = 0 \tag{10.30}$$

where $y_1 = \Phi(x_1)$ and $z_1 = \Psi(x_1)$.

The transversality condition given by Eq. (10.30), together with $y_1 = \Phi(x_1)$ and $z_1 = \Psi(x_1)$, is not enough to evaluate all the constants arising from the general solution of the Euler equations given by Eqs. (10.28) and (10.29). If, however, the boundary point $B(x_1, y_1, z_1)$ can move on a surface $z_1 = \Phi(x_1, y_1)$, then, for independent δx_1 and δy_1, we have

$$[F - y'F_{y'} + (\Phi'_x - z')F_{z'}] = 0 \tag{10.31}$$

$$[F_{y'} + F_{z'}\Phi'_y\Phi_y]_{x = x_1} = 0 \tag{10.32}$$

Equations (10.31) and (10.32), together with $z_1 = \Phi(x_1, y_1)$, are generally enough for the evaluation of the two arbitrary constants resulting from the general solution of the Euler equations given by Eqs. (10.28) and (10.29). Similar conditions may be obtained if the boundary point $A(x_0, y_0, z_0)$ is also variable.

If the form of the functional is

$$\Omega = \int_{x_0}^{x_1} F(x, y, y', y'') \, dx \tag{10.33}$$

we find that the necessary conditions to have an extremum, if the boundaries are fixed, are

$$F_y - \frac{d}{dx} F_{y'} + \frac{d^2}{dx^2} F_{y''} = 0 \tag{10.34}$$

In this case, four arbitrary constants need to be determined from the solution of Eq. (10.34).

When the boundaries are movable, additional expressions are needed. For example, if the end point $B(x_1, y_1)$ is movable and end $A(x_0, y_0)$ is fixed, then the condition $\delta\Omega = 0$ yields

$$\left[F - y'F_{y'} - y''F_{y'} + y'\frac{d}{dx}F_{y'}\right]_{x=x_1}\delta x_1$$
$$+\left[F_{y'} - \frac{d}{dx}F_{y''}\right]_{x=x_1}\delta y_1 + F_{y''}\Big|_{x=x_1}\delta y_1' = 0$$

When δx_1, δy_1, and $\delta y_1'$ are independent, we have

$$\left[F - y'F_{y'} - y''F_{y''} + y'\frac{d}{dx}F_{y''}\right]_{x=x_1} = 0 \qquad (10.36)$$

$$\left[F_{y'} - \frac{d}{dx}F_{y''}\right]_{x=x_1} = 0 \qquad (10.37)$$

$$F_{y''}\big|_{x=x_1} = 0 \qquad (10.38)$$

If $y_1 = \Phi(x_1)$ and $y_1' = \Psi(x_1)$, then Eq. (10.35) yields

$$\left[F - y'F_{y'} - y''F_{y''} + y'\frac{d}{dx}F_{y''} + \left(F_{y'} - \frac{d}{dx}F_{y''}\right)\Phi' + F_{y''}\Psi'\right]_{x=x_1} = 0$$
$$(10.39)$$

Equation (10.39), together with $y_1 = \Phi(x_1)$ and $y_1' = \Psi(x_1)$, is usually sufficient to determine x_1, y_1, and y_1'. Similar procedures may be used when end point $A(x_0, y_0)$ is movable.

10.4 A GENERAL TYPE VARIATIONAL PRINCIPLE FOR ELASTODYNAMICS PROBLEMS

The purpose of the discussion in this section is to present a variational principle that can be used for the solution of elastodynamics problems in one dimension. This variational principle was developed by Thomas Scott Dean [83] and may be considered to be an expanded form of Hamilton's principle. An attempt is also made here to clarify and evaluate this variational principle, correct several errors, and supply steps of explanations and new methods of proof, which are essential in the evaluation and understanding of the variational principle.

The variational principle may be written in the following form:

$$\delta \left[\int_{t_1}^{t_2} \int_0^{\ell} \left(+p\dot{u} - Pu_x - \frac{p^2}{2\rho A} + \frac{P^2}{2EA} - Xu \right) dx\, dt \right.$$

$$\left. - \int_{t_1}^{t_2} \bar{P}_o \bar{u}_0\, dt + \int_{t_1}^{t_2} \bar{P}_\ell \bar{u}_\ell\, dt \right] = 0 \tag{10.40}$$

All terms in Eq. (10.40) represent mechanical energy. Terms 1 and 3 in the first integral are kinetic energy, while terms 2, 4, and 5 in the same integral are potential energy. The last two integrals of the equation represent a state of potential energy at the boundaries.

In Eq. (10.40), ρ is the mass density of the material, u is the displacement of a cross section, P is the normal load at a cross section, X represents body forces, p is the linear momentum per unit length, A is the cross-sectional area, $\bar{P}_0$ is the value of P prescribed at $x = 0$, $\bar{P}_\ell$ is the value of P prescribed at $x = \ell$, and $\bar{u}_0$ and $\bar{u}_\ell$ are the values of u at $x = 0$ and $x = \ell$, respectively. The following restrictions are placed on the variables:

1. The quantities u, P, and p are functions of both space and time. They are allowed to have independent space and time relations.
2. We let u be prescribed at a boundary $x = 0$ or $x = \ell$. All comparison curves must pass through this point, and, consequently, $\delta u = 0$ at this point.
3. $t = t_1$ and $t = t_2$ are the time boundaries for all comparison curves; thus, $\delta u = 0$ at $t = t_1$ and $t = t_2$ for all points in the medium.

If the indicated variation in Eq. (10.40) is performed, the equation can be shown to be equivalent to the conventional formulation for the longitudinal vibration of a medium. If we go ahead and do this, and also perform several mathematical manipulations, such as integration by parts, we find

$$\int_{t_1}^{t_2} \int_0^{\ell} \delta u \left\{ [-\dot{p} + P_x - X] + \delta P \left[-u_x + \frac{P}{AE} \right] + \delta p \left[\dot{u} - \frac{p}{\rho A} \right] \right\} dx\, dt$$

$$+ \int_{t_1}^{t_2} -(P\,\delta u|_0^\ell) - \bar{P}_0\,\delta \bar{u}_0 + \bar{P}_\ell\,\delta \bar{u}_\ell\, dt = 0 \tag{10.41}$$

where δu, δP, and δp are arbitrary. To satisfy Eq. (10.41), the following terms in the first integral must be zero:

$$[-\dot{p} + P_x - X] = 0 \quad \text{or} \quad \dot{p} = P_x - X \tag{10.42}$$

$$\left[-u_x + \frac{P}{AE} \right] = 0 \quad \text{or} \quad u_x = \frac{P}{AE} \tag{10.43}$$

$$\left[+\dot{u} - \frac{p}{\rho A} \right] = 0 \quad \text{or} \quad \dot{u} = \frac{p}{\rho A} \tag{10.44}$$

Equations (10.42), (10.43), and (10.44) are the well-known conventional differential equations, which may be derived from basic mechanics. For example, Eq. (10.42) may be determined by applying Newton's second law of motion to a segment of the medium. Equation (10.43) results from the stress–strain equation of a section of the medium, and Eq. (10.44) is an identity that can be shown by substituting for the value of p; that is, $\dot{u} = p/\rho A = \rho A \dot{u}/\rho A = \dot{u}$, or $\dot{u} = \dot{u}$.

The second integral in Eq. (10.41) contains the boundary condition terms. If $\delta u = 0$ at $x = 0$ and $x = \ell$, then this integral is zero. If $\delta u \neq 0$ at $x = 0$ and $x = \ell$, then we have

$$\int_{t_1}^{t_2} - (P(\ell, t)\, \delta u_\ell + P(0, t)\, \delta u_0 - \bar{P}_0\, \delta \bar{u}_0 + \bar{P}_\ell\, \delta \bar{u}_l)\, dt = 0 \tag{10.45}$$

or

$$\int_{t_1}^{t_2} \delta u_0 [-\bar{P}_0 + P(0, t)] + \delta u_\ell [\bar{P}_\ell - P(\ell, t)]\, dt = 0 \tag{10.46}$$

Since $\delta u_0 \neq 0$ and is independent of $\delta u_\ell \neq 0$, the above condition is satisfied only if

$$\bar{P}_0 = P(0, t) \tag{10.47}$$

$$\bar{P}_\ell = P(\ell, t) \tag{10.48}$$

The above discussion shows that the first integral in Eq. (10.40) may be used to obtain the general differential equation for the elastodynamic problem. The arbitrary constants involved in the solution of the differential equation may be obtained from the end boundary conditions. Equation (10.40), without the end boundary conditions, can also be derived by using Hamilton's principle, which is as follows:

$$\delta \int_{t_1}^{t_2} (T - U)\, dt = 0 \tag{10.49}$$

where T and U are the kinetic and potential energies, respectively. For this case, we have

$$T = \int_0^\ell \left(p\dot{u} - \frac{p^2}{2\rho A} \right) dx \tag{10.50}$$

$$U = \int_0^\ell \left(Pu_x - \frac{p^2}{2EA} + Xu \right) dx \tag{10.51}$$

By substituting Eqs. (10.50) and (10.51) into Eq. (10.49), we find

$$\delta \int_{t_1}^{t_2} \int_0^\ell \left(p\dot{u} - \frac{p^2}{2\rho A} - Pu_x + \frac{p^2}{2EA} - Xu \right) dx \, dt = 0$$

which is identical to the first integral in Eq. (10.40).

The following two examples illustrate the use of the variational principle and also point out some limitations regarding the application of the methodology. In the first example, the exact solution of the differential equation of motion is known, while in the second example, such a solution is very difficult to obtain, and simplified assumptions are made in order to obtain an easier differential equation.

Example 10.2. Determine the longitudinal free frequencies of vibration for a uniform single-span beam that is fixed at both ends. Assume that the system is conservative and that there are no body forces X.

Solution. Since $\bar{u}_0 = \bar{u}_\ell = x = 0$, the variational principle given by Eq. (10.40) reduces to the following equation:

$$\delta \left[\int_{t_1}^{t_2} \int_0^\ell \left(p\dot{u} - Pu_x - \frac{p^2}{2\rho A} + \frac{P^2}{2EA} \right) dx \, dt \right] = 0 \tag{10.52}$$

By substituting $p = \rho A\dot{u}$ and $P = AEu_x$ into Eq. (10.52), we find

$$\delta \left[\int_0^t \int_0^\ell \left(- \rho A\dot{u}^2 + AEu_x^2 + \frac{\rho A\dot{u}^2}{2} - \frac{AEu_x^2}{2} \right) dx \, dt \right] = 0 \tag{10.53}$$

We note here that the functional Ω is of the form

$$\Omega(u(x, t)) = \int_0^t \int_0^\ell F\left(t, x, u, \frac{\partial u}{\partial x}, \frac{\partial u}{\partial t} \right) dx \, dt \tag{10.54}$$

In order to make $\Delta\Omega = 0$, we must satisfy the following equation:

$$F_u - \frac{\partial F_{u'}}{\partial x} - \frac{\partial F_{\dot{u}}}{\partial t} = 0 \qquad (10.55)$$

Here, we have

$$F_u = 0 \qquad (10.56)$$

$$-\frac{\partial}{\partial x}(F_{u'}) = -EAu'' \qquad (10.57)$$

$$-\frac{\partial}{\partial t}(F_{\dot{u}}) = \rho\ddot{u} \qquad (10.58)$$

By substituting Eqs. (10.56), (10.57), and (10.58) into Eq. (10.55), we obtain

$$-EAu'' + \rho\ddot{u} = 0$$

or

$$\frac{\partial^2 u}{\partial x^2} - \frac{\rho}{AE}\frac{\partial^2 u}{\partial t^2} = 0 \qquad (10.59)$$

which is the familiar general differential equation for the free longitudinal vibration of a uniform beam. Its solution yields the longitudinal displacements $u(x, t)$, the free frequencies of vibration, and the corresponding mode shapes.

For the fixed–fixed uniform beam, we can assume that the solution $u(x, t)$ is

$$u(x, t) = f(x)g(t) \qquad (10.60)$$

where $f(x)$ is only a function of x, and $g(t)$ is only a function of time t. On this basis, Eq. (10.59) yields

$$f\ddot{g} - \frac{EA}{\rho}f''g = 0 \qquad (10.61)$$

By separation of variables, we obtain

$$\frac{\ddot{g}}{g} = \frac{EA}{\rho}\frac{f''}{f} = \pm\psi \qquad (10.62)$$

Equation (10.62) yields the following two differential equations:

$$\ddot{g} + \psi g = 0 \tag{10.63}$$

$$a^2 f'' + \psi f = 0 \tag{10.64}$$

where

$$a^2 = \frac{EA}{\rho} \tag{10.65}$$

The solutions of Eqs. (10.63) and (10.64) are

$$g = A \cos \sqrt{\psi} t + B \sin \sqrt{\psi} t \tag{10.66}$$

$$f = C \cos \frac{\sqrt{\psi}}{a} x + D \sin \frac{\sqrt{\psi}}{a} x \tag{10.67}$$

On this basis, Eq. (10.60) yields

$$u(x, t) = (A \cos \sqrt{\psi} t + B \sin \sqrt{\psi} t)\left(C \cos \frac{\sqrt{\psi}}{a} x + D \sin \frac{\sqrt{\psi}}{a} x \right) \tag{10.68}$$

By applying the boundary condition $u = 0$ at $x = 0$, we find $C = 0$. For a nontrivial solution, the boundary condition $u = 0$ at $x = \ell$, where l is the length of the member, yields

$$\sin \frac{\sqrt{\psi}}{a} \ell = 0 \tag{10.69}$$

Equation (10.69) is satisfied if

$$\frac{\sqrt{\psi}}{a} \ell = n \pi$$

or

$$\sqrt{\psi} = \omega_n = \frac{an \pi}{\ell} \qquad n = 1, 2, 3, \ldots$$

Thus, the longitudinal free frequencies of vibration ω_n of the fixed–fixed uniform beam may be determined from the following equation:

$$\omega_n = \frac{n \pi}{\ell} \sqrt{\frac{AE}{\rho}} \qquad n = 1, 2, 3, \ldots \tag{10.70}$$

Example 10.3. Consider a longitudinally vibrating beam with a nonlinear force–strain relation given by the equation

$$P = (E_1 u_x + E_2 u_x^3)A \tag{10.71}$$

where E_1 and E_2 are constants. Find the free longitudinal frequencies of vibration. Assume that there is no damping or body forces.

Solution. The variational principle given by Eq. (10.40) may be written in the following form:

$$\delta \int_{t_1}^{t_2} \int_0^{\ell} \left(-p\dot{u} + Pu_x + \frac{p^2}{2\rho A} - W(P) \right) dx \, dt = 0 \tag{10.72}$$

where $W(P)$ is the *complementary energy density*, which is given by the expression

$$W(P) = \int_0^P u_x \, dP \tag{10.73}$$

From Eq. (10.71), we find

$$dP = (E_1 \, du_x + 3E_2 u_x^2 \, du_x)A \tag{10.74}$$

Thus,

$$W(P) = A \int_0^{u_x} u_x E_1 \, du_x + 3E_2 u_x^3 \, du_x$$

$$= A\left(E_1 \frac{u_x^2}{2} + \frac{3E_2 u_x^4}{4} \right) \tag{10.75}$$

By substituting Eq. (10.75) into Eq. (10.72) and collecting terms, we find

$$\delta \int_{t_1}^{t_2} \int_0^{\ell} \left(-\frac{\rho A \dot{u}^2}{2} + \frac{AE_1 u_x^2}{2} + \frac{AE_2 u_x^4}{4} \right) dx \, dt = 0 \tag{10.76}$$

We note that the functional Ω is of the form

$$\Omega(\dot{u}(x, t)) = \int \int F\left(x, t, u, \frac{\partial u}{\partial x}, \frac{\partial u}{\partial t} \right) dx \, dt \tag{10.77}$$

In order to have an extremum, the following equation must be satisfied:

$$Fu - \frac{\partial F_{u'}}{\partial x} - \frac{\partial F_{\dot{u}}}{\partial t} = 0 \qquad (10.78)$$

We have

$$Fu = 0$$

$$-\frac{\partial}{\partial x}(E_1 u_x + E_2 u_x^3)A = -AE_1 u'' - 3AE_2 u' u''$$

$$-\frac{\partial}{\partial t}(-\rho A \dot{u}) = \rho A \ddot{u}$$

On this basis, Eq. (10.78) yields

$$\rho A \ddot{u} - AE_1 u'' - 3AE_2 u'^2 u'' = 0 \qquad (10.79)$$

or

$$\frac{3E_2}{\rho}\left(\frac{\partial u}{\partial x}\right)^2 \frac{\partial^2 u}{\partial x^2} + \frac{E_1}{\rho}\frac{\partial^2 u}{\partial x^2} = \frac{\partial^2 u}{\partial t^2} \qquad (10.80)$$

which is the general differential equation of motion.
We assume the solution of Eq. (10.80) to be as follows:

$$u = f(x)g(t) \qquad (10.81)$$

where $f(x)$ is a function of x only, and $g(t)$ is a function of t only. By substituting into Eq. (10.80) and separating variables, we find

$$\frac{\ddot{g}}{g} = \left[\frac{E_1}{\rho} + \frac{3E_2}{\rho}(f'^2 g^2)\right]\frac{f''}{f} \qquad (10.82)$$

The solution of Eq. (10.82) appears to be very difficult to solve, since the variables do not separate as desired. However, if the direct method of variational calculus is used, a solution can be obtained with an accuracy that will depend on how accurately $f(x)$ is assumed.
For example, assume that $u(x, t) = f(x)g(t)$, which is the same as Eq. (10.60). From Eq. (10.76), we can write

$$\Omega = \int_0^t \int_0^\ell \left(-\frac{\rho A \dot{u}^2}{2} + \frac{AE_1 u_x^2}{2} + \frac{AE_2 u_x^4}{4}\right) dx\, dt \qquad (10.83)$$

By substituting Eq. (10.60) into Eq. (10.83), we find

$$\Omega = \frac{A}{2} \int_0^{t} \int_0^{\ell} \left(-\rho f^2 \dot{g}^2 + E_1(f')^2 g^2 + \frac{E_2}{2}(f')^4 g^4 \right) dx \, dt \qquad (10.84)$$

or

$$\Omega = \frac{A}{2} \int_0^{t} (-H\dot{g}^2 + Gg^2 + Kg^4) \, dt \qquad (10.85)$$

where

$$H = \int_0^{\ell} \rho f^2 \, dx \qquad (10.86)$$

$$G = \int_0^{\ell} E_1(f')^2 \, dx \qquad (10.87)$$

$$K = \int_0^{\ell} \frac{E_2}{2}(f')^4 \, dx \qquad (10.88)$$

On this basis, the functional Ω is of the form

$$\Omega(g(t)) = \frac{A}{2} \int_0^{t} F(t, g, \dot{g}) \, dt \qquad (10.89)$$

The necessary condition for this functional to have an extremum is the Euler equation; that is,

$$F_g - \frac{\partial F_{\dot{g}}}{\partial t} = 0 \qquad (10.90)$$

This yields

$$2Gg + 4Kg^3 - \frac{\partial}{\partial t}(-2H\dot{g}) = 0$$

or

$$\ddot{g} + \frac{2K}{H}g^3 + \frac{G}{H}g = 0 \qquad (10.91)$$

Equation (10.91) is easier to solve compared to the general differential equation given by Eq. (10.80), and it can be accomplished by using a phase plain plot, provided that K, H, and G can be evaluated. The solution would be exact if the exact function $f(x)$ is known. If the exact expression of $f(x)$ is not known, then an approximate $f(x)$ may be used that satisfies the boundary conditions, and, consequently, approximate values of K, H, and G may be obtained. In other words, the solution will be as good as the assumed expression for $f(x)$.

This example shows that variational calculus can offer a relatively easy method of approximating a solution, which becomes very attractive as the complexity of the general differential equation increases.

10.5   UTILIZATION OF HAMILTON'S VARIATIONAL PRINCIPLE

In this section, Hamilton's variational principle, given by Eq. (10.49), will be used for the solution of various types of vibration problems.

10.5.1   Vibration of Strings and Rods

Consider the freely vibrating elastic string shown in Fig. 10.3. The deflection curve $y(x, t)$ during motion is assumed to be as shown in the same figure. In the deformed configuration, an element ds of the string has the length

$$ds = \sqrt{1 + y_x^2}\, dx \tag{10.92}$$

By applying Taylor's formula, we find

$$\sqrt{1 + y_x^2} \approx 1 + \frac{y_x^2}{2} \tag{10.93}$$

If we assume that the string is perfectly elastic, the potential energy dU of an element dx of the string is proportional to the longitudinal displacement, and it is written as follows:

$$dU = \tfrac{1}{2}Ay_x^2\, dx \tag{10.94}$$

where A is a coefficient that represents the tension force in the string. The

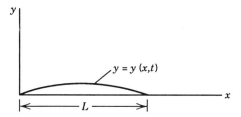

FIGURE 10.3. Vibrating string or rod.

total potential energy U of the string is

$$U = \frac{1}{2} \int_0^{\ell} A y_x^2 \, dx \tag{10.95}$$

The total kinetic energy T of the string is

$$T = \frac{1}{2} m y_t^2 \tag{10.96}$$

where m is the mass per unit length of the string.

By substituting Eqs. (10.95) and (10.96) into Eq. (10.16), we find

$$\Omega = \int_{t_0}^{t_1} \int_0^L \left(\frac{1}{2} m y_t^2 - \frac{1}{2} A y_x^2 \right) dx \, dt \tag{10.97}$$

The functional in Eq. (10.97) is similar to the one given in Item (7) of Appendix L. Thus, the differential equation of motion of the string is the Euler–Lagrange equation, which is written as follows:

$$\frac{\partial}{\partial t} (m\dot{y}) - \frac{\partial}{\partial x} (A y') = 0 \tag{10.98}$$

The deflection curve $y(x, t)$ that satisfies Eq. (10.98) is the function $y(x, t)$ that makes the functional given by Eq. (10.97) have an extremum. If the string is homogeneous, m and A are constant, and Eq. (10.98) yields

$$\frac{\partial^2 y}{\partial t^2} - \frac{A}{m} \frac{\partial^2 y}{\partial x^2} = 0 \tag{10.99}$$

Note that the differential equation of the freely vibrating string is identical to that of the longitudinal and torsional vibration of rods. The assumptions made for the string are that its deflections are small and that the tension in the string remains unaffected by the deflection.

We assume now that the string is subjected to a force $q(x, t)$ that is proportional to the mass of an element of the string. In this case, the work on the element done by the force is

$$mq(x, t) \, dx \tag{10.100}$$

and the functional given by Eq. (10.16) takes the following form:

$$\Omega = \int_{t_0}^{t_1} \int_0^L \left(\frac{1}{2} m y_t^2 - \frac{1}{2} A y_x^2 + mq(x, t) y \right) dx \, dt \tag{10.101}$$

On this basis, the forced vibration may be determined from the following differential equation:

$$\frac{\partial}{\partial t}(my_t) - \frac{\partial}{\partial x}(Ay_x) - mq(x, t) = 0$$

or

$$\frac{\partial^2 y}{\partial t^2} - \frac{A}{m}\frac{\partial^2 y}{\partial x^2} = q(x, t) \tag{10.102}$$

We now consider the free vibration of a rod with deflection $y(x, t)$ and coordinates x, y as shown in Fig. 10.3. The rod's total kinetic energy during vibration is

$$T = \frac{1}{2}\int_0^L my_t^2 \, dx \tag{10.103}$$

where L is its length. We assume here that the rod will not stretch and that its potential energy U is proportional to the square of its curvature. We also assume that the displacements of the rod are small. On this basis, we have

$$U = \frac{1}{2}\int_0^L A\left(\frac{\partial^2 y}{\partial x^2}\right) dx \tag{10.104}$$

and Hamilton's integral, given by Eq. (10.16), yields

$$\Omega = \int_{t_0}^{t_1}\int_0^L \left(\tfrac{1}{2}my_t^2 - \tfrac{1}{2}Ay_{xx}^2\right) dx \, dt \tag{10.105}$$

The function $y(x, t)$ that makes the functional in Eq. (10.105) have an extremum must satisfy the following differential equation:

$$\frac{\partial}{\partial t}\left(my_t + \frac{\partial^2}{\partial x^2}(Ay_{xx})\right) = 0$$

or, for a homogeneous rod,

$$\frac{\partial^2 y}{\partial t^2} + \frac{A}{m}\frac{\partial^4 y}{\partial x^4} = 0 \tag{10.106}$$

10.5.2 Transverse Vibration of Beams

Consider a single-span beam of uniform stiffness EI and length L, and determine the general differential equation of its free undamped vibrating motion. The kinetic energy dT of an element of the beam is

$$dT = \tfrac{1}{2}m\dot{y}^2\, dx \qquad (10.107)$$

where y is the dynamic displacement of the element, and $\dot{y}$ is its velocity. The total kinetic energy T of the beam is

$$T = \int_0^L \frac{m\dot{y}^2}{2}\, dx \qquad (10.108)$$

where m is the mass per unit length of the beam. Note that the displacement $y(x, t)$ of the member is a function of both x and time t.

The potential energy dU of the beam element is

$$dU = \frac{EI}{2}\left(\frac{d^2 y}{dx^2}\right)^2 dx \qquad (10.109)$$

and the total potential energy U of the member is

$$U = \int_0^L \frac{EI}{2}\left(\frac{d^2 y}{dx^2}\right)^2 dx \qquad (10.110)$$

By substituting Eqs. (10.108) and (10.110) into Eq. (10.16), we find

$$\Omega = \int_0^t \int_0^L \left[\frac{m\dot{y}^2}{2} - \frac{EI}{2}\left(\frac{d^2 y}{dx^2}\right)^2\right] dx\, dt \qquad (10.111)$$

Equation (10.111) is represented by a functional of the general form:

$$\Omega = (y(x, t)) = \int_0^t \int_0^L F\left(t, x, y, \frac{\partial y}{\partial x}, \frac{\partial^2 y}{\partial x^2}, \frac{\partial^2 y}{\partial t^2}, \frac{\partial^2 y}{\partial x\, \partial y}\right) dx\, dt \qquad (10.112)$$

Equation (10.112) yields an extremum when the following equation is satisfied:

$$F_y - \frac{\partial f_{y'}}{\partial x} - \frac{\partial F_{\dot{y}}}{\partial t} + \frac{\partial^2 F_{y''}}{\partial x^2} + \frac{\partial^2 F_s}{\partial x\, \partial t} + \frac{\partial^2 F_{\ddot{y}}}{\partial t^2} = 0 \qquad (10.113)$$

where

$$s = \frac{\partial^2 y}{\partial x \, \partial t} \tag{10.114}$$

By substitution, Eq. (10.113) yields

$$\frac{\partial^4 y}{\partial x^4} + \frac{m}{EI} \frac{\partial^2 y}{\partial t^2} = 0 \tag{10.115}$$

which is the familiar general differential equation for the free transverse vibration of beams. Its solution will yield the free frequencies of vibration and the corresponding mode shapes of uniform beams of various boundary conditions.

A direct approach to the variational problem may also be used to determine frequencies of free vibration of beams and their corresponding mode shapes. Consider, for example, a uniform simply supported beam, and assume that a solution $y = f(x)g(t)$ exists, which satisfies the boundary conditions $u = y'' = 0$ at $x = 0$ and $x = L$, and where $f(x)$ is a function only of x, and $g(t)$ varies only with time t. We also assume that

$$f(x) = C \sin \lambda x \tag{10.116}$$

where

$$\lambda = \frac{n\pi}{L} \qquad n = 1, 2, 3, \ldots \tag{10.117}$$

On this basis, we have

$$y = f(x)g(t)$$
$$= (C \sin \lambda x)g \tag{10.118}$$
$$\dot{y} = (C \sin \lambda x)\dot{g} \tag{10.119}$$
$$y'' = (- C\lambda^2 \sin \lambda x)g \tag{10.120}$$

By substituting into Eq. (10.111), we find

$$\Omega = \int_0^t \int_0^L \left(\frac{m}{2} C^2 \sin^2 \lambda x (\dot{g})^2 - \frac{EI}{2} C^2 \lambda^4 \sin^2 \lambda x (g)^2 \right) dx \, dt$$

$$= C^2 \int_0^t \left[\left(\frac{m}{2} (\dot{g})^2 - \frac{EI}{2} \lambda^4 (g)^2 \right) \int_0^L \sin^2 \lambda x \, dx \right] dt$$

or

$$\Omega = C^2 \int_0^{t} \left(\frac{mL}{4} (\dot{g})^2 - \frac{EIL}{4} \lambda^4 (g)^2 \right) dt \tag{10.121}$$

The necessary condition that makes Eq. (10.121) have an extremum is represented by the following Euler equation:

$$F_g - \frac{dF_{\dot{g}}}{dt} = 0 \tag{10.122}$$

Equation (10.122) yields

$$-\frac{EIL\lambda^4}{4} g - \frac{d}{dt} \left(\frac{mL}{2} \dot{g} \right) = 0$$

or

$$\ddot{g} + \frac{EI\lambda^4}{m} g = 0 \tag{10.123}$$

The solution of Eq. (10.123) yields

$$g = A \cos[(EI/m)^{1/2}\lambda^2 t] + B \sin[(EI/m)^{1/2}\lambda^2 t] \tag{10.124}$$

where λ is as shown by Eq. (10.117).

The free frequencies of vibration ω_n of the simply supported beam are as follows:

$$\omega_n = (EI/m)^{1/2} \lambda^2 = (EI/m)^{1/2} \frac{n^2 \pi^2}{L^2}$$

$$= \frac{n^2 \pi}{L^2} \sqrt{\frac{EI}{m}} \qquad n = 1, 2, 3, \ldots \tag{10.125}$$

The corresponding mode shapes are represented by Eq. (10.116). Other beam boundary conditions may be treated in a similar manner.

10.6   BASIC ASPECTS OF RANDOM VIBRATIONS

The kinds of dynamic and vibration responses discussed in the preceding sections and chapters of the text are the ones that are produced by well-defined time-varying excitations, which may be periodic or nonperiodic. Such types of excitations are usually called *deterministic*, because they are defined force functions with predictable time variations. There are, however, situations where the time-varying forces are not known, and the analysis of a structure or a machine subjected to such excitations must be based on predictions. Excitations that involve the element of prediction are called *nondeterministic*, and there is no guarantee that they will actually occur. Often, the term *random* is also used to define such excitations. For example, the excitations that are produced by an earthquake, a blast, or a wind gust are nondeterministic. The prediction of such excitations is usually based on experience or on experimental results, and the design of an engineering system subjected to such excitations depends on how closely an actual situation is predicted.

When the input disturbances on an engineering system are not well defined, probability and statistics may then be used to establish the required methodologies that will help in predicting the dynamic response of such systems. When a random process exhibits a degree of statistical regularity, we say that this process is *stationary*, and its analysis is possible. A stationary random process is also an *ergotic* one if its stationary characteristics do not vary with time.

Statistical theory is based on the concept of probability. The probability of the occurrence of an event is zero when the event cannot possibly occur, and it is taken to be unity when an event is absolutely certain to occur. Consequently, the probability of any event occurring would lie between zero and unity. To illustrate probability, we consider an experiment where an event denoted by Q is certain to occur. When two events a and b are given, the certain event $Q = a + b$ will occur when a, b, or both a and b occur. Events a and b are mutually exclusive when the occurrence of one excludes the occurrence of the other. For example, in the tossing of a coin, the certain event Q to occur is either head or tail. If event a is head and event b is tail, events a and b are mutually exclusive.

The probability of event a is the number $P(a)$ assigned to this event, which obeys the following laws:

$$P(a) \geq 0 \tag{10.126}$$

$$P(Q) = 1 \tag{10.127}$$

If two events a and b are mutually exclusive, we have

$$P(a + b) = P(a) + P(b) \tag{10.128}$$

For a given experiment, the set of all possible outcomes is called the *sample space*.

If there are n mutually exclusive, exhaustive, and equally likely cases, and if m of these events are favorable to event A, then the probability of A is defined as

$$P(A) = \frac{m}{n} \qquad (10.129)$$

In the tossing of a coin, the probability of the event head or tail to occur is $\frac{1}{2}$.

A process is random if it is not possible to predict its final state from its initial state, as in the case of tossing a coin or rolling a die. The process is defined by its sample space, which is the totality of sample points associated with a given experiment. A random variable is a function X that is defined within the sample space. If a number x_k, where $k = 1, 2, 3, \ldots, k$, is associated with a particular outcome of an experiment, then the random variable X can assume all possible values of x_k. Therefore, by assigning a random variable to an experiment, the notation in the discussion that follows is simplified. For example, if a random variable X is assigned to the rolling of a die, the possible values of the outcome are $x_1 = 1$ for one dot, $x_2 = 2$ for two dots, and so on. In this manner, the probability of rolling a five, for example, can be written $P(X = x_5)$.

The *frequency function* $p(x = x_k)$, or simply $p(x_k)$, also known as the *density function*, denotes the probability of the random variable X for a value x_k. For example, the frequency function in the rolling of a die is

$$p(x_1) = p(x_2) = p(x_3) = p(x_4) = p(x_5) = p(x_6) = \tfrac{1}{6} \qquad (10.130)$$

The probability of a random variable being less than or equal to some specified value x_j is called the *distribution function* and is denoted by $P(x \leq x_j)$. Mathematically, the distribution function is written as follows:

$$P(X \leq x_j) = \sum_{k = -\infty}^{j} p(x_k) \qquad (10.131)$$

The frequency and distribution functions for the rolling of a die are shown in Figs. 10.4a and 10.4b, respectively.

Probabilities can be added or subtracted; for example,

$$P(x_1 \leq X \leq x_2) = P(X \leq x_2) - P(X \leq x_1)$$
$$= P(x_2) - P(x_1) \qquad (10.132)$$

In other words, if the distribution function, for example, is represented by the curve in Fig. 10.5a, Eq. (10.132) is represented by the length AB in the figure.

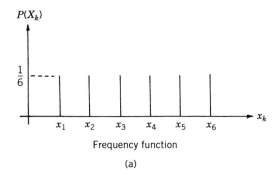

Frequency function

(a)

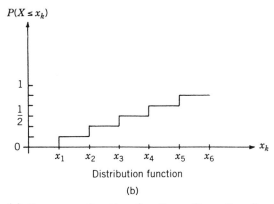

Distribution function

(b)

FIGURE 10.4. (a) Frequency function for the rolling of a die. (b) Distribution function for the rolling of a die.

The probability that X lies between x and $x + dx$ is denoted as $dP(x)$. Therefore, according to Eq. (10.132), we have

$$dP(x) = P(X \leq (x + dx)) - P(X \leq x)$$
$$= P(x + dx) - P(x) \qquad (10.133)$$

For small dx, we have

$$dP(x) = p(x)\,dx \qquad (10.134)$$

and for infinitesimal $dP(x)$ and dx, we have the following expression:

$$p(x) = \frac{dP(x)}{dx} \qquad (10.135)$$

The quantity $p(x)$ is known as the *probability density* and provides an alternative way of describing the probability distribution of a random variable. The

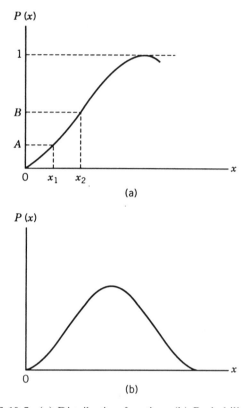

FIGURE 10.5. (a) Distribution function. (b) Probability density.

probability density graph is shown in Fig. 10.5b. When $P(x)$ is known, we can determine $p(x)$. The opposite, however, is also true. Experimentally, $P(x)$ is easier to determine. The reverse process may be carried out by using the following equation:

$$P(x) = \int_{-\infty}^{x} p(z)\, dz \qquad (10.136)$$

When only one random variable is involved, the probability is called *one-dimensional*. A *two-dimensional* probability involves two random variables, say, X and Y, and so on. For a two-dimensional probability, the distribution $P(X \leq x_j, Y \leq y_i)$ is defined as follows:

$$P(X \le x_j, Y \le y_i) = \sum_{k = -\infty}^{j} \sum_{\ell = -\infty}^{i} p(x_k, y_\ell) \qquad (10.137)$$

The frequency function $p(x_k)$ is

$$p(x_k) = \sum_{\ell = -\infty}^{\infty} p(x_k, y_\ell) \qquad (10.138)$$

For a continuous one-dimensional probability, the distribution function may be defined as the probability of a random variable X that assumes a value in the interval between any two values x_1 and x_2; that is,

$$P(x_1 \le X \le x_2) = \int_{x_1}^{x_2} p(x)\, dx \qquad (10.139)$$

where $p(x)$ is the probability frequency function for the continuous case, which is defined as follows:

$$p(x) = \frac{dP(x_1 \le X \le x_2)}{dx} \qquad (10.140)$$

Confidence, or *confidence level*, is a measure of probability, and it may be obtained by reading the curve that provides the distribution function. The statement 95% confidence, for example, means that 95% of a large number of tested components have a life expectancy that exceeds the declared value.

10.7   EXPECTATIONS AND CORRELATION FUNCTIONS

In the utilization of probability and statistics, an important parameter that needs to be determined is the *expected value*, or *mean*, of a random variable X, which is denoted by $E\{X\}$. If, for example, an experiment is repeated n times and the observed outcomes are $x_1, x_2, x_3, \ldots, x_m$, then for each outcome the random variable X assumes a numerical value. Therefore, for the indicated outcomes, the random variable X assumes the values $X(x_1), X(x_2), X(x_3), \ldots, X(x_m)$. If for the n repetitions of the experiment, the symbol n_1 denotes the number of times that $X(x_1)$ occurs, n_2 the number of times that $X(x_2)$ occurs, and so on, the *arithmetic average* X_{av} of the

$X(x_m)$ values, where $m = 1, 2, 3, \ldots$, is given by the following equation:

$$X_{av} = \frac{n_1 X(x_1) + n_2 X(x_2) + n_3 X(x_3) + \cdots + n_m X(x_m)}{n}$$

$$= \frac{1}{n} \sum_{m=1}^{m} n_m X(x_m)$$

(10.141)

where $n = n_1 + n_2 + n_3 + \cdots + n_m$.

If the number of repetitions n approaches infinity, n_1/n approximates the probability $p(x_1)$, n_2/n approximates the probability $p(x_2)$, and so on. The *statistical average*, or expected value $E\{X\}$, of a random variable X is given by the following equation:

$$E\{X\} = \sum_{m=1}^{m} P(x_m) X(x_m)$$

(10.142)

Equation (10.142) is also known as the *first moment* of the random variable X. The expression for the nth moment $E\{X^n\}$ of the random variable X is

$$E\{X^n\} = \sum_{m=1}^{m} p(x_m)[X(x_m)]^n$$

(10.143)

For the continuous case, the analogous expression for the nth moment is

$$E\{X^n\} = \int_{-\infty}^{\infty} x^n p(x)\, dx$$

(10.144)

The second moment $E\{X^2\}$ yields the *mean square value* of X; that is,

$$E\{X^2\} = \bar{X}^2 = \int_{-\infty}^{\infty} x^2 p(x)\, dx$$

(10.145)

The broadness, or spread, of the density function is called the *variance* of the random variable and is denoted by σ^2. The variance σ^2 is defined as follows:

$$\sigma^2 = E([X - E\{X\}]^2)$$

(10.146)

or

$$\sigma^2 = E\{X^2\} - [E\{X\}]^2$$

(10.147)

The probability frequency function discussed earlier provides amplitude variations of a random process, but it is not a unique function of the process. Consequently, two sample functions describing two processes can have the same probability frequency function $p(x)$. Also, a prescribed probability density prescribes only the proportion of time during which a quantity exists in a certain range. If, for example, the displacement in a certain structure exceeds its prescribed level for 3 minutes of time in its lifetime, we need to know the length of time for each occurrence within the total time of 3 minutes, in order to pinpoint more accurately the probable effect on the structure. Hence, it becomes necessary to define a *correlation function* that can be used to distinguish the two processes or pinpoint more accurately probable effects.

Consider, for example, two random processes described by the random variables X and Y at times t_1 and t_2, respectively. The *cross-correlation function* $R(t_1, t_2)$ of these two processes is defined as follows:

$$R(t_1, t_2) = E\{XY\} = \int_{-\infty}^{\infty} \int_{-\infty}^{\infty} xyp(x, y) \, dx \, dy \qquad (10.148)$$

where x and y within the integral signs are the assigned variables of the random variables X and Y, respectively, and $p(x, y)$ is the two-dimensional frequency function.

Equation (10.148) provides a measure of the correlation of the two random processes. The concept of combined probabilities is used to write Eq. (10.148). According to this concept, the *combined probability* is simply the product of the probabilities of the two separate events or values. The two-dimensional probability frequency function $p(x, y)$ could be plotted as a surface above a horizontal xy plane. Note that $p(x, y) \, dx \, dy = p(x)p(y) \, dx \, dy$.

Equation (10.148) may also be written as follows:

$$R_{xy}(t) = \lim_{\tau \to \infty} \frac{1}{2\tau} \int_{-\tau}^{\tau} x(\rho)y(t + \rho) \, d\rho \qquad (10.149)$$

where ρ is a dummy variable of integration, and τ is the period. Equation (10.149) provides the cross-correlation of the random functions $x(t)$ and $y(t)$. The *cross-correlation* $R_{yx}(t)$ of $y(t)$ on $x(t)$ is

$$R_{yx}(t) = R_{xy}(-t) \qquad (10.150)$$

If the random variables X and Y describe the same random processes at their respective times t_1 and t_2, where $t_2 = t_1 + \rho$, and $|\rho|$ is large enough so that X and Y can be considered independent, then the expected value $E\{XY\}$

is given by the following equation:

$$E\{XY\} = E\{X\}E\{Y\} \tag{10.151}$$

If $E\{X\}$ or $E\{Y\}$ is zero, then $E\{XY\}$ is zero, and, consequently, $R\{t_1, t_2\}$ in Eq. (10.148) is zero.

If the dummy variable ρ in Eq. (10.149) approaches zero, the random variables X and Y describe the same process at time $t_1 = t_2$. Therefore, on this basis, Eq. (10.151) may be written

$$R(t_1, t_1) = E\{XY\} = E\{X^2\} \tag{10.152}$$

Equation (10.152) is the *autocorrelation function* of the random process, and it also represents the *second moment* of the random process.

It was pointed out earlier that a random process is ergotic, if all types of ensemble averages, or expected values, are interchangeable with the corresponding time average. For ergotic processes, an ensemble average can be replaced by a time average involving a single sample function $x_i(t)$ of a random $x(t)$. On this basis, Eq. (10.151) yields

$$E\{XY\} = \langle x_i(t_1)x_i(t_2)\rangle \tag{10.153}$$

where $E\{X\}$ and $E\{Y\}$ in Eq. (10.151) are replaced by $x_i(t_1)$ and $x_i(t_2)$, respectively, in Eq. (10.153).

If t_2 in Eq. (10.153) is taken as equal to $t_1 + \rho$, where ρ is a dummy variable, then the autocorrelation function described by Eq. (10.153), or by Eq. (10.152), may be written

$$R_{xx}(t_1) = \lim_{\tau \to \infty} \frac{1}{2\tau} \int_{-\tau}^{\tau} x(\rho)x(t_1 + \rho)\, d\rho \tag{10.154}$$

or, in general,

$$R_{xx}(t) = \lim_{\tau \to \infty} \frac{1}{2\tau} \int_{-\tau}^{\tau} x(\rho)x(t + \rho)\, d\rho \tag{10.155}$$

Equation (10.155) provides a measure of the regularity of a random process. Therefore, the cross-correlation and the autocorrelation functions can be used to distinguish any two random processes that possess similar characteristics. These functions are also used in the following sections of the chapter to carry out power spectra analysis.

10.8 POWER SPECTRA ANALYSIS

As stated earlier in Section 10.6, a random process $x(t)$ is stationary, in the strict sense, if its statistics are not affected by a shift ε of the time origin, which means that $x(t)$ and $x(t + \varepsilon)$ have the same statistics. The power spectrum $\Phi(\omega)$ of a stationary random process $x(t)$ is the Fourier transform of its autocorrelation function.

To prove this concept, let us consider a random $x(t)$. The total energy content of this function is represented by the following integral:

$$\int_{-\tau}^{\tau} x^2(t)\, dt \tag{10.156}$$

Since power is defined as energy per unit of time, the average power P_{av} over the interval $-\tau \le t \le \tau$ is as follows:

$$P_{av} = \frac{1}{2\tau} \int_{-\tau}^{\tau} x^2(t)\, dt \tag{10.157}$$

At $t = 0$, the autocorrelation function given by Eq. (10.155) yields

$$R_{xx}(0) = \lim_{\tau \to \infty} \frac{1}{2\tau} \int_{-\tau}^{\tau} x^2(\rho)\, d\rho \tag{10.158}$$

Equation (10.158) shows that the autocorrelation function can be related to the power of $x(t)$, because the limit of Eq. (10.157) as $\tau \to \infty$ is identical to Eq, (10.158).

By using Eq. (9.20) of Section 9.2.2, the Fourier transform $X(\omega)$ of $x(t)$ may be written

$$X(\omega) = \int_{-\infty}^{\infty} x(t)\, e^{-j\omega t}\, dt \tag{10.159}$$

At a frequency $\omega = \omega_0$, Eq. (10.159) yields

$$X(\omega_0) = A \cos \omega_0 + B \sin \omega_0$$
$$= \sqrt{A^2 + B^2}\, e^{-j[\text{arc tan}(B/A)\, \tan \omega_0]} \tag{10.160}$$

where A and B are the parts of $x(t)$ that travel with frequency ω_0 in the cosine and sine modes, respectively.

From Eq. (10.156), we note that the amount of energy that travels at the frequency ω_0 is simply $A^2 + B^2$. We can verify this result by multiplying Eq. (10.160) by its conjugate. As $\tau \to \infty$, the average power at

ω_0 is

$$P_{av} = \frac{A^2 + B^2}{2\tau} \tag{10.161}$$

By using Eq. (10.161), we can write the average power $\Phi(\omega)$ as a function of the frequency ω in the following way:

$$\Phi(\omega) = \lim_{\tau \to \infty} \frac{1}{2\pi} X(\omega) X^*(\omega) \tag{10.162}$$

where the asterisk $*$ is used to denote the *complex conjugate*. The quantity $\Phi(\omega)$ is known as the *power spectrum* of $x(t)$.

We consider now the autocorrelation function of $x(t)$ that is given by Eq. (10.155), and we write its Fourier traneform as follows:

$$\int_{-\infty}^{\infty} R_{xx}(t) e^{-j\omega t} dt = \lim_{\tau \to \infty} \frac{1}{2\tau} \int_{-\tau}^{\tau} x(\rho) \int_{-\infty}^{\infty} e^{-j\omega t} x(t + \rho) d\rho \, dt \tag{10.163}$$

We change now the variables on the right-hand side of Eq. (10.163) so that $u = t + \rho$ and $du = dt$. On this basis, Eq. (10.163) yields

$$\int_{-\infty}^{\infty} R_{xx}(t) e^{-j\omega t} dt = \lim_{\tau \to \infty} \frac{1}{2\tau} \int_{-\tau}^{\tau} x(\rho) e^{-j\omega \rho} d\rho \, X(\omega) \tag{10.164}$$

where

$$X(\omega) = \int_{-\infty}^{\infty} x(u) e^{-j\omega u} du \tag{10.165}$$

We may rewrite Eq. (10.164) in the following way:

$$R_{xx}(\omega) = \lim_{\tau \to \infty} \frac{1}{2\tau} X(\omega) X^*(\omega) \tag{10.166}$$

where

$$R_{xx}(\omega) = \int_{-\infty}^{\infty} R_{xx}(t) e^{-j\omega t} dt \tag{10.167}$$

$$X^*(\omega) = \int_{-\infty}^{\infty} x(\rho) e^{j\omega \rho} d\rho \tag{10.168}$$

and $X(\omega)$ is as shown by Eq. (10.165). We note here that Eqs. (10.166) and (10.162) are identical, which indicates that the Fourier transform of the

TABLE 10.1. Correlation of $x(t)$, $R_{xx}(t)$, and $\Phi_{xx}(\omega)$ for Various Functions $x(t)$

$x(t)$	$R_{xx}(t)$	$\Phi_{xx}(\omega)$
$Ax(t)$	$\|A\|^2 R_{xx}(t)$	$\|A\|^2 \Phi_{xx}(\omega)$
$\dfrac{dx(t)}{dt}$	$-\dfrac{d^2 R_{xx}(t)}{dt^2}$	$\omega^2 \Phi_{xx}(\omega)$
$\dfrac{d^n x(t)}{dt^n}$	$(-1)^n \dfrac{d^{2n} R_{xx}(t)}{dt^{2n}}$	$\omega^{2n} \Phi_{xx}(\omega)$
$x(t)\, e^{\pm j\omega_0 t}$	$R_{xx}(t)\, e^{\pm j\omega_0 t}$	$\Phi_{xx}(\omega \mp \omega_0)$

autocorrelation function $R_{xx}(t)$ is the power spectrum density $\Phi_{xx}(\omega)$ of the stationary random function $x(t)$.

The autocorrelation function is symmetrical only with respect to the vertical axis, and, on this basis, its Fourier transform $\Phi_{xx}(\omega)$ can be written

$$\Phi_{xx}(\omega) = \int_{-\infty}^{\infty} R_{xx}(t)\, e^{-j\omega t}\, dt$$

$$= \int_{-\infty}^{\infty} R_{xx}(t) \cos \omega t\, dt - j \int_{-\infty}^{\infty} R_{xx}(t) \sin \omega t\, dt \qquad (10.169)$$

$$= 2 \int_{0}^{\infty} R_{xx}(t) \cos \omega t\, dt$$

Note that, because of symmetry, we have

$$\int_{-\infty}^{\infty} R_{xx}(t) \sin \omega t\, dt = 0 \qquad (10.170)$$

Equation (10.169) shows that the power spectrum of a stationary random function $x(t)$ can be evaluated by using only a cosine transformation.

In a similar manner, we can prove that the *cross-power spectrum* $\Phi_{xy}(\omega)$ of two jointly stationary random processes $x(t)$ and $y(t)$ is the Fourier transform of their cross-correlation function $R_{xy}(t)$, where $R_{xy}(t)$ is given by Eq. (10.149). The result is

$$\Phi_{xy}(\omega) = \int_{-\infty}^{\infty} R_{xy}(t)\, e^{-j\omega t}\, dt \qquad (10.171)$$

Table 10.1 provides the correlation of $x(t)$, $R_{xx}(t)$, and $\Phi_{xx}(\omega)$ for various functions $x(t)$. Graphs of various autocorrelation functions and their respective Fourier transforms are shown in Fig. 10.6.

Example 10.4. Determine the power spectrum density $\Phi_{xx}(\omega)$ of a random process $x(t)$, if the autocorrelation function $R_{xx}(t)$ has the form $e^{-2\lambda|\rho|}$.

Solution. The power spectrum density can be obtained by taking the Fourier transform $\Phi_{xx}(\omega)$ of the autocorrelation function $R_{xx}(t)$. It yields

$$\Phi_{xx}(\omega) = \int_{-\infty}^{\infty} e^{-2\lambda|\rho|} e^{-j\omega\rho}\, d\rho$$

$$= \int_{-\infty}^{0} e^{2\lambda\rho} e^{-j\omega\rho}\, d\rho - \int_{0}^{\infty} e^{-2\lambda\rho} e^{-j\omega\rho}\, d\rho \qquad (10.172)$$

$$= \frac{4\lambda}{4\lambda^2 + \omega^2}$$

Example 10.5. Determine the autocorrelation function of a random process that has the power spectrum density shown in Fig. 10.7.

Solution. The autocorrelation function $R_{xx}(t)$ may be determined by taking the inverse Fourier transform of $\Phi_{xx}(\omega)$. Equation (9.21) of Section 9.2.2 may be used for this purpose. The result is

$$R_{xx}(t) = \frac{1}{2\pi} \int_{-\omega_0}^{\omega_0} \Phi_0\, e^{j\omega\rho}\, d\omega$$

$$= \frac{\omega_0 \Phi_0}{\pi} \cdot \frac{\sin \omega_0 \rho}{\omega_0 \rho} \qquad (10.173)$$

10.9 RESPONSE OF SYSTEMS SUBJECTED TO RANDOM EXCITATIONS

We extend the discussion here in order to incorporate the determination of the dynamic response of structural or mechanical systems that are subjected to random excitations. The required equations for such responses are first developed and then used for the solution of single-degree-of-freedom systems. Many structural and mechanical systems, as discussed in preceding chapters of the text, can be idealized as single-degree-of-freedom systems.

The development of the required equations may be initiated by considering the block diagram of a system, such as the one shown in Fig. 10.8. In this diagram, $H(\omega)$ represents the impulse response of the system, $x(t)$ is the random force function applied to the system, and $y(t)$ is the random

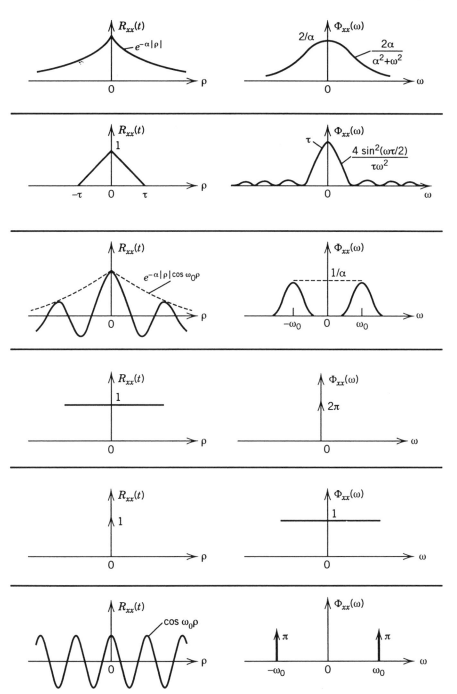

FIGURE 10.6. Graphs of various autocorrelation functions and their respective Fourier transforms.

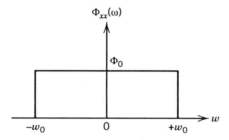

FIGURE 10.7. Power spectrum density of a random process.

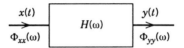

FIGURE 10.8. Block diagram of a system.

output function. The power spectrum of the input is $\Phi_{xx}(\omega)$, and $\Phi_{yy}(\omega)$ is the power spectrum of the output.

By using Eq. (9.96), the output $y(t)$ may be expressed as follows:

$$y(t) = \int_{-\infty}^{\infty} h(\rho)x(t - \rho)\, d\rho \qquad (10.174)$$

where $h(\rho)$ is the time domain of the response $H(\omega)$ of the system. The output $y(t)$ is a stationary random process if the input $x(t)$ is a random stationary force.

If we multiply the conjugate of Eq. (10.174) by $x(t_1)$, we have

$$x(t_1)y^*(t) = \int_{-\infty}^{\infty} x(t_1)x^*(t - \rho)h^*(\rho)\, d\rho \qquad (10.175)$$

The expected value of Eq. (10.175) is

$$E\{x(t_1)y^*(t)\} = \int_{-\infty}^{\infty} E\{x(t_1)x^*(t - \rho)h^*(\rho)\}\, d\rho \qquad (10.176)$$

By using the result obtained in Eq. (10.148), we find that, at $t = t_2$, Eq. (10.176) yields the following expression:

$$R_{xy}(t_1, t_2) = \int_{-\infty}^{\infty} R_{xx}(t_1, t_2 - \rho)h^*(\rho)\,d\rho$$

$$= R_{xx}(t_1, t_2) * h^*(t_2)$$

(10.177)

Equation (10.177) shows that the autocorrelation function $R_{xx}(t_1, t_2)$, considered as a function of time t_2, is *convolved* with the impulse response $h^*(t_2)$, while the variable t_1 is treated as a parameter. In addition, Eq. (10.177) yields the cross-correlation function R_{xy} of the input $x(t)$ and output $y(t)$.

The autocorrelation function of the output $y(t_1, t_2)$ can be obtained by multiplying Eq. (10.174) by $y^*(t_2)$ and proceeding in a similar manner. This procedure yields

$$R_{yy}(t_1, t_2) = \int_{-\infty}^{\infty} R_{xy}(t_1 - \rho)h(\rho)\,d\rho$$

$$= R_{yx}(t_1, t_2) * h(t_1)$$

(10.178)

Because the process is stationary, we can state that $R_{xx}(t_1, t_2 - \rho) = R_{xx}(\beta + \rho)$, where $\beta = t_1 - t_2$. Under this condition, Eq. (10.177) yields

$$R_{xy}(t_1, t_2) = \int_{-\infty}^{\infty} R_{xx}(\beta + \rho)h^*(\rho)\,d\rho$$

(10.179)

We note that Eq. (10.179) is dependent only on β, and it represents a convolution integral. Therefore,

$$R_{xy}(\beta) = R_{xx}(\beta) * h^*(\beta)$$

(10.180)

By following a similar approach, Eq. (10.178) yields

$$R_{yy}(\beta) = R_{yx}(\beta) * h(\beta)$$

(10.181)

By using Eqs. (10.150), (10.180), and (10.181), we can write the following expression:

$$R_{yy}(\beta) = R_{xx}(\beta) * h^*(-\beta) * h(\beta)$$

(10.182)

It was shown earlier that the power spectrum density of a function is the Fourier transform of the autocorrelation function. By using this finding, the following equation is written:

$$\Phi_{yy}(\omega) = \int_{-\infty}^{\infty} R_{yy}(\beta) e^{-j\omega\beta} d\beta \tag{10.183}$$

Therefore, Eqs. (10.180) and (10.181) may be used to write the following equations:

$$\Phi_{xy}(\omega) = \Phi_{xx}(\omega)H^*(\omega) \tag{10.184}$$

$$\Phi_{yy}(\omega) = \Phi_{yx}(\omega)H(\omega) \tag{10.185}$$

The frequency domain of Eq. (10.182) is

$$\Phi_{yy}(\omega) = \Phi_{xx}(\omega)H^*(-\omega)H(\omega) \tag{10.186}$$

or

$$\Phi_{yy}(\omega) = \Phi_{xx}(\omega)|H(\omega)|^2 \tag{10.187}$$

By using the definition given by Eq. (10.152) and Eq. (10.187), we can write the expected value of the input $x(t)$ as follows:

$$E\{|x(t)|^2\} = R_{xx}(0) = \frac{1}{2\pi} \int_{-\infty}^{\infty} \Phi_{xx}(\omega)|H(\omega)|^2 d\omega \tag{10.188}$$

In a similar manner, the expected value of the output $y(t)$ may be written

$$E\{|y(t)|^2\} = R_{yy}(0) = \frac{1}{2\pi} \int_{-\infty}^{\infty} \Phi_{yy}(\omega)|H(\omega)|^2 d\omega$$

$$= \frac{1}{2\pi} \int_{-\infty}^{\infty} \Phi(\omega)|H(\omega)|^2 d\omega \tag{10.189}$$

Equation (10.189) may be used to determine the expected value of the output of a linear system subjected to a random excitation $x(t)$. The procedure is illustrated by the following example.

Example 10.6. Determine the expected value $E\{|y(t)|^2\}$ of the output of the one-degree-of-freedom spring–mass system in Fig. 9.3, which is excited by the random dynamic force $F(t) = F_0(t)f(t) = F_0 \sin \omega_f t$.

Solution. We can determine the autocorrelation function of this random process by using Eq. (10.154). Application of this equation yields

$$R_{ff}(t) = \lim_{\tau \to \infty} \frac{1}{2\pi} \int_{-\tau}^{\tau} F_0^2 \sin \omega_f \rho \sin \omega_f(t + \rho)\, d\rho$$

$$= \frac{F_0^2}{2} \cos \omega_f \rho \tag{10.190}$$

From Fig. 10.6, we note that the power spectrum density $\Phi_{ff}(\omega)$ may be written

$$\Phi_{ff}(\omega) = \frac{\pi F_0^2}{2} \delta(\omega - \omega_f) \tag{10.191}$$

where $\delta(\omega)$ is the Dirac delta function in the frequency domain.

By using Eqs. (9.60) and (9.61), the transfer function $H(s)$ of the single-degree spring–mass system may be written as follows:

$$H(s) = \frac{1}{mZ(s)} \tag{10.192}$$

where

$$Z(s) = s^2 + 2\omega\zeta s + \omega^2 \tag{10.193}$$

With $s = j\omega_f$, Eq. (10.192) yields

$$H(j\omega_f) = \frac{1}{m(\omega^2 - \omega_f^2 + 2j\omega\omega_f\zeta)} \tag{10.194}$$

Equation (10.194) yields the impulse response of the spring–mass system.

By considering the right-hand side of Eq. (10.194) and multiplying the numerator and denominator by the conjugate of the denominator, we obtain the following expression for the absolute value of $|H(\omega_f)|$:

$$|H(\omega_f)| = \frac{1}{m[(\omega^2 - \omega_f^2)^2 + (2\omega\omega_f\zeta)^2]^{1/2}} \tag{10.195}$$

By using Eqs. (10.189), (10.191), and (10.195), the expression for the expected value $E\{|y(t)|^2\}$ of the output $y(t)$ may be written

$$E\{|y(t)|^2\} = \frac{F_0^2}{4m^2} \int_{-\infty}^{\infty} \frac{\delta(\omega - \omega_f)\, d\omega_f}{(\omega^2 - \omega_f^2)^2 + (2\omega\omega_f\zeta)^2} \tag{10.196}$$

The integral in Eq. (10.196) may be evaluated by using the following property of the delta function:

$$\int_{-\infty}^{\infty} \delta(\omega - \omega_f) g(\omega_f) \, d\omega_f = g(\omega) \tag{10.197}$$

On this basis, Eq. (10.196) yields

$$E\{|y(t)|^2\} = \frac{F_0^2}{4} \cdot \frac{1}{4m^2\omega^4\zeta^2}$$

$$= \frac{F_0^2}{16m^2\omega^4\zeta^2} \tag{10.198}$$

Example 10.7. For the single-degree spring–mass system in Fig. 9.3, determine the expected value of the output $y(t)$ when the spring–mass system is excited by a white noise.

Solution. Here, a white noise is defined as any random process whose power spectrum is constant. Under this condition, the power spectrum is independent of the frequency, and, consequently, $\Phi_{ff}(\omega) = \Phi_{ff} = $ constant. Remembering that the autocorrelation function $R_{ff}(t)$ is the inverse Fourier transform of the power spectrum function, we find

$$R_{ff}(t) = \int_{-\infty}^{\infty} \Phi_{ff} e^{j\omega\rho} \, d\omega$$

$$= \Phi_{ff} \delta(\rho) \tag{10.199}$$

Equation (10.199) shows that the autocorrelation function for white noise is an impulse, which indicates that correlation exists only when $\rho = 0$. This observation suggests that, if the single-degree spring–mass system is excited by a white noise, then Eq. (10.189) can be used to evaluate the expected value $E\{|y(t)|^2\}$ of the output $y(t)$. On this basis, we have

$$E\{|y(t)|^2\} = \frac{1}{2\pi} \int_{-\infty}^{\infty} \Phi_{ff} |H(\omega_f)|^2 \, d\omega_f \tag{10.200}$$

By using Eqs. (10.195) and (10.200), we find

$$E\{|y(t)|^2\} = \frac{\Phi_{ff}}{2\pi m^2} \int_{-\infty}^{\infty} \frac{d\omega_f}{(\omega^2 - \omega_f^2)^2 + (2\omega\omega_f\zeta)^2} \tag{10.201}$$

The integral in Eq. (10.201) may be evaluated by using the method of residues. Thus, if the integral in Eq. (10.201) is treated as a function of the complex variable z, which means that the real variable ω_f is replaced by the complex variable z, the result is

$$E\{|y(t)|^2\} = \frac{\Phi_{ff}}{2\pi m^2} \int_{-\infty}^{\infty} \frac{dz}{(\omega^2 - z^2)^2 + (2\omega z \zeta)^2} \qquad (10.202)$$

The evaluation of the integral in Eq. (10.202) may be carried out by using the following equation:

$$E\{|y(t)|^2\} = 2\pi j \sum [\text{of residues}] \qquad (10.203)$$

The integrand in Eq. (10.202) has two simple poles in the upper half-plane:

$$z_1 = \omega\sqrt{1 - \zeta^2} - j\zeta\omega \qquad (10.204)$$
$$z_2 = -\omega\sqrt{1 - \zeta^2} + j\zeta\omega \qquad (10.205)$$

By using Eqs. (10.203), (10.204), and (10.205) and proceeding in a manner similar to the one used for partial fractions, Eq. (10.202) may be written as follows:

$$E\{|y(t)|^2\} = \frac{j\Phi_{ff}}{m^2}[(z - z_1)f(z)|_{z = z_1} + (z - z_2)f(z)|_{z = z_2}] \qquad (10.206)$$

where

$$f(z) = \frac{1}{(\omega^2 - z^2)^2 + (2\omega z \zeta)^2} \qquad (10.207)$$

Evaluation of Eq. (10.206) yields

$$E\{|y(t)|^2\} = \frac{\Phi_{ff}}{2\zeta\omega^2 m^2} \qquad (10.208)$$

or

$$E\{|y(t)|^2\} = \frac{\omega\Phi_{ff}}{2\zeta k^2} \qquad (10.209)$$

Equation (10.209) provides the expected value of the output $y(t)$, and it is directly proportional to the natural frequency of vibration of the spring–mass system and inversely proportional to the damping ratio ζ. In order to design a structural or mechanical system that is idealized as a single-degree spring–mass system and subjected to a given input with power spectrum density function $\Phi_{ff}(\omega_f)$, the quantities ω and ζ of the system may be evaluated so that optimum performance is obtained.

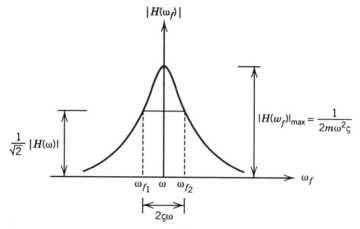

FIGURE 10.9. Plot of $|H(\omega_f)|$ versus ω_f.

As an illustration, let us consider again the one-degree spring–mass system in Fig. 9.3. The absolute value of the impulse response $|H(\omega_f)|$ is given by Eq. (10.195). If this equation is plotted for various values of ω_f, the shape of its plot would be as shown in Fig. 10.9. The peak value of $|H(\omega_f)|$, denoted as $|H(\omega_f)|_{\max}$, occurs at the resonant frequency $\omega = \omega_f$ and has the following value:

$$|H(\omega_f)|_{\max} = \frac{1}{2m\omega^2\zeta}$$

$$= \frac{1}{2k\zeta} \tag{10.210}$$

The frequencies $\omega = \omega_{f_1}$ and $\omega = \omega_{f_2}$, which satisfy the equation

$$|H(\omega_{f_1})|^2 = |H(\omega_{f_2})|^2 = \tfrac{1}{2}|H(\omega)|^2 \tag{10.211}$$

define the location of the so-called half-power points in Fig. 10.9. The difference $\omega_{f_2} - \omega_{f_1}$ in the same figure is the bandwidth of the complex frequency response function $|H(\omega_f)|$ of the system. In the case of viscous damping, this bandwidth is equal to $2\omega\zeta$.

We assume now that the spring–mass system is lightly damped, and that the power spectrum density function $\Phi_{ff}(\omega_f)$ of the exciting force is as shown in Fig. 10.10. In the same figure, the frequency response curve $|H(\omega_f)|$ of the system is also shown. This figure shows that the contribution of the power spectrum to the response of the system is large in the neighborhood of the natural frequency ω of the system and very small in the region outside

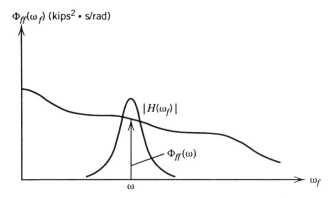

FIGURE 10.10. Plots of $\Phi_{ff}(\omega_f)$ and $|H(\omega_f)|$.

this frequency. With this in mind, we can modify Eq. (10.189) as shown below without introducing appreciable error in the final results. This modification is

$$E\{|y(t)|^2\} = \frac{\Phi_{ff}(\omega)}{2\pi m^2} \int_{-\infty}^{\infty} |H(\omega_f)|^2 \, d\omega_f \qquad (10.212)$$

The integral in Eq. (10.212) is similar to the one in Eq. (10.201), and, consequently, Eq. (10.212) yields

$$E\{|y(t)|^2\} = \frac{\omega \Phi_{ff}(\omega)}{2\zeta k^2} \qquad (10.213)$$

The above equation shows that the power spectrum can be expressed as a function of the natural frequency of the spring–mass system. Consequently, for a given random excitation, it is possible to design a system for a desired response by varying its natural frequency ω.

Assume that $\omega = 6.45$ rps, $k = 4.57$ kips/in., and $\zeta = 0.1$, as shown in Fig. 10.11. If $\Phi_{ff}(\omega) = 0.60$ kip$^2 \cdot$ s/rad, the expected value $E\{|y(t)|^2\}$ of the displacement $y(t)$ can be determined by using Eq. (10.213). This equation yields

$$E\{|y(t)|^2\} = \frac{(6.45)(0.60)}{(2)(0.1)(4.57)^2} = 0.93 \text{ in.}^2 \qquad (10.214)$$

The displacement value in Eq. (10.214) represents the mean square value of the expected displacement of the mass m of the spring–mass system.

The significance of the result given by Eq. (10.214) becomes more meaningful for practical applications, if it is expressed in terms of a probability density function. Such a function may be used to determine the probability

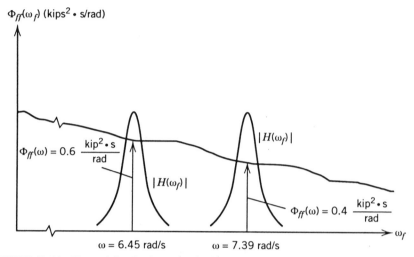

FIGURE 10.11. Plots of $\Phi_{ff}(\omega_f)$ and $|H(\omega_f)|$ with assigned values for ω and $\Phi_{ff}(\omega)$.

that an event, such as the one given by Eq. (10.214), can actually happen. The following section discusses this particular point.

10.10   UTILIZATION OF PROBABILITY IN DESIGN PROCESSES

Random excitations, such as the ones produced by an earthquake or a blast, are approximately of normal distribution. Thus, if a linear system is subjected to such excitations, its response $y(t)$ would also be of normal distribution. We use here a special case of the central limit theorem [80] to discuss the normal probability function. For example, if $X_1, X_2, X_3, \ldots, X_n$ are n independent random variables, and if each random variable is distributed in accordance with a given probability density function, then the probability density $p(y)$ of the composite random variable Y, where

$$Y = X_1 + X_2 + X_3 + \cdots + X_n \tag{10.215}$$

will approach the Gaussian or normal density function if n is large. On this basis, the probability density function $p(y)$ will have the following form:

$$p(y) = \frac{1}{\sqrt{2\pi\sigma^2}} e^{-(y - a)^2/2\sigma^2} \tag{10.216}$$

In Eq. (10.216), we have $a = \bar{Y}$, where $\bar{Y}$ is the mean value of the process.

If we use earthquake as an example, we note that earthquake is a physical process that consists of a summation of a large number of individual similar events. The response $y(x)$ produced by such random excitation is also Gaussian, with zero mean and variance σ^2 that is as follows:

$$\sigma^2 = E\{|y(t)|^2\} \tag{10.217}$$

For the case represented by Eq. (10.214), we have

$$\sigma^2 = E\{|y(t)|^2\} = 0.93 \text{ in.} \tag{10.218}$$

$$a = \bar{Y} = 0 \tag{10.219}$$

Therefore, by using Eq. (10.216), the normal distribution function $p(y)$ is

$$p(y) = \frac{1}{\sqrt{2\pi(0.93)}} e^{-y^2/2(0.93)} \tag{10.220}$$

Assume that the displacement $y(t)$ of the mass m of the single-degree spring–mass system discussed in the preceding section is not permitted to exceed a certain value, say, 2.0 in. The probability that the absolute value of $y(t)$ exceeds 2.0 in. can be determined by using Eq. (10.139). Application of this equation yields

$$P(|y(t)| > 2) = \int_{-\infty}^{-2} p(y) \, dy + \int_{2}^{\infty} p(y) \, dy \tag{10.221}$$

where $p(y)$ is as shown by Eq. (10.220).

By substituting Eq. (10.220) into Eq. (10.221) and making note of the symmetry involved in Eq. (10.221), we find

$$P(|y(t)| > 2) = \frac{2}{\sqrt{2\pi(0.93)}} \int_{2}^{\infty} e^{-y^2/2(0.93)} \, dy \tag{10.222}$$

We now make the following transformations:

$$\frac{t^2}{2} = \frac{y^2}{2(0.93)}$$

$$t = \frac{y}{\sqrt{0.93}} \tag{10.223}$$

$$dt = \frac{dy}{\sqrt{0.93}} \tag{10.224}$$

On this basis, Eq. (10.222) may be written as follows:

$$P(|y(t)| > 2) = \frac{2}{2\pi} \int_{t = 2/\sqrt{0.93}}^{\infty} e^{-t^2/2} \, dt \tag{10.225}$$

The evaluation of the integral in Eq. (10.225) may be obtained by using mathematical tables; see, for example, ref. 84. Or $P(|y(t)| > 2)$ of Eq. (10.225) may be evaluated by using the following expression:

$$P(|y(t)| > 2) = \frac{2}{2\pi} e^{-t^2} \left(\frac{\sqrt{2}}{2t} - \frac{2\sqrt{2}}{2^2 t^3} + \frac{(1)(3)(4\sqrt{2})}{2^3 t^7} - \cdots \right) \tag{10.226}$$

Equation (10.226) is obtained by using repetitive integration by parts. By using either mathematical tables or Eq. (10.226), the result is

$$P(|y(t)| > 2) = 0.0384 \tag{10.227}$$

Equation (10.227) provides the probability that the displacement of 2.0 in. will be exceeded.

If the probability given by Eq. (10.227) is considered to be large, then a smaller one may be obtained by redesigning the spring–mass system. If the parameters are now changed so that $\omega = 7.39$ rps, $k = 6.0$ kips/in., and $\Phi_{ff} = 0.4$ (see also Fig. 10.11 and proceed as in the preceding section), we find that Eq. (10.213) yields

$$E\{|y(t)|^2\} = \frac{(7.39)(0.4)}{(2)(0.1)(6.0)^2} = 0.41 \text{ in.}^2 \tag{10.228}$$

On this basis, Eq. (10.221) yields

$$P(|y(t)| > 2) = 0.00171 \tag{10.229}$$

which is a much smaller probability than the one given by Eq. (10.227).

Remember that the single-degree spring–mass system used in the above examples can be the product of a large structure, such as a single-story building, or the idealization of structural and mechanical components subjected to random dynamic excitations. By changing the sizes of such structural and mechanical components, or changing other parameters that affect their performance, we can obtain different values of ω and k and, consequently, different probabilities that certain events will happen—that is, exceed an assigned displacement.

The above procedure can also be used for systems with two or more degrees of freedom, which can be accomplished by determining the power spectrum density functions of the excitations and the frequency response of the system. On this basis, the expected values of output displacements can be derived in a manner similar to the one used to derive Eq. (10.189). The derivation of these equations is beyond the scope of the material included in this chapter.

10.11 ACOUSTIC SPECTRA ANALYSIS FOR CONCRETE MATERIAL RESPONSE

Conventional approaches used to determine concrete material behavior and properties are usually based on experimental observations, mostly static, and the results obtained are used as a guide for the design of concrete structures. Although numerous investigations are performed on concrete, still more needs to be learned about it, particularly when it is under the action of deterministic and nondeterministic dynamic excitations, by including fatigue. During the latter part of the 1960s and early 1970s, a strong interest developed in using acoustic emission techniques to determine crack propagation, imperfections, and other similar material characteristics for metals and metal structures. Research on this subject continues today and useful information has been obtained for various kinds of materials and their applications.

In this section, we discuss a methodology developed by the author [81, 85] during the late 1960s and early 1970s, which uses acoustic spectra analysis to determine material and strength characteristics of concrete mixtures. The methodology, however, is general and can be used equally well for other types of materials. This methodology involved two phases of work. The first phase included the development of an experimental setup and recording system that is capable of recording the acoustic, or sound, emissions from cracks in the concrete material as it was gradually loaded to failure. The second phase involved the development of a methodology that relates the acoustic emissions to the general material behavior of concrete mixtures, as well as their relationship to specific concrete properties. This undertaking was sponsored by a 3-year grant by the Ohio Department of Transportation, with partial support by the U.S. Department of Highways.

10.11.1 Experimental Setup and Recording System

The basic experimental setup and recording system to record the acoustic emissions consists of (a) a load application system, (b) a microphone transducer system, (c) differential amplifiers, (d) a magnetic tape amplifier system, and (e) oscilloscopes. This system is put together in such a way that it can operate within the experimental specifications, and each instrument is calibrated to operate with maximum efficiency when it is used in cascade

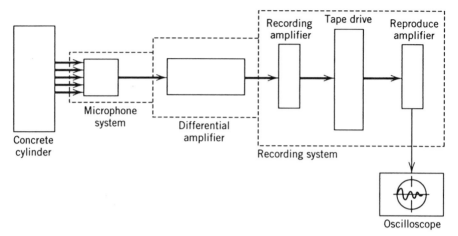

FIGURE 10.12. Block diagram of recording system.

with the other instruments. References 81 and 85 provide complete details, with illustrations, of the experimental setup and recording system.

The tested specimens are standard concrete cylinders 12 in. long and 6 in. in diameter, and they represent specific concrete mixtures that are prepared and cured in accordance with the ASTM specifications. The moisture room, where the curing of the concrete cylinders took place, maintained a nearly constant temperature of 71°F (21.7°C) and 100% relative humidity.

A block diagram of the recording system is shown in Fig. 10.12. This system is used to record the waveforms of the acoustic emissions that are produced by the concrete cylinders as they are loaded gradually to failure. Each instrument of the recording system was calibrated and matched, so that it operates with maximum efficiency when used in conjunction with other instruments. Similar setups can easily be put together today to perform the same operations with excellent accuracy.

10.11.2 Recording of the Acoustic Emissions

If a concrete cylinder made up of a certain concrete mixture is loaded gradually to failure, the energy stored in the material is represented by the area under the stress–strain curve. The characteristics of the mathematical curve by which this energy is represented, as well as the amount of energy stored in the material, are commonly used to define what is known as material properties and strength of concrete.

At failure, the big bulk of the stored energy is suddenly released. However, energy releases are occurring repeatedly from the very early stages of loading because of microfailures and secondary failure buildup before major failures occur. Such energy releases are associated with sound emissions of

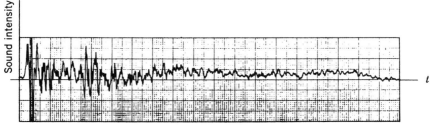

Typical experimental acoustic output

FIGURE 10.13. Strip recorder plot of a typical experimental acoustic emission output. Cylinder number 13C-14-014N.

various intensities, which are believed to depend on the amount of energy released and, consequently, on the material properties and strength characteristics of the mixture used to prepare the concrete cylinders. These are the acoustic emissions that are recorded by the recording system in Fig. 10.12. The units used to designate acoustic power are watts.

During the time the concrete cylinder is loaded to failure by the application of a gradually increasing axial compressive force, the intensities of the acoustic emissions are picked up as input by the microphone transducer system. The output of the microphone system is volts, and, therefore, the volt changes of this system represent the intensity variations of the acoustic emissions. This signal is picked up by the differential amplifier. The output current of the differential amplifier is the input of the recording amplifier. The output voltage of the recording amplifier is recorded on magnetic tape at the speed of 60 in./s. The waveforms of this signal are also displayed on the screen of an oscilloscope for visual observations. It can be played back at lower speeds through the reproduce amplifier. This output, which is in volts, becomes the input of the strip recorder. The output of the strip recorder can be displayed on paper at various speeds for visual observation and analysis.

A typical strip recorder output for a cylinder during ultimate failure time is shown in Fig. 10.13. The acoustic emissions represented by this signal are of random type, or nondeterministic, and, thus, mathematical theories based on random analysis, as discussed in preceding sections, may be used for their analysis. Based on observations and analysis, provisions were made so that the effects of environmental and instrumentation noise were negligible. This is clearly illustrated by the quality of the acoustic emission output shown in Fig. 10.13.

The results that follow are based on the study of the acoustic emissions recorded from energy releases near the ultimate and at the ultimate failure of the concrete cylinder. This should represent well over 90% of the total energy released by the material. The instrumentation, however, can be modified to take into consideration all energy releases starting from zero axial loading.

10.11.3   Method of Analysis of the Acoustic Emissions

Three mathematical methods of analysis were tested to determine which one gives the most satisfactory results for this problem. The first one is based on transform calculus and time series analysis. The second one is based on fast Fourier transforms, and the third one uses the optics theory of spectrum analysis. Preliminary results were obtained with all three methods, and the results were compared. All three methods proved successful, but the transform calculus approach appeared to be in better agreement with the instrumentation set up that was available to the researcher.

The mathematical model was solved by preparing FORTRAN computer programs that were used as an interface between the digitized acoustic emission output of the recording system and the digital computer in use. Four computer programs were prepared and tested, and the results were compared. The one that proved to be most convenient for this type of work is the one based on transform calculus and time series analysis. It incorporates four subroutines. The first one carries out arithmetical operations and transformations on the series. The second subroutine generates a one-sided autocorrelation function. The third subroutine is a Fourier transformation whose purpose is to perform a one-sided Fourier sine or cosine transformation of the data series, and the fourth subroutine is a filter that eliminates the contribution from aliasing the folding.

As was pointed out earlier, the acoustic emissions recorded from tested cylinders representing various types of concrete mixtures are actually releases of energy stored in the cylinders by the application of the axial compressive loading. These acoustic emissions should represent the nonlinear and nonisotropic characteristics of the material, as well as its dynamic characteristics, because these emissions are the result of smaller and larger burst-type failures. There are various ways in which such acoustic emissions can be used to determine important material properties and characteristics of concrete, and a brief discussion of this subject is given later. The purpose here is limited to the development of a procedure that can be used to evaluate the overall material performance and quality of concrete mixtures, by investigating the acoustic power spectra of the recorded intensities of the acoustic emissions from gradually loaded concrete cylinders.

The acoustic emissions from the loaded concrete cylinders are recorded on analog tapes. The analog tapes are then converted into digital tapes. A computer program, prepared as discussed above, is used to determine the acoustic power spectrum of the recorded acoustic emissions. For example, by following this procedure, the experimental acoustic output, the autocorrelation function, and the acoustic power spectrum for concrete cylinder Number 13C-14-014N are shown in Fig. 10.14. The actual computer output of the acoustic power spectrum for this cylinder, showing in detail the random

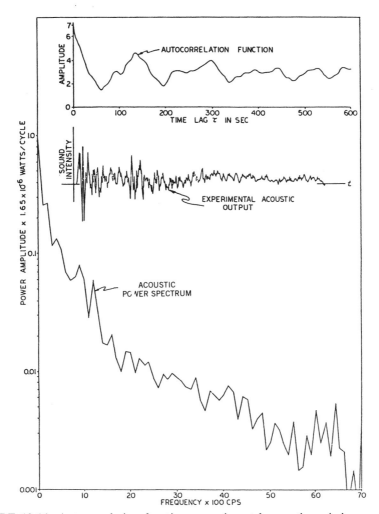

FIGURE 10.14. Autocorrelation function, experimental acoustic emissions output, and acoustic power spectrum for cylinder 13C-14-014N.

aspects of the acoustic power amplitude, is shown in Fig. 10.15. Hundreds of cylinders are tested in this manner, and the results shown in Figs. 10.13 through 10.15 are typical and very consistent.

10.11.4 Histogram Analysis of Acoustic Power Spectra

The investigation of the overall material performance and quality of concrete mixtures, as was stated in the preceding subsection, can be facilitated by introducing a histogram analysis of the acoustic power spectrum shown in Fig. 10.15. On this basis, the acoustic power spectra, such as the one in Fig.

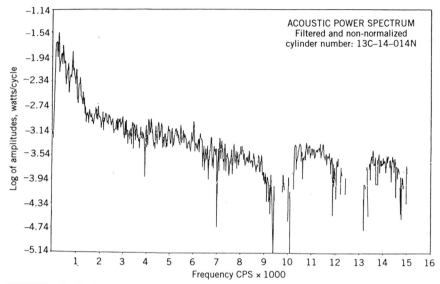

FIGURE 10.15. Actual computer output of acoustic power spectrum of cylinder 13C-14-014N.

10.15, are reduced into bar charts of judiciously selected frequency bandwidth, and the power content within each bandwidth is determined by introducing a summation subroutine in the computer program. For example, by selecting a frequency bandwidth of 1000 cycles, the histogram of the acoustic power spectrum in Fig. 10.15 is shown in Fig. 10.16. Each bar in this figure represents the total power content in each band of 1000 frequencies.

A more convenient way to plot the histograms, particularly when a large number of concrete cylinders and batches of various concrete mixtures are compared, is to use histograms of energy accumulation. For example, for cylinder 13C-14-014N in Table 10.2, one can start with the accumulation of 0.006404 of the bandwidth between 14000 and 14986, which is the first point of coordinates 15000 and 0.006404 in Fig. 10.17. The accumulation 0.006404 + 0.005386 = 0.011790, which is the sum for the bandwidths 14000 to 14986 and 13000 to 13986, is the second point of the curve in the same figure. Proceeding in the same manner, we obtain the complete histogram of energy accumulation as shown. The experimental acoustic output and the autocorrelation function are also shown in the same figure.

The histograms of the three cylinders in Table 10.2, which are made out of the same batch of a concrete mixture, are shown in Fig. 10.18. The ultimate compressive load of each cylinder and the average ultimate static stress are also shown in the same figure.

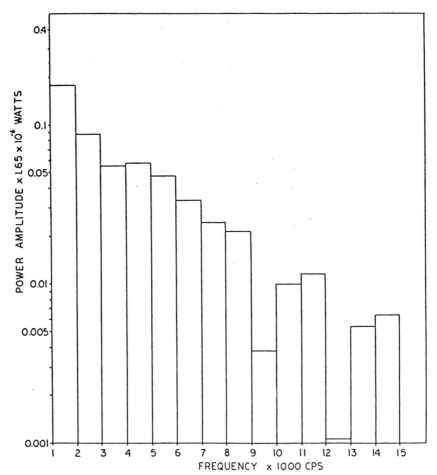

FIGURE 10.16. Histogram of nonnormalized acoustic power spectrum for cylinder 13C-14-014N.

10.11.5 Discussion of the Results

In this subsection, the acoustic experimentation and the method of analysis described in the preceding subsection are applied to specific concrete mixtures. They are (a) Class C concrete mixtures as described in the 1967 Ohio State Department of Transportation specifications, (b) poured-in-place concrete mixtures that are used in the construction of public buildings in the Akron area, and (c) a small number of commercial concrete blocks. Most of the work, however, was concentrated on the Class C mixtures, where certain elements of variation were allowed regarding their mixture, in order to obtain some appreciable variation in their corresponding ultimate static strength.

TABLE 10.2. Frequency Bandwidths for the Three Cylinders of Batch 13

Frequency Bandwidth (cycles)	Cylinder Number		
	13A-15-015N	13B-16-016N	13C-14-014N
0–986	1.025561	1.015256	0.763277
1000–1986	0.312859	0.297045	0.177551
2000–2986	0.119303	0.116368	0.086478
3000–3986	0.099446	0.069696	0.055307
4000–4986	0.075836	0.055512	0.056262
5000–5986	0.053237	0.050575	0.047295
6000–6986	0.036770	0.040788	0.033478
7000–7986	0.030064	0.030095	0.024225
8000–8986	0.012022	0.014790	0.021406
9000–9986	0.011144	0.015356	0.003799
10000–10986	0.006577	0.005842	0.009939
11000–11986	0.008872	0.003926	0.011473
12000–12986	0.004157	0.008088	0.001055
13000–13986	0.006596	0.001426	0.005386
14000–14986	0.003309	0.006880	0.006404
15000–15986	0.005643	0.003183	0.000279

Remarks: Band power spectra with bandwidth of about 1000 cycles; nonnormalized spectra.

The histograms of energy accumulation for eight batches of Class C mixture, are shown in Fig. 10.19. Each batch involved three cylinders. The average histogram of the three cylinders represents the histogram of energy accumulation of each batch. Comparisons between the histograms of the cylinders in each batch were also made, and the correlation was very interesting. The comparisons show that concrete cylinders made out of the same batch do not always accumulate equal amounts of acoustic energy. In fact, it was observed that, in some cases, the variation of acoustic energy was rather large. This was more apparent for the cylinders that came out of batches of lower ultimate strengths. All histograms, however, are characterized by a similar schematic pattern.

The histograms of an additional nine concrete batches are shown in Figs. 10.20 and 10.21. The ultimate static loads and ultimate average compressive static stresses shown in the figures represent the average of the three cylinders in each batch. Note that the variations of these values between batches are not very large.

The histograms shown in Fig. 10.22 are those of cylinders from public buildings that are prepared in the open space, during construction of the building, and under fair to bad weather conditions. However, the curing of the cylinders was performed in the moisture room under almost ideal conditions. It is interesting to note here that the cylinders prepared under bad weather conditions are the three ones in the figure which give histograms of

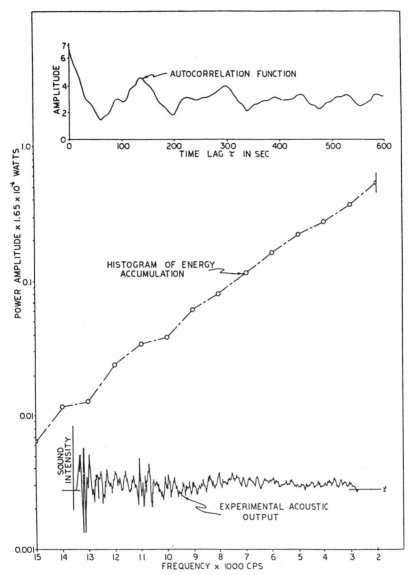

FIGURE 10.17. Autocorrelation function, histogram of energy accumulation, and experimental acoustic emission output for cylinder 13C-14-014N.

much lower energy accumulations. These variations are much greater than the ones observed when the acoustic results of the average compressive stresses between batches are compared.

The histograms of energy accumulation for three commercially available concrete blocks are shown in Fig. 10.23. The agreement here appears to be

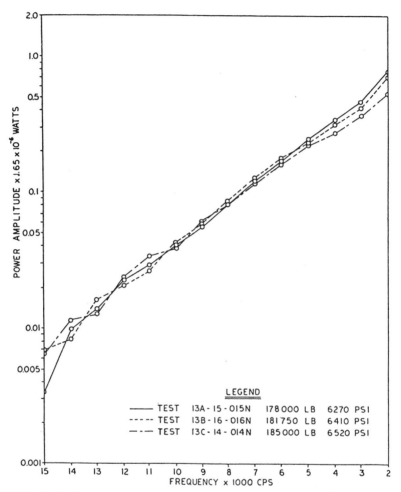

FIGURE 10.18. Histogram of energy accumulation of nonnormalized spectra for three cylinders of batch 13.

excellent. However, it cannot be conclusive because only these three speci-mens were tested. On the other hand, several hundred concrete cylinders that were tested and analyzed by this method all gave results similar to the ones in Figs. 10.19 through 10.21.

It should be stated here that the ultimate stress for the concrete blocks shown in Fig. 10.23 was calculated by dividing the ultimate compressive load shown in the figure by the gross cross-sectional area of the block. The ultimate compressive stress would be much higher if the net solid portion of the concrete block was used. This, however, should not affect appreciably

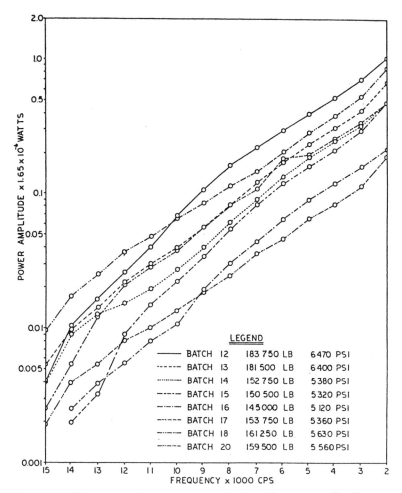

FIGURE 10.19. Histograms of energy accumulation of nonnormalized spectra of eight batches by batch average.

the comparisons made here, because the cavity portions of the concrete blocks are approximately equal.

The above observations, as well as additional observations made during experimentation and analysis, suggest that the energy releases from concrete in the form of acoustic emissions of various intensities represent material properties and strength characteristics of concrete. It was clearly observed that concrete material that satisfies given static strength criteria does not necessarily satisfy appropriate dynamic requirements for structural integrity within the specified static levels. This is an important observation, because many concrete structures are subjected to severe deterministic and nondeterministic dynamic loadings.

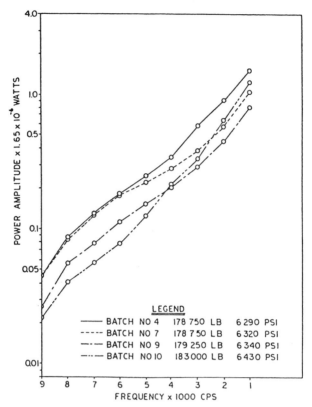

FIGURE 10.20. Histograms of energy accumulation of nonnormalized spectra for four batches by batch average.

During experimentation, it also was observed that poor quality concrete produced very low levels of acoustic energy. This statement is supported by the experimental results of more than 500 concrete cylinders.

The testing of about forty 7-day concrete cylinders by the acoustic emissions method revealed that, although better than 60% of static strength is attained, the dynamic behavior of these cylinders was very poor. This indicates that longer curing time is required for concrete to start developing corresponding levels of dynamic resistance.

It should be pointed out, however, that the results of this research were intended to mark the beginning of a series of problems that could be solved by this methodology. Preliminary research by the author has shown that some of the problems that can be examined by this method are as follows:

1. Prediction of 28-day concrete strength from the testing of 1- to 2-day concrete specimens.

2. Determination of the static and dynamic stress levels at which material

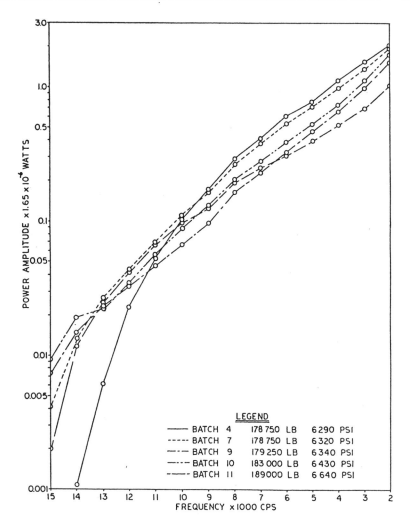

FIGURE 10.21. Histograms of energy accumulation of nonnormalized spectra for five batches by batch average.

deterioration can be tolerated, without endangering the structural integrity of concrete mixtures.

3. General evaluation of various concrete mixtures in terms of both static and dynamic characteristics and strengths.

4. Determination of fatigue properties of concrete.

5. Evaluation of the effects of aggregate constituency on concrete mixture quality and strength. This could also be used for the development of new concrete mixtures of specified characteristics for modern technology demands.

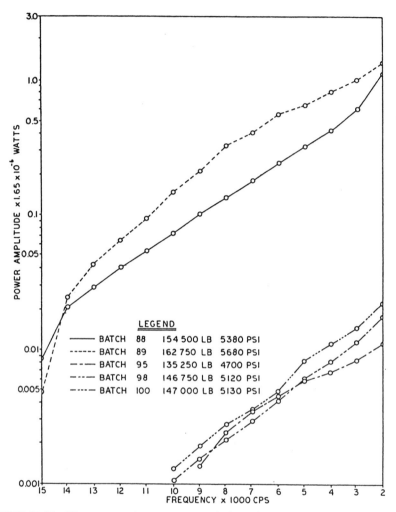

FIGURE 10.22. Histograms of energy accumulation of nonnormalized spectra under varying weather conditions by batch average.

6. Composite materials other than concrete can also be analyzed by acoustic spectra analysis.

PROBLEMS

10.1 By using the appropriate Euler equation, examine the boundary conditions given below and determine the value of a which makes the functional have an extremum.

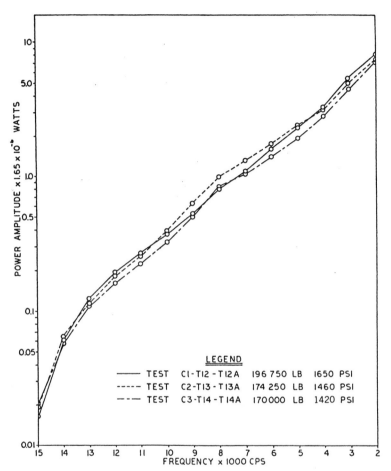

FIGURE 10.23. Histograms of energy accumulation of nonnormalized spectra for three concrete blocks.

$$\Omega(y(x)) = \int_0^1 (y^2 + x^2 y') \, dx$$

$$y(0) = 0 \qquad y(1) = a \qquad y(1) = 1$$

10.2 Which curves can make the functional given below have an extremum?

$$\Omega(y(x)) = \int_0^1 [(y')^2 + 12xy] \, dx$$

$$y(0) = 0 \qquad y(1) = 1$$

10.3 Find a curve that joins two given points A and B, so that when a particle starts at point A and moves along this curve, it reaches point B in the shortest time. Neglect friction and resistance of the medium.

10.4 For the functional given below, the boundary conditions are $y(0) = 1$, $y'(0) = 0$, $y(\pi/2) = 0$, and $y'(\pi/2) = -1$. Determine the extrema of this functional.

$$\Omega(y(x)) = \int_0^{\pi/2} [(y'')^2 - y^2 + x^2]\, dx$$

10.5 Determine the necessary conditions and the function $y(x)$ that cause the functional given below to have an extremum:

$$\Omega(y(x)) = \int_{-L}^{L} \left(\frac{\mu(y'')^2}{2} + \rho y \right) dx$$

where μ and ρ are constants. The boundary conditions are $y(-L) = 0$, $y'(-L) = 0$, $y(L) = 0$, and $y'(L) = 0$.

10.6 Find the extrema of the following functional:

$$\Omega(y(x), z(x)) = \int_0^{\pi/2} [(y')^2 + (z')^2 + 2yz]\, dx$$

The boundary conditions are $y(0) = 0$, $y(\pi/2) = 1$, $z(0) = 0$, and $z(\pi/2) = -1$.

10.7 If a functional is of the form

$$\Omega = \int_0^{x_1} \frac{\sqrt{1 - (y')^2}}{y}\, dx$$

where $y(0) = 0$ and $y_1 = x_1 - 5$, examine the extrema of this functional.

10.8 Determine the longitudinal free frequencies of vibration of a uniform cantilever beam. Use a direct method of variational calculus and assume a solution $u(x, t) = f(x)g(t)$ that satisfies the boundary conditions $u = 0$ at $x = 0$ and $u' = 0$ at $x = L$, where L is the length of the member. Neglect damping.

10.9 Repeat Problem 10.8 by solving the general differential equation of motion, and compare the results.

10.10 Determine the transverse free frequencies of vibration for a uniform simply supported beam of length L, by using a direct method of variational calculus and assuming that $y(x, t) = f(x)g(t)$ with boundary conditions $y = 0$ and $y'' = 0$ at $x = 0$ and $x = L$, respectively. Neglect damping.

10.11 Determine the differential equation of motion for the free vibration of a wedge-shaped cantilever beam. Neglect damping.

10.12 Determine the longitudinal free frequencies of vibration of a wedge-shaped cantilever beam by using a direct method of variational calculus and $u(x, t) = f(x)g(t)$. At $x = 0$, the depth of the beam is equal to $2b$, and it is equal to $2a$ at $x = L$, where L is the length of the member; the uniform width of the member is z. Note that $a > b$ and neglect damping.

10.13 Determine the transverse free frequencies of vibration of the wedge-shaped cantilever beam of Problem 10.12, by using a direct method of variational calculus. Assume that $y(x, t) = f(x)g(t)$ with boundary conditions $y = y' = 0$ at $x = 0$ and $y'' = y''' = 0$ at $x = L$.

10.14 The homogeneous elastic string in Fig. 10.3 is subjected to a harmonic force $q(x, t) = q \cos \omega_f t$, which is uniformly distributed along the length of the string. Determine its differential equation of motion by applying Hamilton's principle. Neglect damping.

10.15 Repeat Problem 10.14 with $q(x, t) = q \sin \omega_f t$.

10.16 By using principles of the calculus of variation, determine the undamped transverse free frequencies of vibration of a uniform steel rod with free–free ends and its corresponding mode shapes.

10.17 Determine the transverse free frequencies of vibration and the corresponding mode shapes of the steel beams in Fig. P5.17, by following the procedure discussed in Section 10.5.2. The stiffness $EI = 30 \times 10^6 \text{ kpi} \cdot \text{in.}^2$ and the weight w per unit length is 0.40 kip/ft.

10.18 If the periodic function $x(t)$, with period $T = 2\pi/\omega_0$, is expressed as a series of harmonically varying terms of the form

$$x(t) = a_0 + \sum_{1}^{\infty} (a_n \cos n\omega_0 t + b_n \sin n\omega_0 t)$$

where ω_0 is the fundamental frequency, determine the mean square value of $x(t)$.

10.19 For the autocorrelation functions $R_{xx}(t)$ given in Fig. 10.6, determine in each case the corresponding power spectra $\Phi_{xx}(\omega)$.

10.20 For the power spectra $\Phi_{xx}(\omega)$ given in Fig. 10.6, determine in each case the corresponding autocorrelation function $R_{xx}(t)$.

10.21 The expected value of the input for a general system is given by Eq. (10.188). By following the procedure used in Section 10.9 to derive Eq. (10.188), verify Eq. (10.189), which represents the expected value of the output.

10.22 By using the power spectra $\Phi_{xx}(\omega)$ derived in Problem 10.20, determine in each case the expected value $E\{|y(t)|^2\}$ of the output for a single-degree spring–mass system.

10.23 For the spring–mass system in Fig. 9.3, assume that $k = 4.57$ kips/in., $\omega = 6.45$ rps, $\zeta = 0.1$, and the power spectrum $\Phi_{xx}(\omega)$ of the exciting force $F(t)$ is $\Phi_{xx}(\omega) = 0.8$ kip$^2 \cdot$ s/rad. Determine the probability that the displacement of the mass of the system will not exceed 2.0 in.

10.24 Design the system in Problem 10.23 so that the probability to exceed the displacement of 2.0 in. is somewhere between 0.0002 and 0.0001.

10.25 Repeat Problem 10.23 for $\Phi_{xx}(\omega) = 0.4$ kip$^2 \cdot$ s/rad.

10.26 Repeat Problem 10.23 for a probability that the displacement of the mass of the spring–mass system will not exceed 1.0 in.

10.27 For the spring–mass system in Fig. 9.3, we have $k = 6.0$ kips/in., $\omega = 7.39$ rps, $\zeta = 0.1$, and the power spectrum $\Phi_{xx}(\omega)$ of the exciting force $F(t)$ is $\Phi_{xx}(\omega) = 0.5$ kip$^2 \cdot$ s/rad. Determine the probability that the displacement of the mass of the system will not exceed 2.0 in.

11 Dimensional and Model Analysis

11.1 INTRODUCTION

The purpose of this chapter is to discuss fundamental principles regarding the preparation of scale models that can be used to predict the response of a prototype in the required respect. In model analysis, a thorough understanding of the principles of dimensional analysis is essential, because it provides the information required to establish similarity between model and prototype and to construct a model that will interpret correctly the response of the prototype.

The discussion in this chapter is initiated by a brief review of dimensional analysis and then continues with the Pi theorem and the prediction equation. Preparation of scale models, distortion considerations, and applications to practical engineering problems are also included.

11.2 PRINCIPLES OF DIMENSIONAL ANALYSIS

The principle of dimensional analysis was first published by Dupré in 1869 [86], and it was further developed by E. Buckingham and Lord Rayleigh [87]. As a powerful analytical tool, its development is based on the idea that absolute numerical equality of quantities may exist only when the quantities are similar qualitatively. The ratio of the magnitudes of two like quantities is also independent of the units used in their measurement, provided that the same units are used for their evaluation. The results obtained from a dimensional analysis are only qualitative. Qualitative results and accurate prediction equations could be obtained if they are combined with experimental procedures. In engineering, an important use of dimensional analysis is its application to model design. This part requires thorough understanding of the significance of the general prediction equation, and the principles involved in deriving this equation.

For scientific or engineering purposes, combinations of measurable quantities such as force–length–time, where force, mass, length, and time are interrelated by Newton's second law of motion, are usually considered as sets of fundamental or basic quantities. The reason is that many other quantities, such as moment, velocity, and acceleration, can be expressed in

683

terms of these basic quantities. Other combinations of basic quantities can be used if a sufficient number of independent quantities is included in the combination. If, for example, mass, length, and time are required to express all the quantities involved in a phenomenon, a different set would, in general, require at least three basic quantities to replace mass, length, and time. In engineering, force, length, and time (F, L, and T) or mass, length, and time (M, L, and T), are usually used for convenience.

With reference to our standard measuring technique, it may be proved that any measurable quantity Q can be expressed dimensionally in the form

$$Q = F^{k_1} L^{k_1} T^{k_3} \tag{11.1}$$

The validity of this dimensional equation comes from the idea that absolute numerical equality of quantities can only exist if they are similar qualitatively, and that if the same units are used to evaluate two like quantities, the ratio of their magnitudes is independent of the units used. On this basis, the principle may be extended to obtain a relationship between any measurable phenomenon Q and the factors $Q_1, Q_2, Q_3, \ldots, Q_n$ influencing or causing this phenomenon. Such a relationship may be proved to be of the form

$$Q = C_Q Q_1^{k_1} Q_2^{k_2} Q_3^{k_3} \cdots Q_n^{k_n} \tag{11.2}$$

which agrees dimensionally with Eq. (11.1). The coefficient C_Q in Eq. (11.2) may be determined either experimentally or theoretically by solving the appropriate mathematical problem. On the other hand, some (or all) of the exponents $k_1, k_2, \ldots, k_n$ may, in many cases, be determined directly, while in other cases, they will have to be determined from experimental data.

As an illustration, if the phenomenon Q under investigation is the axial buckling load P_{cr} of a simply supported beam, and the factors $Q_1, Q_2, Q_3, \ldots, Q_n$ influencing it are the length ℓ of the beam and its flexural stiffness (EI), then Eq. (11.2) becomes

$$P_{cr} = C_Q \ell^{k_1} (EI)^{k_2} \tag{11.3}$$

The exponents k_1 and k_2, in this case, may be determined directly by dimensional analysis. If Eq. (11.3) is dimensionally homogeneous, the corresponding dimensional equation with respect to some convenient basic quantities, in this case force (F) and length (L), may be written

$$L^0 F^1 = (L)^{k_1} (FL^2)^{k_2} \tag{11.4}$$

or

$$L^0 F^1 = L^{k_1 + 2k_2} F^{k_2} \tag{11.5}$$

To satisfy dimensional homogeneity, the exponents of F and L on each side of Eq. (11.5) should be the same; that is,

$$0 = k_1 + 2k_2 \tag{11.6}$$

and

$$1 = k_2 \tag{11.7}$$

Equations (11.6) and (11.7) yield $k_1 = -2$ and $k_2 = 1$. Hence, from Eq. (11.3), we find

$$P_{cr} = C_Q \frac{EI}{\ell^2} \tag{11.8}$$

From the mechanics of solids, it becomes obvious that $C_0 = \pi^2$, and, consequently,

$$P_{cr} = \frac{\pi^2 EI}{\ell^2} \tag{11.9}$$

Equation (11.9) represents the standard form of Euler's buckling equation.

Now consider the transverse natural frequency response of a simply supported beam of uniform mass per unit length. The natural frequencies ω of this beam are known to depend on its length ℓ, its mass m per unit length, and its flexural stiffness EI. Hence, Eq. (11.2) yields

$$\omega = C_Q \ell^{k_1} m^{k_2} (EI)^{k_3} \tag{11.10}$$

The corresponding dimensional equation is

$$F^0 L^0 T^{-1} = (L)^{k_1} (FT^2 L^{-2})^{k_2} (FL^2)^{k_3} \tag{11.11}$$

or

$$F^0 L^0 T^{-1} = F^{k_2 + k_3} L^{k_1 - 2k_2 + 2k_3} T^{2k_2} \tag{11.12}$$

The auxiliary equations are

$$k_2 + k_3 = 0 \tag{11.13}$$

$$k_1 - 2k_2 + 2k_3 = 0 \tag{11.14}$$

$$2k_2 = -1 \tag{11.15}$$

Simultaneous solution of Eqs. (11.13), (11.14), and (11.15) yields

$$k_1 = -2 \qquad k_2 = -\tfrac{1}{2} \qquad k_3 = \tfrac{1}{2} \tag{11.16}$$

Hence, from Eq. (11.10), we find

$$\omega + C_Q \frac{1}{\ell^2} \sqrt{\frac{EI}{m}} \tag{11.17}$$

It is easily recognized that Eq. (11.17) provides the free frequencies of vibration of the simply supported beam, where $C_Q = n^2 \pi^2$ ($n = 1, 2, 3, \ldots$).

The above cases, however, are simple, and thorough analytical treatment may be found in the literature. If the coefficient C_Q is not known, it would have to be determined experimentally. The same applies to cases where the number of k_n ($n = 1, 2, 3, \ldots$) exponents exceeds the number of auxiliary equations available for their determination. This may be demonstrated by considering a uniform cantilevered beam loaded at the free end by a concentrated load P. If its deflection δ at the free end is the argument, then Eq. (11.2) yields

$$\delta = C_Q p^{k_1} \ell^{k_2} (EI)^{k_3} \tag{11.18}$$

where ℓ is the unsupported length of the member, and EI is its flexural stiffness.

The corresponding dimensional equation is

$$F^0 L^1 = (F)^{k_1} (L)^{k_2} (EI)^{k_3} \tag{11.19}$$

Equation (11.19) yields the following auxiliary equations:

$$0 = k_1 + k_3 \tag{11.20}$$

$$1 = k_2 + 2k_3 \tag{11.21}$$

Equations (11.20) and (11.21) are not sufficient to solve for all three unknowns. By expressing k_2 and k_3 in terms of k_1, we find

$$k_2 = 1 + 2k_1 \quad \text{and} \quad k_3 = -k_1 \tag{11.22}$$

On this basis, Eq. (11.18) yields

$$\delta = C_Q P^{k_1} \ell^{1 + 2k_1} (EI)^{-k_1} \tag{11.23}$$

or

$$\delta = C_Q \ell^{1 + 2k_1} \left(\frac{P}{EI} \right)^{k_1} \tag{11.24}$$

In this case, C_Q and k_1 should be determined experimentally if an expression for the deflection δ at the free end of the cantilever beam is the requirement. Since the analytical solution of this problem exists, giving $C_Q = \frac{1}{3}$ and $k_1 = 1$, experimental results should yield approximately the same values for C_Q and k_1, provided that the beam is not stressed beyond its elastic limit.

11.3 THE PI THEOREM AND PREDICTION EQUATION

This theorem was developed by E. Buckingham [88] about 1914 and provides a way by which any physical relation between the dimensional quantities may be formulated as a relation between nondimensional quantities. For this reason, dimensional theory becomes a powerful tool for the investigation of engineering problems.

Consider, for example, a dimensional quantity Q (a physical phenomenon), and suppose that the factors or dimensional quantities influencing it are $Q_1, Q_2, Q_3, \ldots, Q_n$. The relation between Q and $Q_1, Q_2, Q_3, \ldots, Q_n$ may be written as follows:

$$Q = f(Q_1, Q_2, Q_3, \ldots, Q_n) \tag{11.25}$$

where Q is the dependent quantity, and Q_n $(n = 1, 2, 3, \ldots, n)$ are the independent quantities. Now assume that the physical law represented by this function is independent of the choice of the system of units, and that the first j dimensional quantities $(j \leq n)$ have independent dimensions. In other words, j is the largest number of parameters with independent dimensions, usually not more than three for engineering quantities. On this basis, the remaining $Q, Q_j, Q_{j+1}, \ldots, Q_n$ quantities could be expressed in terms of the $Q_1, Q_2, Q_3, \ldots, Q_j$ parameters.

According to the Pi theorem, the number of nondimensional and independent quantities required to obtain a relation among the variables of any physical phenomenon is equal to the total number of dimensional quantities $(n + 1)$, reduced by the number of parameters (j) with independent dimensions. In engineering, the dimensions F, L, T, or M, L, T, in which these quantities may be measured (not more than three), are usually considered to represent j. Hence, the relation between the $n + 1$ dimensional quantities $Q, Q_1, Q_2, Q_3, \ldots, Q_n$ may be reduced to a relation between $n + 1 - j$ nondimensional quantities, usually designated by $\pi, \pi_1, \pi_2, \ldots, \pi_s$ $(s = n - j)$, which are nondimensional combinations of the $n + 1$ dimensional quantities.

Hence, any physical phenomenon may be expressed as a relation between nondimensional quantities in the following form:

$$\pi = f(\pi_1, \pi_2, \ldots, \pi_s) \tag{11.26}$$

The dependent variable, designated by Q in Eq. (11.25), is included in the nondimensional term π on the left-hand side of Eq. (11.26). In cases where $j = n$ and the number of basic units equals the number of characteristic parameters with independent dimensions, a nondimensional combination of the dimensional quantities $Q_1, Q_2, Q_3, \ldots, Q_n$ cannot be formed, but their relation may be represented by Eq. (11.2). The exponents $k_1, k_2, \ldots, k_n$ ($n = j$) may be determined by dimensional analysis using the dimensional equation (11.1), or any other appropriate dimensional equation. The constant C_Q may be determined either experimentally or theoretically by solving the appropriate mathematical problem.

Equation (11.26) is known as the *general prediction equation*, and it serves as a basis for the development of the theory of models. The Pi terms in this equation should always be nondimensional and independent. For the particular problem at hand, they may be determined most conveniently by using dimensional analysis. The difficult and most important task is the formulation of the relation given by Eq. (11.25). In other words, it is important to determine precisely all the parameters affecting the physical phenomenon under investigation. This part may be facilitated by referring to known differential equations or other sources indicating the parameters that affect the physical quantity, or phenomenon, under investigation. This is justified on the basis that every system of equations mathematically describing the phenomenon may be formulated as a relation between nondimensional quantities. When the controlling factors influencing the phenomenon are accurately established, a relation such as the one in Eq. (11.25) can be written. Then, by applying dimensional analysis, the Pi terms in Eq. (11.26) can be determined relatively easily. The following section illustrates the application of the prediction equation.

11.4   APPLICATION OF THE PREDICTION EQUATION

Consider the case in which one is required to determine the free frequency response of the two-degree spring–weight system in Fig. 11.1. The springs of constants K_1 and K_2 are assumed to be weightless. From Section 2.8, Eq. (2.130), we note that the free undamped frequencies ω of the spring–mass system are dependent on the weights W_1 and W_2, the spring constants K_1 and K_2, and the acceleration of gravity g. Thus, Eq. (11.25) may be written as follows:

$$\omega = f(W_1, W_2, K_1, K_2, g) \tag{11.27}$$

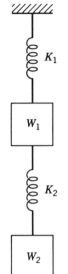

FIGURE 11.1. Two-degree spring–weight system.

The requirement is that the quantities ω, W_1, W_2, K_1, K_2, and g should form nondimensional Pi products that can be arranged in the form given by Eq. (11.26). Such nondimensional products involving these variables should be of the following form:

$$\pi = (\omega)^{k_1}(W_1)^{k_2}(W_2)^{k_3}(K_1)^{k_4}(K_2)^{k_5}(g)^{k_6} \tag{11.28}$$

By choosing force (F), length (L), and time (T) as the basic dimensions, Eq. (11.28) may be written in dimensional form as shown below:

$$\pi = (T^{-1})^{k_1}(F)^{k_2}(F)^{k_3}(FL^{-1})^{k_4}(FL^{-1})^{k_5}(LT^{-2})^{k_6} \tag{11.29}$$

Since π is nondimensional, the exponents of F, L, and T should satisfy the following auxiliary equations:

$$k_2 + k_3 + k_4 + k_5 = 0 \tag{11.30}$$

$$-k_4 - k_5 + k_6 = 0 \tag{11.31}$$

$$-k_1 - 2k_6 = 0 \tag{11.32}$$

The same auxiliary equations may be obtained by applying the algebraic approach. By this method, a dimensional display matrix of the quantities involved may be arranged as follows:

	ω^{k_1}	$W_1^{k_2}$	$W_2^{k_3}$	$K_1^{k_4}$	$K_2^{k_5}$	g^{k_6}	
F	0	1	1	1	1	0	
L	0	0	0	-1	-1	1	(11.33)
T	-1	0	0	0	0	-2	

where each column of the matrix of Eq. (11.33) incorporates the exponents of the dimensional expression, in terms of F, L, and T, for the quantity displayed at the top of the column. The number of nondimensional Pi products that can be formed will be equal to the total number of quantities involved, minus the highest rank of the display dimensional matrix. In this case, the total number of quantities is six, and the rank of the matrix is three. Thus, three nondimensional Pi products can be formed.

The *rank* of the display matrix, which is usually defined to be equal to the number of its rows, is equal to the highest order of a nonzero determinant that can be obtained from the display matrix. In Eq. (11.33), the highest rank of the matrix is three, and it is given by the nonzero third-order determinant; that is,

$$\begin{vmatrix} 0 & 1 & 0 \\ 0 & -1 & 1 \\ -1 & 0 & -2 \end{vmatrix} = -1 \neq 0 \qquad (11.34)$$

The elements of the determinant in Eq. (11.34) are the elements of the columns corresponding to the quantities ω, K_1, and g in Eq. (11.33). Thus, by rearranging Eq. (11.33) so that the columns of the quantities ω, K_1, and g are the last three columns of the dimensional display matrix, the auxiliary equations may be obtained by equating to zero its three rows; that is,

$$k_2 + k_3 + k_5 + k_4 = 0 \qquad (11.35)$$

$$-k_5 - k_4 + k_6 = 0 \qquad (11.36)$$

$$-k_1 - 2k_6 = 0 \qquad (11.37)$$

Equations (11.35), (11.36), and (11.37) are identical to the ones given by Eqs (11.30), (11.31), and (11.32), respectively.

It should be noted that there are six unknowns in Eqs. (11.35) to (11.37), and there are only three equations available to solve for the six unknown exponents $k_1, k_2, \ldots, k_6$. Mathematically, the system of these three equations is indeterminate, and it possesses an infinite number of solutions. Since the coefficients of k_1, k_4, and k_6 form a nonzero determinant, arbitrary values may be assigned for the remaining three unknowns k_2, k_3, and k_5. Then, by solving simultaneously the system of Eqs. (11.35) through (11.37), the values for k_1, k_4, and k_6 are obtained. The same procedure could be used by assigning arbitrary values to k_2, k_3, and k_4, because the coefficients of k_1, k_5, and k_6 form a nonzero determinant. Usually, many combinations are possible.

By solving simultaneously Eqs. (11.35) through (11.37) for k_1, k_4, and k_6, we find

$$k_1 = 2k_2 + 2k_3 \tag{11.38}$$

$$k_4 = -k_2 - k_3 - k_5 \tag{11.39}$$

$$k_6 = -k_2 - k_3 \tag{11.40}$$

By assuming $k_2 = 1$ and $k_3 = k_5 = 0$, Eqs. (11.38) through (11.40) yield

$$k_1 = 2 \quad k_4 = -1 \quad k_6 = -1 \tag{11.41}$$

With $k_3 = 1$ and $k_2 = k_5 = 0$, we have

$$k_1 = 2 \quad k_4 = -1 \quad k_6 = -1 \tag{11.42}$$

Similarly, with $k_5 = 1$ and $k_2 = k_3 = 0$, we obtain

$$k_1 = 0 \quad k_4 = -1 \quad k_6 = 0 \tag{11.42a}$$

The above solutions may be arranged into a matrix form as follows:

	$W_1^{k_2}$	$W_2^{k_3}$	$K_2^{k_5}$	ω^{k_1}	$K_1^{k_4}$	g^{k_6}
π	1	0	0	2	-1	-1
π_1	0	1	0	2	-1	-1
π_2	0	0	1	0	-1	0

$$\tag{11.43}$$

The three Pi products are represented by the three rows of the matrix of solutions given by Eq. (11.43), and they are as follows:

$$\pi = \frac{W_1 \omega^2}{K_1 g} \quad \pi_1 = \frac{W_2 \omega^2}{K_1 g} \quad \pi_2 = \frac{K_2}{K_1} \tag{11.44}$$

The products given by Eq. (11.44) are dimensionless, and, with the help of Eq. (11.26), they are written in the following form:

$$\pi = f(\pi_1, \pi_2) \tag{11.45}$$

or

$$\frac{W_1\omega^2}{K_1 g} = f\left(\frac{W_2\omega^2}{K_1 g}, \frac{K_2}{K_1}\right) \tag{11.46}$$

The expression given by Eq. (11.46) relates the nondimensional products involving the six quantities ω, W_1, W_2, K_1, K_2, and g. Equation (11.46), if preferred, may be altered and simplified by changing the individual Pi products. In this case, the dependent quantity ω appears only in the term π. Hence, the Pi terms in the expressions of Eq. (11.44) are changed as follows:

$$\pi = \pi' = \frac{W_1\omega^2}{K_1 g} \qquad \pi_1' = \frac{\pi}{\pi_1} = \frac{W_1}{W_2} \qquad \pi_2 = \pi_2' = \frac{K_2}{K_1} \tag{11.47}$$

The new Pi products yield the following expression:

$$\frac{W_1\omega^2}{K_1 g} = f\left(\frac{W_1}{W_2}, \frac{K_2}{K_1}\right) \tag{11.48}$$

This equation relates the dependent quantity ω to the independent quantities W_1, W_2, K_1, K_2, and g, in terms of dimensionless Pi products. Hence, it is the prediction equation of the system, and it may be used to investigate the system's frequency response.

Other modifications are possible by again changing the Pi products. The conditions are that the products remain dimensionless and independent.

Usually, the highest rank of the dimensional display matrix is equal to the number of rows contained in the matrix. In exceptional cases, its highest rank may be less, and the matrix is said to be singular. The dimensionless Pi products will always be equal to the total number of quantities involved minus the highest rank that can be obtained from the dimensional matrix. If the rank of the matrix is less than the number of its rows, it will, in general, be sufficient to consider only the rows that define the rank of the matrix and to delete the remaining ones. The procedure for determining the nondimensional products may then be carried out by writing only the auxiliary equations corresponding to these rows and proceeding as described previously.

11.5 MODEL ANALYSIS

The main objective in the theory of models is to develop ways by which the original structure, called the *prototype*, may be investigated by using models of some desirable scale. The preparation of such scale models requires a good understanding of the basic theory associated with the phenomenon under investigation. Models have been used fairly often in the past, particularly in the area of hydraulic structures and machines and in the static and dynamic testing of airplane structures and components, where complicated prototypes can be analyzed satisfactorily by model testing.

A general procedure for the preparation of models may be established by extending the general equation represented by Eq. (11.26). This equation is rewritten below for convenience:

$$\pi = f(\pi_1, \pi_2, \pi_3, \ldots, \pi_s) \tag{11.49}$$

The quantities affecting a phenomenon may be incorporated in Eq. (11.49) in terms of dimensionless Pi products. This equation is general, and any system that is affected by the same quantities may be represented by the same equation. Thus, by using the subscript m to denote model, a similar equation may be written for a model as follows:

$$\pi_m = f(\pi_{1m}, \pi_{2m}, \pi_{3m}, \ldots, \pi_{sm}) \tag{11.50}$$

Equations (11.49) and (11.50) incorporate the same quantities. If the model is designed to be built and to operate so that

$$\pi_1 = \pi_{1m}, \quad \pi_2 = \pi_{2m}, \quad \ldots, \quad \pi_s = \pi_{sm} \tag{11.51}$$

then we have

$$\pi = \pi_m \tag{11.52}$$

The expressions given by Eq. (11.51) are the design conditions, and Eq. (11.52), which involves the dependent variable, is the prediction equation. The model is usually different in size when it is compared to the prototype, and some scale length n, representing the ratio of a pertinent length or distance of the prototype to the corresponding length or distance of the model, would have to be used. Thus, the scaling relation can be written

$$L = nL_m \tag{11.53}$$

where L and L_m are the lengths in the prototype and model, respectively.

When the significant characteristics of the prototype are truly reproduced in the model, and when all the factors influencing the phenomenon under investigation are incorporated in Eq. (11.50), then, by satisfying all the design conditions represented by Eq. (11.51), a true model can be obtained that will represent exactly the response of the prototype. An adequate model may be obtained if accurate predictions can be made for only one characteristic of the prototype, while other characteristics can be violated. *The model is distorted* when a design condition is not completely satisfied, making it necessary to introduce a correction in the prediction equation. Distortion may also be introduced willingly into the model, provided that such distortion improves practical or other requirements of the problem. Dissimilar models may also be obtained. Such models do not provide resemblance to the prototype but, through suitable analogies, could give accurate predictions of the prototype response. Soap film and electrical analogies are examples of such investigations.

The following section illustrates some basic principles regarding the preparation of scale models.

11.6   PREPARATION OF SCALE MODELS

Consider the uniform simply supported beam loaded as shown in Fig. 11.2. The constant width b and constant depth d are 10 and 20 in., respectively. It is required to study its static deflection response under the action of the indicated loading, by using a model of length $\ell_m = 3$ ft (i.e., $n = 10$). Neglect the weight of the beam.

The first step in the analysis is to derive the general prediction equation and establish the design conditions that the model should satisfy.

The transverse deflection y of any point of the neutral axis of the beam is assumed to depend on the distances c_1 and c_2 (Fig. 11.2), the width b and depth d of the beam, the modulus of elasticity E, the coordinate x of the deflection, the W_1 and W_2 concentrated loads, and the length ℓ of the beam. Hence, we have

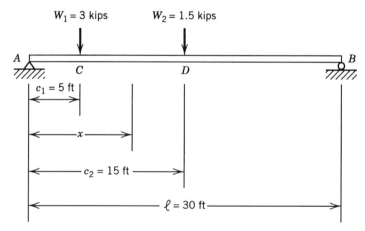

FIGURE 11.2. Simply supported beam of uniform stiffness.

$$y = f(c_1, c_2, b, d, E, x, W_1, W_2, \ell) \tag{11.54}$$

The dimensional display matrix is written as follows:

	y^{k_1}	$c_1^{k_2}$	$c_2^{k_3}$	b^{k_4}	d^{k_5}	E^{k_6}	x^{k_7}	$W_1^{k_8}$	$W_2^{k_9}$	$\ell^{k_{10}}$	
F	0	0	0	0	0	1	0	1	1	0	
L	1	1	1	1	1	-2	1	0	0	1	(11.55)

The highest rank of this matrix is two, since the last two columns form a nonzero determinant. Hence, eight nondimensional Pi products relating the ten dimensional quantities may be formed. The auxiliary equations are

$$k_6 + k_8 + k_9 = 0 \tag{11.56}$$

$$k_1 + k_2 + k_3 + k_5 - 2k_6 + k_7 + k_{10} = 0 \tag{11.57}$$

By solving Eqs. (11.56) and (11.57) simultaneously for k_9 and k_{10}, we find

$$k_9 = -k_6 - k_8 \tag{11.58}$$

$$k_{10} = -k_1 - k_2 - k_3 - k_4 - k_5 + 2k_6 - k_7 \tag{11.59}$$

By using theee expressions and assigning arbitrary values for $k_1, k_2, \ldots, k_8$, the following matrix of solutions is obtained:

	y^{k_1}	$c_1^{k_2}$	$c_2^{k_3}$	b^{k_4}	d^{k_5}	E^{k_6}	x^{k_7}	$W_1^{k_8}$	$W_2^{k_9}$	$\ell^{k_{10}}$	
π	1	0	0	0	0	0	0	0	0	-1	
π_1	0	1	0	0	0	0	0	0	0	-1	
π_2	0	0	1	0	0	0	0	0	0	-1	
π_3	0	0	0	1	0	0	0	0	0	-1	(11.60)
π_4	0	0	0	0	1	0	0	0	0	-1	
π_5	0	0	0	0	0	1	0	0	-1	2	
π_6	0	0	0	0	0	0	1	0	0	-1	
π_7	0	0	0	0	0	0	0	1	-1	0	

Thus, we have

$$\pi = \frac{y}{\ell} \tag{11.61}$$

$$\pi_1 = \frac{c_1}{\ell} \tag{11.62}$$

$$\pi_2 = \frac{c_2}{\ell} \tag{11.63}$$

$$\pi_3 = \frac{b}{\ell} \tag{11.64}$$

$$\pi_4 = \frac{d}{\ell} \tag{11.65}$$

$$\pi_5 = \frac{E\ell^2}{W_2} \tag{11.66}$$

$$\pi_6 = \frac{x}{\ell} \tag{11.67}$$

$$\pi_7 = \frac{W_1}{W_2} \tag{11.68}$$

By changing the term π_5 to $\pi_5^1 = 1/\pi_5 = W_2/E\ell^2$, the general equation may be written

$$\pi = f(\pi_1, \pi_2, \ldots, \pi_7) \tag{11.69}$$

or

$$\frac{y}{\ell} = f\left(\frac{c_1}{\ell}, \frac{c_2}{\ell}, \frac{b}{\ell}, \frac{d}{\ell}, \frac{W_2}{E\ell^2}, \frac{x}{\ell}, \frac{W_1}{W_2}\right) \tag{11.70}$$

For the model, the general equation may be written

$$\frac{y_m}{\ell_m} = f\left(\frac{c_{1m}}{\ell_m}, \frac{c_{2m}}{\ell_m}, \frac{b_m}{\ell_m}, \frac{d_m}{\ell_m}, \frac{W_{2m}}{E_m\ell_m^2}, \frac{x_m}{\ell_m}, \frac{W_{1m}}{W_{2m}}\right) \tag{11.71}$$

Hence, the design conditions are

$$\left.\begin{array}{cc} \dfrac{c_{1m}}{\ell_m} = \dfrac{c_1}{\ell} & c_{1m} = \dfrac{c_1}{10} \\[3mm] \dfrac{c_{2m}}{\ell_m} = \dfrac{c_2}{\ell} & c_{2m} = \dfrac{c_2}{10} \end{array}\right\} \tag{11.72}$$

$$\left.\begin{array}{cc} \dfrac{b_m}{\ell_m} = \dfrac{b}{\ell} & b_m = \dfrac{b}{10} \\[3mm] \dfrac{d_m}{\ell_m} = \dfrac{d}{\ell} & d_m = \dfrac{d}{10} \end{array}\right\} \tag{11.73}$$

$$\left.\begin{array}{l} \dfrac{W_{2m}}{E_m \ell_m^2} = \dfrac{W_2}{E\ell^2} \qquad W_{2m} = \dfrac{W_2 E_m}{10^2 E} \\[12pt] \dfrac{W_{1m}}{W_{2m}} = \dfrac{W_1}{W_2} \end{array}\right\} \qquad (11.74)$$

$$\dfrac{x_m}{\ell_m} = \dfrac{x}{\ell} \qquad x_m = \dfrac{x}{10} \qquad (11.75)$$

By observation, the above design conditions indicate that the conditions given by Eq. (11.73) require the model to be geometrically similar to the prototype, while the conditions given by Eq. (11.72) require the loads to be placed at corresponding points in model and prototype. The conditions given by Eq. (11.74) establish the magnitude and ratio of loads to be used in the model, and the conditions represented by Eq. (11.75) indicate that the deflections y should be measured at geometrically similar points in model and prototype. Hence, the model should have a length $\ell_m = 3$ ft, a width $b_m = 1$ in., and a depth $d_m = 2$ in. By retaining the same modulus of elasticity E for model and prototype, the loads on the model should be as follows:

$$W_{1m} = \dfrac{W_1}{10^2} = 30 \text{ lb} \qquad (11.76)$$

$$W_{2m} = \dfrac{W_2}{10^2} = 15 \text{ lb} \qquad (11.77)$$

$$\dfrac{W_{1m}}{W_{2m}} = \dfrac{W_1}{W_2} = 2 \qquad (11.78)$$

The displacement y, say, at $x = 20$ ft, will be measured at $x_m = x/10 = 2$ ft in the model. Thus, the general prediction equation for the response of the prototype, using the indicated model, becomes

$$\dfrac{y_m}{\ell_m} = \dfrac{y}{\ell} \qquad (11.79)$$

or

$$y = 10 y_m \qquad (11.80)$$

The displacements y of the prototype can adequately be predicted by Eq. (11.80), provided that the material is not stressed beyond its elastic limit.

If the W_1 and W_2 loads are weights securely attached to the weightless beam (prototype) at the indicated points, the obtained model may also be used to determine experimentally the static deflections of the prototype under the load concentration points. If the static deflection curve is assumed to be proportional to the dynamic one corresponding to the free undamped fundamental frequency of the prototype, then this frequency can be determined from the expression

$$\omega^2 = g \frac{\sum\limits_{i=1}^{n} W_i y_i}{\sum\limits_{i=1}^{n} W_i y_i^2} \tag{11.81}$$

Thus, model analysis can be used to determine experimentally quantities that are required in theoretical investigations. Other problems can be worked out in a similar manner.

11.7   DISTORTION CONSIDERATIONS

If the design conditions given by the expressions in Eq. (11.51) are not all satisfied, then the prediction equation as shown by Eq. (11.52) will yield $\pi \neq \pi_m$. In such a case, it becomes necessary either to introduce a prediction factor ν, so that

$$\pi = \nu \pi_m \qquad \nu = \frac{\pi}{\pi_m} \tag{11.82}$$

or to distort additional Pi terms so that

$$\pi = \pi_m \tag{11.83}$$

The distortion may be a distortion of dimensions (length, width, depth); a distortion of configuration (shape); a distortion of loading (loads too large or too small); or a distortion of material properties, manufacturing inaccuracies in building the model, and so on. At any rate, if the distortion is significant and cannot be disregarded, its effect on the prediction equation should be evaluated either by experimental means or by known theoretical relations involving the quantity (Pi term) or quantities (Pi terms) participating in the distortion.

If, in the previous problem, for example, it becomes necessary to distort the width b of the beam, then b_m/ℓ_m in Eqs. (11.70) and (11.71) will be different from b/ℓ. We can express this observation as follows:

$$\frac{b_m}{\ell_m} = \alpha \frac{b}{\ell} \tag{11.84}$$

where the parameter α is known as the distortion factor. The prediction factor ν can be determined experimentally by using Eq. (11.70). In other words, all quantities (Pi products) in this equation may be kept constant—if possible, at the same values as those of the prototype—with the exception of b/ℓ and y/ℓ. Then values of y/ℓ may be obtained for a range of values of b/ℓ, which is analogous to obtaining values of α larger and smaller than 1 in Eq. (11.84). The log plot of y/ℓ versus b/ℓ is then drawn, and the prediction factor ν is determined from the following expression:

$$\nu = \alpha^{-\theta} \tag{11.85}$$

where θ is the slope of the line in the log plot. This will yield a constant value for the slope θ if the plot is a straight line, that is, if b/ℓ satisfies the requirements for combination as a product. If the plot is a curve, the value of y/ℓ corresponding to the correct value of b/ℓ for a true model can be taken from the curve with reasonable accuracy, particularly if the curve is smooth.

Theoretical observations can also be used to determine the prediction factor ν, if the nature of the quantity to be distorted is known. If, in the previous problem, the term b_m/ℓ_m is distorted so that

$$\frac{b_m}{\ell_m} = \alpha \frac{b}{\ell} \tag{11.86}$$

the prediction factor ν can be determined by examining the general differential equation of an elastic line that is given by the following expression:

$$\frac{d^2y}{dx^2} = \frac{M}{EI} \tag{11.87}$$

It is obvious, after integration, that the deflection y due to bending depends on the length ℓ of the beam and the moment of inertia I involving the width b of the beam. If the length ℓ is not distorted, then the dimension b of the cross section is distorted, which, in turn, affects the moment of inertia I. On this basis,

$$I_m = \frac{\alpha b}{n} \cdot \frac{d^3}{n^3} \cdot \frac{1}{12} = \frac{\alpha}{n^4} \cdot \frac{bd^3}{12} = \frac{\alpha}{n^4} I \tag{11.88}$$

In this case, the distortion factor ν is equal to α, provided that the design conditions for moment are satisfied. This, however, is true for the preceding problem, because the loads are placed at corresponding points in the model and prototype.

The aim, in general, is to prepare an undistorted model. If it is advisable, or required, to distort the model, theoretical means should first be investigated in determining (or compensating for) the effects of the distortion. When this is impossible, or very difficult to do, experimental means should be used to determine the prediction factor ν in Eq. (11.82). This procedure usually requires more than one model to be constructed, which explains the reason why theoretical means in correcting for the distortion should first be investigated carefully.

11.8   RING REINFORCED CORRUGATED CYLINDRICAL SHELLS

Consider the investigation of the buckling response of a ring reinforced corrugated cylindrical shell, such as the one shown in Fig. 11.3. At the ends, the shell is acted on by a compressive load uniformly distributed along the geometric center circumference of the corrugation. We wish to determine the design conditions and the prediction equation, by assuming that there are no manufacturing or testing constraints. Such structures are used as intertank connections in space vehicles. A list of the factors influencing the buckling load q per unit length of circumference of the corrugated shell is as follows:

q	Buckling load per unit length of circumference
R	Radius to centerline of corrugated skin
t	Thickness of shell skin
L_R	Ring spacing
A_s	Cross-sectional area of skin
ℓ	Overall length of shell
A_R	Area of ring
h	Depth of corrugation measured from centerline of inner cap to outer cap, along a radial line
b	Width of outer or inner cap
s	Distance between center lines of corrugation along the geometric center circumference
c	Distance from centroid of ring frame to skin centerline
$(EI)_R$	Bending rigidity of ring frame
E_R	Modulus of elasticity of the ring
$(GI)_t$	Torsional rigidity of ring frame
E_s	Modulus of elasticity of skin
β	Angle between a radial line through corner of inner cap and inclined portion of corrugation
μ	Poisson's ratio

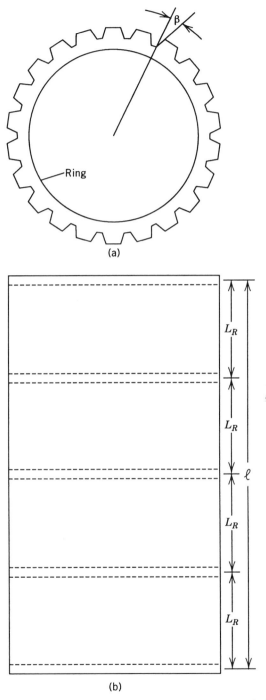

(a)

(b)

FIGURE 11.3. Ring reinforced corrugated cylindrical shell.

The dimensional display matrix is as follows:

	Load	Geometry									
	q^{k_1}	R^{k_2}	t^{k_3}	$L_R^{k_4}$	$A_s^{k_5}$	ℓ^{k_6}	$A_R^{k_7}$	h^{k_8}	b^{k_9}	$s^{k_{10}}$	$c^{k_{11}}$
F	1	0	0	0	0	0	0	0	0	0	0
L	−1	1	1	1	2	1	2	1	1	1	1

Rigidity				Dimensionless Constants		
$(EI)_R^{k_{12}}$	$E_R^{k_{15}}$	$(GI)_t^{k_{13}}$	$E_s^{k_{14}}$	β	μ	(11.89)
1	1	1	1			
2	−2	2	−2			

The rank of the matrix is two, since its last two columns, which are the columns corresponding to the exponents k_{13} and k_{14}, yield a nonzero determinant. The variables in the matrix are arranged into four groups, namely, load, geometry, rigidity, and dimensionless constants. The constants β and μ are dimensionless and may be left out of the display matrix.

The auxiliary equations are

$$k_1 + k_{12} + k_{15} + k_{13} + k_{14} = 0 \tag{11.90}$$

$$-k_1 + k_2 + k_3 + k_4 + 2k_5 + k_6 + 2k_7 + k_8 + k_9 + k_{10} + k_{11}$$
$$+ 2k_{12} - 2k_{15} + 2k_{13} - 2k_{14} = 0 \tag{11.91}$$

By solving Eqs. (11.90) and (11.91) simultaneously for k_{13} and k_{14}, we find

$$k_{13} = -\frac{k_1}{4} - \frac{k_2}{4} - \frac{k_3}{4} - \frac{k_4}{4} - \frac{k_5}{4} - \frac{k_6}{4} - \frac{k_7}{4} - \frac{k_8}{4} - \frac{k_9}{4} - \frac{k_{10}}{4} - \frac{k_{11}}{4} - k_{12} \tag{11.92}$$

$$k_{14} = -\frac{3k_1}{4} + \frac{k_2}{4} + \frac{k_3}{4} + \frac{k_4}{4} + \frac{k_5}{2} + \frac{k_6}{4} + \frac{k_7}{2} + \frac{k_8}{4} + \frac{k_9}{4} + \frac{k_{10}}{4} + \frac{k_{11}}{4} - k_{15} \tag{11.93}$$

By assigning arbitrary values for $k_1, k_2, \ldots, k_{12}$ and k_{15}, the following matrix of solutions is obtained:

	q^{k_1}	R^{k_2}	t^{k_3}	$L_R^{k_4}$	$A_s^{k_5}$	ℓ^{k_6}	$A_R^{k_7}$	h^{k_8}	b^{k_9}	$s^{k_{10}}$	$c^{k_{11}}$	$(EI)_R^{k_{12}}$	$E_R^{K_{15}}$	$(GI)_t^{k_{13}}$	$E_s^{k_{14}}$
π	1	0	0	0	0	0	0	0	0	0	0	0	0		
π_1	0	1	0	0	0	0	0	0	0	0	0	0	0		
π_2	0	0	1	0	0	0	0	0	0	0	0	0	0		
π_3	0	0	0	1	0	0	0	0	0	0	0	0	0		
π_4	0	0	0	0	1	0	0	0	0	0	0	0	0		
π_5	0	0	0	0	0	1	0	0	0	0	0	0	0		
π_6	0	0	0	0	0	0	1	0	0	0	0	0	0		
π_7	0	0	0	0	0	0	0	1	0	0	0	0	0		
π_8	0	0	0	0	0	0	0	0	1	0	0	0	0		
π_9	0	0	0	0	0	0	0	0	0	1	0	0	0		
π_{10}	0	0	0	0	0	0	0	0	0	0	1	0	0		
π_{11}	0	0	0	0	0	0	0	0	0	0	0	1	0		
π_{12}	0	0	0	0	0	0	0	0	0	0	0	0	1		

$$(11.94)$$

By multiplying the elements of the above matrix by a factor of 2 or 4, the nondimensional Pi products are as follows:

$$\pi = \frac{q^4}{(GI)_t E_s^3} \qquad \pi_8 = \frac{b^4 E_s}{(GI)_t}$$

$$\pi_1 = \frac{R^4 E_s}{(GI)_t} \qquad \pi_9 = \frac{s^4 E_s}{(GI)_t}$$

$$\pi_2 = \frac{t^4 E_s}{(GI)_t} \qquad \pi_{10} = \frac{c^4 E_s}{(GI)_t}$$

$$\pi_3 = \frac{L_R^4 E_s}{(GI)_t} \qquad \pi_{11} = \frac{(EI)_R}{(GI)_t}$$

$$\pi_4 = \frac{A_s^2 E_s}{(GI)_t} \qquad \pi_{12} = \frac{E_R}{E_s}$$

$$\pi_5 = \frac{\ell^4 E_s}{(GI)_t} \qquad \pi_{13} = \beta$$

$$\pi_6 = \frac{A_R^2 E_s}{(GI)_t} \qquad \pi_{14} = \mu$$

$$\pi_7 = \frac{h^4 E_s}{(GI)_t}$$

The above Pi products can be simplified in the following way:

$$\pi' = \left(\frac{\pi}{\pi_5}\right)^{1/4} = \frac{q}{E_s\ell} \qquad \pi'_7 = \left(\frac{\pi_7}{\pi_5}\right)^{1/4} = \frac{h}{\ell}$$

$$\pi'_1 = \left(\frac{\pi_1}{\pi_5}\right)^{1/4} = \frac{R}{\ell} \qquad \pi'_8 = \left(\frac{\pi_8}{\pi_5}\right)^{1/4} = \frac{b}{\ell}$$

$$\pi'_2 = \left(\frac{\pi_2}{\pi_5}\right)^{1/4} = \frac{t}{\ell} \qquad \pi'_9 = \left(\frac{\pi_9}{\pi_5}\right)^{1/4} = \frac{s}{\ell}$$

$$\pi'_3 = \left(\frac{\pi_3}{\pi_5}\right)^{1/4} = \frac{L_R}{\ell} \qquad \pi'_{10} = \left(\frac{\pi_{10}}{\pi_5}\right)^{1/4} = \frac{c}{\ell}$$

$$\pi'_4 = \left(\frac{\pi_4}{\pi_5}\right)^{1/2} = \frac{A_s}{\ell^2} \qquad \pi'_{11} = \pi_{11} = \frac{(EI)_R}{(GI)_t}$$

$$\pi'_5 = \frac{\pi_{11}}{\pi_5} = \frac{(EI)_R}{E_s\ell^4} \qquad \pi'_{12} = \pi_{12} = \frac{E_R}{E_s}$$

$$\pi'_6 = \left(\frac{\pi_6}{\pi_5}\right)^{1/2} = \frac{A_R}{\ell^2} \qquad \pi'_{13} = \pi_{13} = \beta$$

$$\pi'_{14} = \pi_{14} = \mu$$

Hence, the general equation for the prototype is

$$\frac{q}{E_s\ell} = f\left(\frac{R}{\ell}, \frac{t}{\ell}, \frac{L_R}{\ell}, \frac{A_s}{\ell^2}, \frac{(EI)_R}{E_s\ell^4}, \frac{A_R}{\ell^2}, \frac{h}{\ell}, \frac{b}{\ell}, \frac{s}{\ell}, \frac{c}{\ell}, \frac{(EI)_R}{(GI)_t}, \frac{E_R}{E_s}, \beta, \mu\right) \qquad (11.95)$$

A similar expression for the model is written as follows:

$$\frac{q_m}{E_{sm}\ell_m} = f\left(\frac{R_m}{\ell_m}, \frac{t_m}{\ell_m}, \frac{L_{Rm}}{\ell_m}, \frac{A_{sm}}{\ell_m^2}, \frac{(EI)_{Rm}}{E_{sm}\ell_m^4}, \frac{A_{Rm}}{\ell_m^2}, \frac{h_m}{\ell_m}, \frac{b_m}{\ell_m}, \right.$$

$$\left. \frac{s_m}{\ell_m}, \frac{c_m}{\ell_m}, \frac{(EI)_{Rm}}{(GI)_{tm}}, \frac{E_{Rm}}{E_{sm}}, \beta_m, \mu_m\right) \qquad (11.96)$$

By introducing a length scale n, that is,

$$\ell = n\ell_m \qquad (11.97)$$

the design conditions are as follows:

$$\frac{R}{\ell} = \frac{R_m}{\ell_m} \qquad R_m = \frac{R}{n} \tag{11.98}$$

$$\frac{t}{\ell} = \frac{t_m}{\ell_m} \qquad t_m = \frac{t}{n} \tag{11.99}$$

$$\frac{L_R}{\ell} = \frac{L_{Rm}}{\ell_m} \qquad L_{Rm} = \frac{L_R}{n} \tag{11.100}$$

$$\frac{A_s}{\ell^2} = \frac{A_{sm}}{\ell_m^2} \qquad A_{sm} = \frac{A_s}{n^2} \tag{11.101}$$

$$\frac{(EI)_R}{E_s\ell^4} = \frac{(EI)_{Rm}}{E_{sm}\ell_m^4} \qquad (EI)_m = \frac{E_{sm}(EI)_R}{n^4 E_s} \tag{11.102}$$

$$\frac{A_R}{\ell^2} = \frac{A_{Rm}}{\ell_m^2} \qquad A_{Rm} = \frac{A_R}{n^2} \tag{11.103}$$

$$\frac{h}{\ell} = \frac{h_m}{\ell_m} \qquad h_m = \frac{h}{n} \tag{11.104}$$

$$\frac{b}{\ell} = \frac{b_m}{\ell_m} \qquad b_m = \frac{b}{n} \tag{11.105}$$

$$\frac{s}{\ell} = \frac{s_m}{\ell_m} \qquad s_m = \frac{s}{n} \tag{11.106}$$

$$\frac{c}{\ell} = \frac{c_m}{\ell_m} \qquad c_m = \frac{c}{n} \tag{11.107}$$

$$\frac{(EI)_R}{(GI)_t} = \frac{(EI)_{Rm}}{(GI)_{tm}} \tag{11.108}$$

$$\frac{E_R}{E_s} = \frac{E_{Rm}}{E_{sm}} \tag{11.109}$$

$$\beta = \beta_m \tag{11.110}$$

$$\mu = \mu_m \tag{11.111}$$

When the above design conditions are satisfied, the prediction equation becomes

$$\frac{q}{E_s\ell} = \frac{q_m}{E_{sm}\ell_m}$$

or

$$q = nq_m \frac{E_s}{E_{sm}} \tag{11.112}$$

By inspection, it may be observed that the above design conditions can be satisfied reasonably well. Considerable thought, however, should be given to the size of the model that can be prepared. Large reductions in controlling dimensions of the prototype (such as skin thicknesses) may introduce significant changes in the structural behavior of the model and additional undesirable effects during manufacturing, handling, and operation of the model.

In practice, the torsional rigidity $(GI)_t$ of corrugated shells is considered small, and it is often neglected in their design. Thus, the design condition given by Eq. (11.108) could be further investigated to determine the validity of this assumption.

In conclusion, it may be stated that models may be used efficiently in the design and analysis of complicated modern structures and machines. If all significant parameters of the prototype are passed into the model by properly satisfying the appropriate design conditions, then the model should predict exactly the response of the prototype in the desired respect. In practice, however, this is not usually the case, and deviations from ideal results often occur. Such deviations are treated as distortions, and they are corrected if their effects on the overall behavior of the model are appreciable. This can be investigated by using models of different scales (i.e., $n = 4, 8, 12$) and establishing proper correlation of response between the models in the desired respect.

PROBLEMS

11.1 By using dimensional analysis, as discussed in Section 11.2, derive in each case an expression that will provide the axial buckling load P_{cr} for the uniform beam cases shown in Fig. P4.17. Discuss the results.

11.2 Repeat Problem 11.1 by investigating the free transverse frequencies of vibration of the uniform beams. Discuss the results.

11.3 Repeat Problem 11.1 by investigating the longitudinal free frequencies of vibration of the uniform beams.

11.4 By following the methodology discussed in Sections 11.3 and 11.4, determine in each case the prediction equation regarding the free vibration of the spring–mass systems in Fig. P1.17.

11.5 Repeat Problem 11.4 by determining in each case the forced vibration response of the indicated spring–mass systems. Assume that $F(t) = F \sin \omega_f t$, where ω_f is the frequency of the exciting force $F(t)$.

11.6 By following the procedures discussed in Sections 11.5 and 11.6, design in each case a scale model of scale length $n = 10$ that can be used to determine experimentally the free frequencies of vibration of the uniform beams shown in Fig. P4.3. Assume that each beam has a rectangular cross section of width b and depth h.

11.7 By following the procedures discussed in Sections 11.5 and 11.6, design in each case a scale model of scale length $n = 10$, for the beam cases in Fig. P4.29, that can be used to determine experimentally the maximum dynamic deflection produced by the applying dynamic load. Assume that the weight w per unit length of each beam is 0.8 kip/ft, $E = 30 \times 10^6$ psi, $I = 1500$ in.^{4}, $F = 30$ kips, and $\omega_f = 30$ rps. Each beam has a rectangular cross section of width b and depth h.

APPENDIX A
Magnification Factors for Undamped, One-Degree Spring–Mass Systems Subjected to Various Types of Dynamic Excitations

Force Function	Magnification Factor Γ
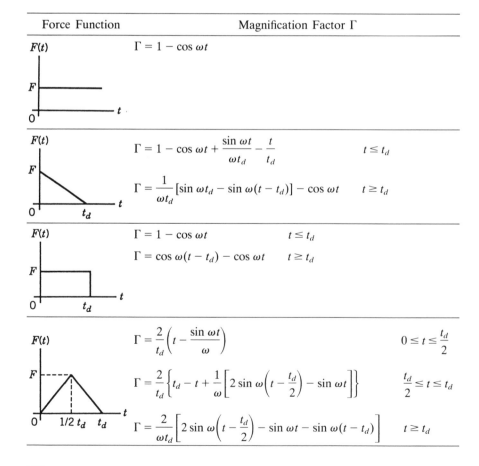	

For the first force function (step):
$$\Gamma = 1 - \cos \omega t$$

For the second force function (decaying ramp):
$$\Gamma = 1 - \cos \omega t + \frac{\sin \omega t}{\omega t_d} - \frac{t}{t_d} \qquad t \le t_d$$
$$\Gamma = \frac{1}{\omega t_d}[\sin \omega t_d - \sin \omega (t - t_d)] - \cos \omega t \qquad t \ge t_d$$

For the third force function (rectangular pulse):
$$\Gamma = 1 - \cos \omega t \qquad t \le t_d$$
$$\Gamma = \cos \omega (t - t_d) - \cos \omega t \qquad t \ge t_d$$

For the fourth force function (triangular pulse):
$$\Gamma = \frac{2}{t_d}\left(t - \frac{\sin \omega t}{\omega}\right) \qquad 0 \le t \le \frac{t_d}{2}$$
$$\Gamma = \frac{2}{t_d}\left\{t_d - t + \frac{1}{\omega}\left[2 \sin \omega\left(t - \frac{t_d}{2}\right) - \sin \omega t\right]\right\} \qquad \frac{t_d}{2} \le t \le t_d$$
$$\Gamma = \frac{2}{\omega t_d}\left[2 \sin \omega\left(t - \frac{t_d}{2}\right) - \sin \omega t - \sin \omega (t - t_d)\right] \qquad t \ge t_d$$

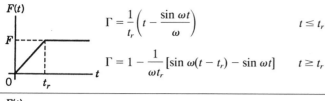

$$\Gamma = \frac{1}{t_r}\left(t - \frac{\sin \omega t}{\omega}\right) \qquad t \leq t_r$$

$$\Gamma = 1 - \frac{1}{\omega t_r}[\sin \omega(t - t_r) - \sin \omega t] \qquad t \geq t_r$$

$F(t)$

$F(t) = F \sin \omega_f t$

$$\Gamma = \frac{1}{(1 - \omega_f^2/\omega^2)}\left(\sin \omega_f t - \frac{\omega_f}{\omega}\sin \omega t\right)$$

where ω_f = frequency of force function
ω = natural frequency of the system

Note: $\Gamma = y(t)/y_{st}$, where $y_{st} = F/k$. In the above expressions, ω is the natural frequency of the one-degree-of-freedom system (except as otherwise stated).

APPENDIX B
Computer Program Prepared for the Determination of the Roots λL_1 That Satisfy Eq. (4.52)

```
        integer nloop
        real  c1,c2,c3,fy2,x,x_int,xa,xb,ya,yb
        read  (*,*)x_int
        open  (1,file='vib.res',status='unknown')
12      nloop=0
        x=x_int
        if (x.eq.0) then
            write  (*,*)'x=0  !!'
            stop
            endif
100     c1=1./tanh(x)-1./tan(x)
        c2=1./tanh(2.*x)-1./tan(2.*x)
        c3=1./tanh(1.25*x)-1./tan(1.25*x)
        fy2=1./sinh(2.*x)-1./sin(2.*x)
        if (nloop.eq.0)then
            xb=1.2*x_int
            yb=1.
            else
                xb=xa
                yb=ya
                endif
        xa=x
        ya=(c1+c2)*(c2+c3)-fy2*fy2
        if(abs(ya).gt.1.e-6)then
          nloop=nloop+1
          x=xa-ya/(yb-ya)*(xb-xa)
          goto 100
          endif
        write  (1,1)nloop,x_int,ya
1       format (//,'Trial loops=',i4,' Initial
        entry=',f7.3,
   @            '         Tolerance= ',e8.2,/)
```

```
        write (1,*)'   nL1='              sital=',cl
        write (1,*)'   nL2=',2*x,'        sita2=',c2
        write(1,*)'    nL3=',1.25*x,'      sita3=',c3
c       x_int=x_int+1.
c       if (x_int.lt.6.1)goto 12
         close (1)
         stop
         end
```

APPENDIX C
Basic Rules of Matrix Algebra

Consider the system of linear equations

$$
\begin{aligned}
y_1 &= a_{11}x_1 + a_{12}x_2 + a_{13}x_3 + \cdots + a_{1n}x_n \\
y_2 &= a_{21}x_1 + a_{22}x_2 + a_{23}x_3 + \cdots + a_{2n}x_n \\
y_3 &= a_{31}x_1 + a_{32}x_2 + a_{33}x_3 + \cdots + a_{3n}x_n \\
&\quad \vdots \\
y_m &= a_{m1}x_1 + a_{m2}x_2 + a_{m3}x_3 + \cdots + a_{mn}x_n
\end{aligned}
\tag{C.1}
$$

The above equations may be expressed in matrix form as

$$
\begin{bmatrix}
y_1 \\ y_2 \\ y_3 \\ \vdots \\ y_m
\end{bmatrix}
=
\begin{bmatrix}
a_{11} & a_{12} & a_{13} & \cdots & a_{1n} \\
a_{21} & a_{22} & a_{23} & \cdots & a_{2n} \\
a_{31} & a_{32} & a_{33} & \cdots & a_{3n} \\
\vdots & \vdots & \vdots & \ddots & \vdots \\
a_{m1} & a_{m2} & a_{m3} & \cdots & a_{mn}
\end{bmatrix}
\begin{bmatrix}
x_1 \\ x_2 \\ x_3 \\ \vdots \\ x_n
\end{bmatrix}
\tag{C.2}
$$

With

$$
\{y\} = \begin{bmatrix} y_1 \\ y_2 \\ y_3 \\ \vdots \\ y_m \end{bmatrix}
\qquad
\{x\} = \begin{bmatrix} x_1 \\ x_2 \\ x_3 \\ \vdots \\ x_n \end{bmatrix}
$$

and

$$[^m_n M] = \begin{bmatrix} a_{11} & a_{12} & a_{13} & \cdots & a_{1n} \\ a_{21} & a_{22} & a_{23} & \cdots & a_{2n} \\ a_{31} & a_{32} & a_{33} & \cdots & a_{3n} \\ \vdots & \vdots & \vdots & \ddots & \vdots \\ a_{m1} & a_{m2} & a_{m3} & \cdots & a_{mn} \end{bmatrix}$$

Equation (C.2) may be written in a more compact form as

$$\{y\} = [^m_n M]\{x\} \tag{C.3}$$

The matrix $[^m_n M]$ is of order (m, n), consisting of m rows and n columns. The elements of this matrix are the a_{ij} coefficients of Eq. (C.1). The column vectors $\{y\}$ and $\{x\}$ are also matrices of order $(m, 1)$ and $(n, 1)$, respectively. A row vector consists of one row and n columns. The order of such vectors is $(1, n)$. When the number of rows of a matrix equals the number of its columns, the result is a *square matrix* of order (m, m).

A matrix whose elements are all zero is called a *null matrix*. The components of a null vector, row or column, are also zero. When the elements of a square matrix are all zero except those of the main diagonal, the resulting matrix is known as a *diagonal matrix*. For example, the square matrix

$$[^m_m M] = \begin{bmatrix} a_{11} & 0 & 0 & \cdots & 0 \\ 0 & a_{22} & 0 & \cdots & 0 \\ 0 & 0 & a_{33} & \cdots & 0 \\ \vdots & \vdots & \vdots & \ddots & \vdots \\ 0 & 0 & 0 & \cdots & a_{mn} \end{bmatrix}$$

is a diagonal matrix. A *unit matrix* is a diagonal matrix whose diagonal elements are all equal to unity. For example,

$$[^m_m M] = \begin{bmatrix} 1 & 0 & 0 & \cdots & 0 \\ 0 & 1 & 0 & \cdots & 0 \\ 0 & 0 & 1 & \cdots & 0 \\ \vdots & \vdots & \vdots & \ddots & \vdots \\ 0 & 0 & 0 & \cdots & 1 \end{bmatrix}$$

is a *unit matrix*..

A *square matrix is symmetrical* about its main diagonal, when the elements satisfy the relation $a_{ij} = a_{ji}$. If it is symmetric about its cross diagonal, it is called a *cross-symmetric matrix*; that is,

$$[^3\overset{3}{M}] = \begin{bmatrix} 1 & 2 & 4 \\ 2 & 3 & 6 \\ 4 & 6 & 7 \end{bmatrix}$$

is a symmetric matrix and

$$[^3\overset{3}{M}] = \begin{bmatrix} 4 & 6 & 7 \\ 2 & 3 & 6 \\ 1 & 2 & 4 \end{bmatrix}$$

is a cross-symmetric matrix.

The *transpose* of a matrix $[M]$ is the matrix $[M']$ that is formed by writing the ith row of $[M]$ as the ith column of $[M']$. For example, the transpose of the matrix

$$[^2\overset{3}{M}] = \begin{bmatrix} a_{11} & a_{12} & a_{13} \\ a_{21} & a_{22} & a_{23} \end{bmatrix}$$

is the matrix

$$[^3\overset{3}{M'}] = \begin{bmatrix} a_{11} & a_{21} \\ a_{12} & a_{22} \\ a_{13} & a_{23} \end{bmatrix}$$

The transpose of $[M']$ is again $[M]$. The transpose of a square matrix that is symmetrical about its main diagonal is the same as the square symmetric matrix.

Matrices may be added or subtracted by adding or subtracting algebraically their corresponding elements. For example,

$$[^4\overset{3}{M}] = \begin{bmatrix} a_{11} & a_{12} & a_{13} \\ a_{21} & a_{22} & a_{23} \\ a_{31} & a_{32} & a_{33} \\ a_{41} & a_{42} & a_{43} \end{bmatrix}$$

and

$$[^4_3N] = \begin{bmatrix} b_{11} & b_{12} & b_{13} \\ b_{21} & b_{22} & b_{23} \\ b_{31} & b_{32} & b_{33} \\ b_{41} & b_{42} & b_{43} \end{bmatrix}$$

may be added or subtracted to form the matrix

$$[^4_3C] = [^4_3M] \pm [^4_3N] = \begin{bmatrix} a_{11} \pm b_{11} & a_{12} \pm b_{12} & a_{13} \pm b_{13} \\ a_{21} \pm b_{21} & a_{22} \pm b_{22} & a_{23} \pm b_{23} \\ a_{31} \pm b_{31} & a_{32} \pm b_{32} & a_{33} \pm b_{33} \\ a_{41} \pm b_{41} & a_{42} \pm b_{42} & a_{43} \pm b_{43} \end{bmatrix}$$

$$= \begin{bmatrix} c_{11} & c_{12} & c_{13} \\ c_{21} & c_{22} & c_{23} \\ c_{31} & c_{32} & c_{33} \\ c_{41} & c_{42} & c_{43} \end{bmatrix}$$

One may easily verify that

$$[M] + [N] = [N] + [M]$$

and

$$([M] + [N])' = [M'] + [N']$$

Also,

$$[A] - ([B] + [C]) = ([A] - [B]) - [C]$$

Multiplication of a matrix $[M]$ by a scalar quantity β yields the matrix $[N]$, consisting of the elements of $[M]$ multiplied individually by the scalar quantity β.

Two matrices,

$$[^mM] = \begin{bmatrix} a_{11} & a_{12} & a_{13} & \cdots a_{1n} \\ a_{21} & a_{22} & a_{23} & \cdots a_{2n} \\ \vdots & \vdots & \vdots & \ddots \vdots \\ a_{i1} & a_{i2} & a_{i3} & \cdots a_{in} \\ \vdots & \vdots & \vdots & \ddots \vdots \\ a_{m1} & a_{m2} & a_{m3} & \cdots a_{mn} \end{bmatrix}$$

and

$$[^n\overset{p}{N}] = \begin{bmatrix} b_{11} & b_{12} & b_{13} & \cdots & b_{1p} \\ b_{21} & b_{22} & b_{23} & \cdots & b_{2p} \\ \vdots & \vdots & \vdots & \ddots & \vdots \\ b_{n1} & b_{n2} & b_{n3} & \cdots & b_{np} \end{bmatrix}$$

where the number of columns of $[^m\overset{n}{M}]$ is equal to the number of rows of $[^n\overset{p}{N}]$, may be multiplied together to form a new matrix $[^m\overset{p}{C}]$ of order (m, p). The elements c_{i1} of the new matrix are obtained by multiplying each element in row i of $[M]$ by the corresponding element in column 1 of $[N]$ and summing up the products; that is,

$$c_{i1} = a_{i1}b_{11} + a_{i2}b_{21} + a_{i3}b_{31} + \cdots + a_{in}b_{n1}$$

In a similar manner,

$$c_{i2} = a_{i1}b_{12} + a_{i2}b_{22} + a_{i3}b_{32} + \cdots + a_{in}b_{n2}$$
$$c_{i3} = a_{i1}b_{13} + a_{i2}b_{23} + a_{i3}b_{33} + \cdots + a_{in}b_{n3}$$
$$\vdots$$
$$c_{in} = a_{i1}b_{1p} + a_{i2}b_{2p} + a_{i3}b_{3p} + \cdots + a_{in}b_{np}$$

The elements of any other row of $[^m\overset{p}{C}]$ may be found in a similar manner.

Consider, for example, the matrices

$$[^2\overset{3}{M}] = \begin{bmatrix} 4 & -2 & 0 \\ 3 & 1 & 5 \end{bmatrix} \quad \text{and} \quad [^3\overset{3}{N}] = \begin{bmatrix} 3 & 2 & -1 \\ 0 & 4 & 0 \\ 1 & -2 & 1 \end{bmatrix}$$

Their product $[^2\overset{3}{M}][^3\overset{3}{M}]$ is the matrix

$$[^2\overset{3}{C}] = [^2\overset{3}{M}][^3\overset{3}{N}] = \begin{bmatrix} 12 & 0 & -4 \\ 14 & 0 & 2 \end{bmatrix}$$

When many matrices are to be multiplied together, it is convenient to follow the multiplication procedure shown in the following example.

Suppose that it is required to multiply the matrices

$$[^3\overset{2}{K}] = \begin{bmatrix} 1 & -3 \\ 4 & 0 \\ -3 & 5 \end{bmatrix} \qquad [^2\overset{3}{M}] = \begin{bmatrix} 4 & -2 & 0 \\ 3 & 1 & 5 \end{bmatrix}$$

and

$$[^3\overset{3}{N}] = \begin{bmatrix} 3 & 2 & -1 \\ 0 & 4 & 0 \\ 1 & -2 & 1 \end{bmatrix}$$

In order to determine the product

$$[^3\overset{3}{D}] = [^3\overset{2}{K}][^2\overset{3}{M}][^3\overset{3}{N}]$$

the multiplication arrangement is as follows:

$$\begin{bmatrix} 3 & 2 & -1 \\ 0 & 4 & 0 \\ 1 & -2 & 1 \end{bmatrix} = [^3\overset{3}{N}]$$

$$[^2\overset{3}{M}] = \begin{bmatrix} 4 & -2 & 0 \\ 3 & 1 & 5 \end{bmatrix} \begin{bmatrix} 12 & 0 & -4 \\ 14 & 0 & 2 \end{bmatrix} = [^2\overset{3}{M}][^3\overset{3}{N}] = [^2\overset{3}{C}] \qquad (C.4)$$

$$[^3\overset{2}{K}] = \begin{bmatrix} 1 & -3 \\ 4 & 0 \\ -3 & 5 \end{bmatrix} \begin{bmatrix} -30 & 0 & -10 \\ 48 & 0 & -16 \\ 34 & 0 & -56 \end{bmatrix} = [^3\overset{2}{K}][^2\overset{3}{C}] = [^3\overset{3}{D}]$$

In the above arrangement, the matrices $[M]$ and $[N]$ are first multiplied together to find the product

$$[^2\overset{3}{C}] = [^2\overset{3}{M}][^3\overset{3}{N}]$$

Then matrices $[^3\overset{2}{K}]$ and $[^2\overset{3}{C}]$ are multiplied, yielding

$$[^3\overset{3}{D}] = [^3\overset{2}{K}][^2\overset{3}{C}] = [^3\overset{2}{K}][^2\overset{3}{M}][^3\overset{3}{N}] = \begin{bmatrix} -30 & 0 & -10 \\ 48 & 0 & -16 \\ 34 & 0 & -56 \end{bmatrix}$$

In the above multiplication, it may be noted that the number of columns of $[M]$ is equal to the number of rows of $[N]$. Also, the number of columns of $[K]$ is equal to the number of rows of $[C]$. This condition is always a requirement for multiplication of matrices.

The commutative law of multiplication is not valid for matrices. In general,

$$[M][N] \neq [N][M]$$

The distributive law

$$[M]([N] + [K]) = [M][N] + [M][K]$$

or

$$([N] + [K])[M] = [N][M] + [K][M]$$

for matrices $[N]$ and $[K]$ of the same order, as well as the associative law

$$[M]([N][K]) = ([M][N])[K]$$

are valid for matrices.

In calculating the transpose of the product of matrices, the following relation may be used:

$$([A][B] \cdots [M][N])' = [N'][M'] \cdots [B'][A'] \tag{C.5}$$

The matrix product of two symmetric matrices does not, in general, yield a symmetric matrix.

The inverse of a square matrix $[M]$ is designated by the symbol $[M^{-1}]$, and it is defined as

$$[M][M^{-1}] = [I] \tag{C.6}$$

Equation (C.3), with $m = n$, yields

$$\{y\} = [{}^{n}\overset{n}{M}]\{x\} \tag{C.7}$$

One application of Eq. (C.6) is to express $\{x\}$ in terms of $\{y\}$ in Eq. (C.7).

For example, by premultiplying both sides of Eq. (C.7) by $[M^{-1}]$, we have

$$[M^{-1}]\{y\} = [M^{-1}][M]\{x\} \tag{C.8}$$

Substituting Eq. (C.6) into Eq. (C.8) yields

$$[M^{-1}]\{y\} = \{x\} \tag{C.9}$$

which expresses $\{x\}$ in terms of $\{y\}$. In this discussion, the term premultiplication of a matrix $[M]$ by a matrix $[N]$ means the product $[N][M]$. The term postmultiplication of $[M]$ by $[N]$ means the product $[M][N]$.

Many approaches have been suggested in the literature for the calculation of the inverse $[M^{-1}]$ of a square matrix $[M]$. The one used here is the expression

$$[M^{-1}] = \frac{[N]}{[M]} \tag{C.10}$$

The elements N_{ki} of the matrix $[N]$ are the subdeterminants, or minors, of the matrix $[M]$ formed by deleting row k and column i. The plus or minus sign by which the N_{ki} elements should be multiplied can be obtained from the identity

$$(-1)^{i+k} \qquad i = 1, 2, 3, \ldots \quad k = 1, 2, 3, \ldots \tag{C.11}$$

where k and i are the deleted row and column, respectively. The elements of the determinant $[M]$ are the elements of the matrix $[M]$.

Consider, for example, the matrix

$$[{}^{3}M] = \begin{bmatrix} a_{11} & a_{12} & a_{13} \\ a_{21} & a_{22} & a_{23} \\ a_{31} & a_{32} & a_{33} \end{bmatrix} = \begin{bmatrix} 1 & 2 & 1 \\ 1 & 1 & 2 \\ 3 & 1 & 2 \end{bmatrix}$$

By applying Eq. (C.10), we find

$$[M^{-1}] = \frac{\begin{bmatrix} \begin{vmatrix} a_{22} & a_{23} \\ a_{32} & a_{33} \end{vmatrix} & -\begin{vmatrix} a_{12} & a_{13} \\ a_{32} & a_{33} \end{vmatrix} & \begin{vmatrix} a_{12} & a_{13} \\ a_{22} & a_{23} \end{vmatrix} \\ -\begin{vmatrix} a_{21} & a_{23} \\ a_{31} & a_{33} \end{vmatrix} & \begin{vmatrix} a_{11} & a_{13} \\ a_{31} & a_{33} \end{vmatrix} & -\begin{vmatrix} a_{11} & a_{13} \\ a_{21} & a_{23} \end{vmatrix} \\ \begin{vmatrix} a_{21} & a_{22} \\ a_{31} & a_{32} \end{vmatrix} & -\begin{vmatrix} a_{11} & a_{12} \\ a_{31} & a_{32} \end{vmatrix} & \begin{vmatrix} a_{11} & a_{12} \\ a_{21} & a_{22} \end{vmatrix} \end{bmatrix}}{\begin{vmatrix} a_{11} & a_{12} & a_{13} \\ a_{21} & a_{22} & a_{23} \\ a_{31} & a_{32} & a_{33} \end{vmatrix}}$$

or

$$[M^{-1}] = \frac{\begin{bmatrix} 0 & -3 & 3 \\ 4 & -1 & -1 \\ -2 & 5 & -1 \end{bmatrix}}{\begin{vmatrix} 1 & 2 & 1 \\ 1 & 1 & 2 \\ 3 & 1 & 2 \end{vmatrix}} = \frac{1}{6} \begin{bmatrix} 0 & -3 & 3 \\ 4 & -1 & -1 \\ -2 & 5 & -1 \end{bmatrix} = \begin{bmatrix} 0 & -\frac{1}{2} & \frac{1}{2} \\ \frac{2}{3} & -\frac{1}{6} & -\frac{1}{6} \\ -\frac{1}{3} & \frac{5}{6} & -\frac{1}{6} \end{bmatrix}$$

Note that

$$[M][M^{-1}] = \begin{vmatrix} 1 & 2 & 1 \\ 1 & 1 & 2 \\ 3 & 1 & 2 \end{vmatrix} \begin{bmatrix} 0 & -\frac{1}{2} & \frac{1}{2} \\ \frac{2}{3} & -\frac{1}{6} & -\frac{1}{6} \\ -\frac{1}{3} & \frac{5}{6} & -\frac{1}{6} \end{bmatrix} = \begin{bmatrix} 1 & 0 & 0 \\ 0 & 1 & 0 \\ 0 & 0 & 1 \end{bmatrix}$$

which verifies the definition for the inverse of a square matrix given by Eq. (C.6).

The inverse of a unit matrix is also a unit matrix. If $a_{11}, a_{22}, a_{33}, \ldots, a_{nn}$ are the elements of the main diagonal of a diagonal matrix, its inverse is also a diagonal matrix with corresponding elements equal to $1/a_{11}$, $1/a_{22}$, $1/a_{33}, \ldots, 1/a_{nn}$. If $[M]$ is a symmetric matrix, its inverse is also symmetric. The inverse of the transpose of a matrix $[M]$ is equal to the transpose of its inverse; that is,

$$[(M')^{-1}] = [(M^{-1})'] \tag{C.12}$$

The inverse of the product of two square matrices may be found by determining the corresponding inverse of each matrix, if it exists, and multiplying them together in reverse order. Suppose that $[C]$ is the product

$$[C] = [M][N] \tag{C.13}$$

Then

$$[C^{-1}] = [N^{-1}][M^{-1}] \tag{C.14}$$

provided that $[N^{-1}]$ and $[M^{-1}]$ exist. If the inverse of a square matrix does not exist, the square matrix is called *singular*. This means that a row or column of such a matrix is a linear combination of other rows or columns of the matrix.

In many cases, it becomes advantageous for computational purposes to partition a matrix $[M]$ into a number of smaller submatrices. Consider, for example, a matrix $[^mM^n]$ of order (m, n). This matrix may be partitioned so that

$$[^m M^n] = \left[\begin{array}{c|c|c|c}
\overset{n_1}{[^{m_1}M_1]} & \overset{n_2}{[^{m_1}M_2]} & \overset{n_3}{[^{m_1}M_3]} & \overset{n_4}{[^{m_1}M_4]} \\
\hline
\overset{n_1}{[^{m_2}M_5]} & \overset{n_2}{[^{m_2}M_6]} & \overset{n_3}{[^{m_2}M_7]} & \overset{n_4}{[^{m_2}M_8]} \\
\hline
\overset{n_1}{[^{m_3}M_9]} & \overset{n_2}{[^{m_3}M_{10}]} & \overset{n_3}{[^{m_3}M_{11}]} & \overset{n_4}{[^{m_3}M_{12}]}
\end{array} \right] \qquad (C.15)$$

where

$$m = m_1 + m_2 + m_3 \quad \text{and} \quad n = n_1 + n_2 + n_3 + n_4$$

The multiplication of partitioned matrices is carried out in the same way as for regular matrices, but the submatrices are treated as if they were elements. If $[M]$ and $[N]$ are partitioned matrices that are supposed to be multiplied together, the partitioning of $[M]$ with respect to columns should be made in the same way as the partitioning of $[N]$ with respect to rows, in order to be able to perform the multiplication.

Suppose that it is required to determine the product

$$[C] = [M][N]$$

where

$$[M] = \begin{bmatrix} 2 & 0 & 2 & 0 \\ 0 & 1 & 0 & 1 \\ 1 & 1 & 1 & 2 \end{bmatrix}$$

and

$$[N] = \begin{bmatrix} 1 & 0 & 0 \\ 0 & 2 & 1 \\ 1 & 0 & 1 \\ 0 & 1 & 2 \end{bmatrix}$$

Matrices $[M]$ and $[N]$ may be partitioned as follows:

$$[M] = \left[\begin{array}{ccc|c} 2 & 0 & 2 & 0 \\ 0 & 1 & 0 & 1 \\ 1 & 1 & 1 & 2 \end{array} \right]$$

$$[N] = \begin{bmatrix} 1 & 0 & 0 \\ 0 & 2 & 1 \\ 1 & 0 & 1 \\ \hline 0 & 1 & 2 \end{bmatrix}$$

Multiplication of $[M]$ by $[N]$ yields

$$[C] = \begin{bmatrix} 2 & 0 & 2 \\ 0 & 1 & 0 \\ 1 & 1 & 1 \end{bmatrix} \begin{bmatrix} 1 & 0 & 0 \\ 0 & 2 & 1 \\ 1 & 0 & 1 \end{bmatrix} + \begin{bmatrix} 0 \\ 1 \\ 2 \end{bmatrix} [0 \quad 1 \quad 2]$$

$$= \begin{bmatrix} 4 & 0 & 2 \\ 0 & 2 & 1 \\ 2 & 2 & 2 \end{bmatrix} + \begin{bmatrix} 0 & 0 & 0 \\ 0 & 1 & 2 \\ 0 & 2 & 4 \end{bmatrix} = \begin{bmatrix} 4 & 0 & 2 \\ 0 & 3 & 3 \\ 2 & 4 & 6 \end{bmatrix}$$

Matrices

$$[M] = \begin{bmatrix} 1 & 0 & 0 & 0 & 0 & 0 \\ 0 & 3 & 0 & 0 & 0 & 0 \\ \hline 0 & 0 & 2 & 0 & 0 & 0 \\ 0 & 0 & 0 & 1 & 0 & 0 \\ \hline 0 & 0 & 0 & 0 & 4 & 0 \\ 0 & 0 & 0 & 0 & 0 & 2 \end{bmatrix}$$

$$[N] = \begin{bmatrix} 1 & 0 & 0 & 0 \\ 0 & 2 & 0 & 0 \\ \hline 0 & 0 & 5 & 0 \\ 0 & 0 & 0 & 4 \\ \hline 0 & 0 & 1 & 0 \\ 0 & 0 & 0 & 1 \end{bmatrix}$$

are partitioned as shown by the dashed lines. Their product $[C] = [M][N]$ is

$$[C] = \begin{bmatrix} \begin{bmatrix} 1 & 0 \\ 0 & 3 \end{bmatrix}\begin{bmatrix} 1 & 0 \\ 0 & 2 \end{bmatrix} & [0] & \\ [0] & \begin{bmatrix} 2 & 0 \\ 0 & 1 \end{bmatrix}\begin{bmatrix} 5 & 0 \\ 0 & 4 \end{bmatrix} \\ [0] & \begin{bmatrix} 4 & 0 \\ 0 & 2 \end{bmatrix}\begin{bmatrix} 1 & 0 \\ 0 & 1 \end{bmatrix} \end{bmatrix} = \begin{bmatrix} 1 & 0 & 0 & 0 \\ 0 & 6 & 0 & 0 \\ 0 & 0 & 10 & 0 \\ 0 & 0 & 0 & 4 \\ 0 & 0 & 4 & 0 \\ 0 & 0 & 0 & 2 \end{bmatrix}$$

Since determinants have already been introduced in the discussion regarding the inverse of a square matrix, it is appropriate at this point to incorporate some of the rules associated with their evaluation. These rules permit one to interchange the rows and columns of a determinant without altering its value; that is,

$$|M| = |M'| \tag{C.16}$$

The sign of a determinant is altered if two rows or two columns are interchanged. A determinant remains unaltered if a column or a row is changed by subtracting from or adding to its elements the corresponding elements of any other column or row. If there exists a common factor β between the elements of a column or row of a determinant, then β may be placed outside the determinant as a multiplier. A determinant is zero when at least one of its rows or columns is a linear combination of the other rows or columns. Products of square matrices satisfy relations of the form

$$\|[M][N][L]\| = \|[N][M][L]\| = \|[L]\|\|[M]\|\|[N]\| \tag{C.17}$$

The elements of a matrix may be a function of some parameter x. In this case, differentiation and integration of a matrix $[M]$ may be defined as follows:

$$\frac{d}{dx}[M] = \begin{bmatrix} \dfrac{da_{11}}{dx} & \dfrac{da_{12}}{dx} & \dfrac{da_{13}}{dx} & \cdots & \dfrac{da_{1n}}{dx} \\ \dfrac{da_{21}}{dx} & \dfrac{da_{22}}{dx} & \dfrac{da_{23}}{dx} & \cdots & \dfrac{da_{2n}}{dx} \\ \vdots & \vdots & \vdots & \ddots & \vdots \\ \dfrac{da_{n1}}{dx} & \dfrac{da_{n2}}{dx} & \dfrac{da_{n3}}{dx} & \cdots & \dfrac{da_{nn}}{dx} \end{bmatrix} \tag{C.18}$$

$$\int [M]\, dx = \begin{bmatrix} \int a_{11}\, dx & \int a_{12}\, dx & \int a_{13}\, dx & \cdots & \int a_{1n}\, dx \\ \int a_{21}\, dx & \int a_{22}\, dx & 2\int a_{23}\, dx & \cdots & \int a_{2n}\, dx \\ \vdots & \vdots & \vdots & \ddots & \vdots \\ \int a_{n1}\, dx & \int a_{n2}\, dx & \int a_{n3}\, dx & \cdots & \int a_{nn}\, dx \end{bmatrix} \quad (C.19)$$

$$\frac{d}{dx}\{y\} = \left\{ \frac{dy_1}{dx} \quad \frac{dy_2}{dx} \quad \frac{dy_3}{dx} \cdots \frac{dy_n}{dx} \right\} \quad (C.20)$$

$$\int \{y\}\, dx = \left\{ \int y_1\, dx \int y_2\, dx \int y_3\, dx \cdots \int y_n\, dx \right\} \quad (C.21)$$

APPENDIX D
The Superposition Principle

In many cases, the response of structural systems is described by a linear equation. Such linear structures are often, but not always, assumed to obey the principle of superposition, which may be stated as follows:

With respect to a coordinate system of axes, components of stress at a point in an elastic body, or its displacements, may be added algebraically in any sequence.

For example, if a member is acted on by forces $F_1, F_2, F_3, \ldots, F_n$, producing a total displacement y at a point C of this member, then

$$y = y_1 + y_2 + y_3 + \cdots + y_n \qquad \text{(D.1)}$$

where $y_1, y_2, y_3, \ldots, y_n$ are the displacements produced individually by the application of $F_1, F_2, F_3, \ldots, F_n$, respectively. The forces $F_1, F_2, F_3, \ldots, F_n$ may be applied to the structure in any sequence, and the corresponding displacements $y_1, y_2, y_3, \ldots, y_n$ may be added algebraically in any order.

This is one of the most important principles in the linear theory of mechanics, and many of the methods developed for the study of such systems are based, at least in part, on the assumption that this principle is valid. The application of this principle to structural problems presupposes that all members of the structural system are constructed of linearly elastic, continuous, homogeneous, and isotropic material, and that the deformations are small so that the geometry of the loading system is not appreciably altered.

The principle of superposition is not valid when the deformation of a structural system resulting from any one of the forces is affected by the deformation due to one of the other forces. For example, in the case of slender flexible members subjected to large axial loads, experience has shown that axial forces may alter appreciably the shape of the elastic curve. Under these conditions, the transverse displacements are significant and one cannot ignore their influence on the equilibrium conditions for a deformed body. As a result, deflections are no longer proportional to the transverse loading, and the principle of superposition is not directly applicable.

APPENDIX E
Betti's Law and Maxwell's Law of Reciprocal Deflections

The laws of Betti and Maxwell have been used extensively in the analysis of the vibration and dynamic responses of structural and mechanical systems. They are general in concept, but their application to structural and mechanical problems is predicated on the assumption that the material obeys Hoode's law and that the superposition principle is valid. Both laws are similar in concept, but Maxwell's law, although discovered somewhat earlier, may be thought of as a special case of the more general Betti's law.

A brief discussion of these concepts may be initiated by considering the two-span continuous beam in Fig. E.1a and assuming that two systems of forces are independently applied to the beam. The system consisting of the forces $M_A, P_1, P_2, \ldots, P_n$ is first applied to the beam and produces the displacements $\theta_C, \delta_1, \delta_2, \ldots, \delta_n$ in the direction of the other force system as it is shown in the figure. Then, the system consisting of the forces $M_C, F_1, F_2, \ldots, F_n$ is applied, producing the displacements θ_A, $\Delta_1, \Delta_2, \ldots, \Delta_n$ in the direction of the first system of forces. Betti's law, in mathematical terms, may be expressed as

$$M_A \theta_A + P_1 \Delta_1 + P_2 \Delta_2 + \cdots + P_n \Delta_n = M_C \theta_C + F_1 \delta_1 + F_2 \delta_2 + \cdots + F_n \delta_n$$

(E.1)

In words, this law is stated as follows:

If a linearly elastic body is subjected independently to two different systems of forces, the work done by the first system of forces moving through the displacements produced by the second system of forces is equal to the work done by the second system of forces moving though the displacements produced by the first system of forces.

If the two force systems consist of only one force each, say, P and F, then,

726

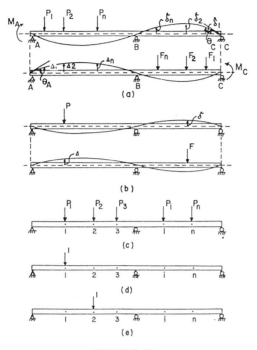

FIGURE E.1

by virtue of the displacements shown in Fig. E.1b,

$$PΔ = Fδ \qquad (E.2)$$

Furthermore, when $P = F$, it follows that

$$Δ = δ \qquad (E.3)$$

The above relation is known as *Maxwell's law of reciprocal deflections*. In words, this law may be stated as follows:

For a linearly elastic body, subjected to a force P that is first applied at a point M and then at another point N, the displacement at M in the direction of the force, when the force is acting at N, is equal to the displacement at N in the direction of the force, when the force is acting at M.

In Fig. E.1a, when only the moments $M_A = M_C = M$ are applied, then

$$\theta_A = \theta_C \tag{E.4}$$

If, in the same figure, only the force P_1 and the moment M_C are applied, then

$$P_1 \Delta_1 = M_C \theta_C \tag{E.5}$$

When the numerical value of P_1 is made equal to M_C, then

$$\Delta_1 = \theta_C \tag{E.6}$$

In this case, Δ_1 may be interpreted as a displacement per unit length.

The law of reciprocal deflections has proved very useful for the computation of flexibility and stiffness coefficients. These coefficients are extensively used in this text.

Suppose that the continuous beam in Fig. E.1c is acted on by forces $P_1, P_2, P_3, \ldots, P_i, \ldots, P_n$ at points $1, 2, 3, \ldots, i, \ldots, n$, respectively. In addition, assume that the total deflections at the corresponding points are $y_1, y_2, y_3, \ldots, y_i, \ldots, y_n$. Each of these deflections can be expressed in terms of the applied loads by using flexibility coefficients. These coefficients are designated by the symbol a_{ij}, and they are defined as the deflection at point i due to a unit load at point j. If, in Fig. E.1b, $P = F = 1$, then, when the unit load is at point i, the flexibility coefficients for points i and j are, respectively, a_{ii} and a_{ji}. When the unit load is at j, the flexibility coefficients for the same points are a_{ij} and a_{jj}. According to Maxwell's law,

$$a_{ij} = a_{ji} \tag{E.7}$$

Therefore, only one of these coefficients needs to be computed.

Returning now to the problem in Fig. E.1c and applying a unit load at point 1, as shown in Fig. E.1d, the flexibility coefficients at points $1, 2, 3, \ldots, i, \ldots, n$ are, respectively, $a_{11}, a_{21}, a_{31}, \ldots, a_{i1}, \ldots, a_{n1}$. For a unit load at point 2 in Fig. E.1e, these coefficients are $a_{12}, a_{22}, a_{32}, \ldots, a_{i2}, \ldots a_{n2}$. In a similar manner, the remaining coefficients may be obtained by applying the unit load at points $3, 4, \ldots, i, \ldots, n$.

The total deflections $y_1, y_2, y_3, \ldots, y_i, \ldots, y_n$ may be written in terms of flexibility coefficients and the applied loads, by making use of the principle of superposition. These expressions are as follows:

$$y_1 = a_{11}P_1 + a_{12}P_2 + a_{13}P_3 + \cdots + a_{1i}P_i + \cdots + a_{1n}P_n \qquad (E.8)$$

$$y_2 = a_{21}P_1 + a_{22}P_2 + a_{23}P_3 + \cdots + a_{2i}P_i + \cdots + a_{2n}P_n \qquad (E.9)$$

$$y_3 = a_{31}P_1 + a_{32}P_2 + a_{33}P_3 + \cdots + a_{3i}P_i + \cdots + a_{3n}P_n \qquad (E.10)$$

$$\vdots$$

$$y_i = a_{i1}P_1 + a_{i2}P_2 + a_{i3}P_3 + \cdots + a_{ii}P_i + \cdots + a_{in}P_n \qquad (E.11)$$

$$\vdots$$

$$y_n = a_{n1}P_1 + a_{n2}P_2 + a_{n3}P_3 + \cdots + a_{ni}P_i + \cdots + a_{nn}P_n \qquad (E.12)$$

Proofs for the above laws are omitted here, because they are beyond the scope of this discussion. Such proofs may easily be found in the literature [8].

APPENDIX F
Stiffness Coefficients

In Eqs. (E.8) through (E.12), the displacements $y_1, y_2, y_3, \ldots, y_n$ were expressed in terms of the loads $P_1, P_2, P_3, \ldots, P_n$ by using flexibility coefficients. For many structural problems, such as frame problems, the computation of flexibility coefficients becomes very laborious. In this case, however, the amount of labor is greatly reduced if stiffness coefficients are used. These coefficients, denoted by the symbol k_{ij}, are used to express the applied loads in terms of the total deflections produced in the directions of the applied loads. They are defined as the force required to be applied at point j in order to produce a deflection equal to unity in the direction of the force, while point i is restrained against translation. In other words, the k_{ij} coefficients represent a force system that is capable of translating point j by an amount equal to unity while preventing the translation of point i.

Consider, for example, the simply supported beam in Fig. F.1a, and assume that forces $P_1, P_2, P_3, \ldots, P_i, \ldots, P_n$ are acting at points $1, 2, 3, \ldots, i, \ldots, n$, respectively. In the same figure, the total deflections under the load concentration points are designated by $y_1, y_2, y_3, \ldots, y_i, \ldots, y_n$. For a unit displacement at point 1, the stiffness coefficients are represented by the force system in Fig. F.1b, which maintains a displacement equal to unity at point 1 and zero displacements at points $2, 3, \ldots, i, \ldots, n$. In Fig. F.1c, the stiffness coefficients for a unit displacement at point 2 are shown. In a similar manner, the k_{ij} coefficients for a unit displacement at points $3, 4, \ldots, i, \ldots, n$ may be obtained.

The following equations may be used to express the forces in Fig. F.1a in terms of the total displacements $y_1, y_2, y_3, \ldots, y_i, \ldots, y_n$ shown in the same figure; that is,

$$P_1 = k_{11}y_1 + k_{12}y_2 + k_{13}y_3 + \cdots + k_{1i}y_i + \cdots + k_{1n}y_n \qquad \text{(F.1)}$$

$$P_2 = k_{21}y_1 + k_{22}y_2 + k_{23}y_3 + \cdots + k_{2i}y_i + \cdots + k_{2n}y_n \qquad \text{(F.2)}$$

$$P_3 = k_{31}y_1 + k_{32}y_2 + k_{33}y_3 + \cdots + k_{3i}y_i + \cdots + k_{3n}y_n \qquad \text{(F.3)}$$

$$\vdots$$

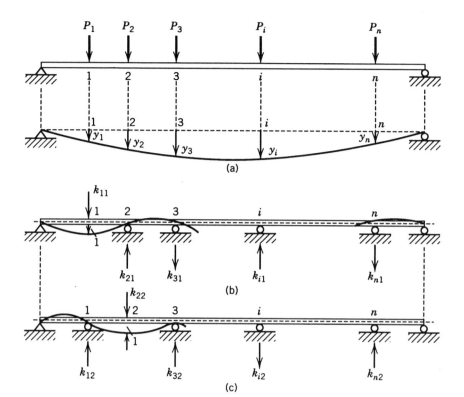

FIGURE F.1

$$P_i = k_{i1}y_1 + k_{i2}y_2 + k_{i3}y_3 + \cdots + k_{ii}y_i + \cdots + k_{in}y_n \qquad (\text{F.4})$$
$$\vdots$$
$$P_n = k_{n1}y_1 + k_{n2}y_2 + k_{n3}y_3 + \cdots + k_{ni}y_i + \cdots + k_{nn}y_n \qquad (\text{F.5})$$

In matrix notation, the above equations are written as follows:

$$
\begin{bmatrix} P_1 \\ P_2 \\ P_3 \\ \vdots \\ P_i \\ \vdots \\ P_n \end{bmatrix}
=
\begin{bmatrix}
k_{11} & k_{12} & k_{13} & \cdots & k_{1i} & \cdots & k_{1n} \\
k_{21} & k_{22} & k_{23} & \cdots & k_{2i} & \cdots & k_{2n} \\
k_{31} & k_{32} & k_{33} & \cdots & k_{3i} & \cdots & k_{3n} \\
\vdots & \vdots & \vdots & & \vdots & & \vdots \\
k_{i1} & k_{i2} & k_{i3} & \cdots & k_{ii} & \cdots & k_{in} \\
\vdots & \vdots & \vdots & & \vdots & & \vdots \\
k_{n1} & k_{n2} & k_{n3} & \cdots & k_{ni} & \cdots & k_{nn}
\end{bmatrix}
\begin{bmatrix} y_1 \\ y_2 \\ y_3 \\ \vdots \\ y_i \\ \vdots \\ y_n \end{bmatrix}
\qquad (\text{F.6})
$$

or

$$\{P\} = [N]\{y\} \tag{F.7}$$

The square matrix

$$[N] = \begin{bmatrix} k_{11} & k_{12} & k_{13} & \cdots & k_{1i} & \cdots & k_{1n} \\ k_{21} & k_{22} & k_{23} & \cdots & k_{2i} & \cdots & k_{2n} \\ k_{31} & k_{32} & k_{33} & \cdots & k_{3i} & \cdots & k_{3n} \\ \vdots & \vdots & \vdots & & \vdots & & \vdots \\ k_{i1} & k_{i2} & k_{i3} & \cdots & k_{ii} & \cdots & k_{in} \\ \vdots & \vdots & \vdots & & \vdots & & \vdots \\ k_{n1} & k_{n2} & k_{n3} & \cdots & k_{ni} & \cdots & k_{nn} \end{bmatrix} \tag{F.8}$$

is known as the stiffness matrix whose elements are the stiffness coefficients k_{ij}. The computation of these coefficients may be carrried out by using known methods of structural analysis and structural mechanics. In performing these computations, it should be remembered that Maxwell's law of reciprocity is valid for stiffness coefficients; that is,

$$k_{ij} = k_{ji} \tag{F.9}$$

In the analysis of structural and mechanical systems for dynamic and vibration response, the use of flexibility coefficients may offer definite advantages over stiffness coefficients, although their computation may be more tedious. In this case, considerable time may be saved by first calculating the stiffness coefficients and then writing the stiffness matrix $[N]$. The inverse of the stiffness matrix, designated by $[N^{-1}]$, will yield the flexibility matrix $[M]$ and, thus, the magnitude of the flexibility coefficients.

On this basis, the flexibility matrix equation can be written

$$\{y\} = [N^{-1}]\{P\} \tag{F.10}$$

The inverse $[N^{-1}]$ of the stiffness matrix $[N]$ always exists and it is unique, provided that the matrix $[N]$ is nonsingular. The inverse of a square matrix can be calculated by applying Eq. (C.10) in Appendix C. Computer programs, however, are usually available for this purpose.

APPENDIX G
Linear Algebraic Equations and Eigenvalues

Consider the following set of n linear algebraic equations:

$$y_1 = a_{11}x_1 + a_{12}x_2 + a_{13}x_3 + \cdots + a_{1n}x_n \tag{G.1}$$

$$y_2 = a_{21}x_1 + a_{22}x_2 + a_{23}x_3 + \cdots + a_{2n}x_n \tag{G.2}$$

$$y_3 = a_{31}x_1 + a_{32}x_2 + a_{33}x_3 + \cdots + a_{3n}x_n \tag{G.3}$$

$$y_n = a_{n1}x_1 + a_{n2}x_2 + a_{n3}x_3 + \cdots + a_{nn}x_n \tag{G.4}$$

The values of $x_1, x_2, x_3, \ldots, x_n$ in the above equations are determined by using Cramer's rule, and they are as follows:

$$x_1 = \frac{1}{|M|} \begin{bmatrix} y_1 & a_{12} & a_{13} & \cdots & a_{1n} \\ y_2 & a_{22} & a_{23} & \cdots & a_{2n} \\ y_3 & a_{32} & a_{33} & & a_{3n} \\ \vdots & \vdots & \vdots & \ddots & \vdots \\ y_n & a_{n2} & a_{n3} & \cdots & a_{nn} \end{bmatrix} \tag{G.5}$$

$$x_2 = \frac{1}{|M|} \begin{bmatrix} a_{11} & y_1 & a_{13} & \cdots & a_{1n} \\ a_{21} & y_2 & a_{23} & \cdots & a_{2n} \\ a_{31} & y_3 & a_{33} & \cdots & a_{3n} \\ \vdots & \vdots & \vdots & \ddots & \vdots \\ a_{n1} & y_n & a_{n3} & \cdots & a_{nn} \end{bmatrix} \tag{G.6}$$

$$x_n = \frac{1}{|M|} \begin{bmatrix} a_{11} & a_{12} & a_{13} & \cdots & y_1 \\ a_{21} & a_{22} & a_{23} & \cdots & y_2 \\ a_{31} & a_{32} & a_{33} & & y_3 \\ \vdots & \vdots & \vdots & \ddots & \vdots \\ a_{n1} & a_{n2} & a_{n3} & \cdots & y_n \end{bmatrix} \tag{G.7}$$

where

$$|M| = \begin{bmatrix} a_{11} & a_{12} & a_{13} & \cdots & a_{1n} \\ a_{21} & a_{22} & a_{23} & \cdots & a_{2n} \\ a_{31} & a_{32} & a_{33} & \cdots & a_{3n} \\ \vdots & \vdots & \vdots & \ddots & \vdots \\ a_{n1} & a_{n2} & a_{n3} & \cdots & a_{nn} \end{bmatrix} \qquad (G.8)$$

is the determinant of the coefficients on the right-hand side of Eqs. (G.1) through (G.4). The above solutions are valid, provided that $|M|$ is different from zero. If $|M|$ is different from zero and $y_1 = y_2 = y_3 = \cdots = y_n = 0$, then $x_1 = x_2 = x_3 = \cdots = x_n = 0$. In this case, the above system of equations is said to be homogeneous.

For a set of n homogeneous equations, a nontrivial solution, that is, other than $x_1 = x_2 = x_3 = \cdots = x_n = 0$, can be obtained only if the determinant $|M|$ is zero. In this case, only relative magnitudes between $x_1, x_2, x_3, \ldots, x_n$ can be found, which indicates that there is at least one equation linearly dependent on the others.

Equations (G.1) through (G.4) can also be written in matrix form as follows:

$$\{y\} = [M]\{x\} \qquad (G.9)$$

where the elements of $[M]$ are the elements of the determinant in Eq. (G.8). If the column vector $\{y\}$ could be expressed as a multiple of the column vector $\{x\}$, that is,

$$\{y\} = \lambda\{x\} \qquad (G.10)$$

then Eq. (G.9) yields

$$\lambda\{x\} = [M]\{x\} \qquad (G.11)$$

or, by expanding Eq. (G.11), we have

$$\lambda x_1 = a_{11}x_1 + a_{12}x_2 + a_{13}x_3 + \cdots + a_{1n}x_n \qquad (G.12)$$

$$\lambda x_2 = a_{21}x_1 + a_{22}x_2 + a_{23}x_3 + \cdots + a_{2n}x_n \qquad (G.13)$$

$$\lambda x_3 = a_{31}x_1 + a_{32}x_2 + a_{33}x_3 + \cdots + a_{3n}x_n \qquad (G.14)$$

$$\vdots$$

$$\lambda x_n = a_{n1}x_1 + a_{n2}x_2 + a_{n3}x_3 + \cdots + a_{nn}x_n \qquad (G.15)$$

By rearranging terms, Eqs. (G.12) through (G.15) may be written as follows:

$$(a_{11} - \lambda)x_1 + a_{12}x_2 + a_{13}x_3 + \cdots + a_{1n}x_n = 0 \qquad (G.16)$$

$$a_{21}x_1 + (a_{22} - \lambda)x_2 + a_{23}x_3 + \cdots + a_{2n}x_n = 0 \qquad (G.17)$$

$$a_{31}x_1 + a_{32}x_2 + (a_{33} - \lambda)x_3 + \cdots + a_{3n}x_n = 0 \qquad (G.18)$$

$$\vdots$$

$$a_{n1}x_1 + a_{n2}x_2 + a_{n3}x_3 + \cdots + (a_{nn} - \lambda)x_n = 0 \qquad (G.19)$$

For a nontrivial solution, the determinant of the coefficients in Eqs. (G.16) through (G.19) must be zero. This yields

$$\begin{vmatrix} (a_{11} - \lambda) & a_{12} & a_{13} & \cdots & a_{1n} \\ a_{21} & (a_{22} - \lambda) & a_{23} & \cdots & a_{2n} \\ a_{31} & a_{32} & (a_{33} - \lambda) & \cdots & a_{3n} \\ \vdots & \vdots & \vdots & \ddots & \vdots \\ a_{n1} & a_{n2} & a_{n3} & \cdots & (a_{nn} - \lambda) \end{vmatrix} = 0 \qquad (G.20)$$

or, in a more compact form,

$$|([M] - \lambda[I])| = 0 \qquad (G.21)$$

By expanding the determinant in Eq. (G.20), an algebraic equation in λ may be written, which will be of the following form:

$$\lambda^n + c_1\lambda^{n-1} + c_2\lambda^{n-2} + \cdots + c_{n-1}\lambda + c_n = 0 \qquad (G.22)$$

This is known as the *characteristic equation* of the matrix $[M]$. The n roots of λ, that is, $\lambda_1, \lambda_2, \lambda_3, \ldots, \lambda_n$, which satisfy Eq. (G.22), are the *eigenvalues* of $[M]$. For each eigenvalue, there corresponds an *eigenvector* that can be determined by using Eqs. (G.16) through (G.19).

Consider, for example, the eigenvalue $\lambda = \lambda_1$. By substituting $\lambda = \lambda_1$ in Eqs. (G.16) through (G.19) and solving the resulting system of n equations simultaneously, by assuming first that $x_1 = 1$, a set of values $x_1 = 1$, $x_2, x_3, \ldots, x_n$ is obtained. This set of values, that is, the vector $\{x_1 = 1, x_2, x_3, \ldots, x_n\}$, is the eigenvector corresponding to the eigenvalue $\lambda = \lambda_1$. The remaining eigenvectors may be found in a similar manner.

Computational mistakes in the calculation of eigenvalues of a matrix $[M]$, or simply checks on the accuracy of the solution for an eigenvalue problem, may be identified by using two important conditions that the eigenvalues must fulfill. These conditions are as follows:

$$\sum_{i=1}^{n} \lambda_i = \text{Tr}[M] \qquad (G.23)$$

where $Tr[M]$ is the *trace*, or *spur*, of the matrix $[M]$, and it is defined as

$$Tr[M] = \sum_{i-1}^{n} a_{ii} \qquad (a_{ii} = a_{11}, a_{22}, \ldots, a_{nn}) \qquad (G.24)$$

The second condition is

$$\prod_i \lambda_i = |M| \qquad (G.25)$$

where $\prod_i \lambda_i$ is the product of all eigenvalues.

As an illustration regarding the above discussion, consider the following set of two equations:

$$6x_1 + 2x_2 = \lambda x_1 \qquad (G.26)$$
$$3x_1 - 2x_2 = \lambda x_2 \qquad (G.27)$$

The matrix $[M]$ of the coefficients is

$$[M] = \begin{bmatrix} 6 & 2 \\ 3 & -2 \end{bmatrix} \qquad (G.28)$$

The eigenvalues of this matrix are calculated by using Eq. (G.20); that is,

$$\begin{vmatrix} (6 - \lambda) & 2 \\ 3 & -(2 + \lambda) \end{vmatrix} = 0 \qquad (G.29)$$

Expansion of this determinant yields the characteristic equation

$$\lambda^2 - 4\lambda - 18 = 0$$

The roots $\lambda_1 = 6.7$ and $\lambda_2 = -2.7$ of this equation are the eigenvalues of the matrix $[M]$. It is easy to verify that these roots satisfy the conditions given by Eqs. (G.23) and (G.25).

The eigenvector corresponding to $\lambda = \lambda_1 = 6.7$ is found from Eqs. (G.26) and (G.27) by substituting $\lambda = 6.7$ and making $x_1 = 1$. This gives $x_1 = 1$ and $x_2 = 0.35$. Thus, the eigenvector is {1   0.35}. For the eigenvaiue $\lambda = \lambda_2 = -2.7$, the eigenvector is {1   -4.35}.

The above discussion is not complete unless we state certain principles regarding matrices, which are known as the *Cayley–Hamilton theorem* [89, 90]. According to this theorem, a matrix $[M]$ satisfies its own characteristic equation; that is, if its characteristic equation is

$$\lambda^n + c_1\lambda^{n-1} + c_2\lambda^{n-2} + \cdots + c_{n-1}\lambda + c_n = 0 \qquad (G.30)$$

then

$$[M]^n + c_1[M]^{n-1} + c_2[M]^{n-2} + \cdots + c_{n-1}[M] + c_n[I] = 0 \quad (G.31)$$

In addition, a function $f([M])$ of a matrix $[M]$, of order (n, n), may be replaced by a polynomial $P([M])$ in $[M]$, of order $n - 1$; that is,

$$f([M]) = P([M]) = a_0[I] + a_1[M] + a_2[M]^2 + \cdots + a_{n-1}[M]^{n-1}$$
$$(G.32)$$

The constants $a_0, a_1, a_2, \ldots, a_{n-1}$ in the above equation can be determined from the equation

$$f(\lambda_i) = a_0 + a_1\lambda_i + a_2\lambda_i^2 + \cdots + a_{n-1}\lambda_i^{n-1} = P(\lambda_i) \quad i = 1, 2, 3, \ldots, n$$
$$(G.33)$$

because the eigenvalues of $[M]$ satisfy Eq. (G.32). In this manner, for n distinct eigenvalues, there correspond n identities of the form given by Eq. (G.33), which can be solved simultaneously for the n unknowns $a_0, a_1, a_2, \ldots, a_{n-1}$.

Equation (G.33) can only be used when the n eigenvalues are real and distinct, that is, when they are different from each other. A real symmetric matrix, of course, will always produce real eigenvalues. In some cases, however, the n eigenvalues are not all distinct. For example, λ_4 could be equal to λ_5 and λ_6. In this case, the $n - 2$ distinct eigenvalues will yield $n - 2$ equations of the form given by Eq. (G.33). The remaining two equations may be found by writing the first and second derivatives of Eq. (G.33) with respect to λ_4, for $i = 4$. In other words, the two additional equations are the first and second derivatives of $f(\lambda_4)$ with respect to λ_4, provided that $f(\lambda)$ and the polynomial $P(\lambda)$ possess the same first two derivatives for $\lambda = \lambda_4$. In this manner, there will be n equations that can be solved for the n unknowns. If $\lambda_i = \lambda_{i+1} = \lambda_{i+2} = \cdots = \lambda_{i+k}$, $k < n$, then k derivatives with respect to λ_i should be written.

APPENDIX H
Graphs of Maximum Magnification Factors and Times of Maximum Response for Undamped Single-Degree-of-Freedom Spring–Mass Systems

Each of the graphs in Figs. H.1 through H.4 contains a plot of the maximum magnification factor Γ_{max} versus the ratio of t_d/τ for single-degree undamped spring–mass systems, in addition to a second plot that gives the variation of the ratio t_m/τ versus the ratio t_d/τ, where t_m is the time of maximum response, τ is the period of vibration of the spring–mass system, and t_d is the time duration of the dynamic force. These graphs emphasize the great importance of the ratio t_d/τ regarding the dynamic response of structural and mechanical components. Examination of the graphs indicates that the maximum response is greatly dependent on the relative values of t_d and τ. This information helps the design engineer to mimimize the effect of the applied dynamic forces and to design a more efficient system. These graphs are also very convenient, because you only need to calculate τ in order to determine Γ_{max} and t_m. It should be noted that the values of Γ_{max} and t_m obtained from the graphs are applicable for displacements, stresses, moments, shear forces, and so on.

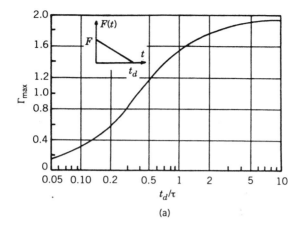

(a)

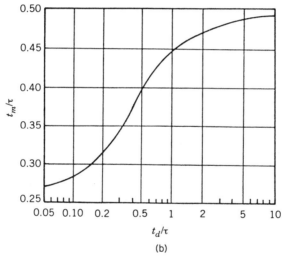

(b)

FIGURE H.1   (From ref. 9.)

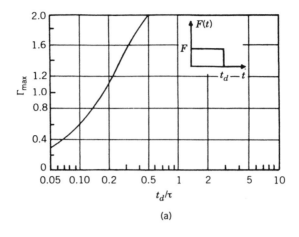

(a)

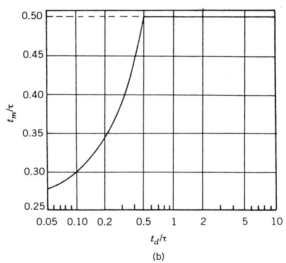

(b)

FIGURE H.2   (From ref. 9.)

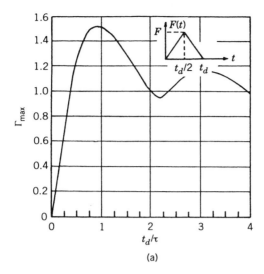

(a)

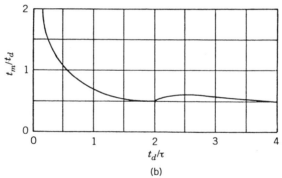

(b)

FIGURE H.3 (From ref. 9.)

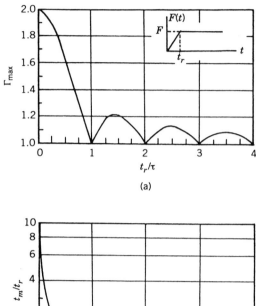

(a)

(b)

FIGURE H.4   (From ref. 9.)

APPENDIX I
Sample Calculations of Reduced Modulus E_r

Calculation of the reduced modulus E_r is based on the idea of assuming a stress distribution through the particular cross section under consideration. The stress distribution is assumed to have the same distribution as the stress–strain diagram of the material, whereby the strain in the outside fiber is proportional to the height of the beam. The area under the stress–strain diagram is calculated, and the moment of this area about the vertical (stress) axis is taken. The moment required to produce this stress distribution is calculated and compared to the actual moment. If they do not match, then a larger or smaller strain is assumed and the calculations are repeated. The following example illustrates the procedure.

In Fig. (I.1a), a variable stiffness cantilever beam with a concentrated force of 60,000 lb at the free end is used as an example. The reduced modulus at $x = 0$, which is the fixed end of the member, is calculated. The actual moment M_x is

$$M_x = (120)(60,000) = 7.20 \times 10^6 \, \text{lb} \cdot \text{in.}$$

The depth $h = 10.0$ in., and $I_A = 500.0$ in.^{4}. We now use Fig. I.1b and assume ε_1 as follows:

$$\varepsilon_1 = 5.425 \times 10^{-3} \, \text{in./in.}$$

$$\Delta = 2\varepsilon_1 = 10.850 \times 10^{-3}$$

The moment of the area of the stress–strain diagram about the origin O is calculated, and it is found to be equal to 1.4127 lb/in. The reduced modulus E_r is

$$E_r = \frac{12}{\Delta^3} \, (\text{moment of the area})$$

$$= \frac{12}{(10.850 \times 10^{-3})^3} \, (1.4127)$$

$$= 13.272 \times 10^6 \, \text{psi}$$

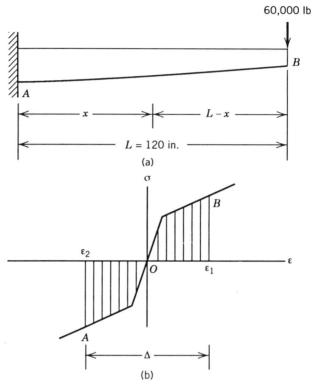

FIGURE I.1. (a) Variable stiffness cantilever beam loaded by a concentrated load at the free end. (b) Bilinear stress–strain curve approximation of Monel.

The radius of curvature r is

$$r = \frac{h}{\Delta} = \frac{10}{10.850 \times 10^{-3}}$$

$$= 921.66 \text{ in.}$$

The required moment M_{req} is

$$M_{\text{req}} = \frac{E_r I_A}{r} = \frac{(13.272 \times 10^6)(500)}{921.66}$$

$$= 7.20 \times 10^6 \text{ lb} \cdot \text{in.}$$

Therefore,

$$\text{required moment } M_{\text{req}} = \text{actual moment } M_x$$

and no further repetition of the procedure is required.

APPENDIX J
Laplace Transforms of Selected Time Functions

Function $f(t)$	Laplace Transform $\mathscr{L}f(t) = \mathscr{L}(s)$
1	$\dfrac{1}{s}$
a	$\dfrac{a}{s}$
t	$\dfrac{1}{s^2}$
$\delta(t)$	1
Ae^{-at}	$\dfrac{A}{(s+a)}$
te^{at}	$\dfrac{1}{(s-a)^2}$
$\dfrac{e^{at}-e^{bt}}{a-b}$	$\dfrac{1}{(s-a)(s-b)}$
$\dfrac{ae^{at}-be^{bt}}{a-b}$	$\dfrac{s}{(s-a)(s-b)}$
$\dfrac{t^{a-1}}{(a-1)!}$	$\dfrac{1}{s^a}$
$\dfrac{e^{-a^2/4t}}{(\pi t)^{1/2}}$	$\dfrac{e^{-a\sqrt{s}}}{\sqrt{s}}$
$\dfrac{\sin \omega t}{\omega}$	$\dfrac{1}{s^2+\omega^2}$
$\cos \omega t$	$\dfrac{s}{s^2+\omega^2}$
$\dfrac{1}{\omega}\sinh \omega t$	$\dfrac{1}{s^2-\omega^2}$

Function $f(t)$	Laplace Transform $\mathscr{L}f(t) = \mathscr{L}(s)$
$\cosh \omega t$	$\dfrac{s}{s^2 - \omega^2}$
$t \cos \omega t$	$\dfrac{s^2 - \omega^2}{(s^2 + \omega^2)^2}$
$\dfrac{1}{t}(e^{bt} - e^{at})$	$\log\left(\dfrac{s-a}{s-b}\right)$
$\dfrac{1}{t}\sin kt$	$\arctan \dfrac{k}{s}$
$\dfrac{e^{-a^2/4t}}{2(\pi t)^{3/2}}$	$\dfrac{e^{-a\sqrt{s}}}{\pi a}$
$\dfrac{df(t)}{dt}$	$sf(s) - f(0)$
$\dfrac{d^n f(t)}{dt^n}$	$s^n F(s) - \displaystyle\sum_{k=1}^{n} s^{n-k}\dfrac{d^{k-1}f(0)}{dt^{k-1}}$

Appendix K
NLFinite.For Computer Program and Output

```
C NLFINITE.FOR (GEOMETRIC NONLlNEAR PR0BS)
C REVISED 10-24-90 (IF STATEMENT IN JCBI REVISED
C MODIFICATION OF PROGRAM FlNITEL.FOR VIBRATION
C ANALYSIS OF BEAMS, RODS, AND PLANE FRAMES USING BEAM
C ELEMENTS WITH AN AXIAL PRETENSION LOAD. SUBROUTINES
C JCBI, DECOMP, MATINV, MATMPY, AND SEARCH.
C DEVICE * IN READ AND WRITE STATEMENTS IS THE CONSOLE.
C DEVICE 2 IN WRITE STATEMENTS IS THE PRINTER.
C * IN THE PLACE OF A FORMAT STATEMENT NUMBER MEANS
  FREE FORMAT
      IMPLICIT REAL * 8 (A-H,O-Z)
      REAL * 8 L.IA,KEL,MEL,KEG,MEG
      INTEGER SUB,ROWSUB,COLSUB,B,Z.EN,CFIX
      DIMENSION KEL(6,6,8),MEL(6,6,8),KEG(6,6,8),
      MEG(6,6,8),
      $RT(6,6),
      $R(6,6), TK(6,6), TM(6,6), SK(27,27), SM(27,27),
      RSK(27,27),
      $RSM(27,27),E(8),A(8),X(9),Y(9),GAMMA(8),IA(8)
      DIMENSION JNM(8,2),CFIX(27),SUB(6)
      OPEN (UNIT=2,FILE='PRN')
C READ IN PROBLEM DATA AS INDICATED BY MESSAGES ON
  CONSOLE. DATA
C REAO IN ARE PRINTED OUT. (PROGRAM STATEMENTS 2
  THROUGH 40)
      WRITE(2,1)
    1 FORMAT(`NLFINITE.FOR; FULL KE+KG MATRICES,REVISED
      10-24-90')
      WRITE(*,2)
    2 FORMAT(/,' ENTER THE NUMBER OF BEAM ELEMENTS
      (I1)',/)
      READ(*,3)NUMEL
    3 FORMAT(I1)
      DO 4 I=1,NUMEL
      WRITE(*,5)I
```

```
 4 READ(*,6)A(I)
 5 FORMAT(/,' ENTER A(',I1,')  (F20.0)',/)
 6 FORMAT(F20.0)
   WRITE(2,7)
 7 FORMAT(6X,'THE AREA ARRAY A IS:',/)
   DO 8 I=1,NUMEL
 8 WRITE(2,9)I,A(I)
 9 FORMAT(6X,'A(',I1,') = ',E14.7)
   DO 10 I=1,NUMEL
   WRITE(*,11)I
10 READ(*,12)E(I)
11 FORMAT(/,' ENTER E(',I1,')  (F20.0)',/)
12 FORMAT(F20.0)
   WRITE(2,13)
13 FORMAT(/6X,'THE ELASTICITY ARRAY E IS;',/)
   DO 14 I=1,NUMEL
14 WRITE(2,15)I,E(I)
15 FORMAT(6X,'E(',I1,') = ',E14.7)
   DO 16 I=1 ,NUMEL
   WRITE(*,17)I
16 READ(*,18)IA(I)
17 FORMAT(/,'ENTER IA(',I1,')  (F20.0)',/)
18 FORMAT(F20.0)
   WRITE(2,19)
19 FORMAT(/,6X, 'THE MOMENT OF INERTIA ARRAY IA
   IS; ',/)
   DO 20 I=1 ,NUMEL
20 WRITE(2,21)I,IA(I)
21 FORMAT(6X,'IA(',I1,') = ',E14.7)
   DO 82 I=1, NUMEL
   WRITE(*,83)I
82 READ(*,84)GAMMA(I)
83 FORMAT(/,' ENTER GAMMA(',I1,')  (F20.0)',/)
84 FORMAT(F20.0)
   WRITE(2,85)
85 FORMAT(/,6X, 'THE GAMMA ARRAY IS;',/)
   DO 86 I=1, NUMEL
86 WRITE(2,87)I,GAMMA(I)
87 FORMAT(6X, 'GAMMA(',I1,') = ',E14.7)
   WRITE(*,88)
88 FORMAT(' ','ENTER THE AXIAL TENSION PRELOAD
   (PLOAD)',/ )
   READ(*,89)PLOAD
89 FORMAT(F20.0)
   WRITE(2,90)
90 FORMAT(/,6X,'THE AXIAL PRETENSION LOAD IS; ',/)
   WRITE(2,91)PLOAD
91 FORMAT(F20.0)
   WRITE(*,22)
```

```
   22 FORMAT(/,'ENTER THE NUMBER OF JOINTS, NJTS
      (I1)',/)
      READ(*,23)NJTS
   23 FORMAT(I1)
      DO 24 I=1,NUMEL
      DO 24 J=1,2
      WRITE(*,25)I,J
   24 READ(*,26)JNM(I,J)
   25 FORMAT(/,' ENTER JNM(',I1, ',' ,I1,') (I1)',/)
   26 FORMAT(I1)
      WRITE(2,27)
   27 FORMAT(/,6X, 'THE JOINT-NUMBER MATRIX IS; ',/)
      DO 28, I=1, NUMEL
   28 WRITE(2,29)JNM(I,1),JNM(I,2)
   29 FORMAT(10X, I5,I4)
      DO 30 I=1,NJTS
      WRITE(*,31 )I,I
   30 READ(*,*)X(I),Y(I)
   31 FORMAT(/ ' ENTER JOINT COORD. X(', I1,'),Y(',I1,')
      (2F20.0)',/)
      WRITE(2,33)
   33 FORMAT(/,6X,'THE JOINT COORDINATES ARE; ',/)
      DO 34 I=1,NJTS
   34 WRITE(2,35)I,X(I),I,Y(I)
   35 FORMAT (6X,'X(',I1,') =',E14.7,5X,'Y(',I1,') =
      ',E14.7)
      WRITE(*,36)
   36 FORMAT( /,' ENTER THE NUMBER OF FIXED COORDINATES
      (12)',/)
      READ(*,37)NB
   37 FORMAT(I2)
      IF(N8.EQ.0)GO TO 94
      DO 38 I=1,NB
      WRITE(*,39)I
   38 READ(*,40)CFIX(I)
   39 FORMAT( /,' ENTER CFIX(',I2,') (I2)',/)
   40 FORMAT(I2)
      WRITE(2,41)
   41 FORMAT(/,6X,'ARRAY CFIX IS;',/)
      DO 42 I=1,NB
   42 WRITE(2,43)I,CFIX(I)
   43 FORMAT(6X,'CFIX(',I2,')=',I2)
C GENERATE NULL 3-DIMENSIONAL ARRAYS KEL AND MEL.
  PLANES OF KEL
C AND MEL WILL LATER CONTAIN THE LOCAL ELEMENT
  STIFFNESS ANO MASS
C MATRICES, RESPECTIVELY .
   94 DO 44 I=1,6
      DO 44 J=1,6
```

```
      DO 44 M=1,NUMEL
      KEL(I,J,M)=0.
   44 MEL(I,J,M)=O.
C GENERATE NULL MATRICES R ANO RT WHICH WILL LATER
  BECOME THE
C TRANSFORMATION MATRIX ANO ITS TRANSPOSE,
  RESPECTIVELY.
      DO 45 I=1,6
      DO 45 J=1,6
      R(I,J)=0.0
   45 RT(I,J)=0.0
C GENERATE THE LOCAL ELEMENT STIFFNESS MATRICES AND
  STORE IN THE
C 3-DIMENSIONAL STIFFNESS ARRAY KEL (SEE FIG. 6.2 FOR
  THE
C EQUATION USED). EACH PLANE IN THE 3-DIM. ARRAY IS ONE
  ELEMENT
C STIFFNESS MATRIX.
      DO 100 EN=1,NUMEL
      IC=JNM(EN,1)
      ID=JNM(EN,2)
      L=DSORT((X(ID)-X(IC))**2+(Y(ID)-Y(IC))**2)
      QUOT=IA(EN)/A(EN)
      R1=DSORT(QUOT)
      F=E(EN)*IA(EN)/L
      P=F/R1**2
      Q=4.*P*R1**2
      S=3.*Q/(2.*L)
      T=S*2./L
      SINA=(Y(ID)-Y(IC))/L
      COSA=(X(ID)-X(IC))/L
      KEL(1,1,EN)=P
      KEL(1,4,EN)=-P
      KEL(2,2,EN)=T+(6.*PLOAD/(5.*L))
      KEL(2,3,EN)=S+.1*PLOAD
      KEL(2,5,EN)=-T-(6.*PLOAD/(5.*L))
      KEL(2,6,EN)=S+.1*PLOAD
      KEL(3,3,EN)=Q+(2.*PLOAD*L/15.)
      KEL(3,5,EN)=-S-.1*PLOAD
      KEL(3,6,EN)=Q/2.-(PLOAD*L/30.)
      KEL(4,4,EN)=P
      KEL(5,5,EN)=T+(6.*PLOAD/(5.*L))
      KEL(5,6,EN)=-S-.1*PLOAD
      KEL(6,6,EN)=Q+(2.*PLOAD*L/15.)
      DO 46 I=2,6
      IM1=I-1
      DO 46 J=1, IM1
   46 KEL(I,J,EN)=KEL(J,I.EN).
```

```
  GENERATE THE LOCAL ELEMENT MASS MATRICES AND STORE
  THEM IN THE
C 3-DIMENSIONAL MASS ARRAY MEL (SEE FIG. 6.2 FOR THE
  EQUATION
C USED). EACH PLANE Of THE 3-DIM. ARRAY MEL CONTAINS
  ONE LOCAL
C ELEMENT MASS MATRIX.
      F=GAMMA(EN)=L/420.
      P=70.*F
      P2=2.*P
      Q=156.*F
      S=22.*L*F
      T=54.*F
      U=4.*L*L*F
      V=13.*L*F
      W=3.*L*L*F
      MEL(1,1,EN)=P2
      MEL(1,4,EN)=p
      MEL(2,2,EN)=Q
      MEL(2,3,EN)=S
      MEL(2,5,EN)=T
      MEL(2,6,EN)=-V
      MEL(3,3,EN)=U
      MEL(3,5,EN)=V
      MEL(3,6,EN)=-W
      MEL(4,4,EN)=P2
      MEL(5,5,EN)=Q
      MEL(5,6,EN)=-S
      MEL(6,6,EN)=U
      DO 47 I=2,6
      IM1=I-1
      DO 47 J=1.IM1
   47 MEL(I,J,EN)=MEL(J,I,EN)
C GENERATE THE TRANSFORMATION MATRIX R ANO ITS
  TRANSPOSE RT.
      R(1,1)=COSA
      R(1,2)=SINA
      R(2,1)=-SINA
      R(2,2)=COSA
      R(3,3)=1.
      R(4.4)=COSA
      R(4,5)=SINA
      R(5,4)=-SINA
      R(5,5)=COSA
      R(6,6)=1.
      DO 48 I=1,3
      DO 48 J=1,3
      RT(I,J)=R(J,I)
   48 RT(I+3,J+3)=R(J+3,I+3)
```

```
C DETERMINE THE ELEMENT STIFFNESS MATRICES IN THE
  GLOBAL
C COORDINATE SYSTEM AND STORE THEM IN THE 3-DIM.
C STIFFNESS ARRAY KEG. EACH PLANE OF THE 3-DIM. ARRAY
  CONTAINS
C ONE GLOBAL ELEMENT STIFFNESS MATRIX.
      DO 95 I=1,6
      DO 95 J=1,6
      TK(I,J)=0.0
      DO 95 K=1,6
   95 TK(I,J)=TK(I,J)+KEL(I,K,EN)=R(K,J)
      DO 96 I=1,6
      DO 96 J=1,6
      KEG(I,J,EN)=0.0
      DO 96 K=1,6
   96 KEG(I,J,EN)=KEG(I,J,EN)+RT(I,K)=TK(K,J)
C DETERMINE THE ELEMENT MASS MATRICES IN THE GLOBAL
  SYSTEM
C AND STORE THEM IN THE 3-DIM. MASS ARRAY MEG. EACH
  PLANE
C OF THE 3-DIM. ARRAY CONTAINS ONE GLOBAL ELEMENT MASS
  MATRIX.
      DO 97 I=1,6
      DO 97 J=1,6
      TM(I,J)=0.0
      DO 97 K=1,6
   97 TM(I,J)=TM(I,J)+MEL(I,K,EN)=R(K,J)
      DO 98 I=1,6
      DO 98 J=1,6
      MEG(I,J,EN)=0.0
      DO 98 K=1,6
   98 MEG(I,J,EN)=MEG(I,J,EN)+RT(I,K)=TM(K,J)
  100 CONTINUE
C GENERATE NULL MATRICES SK ANO SM WHICH WILL BECOME
C THE SYSTEM STIFFNESS ANO MASS MATRICES, RESPECTIVELY.
      N=NJTS*3
      DO 49 I=1,N
      DO 49 J=1,N
      SK(I,J)=0.
   49 SM(I,J)=0.
C ASSEMBLE THE STIFFNESS ANO MASS MATRICES.
      DO 51 I=1,NUMEL
      DO 50 J=1,2
      DO 50 M=1,3
      J1=J*3-M+1
   50 SUB(J1)=3*JNM(I,J)-M+1
      DO 51 B=1,6
      DO 51 Z=1,6
      ROWSUB=SUB(B)
```

```
      COLSUB=SUB(Z)
      SK(ROWSUB,COLSUB)=SK(ROWSUB,COLSUB)+KEG(B,Z,I)
   51 SM(ROWSUB,COLSUB)=SM(ROWSUB,COLSUB)+MEG(B,Z,I)
C CALCULATE THE NUMBER OF DEGREES OF FREEDOM AND REMOVE
  ROWS AND
C COLUMNS FROM THE SYSTEM STIFFNESS AND MASS MATRICES.
      NF=N-NB
      IF(NB .EQ. 0)GO TO 69
      NA=1
      KL=N-1
   62 JC=1
   63 IF(JC .EQ. CFIX(NA))GO TO 64
      JC=JC+1
      IF(JC .EQ. N)GO TO 68
   GO TO 63
   64 DO 65 I=1,N
      DO 65 J=JC,KL
      SK(I,J)=SK(I,J+1)
      SM(I,J)=SM(I,J+1)
   65 CONTINUE
      DO 66 J=1,N
      DO 66 I=JC,KL
      SK(I,J)=SK(I+1,J)
      SM(I,J)=SM(I+1,J)
   66 CONTINUE
      IF(NA .EQ. NB)GO TO 68
      NA=NA+1
      DO 67 I=NA,NB
   67 CFIX(I)=CFIX(I)-1
      GO TO 62
   68 CONTINUE
C ASSIGN REDUCED STIFFNESS AND MASS MATRIX ELEMENTS TO
  ARRAY
C NAMES RSK AND RSM, RESPECTIVELY.
   69 DO 70 I=1,NF
      DO 70 J=1,NF
      RSK(I,J)=SK(I,J)
   70 RSM(I,J)=SM(I,J)
C WRITE OUT THE REOUCED STIFFNESS AND MASS MATRICES
  OBTAINED
C FROM THE BOUNDARY CONDITIONS.
      WRITE(2,71)
   71 FORMAT(/,' THE REDUCED SYSTEM STIFFNESS MATRIX
      IS: ',/)
      WRITE(2,72) ((RSK(I,J),J=1,NF),I=1,NF)
   72 FORMAT(' ',6E11.4/)
      WRITE(2,73)
   73 FORMAT(/,' THE REDUCED SYSTEM MASS MATRIX IS: ',/)
      WRITE(2,74) ((RSM(I,J),J=1,NF),I=1,NF)
```

```
   74 FORMAT(' ',6E11.4/)
      WRITE(2,200)
  200 FORMAT(//,' ')
C CALL SUBPROGRAM JCBI TO CALCULATE FREQUENCIES AND
  MODE SHAPES
      CALL JCBI(NF,RSK,RSM)
      STOP
      END
C LIBRARY. FOR
C SUBROUTINES JCBI, DECOMP, MATINV, MATMPY, AND SEARCH
  AS REQUI
C BY FINITEL.FOR AND TRUSS.FOR
      SUBROUTINE JCBI(N,K,M)
      IMPLICIT REAL*8 (A-H,O-Z)
      REAL*8 K,M,L,LT,LINV, LINVTR,RT,A, OMEGA,PROD, AV,
      DIFF,RAD
      REAL*8 COSINE,SINE,Q.PROD1
      DIMENSION K(27,27),RT(27,27),A(27,27)
      DIMENSION OMEGA(27),M(27,27),L(27,27),LT(27,27)
      DIMENSION LINV(27,27),LINVTR(27,27),PROD(27,27)
      CALL DECOMP(M,N,L,LT)
      CALL MATINV(L,LINV,N)
      DO 204 I=1,N
      DO 204 J=1,N
  204 LINVTR(I,J)=LINV(J,I)
      CALL MATMPY(N,K,LINVTR,PROD)
      CALL MATMPY(N,LINV,PROD,A)
      DO 14 I=1,N
      DO 13 J=1,N
      RT(I,J)=0.0
   13 CONTINUE
      RT(I,I)=1.0
   14 CONTINUE
      NSWEEP=0
   15 NRSKIP=0
      NMIN1=N-1
      DO 25 I=1,NMIN1
      IP1=I+1
      DO 24 J=IP1,N
      AV=0.5*(A(I,J)+A(J,I))
      DIFF*DELTA(T,T)=DELTA(T,T)
      RAD=DSORT(DIFF*DIFF+4.*AV*AV)
      IF(RAD .EQ .0 .0)GO TO 20
      IF(DIFF .LT .0 .0)GO TO 18
      IF(DABS(A(I,I)).EQ.DABS(A(I,I)+100.*DABS(AV))
      GO TO 16
      GO TO 17
   16 IF(DABS(A(J,J)).EQ.DABS(A(J,J))+100.0*DABS(AV))GO
      TO 20
```

```
   17 COSINE=DSORT((RAD+DIFF)/(2.0*RAD))
      SINE=AV/(RAD*COSINE)
      GO TO 19
   18 SINE=DSQRT((RAD-DIFF)/(2.0*RAD))
      IF(AV.LT.0.0)SINE=-SINE
      COSINE=AV/(RAD*SINE)
C  REVISION OF IF STATEMENT FROM ORIGINAL PROGRAM
   19 DBS=DABS(SINE)
      IF(DBS.GT.1.OE-16)GO TO 21
   20 NRSKIP=NRSKIP+1
      GO TO 24
   21 DO 22 L1=1,N
      Q=A(I,L1)
      A(I,L1)*COSINE*Q+SINE*A(J,L1)
      A(J,L1)*-SINE*Q+COSINE*A(J,L1)
   22 CONTINUE
      DO 23 L1=1,N
      Q=A(L1,I)
      A(L1,I)=COSINE*Q+SINE*A(L1,J)
      A(L1,J)=-SINE*Q+COSINE*A(L1,J)
      Q=RT(L1,I)
      RT(L1,I)=COSINE*Q+SINE*RT(L1,J)
      RT(L1,J)=-SINE*Q+COSINE*RT(L1,J)
   23 CONTINUE
   24 CONTINUE
   25 CONTINUE
C  KEEP A TALLY OF THE NUMEER OF SWEEPS.
      NSWEEP*NSWEEP+1
      IF(NSWEEP.GT.100)GO TO 33
      WRITE(2,26)NRSKIP,NSWEEP
   26 FORMAT(' ',5X,'THERE WERE ',I2,
     $' ROTATIONS SKIPPED ON SWEEP NUMBER ',I2)
      IF(NRSKIP.LT.N*(N-1)/2)GO TO 15
      PROD1=0.0
      DO 27 J=1,N
      PROD1=PROD1+RT(J,1)*RT(J,N)
   27 CONTINUE
      WRITE(2,28)
   28 FORMAT(/, ' ',5X, 'THE SCALAR PRODUCT OF THE FIRST
      AND LAST')
      WRITE(2,29 )PROD1
   29 FORMAT(' ',5X, 'EIGENVECTORS OF THE TRANSFORMED
      MATRIX IS ',
     $F19.17/)
      CALL MATMPY(N,LINVTR,RT,PROD)
      DO 30 I=1,N
      DO 30 J=1,N
   30 RT(I,J)=PROD(I,J)
      DO 42 J=1,N
```

```
          SUM=0.0
          DO 31 I=1,N
    31    SUM=SUM+DABS(RT(I,J))
          AV=SUM/N
          QUOT=DABS(RT(1,J))/AV
          IF(QUOT.LT.0.000001)GO TO 40
          DO 32 I=2,N
    32    RT(I,J)=RT(I,J)/RT(1,J)
          RT(1,J)=1.000
          GO TO 42
    40    CALL SEARCH(RT,J,II,N)
          BIG=RT(II,J)
          DO 41 I=1, N
    41    RT(I,J)=RT(I,J)/BIG
    42    CONTINUE
          DO 110 I=1,N
          IF(A(I,I).LE.0.0)GO TO 43
          OMEGA(I)=DSQRT(A(I,I))
          GO TO 110
    43    OMEGA(I)=0.0
   110    CONTINUE
    33    WRITE(2,34)NSWEEP
    34    FORMAT(/,' ',5X,'THERE WERE ',I3,' SWEEPS
          PERFORMED.',
          $/,5X,' THE EIGENVALUES ANO EIGENVECTORS FOLLOW:')
          DO 39 JJ=1,N
          J=N-JJ+1
          WRITE(2,35)JJ,A(J,J)
    35    FORMAT(/,' ',5X,'LAMBDA (',I2,') = ',F20.4)
          WRITE(2,111)JJ,OMEGA(J)
   111    FORMAT(' ',5X,'OMEGA(',I2,') = ',F20.4,' RAD/S')
          WRITE(2,36)
    36    FORMAT(/,' ',5X,'THE ASSOCIATED EIGENVECTOR IS:')
          DO 37 I=1,N
    37    WRITE(2,38)RT(I,J)
    38    FORMAT(' ',5X,D17.10)
    39    CONTINUE
          RETURN
          END
C
C
C
          SUBROUTINE DECOMP(A,N,L,LT)
          IMPLICIT REAL*8(A-H,O-Z)
          DIMENSION A(27,27)
          REAL*8 L(27,27),LT(27,27)
          DO 9 J=1,N
          IF(J.EQ.1)GO TO 7
          JM1=J-1
```

```
      DO 6 I=J,N
      IF(I.NE.J)GO TO 4
      SUM=0.0
      DO 3 K=1,JM1
    3 SUM=SUM+L(I,K)*L(J,K)
      L(J,J)=DSQRT(A(J,J)-SUM)
      GO TO 6
    4 SUM=0.0
      DO 5 K=1,JM1
    5 SUM=SUM+L(I,K)*L(J,K)
      L(I,J)=(A(I,J)-SUM)/L(J,J)
    6 CONTINUE
      GO TO 9
    7 L(1,1)=DSQRT(A(1,1))
      DO 8 I=2,N
    8 L(I,1)=A(I,1)/L(1.1)
    9 CONTINUE
C FILL IN ZERO VALUES OF MATRIX L
      DO 11 J=2,N
      JM1=J-1
      DO 11 I=1,JM1
   11 L(I,J)=0.0
C ASSIGN VALUES TO THE UPPER TRIANGULAR MATRIX LT
      DO 12 I=1,N
      DO 12 J=1,N
   12 T(I,J)=L(J,I)
      RETURN
      END



C
C
C
      SUBROUTINE MATINV(B,A,N)
C MATRIX INVERSION USING GAUSS-JORDAN REDUCTION AND
  PARTIAL
C PIVOTING. MATRIX B IS THE MATRIX TO BE INVERTED AND A
  IS
C THE INVERTED MATRIX.
      IMPLICIT REAL*8(A-H,O-Z)
      DIMENSION B(27,27),A(27,27),INTER(27,2)
      DO 2 I=1,N
      DO 2 J=1,N
    2 A(I,J)=B(I,J)
C CYCLE PIVOT ROW NUMBER FROM 1 TO N
      DO 12 K=1,N
      JJ=K
      IF(K.EQ.N)GO TO 6
      KP1=K+1
```

```
        BIG=DABS(A(K,K))
C SEARCH FOR LARGEST PIVOT ELEMENT
        DO 5 I=KP1,N
        AB=DABS(A(I,K))
        IF(BIG-AB)4,5,5
     4  BIG=AB
        JJ=I
     5  CONTINUE
C MAKE DECISION ON NECESSITY OF ROW INTERCHNGE AND
C STORE THE NUMBER OF THE TWO ROWS INTERCHANGED DURING
  KTH
C REDUCTION. IF NO INTERCHANGE, BOTH NUMBERS STORED
  EQUAL K
     6  INTER(K,1)=K
        INTER(K,2)=JJ
        IF(JJ-K)7,9,7
     7  DO 8 J=1,N
        TEMP=A(JJ,J)
        A(JJ,J)=A(K,J)
     8  A(K,J)=TEMP
C CALCULATE ELEMENTS OF REDUCED MATRIX
C FIRST CALCULATE NEW ELEMENTS OF PIVOT ROW
     9  DO 10 J=1,N
        IF(J.EQ.K)GO TO 10
        A(K,J)=A(K,J)/A(K,K)
    10  CONTINUE
C CALCULATE ELEMENT REPLACING PIVOT ELEMENT
        A(K,K)=1./A(K,K)
C CALCULATE NEW ELEMENTS NOT IN PIVOT ROW OR COLUMN
        DO 11 I=1,N
        IF(I.EQ.K)GO TO 11
        DO 110 J=1,N
        IF(J.EQ.K)GO TO 110
        A(I,J)=A(I,J)-A(K,J)*A(I,K)
   110  CONTINUE
    11  CONTINUE
C CALCULATE NEW ELEMENTS FOR PIVOT COLUMN-EXCEPT PIVOT
  ELEMENT
        DO 120 I=1,N
        IF(I.EO.K)GO TO 120
        A(I,K)=-A(I,K)*A(K,K)
   120  CONTINUE
    12  CONTINUE
C REARRANGE COLUMNS Of FINAL MATRIX OBTAINED
        DO 13 L=1,N
        K=N-L+1
        KROW=INTER(K,1)
        IROW=INTER(K,2)
        IF(KROW.EQ.IROW)GO TO 13
```

```
        DO 130 I=1,N
        TEMP=A(I,IROW)
        A(I,IROW)=A(I.KROW)
        A(I,KROW)=TEMP
  130   CONTINUE
   13   CONTINUE
        RETURN
        END
C
        SUBROUTTNE MATMPY(N,A,B,C)
        IMPLICIT REAL*8(A-H,O-Z)
C C IS THE PRODUCT MATRIX OF A AND B
        DIMENSION A(27,27),B(27,27),C(27,27)
        DO 2 I=1,N
        DO 2 J=1,N
        C(I,J)=0.0
        DO 2 K=1,N
    2   C(I,J)=C(I,J)+A(I,K)*8(K,J)
        RETURN
        END
C
C
        SUBROUTINE SEARCH(RT,J,II,N)
C THIS SUBROUTINE SEARCHES THE JTH COLUMN OF THE MATRIX
  RT
C FOR THE LARGEST EIGENVECTOR COMPONENT. ITS ROW NUMBER
  IS
C ASSIGNED TO THE NAME II.
        IMPLICIT REAL*8(A-H,O-Z)
        DIMENSION RT(27,27)
        II=1
        BIG=DABS(RT(1,J))
        DO 3 I=2,N
        AB=DABS(RT(I,J))
        IF(BIG-AB)2,3,3
    2   BIG=AB
        II=I
    3   CONTINUE
        RETURN
        END


NLFINITE.FOR; FULL KE+KG MATRICES,REVISED 10-24-90
        THE AREA ARRAY A IS:
        A(1)=0.4800000E+02
        THE ELASTICITY ARRAY E IS;
        E(1)=0.3000000E+08
        THE MOMENT OF INERTIA ARRAY IA IS;
        IA(1)=0.1000000E+04
```

```
          THE GAMMA ARRAY IS;
          GAMMA(1)=0.3525000E-01
          THE AXIAL PRETENSION LOAD IS;
          0.
          THE JOINT-NUMBER MATRIX IS;
          1 2
          THE JOINT COORDINATES ARE;
          X(1)=0.00000000E+00 Y(1)=0.0000000E+00
          X(2)=0.10000000E+03 Y(2)=0.0000000E+00
THE REDUCED SYSTEM STIFFNESS MATRIX IS:
 0.1440E+08 0.0000E+00 0.0000E+00
 0.0000E+00 0.3600E+06 0.1800E+08
 0.0000E+00 0.1800E+08 0.1200E+10
-0.1440E+08 0.0000E+00 0.0000E+00
 0.0000E+00-0.3600E+06-0.1800E+08
 0.0000E+00 0.1800E+08 0.6000E+09
                      -0.1440E+08 0.0000E+00 0.0000E+00
                       0.0000E+00-0.3600E+06 0.1800E+08
                       0.0000E+00-0.1800E+08 0.6000E+09
                       0.1440E+08 0.0000E+00 0.0000E+00
                       0.0000E+00 0.3600E+06-0.1800E+08
                       0.0000E+00-0.1800E+08 0.1200E+10
THE REDUCED SYSTEM MASS MATRIX IS:
0.1175E+01 0.0000E+00 0.0000E+00
0.0000E+00 0.1309E+01 0.1846E+02
0.0000E+00 0.1846E+02 0.3357E+03
0.5875E+00 0.0000E+00 0.0000E+00
0.0000E+00 0.4532E+00 0.1091E+02
0.0000E+00-0.1091E+02-0.2518E+03
                       0.5875E+00 0.0000E+00 0.0000E+00
                       0.0000E+00 0.4532E+00-0.1091E+02
                       0.0000E+00 0.1091E+02-0.2518E+03
                       0.1175E+01 0.0000E+00 0.0000E+00
                       0.0000E+00 0.1309E+01-0.1846E+02
                       0.0000E+00-0.1846E+02 0.3357E+03


          THERE WERE   5 ROTATIONS SKIPPED ON SWEEP NUMBER 1
          THERE WERE   9 ROTATIONS SKIPPED ON SWEEP NUMBER 2
          THERE WERE   9 ROTATIONS SKIPPED ON SWEEP NUMBER 3
          THERE WERE  14 ROTATIONS SKIPPED ON SWEEP NUMBER 4

     THERE WERE 15 ROTATIONS SKIPPED ON SWEEP NUMBER 5

     THE SCALAR PRODUCT OF THE FIRST AND LAST
     EIGENVECTORS OF THE TRANSFORMED MATRIX
     IS 0.00000000000000000

     THERE WERE 5 SWEEPS PERFORMED.
```

THE EIGENVALUES AND EIGENVECTORS FOLLOW:

LAMBDA (1) = 0.0000
OMEGA (1) = 0.0000 RAD/S

THE ASSOCIATED EIGENVECTOR IS:
0.1000000000D+01
0.0000000000D+00
0.0000000000D+00
0.1000000000D+01
0.0000000000D+00
0.0000000000D+00

LAMBDA (2) = 0.0000
OMEGA (2) = 0.0000 RAD/S

THE ASSOCIATED EIGENVECTOR IS:
0.0000000000D+00
0.9334669755D+00
0.6653302446D-03
0.0000000000D+00
0.1000000000D+01
0.6653302446D-03

LAMBDA (3) = 0.0000
OMEGA (3) = 0.0000 RAD/S

THE ASSOCSATED EIGENVECTOR IS:
 0.0000000000D+00
 0.1000000000D+01
-0.1977319320D-01
 0.0000000000D+00
-0.9773193204D+00
-0.1977319320D-01

LAMBDA (4) = 6127659.5745
OMEGA (4) = 2475.4110 RAD/S

THE ASSOCIATED EIGENVECTOR IS:
 0.0000000000D+00
 0.1000000000D+01
-0.6000000000D-01
 0.0000000000D+00
 0.1000000000D+01
 0.6000000000D-01

LAMBDA (5) = 49021276.5957
OMEGA (5) = 7001.5196 RAD/S

THE ASSOCIATED EIGENVECTOR IS:
 0.1000000000D+01
 0.0000000000D+00
 0.0000000000D+00
-0.1000000000D+01
 0.0000000000D+00
 0.0000000000D+00

LAMBDA (6) = 71489361.7021
OMEGA (6) = 8455.1382 RAD/S

THE ASSOCIATED EIGENVECTOR IS:
 0.00000000000+00
 0.1000000000D+01
-0.1200000000D+00
 0.0000000000D+00
-0.1000000000D+01
-0.1200000000D+00

APPENDIX L
Necessary Conditions for Functionals with Fixed Boundaries

Functionals with Fixed Boundaries	Necessary Conditions
1. $\Omega(y(x)) = \int_{x_0}^{x_1} F(x, y, y') \, dx$ $y(x_0) = y_0, \quad y(x_1) = y_1$	1. $F_y - \dfrac{d}{dx} F_{y'} = 0 \quad$ (Euler equation)
2. $\Omega(y_1(x), y_2(x)) = \int_{x_0}^{x_1} F(x, y_1, y_2, y_1', y_2') \, dx$ $y_1(x_0) = y_{10}, \quad y_2(x_0) = y_{20}, \quad y_1(x_1) = y_{11}, \quad y_2(x_1) = y_{21}$	2. $F_{y_1} - \dfrac{d}{dx} F_{y_1'} = 0, \quad F_{y_2} - \dfrac{d}{dx} F_{y_2'} = 0$
3. $\Omega(y_1, y_2, \ldots, y_n) = \int_{x_0}^{x_1} F(x, y_1, y_2, \ldots, y_n, y_1', y_2', \ldots, y_n') \, dx$ $y_1(x_0) = y_{10}, \quad y_2(x_0) = y_{20}, \ldots, \quad y_n(x_0) = y_{n0}$ $y_1(x_1) = y_{11}, \quad y_2(x_1) = y_{21}, \ldots, \quad y_n(x_1) = y_{n1}$	3. $F_{y_i} - \dfrac{d}{dx} F_{y_i'} = 0 \quad i = 1, 2, 3, \ldots, n$
4. $\Omega(y(x)) = \int_{x_0}^{x_1} f(x, y, y', y'', \ldots, y^{(n)}) \, dx$ $y(x_0) = y_0, \quad y'(x_0) = y_0', \ldots, y^{(n-1)}(x_0) = y_0^{(n-1)}$ $y(x_1) = y_1, \quad y'(x_1) = y_1', \ldots, y^{(n-1)}(x_1) = y_1^{(n-1)}$	4. $F_y - \dfrac{d}{dx} F_{y'} + \dfrac{d^2}{dx^2} F_{y''} + \cdots + (-1)^{(n)} \dfrac{d^n}{dx^n} F_y^{(n)} = 0$ (Euler–Poisson equation)

Functionals with Fixed Boundaries (cont'd)

Necessary Conditions (cont'd)

5. $\Omega(y(x), z(x)) = \int_{x_0}^{x_1} F(x, y, y', \ldots, y^{(n)}, z, z', \ldots, z^{(n)}) \, dz$

5. $F_y - \dfrac{d}{dx} F_{y'} + \cdots + (-1)^n \dfrac{d^n}{dx^n} F_y^{(n)} = 0$

$F_z - \dfrac{d}{dx} F_{z'} + \cdots + (-1)^m \dfrac{d^m}{dx^m} F_z^{(m)} = 0$

6. $\Omega(y_1, y_2, \ldots, y_m) = \int_{x_0}^{x_1} F(x, y, y', \ldots, y_1^{(n_1)}, y_2, y_2', \ldots, y_2^{(n_2)}, \ldots, y_m, y_m', \ldots, y_m^{(n_m)}) \, dx$

6. $F_{y_i} - \dfrac{d}{dx} F_{y_i'} + \cdots + (-1)^{n_i} \dfrac{D^{n_i}}{dx^{n_i}} F_{y_i}^{(n_i)} = 0 \qquad i = 1, 2, \ldots, m$

7. $\Omega(z(x,y)) = \iint_D F\left(x, y, z, \dfrac{\partial z}{\partial x}, \dfrac{\partial z}{\partial y}\right) dx \, dy$

7. $F_x - \dfrac{d}{dx}\{F_p\} - \dfrac{\partial}{\partial y}\{F_q\} = 0, \quad p = \dfrac{\partial z}{\partial x}, \quad q = \dfrac{\partial z}{\partial y}$

(The second and third terms are total partial derivatives with respect to x and y, respectively.)

8. $\Omega(z(x_1, x_2, \ldots, x_n)) = \iint_D \cdots \int F(x_1, x_2, \ldots, x_n, z, p_1, p_2, \ldots, p_n) \, dx_1 \, dx_2 \cdots dx_n$

where $p_i = \dfrac{\partial z}{\partial x_i}$

8. $F_z - \displaystyle\sum_{i=1}^{n} \dfrac{\partial}{\partial x_i}\{F_{p_i}\} = 0$

9. $\Omega = \iint_D \left[\left(\dfrac{\partial^2 z}{\partial x^2}\right)^2 + \left(\dfrac{\partial^2 z}{\partial y^2}\right)^2 + 2\left(\dfrac{\partial^2 z}{\partial x \, \partial y}\right)^2 \right] dx \, dy$

9. $\dfrac{\partial^4 z}{\partial x^4} + 2\dfrac{\partial^4 z}{\partial x^2 \, \partial y^2} + \dfrac{\partial^4 z}{\partial y^4} = 0 \qquad$ (biharmonic equation)

References

1. D. G. Fertis, *Dynamics and Vibration of Structures*, John Wiley & Sons, New York, 1973.
2. D. G. Fertis, *Nonlinear Mechanics*, CRC Press, Boca Raton, FL, 1993.
3. F. C. Moon, *Chaotic and Fractal Dynamics*, John Wiley & Sons, New York, 1992.
4. G. L. Baker and J. P. Gollub, *Chaotic Dynamics*, Cambridge University Press, Cambridge, UK, 1990.
5. M. Barnsley, *Fractals Everywhere*, Academic Press, Orlando, FL, 1988.
6. M. S. El Naschie, *Stress, Stability and Chaos*, McGraw-Hill, New York, 1990.
7. D. G. Fertis, *Dynamics and Vibration of Structures*, rev. ed., Robert E. Krieger Publishing Co., Malabar, FL, 1984.
8. H. I. Laursen, *Structural Analysis*, McGraw-Hill, New York, 1978.
9. U.S. Army Corps of Engineers, *Design of Structures to Resist the Effects of Atomic Weapons*.
    - Manual EM-1110-345-413, *Weapon Effect Data*. Published July 1, 1959.
    - Manual EM-1110-345-415, *Principles of Dynamic Analysis and Design*. Published March 15, 1957.
    - Manual EM-1110-345-416, *Structural Elements Subjected to Dynamics Loads*. Published March 15, 1957.
10. D. G. Fertis, "Safety of Long-Span Bridges Based on Dynamic Response", in *Proceedings of the 87th Structures Congress of ASCE*, Orlando, FL, 1987, pp. 449–468.
11. R. V. Churchill, *Fourier Series and Boundary Value Problems*, McGraw-Hill, New York, 1941, pp. 70–94.
12. D. G. Fertis and M. E. Keene, "Elastic and Inelastic Analysis of Nonprismatic Members", *Journal of Structural Engineering*, ASCE, Vol. 116, No. 2, pp. 475–489, 1990.
13. D. G. Fertis and R. Taneja, "Equivalent Systems for Inelastic Analysis of Prismatic and Nonprismatic Members," *Journal of Structural Engineering*, ASCE, Vol. 117, No. 2, pp. 473–488, 1991.
14. D. G. Fertis and C. T. Lee, "Inelastic Analysis of Flexible Bars Using Simplified Nonlinear Equivalent Systems," *International Journal of Computers and Structures*, Vol. 41, No. 5, pp. 947–958, 1991.
15. *Steel Construction Manual*, 6th ed., AISC, New York, 1964.
16. S. Timoshenko and S. Woinosky-Krieger, *Theory of Plates and Shells*, 2nd ed., McGraw-Hill, New York, 1959.

17. D. G. Fertis and C. T. Lee, "Nonlinear Vibration of Axially Restrained Elastically Supported Beams," *Proceedings of the 1992 Workshop on Dynamics, ASME, PD-Vol. 44, Dynamics and Vibration*, Book No. G00652, 1992, pp. 13–19.

18. D. G. Fertis and C. T. Lee, "Nonlinear Vibration and Instabilities of Elastically Supported Beams with Axial Restraints," *International Journal of Sound and Vibration*, Vol. 165, No. 1, pp. 123–135, 1993.

19. D. G. Fertis, "Inelastlc Vibration of Prismatic and Nonprismatic Members," in *Proceedings of the Second European Joint Conference on Engineering Systems Design and Analysis*, July 4–7, 1994, London, England.

20. A. N. Kounadis, "The Existence of Regions of Divergence Instability for Nonconservative Systems Under Follower Forces," *International Journal of Solids and Structures*, Vol. 19, No. 8, pp. 725–733, 1983.

21. A. N. Kounadis, "Divergence and Flutter Instability of Elastically Restrained Structures Under Follower Forces," *International Journal of Engineering Science*, Vol. 19, pp. 553–562, 1981.

22. W. H. Press, B. P. Flannery, S. A. Tecikolsky, and W. T. Vetterling, *Numerical Recipes*, Cambridge University Press, Cambridge, UK, 1989, pp. 243–248.

23. D. G. Fertis and E. C. Zobel, *Transverse Vibration Theory*, Ronald Press, New York, 1961.

24. D. G. Fertis, "Dynamic Hinge Concept for Beam Vibrations," *Journal of the Structural Division, Proceedings of ASCE*, Paper 4689, pp. 365–380, February 1966.

25. D. G. Fertis and G. E. Chidiac, "Dynamically Equivalent Systems for Beams and Frames," in *Proceedings of the Twenty-first Midwestern Mechanics Conference*, Vol. 15, pp. 399–401, 1989, Michigan Technological University, Houghton, Michigan.

26. Lord Rayleigh, *Theory of Sound*, Dover Publications, New York, 1945 (reprint).

27. J. M. Gere and S. P. Timoshenko, *Mechanics of Materials*, 3rd ed., PWS-KENT Publishing Co., Boston, 1990, pp. 554–561.

28. N. O. Myklestad, "A New Method of Calculating Natural Modes of Uncoupled Bending Vibration of Airplane Wings and Other Types of Beams," *Journal of Aeronautical Sciences*, pp. 153–162, April 1944.

29. M. A. Prohl, "A General Method for Calculating Critical Speeds of Flexible Rotors," *Transactions of the ASME*, Vol. A-142, September 1945.

30. E. C. Pestel and F. A. Leckie, *Matrix Methods in Elastomechanics*, McGraw-Hill, New York, 1963.

31. J. H. Argyris, "Energy Theorems in Structural Analysis," Aircraft Engineering, Vol. 27, 1955 (see also subsequent issues).

32. R. Courant, "Variational Method for the Solution of Problems of Equilibrium and Vibrations," *Bulletin of the American Mathematical Society*, Vol. 49, 1943.

33. M. J. Turner, R. W. Clough, H. L. Martin, and L. J. Topp, "Stiffness and Deflection Analysis of Complex Structures," *Journal of the Aeronautical Sciences*, Vol. 23, No. 9, 1956.

34. R. Clough and J. Penzien, *Dynamics of Structures*, McGraw-Hill, New York, 1975.

35. J. H. Argyris and S. Kelsey, *Energy Theorems in Structural Analysis*, Butterworth, London, 1960.

36. *Proceedings of the Matrix Method in Structural Mechanics*, AFFDL-TR-66-80, Wright–Patterson Air Force Base, Ohio, 1966.

37. K. J. Bathe, *Finite Eiement Procedures in Engineering Analysis*, Prentice-Hall, Englewood Cliffs, NJ, 1982.

38. K. H. Huebner and E. A. Thornton, *The Finite Element Method for Engineers*, John Wiley & Sons, New York, 1983.

39. R. D. Cook, "Concepts and Applications of Finite Element Analysis," John Wiley & Sons, Inc., New York, 1981.

40. H. C. Martin, "On the Derivation of Stiffness Matrices for the Analysis of Large Deflection and Stability Problems," in *Proceedings of the ?? Matrix Method in Structural Mechanics*, AFFDL-TR-66-80, Wright–Patterson Air Force Base, Ohio, 1966.

41. H. C. Martin and G. F. Carey, *Introduction to Finite Element Analysis Theory and Application*, McGraw-Hill, New York, 1973.

42. P. Bellini, *Advanced Concepts of Matrix Structural Mechanics*, Cleveland State University, 1985.

43. T. Y. Yang, *Finite Element Structural Analysis*, Prentice-Hall, Englewood Cliffs, NJ, 1986.

44. P. A. Bosela, *Development of Finite Element Method for the Dynamic Analysis of a Pre-loaded Beam in Space*, Ph.D. Dissertation, The University of Akron, Akron, OH, 1991.

45. P. A. Bosela, F. J. Shaker, and D. G. Fertis, "Dynamic Analysis of Space-Related Linear and Nonlinear Structures," *International Journal of Computers and Structures*, Vol. 44, No. 5, pp. 1145–1148, 1992.

46. P. A. Bosela, D. G. Fertis, and F. J. Shaker, "Grounding of Space Structures," *International Journal of Computers and Structures*, Vol. 45, No. 1, pp. 143–153, 1992.

47. P. A. Bosela, D. G. Fertis, and F. J. Shaker, "A New Pre-loaded Beam Geometric Stiffness Matrix with Full Rigid Body Capabilities," *International Journal of Computers and Structures*, Vol. 45, No. 1, pp. 155–163, 1992.

48. A. Collar and A. Simpson, *Matrices and Engineering Dynamics*, Halsted Press, New York, 1987.

49. J. H. Argyris and S. Symeonidis, "Nonlinear Finite Element Analysis of Elastic Systems Under Nonconservative Loading–Natural Formulation. Part 1. Quasi-static Problems," *Computer Methods in Applied Mechanics and Engineering*, Vol. 26, pp. 75–123, 1981.

50. M. Paz and L. Dung, "Power Series Expansion of the General Stiffness Matrix for Beam Elements," *International Journal for Numerical Methods in Engineering*, Vol. 9, 1975.

51. M. Paz and L. Dung, "Power Series Expansion of the General Stiffness Matrix

Including Rotatory Inertia and Shear Deformation, *Shock and Vibration Bulletin*, Vol. 46, Part 2, pp. 181–184, August 1976.

52. R. Craig, *Structural Dynamics — An Introduction to Computer Methods*, John Wiley & Sons, New York, 1981.

53. M. L. James, G. M. Smith, J. C. Wolford, and P. W. Whaley, *Vibration of Mechanical and Structural Systems with Microcomputer Applications*, Harper & Row Publishers, New York, 1989.

54. W. Weaver and P. R. Johnston, *Structural Dynamics by Finite Elements*, Prentice-Hall, Englewood Cliffs, NJ, 1987.

55. R. Craig and M. Bampton, "Coupling of Substructures for Dynamic Analysis," *AIAA Journal*, Vol. 6, No. 7, July 1968.

56. W. Hurty, "Dynamic Analysis of Structural Systems Using Component Modes," *AIAA Journal*, Vol. 3, No. 4, April 1965.

57. D. Kammer and M. Triller, "Ranking the Dynamic Importance of Fixed Interface Modes Using a Generalization of Effective Mass," *International Journal of Analytical and Experimental Modal Analysis*, in press.

58. S. Rubin, "Improved Component-Mode Representation for Structural Dynamic Analysis," *AIAA Journal*, Vol. 13, No. 8, August 1975.

59. R. MacNeal, *A Hybrid Method of Component Mode Synthesis, Computers and Structures*, Vol. 1, Pergamon Press, London, 1971.

60. R. Craig and T. J. Sue, "A Review of Model Reduction Methods for Structural Control Design," in *Dynamics of Flexible Structures in Space*, Kirk and Junkings, Editors, Computational Mechanics Publications, Southampton, 1990, Proceedings of the First International Conference, Cranfield, UK, 15–18 May 1990.

61. P. A. Blelloch and K. S. Carney, "Modal Selection in Structural Dynamics," in *Proceedings of the 7th International Modal Analysis Conference*, Schenectady, NY, 1989, pp. 742–749.

62. P. A. Blelloch and K. S. Carney, "Selection of Component Modes," in *30th Structures, Structural Dynamics, and Materials Conference*, AIAA, New York, 1989.

63. J. H. Hodgson, *Earthquake and Earth Structure*, Prentice-Hall, Englewood Cliffs, NJ, 1964.

64. D. G. Fertis and A. O. Afonta, "Equivalent Systems for Large Deformation of Beams of Any Stiffness Variation," *European Journal of Mechanics*, *A/Solids*, Vol. 10, No. 3, pp. 265–293, 1991.

65. J. L. Lagrange, "Sur la Force de Resorts Plies," *Member of Academy of Berlin*, 1970.

66. L. Euler, *Methodus Inveniendi Lineas Curvas*, 1744.

67. Plana, "Equation de la Courbe Fermee par Une Lame Elastique", *Memoirs of the Royal Society, Turin*, 1809.

68. R. Frisch-Fay, *Flexible Bars*, Butterworths and Co., Publishers, Washington, DC, 1962.

69. D. G. Fertis and A. O. Afonta, "Small Vibrations of Flexible Bars by Using the Finite Element Method with Equivalent Uniform Stiffness and Mass Methodology," *Journal of Sound and Vibration*, Vol. 163, No. 2, pp. 343–358, 1993.

70. A. K. Gupta, "Exact Solution for a Variable Stiffness Beam," *Journal of Structural Engineering, ASCE*, Vol. 111, No. 1, January 1985.

71. S. Timoshenko, *Strength of Materials, Part II, Advanced Theory and Problems*, 3rd ed., Robert E. Krieger Publishing Co., Huntington, NY, 1976.

72. M. A. Biot, "Analytical and Experimental Methods in Engineering Seismology," *Transactions of ASCE*, Vol. 108, pp. 365–408, 1943.

73. F. E. Richart, R. D. Woods, and J. R. Hall Jr., "Vibration of Soils and Foundations," Prentice-Hall, Englewood Cliffs, NJ, 1970.

74. E. D'Appolonia, "Dynamic Loadings," Presented at the ASCE Special Conference on Placement and Improvement of Soil to Support Structures," M.I.T., August 1968.

75. G. N. Bycroft, "Forced Vibration of a Rigid Circular Plate on a Semi-infinite Elastic Space and on an Elastic Stratum," *Philosophical Transactions of the Royal Society of London, Series A*, Vol. 248, pp. 327–368, 1956.

76. T. K. Hsieh, "Foundation Vibrations," *Proceedings of the Institution of Civil Engineering*, Vol. 22, pp. 211–226, 1962.

77. L. Toriumi, "Vibrations in Foundations of Machines," *Technical Report of Osaka University*, Vol. 5, No. 146, pp. 103–126, March 1955.

78. D. D. Barkan, *Dynamics of Bases and Foundations*, translated from the Russian by L. Drashevka, translation edited by G. P. Tsebotarioff, McGraw-Hill, New York, 1962.

79. S. H. Crandall et al., *Random Vibration*, Vols. 1 and 2, M.I.T. Press, Cambridge, MA, 1963.

80. A. Papoulis, *Probability, Random Variables, and Stochastic Processes*, McGraw-Hill, New York, 1965.

81. D. G. Fertis, "Material and Strength Characteristics of Concrete Mixtures by Acoustic Spectra Analysis," Interim Research Report No. APS-201A, Institute of Science and Engineering Research, The University of Akron, Akron, OH, 1971.

82. L. E. Elsgolc, *Calculus of Variation*, Addison-Wesley Publishing Co., Reading, MA, 1962.

83. T. S. Dean, *Calculus of Variation*, Ph.D. Dissertation, University of Texas, Texas University Microfilms, 1964.

84. U.S. Department of Commerce, National Bureau of Standards, *Handbook of Mathematical Functions with Formulas, Graphs, and Mathematical Tables*, Applied Mathematics Series 5, June 1964.

85. D. G. Fertis, "Concrete Material Response by Acoustic Spectra Analysis," *Journal of the Structural Division, ASCE*, Vol. 102, No. ST2, pp. 387–400, 1976.

86. X. Dupré, *Theorie Mechanique de la Chaleur*, Paris, 1869.

87. E. Buckingham and Lord Rayleigh.

88. E. Buckingham (1914) reference.

89. R. A. Frazer and W. J. Duncan, *Elementary Matrices*, Cambridge University Press, Cambridge, UK, 1938.

90. F. R. Gantmacher, *The Theory of Matrices*, Vols. I and II, Chelsey Publishing Co., New York, 1959.

91. R. W. Clough, "The Finite Element Method in Plane Stress Analysis," *Proceedings of the Second Conference on Electronic Computations*, American Society of Civil Engineers, pp. 345–377, New York, 1960.

Bibliography

J. H. Argyris and S. Kelsey, "Structural Analysis by the Matrix Force Method with Application to Aircraft Wings," *Jahrbuch der Wissenschaftliche Gesellschaft für Luftfahrt*, pp. 78–98, 1956.

J. R. Brauer, *What Every Engineer Should Know About Finite Element Analysis*, Marcel Dekker, New York, 1988.

D. G. Fertis, "Vibration of Beams and Frames by Using Dynamically Equivalent Systems," *Proceedings of the Structural Dynamics and Vibration Symposium*, Energy-Sources Technology Conference and Exhibition, ASME, January 23–26, 1994, New Orleans, LA.

D. G. Fertis and H. Cunningham, "Equivalent Systems for Shear Deflections of Variable Thickness Members," in *Industrial Mathematics*, Vol. 12, Part 1, The Industrial Mathematics Society, Detroit, MI, 1962.

M. Paz, *Structural Dynamics Theory and Applications*, Van Nostrand Reinhold Co., New York, 1980.

H. Saunders, "Stiffness Matrix of a Beam-Column Including Shear Deformations," *Shock and Vibration Bulletin*, Vol. 40, Part 4, pp. 187–196, December 1969.

Answers to Selected Problems

CHAPTER 1

1.1 $\ddot{x} + \omega^2 y = 0$

1.12c $m_1 \ddot{y}_1 - c(\dot{y}_2 - \dot{y}_1) - k_2(y_2 - y_1) + 2k_1 y_1 = 0$

         $m_2 \ddot{y}_2 - c(\dot{y}_2 - \dot{y}_1) + k_2(y_2 - y_1) = F \sin \omega_f t$

1.15c $m\ddot{y} + c\dot{y} + k_e y = 0$

$$\frac{1}{k_e} = \frac{1}{k_1} + \frac{1}{k_2} + \frac{1}{k_3}$$

1.17b $m_1 \ddot{y}_1 + k_1 y_1 - k_2(y_2 - y_1) = F(t)$

         $m_2 \ddot{y}_2 + c\dot{y}_2 + k_2(y_2 - y_1) - k_3 y_3 = 0$

1.19b $J_1 \ddot{\varphi}_1 + c_1 \dot{\varphi}_1 + k_{t_1} \varphi_1 - c_2(\dot{\varphi}_2 - \dot{\varphi}_1) - k_{t_2}(\varphi_2 - \varphi_1) = 0$

         $J_2 \ddot{\varphi}_2 + c_2(\dot{\varphi}_2 - \dot{\varphi}_1) + k_{t_2}(\varphi_2 - \varphi_1) = T(t)$

1.22a $k_A = 1{,}220.7$ lb/in. $k_B = 2{,}893.5$ lb/in.

1.22b $k_A = 78{,}125$ lb/in.

1.22c $k_A = 58{,}593.75$ lb/in.

1.24 $k_{t_e} = 1{,}500{,}000$ in.lb/rad

CHAPTER 2

2.3 $m = 2.702$ kg

2.5 $f = 4.574$ Hz

2.8 $c_c = 0.6435$ lb $\cdot$ sec/in.
        $\zeta = 0.093$
        $\omega = 24.86$ rps
        $\omega_d = 24.75$ rps

2.9 $c = 7545.70$ kg $\cdot$ s/m

2.12 Using Eq. (2.60): $\delta = 0.3796$
Using Eq. (2.61): $\delta = 0.3789$
$y_1/y_2 = 1.4607$

2.13 $\zeta = 0.457$; $\delta = 0.320$; $Y_1/Y_2 = 24.57$

2.15 $\zeta = 0.0297$
$\delta = 0.1867$
$y_1/y_2 = 1.2052$

2.17 $x_{\text{second cycle}} = 8.93$ in.; $x_{\text{fifth cycle}} = 7.33$ in.

2.19 $\zeta_e = 0.2865$ $c_e = 2.8576$ lb · sec/in.

2.21 $J_{\text{disk}} = 0.2488$ kg · m

2.22 $f = (d^2/8\pi)\sqrt{\pi G/JL}$

2.25 $d = 0.0316$ m
$= 3.16$ cm

2.26 $\omega_1 = 0.458\sqrt{k/m}$ $\omega_2 = 2.189\sqrt{k/m}$
$Y_1^{(1)} = 1.0000$ $Y_1^{(2)} = 1.0000$
$Y_2^{(1)} = 1.2658$ $Y_2^{(2)} = -0.2639$

2.29 $\omega_1 = 0.4451\sqrt{k/m}$; $\omega_2 = 1.2470\sqrt{k/m}$; $\omega_3 = 1.8019\sqrt{k/m}$

MODE SHAPES

First Mode	Second Mode	Third Mode
$y_1^{(1)} = 0.4450$	$y_1^{(2)} = -1.247$	$y_1^{(3)} = 1.802$
$y_2^{(1)} = 0.8019$	$y_2^{(2)} = -0.555$	$y_2^{(3)} = -2.247$
$y_3^{(1)} = 1.000$	$y_3^{(2)} = 1.000$	$y_3^{(3)} = 1.000$

2.35 $\omega_1 = 47.279$ rps $\omega_2 = 103.207$ rps
$\Phi_1^{(1)} = 1.000$ $\Phi_1^{(2)} = 1.000$
$\Phi_2^{(1)} = 1.884$ $\Phi_2^{(2)} = -0.809$

2.38b The frequency equation is

$$\sinh \lambda L[EI\lambda^3 \cos \lambda L - k \sinh \lambda L]$$
$$+ \sin \lambda L[EI\lambda^3 \cosh \lambda L + k \sinh \lambda L] = 0$$

$$\lambda_1 L \approx 2.455 \qquad \omega_1 = \frac{\lambda_1^2}{L^2}\sqrt{\frac{EI}{m}} = 22.38 \text{ rps}$$

$$\lambda_2 L = 5.49 \qquad \omega_2 = \frac{\lambda_2^2}{L^2}\sqrt{\frac{EI}{m}} = 111.92 \text{ rps}$$

2.39 $\omega_1 = 3.53/L^2\sqrt{EI/m}$; $\omega_2 = 22.00/L^2\sqrt{EI/m}$; $\omega_3 = 61.70/L^2\sqrt{EI/m}$

2.40 $\omega_n = n\pi c/L$ $n = 0, 1, 2, \ldots$;

$U_n(x) = A_n \cos n\pi x/L$ $n = 0, 1, 2, \ldots$

2.41 $\omega_n = (2n + 1)\pi c/2L$ $n = 0, 1, 2, \ldots$;

$U_n(x) = A_n \sin (2n + 1)\pi x/2L$ $n = 0, 1, 2, \ldots$

2.43 $\omega_n = (2n + 1)\pi a/2L$ $n = 0, 1, 2, \ldots$;

$\Phi(x) = B_n \sin (2n + 1)\pi x/2L$ $n = 0, 1, 2, \ldots$

2.44 $\omega_n = n\pi a/L$ $n = 0, 1, 2, \ldots$;

$\Phi(x) = B_n \cos n\pi x/L$ $n = 0, 1, 2, \ldots$

2.46 $c = 196.57(10)^3$ in./sec

CHAPTER 3

3.1a $y_{max} = 0.457$ cm

3.1b $y_{max} = 0.739$ cm

3.4 $(\omega_f)_{resonance} = 10.85$ rps
$Y_{resonance} = 0.03223$ m
$\Phi_{resonance} = 90°$

3.5b $y_{max} = Y = 6.11$ mm

3.7b $\omega_f = 8$ rps $y_{max} = 3.0$ mm
$\omega_f = 12$ rps $y_{max} = 4.0$ mm
$\omega_f = 15$ rps $y_{max} = 5.0$ mm
$\omega_f = 20$ rps $y_{max} = 7.0$ mm
$\omega_f = 25$ rps $y_{max} = 14.5$ mm
$\omega_f = \omega = 26.2$ rps $y_{max} = 16.0$ mm
$\omega_f = 30$ rps $y_{max} = 9.0$ mm

3.8 $X = 0.3651$ m; $4F_f < \pi F$

3.10 $X = 0.3719$ m

3.13a   With force function as shown in Fig. A3.13b.
(a) $0 \le t \le t_d$ $\mu = \zeta\omega$

$$y(t) = \frac{Fe^{-\mu t}}{m\omega_d} \left(\frac{e^{\mu t}\omega_d - \mu\sin(\omega_d t) - \omega_d\cos(\omega_d t)}{\mu^2 + \omega_d^2} \right)$$

$t_d \le t$

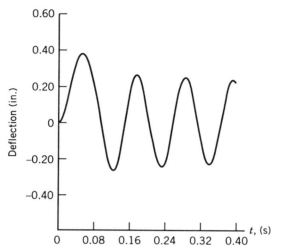

FIGURE A3.13b. Plot of $y(t)$ versus time t for problem in Fig. P3.13a with force function as shown in Fig. P3.13b.

$$y(t) = \frac{Fe^{-\mu(t + t_d)}}{m\omega_d(\mu^2 + \omega_d^2)} \{\omega_d[e^{\mu t_d} \cos(\omega_d t) \cos \omega_d(t + t_d)]$$

$$+ \mu[e^{\mu t_d} \sin(\omega_d t) \sin \omega_d(t + t_d)]\}$$

(b) For $c = 0.1$ kip $\cdot$ s/in., the plot of $y(t)$ versus time t is shown in Fig. A3.13b.

3.14d (a) $y(t) = e^{-4.144t}(0.29436 \cos 41.23t + 0.3183 \sin 41.23t)$
$t \geq t_d$
(b) The plot of $y(t)$ versus t is shown in Fig. A3.14.
(c) $y_{max} = 0.3956$ in.; $t_{max} = 0.061$ s

3.15a $y_{max} = Y_{st}\Gamma_{max} = 0.3876$ in. See also Appendix A for general expression for Γ.

3.17 $y(t) = \frac{F}{k}(1 - \cos \omega t)$

3.18 $y(t) = 0.08 - 0.008386e^{-3t}[3 \sin 9.5394t + 9.5394 \cos 9.5394t]$

3.23 $y(t) = 0.076971 \sin[12t - 0.5449] + 0.035 \sin 12t$
$y(0.2) = 0.0975$ m

3.24a $y_{1max} = 4.595$ in. $\phi_1 = 42.01°$; $y_{2max} = 7.319$ in. $\phi_2 = 41.88°$

3.29 $y_{max} = 0.3989$ in.

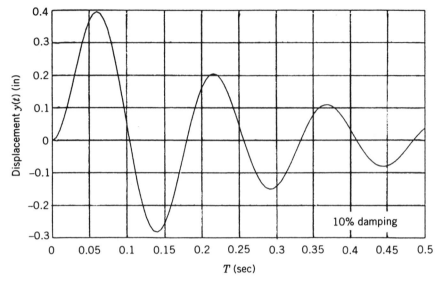

FIGURE A3.14. Plot of $y(t)$ versus time t for Problem P3.14.

3.35a $F(t) = \dfrac{200}{\pi}(\sin 10\,\pi t + \tfrac{1}{3}\sin 30\,\pi t + \tfrac{1}{5}\sin 50\,\pi t + \cdots)$

n	y_{st} (in.)	ω_f/ω	Γ_n	ϕ_n^0	$y(t) = y_{st}\Gamma\sin(\omega_f t - \phi_n)$
1	0.6366	1.5661	0.6844	173.85	$0.4357\sin(10\,\pi t - 173.85°)$
3	0.2122	4.6983	0.0474	178.72	$0.0101\sin(30\,\pi t - 178.72°)$
5	0.1273	7.8305	0.0166	179.26	$0.0021\sin(50\,\pi t - 179.26°)$

3.42d $y_{max} = 3.40$ in.
    $(y_{beam})_{max} = 1.95$ in.
    $\sigma_{max} = 31.72$ ksi

3.42e $y_{max} = 0.874$ in.; $(y_{beam})_{max} = 0.207$ in.; $\sigma_{max} = 9.54$ ksi

3.43e $y_{max} = 0.881$ in.; $\sigma_{max} = 8.37$ ksi

3.47e $y_{max} = 0.760$ in.; $\sigma_{max} = 7.22$ ksi

3.49a $x_{max} = 0.2803$ in.; $\sigma_{max} = 3.89$ ksi   (first column); $\sigma_{max} = 7.79$ ksi
    (second and third columns)

3.59 $y_{max} = 5.9886$ in.   @ $t = 0.24$ sec.

CHAPTER 4

4.3a $\omega_1 = 61.22$ rps
$\omega_2 = 121.80$ rps
$\omega_3 = 239.38$ rps

4.3b Frequency equation is

$$\theta_1(\theta_1 + \theta_2) - \psi_1^2 = 0$$

when $EI = 60 \times 10^6$ k-in^2, we have

$$\omega_1 = 51.8 \text{ rps}, \quad \omega_2 = 78.32 \text{ rps}, \quad \omega_3 = 102.60 \text{ rps},$$

4.25 $\omega_1 = \omega_2 = 2.177$ rps.

4.26 For $k_v = 100$ lb/in and $P = 0$
$\omega_1 = 2.177$ rps, $\quad \omega_2 = 3.758$ rps
For $kv = 100$ lb/in and $P = 1000$ lb
$\omega_1 = 2.177$ rps, $\omega_2 = 3.362$ rps
For $k_v = 100$ lb/in and $P = 2500$ lb
$\omega_1 = 2.177$ rps, $\omega_2 = 2.658$ rps

4.27 For $k_v = 1000$ lb/in and $P = 0$
$\omega_1 = 6.876$ rps, $\omega_2 = 11.882$ rps
For $k_v = 1000$ lb/in and $P = 15000$ lb
$\omega_1 = 6.876$ rps, $\omega_2 = 9.941$ rps
For $k_v = 1000$ lb/in and $P = 33501$ lb
$\omega_1 = \omega_2 = 6.876$ rps

4.28 For $k_v = 5000$ lb/in and $P = 0$
$\omega_1 = 11.882$ rps, $\omega_2 = 20.572$ rps
For $k_v = 5000$ lb/in and $P = 95564$ lb
$\omega_1 = \omega_2 = 11.882$ rps

4.29c $y = Y \cos \omega_f t$

$$Y = -9.6678 \cosh(0.007953x) + 7.0274 \sinh(0.007953x)$$
$$+ 10.5939 \cos(0.007953x) - 7.0274 \sin(0.007953x)$$
$$- 0.9261$$

4.33 The length L_e of the exact dynamically equivalent system with simply supported ends is

$$L_e = L/10$$

4.34 $f_1 = 3.56/L^2 \sqrt{EI/m}$ Hz.

4.35 $f_2 = 9.82/L^2 \sqrt{EI/m}$ Hz.

4.36 $f_3 = 19.2/L^2 \sqrt{EI/m}$ Hz.

4.37 $\omega_{10} = (10)^2 \pi^2/L^2 \sqrt{EI/m}$ rps

4.39   The location of node points and dynamic hinge points is the same. They are located at $L/5$, $2L/5$, $3L/5$, and $4L/5$.

4.41   The location of node points coincides with the location of the dynamic hinge points. They are located at $0.277L$, $0.409L$, $0.591L$, and $0.773L$.

4.43a Node points and dynamic hinge points coincide.

4.43b Follow the procedure of Example 4.7.

4.43c Node points and dynamic hinge points coincide.

4.46c Node points and dynamic hinge points coincide, and $k_H = \infty$. i.e., for the third mode, the dynamically equivalent system would be a simply supported beam of length $L_e = 0.282L$, or a beam fixed at the one end and simply supported at the other end of length $L_e = 0.359L$.

4.50 $\omega = 40.65$ rps; $f = \omega/2\pi = 6.47$ Hz

CHAPTER 5

5.2 $f = 3.77\sqrt{gEI/wL^4}$ in units of hertz

5.5 $\omega = 35.52$ rps

5.9d $f_1 = 3.21$ Hz.
           Mode shape: $\beta_1 = 4.9176$
                          $\beta_2 = 9.1139$
                          $\beta_3 = 3.5659$
                          $\beta_4 = -1.4061$
                          $\beta_5 = -2.2244$
                          $\beta_6 = -1.0000$

5.10 $f_1 = 5.15$ Hz.

5.14 $f_1 = 3.15$ Hz, $f_2 = 12.91$ Hz, $f_3 = 29.19$ Hz; modes shapes are shown in Fig. A5.14.

5.15 $f_1 = 2.1157$ Hz, $f_2 = 6.1105$ Hz, $f_3 = 8.488$ Hz; modes shapes are shown in Fig. A5.15.

5.16 $f_1 = 2.70$ Hz, $f_2 = 7.12$ Hz, $f_3 = 10.28$ Hz; modes shapes are shown in Fig. A5.16.

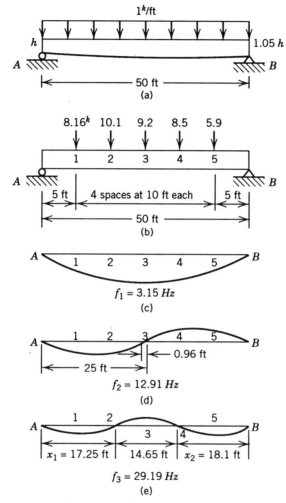

FIGURE A5.14. (a) Original variable stiffness beam. (b) Equivalent uniform stiffness system. (c) First mode shape. (d) Second mode shape. (e) Third mode shape. Five lumped masses are used.

5.17e $\omega_1 = 35.52$ rps
$\omega_2 = 127.84$ rps

5.18 $\omega = 16.65$ rps

5.30a $\omega_1 = 42.49$ rps $\omega_2 = 94.13$ rps

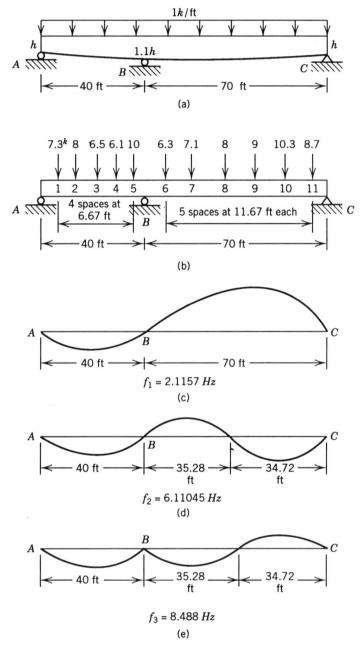

FIGURE A5.15. (a) Statically indeterminate variable stiffness beam. (b) Equivalent system of uniform stiffness. (c) First mode shape. (d) Second mode shape. (e) Third mode shape.

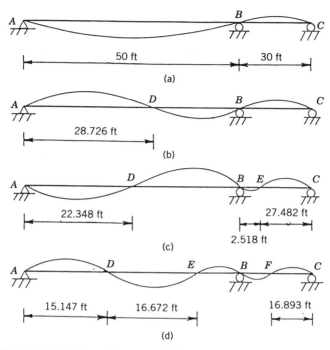

FIGURE A5.16. (a) First, (b) second, (c) third, and (d) fourth mode shapes.

CHAPTER 6

6.1 $\quad K_{3,5} = \dfrac{-d(65a^4 - 42a^2b^2 + 135b^4)}{175a^2b^3}$

6.2 $\quad [M] = \dfrac{\rho AL}{420} \begin{bmatrix} 156 & 54 & 22L & -13L \\ 54 & 156 & 13L & -22L \\ 22L & 13L & 4L^2 & -3L^2 \\ -13L & -22 & -3L^2 & 4L^2 \end{bmatrix}$

6.3 Determinant minors along diagonal are 3.21×10^9, 7.65×10^8, 2.96×10^7, 8.23×10^3. Since they are all positive, matrix is positive definite.

6.4 $\quad R = [0\ \ -27\ -360\ \ 0\ -180\ -720\ \ 0\ -153\ \ 1440]^T$

6.5 $\quad R = [2.88\ \ 0\ -92.16\ \ 5.04\ -21.6\ -99.48$
$\qquad 2.16\ -15.696\ -722.354$
$\qquad 0\ -17.424\ \ 815.616\ \ 0\ -21.6\ \ 1036.8\ \ 0$
$\qquad 0\ \ 0]^T$

6.6 $K_A =$

$$
\begin{bmatrix}
0.94E8 & 0 & 0 & -0.47E8 & 0 & 0 & 0 & 0 \\
0 & 0.138E8 & 0 & 0 & 0.207E9 & 0 & 0 & 0 \\
0 & 0 & 0.1656E11 & 0 & 0.414E10 & 0 & 0 & 0 \\
-0.47E8 & 0 & 0 & 0.141E9 & 0 & -0.94E8 & 0 & 0 \\
0 & 0.207E9 & 0.414E10 & 0 & 0.2484E11 & 0 & -0.828E9 & 0.828E10 \\
0 & 0 & 0 & -0.94E8 & 0 & 0.94E8 & 0 & 0 \\
0 & 0 & 0 & 0 & -0.828E9 & 0 & 0.552E8 & -0.828E9 \\
0 & 0 & 0 & 0 & 0.828E10 & 0 & -0.828E9 & 0.1656E11
\end{bmatrix}
$$

6.7

$$
\begin{bmatrix}
0.288E8 & 0 & 0 & -0.288E8 & 0 & 0 & 0 & 0 & 0 \\
0 & 0.576E7 & 0.84E8 & 0 & -0.576E7 & 0.84E8 & 0 & 0 & 0 \\
0 & 0.84E8 & 0.32E10 & 0 & -0.84E8 & 0.1E10 & 0 & 0 & 0 \\
-0.288E8 & 0 & 0 & 0.576E8 & 0 & 0 & -0.288E8 & 0 & 0 \\
0 & -0.576E7 & -0.84E8 & 0 & 0.1152E8 & 0 & 0 & -0.576E7 & 0.84E8 \\
0 & 0.84E8 & 0.1E10 & 0 & 0 & 0.64E10 & 0 & -0.84E8 & 0.1E10 \\
0 & 0 & 0 & -0.288E8 & 0 & 0 & 0.288E8 & 0 & 0 \\
0 & 0 & 0 & 0 & -0.576E7 & -0.84E8 & 0 & 0.576E7 & -0.84E8 \\
0 & 0 & 0 & 0 & 0.84E8 & 0.1E10 & 0 & -0.84E8 & 0.32E10
\end{bmatrix}
$$

$$
\times
\begin{bmatrix}
1 & 0 & 0 \\
0 & 1 & -50 \\
0 & 0 & 1 \\
1 & 0 & 0 \\
0 & 1 & 0 \\
0 & 0 & 1 \\
1 & 0 & 0 \\
0 & 1 & 50 \\
0 & 0 & 1
\end{bmatrix}
=
\begin{bmatrix}
0 \\
-120000000 \\
0 \\
0 \\
0 \\
0 \\
0 \\
120000000 \\
0
\end{bmatrix}
$$

Pseudoforces exist; lacks rigid-body rotation capability.

6.8

$$
\begin{bmatrix}
0.3754E8 & -0.1697E8 & 0.1697E8 \\
-0.1697E8 & 0.1697E8 & -0.1697E8 \\
0.1697E8 & -0.1697E8 & 0.2734E8
\end{bmatrix}
$$

6.9

$$
\begin{bmatrix}
0.4114E10 & 0 & 0.8571E9 & 0.432E8 & -0.576E8 & 0.12E10 \\
0 & 0.399E8 & -0.7071E8 & -0.1933E8 & 0.1461E8 & -0.7071E8 \\
0.8571E9 & -0.7071E8 & 0.4543E10 & 0.7071E8 & 0.7071E8 & 0.1414E10 \\
0.432E8 & -0.1933E8 & 0.7071E8 & 0.388E8 & -0.2172E7 & 0.1139E9 \\
-0.576E8 & 0.1461E8 & 0.7071E8 & -0.2172E7 & 0.3154E8 & 0.1311E8 \\
0.12E10 & -0.7071E8 & 0.1414E10 & 0.1139E9 & 0.1311E8 & 0.5228E10
\end{bmatrix}
$$

6.10

$$[L] = \begin{bmatrix} 12.490 & 0 & 0 & 0 \\ 105.68 & 56.848 & 0 & 0 \\ 4.3235 & 5.6683 & 10.256 & 0 \\ -62.45 & -73.886 & -61.543 & 35.402 \end{bmatrix}$$

6.11 $\omega_1 = 704.3951$ rps

$$\phi_1 = \begin{bmatrix} 1 \\ 0 \\ -1.9396 \\ 0 \\ -175.48 \\ -4.0894 \end{bmatrix}$$

$\omega_2 = 2734.6631$ rps

$$\phi_2 = \begin{bmatrix} 0 \\ 0.70711 \\ 0 \\ 1 \\ 0 \\ 0 \end{bmatrix}$$

$\omega_3 = 4096.0364$ rps

$$\phi_3 = \begin{bmatrix} 1 \\ 0 \\ -0.71901 \\ 0 \\ 9.9328 \\ 0.73190 \end{bmatrix}$$

$\omega_4 = 8522.6278$ rps

$$\phi_4 = \begin{bmatrix} 1 \\ 0 \\ 0.43297 \\ 0 \\ -12.467 \\ -1.6157 \end{bmatrix}$$

$\omega_5 = 9553.2442$ rps

$$\phi_5 = \begin{bmatrix} 1 \\ -0.70711 \\ 0 \\ 1 \\ 0 \\ 0 \end{bmatrix}$$

$\omega_6 = 18889.4449$ rps

$$\phi_6 = \begin{bmatrix} 1 \\ 0 \\ 1.0998 \\ 0 \\ 11.843 \\ 2.4235 \end{bmatrix}$$

6.12 $\omega_1 = 740.8074$ rps

$$\phi_1 = \begin{bmatrix} 1 \\ 0 \\ 34.599 \\ 0 \\ 0 \\ -1 \end{bmatrix}$$

$\omega_2 = 2734.6631$ rps

$$\phi_2 = \begin{bmatrix} 0 \\ 0.70711 \\ 0 \\ 0 \\ 1 \\ 0 \end{bmatrix}$$

$\omega_3 = 3394.1125$ rps

$$\phi_3 = \begin{bmatrix} 1 \\ 0 \\ 0 \\ -1 \\ 0 \\ 1 \end{bmatrix}$$

$\omega_4 = 7471.8702$ rps

$$\phi_4 = \begin{bmatrix} 1 \\ 0 \\ -5.3724 \\ 0 \\ 0 \\ -1 \end{bmatrix}$$

$\omega_5 = 9553.2442$ rps

$$\phi_5 = \begin{bmatrix} 0 \\ -0.70711 \\ 0 \\ 0 \\ 1 \\ 0 \end{bmatrix}$$

$\omega_6 = 13148.1619$ rps

$$\phi_6 = \begin{bmatrix} 1 \\ 0 \\ 0 \\ 1 \\ 0 \\ 1 \end{bmatrix}$$

6.13

$$\omega_1 = 0, \quad \phi_1 = \begin{bmatrix} 1 \\ 0 \\ 0 \\ 1 \\ 0 \\ 0 \\ 1 \\ 0 \\ 0 \end{bmatrix} \qquad \omega_2 = 0, \quad \phi_2 = \begin{bmatrix} 0 \\ 1 \\ 0 \\ 0 \\ 1 \\ 0 \\ 0 \\ 1 \\ 0 \end{bmatrix}$$

$$\omega_3 = 1935.0943 \qquad \omega_4 = 4505.4561 \qquad \omega_5 = 7001.5196$$

$$\phi_3 = \begin{bmatrix} 0 \\ -1 \\ 0.012196 \\ 0 \\ 0 \\ 0.026482 \\ 0 \\ 1 \\ 0.012196 \end{bmatrix} \qquad \phi_4 = \begin{bmatrix} 0 \\ 1 \\ -0.036947 \\ 0 \\ -0.69211 \\ 0 \\ 0 \\ 1 \\ 0.036947 \end{bmatrix} \qquad \phi_5 = \begin{bmatrix} 1 \\ 0 \\ 0 \\ 0 \\ 0 \\ 0 \\ -1 \\ 0 \\ 0 \end{bmatrix}$$

$$\omega_6 = 8986.086 \qquad \omega_7 = 14003.0392 \qquad \omega_8 = 19070.943 \qquad \omega_9 = 28463.028$$

$$\phi_6 = \begin{bmatrix} 0 \\ 1 \\ -0.081804 \\ 0 \\ 0 \\ 0.077177 \\ 0 \\ -1 \\ -0.081804 \end{bmatrix} \qquad \phi_7 = \begin{bmatrix} 1 \\ 0 \\ 0 \\ -1 \\ 0 \\ 0 \\ 1 \\ 0 \\ 0 \end{bmatrix} \qquad \phi_8 = \begin{bmatrix} 0 \\ 1 \\ -0.16761 \\ 0 \\ 0.39677 \\ 0 \\ 0 \\ 1 \\ 0.16761 \end{bmatrix} \qquad \phi_9 = \begin{bmatrix} 0 \\ 1 \\ -0.21030 \\ 0 \\ 0 \\ -0.10646 \\ 0 \\ -1 \\ -0.21030 \end{bmatrix}$$

6.14

Number	Frequency (rps)	Number	Frequency (rps)
1	764.6948	7	15374.3131
2	2682.8899	8	21579.7091
3	3070.0546	9	22232.6050
4	7006.1899	10	34125.5239
5	8463.5518	11	51125.0480
6	13576.4502	12	62215.1107

6.15

Number	Frequency (rps)	Number	Frequency (rps)
1	1.4142	6	5878.7759
2	6.3244	7	11757.5510
3	10.9544	8	13592.6270
4	1736.9434	9	21715.6853
5	5435.9446		

6.16 $\omega_1 = 195.6864$ rps $\qquad \omega_2 = 733.078$ rps $\qquad \omega_5 = 908.4975$ rps

$$\phi_1 = \begin{bmatrix} 1 \\ 4.3842 \\ 0.65972 \\ 0.010722 \\ 4.5885 \end{bmatrix} \qquad \phi_2 = \begin{bmatrix} 1 \\ 0.097950 \\ -0.69251 \\ 0.43265 \\ 0.018529 \end{bmatrix} \qquad \phi_3 = \begin{bmatrix} 1 \\ 0.80650 \\ -3.1966 \\ -8.1145 \\ 1.0279 \end{bmatrix}$$

6.17 $\omega_1 = \quad 0 \qquad\qquad \omega_2 = 0 \qquad\qquad \omega_3 = 0$

$$\phi_1 = \begin{bmatrix} 0 \\ 1 \\ -0.02 \\ 0 \\ 0 \\ -0.02 \\ 0 \\ -1 \\ -0.02 \end{bmatrix} \qquad \phi_2 = \begin{bmatrix} 1 \\ 0 \\ 0 \\ 1 \\ 0 \\ 0 \\ 1 \\ 0 \\ 0 \end{bmatrix} \qquad \phi_3 = \begin{bmatrix} 0 \\ 1 \\ 0 \\ 0 \\ 1 \\ 0 \\ 0 \\ 1 \\ 0 \end{bmatrix}$$

$\omega_4 = 4505.4561$ rps $\qquad \omega_5 = 7001.5196$ rps $\qquad \omega_6 = 8662.2412$ rps

$$\phi_4 = \begin{bmatrix} 0 \\ 1 \\ -0.036947 \\ 0 \\ -0.69211 \\ 0 \\ 0 \\ 1 \\ 0.036947 \end{bmatrix} \qquad \phi_5 = \begin{bmatrix} 1 \\ 0 \\ 0 \\ 0 \\ 0 \\ 0 \\ -1 \\ 0 \\ 0 \end{bmatrix} \qquad \phi_6 = \begin{bmatrix} 0 \\ 1 \\ -0.087083 \\ 0 \\ 0 \\ 0.079375 \\ 0 \\ -1 \\ -0.087083 \end{bmatrix}$$

$\omega_7 = 14003.0392$ rps $\qquad \omega_8 = 19070.943$ rps $\qquad \omega_9 = 28269.7425$ rps

$$\phi_7 = \begin{bmatrix} 1 \\ 0 \\ 0 \\ -1 \\ 0 \\ 0 \\ 1 \\ 0 \\ 0 \end{bmatrix} \qquad \phi_8 = \begin{bmatrix} 0 \\ 1 \\ -0.16761 \\ 0 \\ 0.39677 \\ 0 \\ 0 \\ 1 \\ 0.16761 \end{bmatrix} \qquad \phi_9 = \begin{bmatrix} 1 \\ 1 \\ -0.21221 \\ 0 \\ 0 \\ -0.10832 \\ 0 \\ -1 \\ -0.21221 \end{bmatrix}$$

6.18 $\omega_1 = 1.4142$ rps $\qquad \omega_2 = 1.8257$ rps $\qquad \omega_3 = 5.0$ rps

$$\phi_1 = \begin{bmatrix} 1 \\ 1 \\ 1 \end{bmatrix} \qquad \phi_2 = \begin{bmatrix} 1 \\ -0.3333 \\ -1 \end{bmatrix} \qquad \phi_3 = \begin{bmatrix} 1 \\ -2 \\ 4 \end{bmatrix}$$

6.19

$$\phi_1 = \begin{bmatrix} \dfrac{\sqrt{6}}{6} \\[2mm] \dfrac{\sqrt{6}}{6} \\[2mm] \dfrac{\sqrt{6}}{6} \end{bmatrix} \qquad \phi_2 = \begin{bmatrix} \dfrac{\sqrt{30}}{10} \\[2mm] -\dfrac{\sqrt{30}}{30} \\[2mm] -\dfrac{\sqrt{30}}{10} \end{bmatrix} \qquad \phi_3 = \begin{bmatrix} \dfrac{\sqrt{30}}{30} \\[2mm] -\dfrac{\sqrt{30}}{15} \\[2mm] 2\dfrac{\sqrt{30}}{15} \end{bmatrix}$$

6.20 $\omega_1 = 54.0206$ rps $\omega_2 = 341.3405$ rps $\omega_3 = 786.2903$ rps

$$\phi_1 = \begin{bmatrix} 0 \\ 0.34015 \\ 0.0096957 \\ 0 \\ 1 \\ 0.011452 \end{bmatrix} \qquad \phi_2 = \begin{bmatrix} 0 \\ -0.72158 \\ 0.0036448 \\ 0 \\ 1 \\ 0.040079 \end{bmatrix} \qquad \phi_3 = \begin{bmatrix} 1 \\ 0 \\ 0 \\ 1.4142 \\ 0 \\ 0 \end{bmatrix}$$

$\omega_4 = 1152.3654$ rps $\omega_5 = 2746.8187$ rps $\omega_6 = 3343.4041$ rps

$$\phi_4 = \begin{bmatrix} 0 \\ 0.10145 \\ -0.063731 \\ 0 \\ 1 \\ 0.080341 \end{bmatrix} \qquad \phi_5 = \begin{bmatrix} 1 \\ 0 \\ 0 \\ -1.4142 \\ 0 \\ 0 \end{bmatrix} \qquad \phi_6 = \begin{bmatrix} 0 \\ 0.25316 \\ 0.043361 \\ 0 \\ 1 \\ 0.16107 \end{bmatrix}$$

6.21

$$\begin{bmatrix} 1 & 0 & 0 & 0 \\ 8.4615 & 1 & 0 & 0 \\ 0.34615 & 0.1 & 1 & 0 \\ -5 & -1.3 & -6 & 1 \end{bmatrix} \begin{bmatrix} 156 & 0 & 0 & 0 \\ 0 & 3230.8 & 0 & 0 \\ 0 & 0 & 105 & 0 \\ 0 & 0 & 0 & 1260.1 \end{bmatrix} \begin{bmatrix} 1 & 8.4615 & 0.34615 & -5 \\ 0 & 1 & 0.1 & -1.3 \\ 0 & 0 & 1 & -6 \\ 0 & 0 & 0 & 1 \end{bmatrix}$$

$\qquad\qquad\qquad L \qquad\qquad\qquad\qquad\quad D \qquad\qquad\qquad\qquad\quad L_T$

6.22

t	u_1	u_2	u_3
0.2	6.02391×10^{-6}	6.20463×10^{-4}	0.0961658
0.4	6.98898×10^{-5}	0.00478908	0.369960
0.6	4.06840×10^{-4}	0.0187677	0.779702

6.23

t	x_1	x_2	x_3	u_1	u_2	u_3
0.2	0.057356	−0.0766031	0.101570	2.98181×10^{-6}	2.22763×10^{-4}	0.0986778
0.4	0.224866	−0.296311	0.386302	2.91695×10^{-5}	0.00342941	0.379166
0.6	0.489219	−0.630155	0.798193	2.25204×10^{-4}	0.016497	0.797483

CHAPTER 7

7.4

Level	Column	V (kips)	m (kip · in.)	S (in³.)	σ (ksi)
1	Left	38.06	3425.4	64.7	52.9
	Right	10.33	1859.4	62.7	29.7
2	Left	23.00	1656.0	64.7	25.6
	Right	18.05	1230.0	51.9	25.0
3	Left	5.05	363.6	26.5	13.7
	Right	5.05	363.6	26.5	13.7

7.5 $x_{1\text{max}} = 1.1445$ in.; $x_{2\text{max}} = 1.2827$ in.
First floor: $\sigma_{1\text{max}} = 22.34$ ksi; second floor: $\sigma_2 = 6.27$ ksi

7.6 $x_{1\text{max}} = 1.2906$ in.; $x_{2\text{max}} = 1.4508$ in.; $\sigma_{1\text{max}} = 25.20$ ksi; $\sigma_{2\text{max}} = 6.97$ ksi

7.7b $x_{1\text{max}} = 0.6314$ in.; $x_{2\text{max}} = 0.9098$ in.
Maximum Stresses in first story columns:
$\sigma_{\text{first column}} = 10.68$ ksi
$\sigma_{\text{second column}} = 21.36$ ksi
Maximum Stresses in second story columns:
$\sigma_{\text{first column}} = \sigma_{\text{second column}} = 24.10$ ksi

7.8b $x_{1\text{max}} = 0.8612$ in.; $x_{2\text{max}} = 1.2148$ in.
$\sigma_{\text{max}} = 15.33$ ksi (second-story columns); $\sigma_{\text{max}} = 14.35$ ksi (first-story column with hinged end); $\sigma_{\text{max}} = 28.71$ ksi (first-story column with fixed end)

7.11 $y_{\text{center}} = 2.412$ in. (using the contributions of the first four modes; the even modes contribute zero). The contribution of the first mode predominates, and it is equal to 2.4104 in.

7.17 Maximum response occurs at the center of the plate. $w_{\text{max}} = 0.834$ in.; $(\sigma_x)_{\text{max}} = 31,730$ psi; $(\sigma_y)_{\text{max}} = 58,328$ psi.

CHAPTER 8

8.1 $\Delta_B = 376.80$ in.; $\theta_B = 75.53°$; $\delta_B = 682.30$ in.

8.2   For $n = 1$: $\Delta_B = 414.94$ in.; $\theta_B = 71.96°$; $\delta_B = 732.38$ in. For $n = 2$: $\Delta_B = 78.42$ in.; $\theta_B = 36.14°$; $\delta_B = 334.37$ in.

8.3 $\Delta_B = 483.11$ in.; $\theta_B = 81.70°$; $\delta_B = 747.33$ in.

8.5   The values obtained by solving the initial system by pseudolinear analysis are $\Delta_A = 300.64$ in.; $\theta_A = 59.71°$; $\delta_A = 653.57$ in.

8.6   The direct solution by using pseudolinear analysis yielded $\Delta_A = 137.82$ in.; $\theta_A = 39.89°$; $\delta_A = 466.20$ in.

8.7   The direct solution of the original system by using pseudolinear analysis yielded $\Delta_B = 159.14$ in.; $\theta_B = 43.48°$; $\delta_B = 493.82$ in.

8.8   The direct solution of the original system by pseudolinear analysis yielded:
For $n = 1$: $\Delta_B = 562.15$ in.; $\theta_B = 75.73°$; $\delta_B = 843.02$ in.
For $n = 3$: $\Delta_B = 39.78$ in.; $\theta_B = 22.88°$; $\delta_B = 250.37$ in.

8.10 $R_A = 38.4$ kips $\uparrow$ $M_A = 427.0$ kips $\cdot$ ft

8.11 $R_B = 50.31$ kips

8.12 $f_1 = 3.73$ Hz; $f_2 = 15.39$ Hz; $f_3 = 34.77$ Hz

8.14 $f_1 = 5.75$ Hz; $f_2 = 31.43$ Hz; $f_3 = 65.60$ Hz

8.20 $f_1 = 10.25$ Hz

8.23 $f_1 = 8.85$ Hz; $f_2 = 17.72$ Hz; $f_3 = 34.86$ Hz

8.24 $f_1 = 6.75$ Hz; $f_2 = 13.33$ Hz; $f_3 = 25.96$ Hz

8.25 $f_1 = 13.04$ Hz; $f_2 = 26.64$ Hz; $f_3 = 56.54$ Hz

8.29 $\delta_B = 7.79$ in.

8.32 $\delta_B = 20.06$ in.

CHAPTER 9

9.1a
$$c_n = -\frac{A}{jn\pi}$$

$$f(t) = \sum_{n=-\infty}^{\infty} -\frac{A}{jn\pi} e^{jn\omega_0 t}$$

$$j\pi c_n = -\frac{A}{n}$$

9.1b

$$F(\omega) = \frac{2A\sin(5\omega)}{\omega}$$

The above frequency spectrum is indeterminate at $\omega = 0$.

$$\lim_{\omega \to 0} F(\omega) = 10A$$

9.2

$$f(t) = \frac{1}{2\pi} \int_{\omega = -\infty}^{\infty} \frac{e^{j\omega t}}{1 + j\omega} \, d\omega$$

which is the Fourier integral

$$|F(\omega)| = \frac{1}{\sqrt{1 + \omega^2}}$$

9.3

$$F(\omega) = \frac{\omega_0}{(j\omega + 1/\tau)^2 + \omega_0^2}$$

τ may be assumed as equal to $2\pi/\omega_0$.

9.6

$$H(s) = \frac{1}{s^2 + 2\omega\zeta s + \omega^2} \qquad \text{(transfer function)}$$

$$Z(s) = s^2 + 2\omega\zeta s + \omega^2 \qquad \text{(impedance)}$$

The steady state and transient responses $y_p(t)$ and $y_h(t)$, respectively, are

$$y_p(t) = \frac{(F/k)\sin(\omega_f t - \phi)}{\{[1 - (\omega_f/\omega)^2]^2 + (2\zeta\omega_f/\omega)^2\}^{1/2}}$$

$$y_h(t) = e^{-\omega\zeta t}\left[\left(\frac{2(F/k)\,\omega_f\zeta}{\omega\{[1 - (\omega_f/\omega)^2]^2 + (2\omega_f\zeta/\omega)^2\}} + y_0\right)\cos\omega(\sqrt{1 - \zeta^2})t\right.$$

$$+ \left(\frac{(F/k)\,\omega_f[2\zeta^2 - 1 + (\omega_f/\omega)^2]}{\omega\sqrt{1 - \zeta^2}\{[1 - (\omega_f/\omega)^2]^2 + (2\omega_f\zeta/\omega)^2\}}\right.$$

$$\left.\left. + \frac{\omega\zeta y_0 + \dot{y}_0}{\omega\sqrt{1 - \zeta^2}}\right)\sin\omega(\sqrt{1 - \zeta^2})t\right]$$

Total response $y(t) = y_p(t) + y_h(t)$

9.10   The roots s_1, s_2, s_3, and s_4 of the quartic equation are

$$s_1 = \frac{-J - K + \sqrt{(J + K)^2 - 4(L + M)}}{2}$$

$$s_2 = \frac{-J - K - \sqrt{(J + K)^2 - 4(L + M)}}{2}$$

$$s_3 = \frac{-J + K + \sqrt{(J - K)^2 - 4(L - M)}}{2}$$

$$s_4 = \frac{-J + K - \sqrt{(J - K)^2 - 4(L - M)}}{2}$$

where

$$J = \frac{c(2m_2 + m_1)}{2m_1 m_2}$$

$$K = \sqrt{\frac{c^2(2m_2 + m_1)^2}{4m_1^2 m_2^2} + Z_{\text{real}} - \frac{(k_1 + k_2)m_2 + c^2 + k_2 m_1}{m_1 m_2}}$$

$$L = \frac{Z_{\text{real}}}{2}$$

$$M = \sqrt{\frac{Z_{\text{real}}^2}{4} - \frac{k_1 k_2}{m_1 m_2}}$$

$$Z_{\text{real}} = A + B$$

A and B are given by Eqs. (a) and (b), respectively, of Section 2.9.

$$y_1(t) = A_1 e^{s_1 t} + B_1 e^{s_2 t} + C_1 e^{s_3 t} + D_1 e^{s_4 t} + E_1 e^{\zeta \omega_f} + E_1^* e^{-\zeta \omega_f}$$

$$y_2(t) = A_2 e^{s_1 t} + B_2 e^{s_2 t} + C_2 e^{s_3 t} + D_2 e^{s_4 t} + E_2 e^{\zeta \omega_f} + E_2^* e^{-\zeta \omega_f}$$

The constants A_1, B_1, . . . , E_1^*, A_2, B_2, . . . , E_2^* may be determined by using partial fractions as discussed in Sections 9.4 and 9.5. When these constants are in numerical form, appropriate mathematical manipulations may be performed and conclusions may be formulated.

9.12   The expressions for $y_1(t)$ and $y_2(t)$ incorporate both transient and steady-state responses. They are

$$y_1(t) = e^{-2.1178t}[0.108951 \sin(6.3742t + 0.76858\pi)]$$

$$+ e^{-2.8826t}[0.0911028 \sin(2.3592t + 0.596127\,\pi)]$$

$$+ 0.00199849 \sin(40t + 0.564895\,\pi)$$

$$+ 0.174659 \sin(15t + 1.625065\,\pi)$$

$$y_2(t) = e^{-2.1178t}[0.0932766 \sin(6.3742t + 0.31807\,\pi)]$$

$$+ e^{-2.8826t}[0.4457576 \sin(2.3592t + 1.968582\,\pi)]$$

$$+ 0.0153643 \sin(40t + 1.0390287\,\pi)$$

$$+ 0.0579752 \sin(15t + 1.1903923\,\pi)$$

9.14 For $G = 100,000$ psi, the results are given in Table 9.1.

$$x_g(t) = 1.7434 \times 10^{-3} + e^{-8.1861t} \times 10^{-3}$$

$$\times [-1.7726 \cos(79.2962t) - 0.2332 \sin(79.2962t)]$$

$$+ e^{-123.3998t} \times 10^{-3}[0.02928 \cos(189.303t) + 0.03976 \sin(189.303t)]$$

$$\psi(t) = 0.8132 \times 10^{-5} + e^{-8.1861t} \times 10^{-5}$$

$$\times [-0.7974 \cos(179.2962t) + 0.04318 \sin(79.2962t)]$$

$$+ e^{-123.3998t} \times 10^{-5}[-0.01573 \cos(189.303t) - 0.06272 \sin(189.303t)]$$

CHAPTER 10

10.1 An extremum can occur only along the curve $y = x^3$.

10.3 The curve is a cycloid.

10.4 An extremum can be taken only along the curve $y = \cos x$

10.5 Euler–Poisson equation: $\rho + \dfrac{d^2}{dx^2}(\mu y'') = 0$

or $\quad y'''' = \dfrac{\rho}{\mu}$

$$y(x) = \frac{\rho}{24\mu}(x^2 - L^2)^2$$

Represents the buckled axis of a fixed–fixed homogeneous cylindrical beam.

10.6 $y = \sin x;\ z = -\sin x$

10.7 Extrema can occur only along the path

$$y = \sqrt{10x - x^2} \quad \text{or} \quad y = -\sqrt{10x - x^2}$$

10.8

$$\omega_n = \frac{(2n - 1)\pi}{2L} \sqrt{\frac{EA}{\rho}} \qquad n = 1, 2, 3, \ldots$$

10.10

$$\omega_n = \frac{n\pi}{2L} \sqrt{\frac{EI}{\rho}} \qquad n = 1, 2, 3, \ldots$$

where ρ is the mass per unit length

10.11

$$\rho A \ddot{u} = A E u'' + E u' \frac{\partial A}{\partial x}$$

10.12
$$\omega_n \approx \sqrt{\frac{(EM/2)(\omega_n^2 L^2 - 1) + EN\omega_n^2 L}{(\rho M/2\omega_n^2)(\omega_n^2 L^2 + 1) + \rho NL}} \qquad n = 1, 2, 3, \ldots$$

where $\omega_n = (2n - 1)\pi/2L$ for $n = 1, 2, 3, \ldots$; $M = -2[(a - b)/L]z$; and $N = 2az$

10.13 $\omega_n \approx \sqrt{H/G}$

where

$$H = \frac{\omega^4 E z}{24} \left(\frac{3M^3}{8\omega^4}(1 - \omega^2 L^2) - \frac{3NM^2 L}{4\omega^3} + \frac{M^3 L^4}{8} + \frac{NM^2 L^3}{2} \right.$$
$$\left. + \frac{3MN^2}{4\omega^2}(\omega^2 L^2 - 1) + \frac{N^3 L}{2} \right)$$

$$G = \frac{\rho z}{2} \left[M \left(\frac{3L^2}{4} - \frac{1}{4\omega^2} - \frac{2L(-1)^{n+1}}{\omega} + \frac{2}{\omega^2} \right) \right.$$
$$\left. + N \left(\frac{3L}{2} - \frac{2(-1)^{n+1}}{\omega} \right) \right]$$

$$M = \frac{2(a - b)}{L} \qquad N = 2a \qquad \omega = \frac{(2n - 1)\,\pi}{2L}$$

10.18 Mean square value $= \dfrac{1}{T} \displaystyle\int_{-T/2}^{T/2} x^2(t)\, dt$

$$= a_0^2 + \tfrac{1}{2} \sum_1^\infty (a_n^2 + b_n^2)$$

This is a form of Parceval's theorem.

INDEX